Joseph John Bevelacqua

Health Physics

Related Titles

Joseph John Bevelacqua

Basic Health Physics: Problems and Solutions

Second Edition

2010

ISBN: 978-3-527-40823-8

Joseph John Bevelacqua

Contemporary Health Physics: Problems and Solutions

Second Edition

2009

ISBN: 978-3-527-40824-5

Joseph John Bevelacqua

Health Physics in the 21st Century

2008

ISBN: 978-3-527-40822-1

Joseph John Bevelacqua

Health Physics

Radiation-Generating Devices, Characteristics, and Hazards

WILEY-VCH
Verlag GmbH & Co. KGaA

Author

Dr. Joseph John Bevelacqua
Bevelacqua Resources
343 Adair Drive
Richland, WA
USA

Cover Illustration:
Normalized absorbed dose distribution from an array of internal radiation-generating devices generating a spectrum of protons in water. This figure was initially published in Bevelacqua, J. J. (2010). Feasibility of Using Internal Radiation-Generating Devices in Radiotherapy. *Health Physics*, **98**, 614.

Library of Congress Card No.: applied for

British Library Cataloguing-in-Publication Data
A catalogue record for this book is available from the British Library.

Bibliographic information published by the Deutsche Nationalbibliothek
The Deutsche Nationalbibliothek lists this publication in the Deutsche Nationalbibliografie; detailed bibliographic data are available on the Internet at <http://dnb.d-nb.de>.

Print ISBN: 978-3-527-41183-2
ePDF ISBN: 978-3-527-69433-4
ePub ISBN: 978-3-527-69434-1
Mobi ISBN: 978-3-527-69432-7
oBook ISBN: 978-3-527-69435-8

Typesetting SPi Global, Chennai, India
Printing and Binding Markono Print Media Pte Ltd, Singapore

Printed on acid-free paper

This book is dedicated to my wife Terry

and

Sammy, Chelsea, Molly, and Eli

and

Anthony, Stayce, Lucy, Anna, Samuel, Matthew, and Henry

and

Jeffrey, David, and Hannah

and

Megan, Marlando, Isaiah, and Annabelle

and

Peter and Jessica

and

Michael, Tara, Lauren, Janelle, and Lucas

and

Karen, Adam, and Hemma

Contents

Preface

Health Physics: Radiation-Generating Devices, Characteristics, and Hazards addresses emerging radiation protection topics that are judged by the author to be relevant in the upcoming decades, but were not the primary focus in *Health Physics in the 21st Century.* The selection of topics represents the author's judgment regarding the importance of these emerging and evolving areas, which are significantly influenced by his experience, educational background, research interests, and national and international events that have health physics implications. *Health Physics: Radiation-Generating Devices, Characteristics, and Hazards* encompasses emerging radiation-generating technologies, advances in existing technology, applications of existing technology to new areas, and postulated new technologies and devices.

The text covers discussions of technology that will affect the world's population as the twenty-first century proceeds. Topics include the nuclear fuel cycle and the proliferation of nuclear materials and associated technologies. Laser isotope separation and advanced centrifuge technologies have the potential for efficient uranium enrichment and the production of highly enriched uranium.

Expansion of nuclear power technology to less developed nations with limited technical and operational experience increases the potential for nuclear events and accidents. The 2011 Fukushima Daiichi accident highlighted the fact that even advanced nations are vulnerable to nuclear accidents, and the licensing basis of nuclear power facilities must be carefully examined to ensure that these facilities are capable of protecting their fission product barriers during natural and man-made events. Degradation of fission product barriers facilitates the release of radioactive material into the environment and has the potential for significant environmental impacts and economic disruption. Power reactor accidents are not the only source of human and environmental disruption related to the release of radioactive material.

Associated with the proliferation of nuclear technology is the clandestine development of nuclear weapons, improvised nuclear devices (INDs), and radiological dispersal devices (RDDs). These devices can be utilized for terrorist purposes and have the potential for significant harm. The use of stolen nuclear weapons or INDs would produce mass casualties and widespread destruction and result in contamination around the detonation site. RDDs are a lower-level threat, but their use would create significant psychological harm and economic disruption.

On a more positive note, nuclear materials and techniques are advancing medical imaging and therapy procedures. New techniques that deliver targeted dose to the tumor site while minimizing the absorbed dose to healthy tissue enhance the efficacy of treatments and minimize negative side effects. These techniques should enhance a patient's quality of life following treatment.

Expansion in the use of nuclear materials also affects the radiation dose received by the public. As noted in NCRP 160, the expanded use of nuclear medicine techniques has significantly increased public doses. Nuclear materials have also inadvertently found their way into consumer products through a variety of sources including recycled metals.

Twenty-first century technologies are also facilitating the entry of private firms to develop orbital transport vehicles. These vehicles initially focus on low earth orbit, but may eventually permit travel beyond the orbital trajectory. This technology will expose the public to new sources of radiation as they leave the protective electromagnetic shield provided by the earth and the shielding afforded by the atmosphere.

The public will initially have the opportunity for low earth orbit and suborbital flights where they have the potential for increased exposure to cosmic rays and solar particle events. Their exposure to protons and heavy ions presents new challenges for radiation protection professionals.

The increased use and application of nuclear materials and technology also affect nuclear regulations. In addition to the Fukushima Daiichi accident, low earth orbit activities involving public passengers, additional medical treatment methodologies, and unforeseen events will likely influence regulatory involvement and rulemaking. International regulations and the harmonization of national nuclear regulations are other areas that will receive additional emphasis in the forthcoming decades. These and many more topics are addressed in this book.

The topics selected for inclusion in this text are based on near-term technologies and their extrapolation into the future and cutting-edge technologies. Some areas involve incremental steps in existing health physics knowledge including aspects of Generation III and IV fission reactors. Other topics, such as uranium enrichment using laser isotope separation and cancer therapy using internal radiation-generating devices, require the development of concepts that may be relatively new to some health physicists. The extent to which public space travel becomes practical is uncertain and depends on technology development, demonstration of flight safety, economic viability, public interest and support, and regulatory involvement.

Paradigm shifts in thinking are necessary. For example, health physicists are currently trained to accept current regulatory practices (e.g., adequacy of reactor designs and appropriateness of existing emergency planning zones) as providing a bounding, safe framework for public protection following a power reactor accident. However, the Fukushima Daiichi accident challenged these paradigms and suggested that a number of basic design assumptions require challenge to ensure their adequacy. Emerging technologies also require independent thinking and a degree of open-mindedness that is often inhibited by regulatory practices, litigation concerns, and lack of confidence in the future.

As a means to facilitate the transition to new concepts, over 300 problems with solutions are provided. These problems are an integral part of the text, and they serve to integrate and amplify the chapter and appendix information. Readers are encouraged to carefully work each problem to maximize the benefit of this text.

This book is primarily intended for upper level undergraduate and graduate level health physics courses. *Health Physics: Radiation-Generating Devices, Characteristics, and Hazards* is also written for advanced undergraduate and graduate science and engineering courses. It will also be a useful reference for scientists and engineers participating in evolving nuclear technology areas including advanced fuel cycles, laser isotope separation, nuclear proliferation, and Generation IV fission reactors. *Health Physics: Radiation-Generating Devices, Characteristics, and Hazards* has applicability for studies involving nuclear power accidents, terrorist events utilizing INDs and RDDs, advanced nuclear medicine imaging and therapy approaches, public involvement in nuclear licensing, regulatory challenges, and establishing radiological standards and criteria for normal operations and major accident events. The book also is pertinent to the various health physics certification boards (e.g., the American Board of Health Physics) in developing examination questions.

The author offers his best wishes to health physicists as we meet the radiation protection challenges that will unfold in the twenty-first century.

Good luck.
Bonne chance.
Viel Glück.
Удачи.
Buena suerte.
Buona fortuna.

Richland, WA, USA
25 May 2015

Joseph John Bevelacqua, PhD, CHP, RRPT
Bevelacqua Resources

Acknowledgments

Many individuals and organizations assisted the author in the development of this book. Assistance included discussions, information exchanges, and kind guidance. The author apologizes in advance to any individual or organization that was inadvertently omitted and is pleased to acknowledge the support of the following individuals and organizations:

- American Nuclear Society
- American Physical Society
- Argonne National Laboratory
- Brookhaven National Laboratory, National Nuclear Data Center
- Dr Lowell Charlton, Oak Ridge National Laboratory
- Dr Mohan Doss, Fox Chase Cancer Center, Philadelphia, PA
- European Organization for Nuclear Research (CERN)
- Fermi National Accelerator Laboratory
- GE Healthcare
- Health Physics Society
- Professor Franck Guarnieri, Ecole des Mines de Paris
- International Atomic Energy Agency
- Japan Atomic Energy Research Institute
- Organisation for Economic Co-operation and Development
- Professor Dale Kunz, University of Colorado
- Professor Don Robson, Florida State University
- Dr Eric Loewen, GE Hitachi Nuclear Energy
- Lawrence Berkeley National Laboratory
- Lawrence Livermore National Laboratory
- Los Alamos National Laboratory
- National Aeronautics and Space Administration
- National Research Council of Canada
- Dr Robert C. Nelson, Research Reactor Safety Analysis Services
- Oak Ridge National Laboratory
- Oak Ridge National Laboratory's Radiation Safety Information Computational Center
- Dr John Parmentola, General Atomics
- Professor M. L. Raghavan, University of Iowa

- Research Center for Charged Particle Data, National Institute of Radiological Sciences, Chiba, Japan
- Dr Paul Rittman
- Dr Joseph Shonka
- Stanford Linear Accelerator Center
- Dr Igor Tarasov, Michigan State University
- University of Wisconsin
- US Department of Energy
- US Department of Homeland Security
- US Environmental Protection Agency
- US Nuclear Regulatory Commission
- Dr Keith Woodard, ABS Consulting, Inc.
- Waste Isolation Pilot Plant.

Since one of the purposes of this text is to support the technical basis for evolving American Board of Health Physics certification examinations, a portion of the problems were derived from questions that appeared on previous examinations. As a prior panel member, vice chair, and chair of the Part II Examination Panel, I would like to thank my panel and all others whose exam questions have been consulted in formulating questions for this textbook.

The author is also fortunate to have worked with colleagues, students, mentors, and teachers who have shared their wisdom and knowledge, provided encouragement, or otherwise influenced the content of this text. The following individuals are acknowledged for their assistance during the author's career: Dick Amato, John Auxier, Lee Booth, Ed Carr, Paul Dirac, Bill Halliday, Tom Hess, Gordon Lodde, Bob Nelson, John Philpott, Lew Pitchford, John Poston, John Rawlings, Don Robson, Bob Rogan, Mike Slobodien, Jim Tarpinian, Jim Turner, and George Vargo. Sadly, a number of these colleagues are now deceased. The continuing encouragement of my best friend and wife, Terry, is gratefully acknowledged.

I would also like to thank the staff of Wiley-VCH with whom I have enjoyed working, particularly Dr Ulrike Fuchs, Sarah Tilley Keegan, Hans-Jochen Schmitt, Dr Heike Nöthe, Dr Martin Preuss, Anja Tschörtner, Stefanie Volk, and Ulrike Werner. The advice and encouragement of George Telecki of John Wiley & Sons, Inc. is also acknowledged. Sujatha Krishna and the staff at SPi Global are acknowledged for preparation of the book's proofs.

A Note on Units

Although traditional English units are a source of comfort to the author and many applied health physicists in the United States, this text uses the International System of Units (SI). As US regulations are harmonized with international recommendations and regulations, there is an evolving transition to SI.

For those readers that feel more comfortable with conventional units, the following conversion factors are provided:

SI unit	Traditional US unit
Bq	2.7×10^{-11} Ci
Gy	100 rad
C/kg air	3881 R
Sv	100 rem

As the reader can attest, the choice of units is often a matter of familiarity and comfort. However, uniformity and clear communication between various scientific and engineering fields and nations suggest the need to adopt SI System of Units.

With the Fukushima Daiichi accident, some health physicists saw a set of unfamiliar units including TBq, PBq, and EBq. For specificity, standard metric prefixes are used in *Health Physics: Radiation-Generating Devices, Characteristics, and Hazards:*

Standard metric prefixes

Metric prefix	Abbreviation	Value
exa	E	10^{18}
peta	P	10^{15}
tera	T	10^{12}
giga	G	10^{9}
mega	M	10^{6}
kilo	k	10^{3}
hecto	h	10^{2}
deka	da	10^{1}

(*continued overleaf*)

Standard metric prefixes		
Metric prefix	**Abbreviation**	**Value**
deci	d	10^{-1}
centi	c	10^{-2}
milli	m	10^{-3}
micro	μ	10^{-6}
nano	n	10^{-9}
pico	p	10^{-12}
femto	f	10^{-15}
atto	a	10^{-18}

Part I
Overview of Health Physics: Radiation-Generating Devices, Characteristics, and Hazards

Health Physics: Radiation-Generating Devices, Characteristics, and Hazards connects twentieth-century and twenty-first-century health physics in selected areas including the nuclear fuel cycle, nuclear accidents, radiological emergencies, nuclear terrorism and related events, nuclear medicine, public issues related to radiation and radioactive materials, and evolving regulatory issues. Specific topics include advanced nuclear reactors, laser uranium enrichment, actinide transformation, advanced medical devices, radiation therapy utilizing exotic particles and heavy ions, nuclear accidents, terrorism involving radioactive dispersal devices and improvised nuclear devices, and evolving regulatory requirements. These topics are active health physics areas. Other topics involving public space travel, harmonization of radiation protection regulations, using antimatter and internal radiation-generating devices in nuclear therapy applications, and implementation of advanced fuel cycles using Generation IV reactors are evolving areas that will more fully emerge as the twenty-first century progresses.

Seven chapters introduce these topics and basic knowledge required to understand the anticipated evolution of the health physics field. Background information is provided in eight appendices to smooth the transition to information needed to comprehend the emerging radiation-generating technologies. The reader should consult these appendices as they are referenced in the main text.

Some topical areas naturally appear in multiple chapters since they are significant and have many aspects. For example, the major nuclear power reactor accidents at Three Mile Island, Chernobyl, and Fukushima Daiichi are addressed throughout the book and not restricted to Chapter 3 that focuses on reactor accidents. This organizational structure is appropriate since these accidents had a significant impact on the nuclear fuel cycle, planning for future nuclear emergencies, public issues associated with the nuclear power debate, and regulatory issues associated with reactor licensing and the selection of design and beyond design basis accidents.

Health Physics: Radiation-Generating Devices, Characteristics, and Hazards, First Edition. Joseph John Bevelacqua.

The nature of this text suggests that its content is continuously evolving. As with any book, it is necessary to eventually freeze the content and focus on consolidation and editing. Text material was finalized in mid-2014 and the addition of new material essentially terminated at that time. Accordingly, some material may have evolved after that date including the ongoing description and development of Generation IV reactors, proposed changes to the 10CFR20 radiation protection regulations in the United States, and emerging advances in nuclear medicine. As warranted, references were added to reflect these evolving topics.

1
Introduction to Twenty-First Century Health Physics

1.1
Overview of Twenty-First Century Health Physics

History has the unfortunate habit of repeating. Significant events of a given classification (e.g., accidents, natural disasters, and conflicts over natural resources) reoccur and are often influenced by available technology. For example, wars continue to be waged, but their scope and destructive power are amplified by technology. The development of nuclear technology and the fabrication of nuclear weapons continue to influence world events and health physics concerns as the twenty-first century unfolds.

1.2
Health Physics Issues, Challenges, and Opportunities

The twentieth-century power reactor accidents at Three Mile Island Unit 2 and Chernobyl Unit 4 revealed weaknesses in the management and regulation of nuclear reactors. Unfortunately, the nuclear accident *hat trick* was achieved in the twenty-first century with the accident involving Fukushima Daiichi Units 1, 2, 3, and 4. This most recent accident reveals additional structural weaknesses in nuclear regulation and management that involve fundamental licensing basis issues. The legacies of Three Mile Island and Chernobyl remain, and final cleanup actions for these sites either are delayed until facility decommissioning or are ongoing. The decade cleanup duration of Three Mile Island is dwarfed by the projected 40–100-year recovery effort for Fukushima Daiichi. Associated with these three accidents are issues involving environmental impacts, stakeholder concerns, regulatory changes, licensing impacts, and financial implications. These issues are addressed in this book and have a profound influence on health physics activities associated with these accidents and the subsequent expansion of nuclear power generation.

In a similar fashion, the terrorist attacks of the twentieth century culminated in the 11 September 2001 events involving the World Trade Center in New York and the US Pentagon. These attacks spawned significant concerns regarding the

Health Physics: Radiation-Generating Devices, Characteristics, and Hazards, First Edition. Joseph John Bevelacqua.

escalation of terrorist events to include a variety attacks including those utilizing radioactive materials and nuclear weapons. Technology has once again opened a door to an escalation of attack profiles that significantly affect the health physics profession.

The nuclear fuel cycle has successfully enriched uranium for reactor fuel and weapons production and reprocessed spent nuclear fuel to recover uranium and plutonium. Historically, the enrichment process required large facilities because diffusion and centrifuge technologies are relatively inefficient processes for uranium enrichment. The advent of advanced centrifuge technology and laser isotope separation makes the uranium enrichment step considerably more efficient and permits smaller facilities to be constructed and operated. These facilities are easier to conceal than the large centrifuge and gaseous diffusion plants. This presents the opportunity for a clandestine enrichment facility to produce weapons-grade uranium. Advanced technologies, particularly laser uranium enrichment, present a twenty-first-century nuclear proliferation concern.

In a similar manner, reprocessing technology has successfully recovered plutonium, and this technology is well known. The expansion of nuclear power facilities offers the possibility for the diversion of spent fuel that could be reprocessed and the recovered plutonium diverted toward weapons production or terrorist purposes.

On a more positive note, nuclear medicine has advanced and improved diagnostic and therapeutic techniques. The capability to localize the absorbed dose has improved, and additional radiation types are being utilized to target tumors. Proton and heavy ion therapy techniques are becoming more common, and the initial studies using antiprotons have been published. The use of nanotechnology and internal radiation-generating devices in cancer therapy applications is in development for the selective delivery of absorbed dose.

The advancement of nuclear medicine techniques increased the average absorbed dose delivered to the public. An increased use of nuclear materials in commercial products and their inadvertent entry into scrap metal used in consumer products offer additional challenges. Public concerns regarding the use of nuclear power generation and the effects of major accidents have been heightened by the Fukushima Daiichi accident and its sensationalism by the media and antinuclear groups.

Public interest and the involvement of stakeholder groups in nuclear licensing have also increased following the Fukushima Daiichi accident. Events involving radioactive materials and their associated media attention suggest that the interest of the public in radiation-generating technologies and radioactive materials will likely increase. The media presents a significant challenge because its perspective is often influenced more by emotion and sensationalism than scientific reasoning and knowledge.

Heightened public concern, media presentations that sensationalize events, increasing political pressure and influence, and active stakeholder involvement in nuclear projects suggest that the twenty-first-century regulatory environment will be dynamic and challenging. These elements affected the US fuel repository at

Yucca Mountain and led to a temporary suspension of construction and operating licenses for new power reactors related to fuel storage environmental concerns and the associated legal issues. There has also been significant regulatory action following the Fukushima Daiichi accident that affects existing plants and those facilities under design and construction. The twenty-first century will likely offer a challenging health physics environment with considerable emphasis on postulated power reactor release scenarios, assumed accident severity, and the definition of credible design basis events.

The twentieth century saw a maturation of the health physics profession and its scientific basis, and the twenty-first century will require additional scientific training for health physics professionals to meet the significant challenges posed by advanced technologies. These challenges include continued debate over the fundamental regulatory assumption regarding the linear-nonthreshold (LNT) dose–response hypothesis, applicability of hormesis to the human species, evaluation of doses to reference plants and animals and their inclusion in environmental assessments and regulations, and the inclusion of occupational dosimetry and environmental doses into assessments of the biological effects of ionizing radiation.

National and international organizations continue to foster sustained development and standardization, but they run the risk of becoming decoupled from applied health physicists over issues such as the LNT hypothesis and environmental protection. Instrumentation advances will permit the enhanced detection of a variety of ionizing radiation types over a wide range of energies, and these detectors will find their incorporation into consumer products such as cell phones and enhance the detection of illicit nuclear materials.

Health Physics: Radiation-Generating Devices, Characteristics, and Hazards reviews emerging and maturing radiation-generating technologies that will affect the health physics profession. It is hoped that this review will foster additional research into these and supporting areas.

Health physics is a dynamic and vital field and has an exciting future. The topics addressed in this text encompass energy generation, medical applications, fuel cycle technologies, consumer applications, public exposures, and national defense. However, significant challenges will likely arise as new technologies expand the use of radioactive materials and radiation-generating devices, failures of existing technology occur, terrorist attacks expand to include radioactive materials or nuclear weapons, and old paradigms fall.

There is an intimate linkage between the health physics profession and the expansion of nuclear technology and nuclear-related events. This linkage will manifest itself in traditional fields and possibly in new areas including the response to public space tourism and nuclear terrorism. Communications with stakeholders and the public are essential to counter misinformation and hysteria that often accompanies media reports of nuclear-related events. The twenty-first-century health physicist must be technically capable and able to communicate information to the public in a commonsense manner that is understandable to a group with limited scientific knowledge. It will be an exciting time, but a time filled with challenges. The following areas are judged by the author to be representative of future health physics challenges, and these topics are further explored in this book:

- Generation IV fission power reactors
- Low earth orbit tourism by the public
- Advanced nuclear fuel cycles incorporating laser uranium enrichment and actinide transmutation
- Radiation therapy using heavy ions, exotic particles, internal radiation-generating devices, and antimatter
- Public radiation exposure
- Radioactive dispersal and improvised nuclear devices
- Nuclear accidents
- Evolving regulatory considerations

1.3 Forecast of Possible Future Issues

Table 1.1 summarizes a selected set of twentieth-century and early twenty-first-century events that are used to forecast events that may have health physics relevance. For example, the occurrence of the Three Mile Island and Chernobyl reactor accidents suggested that future accidents are likely and have occurred at Fukushima Daiichi. However, the cause of a future accident is not predicted by the recurrence of these events. An examination of the events is summarized in Table 1.1 suggesting possible causes for a future nuclear event which include natural events such as an earthquake, rare natural phenomena, military action, terrorism, technology failure, management failure, human error, an unrelated industrial accident, economic failure, and social disruption. The 2011 Fukushima Daiichi accident was caused by an earthquake and subsequent tsunami. The predictive power of the aforementioned approach is speculative. However, it does suggest possible twenty-first-century health physics events having the potential for significant environmental releases of radioactive materials and associated public doses.

Given the history of humankind, twenty-first-century wars are likely. With the expansion of the use of nuclear technology, these wars could include a nuclear exchange between nations and a military attack or intentional sabotage of a nuclear facility.

Terrorist events have continued into the twenty-first century including the 11 September 2001 attacks in the United States and the 2004 Madrid and 2005 London transportation bombings. Terrorist attacks on a nuclear facility are possible twenty-first-century radiological events. Other terrorist events with radiological consequences include the use of nuclear weapons, intentional dispersal of radioactive materials into a populated area, intentional contamination of water supplies, and contamination of food supplies.

Table 1.1 Selected significant twentieth-century and early twenty-first-century events.

Event	Event type	Possible twenty-first-century health physics event extrapolation
1906 San Francisco earthquake and fire	Natural event—massive earthquake	Massive earthquake damaging a nuclear facility
1908 Tunguska explosion in Siberia	Unknown cause, possibly a meteorite strike Impact energy equivalent of about 15 MT of TNT[a)]	Rare natural event damaging a nuclear facility
1914–1919 World War I	International armed conflict	Military attack on a nuclear facility
1918 Spanish flu pandemic	Epidemic	Epidemic affects staffing and disrupts nuclear facility operations
1929 stock market crash	Economic disruption	Economic event disrupts nuclear facility operations
1930s to early 1940s Great Depression	Economic collapse	Worldwide economic collapse disrupts nuclear facility operations
1939–1945 World War II	International armed conflict	Military attack on a nuclear facility
1945 nuclear bombing of Hiroshima and Nagasaki, Japan	Nuclear attack	Nuclear exchange between nations or terrorist nuclear event in a major city
1950–1970s Space Race	Development of long-range rockets and space exploration	Nuclear missile attack Public space tourism
1960s political assassinations in the United States	Disruption of government	Social unrest disrupts nuclear facility operations
1965 Northeast US and Canada blackout	Disruption of electrical energy supply	Loss of off-site power for a nuclear facility
1972 Munich Olympics massacre	Terrorist attack	Nuclear terrorism
1976 earthquake hits Tangshan, in northeastern China	Natural event—massive earthquake	Massive earthquake with significant loss of life affects nuclear facility operations
1979 Three Mile Island nuclear accident	Power reactor accident with minimal release of radioactive material	Major power reactor accident
1981 Israeli military successfully attacks and destroys the Osirak nuclear reactor in Iraq	Military attack on a nuclear power facility	Major power reactor accident following a military attack
1984 massive poison gas leak in Bhopal, India	Major industrial accident	Major industrial accident affects nuclear facility operations

(Continued Overleaf)

Table 1.1 (*Continued*)

Event	Event type	Possible twenty-first-century health physics event extrapolation
1986 Chernobyl nuclear accident	Power reactor accident with significant release of radioactive material	Major power reactor accident
1986 Space Shuttle Challenger explosion	Technological and management failure	Failure of safety and management systems disrupts nuclear facility operations
1987 Goiania, Brazil, contamination event	^{137}Cs orphan source contaminates homes and individuals, resulting in four fatalities	Radiological dispersal device is utilized in a terrorist attack
1989 Northeast United States, Canada, and Sweden experience a power blackout caused by a solar flare	Disruption of electrical energy supply	Loss of off-site power for a nuclear facility
2001 New York City and Pentagon terrorist attacks	Terrorist attack	Nuclear terrorism including a direct attack on a nuclear facility
2003 Space Shuttle Columbia accident	Technological and management failure	Failure of safety and management systems disrupts nuclear facility operations
2003 Northeastern and Midwestern United States and Ontario, Canada, blackout caused by a solar flare	Disruption of electrical energy supply	Loss of off-site power for a nuclear facility
2004 Madrid commuter train bombing	Terrorist attack	Nuclear terrorism including a direct attack on a nuclear facility
2005 London underground train and double-decker bus bombings	Terrorist attack	Nuclear terrorism including a direct attack on a nuclear facility
2005 Hurricane Katrina floods New Orleans, kills nearly 2000, and damages critical infrastructure	Massive storm disrupts a major city and surrounding areas	Loss of power and critical infrastructure support to a nuclear facility Flooding a nuclear facility
2009 terrorist attack occurred at Fort Hood in Texas. A US Army major and psychiatrist fatally shot 13 people and injured more than 30 others	Insider terrorist attack by a member of the operating organization	Trusted employee becomes a terrorist and sabotages a nuclear reactor to create severe core damage and release of fission products to the environment

Table 1.1 (*Continued*)

Event	Event type	Possible twenty-first-century health physics event extrapolation
2011 Fukushima Daiichi nuclear accident	Massive earthquake and tsunami causes a power reactor accident involving multiple units with a significant release of radioactive material	Major power reactor accident involving multiple units caused by a natural event
2012 Hurricane Sandy storm surge floods New York City and neighboring areas	500-year storm surge disrupts city services	Loss of power and infrastructure support to a nuclear facility Flooding of a nuclear facility
2013 18 m-diameter meteorite explodes over Chelyabinsk, Russia, and injures 1000 people	Impact event corresponding to an energy equivalent of about 1 MT of TNT[a)]	Rare natural event damages a nuclear facility
2013 Asteroid DA14 (5.7×10^8 kg) passes within 1.0×10^4 km of the earth	Astronomical near miss with a 2046 predicted return to earth	Rare natural event damages a significant geographical area including infrastructure and nuclear facilities
2013 Typhoon Haiyan devastates the eastern Philippines	The massive typhoon leads to a death toll in the thousands with hundreds of thousands displaced and critical infrastructure destroyed	Massive typhoon disrupts nuclear facility operations
2013 110 TBq ^{60}Co teletherapy source stolen in Mexico	Theft of radioactive material	Stolen radioactive material is incorporated into a terrorist device
2014 Waste Isolation Pilot Plant waste container undergoes an unanticipated chemical reaction and releases americium and plutonium into the environment	Underground geologic waste repository event	Major accident at a high level waste geologic repository caused by the failure of assumed controls and inadequate oversight
2014 Belgian Doel 4 nuclear reactor's turbine is sabotaged and severely damaged	Sabotage of a nuclear power reactor	Sabotage of a nuclear reactor leading to a major accident with severe core damage and an off-site release
2015 Germanwings Airbus A320 carrying more than 140 passengers intentionally crashed by its copilot	Catastrophic act committed by a trusted employee	Trusted employee sabotages a nuclear reactor to create severe core damage and release of fission products to the environment

a) Megatons of trinitrotoluene.

Many aspects of health physics activities are reactive. These reactive aspects include resolution of audit and inspection findings, response to abnormal and emergency events, and development of procedures and programs to meet regulatory requirements. However, the author prefers a proactive approach that challenges accepted assumptions and established practices to address and anticipate future events. For example, the National Council on Radiation Protection and Measurements assumes that improvised nuclear devices will not exceed 10 kT (trinitrotoluene, TNT equivalent). Given the level of technology, availability of weapons information in the open literature, abundance of raw scientific data, proliferation of nuclear materials, and availability of necessary computational tools, the 10 kT design assumption should be expanded to include larger weapons yields to develop bounding plans, procedures, and resource allocation requirements. The 10 kT limit also appears to exclude the possibility of the theft of an existing device from a nuclear power, transfer of a device from a nuclear power to a terrorist organization, or use of proven scientific resources to develop a higher-yield clandestine device.

Other nuclear scenarios that could present twenty-first-century health physics challenges are developed in subsequent chapters and their associated problems. However, to provide a preview of upcoming topics, a series of general problems are provided in this chapter to illustrate possible twenty-first-century events of significant health physics consequences. These problems are based on the events in Table 1.1. They are low-probability, high-consequence events that are often classified as X factors or black swan events.

In the twentieth century, the causes of the Fukushima Daiichi accident would have been classified as X factors. Unfortunately, my Mark-I Crystal Ball is out of service, but past events are often a guide to future events. Therefore, the Chapter 1 solutions are necessarily general and brief. However, considerable additional detail is provided in the subsequent chapters that more fully characterized the consequences of more probable event types.

One way to minimize the consequences of future radiological events is to constantly challenge assumptions, focus on the mitigation of significant events, and have an informed public that understands the risks and benefits of nuclear technologies. Scientific prediction and mitigation of significant nuclear events have not been completely successful, and we must do a significantly better job in the future. That is not an easy task. I hope that this text will motivate additional improvements to minimize the probability and consequence of future radiological events.

The twenty-first century will be an exciting time for the health physics profession. It is the author's desire that this book contributes in some small measure to the education of twenty-first-century health physicists and their understanding of existing, evolving, and emerging radiation-generating technologies. The author also hopes that this text will foster additional effort to improve upon and further develop the topics of this text.

Problems

1.1 The 1908 Tunguska explosion in Siberia is believed to have been caused by a meteorite. On 30 June 1908, a meteorite exploded about 10 km above the ground in a sparsely populated region. The blast released about 15 MT of energy and leveled about 2000 km^2 of forest. Predict the consequences of a Tunguska-type event that explodes in the air within 1 km of the underground Hanford Tank Farms containing fuel reprocessing waste. List the most likely public effects and required health physics actions resulting from this event.

1.2 In 1984, a huge poison gas leak in Bhopal, India, led to the death of thousands of people. A storage tank containing methyl isocyanate at a pesticide plant leaked gas into the densely populated city of Bhopal. It was one of the worst industrial accidents in history. Predict the consequences of a Bhopal-type event that occurs in proximity of an operating nuclear power reactor. List the most likely effects and health physics consequences of a Bhopal-style event if the gas cloud covered a nuclear power facility for an extended period.

1.3 The North American blackout of 1965 was a significant disruption in the supply of electricity that affected parts of Ontario, Canada, and New England in the United States. Over 30 million people and 207 000 km^2 were without electricity for over 10 h. Predict the consequences of an extended (e.g., several weeks) power blackout event that occurs at a uranium enrichment facility using lasers and UF_6 gas as the working fluid.

1.4 Assume that a terrorist group acquires medical isotopes (i.e., ^{32}P, ^{60}Co, and ^{131}I) and incorporates them into a dirty bomb. What is the relative hazard of these isotopes if the dirty bomb is detonated in a populated area? How do these hazards affect recovery activities?

1.5 A massive solar event has the potential to disrupt the electrical grid for an extended period. If a solar event an order of magnitude larger than the 1859 Carrington event (see Chapter 6) occurred, what is the impact on the capability of a nuclear power reactor to preserve its fission product barriers? Assume the event disrupts the power grid supplying the reactor and its surrounding area for 1 month.

1.6 A limited nuclear exchange occurs between two neighboring nations. Each nation has detonated three, 250 kT ^{239}Pu fission devices over separate, heavily populated targets. You have been requested to advise the population residing outside the immediate blast area. Stakeholders are particularly interested in the radiological effects of fallout. The impacts on the food supply and means to limit the associated effective doses are immediate concerns. From a health physics perspective, what isotopes are of concern, what pathways can these isotopes enter the food chain, and what protective actions can be applied to limit the absorbed dose from these isotopes?

1.7 A terrorist group has stolen sufficient ^{235}U to fabricate a crude nuclear weapon. In the process of constructing the device, the explosive package prematurely detonates, but the weapon does not achieve a significant nuclear yield. What isotopes are of concern? What health physics actions should be implemented to permit reentry into the ground zero area?

1.8 A research team has developed a cancer therapy technique using anti-^{12}C ions. List three challenges to deploying this technology to medical facilities. List three positive aspects of this technology.

References

Artemieva, N. (2013) Russian skyfall. *Nature*, **503**, 202.

Bevelacqua, J.J. (2008) *Health Physics in the 21st Century*, Wiley-VCH Verlag GmbH, Weinheim.

Bevelacqua, J.J. (2009) *Contemporary Health Physics: Problems and Solutions*, 2nd edn, Wiley-VCH Verlag GmbH, Weinheim.

Bevelacqua, J.J. (2010a) *Basic Health Physics: Problems and Solutions*, 2nd edn, Wiley-VCH Verlag GmbH, Weinheim.

Bevelacqua, J.J. (2010b) Radiological dispersion devices and basic radiation science. *Phys. Teach.*, **48**, 302.

Bevelacqua, J.J. (2010c) Feasibility of using internal radiation-generating devices in radiotherapy. *Health Phys.*, **98**, 614.

Bevelacqua, J.J. (2012a) The Fukushima Daiichi accident and its radiological impact on the environment. *Phys. Teach.*, **50**, 368.

Bevelacqua, J.J. (2012b) Applicability of health physics lessons learned from the three mile island unit-2 accident to the Fukushima Daiichi accident. *J. Environ. Radioact.*, **105**, 6.

Bevelacqua, J.J. (2014) Health physics aspects of generation IV reactors. *Int. Nucl. Saf. J.*, **3** (1), 13, http://nuclearsafety.info/international-nuclear-safety-journal/index.php/INSJ/article/view/20/pdf_1 (accessed 05 August 2015).

Brown, P.G. (2013) A 500-kiloton airburst over chelyabinsk and an enhanced hazard from small impactors. *Nature*, **503**, 238.

Brumfiel, G. (2011a) The meltdown that wasn't. *Nature*, **471**, 417.

Brumfiel, G. (2011b) Fukushima set for epic clean-up. *Nature*, **472**, 146.

Brumfiel, G. (2013) Fallout or fear. *Nature*, **493**, 290.

Brumfiel, G. and Cyranoski, D. (2011) Quake sparks nuclear crisis. *Nature*, **471**, 273.

Bushnell, D.M. (2001) *Future Strategic Issues/Future Warfare [Circa 2025]*, National Aeronautic and Space Administration, NASA Langley Research Center, Langley, VA, http://www.standeyo.com/NEWS/13_USA/130729.future-strategic-issues-and-warfare.pdf (accessed 05 August 2015).

Butler, D. (2011a) Radioactivity spreads in Japan. *Nature*, **471**, 355.

Butler, D. (2011b) Fukushima health risks scrutinized. *Nature*, **472**, 13.

Cember, H. and Johnson, T.A. (2008) *Introduction to Health Physics*, 4th edn, McGraw-Hill, New York.

Chernobyl accident special edition. *Nucl. Saf.*, **28**, 1 (1987).

Clifton, D. (1995) *Chronicle of the 20th Century: The Ultimate Record of Our Times*, Simon & Schuster, New York.

Generation IV International Forum (2009) *GIF R&D Outlook for Generation IV Nuclear Energy Systems*, US Department of Energy, Washington, DC.

Glaser, A. (2008) Characteristics of the gas centrifuge for uranium enrichment and their relevance for nuclear weapon proliferation. *Sci. Global Secur.*, **16**, 1.

Glasstone, S. (1982) *Energy Deskbook*, United States Department of Energy, Technical Information Center, Oak Ridge, TN.

Hecht, J. (2011) Laser enrichment returns from the dead. *Laser Focus World*, **47** (10), 18.

ICRP (2007) The 2007 Recommendations of the International Commission on Radiological Protection. ICRP Report No. 103, Elsevier, Amsterdam.

International Atomic Energy Agency (2002) *IAEA Safeguards Glossary 2001 Edition*, International Atomic Energy Agency, Vienna.

International Atomic Energy Agency (2006) *Nuclear Forensics Support Reference Manual*, IAEA Nuclear Security Series No. 2 Technical Guidance, International Atomic Energy Agency, Vienna.

International Atomic Energy Agency (2007) *Status of Small Reactor Designs without On-site Refuelling*. IAEA-TECDOC-1536,

International Atomic Energy Agency, Vienna.

Jones, N. (2013) It could happen one night: catastrophes from the past will strike again – We just do not know when. *Nature*, **493**, 154.

NCRP (2001) Management of Terrorist Events Involving Radioactive Material. NCRP Report No. 138, NCRP, Bethesda, MD.

NCRP (2007) Uncertainties in the Measurement and Dosimetry of External Radiation. NCRP Report No. 158, NCRP, Bethesda, MD.

NCRP (2008) Management of Persons Contaminated with Radionuclides. NCRP Report No. 161, NCRP, Bethesda, MD.

NCRP (2009a) Ionizing Radiation Exposure of the Population of the United States. NCRP Report No. 160, NCRP, Bethesda, MD.

NCRP (2009b) Uncertainties in Internal Radiation Dose Assessment. NCRP Report No. 164, NCRP, Bethesda, MD.

NCRP (2010a) Responding to a Radiological or Nuclear Terrorism Incident: A Guide for Decision Makers. NCRP Report No. 165, NCRP, Bethesda, MD.

NCRP (2010b) Population Monitoring and Radionuclide Decorporation Following a Radiological or Nuclear Incident. NCRP Report No. 166, NCRP, Bethesda, MD.

New York Times (2015) Archives of The New York Times, available at www.nytimes.com/content/help/search/archives/archives.html (accessed 7 August 2015).

Rogovin, M. and Frampton, G.T. (1980) Three Mile Island, A Report to the Commissioners and to the Public. U. S. Nuclear Regulatory Commission, NUREG/CR-1250, Vols. 1–2, U. S. Government Printing Office, Washington, DC.

Tollefson, J. (2013) New York vs. the sea. *Nature*, **494**, 162.

Turner, J.E. (2007) *Atoms, Radiation, and Radiation Protection*, 3rd edn, Wiley-VCH Verlag GmbH, Weinheim.

U.K. Ministry of Defense (2010) *Global Strategic Trends – Out to 2040*, 4th edn, Development, Concepts and Doctrine Centre (DCDC), London.

US Department of Energy (2001) *Generation-IV Roadmap Report of the Fuel Cycle Crosscut Group*, US Department of Energy, Washington, DC.

US Department of Energy (2012) *Reactor and Fuel Cycle Technology Subcommittee Report to the Full Commission - Updated Report*, US Department of Energy, Blue Ribbon Commission on America's Nuclear Future, Washington, DC.

US Department of Energy (2013a) *FY 2012 Annual Report: Addressing the Issues of Today*, US Department of Energy, Office of Nonproliferation and International Security, Washington, DC.

US Department of Energy (2013b) *Strategy for the Management and Disposal of Used Nuclear Fuel and High-Level Radioactive Waste*, US Department of Energy, Washington, DC.

US Department of Energy (2014) *Phase 1-Radiological Release Event at the Waste Isolation Pilot Plant on February 14, 2014*, US Department of Energy, Washington, DC.

U. S. Nuclear Regulatory Commission (1979a) Population Dose and Health Impact of the Accident at the Three Mile Island Nuclear Station. NUREG-0558, U. S. Government Printing Office, Washington, DC.

U. S. Nuclear Regulatory Commission (1979b) Investigation into the March 28, 1979 Three Mile Island Accident by the Office of Inspection and Enforcement. NUREG-0600, U. S. Government Printing Office, Washington, DC.

U. S. Nuclear Regulatory Commission (1979c) NRC Views and Analysis of the Recommendations of the President's Commission on the Accident at Three Mile Island. NUREG-0632, U. S. Government Printing Office, Washington, DC.

U. S. Nuclear Regulatory Commission (1987) Report on the Accident at the Chernobyl Nuclear Power Station. NUREG-1250, Rev. 1, U. S. Government Printing Office, Washington, DC.

U. S. Nuclear Regulatory Commission (1989) Implications of the Accident at Chernobyl for Safety Regulation of Commercial Nuclear Power Plants in the United States. NUREG-1251, Vols. 1 and 2,

U. S. Government Printing Office, Washington, DC.

U. S. Nuclear Regulatory Commission (1992) Summary of Chernobyl Followup Research Activities. NUREG-1422, U. S. Government Printing Office, Washington, DC.

Zhang, S. (2013) Haiyan prompts risk research. *Nature*, **503**, 324.

Part II
Nuclear Fuel Cycle Issues

Part II examines the nuclear fuel cycle and the associated challenges to the operation of its facilities and associated infrastructure. The ramifications of the Fukushima Daiichi accident on facility operations are examined. New technologies for the enrichment of uranium using advanced centrifuges and laser techniques, sophisticated methods to address high-level nuclear waste, advanced reactor designs, and various fuel cycle options are a portion of the topics addressed in this part.

The nuclear fuel cycle is growing in importance, and its relationships to contemporary issues (e.g., climate change, nuclear proliferation, medical isotope production, and environmental protection) are active media topics and research areas. Fuel cycle activities have a growing influence on our daily lives. These activities will increase in importance as fossil fuels are more heavily regulated and nuclear energy emerges as an environmentally friendly energy source.

As used in this part, nuclear proliferation is the spread of nuclear weapons and associated information and technology. Technologies associated with nuclear weapons and developing or acquiring their key components such as weapons-grade uranium and plutonium are aspects of nuclear proliferation.

Health Physics: Radiation-Generating Devices, Characteristics, and Hazards, First Edition. Joseph John Bevelacqua.

2
Nuclear Fuel Cycle

2.1
Overview

The twentieth century developed and partially advanced the nuclear fuel cycle but left a number of open issues for resolution in the twenty-first century. These issues include questions of reactor safety, the storage and disposition of nuclear waste including spent reactor fuel, regulatory challenges, and changes in the strategic approach to fuel cycle operations.

A successful twenty-first-century fuel cycle must deal with the nuclear safety aftermath of the Three Mile Island (TMI), Chernobyl, and Fukushima Daiichi accidents as well as the terrorist events of 11 September 2001. Terrorism and the threat of sabotage at a nuclear power reactor complicate fuel cycle nuclear safety issues.

Nuclear waste remains a cornerstone liability to the continued viability of the nuclear power portion of the fuel cycle. Waste issues encompass a variety of concerns including the storage and disposition of spent nuclear fuel.

Many of the issues associated with the advancement and deployment of nuclear power are intimately linked to the choice of a nuclear fuel cycle. Issues associated with the deployment of nuclear reactors including resource allocation, nuclear proliferation, and waste management are inherent fuel cycle concerns. The fuel cycle provides the infrastructure connecting uranium and thorium derived from mining, through nuclear power production, and to the eventual management of the associated waste.

Advanced fuel cycles include Generation IV power reactor designs. The various stages of power reactor evolution are specified in terms of generations. Generation I reactors were experimental devices used to investigate and verify design concepts. All currently operating reactors in the United States are Generation II designs. These reactors utilize active safety systems. Generation III reactors are similar to Generation II designs but incorporate passive safety systems.

An advanced fuel cycle must necessarily consider the long-term resource potential, energy supply and demand projections, and the potential deployment of advanced reactors including the planned Generation IV reactors. Generation IV technologies are envisioned for deployment in the 2030s and will include a 60-year design lifetime. The Generation IV goals of sustainability, safety

Health Physics: Radiation-Generating Devices, Characteristics, and Hazards, First Edition. Joseph John Bevelacqua.

and reliability, and economic viability require consideration of the entire fuel cycle. Health physics aspects of an advanced fuel cycle are considered in this chapter.

Emerging technologies and changes in deployment strategy affect the operation and inherent structure of the twenty-first-century nuclear fuel cycle. These technologies include the laser enrichment of uranium and use of Generation IV reactors to minimize the quantity of high-level waste. Changes in strategy including a renewed interest in fuel reprocessing are also likely.

This chapter reviews these and other issues and challenges that face the nuclear fuel cycle in the twenty-first century. The health physics challenges will likely be severe as new regulatory requirements emerge and economic considerations place a premium on operational efficiency with limited resources. A detailed consideration of fuel cycle regulatory issues is provided in Chapter 7.

2.2
Basic Fuel Cycle Options

In the twentieth century, light water reactors (LWRs) using uranium dioxide fuel were the dominant focus of the nuclear fuel cycle. Most uranium enrichment facilities were based on gaseous diffusion and gas centrifuge technology using uranium hexafluoride as the working fluid. These choices dictated the chemical conversion facilities required to convert uranium ore into the compounds utilized in the enrichment (UF_6) and reactor (UO_2) fuel cycle stages. The twenty-first century will witness a variety of reactor fuel forms as well as enrichment technologies that may use other uranium compounds. These changes will have a significant influence on the fuel cycle and its required infrastructure.

2.3
Overview of the Twentieth-Century Nuclear Fuel Cycle

The nuclear fuel cycle describes the production, utilization, and disposition of specific nuclear materials. The two primary fuel cycles involve uranium and thorium. Fuel cycles include the successive stages from mining of the uranium or thorium ores to the final disposal of the radioactive wastes derived from the storage or reprocessing of spent nuclear fuel. In its more advanced twentieth-century form, the spent fuel removed from a reactor is reprocessed and ^{235}U and plutonium are recovered for subsequent reuse in new reactor fuel. The twentieth-century fuel cycle incorporated two fuel cycle options. Option 1 is the storage of spent reactor fuel with no reprocessing, and Option 2 is reprocessing with the recovery of ^{235}U and plutonium for use in a subsequent reactor cycle. These options are reviewed in the subsequent discussion.

2.3.1 Uranium Fuel Cycle

Natural uranium consists of three primary isotopes, namely, ^{238}U, ^{235}U, and ^{234}U, whose natural abundances are 99.2739, 0.7204, and 0.0057, respectively. Given their respective half-lives, noted in parenthesis, ^{238}U (4.468×10^9 years), ^{235}U (7.04×10^8 years), and ^{234}U (2.46×10^5 years) contribute 49, 2, and 49% of the specific activity of natural uranium, respectively. Using twentieth-century enrichment technologies, the radiological hazard of a uranium compound increases as the ^{235}U enrichment increases. This occurs because gaseous diffusion and gas centrifuge technologies increase the ^{234}U and ^{235}U enrichments. The increase in ^{234}U enrichment enhances the radiotoxicity of the enriched product relative to the natural feed material. The decay products of these uranium isotopes consist of long decay chains that decay by alpha, beta, and photon radiation types.

Uranium and its decay products are predominantly an internal radiation hazard, and standard internal dosimetry models assess their associated radiological hazard. In addition, the ^{238}U natural decay series and its decay products produce a significant external hazard (2.33 mSv/h beta radiation at 7 mg/cm^2 from an equilibrium thickness of uranium metal). This beta dose rate arises primarily from the decay of ^{234m}Pa.

The radiological hazard of natural uranium is overshadowed by its chemical toxicity. As a heavy metal, uranium is chemically toxic to the kidneys. The radiological hazards become more significant as the ^{235}U enrichment increases. For example, neutron radiation levels from UF_6 increase from about 2 μSv/h for low-enriched material (<5 wt% ^{235}U) to about 40 μSv/h for highly enriched material (>95 wt% ^{235}U). These effective dose rates are applicable to enriched natural uranium. Higher effective dose rates occur for reprocessed uranium. These neutron levels are driven by the $^{19}F(\alpha, n)$ reaction with limited dose derived from other reactions including the $^{19}F(n, 2n)$ reaction.

Option 2 of the theoretical uranium cycle assumes that the spent reactor fuel is reprocessed in order to recover uranium and plutonium for subsequent recycling as reactor fuel. Typically, 99.5% of the available uranium and plutonium is recovered with single-pass recycling.

2.3.1.1 Uranium Ore and Chemical Processing

The average uranium content of US ores, expressed as the oxide U_3O_8, is about 0.2 wt% (i.e., 1000 kg of ore contains the uranium equivalent of 2 kg of U_3O_8). After processing, the raw ore is converted to U_3O_8 or yellowcake. The U_3O_8 is reduced to UO_2 (brown oxide) utilizing hydrogen gas.

Uranium tetrafluoride (UF_4) or green salt is formed by heating uranium dioxide in hydrogen fluoride gas. The tetrafluoride is subsequently converted into uranium hexafluoride using fluorine gas. The hexafluoride form is the working fluid used in the uranium enrichment step for gaseous diffusion, gas centrifuge, and some laser enrichment approaches.

The mining, milling, and chemical conversion steps of the fuel cycle primarily create an internal intake hazard. Worker intakes are minimized using engineering controls or respiratory protection if controls are not practical.

2.3.1.2
Overview of Existing Enrichment Technologies

The next step in the uranium fuel cycle is the enrichment of the ^{235}U content from its nominal value of 0.72% by weight in natural uranium to 3–5 wt% reactor grade material. The traditional methods for enriching uranium are the gaseous diffusion and gas centrifuge technologies.

2.3.1.2.1
Gaseous Diffusion

Isotopic separation by the diffusion process is accomplished by flowing uranium hexafluoride gas through a porous membrane. The various uranium isotopes reach an equilibrium condition after numerous collisions as gas flows into the separation device. With equilibrium established, each UF_6 molecular form has an equal momentum ($\vec{p}$):

$$\vec{p}(^{234}UF_6) = \vec{p}(^{235}UF_6) = \vec{p}(^{238}UF_6) \tag{2.1}$$

Since the product of the molecular mass (m) and velocity (v) is the momentum, the diffusion velocity through a membrane is inversely proportional to the molecular mass.

The different molecular weights of $^{235}UF_6$ and $^{238}UF_6$ and the resulting difference in molecular velocities are used as the basis for separating ^{235}U from ^{238}U. In a mixture of $^{235}UF_6$ and $^{238}UF_6$, the average speed of the lighter $^{235}UF_6$ molecules is greater than that of the heavier $^{238}UF_6$ molecules. When the mixture contacts a porous barrier, the lighter $^{235}UF_6$ molecules strike the barrier and diffuse through it more quickly than the heavier $^{238}UF_6$ molecules. Since the velocity difference is small, the enrichment through each gaseous diffusion chamber or stage is small. Consequently, thousands of stages are required to increase the assay from 0.72 wt% ^{235}U to the desired enrichment of 3–5 wt% for power reactor use. The ^{234}U content is also enriched in a gaseous diffusion facility because the technology is based on molecular mass differences. Since the uranium specific activity is increased following enrichment, the product material has a greater radiological hazard than feed material.

A gaseous diffusion stage consists of a motor, compressor, and converter that contains the porous barrier or membrane and a cooler. Uranium hexafluoride is introduced as a gas and flows through the inside of the barrier tube. A portion of the gas, about half, diffuses through the barrier and is fed to the next higher (increased ^{235}U enrichment) stage. The remaining gas that did not diffuse through the tube is fed to the next lower enrichment stage. The diffused or product stream is slightly enriched in ^{235}U, and the gas remaining in the tube is slightly depleted in ^{235}U.

The stages above the location of feed entry are the enriching section, and the ^{235}U concentration exceeds that of the nominal feed concentration. In the stripping section, below the feed point, the concentration of ^{235}U is less than the nominal feed concentration. The enrichment increases (decreases) the further the stage is upstream (downstream) of the feed point.

The feed for each stage in the gaseous diffusion cascade is a mixture of the enriched material from the stage immediately below and the depleted material from the stage immediately above. The cascade operates continuously with the addition of new feed material.

The number of stages in a cascade is a function of a number of variables including the isotopic concentration of the feed material, the desired product and tails concentrations, and the efficiency of the diffusion barrier material. For a typical application of natural uranium feed material, reactor grade fuel product, and a tails assay of 0.2 wt% ^{235}U, about 2000 stages are required.

The number of stages could be altered if the product or tails assay were adjusted. For example, the number of stages is reduced if the ^{235}U content of the tails material is increased. However, this change rejects a larger amount of ^{235}U eliminated as tails material.

In a gaseous diffusion plant, an acute exposure results from a release of uranium hexafluoride from the process equipment. Chronic exposures arise from routine maintenance or processing operations. The radiological hazard varies with the ^{235}U enrichment, and internal doses are the primary concern.

Minor leakage of UF_6 from process equipment leads to surface contamination that can become an airborne hazard. Health physics hazards are minimized by timely maintenance of leaking components and decontamination of affected areas. External radiation concerns are managed by good radiological controls practices and dose reduction measures.

The uranium feed materials for the enrichment process may include small quantities of neptunium and plutonium if reprocessing is incorporated into the fuel cycle. Good radiological controls practices are usually adequate to control the presence of these transuranium contaminants. However, these transuranics (TRUs) represent a significant internal radiation hazard. Their specific activities and annual limits on intake are generally more limiting than the uranium isotopes.

For low ^{235}U enrichments, chemical toxicity remains the controlling hazard. At higher enrichments, radiation effects become the primary concern. Criticality must also be considered in the higher enrichment stages primarily at stages near the top of the cascade.

Most of the chemical compounds encountered in a gaseous diffusion plant, including uranium hexafluoride and uranyl fluoride, are ICRP 30 Class D compounds. Interactions of these materials with the process equipment and the environment can produce Class W compounds during normal and abnormal operations.

In a gaseous diffusion plant, or other facility utilizing uranium hexafluoride, the probability of a criticality is minimized by controlling the process parameters to prevent the solidification of the UF_6. The integrity of the process stages is also maintained to prevent the intrusion of water or moist air.

Radiation monitors located in key locations provide early detection of an accumulation of solidified uranium hexafluoride. For plant components containing uranium solutions or storing uranium compounds, various criticality controls are applied. These controls include geometry and batch control, limitations on the uranium concentrations and enrichment, and administrative or procedural controls.

The primary radiological hazards from a criticality event are neutron and photon radiation to personnel in the immediate vicinity of the event. Timely evacuation of personnel is an effective radiation control measure. Criticality alarms do not prevent an inadvertent criticality, but they facilitate the evacuation of personnel from the immediate area of the event.

2.3.1.2.2
Gas Centrifuge

Gas centrifuge technology utilizes uranium hexafluoride as its working fluid. Consequently, the health physics considerations are similar to those in a gaseous diffusion facility.

As applied to uranium enrichment, a centrifuge is a cylindrical device that rotates about its long axis. Its enrichment capacity increases with the length of the device, the radius of the device, and an increase in its speed. Limits in material properties restrict the available values of these parameters. As noted in Table 2.1, the actual design of a centrifuge depends upon the enrichment desired, the technology level of the group developing the device, and the materials available for device construction.

The centrifugal force (F) imposed on a molecule within a centrifuge rotating at a velocity v is

$$F = \frac{mv^2}{r} \tag{2.2}$$

where m is the molecular mass and r is the radius of its circular path relative to the machine's axis. For a given centrifuge design, heavier molecules (i.e., $^{238}UF_6$) are subjected to a larger force and tend to be moved to a larger radial distance than lighter molecules (i.e., $^{234}UF_6$ and $^{235}UF_6$). This difference in trajectories permits the separation of the heavier and lighter UF_6 molecules. The mass difference of the various UF_6 molecules is the basis for the use of a centrifuge for the enrichment of uranium isotopes.

The working fluid in a gaseous centrifuge is composed of primarily $^{235}UF_6$ and $^{238}UF_6$. Consequently, when normal uranium hexafluoride is centrifuged, material drawn off from the interior region will be somewhat enriched in the lighter ^{235}U isotope. $^{234}UF_6$ is also enriched.

Table 2.1 Characteristics of selected centrifuge designs[a].

Type	Original machine designation	Deployment period	Rotor characteristics				
			Material	Speed (m/s)	Diameter (cm)	Length (m)	Separative power (SWU/year)
—	Zippe	1940–1950s	Aluminum	350	7.4	0.3	0.44
P-1	SNOR/CNOR	1960–1970s	Aluminum	350	10	2.0	2–3
P-2	G-2	1960–1970s	Maraging steel	485	15	1.0	5–6
P-3	4-M[b]	Early 1980s	Maraging steel	485	[c]	2.0	12
P-4	SLM (TC-10)[b]	Late 1980s	Maraging steel	500	15	3.2	21
—	TC-11[b]	Late 1980s	Carbon fiber	600	[c]	3.0	[c]
—	TC-12[b]	1990s	Carbon fiber	620	20[a]	3.0	40
—	TC-21[b]	2000s	Carbon fiber	770	20[a]	5.0	100
—	AC100[d]	2000s	Carbon fiber	900	60[a]	12.0	330

a) Derived from Glaser (2008).
b) URENCO designation.
c) Not provided by Glaser (2008).
d) USEC, American Centrifuge designation.

The uranium hexafluoride feed material is introduced at or near the axis of the device. Since separation is based on centrifugal force, the product withdrawal point is located at a smaller radius than the tails withdrawal location.

In order to obtain the desired ^{235}U enrichment, gas centrifuges are operated in a facility utilizing thousands of machines. Although the centrifuge machines operate in a manner similar to a gaseous diffusion cascade, the degree of ^{235}U enrichment per separation unit is greater for the centrifuge technology. However, twentieth-century centrifuge and gaseous diffusion facilities are large, easily observed structures. Their size provided a natural obstacle to the covert development of enrichment technology and diversion of that material for clandestine purposes (e.g., nuclear weapons production). As the centrifuge efficiency improves, the facility footprint decreases and its proliferation potential increases.

The health physics hazards encountered in a centrifuge facility are similar to those noted previously in the gaseous diffusion discussion. This is expected because the working fluids are the same, the separation basis is molecular mass, and the technologies require thousands of separating units to achieve a commercial, production-scale facility.

2.3.1.3
Nuclear Fuel

Fission reactor fuels usually consist of a mixture of fissile materials (^{233}U, ^{235}U, ^{239}Pu, and ^{241}Pu) and a fertile material (^{232}Th and ^{238}U). These materials are present or derived from naturally occurring uranium and thorium.

Power reactors utilize UO_2 or a combination of UO_2 and PuO_2 mixed oxide (MOX) fuel. MOX fuel derives its Pu content from fuel reprocessing or utilization of recycled nuclear weapons plutonium. Uranium and plutonium oxide fuels have good fission product retention characteristics.

2.3.1.4
Reactor Power Production

In the twentieth century, Generation II and III reactors dominated power production. About one-third of the operating reactors are boiling water reactors (BWRs) and two-thirds are pressurized water reactors (PWRs). The discussion in this section is based on PWR terminology. Specific BWR terminology is outlined in the subsequent discussion.

Reactor power production creates a variety of fission and activation products. These nuclides are produced following the fission of uranium and plutonium. The fission reaction produces intense neutron and photon radiation that escapes from the fuel core that is enclosed within a reactor vessel. Fission products are also produced, and barriers are required to prevent their release to the environment.

Uranium and plutonium oxides are formed into fuel pellets enclosed within a long tube. These tubes are bundled to form fuel assemblies that form the reactor core. The tube and fuel pellets form the first fission product barrier preventing the release of fission products to the environment.

The reactor vessel and its included piping comprise the primary system that provides water to cool the core. This piping and the reactor vessel are the second fission product barrier. A third fission product barrier, the containment building, encloses the primary coolant system to further limit the release of fission products to the environment.

Reactor operations produce radiological conditions that must be controlled. Power operations require the monitoring and control of neutron, alpha, photon, and beta radiation types. During outage periods, maintenance activities involving primary system components are significantly impacted by the presence of fission and activation products.

Power is produced by fission of the various fuel materials. Thermal neutrons produce fission in ^{233}U, ^{235}U, ^{239}Pu, and ^{241}Pu. Fast neutrons cause fission of ^{232}Th and ^{238}U. For example, a Generation II LWR derives most of its power from the thermal fission of ^{235}U and from the ^{239}Pu produced during its operating cycle. The fast fission of ^{238}U also contributes to power production, but its contribution is much less than the contribution from the thermal fission of ^{235}U and ^{239}Pu.

Table 2.2 Composition of spent light water reactor fuel[a].

Fuel constituent	Fuel constituent (wt%)	
	Initial	Final
^{238}U	97	95
^{236}U	—	0.4
^{235}U	3	0.8
^{239}Pu and ^{241}Pu	—	0.65
^{240}Pu and ^{242}Pu	—	0.25
Fission products	—	2.9

a) Derived from Glasstone (1982).

2.3.1.5
Spent Fuel

Spent fuel is the designation for fuel following its discharge from a power reactor. The specific isotopic content of spent fuel is affected by the mode of reactor operation including the duration of the operating cycle and the initial fuel characteristics. Operating cycles are typically 12–24 months in duration. For 3 wt% ^{235}U in the initial fuel assembly, the discharged fuel after a 12-month operating cycle has the isotopic composition summarized in Table 2.2.

Table 2.2 illustrates the depletion of ^{235}U and production of ^{239}Pu in a Generation II LWR during its operating cycle. Spent fuel contains significant ^{235}U and ^{239}Pu that provides an economic and environmental motivation for reprocessing to recover these fissile materials. However, proliferation concerns prevent fuel reprocessing from inclusion in the fuel cycle utilized by many nations including the United States.

The licensing of spent fuel repositories is complicated by the presence of actinides in spent fuel. Remnant neptunium, plutonium, americium, and curium in spent fuel or reprocessing waste solutions affect the licensing basis of fuel repositories. These actinides could be removed by Generation IV technologies. Table 2.3 provides examples of actinides that are important from a fuel cycle and health physics perspective. The half-life, production mode, and decay mode are presented for selected radionuclides.

2.3.1.6
Fuel Reprocessing

Nuclear fuel reprocessing is the treatment of used or spent reactor fuel to recover fissile and fertile materials. Following irradiation, spent fuel assemblies are removed from the reactor and transferred to a water-filled storage pool. After the decay heat sufficiently decreases (typically 12 months or longer), the fuel assemblies are transported to a reprocessing facility. At this facility, the spent fuel assemblies are shredded and immersed in hot, concentrated nitric acid. The acid chemically reacts with the uranium dioxide fuel and included isotopes,

Table 2.3 Key actinides encountered in spent nuclear fuel.

Nuclide	Half-life (year)	Production mode	Dominant decay mode
^{235}U	7.04×10^{8}	Naturally occurring	$^{235}\text{U} \rightarrow {}^{231}\text{Th} + \alpha$
^{238}U	4.47×10^{9}	Naturally occurring	$^{238}\text{U} \rightarrow {}^{234}\text{Th} + \alpha$
^{237}Np	2.14×10^{6}	$^{235}\text{U} + \text{n} \rightarrow {}^{236}\text{U} + \text{n} \rightarrow {}^{237}\text{U} \xrightarrow{\beta^-} {}^{237}\text{Np}$	$^{237}\text{Np} \rightarrow {}^{233}\text{Pa} + \alpha$
^{238}Pu	87.7	$^{237}\text{Np} + \text{n} \rightarrow {}^{238}\text{Np} \xrightarrow{\beta^-} {}^{238}\text{Pu}$	$^{238}\text{Pu} \rightarrow {}^{234}\text{U} + \alpha$
^{239}Pu	24 100	$^{238}\text{U} + \text{n} \rightarrow {}^{239}\text{U} \xrightarrow{\beta^-} {}^{239}\text{Np} \xrightarrow{\beta^-} {}^{239}\text{Pu}$	$^{239}\text{Pu} \rightarrow {}^{235}\text{U} + \alpha$
		$^{237}\text{Np} + \text{n} \rightarrow {}^{238}\text{Np} + \text{n} \rightarrow {}^{239}\text{Np} \xrightarrow{\beta^-} {}^{239}\text{Pu}$	
^{240}Pu	6.56×10^{3}	$^{239}\text{Pu} + \text{n} \rightarrow {}^{240}\text{Pu}$	$^{240}\text{Pu} \rightarrow {}^{236}\text{U} + \alpha$
		$^{239}\text{Np} + \text{n} \rightarrow {}^{240}\text{Np} \xrightarrow{\beta^-} {}^{240}\text{Pu}$	
^{241}Pu	14.29	$^{240}\text{Pu} + \text{n} \rightarrow {}^{241}\text{Pu}$	$^{241}\text{Pu} \xrightarrow{\beta^-} {}^{241}\text{Am}$
^{242}Pu	3.75×10^{5}	$^{241}\text{Pu} + \text{n} \rightarrow {}^{242}\text{Pu}$	$^{242}\text{Pu} \rightarrow {}^{238}\text{U} + \alpha$
		$^{242}\text{Am} \xrightarrow{\beta^-} {}^{242}\text{Pu}$	
^{241}Am	432.7	$^{241}\text{Pu} \xrightarrow{\beta^-} {}^{241}\text{Am}$	$^{241}\text{Am} \rightarrow {}^{237}\text{Np} + \alpha$
^{242m}Am	141	$^{241m}\text{Am} + \text{n} \rightarrow {}^{242}\text{Am}$	$^{242m}\text{Am} \rightarrow {}^{238}\text{Np} + \alpha$
^{243}Am	7.37×10^{3}	$^{242}\text{Pu} + \text{n} \rightarrow {}^{243}\text{Pu} \xrightarrow{\beta^-} {}^{243}\text{Am}$	$^{243}\text{Am} \rightarrow {}^{239}\text{Np} + \alpha$
^{243}Cm	29.1	$^{241}\text{Am} + \text{n} \rightarrow {}^{242}\text{Am} + \text{n} \rightarrow {}^{243}\text{Am} \xrightarrow{\beta^-} {}^{243}\text{Cm}$	$^{243}\text{Cm} \rightarrow {}^{239}\text{Pu} + \alpha$
^{244}Cm	18.1	$^{243}\text{Cm} + \text{n} \rightarrow {}^{244}\text{Cm}$	$^{244}\text{Cm} \rightarrow {}^{240}\text{Pu} + \alpha$
^{245}Cm	8.50×10^{3}	$^{244}\text{Cm} + \text{n} \rightarrow {}^{245}\text{Cm}$	$^{245}\text{Cm} \rightarrow {}^{241}\text{Pu} + \alpha$
^{246}Cm	4.77×10^{3}	$^{245}\text{Cm} + \text{n} \rightarrow {}^{246}\text{Cm}$	$^{246}\text{Cm} \rightarrow {}^{242}\text{Pu} + \alpha$

and uranium and plutonium nitrates are extracted. However, the acid does not dissolve the fuel cladding.

The unique valence states of the uranium, plutonium, and fission product nitrate compounds form the basis for the chemical separation of uranium and plutonium. These include hexapositive(VI), tetrapositive(IV), and tripositive(III) states of uranium and plutonium nitrates. Separation of uranium and plutonium from the nitric acid solution is achieved because the VI and IV valence states of uranium and plutonium are soluble in certain organic solvents. However, the III valence states of uranium, plutonium, and fission product nitrates are not soluble in these solvents.

In 1977, proliferation concerns prompted the United States to postpone the reprocessing of nuclear fuel from LWRs. In addition, no US commercial plutonium breeder reactors are in operation. Without breeder reactors, the fuel cycle does not require a fuel reprocessing step. Therefore, the current US fuel cycle does not include fuel reprocessing.

With the current mix of LWRs and political concerns regarding nuclear proliferation, the US fuel cycle is incomplete and fails to operate in the most efficient or environmentally sound manner. Other nations reprocess fuel, which permits a more efficient use of available uranium and plutonium resources and development of a more environmentally friendly fuel cycle.

As noted previously, two fuel reprocessing options were utilized in the twentieth century. There are two additional fuel reprocessing options. Each option has unique aspects, and the selection of a fuel cycle reprocessing option depends on government policy, public support, political will, and available infrastructure. Given current technology levels, only the first two options are currently in use. Options 3 and 4 require additional technology deployment and are addressed in the subsequent discussion.

2.3.1.6.1
Single-Pass Fuel Cycle

Option 1 is a once-through fuel cycle in which reactor spent fuel is not reprocessed. Most of the reactors supporting this fuel cycle are LWRs and high-temperature gas-cooled reactors (HTGRs). Pebble bed modular reactors (PBMRs) and thorium cycle reactors are also included within this option. The most significant drawback of the once-through fuel cycle is that it generates the most radioactive waste and the unburned ^{235}U and ^{239}Pu are not recovered. In addition, the presence of actinides and long-lived fission products complicates the licensing basis of an Option 1 spent fuel repository.

The fuel cycle in a number of countries including the United States, Canada, Spain, and Sweden is truncated. In Option 1, fuel discharged from a commercial light water power reactor is stored until it can be placed in a long-term geologic repository. The residual fissile material contents in the spent fuel of about 1 wt% ^{235}U and ^{239}Pu as well as fertile ^{238}U material are not recovered and are considered waste.

The short-term (<1000 years) source term of the Option 1 fuel cycle includes fission products. After a few hundred years, most fission products have decayed except for long-half-life radionuclides including ^{90}Sr (28.8 years), ^{93}Zr (1.5×10^6 years), ^{129}I (1.57×10^7 years), ^{135}Cs (2.3×10^6 years), and ^{137}Cs (30.07 years).

As noted in Table 2.3, the long-term (>1000 years) radiotoxicity of the spent fuel is primarily associated with Np, Pu, Am, and Cm. These elements will comprise a major portion of the source term for a geologic repository.

One of the disadvantages of Option 1 is that there is no international agreement for the confinement time required for a geologic repository. Times of $10^4 - 10^6$ years or longer have been proposed.

Option 1 has a number of health physics and radiological safety issues. Criticality is a potential concern because there are significant quantities of residual ^{235}U and ^{239}Pu in the fuel. During pool storage, decay heat from the fission product and actinides presents a driving mechanism for the dispersion of radionuclides. External radiation levels near spent fuel and waste reprocessing solutions are dominated by ^{90}Sr and ^{137}Cs for a few hundred years. Over this period, internal doses are dominated by actinides, particularly ^{239}Pu and ^{241}Am. Following the decay of ^{239}Pu and ^{241}Am, ^{129}I transport to the environment is a release pathway of concern. Release pathways must consider other long-lived radionuclides including ^{99}Tc (2.13×10^5 years), ^{242}Pu (3.75×10^5 years), and ^{237}Np (2.14×10^6 years).

2.3.1.6.2
Partial Uranium and Plutonium Recycle Fuel Cycle

Option 2 involves the partial recycle of uranium and plutonium fissile material. The recycled U and Pu are fabricated into MOX fuel. In the second fuel cycle option, reactors burn MOX fuel using uranium and plutonium recovered from reprocessed LWR fuel. However, residual uranium and plutonium, minor actinides (e.g., Np, Am, and Cm), and fission products remain in the high-level waste.

Reprocessing to recover the residual ^{235}U and produced ^{239}Pu was utilized in the twentieth century as a fuel cycle option. Option 2 is the fuel cycle with reprocessing and is utilized by a number of countries including China, France, India, Japan, Russia, and the United Kingdom. Concerns over nuclear proliferation reduce the popularity of Option 2 and have limited its development. These proliferation concerns are often raised because Option 2 recovers about 99.5–99.9% of the uranium and plutonium from the spent fuel via reprocessing.

In Option 2, mixed plutonium and uranium oxide fuel undergoes a single recycle wherein the discharged fuel is separated into three product streams. Recovered irradiated uranium is placed into interim storage, fission products and minor actinides are converted to a glass waste form for disposal, and plutonium is recovered and used to fabricate MOX fuel for a single return pass through a reactor. The discharged MOX fuel is placed in interim storage awaiting further disposition options. The uranium oxide once-through cycle has been the baseline US approach, and the MOX monorecycle option is in commercial deployment in portions of Asia and Europe.

Option 2 includes a step to vitrify high-level waste, which is an irreversible process. This is neither the most environmentally friendly solution nor one that permits recovery of valuable resources. If the waste has not been reprocessed and vitrified, then the spent fuel should be stored in an engineered facility until a long-term actinide management strategy is developed. In lieu of an actinide management strategy, retrievable storage in an underground facility has merit as an intermediate solution.

Option 2 reduces the actinide content of high-level waste. However, the recycling of uranium and plutonium into a MOX fuel cycle is at best a partial solution to reduce the volume of actinide waste. Recycling of plutonium in LWRs using MOX fuel is not efficient in reducing the minor actinide (MA) composition of high-level waste.

Traditional reprocessing removes uranium and plutonium, which comprises about 30% of the total alpha activity. The remaining 70% of the alpha activity in high-level waste is attributed to Am and Cm with a much smaller Np contribution. Thermal neutron reactions alone will not remove the Np, Am, and Cm activity. In view of this situation, spent LWR MOX fuel should be stored in engineered facilities until fast nuclear reactors can be developed to process the LWR fuel with high actinide content.

Fuel cycle Option 3 requires multipass reprocessing of high burnup spent fuel to remove uranium and TRU elements to produce vitrified waste or process these materials into another acceptable waste form. The minor actinides would be eventually burned in a fast reactor (Option 4).

The development of fuel types containing increased plutonium and minor actinide (PMA) fractions is a significant challenge. However, advanced Generation IV reactor concepts and fuel cycles provide the opportunity to explore new fuel types that facilitate recycling of fuel over multiple cycles. Additional fuel cycle stages could include accelerator systems to burn the minor actinides and long-lived fission products as well as recover valuable isotopes using laser technology. These Generation IV reactors and accelerator systems are addressed in Sections 2.4.2.2 and 2.7, respectively.

2.3.1.6.3
Uranium and Plutonium Recycle Fuel Cycle

The third option involves recycling uranium and plutonium. In Option 3, fast reactors support LWRs with excess fissile material being recycled to additional LWRs or PBMRs. Less waste is generated than in Option 2, but the waste products are similar and the minor actinides remain. Option 3 utilizes multiple fuel reprocessing passes.

2.3.1.6.4
Full Recycle Fuel Cycles

Option 4 recycles actinides and fissile material, and a number of operating modes are envisioned. These include LWRs supported by fast reactors or molten salt reactors. Option 4 generates less waste than Option 3. The Option 4 waste contains minimal minor actinides and some fission products.

The key aspect of these fuel cycle options is to reduce the decay heat from ^{137}Cs, ^{90}Sr, and actinides. If the actinides are transmuted, the long-term decay heat issues are minimized. There are a number of options for cesium and strontium management including their separation and storage until the decay heat is reduced. By removing the cesium and strontium and minor actinides, the licensing requirements for waste storage facilities are significantly reduced.

2.3.1.7
Radioactive Waste

One of the consequences of power production in a nuclear reactor and subsequent spent fuel reprocessing is the generation of radioactive waste or effluents that may be released from the facility. A summary of these waste and effluent isotopes typically associated with the uranium fuel cycle is summarized in Table 2.4. This listing

Table 2.4 Radioactive waste and effluent isotopes from the nuclear fuel cycle[a].

Fuel cycle stage	Waste form		
	Gas	Liquid	Solid
Mining and milling	^{222}Rn and daughters (^{214}Pb, ^{214}Bi, ^{214}Po, and ^{218}Po)	—	U, ^{210}Pb, ^{226}Ra, and ^{230}Th
Refining	—	^{226}Ra, ^{234}Pa, ^{234}Th, and ^{238}U	—
Fuel fabrication	—	—	Th, U, and Pu
Reactor operation	^{41}Ar, ^{85}Kr, ^{87}Kr, ^{89}Kr, ^{129}I, ^{131}I, ^{133}Xe, ^{135}Xe, and ^{138}Xe	^{3}H	^{58}Co, ^{60}Co, ^{59}Fe, and ^{51}Cr
Chemical processing	^{3}H, ^{85}Kr, ^{129}I, ^{131}I, and ^{133}Xe	Fission products dissolved in acid solutions	Np, Pu, Am, Cm, and fission products

a) Cember and Johnson (2008).

is not complete but is indicative of the isotopes that will be encountered in the various fuel cycle activities. The isolation of radioactive waste to prevent public access and potential releases to the environment are key health physics concerns.

2.3.1.7.1
Low-Level Radioactive Waste

Radioactive wastes are loosely characterized as either high level or low level. Low-level wastes include contaminated articles of disposable protective clothing, spent ion exchange resins, trash, animal carcasses, or other items commonly used in a nuclear power facility and medical, research, or industrial environments. High-level waste is addressed in Section 2.3.1.7.3.

2.3.1.7.2
Transuranic Waste

Another broad waste category associated with the nuclear fuel cycle is TRU wastes. TRU wastes contain appreciable quantities of elements heavier than uranium (e.g., neptunium, plutonium, americium, and curium).

2.3.1.7.3
High-Level Waste

High-level wastes include spent nuclear fuel and the wastes associated with their reprocessing. Both of these waste streams contain fission products with their associated beta–gamma activity and actinides. As part of spent fuel reprocessing, fissile material is extracted. The remaining waste contains fission products, minor actinides, and residual uranium and plutonium. These waste solutions remain a long-term hazard unless the minor actinides and residual plutonium are removed.

2.3.2
Thorium Fuel Cycle

Thorium is periodically proposed as a new energy source for power production because it is three to four times more abundant than uranium. It has been proposed as a Generation III reactor fuel as well Generation IV fuel in molten salt reactors. Small modular reactors (SMRs), including high-temperature gas-cooled reactors, are also being developed with thorium fuel.

Natural thorium is dominated by ^{232}Th that has a relatively small fast fission cross-section relatively to the thermal fission of ^{235}U or ^{239}Pu. However, when irradiated by neutrons, ^{232}Th is converted into fissile ^{233}U, which has a large thermal fission cross-section and a half-life of 1.592×10^5 years:

$$^{232}\mathrm{Th} + \mathrm{n} \rightarrow {}^{233}\mathrm{Th} \xrightarrow{\beta^-} {}^{233}\mathrm{Pa} \xrightarrow{\beta^-} {}^{233}\mathrm{U} \tag{2.3}$$

An additional reaction product of ^{232}Th irradiation is the production of ^{232}U whose daughters yield a large photon radiation component. Proponents of the thorium fuel cycle argue that spent thorium fuel with its intense photon radiation creates a natural resistance to proliferation. However, processes can be designed to minimize ^{232}U production or chemically separate and rigorously control ^{233}Pa to alter the proliferation potential of ^{233}U production.

The thorium fuel cycle requires reprocessing to recover the fissile ^{233}U. In the early stages of the use of the thorium cycle, some ^{235}U would be used to ensure sufficient fission neutron production to produce an adequate yield of ^{233}U. Reprocessing recovers ^{233}U in a manner analogous to ^{239}Pu recovery in the uranium fuel cycle. ^{233}U represents a proliferation concern because the International Atomic Energy Agency (IAEA) suggests that about 8 kg of ^{233}U is sufficient to construct a nuclear weapon.

Thorium represents an alternative to uranium, but its fuel cycle has not been as widely utilized as the uranium fuel cycle. However, the health physics concerns of the uranium and thorium fuel cycles are similar. The proliferation aspects of the nuclear fuel cycle are addressed in the subsequent discussion.

2.4
Twenty-First-Century Changes and Innovations

The baseline twentieth-century fuel cycle is enhanced by technology currently in existence or likely to be developed in the twenty-first century. Uranium enrichment technology has progressed using advanced centrifuge technology and laser techniques. These technologies offer the potential for a significant cost reduction in enrichment services but also have a negative proliferation aspect.

Significant fuel cycle changes that will likely occur in the twenty-first century include the continued deployment of Generation III reactors and development and licensing of Generation IV and SMRs. Additional innovations are possible in the reprocessing of spent fuel using lasers. Advanced reprocessing technologies also offer the potential for the enhanced recovery of uranium and plutonium, separation of valuable metals and medical isotopes, and destruction of minor actinides and long-lived fission products. These enhancements are discussed in the subsequent sections of this chapter.

2.4.1 Advanced Uranium Enrichment Technologies

The number of operating reactors and their characteristics drives the demand for uranium enrichment services. Even without a significant increase in the number of operating reactors, providers of uranium enrichment services are driven to lower costs and improved efficiency. Economic considerations have led to the transition from gaseous diffusion to centrifuge enrichment.

Improved centrifuges continue to evolve, and laser methods have achieved their initial commercial licensing. Additional techniques may also emerge, but this chapter focuses on advanced centrifuge and laser enrichment techniques.

2.4.1.1 Advanced Centrifuge Technology

The gas centrifuge for uranium enrichment has assumed an increasingly important role in the nuclear fuel cycle. This is attributed to their improving efficiency and economics compared to conventional gaseous diffusion technology. The improved efficiency of gas centrifuge machines is accompanied by nuclear proliferation concerns and the possibility of covert construction or converting a civilian enrichment facility into one that produces highly enriched uranium (HEU).

The proliferation concerns associated with advanced centrifuge machines limits discussion of an operational envelope and associated characteristics. Since available information is limited, only general machine characteristics and operating parameters are provided through reference to open literature sources.

Numerous centrifuge designs have been developed, and these incorporate a variety of rotor materials, lengths, and speeds. Table 2.1 lists estimated design characteristics of historical and advanced centrifuge machines. For consistency with the literature, common shorthand notation (P-1, P-2, etc.) is used to refer to some of these machines.

The P-1 design is based on early Dutch machines designated scientific nuclear orbital rotor (SNOR) and cultivated nuclear orbital rotor (CNOR) developed by an international corporation URENCO. URENCO operates enrichment facilities in Germany, the Netherlands, the United Kingdom, and the United States.

The P-2 machine is a modified version of the German G-2 centrifuge that was a pre-URENCO design. As noted in Table 2.1, it has an improved capability for performing separative work than the P-1 machine. Separative work is defined in Eq. (2.4). The P-3 and P-4 designs represent evolving URENCO machines.

The additional machines listed in Table 2.1 illustrate the increasing enrichment power of centrifuge designs. TC-11, TC-12, and TC-21 are URENCO machines with carbon fiber construction. The estimated separative power of the American Centrifuge (AC100) has significantly more enrichment capability than the URENCO devices.

Publicly available information for other centrifuge machines listed in Table 2.1 is more uncertain. Glaser estimated the characteristics of the advanced centrifuge designs based on known characteristics of older machines and projected improvements.

Table 2.1 provides machine-specific information including characteristics of the rotor (e.g., materials of construction, physical envelope, and operating parameters). The machine's capacity is also provided in terms of separative work units (SWU)/year.

A SWU is a measure of work (W) or separative power expended by an enrichment device to separate a mass m_F of assay (in wt%) x_F into a mass m_P of product with assay x_P and mass m_T of tails with assay x_T:

$$W_\mathrm{SWU} = m_\mathrm{P} V(x_\mathrm{P}) + m_\mathrm{T} V(x_\mathrm{T}) - m_\mathrm{F} V(x_\mathrm{F}) \tag{2.4}$$

where $V(x)$ is the value function defined by the relationship

$$V(x) = (1 - 2x) \ln\left(\frac{1 - x}{x}\right) \tag{2.5}$$

The masses are often expressed in terms of annual throughput with units of kg/year. Another parameter used to describe an enrichment device is the cut (θ) which is the fraction of the feed material that leaves the device as product ($m_\mathrm{P} = \theta m_\mathrm{F}$). Centrifuge cascades typically have cut values in the range of 0.4–0.5.

Machines with larger separative power produce more separative work. A device with a larger separative power will more easily increase the enrichment of an isotope. Machines with larger separative power permit an enrichment facility to be smaller and more efficient. This increased efficiency and smaller size has the potential for an increased proliferation potential and the clandestine use of the facility for the production of weapons-grade materials. A comparison of the various enrichment technologies in terms of their separation factor, number of separating units, energy consumption, and relative capital costs is provided in Table 2.5.

2.4.1.2
Laser Isotope Separation

Laser methods for uranium enrichment are an emerging commercial technology and are projected to be more efficient and economical than the gaseous diffusion and gas centrifuge processes. As noted in Table 2.5, electrical energy consumption

Table 2.5 Comparison of potential commercial enrichment technologies[a].

Enrichment process	Separation factor	Number of equipment units	Energy consumption (kW h/SWU)	Capital costs
Gaseous diffusion	1.004	Thousands	2400	Reference cost
Advanced gas centrifuge	1.3	Hundreds to thousands	~100	Higher than diffusion
Molecular laser isotope separation	2–6	<4	~100	Lower than diffusion
Atomic vapor laser isotope separation	2–6	<4	~100	Lower than diffusion

a) Bevelacqua (2014b).

is expected to be significantly less than used in existing technologies. With its significantly higher enrichment factors, laser technology could recover the residual ^{235}U residing in the tailings from either diffusion or centrifuge plants.

The separative power of a uranium enrichment device is defined as a function of its design and operational parameters and its capability to increase the ^{235}U concentration in the product material. Parameters that define the separative power include the enrichment factor α and the depletion factor β:

$$\alpha = \frac{x_P/(1 - x_P)}{x_F/(1 - x_F)} \tag{2.6}$$

and

$$\beta = \frac{x_F/(1 - x_F)}{x_T/(1 - x_T)} \tag{2.7}$$

Equivalently, the capability of the enrichment device can be defined by its separation factor (S):

$$S = \alpha\beta = \frac{x_P/(1 - x_P)}{x_T/(1 - x_T)} \tag{2.8}$$

Applicable separation factors and associated facility characteristics for various enrichment technologies are summarized in Table 2.5. The parameters summarized in Table 2.5 support the previous discussion regarding the efficiency of laser uranium enrichment technology and its potential for reduced reactor fuel costs.

Laser isotope separation techniques rely on the property that different isotopic species, in either an atomic or a molecular form, exhibit small differences in their atomic or molecular energy level spectra. Equivalent transitions from one energy level to another are isotope specific and require a different energy to induce the transition. Thus, selective excitation is possible, and this property is a critical factor in the viability of laser isotope separation. With laser techniques, the enrichment of the ^{235}U isotope is not accompanied by ^{234}U enrichment. This is a significant change from the gaseous diffusion and gas centrifuge uranium enrichment technologies.

In order to utilize the selective excitation property, an energy source is tuned to the desired excitation energy. Lasers offer a useful tool for this selective excitation.

A laser is a source of radiation that can be designed to operate at a specified frequency and intensity. Therefore, it is possible to preferentially excite one isotopic species via a precisely tuned laser and leave other isotopic species in their ground states.

Two general laser techniques are under evaluation for the enrichment of ^{235}U. One technique involves the use of uranium vapor, and it is based on the selective photoionization of ^{235}U atoms. A second method of laser enrichment is based on the photodisintegration of $^{235}UF_6$ molecules. The molecular laser isotope separation (MLIS) and atomic vapor laser isotope separation (AVLIS) technologies will be briefly addressed. The first commercial deployment of the molecular process is also discussed.

2.4.1.2.1 MLIS

In the molecular process, an infrared (IR) laser is utilized to preferentially excite the $^{235}UF_6$ vibrational energy levels until the molecule dissociates:

$$^{235}UF_6 + \gamma_{IR} \rightarrow {}^{235}UF_6^* + n\gamma_{IR} \rightarrow {}^{235}UF_5 + F \tag{2.9}$$

Multiple infrared absorptions (n) are required to fully dissociate the excited $^{235}UF_6$ molecule. However, the $^{234}UF_6$ and $^{238}UF_6$ molecules are not excited and remain in their ground states.

The excitation process is based upon the inherent assumption that the UF_6 molecules are in their ground states before being illuminated by the laser radiation. Thus, it may be necessary to cool the molecules via flow through an expansion nozzle in order to ensure all molecules reside in their vibrational ground states.

The UF_6 dissociation is enhanced using additional laser types. For example, an ultraviolet laser could be utilized to dissociate the vibrationally excited $^{235}UF_6^*$ molecule

$$^{235}UF_6^* + \gamma_{UV} \rightarrow {}^{235}UF_5 + F \tag{2.10}$$

Once formed, the $^{235}UF_5$ molecule precipitates as a solid and is collected. The unaffected $^{234}UF_6$ and $^{238}UF_6$ gas flows through the enrichment device and is separated from the solid $^{235}UF_5$ product.

The health physics concerns associated with gaseous diffusion and gas centrifuge noted previously apply to the molecular separation process. The use of laser components with high-voltage power supplies introduces X-ray and non-ionizing radiation hazards that need to be addressed. Other health physics issues associated with laser technology are addressed in the subsequent discussion.

2.4.1.2.2
SILEX

The first commercialization of the MLIS process is a joint Australian-General Electric-Hitachi venture located in Wilmington, North Carolina. The separation of isotopes by laser excitation (SILEX) is under the regulatory jurisdiction of the US Nuclear Regulatory Commission.

SILEX technology is a proprietary commercial uranium enrichment process. The subsequent discussion is derived from regulatory documentation and open literature sources.

SILEX is a variation of the MLIS process described previously. It is based on the selective excitation of $^{235}UF_6$ using 16 μm infrared laser radiation, which pumps energy into one of the $^{235}U_5$ + F bonds. The IR laser creates a $^{235}UF_5$ + F excited molecular state with the uranium pentafluoride–fluorine bond weakened by the absorption of IR radiation. A second laser adds sufficient energy to initiate a photochemical reaction that severs this bond to create a new $^{235}UF_5$ molecule and an unbound fluorine atom. The $^{235}UF_5$ particulate separates from the UF_6 gas, which forms the physical basis for ^{235}U enrichment.

Hecht notes one possible laser configuration uses a pulsed CO_2 laser. High-pressure para-hydrogen cells convert the 10.8 μm output to produce the desired 16 μm infrared radiation. SILEX is projected to increase the ^{235}U concentration by a factor of 2–20. However, licensing basis documents issued by the NRC only authorize enrichment of ^{235}U to 8 wt%.

As noted in Table 2.5, MLIS enrichment factors are higher than gas centrifuge (1.3) or gaseous diffusion (1.004). The higher enrichment factors projected for SILEX reduce costs, which would provide a significant economic advantage over the diffusion and centrifuge technologies. SILEX health physics issues are similar to those encountered in an MLIS facility and are reviewed in Section 2.4.1.2.4. Possible proliferation concerns are addressed in the subsequent discussion.

2.4.1.2.3
AVLIS

Laser enrichment methods also utilize uranium vapor as the working fluid. The use of metallic uranium affects the conventional fuel cycle chemical conversion requirements both prior to and post enrichment. Since UF_6 is not used as the working fluid, chemical conversion from U_3O_8 to UF_6 is not required. The extent of the chemical conversion changes will be governed by the manner in which this technology is implemented on a production scale.

In the AVLIS process, uranium metal is fed into a vacuum vessel where it is melted and then vaporized. Vaporization could be achieved using a variety of heat sources (e.g., conventional heating elements, sputtering methods, and electron beam impingement). The uranium vapor is illuminated by laser radiation tuned to selectively ionize only the ^{235}U atoms. Collection of the ^{235}U ions is accomplished using electromagnetic (EM) fields that alter the ion's trajectory. The unionized

^{234}U and ^{238}U atoms pass through the EM field region and are separately removed on a tails collector.

The AVLIS enrichment process produces both internal and external radiation hazards. Internal exposure is due to the alpha decay of ^{234}U, ^{235}U, and ^{238}U and their daughter products. In contrast to UF_6-based enrichment technologies, uranium metal oxidizes during maintenance activities. The AVLIS technology produces primarily ICRP 30 Class W material (UO_2, UO_4, and U_3O_8). Class D UO_3 may also be produced during the variety of maintenance activities required in an AVLIS facility. These oxides are an inhalation hazard if mobilized.

An external radiation hazard is created by photons and X-rays generated from the electron beam impingement on uranium metal; from the high-voltage equipment utilized by the laser systems, electron beam heating, and ion collection systems; and by the possibility of an inadvertent criticality event following the enrichment process. Uranium metal and its compounds also present a beta radiation hazard. These hazards and potential controls are summarized in the subsequent discussion.

2.4.1.2.4 Laser Enrichment Health Physics and Other Hazards

There are unique health physics hazards associated with laser enrichment. Unique hazards arise for a variety of reasons that include the use of uranium metal in the AVLIS process, vaporization of uranium, use of lasers to create ions and excited molecules, and collection of ^{235}U ions and $^{235}UF_5$ molecules. This section discusses the specific health physics aspects that are unique to uranium laser enrichment technologies. These aspects include (i) X-ray production, (ii) airborne radioactive material, (iii) nuclear criticality, (iv) laser hazards, (v) electromagnetic hazards, (vi) thermal hazards, and (vii) noise hazards. Reduced radiotoxicity created by the selective enrichment of ^{235}U is also addressed. Traditional uranium enrichment hazards encountered in gaseous diffusion and gas centrifuge facilities that were previously addressed are not specifically repeated.

X-ray Production X-rays are produced from stray currents in the high-voltage power supplies supporting the various laser systems and the AVLIS product collection system. The AVLIS vapor generation system also produces X-rays if it utilizes an electron beam to heat the uranium metal. The collection system is unique to the AVLIS technology that extracts ^{235}U ions using electromagnetic fields.

X-ray hazards are minimized using shielding and locating equipment in low occupancy locations. Shielding should be added to all high-voltage power supplies. The X-ray hazard is mitigated by locating high-voltage equipment in areas not usually occupied or in areas of restricted access. AVLIS vaporization units should also be shielded to minimize the X-ray hazard and have restricted access during enrichment operations to minimize worker doses.

Airborne Radioactive Material Gaseous diffusion and gas centrifuge enrichment facilities have numerous components (e.g., valves, compressors, and instrument lines) that leak and lead to surface contamination and airborne radioactive material. These hazards also exist in laser enrichment facilities. The AVLIS facility vaporizes uranium metal that coats the process vessel with a fine particulate layer. If the AVLIS separation unit is breached during operations (e.g., by a mechanical impingement accident, chemical reaction, thermal excursion, overpressure event, or laser-induced damage), this particulate material is released. Since uranium is pyrophoric, the fine particulate material will ignite and disperse oxidized uranium throughout the enrichment facility. As such, it presents a unique airborne hazard. Airborne uranium is also created if air is rapidly introduced into the AVLIS reaction chamber during maintenance operations.

Maintenance operations should proceed by slowly drawing air into the AVLIS separation unit. A controlled transition from vacuum pressures to normal pressures will minimize the pyrophoric reaction and its contribution to airborne radioactive material.

Airborne hazards are also minimized using traditional health physics controls. Airflow and ventilation systems are designed and located to minimize airborne radioactive materials. Localized ventilation is used to supplement installed systems to minimize airborne activity. In addition, exhaust air should be high-efficiency particulate air (HEPA) filtered and not recirculated from higher airborne concentration areas to lower airborne concentration areas. Alarming air monitors should be installed in areas where uranium dust may be present. These monitors warn personnel to exit process areas and minimize the potential for an intake of uranium. The system design should consider integral glove boxes or other confinement structures to facilitate maintenance and repair operations within the separation unit. These engineering controls provide a needed barrier to limit the internal intake of radioactive materials.

The airborne hazard becomes more significant if reprocessed uranium is used as the feed material. Dust and debris resulting from maintenance operations present disposal problems due to the presence of transuranium elements. This hazard only exists if the uranium feed includes recycled material derived from a fuel cycle that uses reprocessed uranium. The control measures noted previously require careful review to ensure their adequacy if PMAs are present in the feed and product material.

Nuclear Criticality The high separation factor expected in a laser enrichment facility suggests that HEU can be achieved in a single separation unit. This contention is supported by a recent study by the American Physical Society (APS). Higher ^{235}U enrichments increase the likelihood of a criticality event.

Neutron and gamma rays are produced by a criticality event that can occur with either the uranium metal or UF_5 product forms. The proper sizing and arrangement of transfer lines and storage containers minimize the probability of a criticality. Criticality alarms should be installed with detectors properly located in enriched materials storage areas or near enriched material transport areas. Since lethal absorbed doses are produced in a criticality, alarm systems can significantly reduce worker doses if personnel rapidly exit the separation unit area.

Laser Hazards Under normal conditions, laser radiation is directed into the separation unit. However, an abnormal event (e.g., facility accident, beam misalignment, or failure of the optical system) could redirect the laser radiation into occupied areas. This is a concern because high-intensity (Class 4) laser radiation damages the skin and eyes. In addition, reflected laser light can damage the skin and eyes.

A number of controls mitigate the laser radiation hazard. Beam tubes and optical transport systems are designed to keep laser radiation out of occupied or accessible areas. Interlocks and access controls preclude entry into areas having high-intensity laser radiation. In addition, interlocks interrupt the laser power supply, terminate enrichment operations, and eliminate the hazard.

Electromagnetic Hazards High-strength EM fields associated with the AVLIS ^{235}U ion collection system may cause biological injury and require controls to limit their effects. Areas with high-strength EM fields should be interlocked to preclude inadvertent personnel access or subjected to strict access controls.

Thermal Hazards Heat buildup from the uranium AVLIS vaporizer presents a worker safety issue. Thermal insulation and vaporizer cooling water systems should ensure worker habitability conditions are met. However, cooling water systems must be carefully routed to avoid enriched uranium removal systems and preclude an inadvertent criticality event.

Noise Hazards Noise hazards are created by high-energy systems transporting large quantities of matter. These systems include laser and enrichment process support equipment. The laser enrichment facility processes in a few separation units the equivalent material handled by thousands of gaseous diffusion stages or gas centrifuge machines. The smaller laser enrichment facility could magnify the expected noise hazard.

High noise areas should have restricted access. Noise hazards are reduced through the use of shielding or access controls. Personnel protective equipment provides a means to mitigate high noise levels.

Reduced Radiotoxicity The elimination of ^{234}U enrichment is a positive aspect of laser isotope separation technology. The enriched product specific activity is reduced relative to equivalent enrichments from gaseous diffusion and gas centrifuge facilities because only ^{235}U is enriched in a laser enrichment facility.

The $^{235}UF_5$ laser products will have the expected increase in neutron radiation as the ^{235}U enrichment increases. This neutron radiation will not be present in the AVLIS product. Somewhat higher beta radiation levels arise from uranium metal (2.33 mSv/h at 7 mg/cm^2) vice $^{235}UF_6$ (~1.8 mSv/h at 7 mg/cm^2). The beta, gamma, and neutron radiation levels associated with uranium materials require dose management and as low as reasonably achievable (ALARA) planning to ensure that worker doses are properly controlled.

2.4.2
Power Reactors

Nuclear reactors are a key component of the nuclear fuel cycle, and their designs will significantly evolve in the twenty-first century. Generation II and some Generation III systems dominate the near term with Generation IV reactors subsequently emerging to foster a more complete and environmentally friendly fuel cycle.

2.4.2.1
Generation III Reactors

Generation III reactors were developed to minimize design vulnerabilities that became apparent as operating experience with Generation II reactors was accumulated. The Generation III reactors also offer standardized designs that facilitate their licensing, construction, and operation. Generation III designs emphasize passive reactor safety systems.

As illustrated by the TMI Unit 2 and Fukushima Daiichi accidents, Generation II reactors are vulnerable to off-normal operating conditions because these designs rely on active safety systems. Active systems require either electrical energy or mechanical actuation for their effective operation, and these systems are vulnerable to mechanical and electrical failures. The complete loss of power at Fukushima Daiichi clearly illustrated vulnerabilities associated with active safety systems.

A more reliable safety system operates passively and uses inherent physical properties as the basis for their design. These inherent properties include physical phenomena such as gravity or convection as the basis for functionality. No electric power is required for a passive safety system to function.

Examples of Generation III reactors are summarized in Table 2.6. Table 2.6 includes the Advanced Boiling Water Reactor (ABWR), Advanced Pressurized Water Reactor (APR or APWR), Advanced Passive (AP) PWR, European Pressurized Water Reactor (EPR), Economic and Simplified Boiling Water Reactor (ESBWR), Advanced Canadian Deuterium Reactor (ACR), and PWR designs with multiple train safety systems. These designs may require modifications resulting from Fukushima Daiichi regulatory mandates. Other Generation III design concepts include the PBMR and Gas Turbine-Modular Helium Reactor (GT-MHR).

The reactor types noted in Table 2.6 include light water and heavy water reactors. Light water and heavy water reactors operate predominantly with uranium fuel, but operation with uranium and plutonium fuels is also possible. Other Generation III systems have proposed the use of high-temperature gas-cooled systems that utilize a variety of fuel types including HEU and thorium, ^{233}U and thorium, and plutonium and thorium. Thorium fuels have been used most prominently in HTGRs.

A discussion of the operating characteristics of these reactors is presented to illustrate the health physics aspects of Generation III reactors. The health physics

Table 2.6 Generation III reactors[a].

Country	Reactor	Power rating (MWe)	Construction/ operating status	Main features
United States/Japan	ABWR	1300	Commercial operation in Japan since 1996 Under construction in Taiwan	BWR with improved efficiency, simplified operation, and evolutionary design
South Korea	APR-1400	1400	Under construction in South Korea and the United Arab Emirates	PWR with simplified operation, increased reliability, and evolutionary design
United States	AP-1000	1100	Under construction in the United States and China	PWR with passive safety features, 60-year plant life, and capable of operating with a mixed oxide core
Japan	APWR	1500–1700	No units are currently under construction	PWR with hybrid active and passive safety systems and simplified design, construction, and operation
France	EPR	1600	Under construction in China, Finland, and France	PWR with improved safety features, high fuel efficiency, and low projected costs
United States	ESBWR	1500	No units are currently under construction	BWR with short construction time and enhanced safety features
Canada	ACR-1000	1000	No units are currently under construction	Light water reactor with low-enriched fuel and passive safety features
Russia	AES-2006	1150	Under construction in Russia	PWR with four train safety systems with passive backup systems
Japan/France	ATMEA1	1000	No units are currently under construction	PWR with three train safety systems with passive features including a core catcher and hydrogen recombiners

a) Uranium Information Centre (2005), Nuclear News Summary Report (2013), and Blake (2013).

characteristics are not specific to any of the reactor types of Table 2.6 and represent generic descriptions. For simplicity, the Generation III reactors are classified as PWRs, BWRs, ACRs, and HTGRs. Since the Generation III designs are incremental improvements over their Generation II counterparts, only differences between these two design concepts are addressed.

2.4.2.1.1 PWRs

Generation III PWRs incorporate the basic Generation II design but have enhanced safety systems and improved operating characteristics. The operational safety of Generation III PWRs is improved by incorporating passive safety systems; adding new safety systems, structures, and components (SSCs); and improving the reliability of existing SSCs. For example, the control room and information processing systems facilitate operator response to changing plant conditions.

The TMI accident was caused by a valve failure that led to a loss-of-coolant accident (LOCA) and core damage. Improved control room architecture incorporated into the Generation III PWR designs includes Generation II TMI modifications that facilitate the detection of loss-of-coolant conditions by providing containment parameters (e.g., sump level, humidity, temperature, and hydrogen concentration) that are indicative of accident conditions. Generation III designs also have enhanced core cooling water supplies that would mitigate a Fukushima Daiichi-type event. Licensing changes resulting from evaluations of the Fukushima Daiichi accident require supplemental power sources and core cooling water injection capability that further enhance the capability of Generation III designs to preserve the three fission product barriers.

These changes lead to increased safety and improved operability, reliability, and availability. Advanced PWRs can be configured to operate with plutonium using MOX fuel. The use of MOX fuel provides the potential to extend uranium reserves and minimizes the diversion of reprocessed plutonium for illicit purposes.

Improvements in reliability and maintainability in the Generation III PWRs reduce occupational doses. For example, dose savings are achieved by improving steam generator performance that minimizes inspection, surveillance, and repair activities. Improved materials and chemistry controls also minimize steam generator replacement outages.

2.4.2.1.2 BWRs

ABWRs incorporate a variety of innovations including natural circulation in the core region, no recirculation pumps, and passive decay heat removal systems. These safety improvements yield a more compact design than the Generation II BWR. This size reduction decreases the construction time and cost and makes the design more robust and less susceptible to earthquakes. In Generation II BWRs, the control rods are hydraulic. In the ABWR, they are electrohydraulic. Having an additional drive mechanism reduces the probability of failure and enhances the ability to achieve a safe shutdown condition. All major equipment and components are engineered for reliability and ease of maintenance.

Given the improvements in reliability and maintainability, Generation III BWRs should have reduced occupational doses compared to their Generation II counterparts. Minimizing high dose activities involving contaminated systems (e.g., recirculation piping replacement) is a key element of dose reduction at a Generation III BWR.

Both Generation III PWR and BWR units need to demonstrate source term control including minimizing corrosion, system activity, and source term buildup in primary system components. Source term control depends on maintaining rigorous chemistry specifications, ensuring replacement components minimize the generation of activation products, and maintaining good fuel performance with minimal fuel damage.

As noted in the previous section, the Fukushima Daiichi accident illustrates the importance of backup power and core cooling capability. The Fukushima Daiichi design modifications supplement the inherent passive safety features of the Generation III designs. Significant design modifications are intended to minimize the consequences of accidents that result in severe core damage.

Two modifications recommended by regulatory organizations evaluating the Fukushima Daiichi accident are hardened containment vents and filtered vents. A hardened primary containment venting system minimizes the possibility of a hydrogen explosion, and filtered vents mitigate the release of iodine and particulate radioactive material to the environment.

2.4.2.1.3
ACRs

The ACRs incorporate a compact core design, a light water primary cooling system, a heavy water moderator, online refueling, an extended fuel life of three to six times over natural uranium, slightly enriched uranium (1.2 wt% uranium dioxide fuel), and prefabricated structures and systems. Other Generation III ACR options using thorium fuel are also under consideration.

The compact core size in the Generation III ACR has a number of positive health physics implications. Maintenance and surveillance requirements are reduced because there are fewer fuel channels and the core is smaller. These changes result in reduced primary system work and the associated occupational dose. The tritium dose is also reduced since the coolant has been changed from heavy to light water, which minimizes the production reaction:

$$^{2}H + n \rightarrow {}^{3}H + \gamma \tag{2.11}$$

In a Generation II Canadian Deuterium reactor, 30–40% of the anticipated occupational dose is attributed to tritium intakes. Eliminating the heavy water coolant reduces this dose component.

The Generation III ACR designs include large low-temperature and low-pressure water reservoirs surrounding the fuel/coolant channels. These heat sinks provide fission product barrier protection by minimizing core damage and breaching of the reactor vessel.

2.4.2.1.4
HTGRs

Another Generation III reactor type is the HTGR. The HTGR concept has a number of positive characteristics including proliferation resistance, inherent safety, fuel construction that retains fission products, and ability to produce high temperatures for hydrogen production. Generation III HTGRs are inherently safe since fuel damage is inhibited due to the physical structure of the core. HTGRs are not susceptible to meltdown because additional heat removal capabilities are provided. These capabilities include convective airflow through a heat exchanger to achieve residual heat removal. This is an inherent passive safety feature that maintains the fuel fission product barrier.

The HTGR design offers a number of health physics advantages. An enhanced fission product barrier is achieved using a coated fuel concept. The fuel coatings minimize releases of fission products into the gas coolant and subsequently to the environment. Compared to PWRs and BWRs, the HTGR gas coolant produces minimal activation products. Although tritium is produced, limited ^{16}N production has definite benefits from both radiation protection and design simplification perspectives.

2.4.2.1.5
Generation III Safety Systems

The passive safety system philosophy of Generation III reactors is illustrated by reviewing selected aspects of BWR safety systems. Examples of proposed BWR safety systems include the (i) emergency condenser system, (ii) containment cooling condenser, (iii) core flooding system, and (iv) pressure pulse transmitters. These systems are important from a health physics perspective because they mitigate the radiological source term by protecting one or more fission product barriers.

Emergency condensers remove heat from the reactor core when the water level in the reactor pressure vessel (RPV) decreases. This condition can occur during a severe loss-of-coolant accident. The emergency condenser tubes are submerged in the core flooding pool and filled with water during normal operating conditions.

During a severe emergency, the water level in the RPV decreases and water drains from the condenser tubes and flows into the reactor. Steam from the reactor then enters the drained condenser tubes and condenses. The condensed steam flows by gravity into the RPV and maintains core cooling. Emergency condensers operate passively and require no electrical power, control logic, or switching operations. Operation of the emergency condensers maintains the integrity of the fuel and minimizes the radiological source term available for release to the plant/environment. This system was not available at Fukushima Daiichi and could have mitigated that severe event.

The primary containment vessel at a BWR includes a dry well. Containment cooling condensers remove heat from the dry well and transfer it to the water in the fuel storage pool located above the reactor. These systems mitigate a LOCA when steam is released to the BWR dry well increasing its temperature. The containment cooling condensers provide passive heat transfer to limit temperature and pressure increases. Removing containment heat minimizes the potential for a loss of fission product barriers and limits the release of radioactive material to the environment. This system was not part of the Fukushima Daiichi design basis.

The core flooding system provides a passive means of water addition to the RPV. When reactor coolant system pressure drops below a specified value, check valves open and permit the gravity flow of water from the core flooding system to the reactor vessel. These check valves open on differential pressure between the core flooding tank and the primary coolant system. Both pressure- and gravity-induced flows are passive features requiring no electric power or active switching operations. Preserving the fuel/clad fission product barrier minimizes the release of radioactive material from the primary system. Passive core flooding systems are also incorporated into a number of Generation II designs.

Core flooding systems offer limited relief because only the available tank volume is injected into the core and there is no refill capability. These systems do mitigate the event and provide operators time for additional action.

Passive pressure pulse transmitters are small heat exchangers. When reactor water level decreases, pressure increases on the secondary side of the heat exchanger. This pressure increase changes the position of a pilot valve connected to the secondary side of the heat exchanger. The change in valve position initiates action to shut down the reactor and triggers containment isolation without the need for electrical power or logic signals. This passive reactor shutdown feature protects the fuel/clad and primary coolant system fission product barriers. Isolation ensures the integrity of the containment fission product barrier.

Generation III safety systems are also being enhanced by implementing a variety of corrective actions following reviews of the Fukushima Daiichi accident. These actions include the storage of portable electrical generators and core cooling pumps that can be readily installed, hardened vents in BWRs, filtered vents, enhancements to spent fuel pool instrumentation, and enhanced emergency preparedness organizations. Details of these enhancements are provided in Chapter 7.

2.4.2.2
Generation IV Reactors

Operating Generation II and III reactors are basic light water systems. LWRs dominate the current reactor fleet because they were developed in the 1940–1950s as a compact power source for naval vessels and were subsequently scaled to commercial sizes. Problems with Generation II LWR designs were illustrated during the TMI and Fukushima Daiichi accidents. In addition, the light water design does not fulfill the original vision of nuclear power generation that incorporated reprocessing spent fuel to extract the maximum energy from this technology. Reprocessing has been limited by cost and proliferation concerns. The aforementioned Generation III designs improve Generation II safety performance but do not minimize high-level waste, effectively contribute to operating the fuel cycle in an optimum environmental manner, or limit the production of plutonium.

Generation IV reactors strive to minimize these issues. The Generation IV reactors are in the development phase. Currently, Generation IV is used to categorize these evolving designs that incorporate promising and innovative concepts that require significant research to achieve a final design that is capable of being licensed by a regulatory body.

Nuclear reactors produce plutonium that could be diverted to weapons use, but Generation IV systems minimize this possibility. This is accomplished using a fast neutron spectrum.

Reactors are classified into two broad categories: thermal and fast. A thermal reactor is a reactor that primarily operates using the thermal neutron fission of fissile nuclides (e.g., ^{233}U, ^{235}U, and ^{239}Pu). Fast reactors utilize fast neutrons to fission fertile nuclides (e.g., ^{232}Th and ^{238}U). Reactors utilizing fast neutrons also produce fissile materials including ^{233}U and ^{239}Pu. The presence of ^{233}U, ^{235}U, and ^{239}Pu are important considerations when evaluating the proliferation potential of fuel cycle technologies. Nuclear proliferation is addressed in more detail in the subsequent discussion.

Generation IV reactors have the potential for being proliferation resistant and incorporate reprocessing as part of their integrated design concept. To advance this fuel cycle concept, the Generation IV International Forum (GIF) was established in 2000 and included the United States, Argentina, Brazil, Canada, France, Japan, South Korea, South Africa, Switzerland, and the United Kingdom. These nations committed to the joint development of the next generation of nuclear technology. The 10 nations agreed on six Generation IV nuclear reactor technologies for deployment in the 2030 time frame. Some of these reactors operate at higher temperatures than the Generation II and III reactors, four are designated for hydrogen production, their characteristics are summarized in Table 2.7, and the primary activation products are listed in Table 2.8. Additional GIF members include China, Russia, and Euratom.

The six design concepts offer the potential for improved economics, safety, reliability, and proliferation resistance. These designs also maximize the utilization of fissile resources and minimize high-level waste. Generation IV reactors addressed by the GIF include gas-cooled fast reactors (GFRs), lead-cooled fast reactors (LFRs), molten salt epithermal reactors (MSRs), sodium-cooled fast reactors (SFRs), supercritical water-cooled reactors (SCWRs), and very-high-temperature, helium-cooled, graphite-moderated thermal reactors (VHTRs). Lead–bismuth-cooled fast reactors (LBFRs) were previously considered as a Generation IV candidate but were replaced by LFRs. An advantage of the Generation IV design is the capability for full actinide recycling (Option 4) using a closed fuel cycle concept. Open fuel cycles do not incorporate actinide recycling and are addressed in the subsequent discussion.

Table 2.7 Generation IV reactor concept characteristics[a].

Reactor technology	Power rating (MWe)[b]	Operating temperature (°C)[b]	Fuel cycle options	Economic justification
Gas-cooled fast reactors	1200	850	Closed	Electricity and hydrogen production
Lead-cooled fast reactors	20–180, 300–1200, 600–1000	480–800	Closed	Electricity and hydrogen production
Molten salt epithermal reactors	1000	700–800	Closed	Electricity and hydrogen production
Sodium-cooled fast reactors	50–150, 300–600, 600–1500	550	Closed	Electricity production
Supercritical water-cooled reactors (thermal and fast versions)	300–700, 1000–1500	510–625	Open Closed	Electricity production
Very-high-temperature, helium-cooled, graphite-moderated thermal reactors	250–300	900–1000	Open	Electricity and hydrogen production

a) Derived from Generation IV International Forum Report (2009, 2013) http://www.gen-4.org/Technology/roadmap.htm.
b) A range of values is presented by the Generation IV International Forum Report (2009, 2013).

The principal goals of the Generation IV systems are to achieve high levels of safety and reliability, sustainability, proliferation resistance and physical protection, and economic competitiveness. GIF members collaborating in the development of the six reactor concepts are listed in Table 2.9. Collaborating members are required to sign a formal agreement that governs intellectual property rights and associated reactor technology. These arrangements have been signed for the SFR, VHTR, GFR, and SCWR. Limited studies are governed by a memorandum of understanding for the LFR and MSR systems.

The GIF plans to construct an SFR and VHTR in the near future following the development of safety design criteria. These criteria must be met while satisfying the four main Generation IV system goals. An acceptable design must not only achieve the highest safety standards, but the enhanced safety must be clearly communicated to the public. Sustainability includes the long-term viability of these systems as related to fuel design, waste generation, and potential design enhancements.

Table 2.8 Activation products produced in materials unique to Generation IV fission power reactors[a)].

Nuclide	Half-life	Decay mode	Production mode
^{3}H	12.3 years	β^-	GFRs and VHTRs (^{4}He gas coolant): ^{4}He(γ, p)^{3}H and ^{4}He(n, d)^{3}H MSRs (lithium fluoride salt coolant): ^{6}Li(n, α)^{3}H
^{10}Be	1.56×10^{6} years	β^-	MSRs (beryllium fluoride salt coolant): ^{9}Be(n, γ)^{10}Be
^{14}C	5715 years	β^-	GFRs, MSRs, and VHTRs (graphite moderator): ^{14}N(n, p)^{14}C and ^{13}C(n, γ)^{14}C GFRs (gas coolant) and SCWRs (water coolant): ^{17}O(n, α)^{14}C
^{15}O	2.037 min	β^+ γ	GFRs (gas coolant) and SCWRs (water coolant): ^{16}O(n, 2n)^{15}O and ^{16}O(γ, n)^{15}O
^{16}N	7.13 s	β^- γ	GFRs (gas coolant) and SCWR (water coolant): ^{16}O(n, p)^{16}N MSRs (fluoride salt coolant): ^{19}F(n, α)^{16}N
^{17}N	4.174 s	β^- γ n	GFRs (gas coolant) and SCWR (water coolant): ^{17}O(n, p)^{17}N
^{18}F	1.8293 h	β^+ γ	MSRs (fluoride salt coolant): ^{19}F(n, 2n)^{18}F
^{19}O	26.9 s	β^- γ	GFRs (gas coolant) and SCWR (water coolant): ^{18}O(n, γ)^{19}O MSRs (fluoride salt coolant): ^{19}F(n, p)^{19}O
^{20}F	11.1 s	β^- γ	MSRs (fluoride salt coolant): ^{19}F(n, γ)^{20}F SFRs (liquid sodium coolant): ^{23}Na(n, α)^{20}F
^{22}Na	2.604 years	β^+ γ	MSRs (sodium salt coolant) and SFRs (liquid sodium coolant): ^{23}Na(n, 2n)^{22}Na and ^{23}Na(γ, n)^{22}Na
^{23}Ne	37.1 s	β^- γ	MSRs (sodium salt coolant) and SFRs (liquid sodium coolant): ^{23}Na(n, p)^{23}Ne
^{24}Na	14.97 days	β^- γ	MSRs (sodium salt coolant) and SFRs (liquid sodium coolant): ^{23}Na(n, γ)^{24}Na GFRs (in core materials): ^{24}Mg(n, p)^{24}Na
^{25}Na	59.3 s	β^- γ	GFRs (in core materials): ^{25}Mg(n, p)^{25}Na
^{27}Mg	9.45 min	β^- γ	GFRs (in core materials): ^{26}Mg(n, γ)^{27}Mg GFRs (in core materials): ^{30}Si(n, α)^{27}Mg
^{28}Al	2.25 min	β^- γ	GFRs (in core materials) and VHTRs (fuel coating): ^{28}Si(n, p)^{28}Al
^{29}Al	6.5 min	β^- γ	GFRs (in core materials) and VHTRs (fuel coating): ^{29}Si(n, p)^{29}Al
^{31}Si	2.62 h	β^- γ	GFRs (in core materials) and VHTRs (fuel coating): ^{30}Si(n, γ)^{31}Si
^{35}S	87.2 days	β^-	GFRs, MSRs, and VHTRs (graphite moderator): ^{35}Cl(n, p)^{35}S and ^{34}S(n, γ)^{35}S
^{38}Cl	37.2 min	β^- γ	GFRs, MSRs, and VHTRs (graphite moderator): ^{37}Cl(n, γ)^{38}Cl

Table 2.8 *(Continued)*

Nuclide	Half-life	Decay mode	Production mode
^{45}Ca	162.7 days	β^- γ	GFRs (in core materials): ^{48}Ti(n, α)^{45}Ca
^{45}Ti	3.078 h	β^+ γ	GFRs (in core materials): ^{46}Ti(n, 2n)^{45}Ti and ^{46}Ti(γ, n)^{45}Ti
^{46}Sc	83.8 days	β^- γ	GFRs (in core materials): ^{46}Ti(n, p)^{46}Sc GFRs, MSRs, and VHTRs (graphite moderator): ^{45}Sc(n, γ)^{46}Sc
^{47}Ca	4.536 days	β^- γ	GFRs (in core materials): ^{46}Ca(n, γ)^{47}Ca and ^{50}Ti(n, α)^{47}Ca
^{47}Sc	3.349 days	β^- γ	GFRs (in core materials): ^{47}Ti(n, p)^{47}Sc
^{48}Sc	43.7 h	β^- γ	GFRs (in core materials): ^{48}Ti(n, p)^{48}Sc
^{51}Ti	5.76 min	β^- γ	GFRs (in core materials): ^{50}Ti(n, γ)^{51}Ti
^{82}Br	1.471 days	β^- γ	GFRs, MSRs, and VHTRs (graphite moderator): ^{81}Br(n, γ)^{82}Br
^{88}Y	106.63 days	β^+ γ	GFRs (in core materials): ^{89}Y(n, 2n)^{88}Y and ^{89}Y(γ, n)^{88}Y
^{89}Sr	50.61 days	β^- γ	GFRs (in core materials): ^{89}Y(n, p)^{89}Sr GFRs (in core materials), MSRs (coolant component), and VHTRs (fuel coating): ^{92}Zr(n, α)^{89}Sr
^{89m}Y	15.7 s	γ	GFRs (in core materials): ^{89}Y(n, n′)^{89m}Y
^{89}Zr	3.27 days	β^+ γ	GFRs (in core materials), MSRs (coolant component), and VHTRs (fuel coating): ^{90}Zr(n, 2n)^{89}Zr and ^{90}Zr(γ, n)^{89}Zr
^{90}Y	2.669 days	β^-	GFRs (in core materials): ^{89}Y(n, γ)^{90}Y and ^{93}Nb(n, α)^{90}Y GFRs (in core materials), MSRs (coolant component), and VHTRs (fuel coating): ^{90}Zr(n, p)^{90}Y
^{90m}Y	3.19 h	β^- γ	GFRs (in core materials): ^{89}Y(n, γ)^{90m}Y
^{92}Nb	3.5×10^7 years	γ	GFRs (in core materials): ^{93}Nb(n, 2n)^{92}Nb and ^{93}Nb (γ, n) ^{92}Nb
^{92m}Nb	10.13 days	γ	GFRs (in core materials): ^{93}Nb(n, 2n)^{92m}Nb and ^{93}Nb (γ, n) ^{92m}Nb
^{93}Zr	1.5×10^6 years	β^- γ	GFRs (in core materials), MSRs (coolant component), and VHTRs (fuel coating): ^{92}Zr(n, γ)^{93}Zr
^{93m}Nb	16.1 years	γ	GFRs (in core materials): ^{93}Nb(n, n′)^{93m}Nb
^{94}Nb	2.0×10^4 years	β^- γ	GFRs (in core materials): ^{93}Nb(n, γ)^{94}Nb
^{94m}Nb	6.263 min	β^- γ	GFRs (in core materials): ^{93}Nb(n, γ)^{94m}Nb
^{95}Zr	64.02 days	β^- γ	MSRs (coolant component) and VHTRs (fuel coating): ^{94}Zr(n, γ)^{95}Zr

(Continued Overleaf)

Table 2.8 (*Continued*)

Nuclide	Half-life	Decay mode	Production mode
^{97}Zr	16.75 h	β^- γ	MSRs (coolant component) and VHTRs (fuel coating): ^{96}Zr(n, γ)^{97}Zr
^{152}Eu	13.54 years	β^- β^+ γ	GFRs, MSRs, and VHTRs (graphite moderator): ^{151}Eu(n, γ)^{152}Eu
^{203}Pb	2.164 days	γ	LFRs (lead coolant): ^{204}Pb(n, 2n)^{203}Pb and ^{204}Pb(γ, n)^{203}Pb
^{204m}Pb	1.12 h	γ	LFRs (lead coolant): ^{204}Pb(n, n′)^{204m}Pb
^{205}Pb	1.5×10^7 years	ε[b)]	LFRs (lead coolant): ^{204}Pb(n, γ)^{205}Pb
^{209}Pb	3.25 h	β^-	LFRs (lead coolant): ^{208}Pb(n, γ)^{209}Pb
^{210}Pb	22.3 years	β^- γ α	LBFRs (bismuth coolant): ^{209}Bi(n, p)^{209}Pb + n $\rightarrow$ ^{210}Pb LFR(lead coolant): ^{208}Pb(n, γ)^{209}Pb followed by ^{209}Pb(n, γ)^{210}Pb
^{210}Bi	5.01 days	β^- γ α	LBFRs (bismuth coolant): ^{209}Bi(n, γ)^{210}Bi
^{210m}Bi	3.0×10^6 years	α γ	LBFRs (bismuth coolant): ^{209}Bi(n, γ)^{210m}Bi
^{210}Po	138.38 days	α γ	LBFR (bismuth coolant): ^{209}Bi(n, γ)^{210m}Bi $\xrightarrow{\beta^-}$ ^{210}Po

a) See Appendix A for additional fission and activation products and a discussion of decay modes.
b) Electron capture (ε).

Table 2.9 Generation IV industrial forum agreements.

Design	Canada	France	Japan	South Korea	South Africa	Switzerland	United States	Euratom	China	Russia	United Kingdom	Brazil	Argentina
SFR		a)	a)	a)			a)	a)	a)	a)			
VHTR	a)	a)	a)	a)		a)	a)	a)	a)				
GFR		a)	a)			a)		a)					
SCWR	a)		a)					a)		a)			
LFR			b)					b)		b)			
MSR		b)						b)					

a) Signatory of formal agreement.
b) Memorandum of understanding.

The development of the six Generation IV systems is divided into four phases that include viability, performance, demonstration, and commercialization. Design viability and performance require substantial research and development that involve collaboration of GIF signatories. The demonstration and commercialization phases require significant resources and funding commitments. The GIF categorized the six Generation IV designs in terms of these phases.

Over the next decade, four systems (GFR, MSR, SCWR, and VHTR) will be assigned to the viability phase. The LFR system will move from the viability phase to the performance phase, and the SFR design will transition from the performance to the demonstration category.

The basic characteristics of these Generation IV systems are summarized in the subsequent discussion. Before reviewing the Generation IV reactor types, an examination of expected activation products is provided. These activation products are summarized in Table 2.8 and focus on the unique Generation IV specific isotopes associated with fuel, coolant, and primary system components. The production mode for the activation product is also provided. The subsequent discussion outlines the unique Generation IV materials that lead to radionuclides or production modes not normally associated with Generation II and III reactors. The Generation IV reactors also produce the activation products encountered in Generation II reactors. These common fission and activation products include ^{3}H, ^{16}N, ^{54}Mn, ^{59}Fe, ^{58}Co, ^{60}Co, ^{85}Kr, ^{90}Sr, ^{90}Y, ^{95}Zr, ^{131}I, ^{133}Xe, ^{135}Xe, and ^{137}Cs. The properties of selected nuclides are summarized in Appendix A.

Generation IV graphite-moderated reactors produce other isotopes due to impurities in this material. These isotopes include ^{38}Cl [^{37}Cl(n, γ)], ^{46}Sc [^{45}Sc(n, γ)], ^{82}Br [^{81}Br(n, γ)], and ^{152}Eu [^{151}Eu(n, γ)]. The specific isotopes and activity levels produced in graphite-moderated reactors depend on the acceptable impurity content of graphite materials incorporated in the design.

2.4.2.2.1
Gas-Cooled Fast Reactors (GFRs)

The reference Generation IV GFR incorporates a fast neutron spectrum and a helium-cooled reactor core. It uses a closed cycle process that incorporates a direct cycle helium turbine for electricity production. Process heat is utilized for the thermochemical production of hydrogen. With a fast neutron spectrum and full actinide recycle, the GFR minimizes the production of long-lived radionuclides.

Operating in conjunction with a closed fuel cycle, GFRs enhance the utilization of uranium and minimize high-level waste generation. High-level waste is limited by incorporating full actinide recycle as part of the closed fuel cycle. To improve efficiency, the GFR is colocated with other fuel cycle facilities for on-site spent fuel reprocessing and fuel fabrication that incorporates uranium, plutonium, and minor actinides in the fuel matrix. The Option 4 fuel cycle more closely links the power production, reprocessing, and waste disposal options in order to limit the generation of high-level waste.

The GFR concept has a number of characteristics that support health physics design objectives. These include a fuel composition that enhances fission product retention and actinide recycling that permits operation of a closed fuel cycle. GFR fuel incorporates a number of enhancements including advanced coatings and ceramic fuel composites (e.g., SiC, ZrC, TiC, NbC, ZrN, TiN, MgO, and $ZrYO_2$)

to facilitate fission product retention. These fuel characteristics enhance the fuel fission product barrier and limit the probability of a release of radioactive material to the environment.

Full actinide recycle eliminates long-term waste disposal and associated radiation dose concerns. In addition, the closed nature of the fuel cycle limits the occupational doses associated with geologic waste disposal and storage. The extent of these health physics advantages depends on GFR fuel performance and the development of actinide recycling technology. Accordingly, near-term GFR activities focus on the development of severe accident mitigation approaches, demonstration of integrated performance, and design of a small experimental reactor.

2.4.2.2.2
Lead-Cooled Fast Reactors (LFRs)

The LFR system utilizes a fast neutron spectrum, in-vessel steam generators, and a core that is cooled passively through natural convection with a liquid lead coolant. This reactor type is an inherently safe system. With a closed fuel cycle, an LFR has the potential for significant waste volume reduction relative to advanced LWRs. A key advantage of the liquid metal reactor is the potential to recycle essentially all actinides. LFR applications include the generation of electricity, hydrogen production, and desalination of seawater.

The LFR design must demonstrate successful proliferation resistance and economic viability. Economics are improved through simplification including modularization of the design. Proliferation issues are minimized if the design is successful in efficiently recycling actinides, particularly ^{239}Pu.

LFR core lifetimes are projected to approach 15–20 years. The LFR system offers considerable flexibility, and facility options include a 20 MWe fabricated reactor module, a 600 MWe modular design, and a 1200 MWe base load facility.

The LFR concept utilizes a closed fuel cycle with the supporting facilities residing in a central or regional location. Within the closed fuel cycle, LFR facilities provide efficient utilization of uranium resources and management of actinides.

A number of issues must be resolved for the LFR concept to become a commercial viability. Chemistry criteria are needed to facilitate the control of oxygen and ^{210}Pb. The development of fuel and reactor materials and achieving acceptable corrosion properties for these materials are additional issues.

The liquid metal design initially incorporated a LBFR, but a number of design issues suggested a lead coolant provides better performance. One health physics issue associated with the LBFR was the capability of the lead–bismuth eutectic to retain fuel and fission products during all operating conditions.

The limitations regarding fuel and fission product retention in the lead–bismuth coolant are mitigated using lead. According to the GIF, the LFR system has excellent materials capabilities and fission product retention. In addition, the LFR's molten lead coolant is relatively inert which should lead to improved safety performance and reliability.

From a health physics perspective, a lead coolant has several positive features that can minimize the potential for a fission product release. In particular, lead has a high boiling point, has a low vapor pressure, and provides an efficient gamma-ray shielding material. Although these characteristics enhance the fuel fission product barrier, they must be demonstrated under accident conditions including design basis and beyond design basis events.

The lead coolant also enhances reactor safety by contributing to a low core damage probability. This result is supported by lead's heat transfer characteristics, high specific heat and thermal expansion coefficients, and inherent negative reactivity contribution to the LFR core. These characteristics also support good heat transfer from the core to lead coolant and the capability for natural circulation of the reactor coolant during emergency conditions. From a release perspective, lead reduces the risk of a recriticality following a core melt event.

The use of a liquid lead coolant creates a number of operational difficulties that are encountered during routine outage activities. For example, the high-temperature lead coolant presents a challenge during refueling operations. The coolant must remain in a liquid state for refueling to occur. There are engineering solutions that resolve the refueling issue including the use of a cool cover gas to facilitate access to the fuel assemblies. In addition, a number of operational requirements and maintenance activities involving primary system components in a liquid lead environment must be addressed for the LFR design to become viable.

The LFR is primarily envisioned for electricity and hydrogen production and actinide management. Since the LFR system is transitioning into the performance phase, research and development focus on reactor safety and ensuring that the fuel, reactor materials, and associated corrosion control measures perform as anticipated.

Two LFR concepts are currently being advanced. These are the 20 MWe Small Secure Transportable Autonomous Reactor (SSTAR) developed in the United States and the 600 MWe European Lead-cooled System (ELSY) developed in the European Union. Most of the engineering and materials challenges are projected to be addressed by the ELSY design configuration. Separate designs for a small, transportable LFR with a long core life and a moderate-sized power plant will incorporate the operating experience derived from the SSTAR and ELSY demonstration facilities. Larger facilities are dependent on the success of these designs.

2.4.2.2.3 Molten Salt Epithermal Reactors (MSRs)

Molten salt reactors have potential advantages in terms of proliferation resistance attributable to the lower fuel inventory and plutonium buildup and a reduced source term associated with the online separation and removal of fission products. The circulating molten salt fuel is a mixture of zirconium, sodium, and uranium fluorides. Other molten salt options include lithium and beryllium fluoride with dissolved thorium and ^{233}U. The molten salt/fuel flows in channels through the core's graphite moderator. The MSR reference power level is 1000 MWe.

Since the fuel is in a liquid state, fuel processing is performed while the reactor is operating. The produced actinides and fission products form fluorides in the liquid coolant, which remain in the eutectic mixture and are burned as they pass through the core. The core's fast neutron fluence transmutes the minor actinides and some of the fission products.

This chemical environment permits the reactor cycle to be tailored for the destruction (burnup) of minor actinides and plutonium and the removal of fission products. Since the MSR fuel cycle allows full actinide recycling, waste issues are dominated by fission products. However, the MSR concept requires refinement and developing high-temperature structural materials, establishing and demonstrating appropriate fuel characteristics, and resolving nuclear and hydrogen safety issues.

Since the fuel forms a eutectic mixture with the coolant, the MSR design only has two fission product barriers. This is a significant departure from the current safety philosophy based on three fission product barriers. Any primary coolant leakage leads to the release of fuel and fission products to the facility. If leakage occurs in the containment building, it is the only remaining fission product barrier. Leakage outside containment merits special attention, if these areas do not provide a fission product barrier. Therefore, health physics activities at a MSR will be strongly influenced by the ability of the coolant to retain fuel, fission products, activation products, and actinides.

Although there are possible radiation safety issues associated with the MSR's liquid fuel, there are also positive nuclear and radiological safety characteristics. The liquid fuel has an advantage that it is recirculated and fission products can be continuously removed to minimize the source term. Although the removal mechanisms are yet to be completely defined, systems analogous to LWR demineralizers are envisioned.

The MSR design incorporates a unique reactor vessel that includes a hole in its lower head. This hole is plugged with solidified fuel material. The fuel plug remains in a solid state and is cooled by a refrigeration unit. When the plug is solidified, it preserves the reactor vessel fission product barrier's integrity. If the facility loses power during an emergency, the refrigeration unit becomes inoperable. Without power to maintain the solidified material, the plug melts and the fuel drains into underground holding tanks. These tanks are designed to provide a stable, safe shutdown condition that prevents the release of fission products.

Given the level of development required for the MSR design to fully mature, additional health physics issues may emerge. A potential area of concern is the capability of the liquid fuel/coolant to retain fission and activation products and actinides during a severe accident. The capability of safety systems to preserve fission product barriers during design and beyond design basis accidents must be demonstrated. In addition, a number of technical issues are unresolved including demonstrating a viable safety approach, completing a fuel reprocessing flow sheet, characterizing the properties and behavior of the liquid salt coolant, and developing robust reactor materials.

2.4.2.2.4

Sodium-Cooled Fast Reactors (SFRs)

SFRs operate with a fast neutron spectrum and utilize a liquid sodium coolant. The SFR design is associated with a fuel cycle that incorporates full actinide recycling. With design improvements, the SFR also generates electricity. Three SFR concept designs are currently envisioned.

The first design is a large-scale 600–1500 MWe loop-type sodium-cooled reactor using mixed uranium–plutonium oxide fuel. Its fuel cycle is based upon advanced aqueous fuel reprocessing technology. The second design is an intermediate-sized 300–600 MWe pool-type reactor and the third is a small-scale 50–150 MWe modular-type sodium-cooled reactor utilizing uranium–plutonium–minor actinide–zirconium metal alloy fuel. SFRs are supported by a fuel cycle based on fuel reprocessing in facilities that are integrated with the reactor.

The SFR system benefits from operating experience with Generation II sodium-cooled reactors. If the SFR capability to efficiently consume plutonium and other minor actinides is achieved, it significantly reduces the actinide loadings in high-level radioactive waste. These actinide characteristics reduce the SFR's radioactive waste disposal requirements and enhance its nonproliferation potential. Reducing capital cost and improving passive safety system performance under transient conditions are the major challenges for implementing the SFR design concept.

Given existing experience with Generation II sodium-cooled reactors, the Generation IV SFR health physics concerns are well defined. These health physics issues are complicated by the potential for the sodium–water chemical reaction to mobilize fission and activation products. Health physics issues could also arise from the implementation of the closed fuel cycle with full actinide recycle. Experience with twentieth-century reprocessing approaches suggests that waste storage, environmental concerns, maintenance of heavily contaminated equipment, and decommissioning issues merit thorough evaluation.

The SFR is advancing to the demonstration phase. The planned reactors supporting the SFR concept include the Russian BN-800, the French Advanced Sodium Technological Reactor for Industrial Demonstration (ASTRID) with an operation date near 2023, and Japanese and Korean designs that are in development. Research and development are focusing on enhanced safety options.

2.4.2.2.5

Supercritical Water-Cooled Reactors (SCWRs)

The SCWR system is similar to the BWR design. It is primarily designed for efficient electricity production, with an option for actinide management. SCWR designs are based on either a thermal or a fast neutron spectrum. The thermal neutron version uses once-through uranium dioxide fuel and has similar waste management issues associated with a Generation II and III single-pass fuel cycle.

From a health physics perspective, a fuel cycle without actinide recycling is not a desirable Generation IV alternative.

SCWRs have a thermal efficiency about one-third higher than current Generation II and III light water reactors. The plant design is considerably simplified because the coolant does not change phase in the reactor and is directly coupled to the energy conversion equipment. As with other light water systems, the fuel is uranium dioxide. Passive safety features are similar to those utilized in Generation III simplified BWRs.

The fast spectrum version permits actinide recycling using conventional reprocessing technology. However, the fast reactor version must overcome materials development issues. Both SCWR options utilize passive safety systems and operational characteristics similar to those of the Generation III ESBWR. The full actinide recycle version is based on advanced aqueous fuel reprocessing. A fuel reprocessing facility supports individual or multiple SCWRs and is integrated with these reactors for this Generation IV concept to achieve economic viability.

Based on initial design efforts, the SCWR concept has a number of reactor safety issues. First, the design has a tendency to have a positive void reactivity coefficient. A positive reactivity coefficient intensifies the fission reaction during operating conditions (e.g., as the parameter increases the fission rate and reactor power increases). Ideally, a negative coefficient is desired such that an increase in the parameter (e.g., void volume) dampens or shuts down the fission reaction. Therefore, a positive void reactivity coefficient complicates reactor operation and limits the capability to reach a stable configuration during a severe reactor transient. There is also the potential for design basis loss-of-coolant accidents to occur. These two characteristics complicate the advancement of the SCWR. Other SCWR challenges include the production of a viable core design, accurately estimating the heat transfer characteristics, and developing fuel and core structural materials that are corrosion-resistant during the various SCWR normal and transient operating conditions.

The SCWR facility should have health physics issues that are similar to those encountered in Generation II and III BWRs. Additional health physics issues arise if reactor materials fail to achieve the desired lifetime and reliability goals.

2.4.2.2.6
Very-High-Temperature Reactors (VHTRs)

The VHTR is a high-efficiency, graphite-moderated, helium-cooled reactor that operates with a thermal neutron spectrum. It can be utilized for the cogeneration of electricity and hydrogen and to provide process heat for industrial applications. The basic technology for VHTR systems has been established in Generation II HTGRs.

VHTR fuel consists of coated particles using materials such as SiC and ZrC that are formed into pebble elements or prismatic blocks. The design uses once-through uranium fuel or U/Pu fuel. Waste disposal issues associated with long-term spent fuel storage are not resolved by the VHTR's open fuel cycle.

Since the basic technology for VHTR systems has already been established in Generation II HTGRs, the Generation IV design is an evolutionary development. However, the system's goal of operating at 1000°C presents challenges in terms of fuel and materials development and in maintaining reactor safety under transient conditions.

Technology advancements in fuel performance and high-temperature materials development are required for the VHTR to be a viable technology. Shortcomings in either of these areas would potentially weaken the fuel and the primary coolant system fission product barriers. If these issues are resolved, the health physics issues will resemble those at a Generation II HTGR facility.

The prototype design is currently focusing on achieving the desired high outlet temperatures. Developing advanced materials and fuel designs governs the long-term viability of the VHTR.

Other projects are supporting VHTR development. China has resumed construction of its high-temperature demonstration reactor. This reactor has a tentative 2017 operation date, and the United States expressed interest in participating in this venture. In support of the VHTR concept, the US Department of Energy is focusing on fuel and materials development including the graphite used in the high-temperature core structures and steel for the pressure vessel. The United States also supports the VHTR through its Next Generation Nuclear Plant Demonstration Project.

2.4.2.2.7
Radionuclide Impacts

The extent to which the radionuclides of Table 2.8 dominate effective doses at a Generation IV facility ultimately depends on the reactor's operational characteristics. Based on Generation II and III experience, a number of health physics considerations apply to Generation IV systems.

Internal radiation hazards are presented by ^{3}H in the HTO form and ^{14}C as CO_2 particularly during refueling operations and primary system maintenance. The extent of the hazard depends on allowable leakage and primary system performance characteristics. Traditional Generation II and III activation products and fission products (APFPs) including ^{60}Co and ^{131}I also present internal intake hazards.

Submersion hazards result from short-lived radioactive gases (e.g., ^{15}O, ^{16}N, ^{17}N, ^{19}O, and ^{23}Ne). The noble gases produced in the fission process also present a submersion hazard, and these are primarily comprised of isotopes of Kr and Xe.

External hazards exist for a variety of nuclides including the coolant activation products ^{16}N and ^{24}Na. The extent of the external radiation hazard is dependent on the magnitude of the production of fission and activation products that decay via beta and gamma emission. It is likely that ^{58}Co and ^{60}Co will significantly contribute to worker effective doses, which is consistent with Generation II and III operating experience.

Reactor coolant leakage in the MSR, LFR, and SFR designs introduces new hazards that were not routinely encountered in Generation II and III reactors. The leakage of MSR coolant containing entrained fuel and fission products presents a significant source term that is much greater than encountered in Generation II and III LWRs. These coolant activity levels may be comparable to the levels encountered during the TMI-2 and Fukushima Daiichi accidents. High activity levels require changes in operating practices compared to contemporary PWR auxiliary building or BWR reactor building maintenance activities in Generation II and III reactors. The release of fission gases and entrained fission and activation products could require that routine maintenance and spill cleanup be accomplished remotely or using remote handling techniques.

The LFR and SFR coolant activity levels are governed by fuel integrity. However, SFR leakage presents a challenge because the energetic sodium–water reaction has significant potential to disperse radioactive material. LFR leakage must address the toxic characteristics of lead. The consequences of liquid lead and liquid sodium leakage in an industrial environment require controls to mitigate their effects. Radiation work permits and personal protective equipment must address these leakage issues in an operating Generation IV reactor.

Off-site releases of radioactive material from a Generation IV reactor are expected to be similar to those from Generation II and III facilities. Unless the molten salt coolant provides a demonstrated fuel fission product barrier, MSRs present potential health physics issues because there are only two fission product barriers. The MSR source term requires further investigation and characterization in terms of the capability of the coolant to retain fission products. However, noble gas and iodine will be a major portion of the MSR release source term.

Open fuel cycles present additional health physics concerns. The open fuel cycles associated with SCWRs (thermal option) and VHTRs have negative waste storage and associated effective dose impacts. Impacts include the long-term storage of high-level waste with the potential for the release of fission products and actinides to the environment. Closed fuel cycle options have positive nuclear proliferation and waste disposal aspects since actinides are destroyed during reactor operation.

2.4.2.2.8
Hydrogen Production

Hydrogen production for use as an alternative fuel is another application of advanced reactors, and four of the six Generation IV design concepts have hydrogen production as a design goal. Three basic approaches have been advanced for the nuclear energy production of hydrogen. The first (nuclear-assisted steam reforming) introduces natural gas to produce hydrogen. Hot electrolysis is the second approach, and it produces oxygen and hydrogen from water. The third method (thermochemical production) uses a series of chemical reactions and high temperatures to convert water into hydrogen and oxygen. All three processes use reactor heat to drive hydrogen production. Of these three, thermochemical hydrogen production is currently viewed as the most cost-effective method.

The health physics aspects of hydrogen production depend on the reactor design generating process heat. Since only high-temperature reactors are candidates for hydrogen production, an optimum design matches the reactor output and hydrogen generation requirements. In addition, the nuclear reactor and chemical hydrogen production facility must be physically separated. Preliminary design studies suggest that a separation distance of at least a kilometer is necessary to ensure that hydrogen facility accidents do not affect the high-temperature reactor.

Hydrogen explosions occurred at TMI-2 and Fukushima Daiichi Units 1, 3, and 4. The Fukushima Daiichi accident demonstrated that a hydrogen explosion is an effective method to disperse radioactive material. Hydrogen production in significant quantities requires careful examination. Appropriate design and beyond design basis accident evaluations limit the negative impact of hydrogen production on Generation IV reactor safety.

2.4.2.2.9 Deployment of Generation IV Reactors

Generation IV reactors are projected to be deployed in the 2030s. The SFR has the most optimistic deployment outlook which is somewhat expected since there is scalable operating experience from Generation II SFR designs. Deployment dates are contingent on the development of the Generation IV reactor types and resolution of the issues previously identified.

2.4.2.2.10 Generation IV Radiological Design Characteristics

From a radiological perspective, the Generation IV facility design should ensure that effective doses to plant workers and to members of the public are ALARA. This is achieved through the design of SSCs that are reliable, are easily maintained, and do not significantly contribute to the radiological source term. The proposed radiological design characteristics are common to a variety of organizations involved in advanced reactor regulation and standards. This includes the US Nuclear Regulatory Commission and the International Atomic Energy Agency.

With these considerations, Generation IV SSCs should limit their radioactive source term contribution. This entails the reduction in the concentrations of cobalt and nickel for materials in contact with the primary coolant to minimize the production of the ^{58}Co and ^{60}Co activation products. These isotopes are the major sources of radiation exposure during shutdown, maintenance, and inspection activities at Generation II and III LWRs. Exceptions to this design specification may be necessary to enhance component or system reliability and minimize component maintenance. However, the decision to utilize materials that produce ^{58}Co and ^{60}Co activation products must be made in a deliberate manner using ALARA considerations as a guide.

The resulting reduced radiation fields allow operations, maintenance, and inspection activities to proceed in a manner that leads to minimizing effective doses. Effective doses are also maintained ALARA by incorporating the use of robotic technology in maintenance and surveillance tasks in high radiation areas. The design should also accommodate remote and semiremote operation, maintenance, and inspection to reduce the time spent in radiation fields. Reach rods and motor operators should be evaluated for incorporation into valves located in high radiation areas.

Generation IV SSCs should attain optimal reliability and maintainability to reduce the frequency and duration of maintenance requirements. This is particularly true for systems in contact with fluids cooling the reactor core. Adequate equipment spacing and job preparation areas facilitate access for maintenance, repair, and inspection. Modularized components facilitate their replacement or removal to a lower radiation area for repair. These requirements reduce access, repair, and equipment removal times and limit the time spent in radiation fields.

The SSC design should facilitate the physical separation of radioactive and nonradioactive systems. High radiation sources should be located in separate shielded cubicles. In addition, equipment requiring periodic servicing or maintenance (e.g., pumps, valves, and control systems) should be physically separated from sources with higher radioactive material concentrations including tanks and demineralizers.

The accumulation of radioactive materials in equipment and piping should be minimized. This is often accomplished using flushing connections to facilitate the removal of radioactive materials from system components. Locating drains at low points enhances the achievement of this design objective. Piping should be seamless, and the number of fittings minimized to reduce the accumulation of radioactive materials at seams and welds.

Systems that generate radioactive waste should be located close to waste processing areas to minimize the length of piping carrying these materials. The potential for pipe plugging is minimized by routing lines that carry resin slurries vertically. Large-radius bends are used instead of elbows to limit the potential for pipe plugging.

The radiological design considerations are most easily met for water- or gas-cooled reactors, which have considerable Generation II and III operating experience. SFR designs also benefit from Generation II operational experience, but sodium reactors do not have as much operating experience as the light water systems. The MSR and LFR designs have limited operating experience and require significant development to achieve the performance levels currently available in operating light water- and gas-cooled reactors.

2.4.2.2.11
Economic Considerations

At the most basic level, nuclear power plants utilize the fission of uranium, thorium, and plutonium to provide a heat source to boil water and produce steam that drives a turbine generator to produce electricity. Accordingly, nuclear energy competes with other energy sources and is ultimately judged by its safety, cost-effectiveness, and public acceptance. Nuclear power has an added regulatory overhead that adds to its cost profile. Other energy sources such as natural gas do not have the regulatory burden attached to a nuclear power plant. Nuclear power is also associated with radiation and its deleterious effects. In the twentieth century, this association created a climate of fear among a portion of the public that has been reinforced by the accidents at TMI and Chernobyl.

In the first decade of the twenty-first century, nuclear power appeared to be undergoing a renaissance with numerous organizations expressing interest in new Generation III plants. In the United States, new plant designs were certified and a streamlined licensing approach contributed to a positive outlook for nuclear power. In addition, low interest rates promoted investment by nuclear utilities.

These positive conditions began to erode with declining world economic conditions and a major recession in the United States. A combination of increasing capital costs (5–10 billion US dollars per unit), additional regulatory requirements, and eroding public confidence following the 2011 Fukushima Daiichi accident further dampened the outlook for new US and European nuclear power growth. In addition, recently discovered natural gas reserves and low natural gas prices have soured the outlook for a nuclear renaissance in the United States. Instead of growth, a number of Generation II US reactors have been shut down before their license expiration.

Utilities could not justify the continued operation of some nuclear units, because outstanding issue resolution was cost prohibitive or the facility was no longer economically viable. The issues and concerns that led to these decisions include steam generator degradation, loss of containment concrete integrity, public and state government opposition, power generation costs, failure to obtain long-term power contracts, and government mandates for renewable energy.

The Fukushima Daiichi accident also led to a number of nations deciding to eliminate the nuclear option from their future energy plans. Although these conditions are not as favorable as those of the early twenty-first century, the nuclear option still exists in the United States and at least 60 nations have expressed interest in developing nuclear generating capability. Some of the improved global outlook is associated with SMRs.

2.4.2.3 Small Modular Reactors

SMRs offer more distributed generation and the initial costs are projected to be about an order of magnitude less than nuclear base load units having outputs of 1000 MWe or greater. The SMR deployment strategy is based on the premise that these units can be manufactured, shipped, and assembled on-site within about 3 years of being ordered. Conventional base load nuclear reactor construction times are much longer and can take 5–10 or more years.

There are a number of SMRs under development, and these systems could be available as prototype plants before 2020. These reactors have a number of potential applications including electrical generation in developed and developing markets, generation of industrial process heat, desalination, hydrogen production, oil shale recovery, transmission boosting, and district heating.

The IAEA defines a reactor with an output of 700 MWe or greater as a large conventional nuclear reactor. Small nuclear reactors are defined as those producing less power, typically on the order of about 300 MWe or less, and are intended to

Table 2.10 Candidate SMRs for deployment in the 2020–2030 time frame[a)].

Vendor	SMR designation	Reactor type	Power (MW_e)	Refueling interval (year)
Babcock & Wilcox	mPower	PWR	125	5
NuScale Power	NuScale module	PWR	45	2
Westinghouse	IRIS	PWR	335	3–3.5
Toshiba	4S	Sodium-cooled fast reactor	10 50	30
GE-Hitachi	PRISM	Sodium-cooled fast reactor	311	1–2
PBMR, Ltd.	PBMR	Pebble bed modular reactor	165	Online refueling
Hyperion	Hyperion power module	Lead–bismuth-cooled fast reactor	25	7–10

a) Energy Policy Institute Report (2010).

be modular in construction. The modular approach facilitates component standardization and construction.

Although there are a number of viable SMR designs under development, only a small number are expected to become NRC certified and commercially available within the United States in the next 20 years. Table 2.10 provides a summary of SMR vendors that have submitted a letter of intent to certify their reactors with the NRC. The reactor types summarized in Table 2.10 can be grouped into three major categories: PWRs, PBMRs, and advanced reactor concepts including liquid metal and SFRs.

The small PWRs noted in Table 2.10 are generally intended for electricity generation, and their vendors forecast a deployment schedule of 5–10 years. These reactors have direct application in areas where electrical grids have limited capacity and conventional power sources are not available. Accordingly, these PWRs have potential for power production in developing countries or remote areas. However, security concerns must be addressed during transport and emplacement of SMRs.

PBMRs are designed to generate process heat for industrial applications. Advanced reactor concepts including liquid metal and SFRs have emerging fuel recycle applications and have an extended core lifetime. These designs have the longest licensing and deployment schedule of the three groups that is estimated to be 15–25 years.

An illustration of the SMR application for electrical production in remote areas is provided by a Toshiba reactor proposal. Galena, Alaska, is not readily served by conventional sources of energy and is an ideal candidate for a small modular nuclear reactor. The reactor under consideration is Model 4S (supersafe, small, and simple).

Model 4S has a design lifetime of 30 years and is available in 10 and 50 MWe versions. This model is a sodium-cooled, fully sealed, passively safe, and transportable design that is factory fabricated and shipped to the installation site.

The 4S reactor is a totally enclosed unit with the core and primary coolant loops sealed in a cylindrical structure. Radioactive emissions and radiation exposures from the 4S reactor are projected to be minimal. The core is designed to operate with an initial enrichment below 20% to meet nonproliferation recommendations.

The main reactor components are located within the reactor vessel that is surrounded by a second vessel. This second boundary or guard vessel forms an additional boundary for the primary sodium circulated by pumps located above the core. The 4S design contains three heat transport systems: the primary sodium system, a secondary sodium system, and the steam turbine generator system.

A number of positive health physics features are associated with the 4S design. First, the reactor does not require refueling during its useful lifetime. This feature eliminates refueling outages and their associated doses. It also enhances proliferation resistance since the need to store spent fuel at the site is eliminated. The elimination of these outages simplifies the design requirements and minimizes related maintenance. Design simplification and reduced maintenance requirements lead to reductions in radioactive effluents and radiation doses.

A second feature is that the 4S reactor utilizes passive safety systems. Passive safety features include a negative temperature coefficient, a primary sodium coolant system operating at about atmospheric pressure, and a reactor vessel auxiliary cooling system using natural air circulation to provide decay heat removal. The negative temperature coefficient, achieved by the fuel and core design, dampens the fission reaction as the temperature increases.

A third feature is that the reactor and support buildings are modular and transportable. This feature minimizes the time to construct, test, and operate the reactor.

Low maintenance is a fourth positive feature of the 4S. Since electromagnetic pumps circulate primary coolant, there are no mechanical systems within the reactor vessel. Therefore, the potential for equipment failures associated with the primary system is low. The secondary plant systems are modular that facilitates easy replacement with limited maintenance.

A fifth feature is enhanced security. The 4S is located in an underground concrete shaft. This design feature provides a robust fission product barrier and minimizes the radiological impact of a terrorist attack.

A sixth feature is design standardization. Standardization provides a positive influence on many reactor considerations including licensing, fabrication, transportation, safety performance, and cost.

The licensing of the 4S represents a regulatory challenge because it is a new system. As noted in Chapter 7, the US regulatory environment must adapt to such challenges for nuclear reactors to remain a viable source of power. The passive design features noted earlier should make the 4S relatively immune to severe accidents. However, the 4S design requires thorough review and verification during the licensing process.

2.4.2.4
Health Physics Hazards

The essential health physics aspects of Generation II and III reactors are applicable to the Generation IV systems. Under normal operating conditions, activation products including ^{58}Co and ^{60}Co dominate worker effective doses. Assuming the effectiveness of health physics programs, internal doses will not be limiting. Accident releases will be dominated by noble gases and iodine.

It is also expected that maintenance and surveillance activities continue and that normal and outage activities are similar to Generation II and III operating practices. Given these assumptions, the subsequent discussion outlines generic and specific activities of health physics concern at Generation IV reactors.

2.4.2.4.1
Generic Generation IV Health Physics Hazards

The power reactor health physicist deals with a variety of issues. Internal and external dose controls are not unique to the power reactor environment, but their implementation is dependent on the specific reactor environment and its operating characteristics.

A summary of the health physics concerns associated with generic Generation IV power reactor activities is provided in Table 2.11. Examples of these work activities include primary component maintenance during outages and power operations, steam generator surveillance and repair, recirculation pipe replacement, spent fuel pool work activities, refueling operations, containment at power inspections, waste processing operations, component decontamination, and spill cleanup. The activities illustrated in Table 2.11 involve both internal and external exposure pathways. Activation products, fission products, and hot particles (APFPHPs) are common health physics concerns in many of these activities. The presence of fission products depends on fuel integrity.

LFR, MSR, and SFR Generation IV reactors have lead, molten salt, and sodium core coolants, respectively, that will adhere to primary system components. The core coolant contains APFPs. MSR coolants also contain TRU isotopes. These radionuclides present a unique maintenance challenge that may require component removal for conventional maintenance or development of specialized remote repair methods including robotic techniques.

From Generation II and III experience, activation of the core's ^{16}N coolant presents an operational concern in PWR and BWR systems. ^{16}N doses are an operational concern near primary system piping. A similar coolant activation concern exists in Generation IV reactors. Isotopes of concern for the various Generation IV reactors are addressed in Section 2.4.2.2.

Table 2.11 Generation IV power reactor generic work activities and associated health physics hazards.

Work activity	Reactor types	Radiological concern[a]
Primary component maintenance during a refueling or maintenance outage	All	APFPHP[b] personal contaminations APFP direct dose Core coolant adhering to primary system components (LFRs, MSRs, and SFRs)
Primary component maintenance during power operations	All	APFPHP[b] personal contaminations APFP direct dose Core coolant direct dose[c] Core coolant adhering to primary system components (LFRs, MSRs, and SFRs)[c] Neutrons Photons
Steam generator eddy current surveillance and tube repair during an outage	All reactors with steam generators (independent of type)	APFPHP[b] personal contaminations APFP direct dose Core coolant adhering to steam generator tubes in LFRs, MSRs, and SFRs[c]
Recirculation pipe replacement	SCWR systems depending on the final design	APFPHP[b] personal contaminations APFP direct dose
Spent fuel pool activities including fuel rearrangement, control rod replacement, fuel assembly reconstitution, and cleanup activities	All except MSR[d]	APFPHP[b] personal contaminations Criticality APFP direct dose
Refueling operations	All	APFPHP[b] personal contaminations APFP direct dose Core coolant adhering to primary system components (LFRs, MSRs, and SFRs)[c] Core coolant direct dose[c] Tritium Criticality

(*Continued Overleaf*)

Table 2.11 (*Continued*)

Work activity	Reactor types	Radiological concern[a]
Containment at power inspections[e]	All	APFPHP[b] personal contaminations APFP direct dose Noble gases Tritium Neutrons Photons Core coolant direct dose[c]
Online refueling, radioactive waste processing, and actinide recycle	MSR[d]	APFPHP[b] personal contaminations APFP direct dose Core coolant direct dose[c] Neutrons Photons Criticality
Radioactive waste processing	All	APFPHP[b] personal contaminations APFP direct dose
Component decontamination	All	APFPHP[b] personal contaminations APFP direct dose Core coolant direct dose[c] Core coolant adhering to primary system components (LFRs, MSRs, and SFRs)[c]
Spill cleanup	All	APFPHP[b] personal contaminations APFP direct dose Core coolant direct dose[c] Solidified core coolant (MSRs, LFRs, and SFRs)[c]

a) Tritium is a hazard for all activities at heavy water reactors.
b) Activation products, fission products, and hot particles. The fission product activity levels depend on fuel integrity or ability of the MSR coolant to retain fission products and other radioactive materials.
c) Core coolant activation products vary by reactor type and are discussed in the text.
d) MSRs have no fuel fission product barrier since the fuel and coolant form a eutectic mixture. Refueling occurs while the reactor is operating.
e) This is a Generation II and III reactor activity that improved maintenance and outage planning. Operating experience and operating philosophy will determine if it is utilized at Generation IV facilities.

2.4.2.4.2
Specific Health Physics Hazards

The generic descriptions of Table 2.11 provide an overview of the radiation hazards that affect task performance at Generation IV reactors. Knowledge of these generic hazards facilitates the introduction of specific hazards. For specificity, selected tasks and facility conditions are chosen to illustrate specific health physics hazards. These tasks and conditions are the buildup of radioactive material in components such as filters, demineralizers, and waste gas decay tanks; activation of reactor components; fuel damage; reactor coolant system leakage; hot particles; and effluent releases.

2.4.2.4.3
Buildup of Activity in Filters, Demineralizers, and Waste Gas Decay Tanks

The reduction of activity concentrations in radioactive fluids is an important consideration in minimizing worker doses. Filters, demineralizers, and waste gas decay tanks are often used to reduce fluid system activity levels.

Air filters trap airborne radioactive material, liquid filters remove suspended particulates, demineralizers use an ion exchange technique to extract radioactive material from liquid streams, and waste gas decay tanks collect fission gases and iodine removed from the primary coolant. LWR operating experience demonstrated the effectiveness of these components to reduce the release source term. Similar components will be developed to limit the source term for the MSR, LFR, and SFR designs. This is important from a health physics perspective, because all power reactor types benefit from minimizing the activity of radioactive material available for release.

The radioactive materials that accumulate in filters and demineralizers are primarily activation products. Fission products accumulate if fuel damage has occurred. MSR designs must contend with fission products and actinides since the fuel forms a eutectic mixture with the salt coolant. Metal-cooled reactors exhibit complex interactions between the coolant, fission products, and fuel form.

Filters are commonly used to reduce effluent concentrations. A variety of air filter types (e.g., HEPA and charcoal) remove airborne activation products, fission products, and iodine. Liquid filters vary in construction and composition, but most types mechanically remove radioactive material suspended in fluid streams.

Filter performance is affected by the coolant medium and its interaction with radioactive materials. Although similar fission products and actinides are generated in both light water- and metal-cooled reactor types, their chemical interactions with the coolant and release to the containment atmosphere are unique to the specific reactor type. Since most metal research has been performed on liquid sodium coolant, the discussion focuses on that material.

In a severe accident at a sodium-cooled reactor that breaches primary system piping, noble gases are immediately released to the containment. The volatile halogens (iodine and bromine), alkali metals (cesium and rubidium), alkali earths (strontium and barium), and chalcogens (tellurium and selenium) are soluble in liquid sodium metal. Some of these fission products form soluble sodium compounds. These compounds are released from the sodium coolant by vaporization from the liquid surface or though aerosol production via sodium combustion.

Sodium aerosols tend to agglomerate into rather large fluffy particles. In addition, sodium chemically reacts with several fission products and forms aerosol compounds that settle quickly. These aerosols tend to agglomerate into large particles that precipitate close to the release point during atmospheric dispersion. These properties must be considered when analyzing fission product transport and retention in filters.

Demineralizers or equivalent components are expected to be utilized in Generation IV systems to limit the source term. In Generation II reactors, the radiation levels inside demineralizer cubicles associated with spent fuel cleanup systems can exceed the US regulatory criteria for very high radiation areas (5 Gy/h at 1 m from the source). Following fuel damage, demineralizer radiation levels increase dramatically with the release of fission products through the fuel fission product barrier.

MSR demineralizer or equivalent systems are unique because the fuel and salt coolant form a eutectic mixture. APFPs are removed from the coolant as part of the facility's design. Demineralizer loading and changeout are unique aspects of the MSR, and the selection of ion exchange or equivalent media requires careful selection to avoid radiation or physical degradation of the media.

Waste gas decay tanks accumulate fission gases and iodine that are removed from the reactor coolant. The radioactive material is stored and retained in these tanks until it meets the regulatory criteria for release to the environment. These tanks and supporting systems that remove fission gases from the core coolant should be integral Generation IV systems.

The buildup and decay of radioactive material in a system is described in terms of production equations. Production equations describe a variety of physical processes, are important in a number of health physics applications, and are derived in Appendix B. The activity deposited into a filter, demineralizer bed, or waste gas decay tank is described in terms of a production equation.

The buildup of activity of isotope i (A_i) on a filter, in a demineralizer bed, or in a waste gas decay tank is determined from the system properties and isotopes present in the fluid entering these components. As noted previously, additional components and physical mechanisms must be considered with metal and salt coolants, but these systems are also described by production equations. The following discussion assumes a constant rate of production and exponential removal terms. Other phenomena introduced by changes in chemical composition, particle interactions, or chemical reactions are also described by production equations if their associated removal terms are exponential.

Important parameters impacting the buildup of activity of isotope i in filters, demineralizers, or waste gas decay tanks include the concentration of the isotope in the fluid entering the device (C_i), the system flow rate (F), the time the system is operating or processing influent (t_{op}), and the time the system is isolated (t_{decay})

from the influent stream:

$$A_i = \frac{C_i e_i F}{\lambda_i} \left(1 - \exp(-\lambda_i t_{\rm op})\right) \exp(-\lambda_i t_{\rm decay}) \quad (2.12)$$

where e_i is the efficiency of the filter or demineralizer for removal of isotope i and λ_i is the radioactive decay constant of isotope i. Fluids containing multiple isotopes require the application of Eq. (2.12) for each nuclide present in the influent stream. The retention efficiency is typically unity for an intact waste gas decay tank.

The types of radioactive material deposited in filters, demineralizers, and waste gas decay tanks vary with the specific reactor design. Activation products are design specific as noted in Table 2.8. These activation products vary considerably and depend on the coolant type, materials used in the construction of the primary system, fuel type, and the reactor's neutron spectrum (i.e., thermal or fast).

In a similar manner, fission product generation depends on the specific fuel composition and neutron spectrum incorporated into the design. For example, fission products are derived from a variety of fissile nuclides including ^{233}U, ^{235}U, ^{239}Pu, and ^{241}Pu for thermal fission and ^{232}Th and ^{238}U for fast fission.

2.4.2.4.4
Activation of Reactor Components

The activation of reactor components and corrosion products are described in terms of production equations. Corrosion or wear products dissolved or suspended in the primary coolant are subjected to the core's neutron fluence. Activation occurs by a variety of neutron-induced reactions, and the nuclides produced depend on the neutron spectrum and fluence impinging upon the material in the core region. Generation IV activation reactions are illustrated in Table 2.8.

The activity resulting from an activation reaction has the form

$$A_i = N_i \sigma \phi \left(1 - \exp(-\lambda_i t_{\rm irr})\right) \exp(-\lambda_i t_{\rm decay}) \quad (2.13)$$

where N_i are the number of target atoms that are activated, σ is the microscopic cross-section for the activation reaction, ϕ is the neutron fluence rate or flux inducing the activation reaction, $t_{\rm irr}$ is the time the target is irradiated or exposed to the core flux, and $t_{\rm decay}$ is the decay time or time the target was removed from the reactor's core region or activating flux.

Activated material presents an internal as well as external dose concern. In a power reactor environment, external radiation sources are dominated by ^{60}Co in most Generation II reactors and ^{58}Co in late Generation II and many Generation III PWRs and BWRs. External doses are dominated by activation sources that emit beta and gamma radiation types.

Once the activity of a source is determined, its dose rate impact is governed by its geometry. Common source geometries include the point, line, disk, and slab configurations. Relationships for calculating dose rates from these geometries are summarized in Appendix C. Appendix C provides a summary of other important health physics relationships that are useful in external dosimetry applications. Internal dosimetry relationships are provided in Appendix D.

For example, the dose rate at a distance *r* from a small source is obtained from a point source approximation. Hot particles, radiography sources, criticality events, and other small sources are often represented using a point source approximation. The point source approximation is accurate to about 1% whenever the distance from the source is at least three times the largest source dimension.

A second useful relationship encountered in a power reactor environment is the line source approximation. The line source equation is useful when assessing the dose from sample lines or piping carrying primary coolant or other radioactive fluids. Fuel rods, resin columns, and irradiated rods are also accurately approximated using line sources.

The third useful relationship for estimating the dose rate from typical power reactor components is the thin disk source approximation. A disk source provides a reasonable approximation to the dose rate from a radioactive spill or contaminated surface.

Slab sources are useful in approximating the dose rates from contaminated soil, contaminated pools, or demineralizer beds. Dose rates from a spent fuel pool and contaminated concrete floors or walls are obtained with reasonable accuracy with a slab source approximation.

2.4.2.4.5
Fuel Damage

With the exception of MSRs, a nuclear reactor contains three barriers to prevent fission products from escaping from the reactor core to the environment. These barriers are the fuel matrix (e.g., pellet) and associated cladding or coating, the reactor coolant system and included piping, and the containment building. A breach of any of these barriers enhances the probability that radioactive material will be released to the environment.

The robustness of the fuel fission product barrier depends on its specific composition. For example, in PWRs, BWRs, and Canada Deuterium Uranium (CANDU) reactors, the fuel fission product barriers consist of UO_2 pellets enclosed within a stainless steel or zirconium alloy tube. In VHTRs, the fuel is coated in a ceramic, and the fuel fission product barrier is the SiC or ZrC fuel coating and the fuel material matrix.

Fuel damage facilitates the release of fission products contained between the matrix and cladding (gap activity) or between the ceramic coating and fuel and increases the primary coolant activity. Noble gas activity entering the primary coolant is released to the containment atmosphere through leakage paths or removed by gas stripping systems. The detection of these gaseous fission products is an early indication that a failure or mechanical damage to the fuel cladding/coating has occurred. BWRs normally detect fuel failure by detection of fission gases in the off-gas system. However, PWRs normally monitor the primary coolant or letdown line for these fission products. Fission products are also detected by sampling the containment atmosphere for released noble gases (e.g., xenon and krypton) and their daughter products. The analysis of primary coolant samples by gamma spectroscopy is a routine confirmatory action.

Fuel performance in a Generation IV system must be demonstrated. GFR, VHTR, SCWR, and SFR systems derive fuel performance experience based on Generation II and III designs that operated at lower temperatures. The effects of increased temperatures can be significant since chemical reaction and corrosion rates tend to increase as temperatures increase. Fuel performance in these systems must be achieved to ensure a reliable fission product barrier.

There is significantly less data regarding the fuel performance in LFRs and MSRs. The ability of the LFR fuel and lead coolant to retain fission products during normal operations and transient conditions has yet to be demonstrated.

A severe accident involving MSR fuel has a different character than light water uranium dioxide fuel. In a MSR accident in which core cooling capability is lost and the solidified fuel plug melts (Section 2.4.2.2.3), the fuel storage tanks are assumed to provide a barrier equivalent to that of the reactor coolant system. If the fuel storage tank barrier fails, a unique accident condition exists that requires additional evaluation. The accident's severity depends on the capability of the MSR coolant to retain fission products when its temperature is elevated and the capability of the containment to withstand the stress induced by a breached fuel storage tank. If the containment is breached, the release of fission products to the environment depends on the aerosol characteristics generated during the accident.

2.4.2.4.6
Reactor Coolant System Leakage

Since reactors are electromechanical systems, leakage from the primary coolant system is an undesirable but inevitable problem. This leakage occurs in Generation II and III systems and will occur in Generation IV reactors. Generation II and III leakage is well quantified, but some Generation IV systems use metal and salt coolants instead of light water, and their leakage characteristics during operating conditions have yet to be determined.

Value stems, pump seals, value packing, and instrument line connections provide pathways for small leaks to contaminate local areas. This contamination must be controlled in order to limit external and internal doses. In addition to primary system leaks, health physicists must address leakage from the primary to secondary systems for reactors using steam generators.

Leakage of primary coolant from steam generator tubes to the secondary system presents a health physics concern because additional plant areas become contaminated. Since the secondary components are clean systems, the presence of contamination has a negative impact on facility operations and expands areas requiring stringent radiological controls.

Secondary coolant contamination has a number of negative health physics aspects. The secondary activity tends to concentrate in components such as main steam isolation valves and high-pressure turbine piping resulting in surface contamination areas and local hot spots. Secondary ion exchange resins and filters become contaminated which adds to the facility's radiological control requirements and increases the volume of radioactive waste. Steam generator cleanup systems also become contaminated. Contaminated secondary system areas increase health physics survey requirements and associated decontamination activities.

Primary to secondary leakage increases the likelihood of a release of radioactive material (e.g., noble gases and iodine) to the environment. The most likely release pathways are through a secondary system relief valve or the condenser air ejector.

Leakage of metal or salt coolants presents additional challenges beyond those encountered in LWRs. In addition to the presence of fission products and their dispersion by the core's decay heat following an accident, metal and salt coolants present the possibility for chemical reactions and phase transition energies to enhance the dispersion of radioactive material. Although liquid sodium reactions have been studied, less is known about other metal and salt coolants and their reactions with fission products and construction materials encountered as the coolant leaks onto surfaces supporting the reactor coolant system. These reactions and their ability to mobilize fission products govern the health physics consequences of leakage events in LFR and MSR Generation IV systems.

2.4.2.4.7
Hot Particles

The maintenance of pumps, valves, and primary system components and piping create small particles during the process of testing, cutting, grinding, and welding. Operation of valves and pumps leads to wearing of active surfaces that produces small particulate material. Cladding erosion and failures or erosion of control rod surfaces contribute additional particulate matter to the reactor coolant system. This material is often too small to be removed by the reactor coolant system's filters. It passes through the core, and fission neutrons create highly activated, microscopic material commonly called a *hot particle*. Given the nature of the MSR coolant–fuel eutectic, hot particles could present a health physics challenge in these Generation IV reactors.

Hot particles are comprised of combinations of activation products, fuel fragments, and fission products depending upon the integrity of the fuel fission product barrier. Particles may contain either a single or a large number of radioisotopes. Beta radiation is the dominant contributor to the skin dose, but gamma contributions can approach about 30% of the total dose contribution.

The absorbed dose (D) to the skin from a hot particle is given by the relationship

$$D = \frac{t}{S} \sum_i A_i F_i \tag{2.14}$$

where A_i is the particle activity for radionuclide i, F_i is the dose factor for radionuclide i, t is the residence time on the skin, S is the area over which the dose is averaged, and i is the hot particle radionuclide label.

Hot particles also attach to the eye, deposit in the ear, enter the lungs through inhalation, and irradiate the gastrointestinal tract following ingestion. NCRP 130 provides guidance for addressing these specific hot particle conditions.

Equation (2.14) is often evaluated using the VARSKIN computer code described in Appendix E. This code permits the skin dose to be calculated at various depths or volumes.

Following NCRP 130, the skin dose is usually averaged over 10 cm^2 and evaluated at a depth of 7 mg/cm^2 at the depth of the basal cell layer. Since the dose from a point source falls off rapidly as $1/r^2$, the dose from a hot particle is highly localized.

2.4.2.4.8
Environmental Releases

The effluents that characterize a facility depend on the core materials, reactor materials, and specific design aspects of the Generation IV system. Examples of the unique nuclides that will likely appear in a Generation IV facility are summarized in Table 2.8. Additional nuclides common to Generation II, III, and IV systems are provided in Appendix A.

Light water and heavy water reactor effluents are primarily isotopes generated through the activation and fission processes. Although waste gas systems are designed to trap most gaseous effluents, quantities of noble gases, ^{3}H, ^{14}C, and iodine isotopes are available for release. Their release is facilitated by defects in the fuel clad/coating or failure of the MSR coolant to retain fission products.

Isotope production mechanisms are design dependent. For example, tritium arises from the neutron activation of the primary coolant [$^{2}H(n, \gamma)^{3}H$] and from tertiary fission. Tritium production is enhanced in CANDU reactors that use a D_2O coolant. In a PWR, tritium is also produced from neutron capture in ^{10}B used for reactivity control [^{10}B (n, 2α) ^{3}H] and from neutron capture in ^{6}Li used for primary system chemistry control [$^{6}Li(n, \alpha)^{3}H$]. ^{14}C is usually produced from $^{14}N(n, p)^{14}C$ and the $^{17}O(n, \alpha)^{14}C$ reaction in a CANDU reactor. In GFRs and VHTRs, tritium is produced in the gas coolant reactions $^{4}He(\gamma, p)^{3}H$ and $^{4}He(n, d)^{3}H$. MSR salt coolants produce tritium via $^{6}Li(n, \alpha)^{3}H$.

Liquid effluents include fission and activation products as well as tritium. Tritium is the dominant liquid effluent in PWRs. The quantity of fission products in liquid waste depends on the integrity of the fuel fission product barrier. Liquid waste cleanup systems, including filtration and demineralization, remove most of these radionuclides from the effluent stream. Similar effluents are expected in Generation IV reactors. However, the high-temperature SFR, LFR, and MSR are departures from light water- and gas-cooled reactors, and their liquid release characteristics are not well defined.

Fission product radionuclides generated from binary fission in Generation III reactors include ^{85}Kr, ^{87}Kr, ^{88}Kr, ^{133}Xe, ^{135}Xe, ^{137}Xe, ^{131}I, ^{137}Cs, ^{137m}Ba, ^{141}Ce, ^{144}Ce, ^{103}Ru, ^{106}Ru, ^{103}Rh, ^{106}Rh, ^{90}Sr, and ^{90}Y. Generation IV reactors will produce similar fission products.

Generation IV reactor activation products are produced from a variety of reactions including $^{54}Fe(n, p)^{54}Mn$, $^{58}Fe(n, \gamma)^{59}Fe$, $^{57}Co(n, \gamma)^{58}Co$, $^{58}Ni(n, p)^{58}Co$, $^{59}Co(n, \gamma)^{60}Co$, and $^{94}Zr(n, \gamma)^{95}Zr$. The aforementioned (n, γ) reactions are normally induced by thermal neutrons, and the (n, p) reactions are initiated by fast neutrons. The specific activation products vary with reactor type and generation.

As an illustration, the activation products and associated effluents in gas-cooled reactors are briefly reviewed.

Since gas-cooled reactors have different materials of construction than water-cooled reactors, distinct radionuclides inventories and effluents are expected. As an illustration, the effluents from CO_2 and ^{4}He gas-cooled reactors are outlined in the subsequent discussion.

Advanced CO_2 gas-cooled reactors developed in the United Kingdom are graphite-moderated facilities. Activation of the CO_2 primary coolant produces ^{14}C, ^{16}N, and ^{41}Ar, and activation of the graphite moderator yields ^{3}H, ^{14}C, and ^{35}S. Fission products similar to those noted for PWRs, BWRs, and CANDUs are also produced.

The graphite moderator may contain trace sulfur and chlorine impurities that lead to ^{35}S production through the ^{34}S(n, γ)^{35}S and ^{35}Cl(n, p)^{35}S reactions. Graphite also contains lithium impurities that upon capture of neutrons produce tritium through the ^{6}Li(n, α)^{3}H reaction.

One of the key features affecting the effluent releases in helium-cooled reactors is the concentration of impurities in the graphite moderator. These impurities vary with the type of graphite used in the design. It is expected that a variety of elements will be found in the graphite moderator including boron, cesium, calcium, carbon, chlorine, cobalt, helium, iron, lithium, nickel, nitrogen, niobium, and uranium. The concentrations of these elements directly affect the effluent characteristics. Therefore, identical Generation IV helium-cooled reactors could have different effluent radionuclide characteristics if their graphite specifications are not the same.

Helium-cooled reactor metallic materials are dominated by chromium, iron, and nickel with smaller quantities of cobalt and molybdenum. These elements lead to activation products including ^{55}Fe, ^{59}Ni, ^{60}Co, and ^{63}Ni.

The previous discussion illustrates the uncertainty involved in the production of gas-cooled reactor effluents. The specific design requirements including materials specifications govern the radionuclides produced and their abundance.

2.4.2.4.9
Advanced Reactor ALARA Measures

One of the Generation IV goals is minimizing worker radiation doses. Reactor components are designed to be nearly maintenance-free and minimize the production of activation products. In particular, cobalt alloys are restricted. This minimizes the ^{60}Co source term that dominates in Generation II facilities.

Component arrangement and accessibility are optimized in Generation IV reactors. These features enhance task completion, minimize radiation doses, and facilitate operability testing of reactor components. For example, heat exchangers, tanks, and vessels are designed to minimize the collection of radioactive material and facilitate the removal of any radioactive material collecting within their boundaries. Components are arranged to allow for sufficient room for maintenance, surveillance, and inspection activities.

Passive safety systems are incorporated to preserve fission product barriers and minimize the consequences of events that led to the TMI-2 and Fukushima Daiichi accidents. Design enhancements include additional core cooling inventories associated with safety systems, improved materials to minimize corrosion of components such as steam generator tubes, enhanced control room instrumentation to provide indication of abnormal conditions, enhanced core cooling capability, and improvements in emergency electrical supply capabilities during loss of power events.

2.5 Nuclear Proliferation

The nuclear nonproliferation treaty provides for nations to acquire nuclear technology (e.g., fission reactors, fuel reprocessing facilities, and uranium enrichment systems), and most of these facilities are subject to monitoring. These facilities are monitored and inspected by the IAEA whose goal is to ensure that material is not diverted for military or terrorist purposes.

Nuclear proliferation concerns arise from a number of fuel cycle activities including uranium enrichment, ^{239}Pu production through reactor operation, and spent fuel reprocessing. The advanced centrifuges and laser enrichment technologies have the capability to produce HEU that can be used for nuclear weapons production. Reactor operation produces ^{239}Pu that if extracted through reprocessing can be diverted to weapons production. Each of these fuel cycle steps is examined in terms of its proliferation impact.

2.5.1 Advanced Centrifuge and Laser Enrichment

Traditional uranium enrichment technologies (e.g., gaseous diffusion and gas centrifuge) require facilities that are difficult to hide. These facilities also have noticeable electric power requirements. Advanced centrifuge facilities utilizing higher output machines and laser enrichment techniques require significantly less space and less electric power that further facilitates their clandestine operation. The efficiency of advanced enrichment technologies increases the likelihood of their use to produce HEU which is an integral component of a uranium nuclear weapon.

These advanced uranium enrichment technologies have the potential to lower the fuel and associated generating cost for nuclear power plants but also pose increased proliferation risks. The proliferation risks of a uranium enrichment process increase as the technology becomes more efficient. If the size of an enrichment facility decreases, its construction may no longer be visible through aerial surveillance. In addition, electrical efficiency precludes the necessity for an observable, dedicated power facility and may preclude a large heat signature observable through satellite imaging. Therefore, an extremely efficient uranium enrichment facility could be clandestinely constructed and operated to produce weapons-grade uranium. This would create a significant security risk and an associated nuclear proliferation concern.

2.5.1.1
Advanced Centrifuge

Glaser notes that the production of weapons-grade uranium is accomplished with natural uranium feed material in an advanced centrifuge. Using standard formulas for separative work (Eq. (2.4)), Glaser determined that only about 280 kg of natural uranium feed is needed to produce 1 kg of weapons-grade HEU (≥90 wt%). This proliferation scenario assumes batch recycling, no discarded material, and no mixing occur.

The use of pre-enriched feedstock and batch recycling presents an additional proliferation example. Results published by Glaser predict that more than 100 kg of weapons-grade uranium can be produced in 1 year with 3.5 wt% pre-enriched feed material in an enrichment facility with a capacity of 5000 SWU/year, which is equivalent to about 2000 P-1 centrifuges (see Table 2.1). Without international inspections, an existing uranium enrichment facility could produce weapons-grade material if higher enrichments are used as feed material. This is accomplished using existing enriched material or altering the enrichment cascade to recycle product material to enrich it to higher levels. By repeating this process, weapons-grade material can be produced over time. These examples illustrate the inherent proliferation risk of advanced enrichment technologies.

Detecting the diversion of enriched material or enriching material beyond facility limits is a complex task. Higher enrichments are detected by monitoring process lines or UF_6 cylinders entering the facility as feed material or exiting as product. Operating data from various facility systems could also be used to detect the production or diversion of higher-enriched uranium. These monitoring activities are implemented through international safeguards efforts but would not be applicable for a clandestine facility or a facility that was not open to international inspection.

2.5.1.2
Laser Enrichment

The APS raised concerns regarding the proliferation risk of the SILEX process that uses lasers to enrich uranium. In its 2010 petition to the NRC, the APS argued that the SILEX technology could increase the global risk of a nation clandestinely acquiring ^{235}U of sufficient enrichment to increase their nuclear weapons capability. The APS noted that SILEX is up to 16 times more efficient than centrifuge technology. This efficiency permits SILEX enrichment facilities to be smaller and more easily concealed than other enrichment technologies including gaseous diffusion and gas centrifuge.

In view of the projected separation factors and performance characteristics, a single laser isotope separation unit would likely exceed the 5000 SWU/year criteria to obtain 100 kg of HEU noted by Glaser in Section 2.5.1.1. This means that a single laser isotope separation unit could replace thousands of conventional centrifuges or a hundred more advanced machines. This equivalency illustrates the proliferation potential that results from laser enrichment technology.

The companies operating SILEX argue that it has no greater proliferation potential than a gas centrifuge plant. They also argue that the level of technology required to construct and operate a SILEX facility is well beyond the capabilities of nations that are attempting to acquire nuclear weapons.

The APS did not fully concur with these contentions and raised the additional concern that over time process information will be acquired by nations seeking nuclear weapons. APS concerns also include the commercial availability of baseline SILEX components including the carbon dioxide lasers used in the separation process.

Although the NRC has issued a license for the SILEX process, the concerns noted by the APS illustrate the need for strong controls to limit the dissemination of process-specific information and sale of process-related equipment. Historically, these controls have not been completely successful in limiting the spread of nuclear weapons related technology. The most obvious example of the failure of these controls is the proliferation of centrifuge technology and supporting hardware. Failure to control centrifuge technology has led to the expansion of this enrichment method to nations that have developed or are attempting to develop nuclear weapons capability.

The development of centrifuge enrichment capability by North Korea and Iran focuses worldwide attention on the proliferation issue. Health physicists have a significant role in the proliferation arena since the detection of ^{239}Pu and ^{235}U are tasks well within their capability. The diversion of nuclear material and aspects of nuclear forensics also require significant health physics resources to ensure radioactive materials are detected and their origins identified.

2.5.2
Reactor Plutonium Production

Although Generation IV reactors have a proliferation resistance design goal, experience with the control of centrifuge technology suggests that reactor production of plutonium requires careful monitoring. Reactors produce copious quantities of ^{239}Pu through the neutron capture reaction

$$^{238}\mathrm{U} + \mathrm{n} \rightarrow {}^{239}\mathrm{U} \xrightarrow{\beta^-} {}^{239}\mathrm{Np} \xrightarrow{\beta^-} {}^{239}\mathrm{Pu} \qquad (2.15)$$

About one-third of the power output of a LWR is derived from the fission of ^{239}Pu. In addition, fuel discharged from a LWR (spent fuel) contains significant ^{239}Pu (see Table 2.2). The handling and processing of spent fuel creates a proliferation concern that requires oversight to ensure the ^{239}Pu is not recovered and diverted to illicit weapons production.

The options for a twenty-first-century fuel cycle depend on the acceptance, development, and deployment of new nuclear generating capacity. Although the development of a new generation of nuclear power plants in the United States has lost momentum following the Fukushima Daiichi accident, a global resurgence of nuclear power is emerging.

Any resurgence of US nuclear power depends on a number of considerations including the proliferation resistance of new nuclear power plants and their support facilities (e.g., uranium enrichment, fuel fabrication, and fuel reprocessing). Nuclear power production is intimately linked to proliferation because the technologies used in power production overlap with those used in the production of fissionable material for nuclear weapons.

A 2005 APS report makes a number of recommendations regarding the successful expansion of nuclear power operations in the United States. These recommendations are related to ensuring the proliferation resistance of power reactors and fuel reprocessing activities. Four specific recommendations were provided in the APS report.

A strong research and development program on advanced safeguards technology is the first APS recommendation. The second recommendation calls for making proliferation resistance a high priority in the design and development of future nuclear energy systems. These systems should be open to international inspections and be implemented on a global basis. The third recommendation is to increase international nuclear security and safeguards cooperation. Expansions of US safeguards efforts similar to those in place with Japan and Russia would be warranted. The final recommendation focuses on spent fuel reprocessing, but no specific proliferation position was advocated.

Currently, no option exists in the United States to reprocess spent fuel in order to recover and recycle its valuable constituents. In order to implement reprocessing, several technical decisions need to be made. In addition, political support is required for spent fuel reprocessing to become an accepted national policy.

2.5.3
Fuel Reprocessing

An aqueous chemical method is a logical initial option for fuel reprocessing. It is a proven and mature technology, but its deployment would require many years. Licensing a US reprocessing plant requires overcoming regulatory, environmental, and political challenges. In addition, the plant or group of plants would need a capacity of at least 200 000 tons of spent fuel per year that would make it the largest reprocessing operation in the world.

Opponents would likely argue that the facility is essentially the Plutonium/Uranium Redox Extraction (PUREX) process that was used in the United States in the 1950s to extract plutonium and uranium from low-burnup fuel. The PUREX process recovered plutonium for military applications, and the uranium was re-enriched and fabricated into new fuel. The remaining minor actinides are part of the process waste and are stored in tanks. These PUREX waste storage tanks have degraded and pose an environmental concern. The volume of this liquid waste, their radionuclide content, and potential for leakage to the environment are major objections of PUREX-type processes. These PUREX wastes are considered high-level waste, and there is currently no commercial US facility licensed to accept this material. Legacy reprocessing waste storage tanks are addressed in more detail in Section 3.7.

The basic PUREX process has evolved considerably since its initial use. As currently utilized in France, the United Kingdom, and Japan, the PUREX process focuses on recovering plutonium for the fabrication of MOX fuel. Efforts over the years have reduced processing costs and minimized waste volumes. Currently, minimal liquid waste is produced in the modern PUREX process.

Since there is currently no fuel reprocessing component to the US Nuclear Fuel Cycle, a new twenty-first-century plant can be designed and built to accommodate a number of stakeholder issues. These considerations include the use of advanced instrumentation, detailed process monitoring, proliferation resistance, and a design for physical protection to repel potential terrorist attack scenarios. Each of these considerations will evolve as the twenty-first century progresses.

To overcome the planned proliferation barriers at a twenty-first-century fuel reprocessing plant, the individual or terrorist group would need to expend considerable effort and resources. Technologies incorporated at a new fuel reprocessing plant will include proliferation resistant fissile material characteristics, facilitate a minimum time to detect any intrusion or attempt to divert fissile material, and incorporate significant detection resources.

Individuals or groups that intend to defeat the proliferation measures at a twenty-first-century fuel reprocessing facility would need to (i) defeat the reprocessing facility's design controls, (ii) expend significant financial resources, and (iii) require an extended time to implement proliferation activities. To inhibit the diversion of weapons-grade material, the nonproliferation technologies incorporate at least six characteristics. First, the reprocessing technology should present an extreme technical challenge for the diversion of special nuclear material. Multiple proliferation barriers present an inherent challenge to recover special nuclear material. The significant technical sophistication and materials handling capabilities required to overcome the multiple barriers to proliferation limit the probability of the successful diversion of weapons-grade material.

Second, the reprocessing system design should maximize the economic and workforce resources required to overcome the multiple barriers to proliferation including the use of existing or new facilities. Workforce requirements present an inherent impediment to diverting significant quantities of special nuclear materials.

Third, system design should maximize the time required to overcome the multiple barriers to proliferation. Ideally, the time to overcome the design should exceed the facility lifetime. This time is extended by frequent materials inventory audits, unannounced inspections, and active radiological and visual surveillance of plant areas handling special nuclear materials.

Fourth, the system should be designed to operate using material possessing, chemical, and physical characteristics (e.g., very high dose rates) that negatively affect its utility for use in fabricating nuclear weapons. Although the form of special nuclear material does not prevent diversion, it is an added complication that limits the value of the diverted material.

Fifth, system design should minimize the time following the initiation of diversion for detection resources to reveal any irregularities. The system design should also provide verification that diversion or undeclared production has occurred or is in progress. Detection using a variety of nuclear and radiation instrumentation would add to the effectiveness of system design.

Sixth, systems should be designed to minimize the workforce, technology, and funding required to achieve international detection safeguards requirements. The detection resources should utilize a variety of techniques to maximize the probability of detecting a diversion attempt.

2.5.4 Nuclear Forensics

Since special nuclear and radioactive materials are produced throughout the world, a comprehensive global response is required to effectively stop the diversion and transportation of these materials. If nuclear material is diverted to a clandestine group, its illicit use can be prevented if it is detected during transit. Nuclear material detection reveals its origin through forensic techniques. Forensic analysis includes the characterization of the material to determine its production history and facility of origin. It also provides information that identifies the responsible individuals or groups diverting the material.

The physical and chemical characteristics of a radioactive material (e.g., physical appearance, microstructure, elemental composition, and isotopic composition) provide clues to its origin and history. For example, the isotopic ratios and elemental impurities in natural uranium indicate its geographic origin. The production date of a material can be determined by the presence and activity of its daughter products. Isotopic constituents such as ^{236}U suggest the material was produced in a nuclear reactor. The specific reactor producing special nuclear material is revealed by the spectrum of isotopes present in that material. For example, the percentage of certain fission products is indicative of a specific facility and its unique operating characteristics. Therefore, careful analysis of illicit material reveals its origin and production history. Effective analysis techniques include alpha spectroscopy, electron microscopy, gamma spectroscopy, mass spectrometry, beta spectroscopy, and radiochemical separation techniques.

Although the techniques and procedures involved in nuclear forensics are well established in nuclear-capable countries, an effective nuclear materials interdiction program must involve less advanced nations since their borders could be penetrated to facilitate the transit of these materials. Therefore, personnel from these countries must be trained in nuclear forensic techniques. In addition, these nations must acquire the necessary equipment and facilities for an international program to effectively function. This requires both financial support and resource dedication from nuclear-capable nations.

2.5.5 Nuclear Suppliers

Another approach to combat the illicit procurement of nuclear materials is to restrict the sale of products that support their acquisition. This is a complex task because many of these items also have peaceful uses. Therefore, export control measures must be carefully implemented when establishing items that are restricted for proliferation concerns.

Export control guidelines for nuclear and nuclear-related material must evaluate dual-use items, equipment, tools, software, and technology. These restrictions strive to avoid restricting international trade and cooperation in the peaceful uses of nuclear energy. As with nuclear forensics, export controls require international cooperation to be effectively implemented.

2.6 Twentieth-Century Waste Disposal Options and Solutions

Shallow land burial for low-level radioactive waste and geologic repositories for high-level waste have been the traditional twentieth-century answers for the long-term disposition of fuel cycle waste. These techniques do not address the fundamental issue of minimizing waste generation. Generation IV reactors and transmutation techniques offer the potential to significantly limit the quantity of high-level waste and to reduce the challenges of licensing a high-level waste facility. The objectives and operational characteristics of Generation IV reactors were outlined in the previous discussion.

A number of options have been proposed for the disposal of high-level waste. The most popular options are addressed in the subsequent discussion. A successful option must meet a variety of technical and environmental requirements including minimizing the proliferation risks of deploying the technology, being environmentally acceptable, and being politically acceptable to affected stakeholders.

2.6.1 Boreholes

Deep boreholes have been studied in Sweden, Finland, and Russia, as a possible repository for low-level radioactive waste. If this method were to be utilized for spent nuclear fuel or fuel reprocessing waste, the boreholes would be deep (e.g., several kilometers) and narrow (e.g., diameters typically less than a meter). The high-level waste packages would then be stacked in the borehole and separated by a layer of nonradioactive material (e.g., clay, grout, or concrete).

There is no definitive analysis to prove that radioactive material inserted into a borehole would not enter the environment. In addition, retrieval of the high-level waste is problematic. A number of technical issues remain regarding the borehole approach including (i) the physical integrity of the waste packages under the pressures and temperatures encountered within in the boreholes, (ii) horizontal and vertical migration of radioactive material, (iii) long-term environmental effects, and (iv) effects on animal and plant species in the vicinity of the borehole. Although deep boreholes have applications for small quantities of radioactive waste, this approach is not sufficiently developed as a disposal option for high-level waste.

Deep borehole disposal of high-level waste is periodically revaluated. For example, the US Department of Energy is reviewing this approach for the disposal of cesium and strontium extracted from Hanford's reprocessing waste tanks.

2.6.2 Deep Injection

The issues associated with boreholes are also applicable to the deep injection of liquid high-level waste into rock strata or engineered structures. The inherent physical mobility of liquid waste makes the deep injection method problematic. For example, no proven methods exist to map the minute crevices or cracks that provide a pathway to enhance liquid waste migration. In addition, geologic features change over time, and these alterations greatly complicate a pathway analysis. Deep injection is not currently being pursued by any country. However, the former Union of Soviet Socialist Republics used it in a number of locations.

2.6.3 Rock Melting

A curious approach seals the high-level waste in an underground cavity or borehole. This configuration utilizes the waste's decay heat to melt the surrounding rock. Rock melting assumes that the raw waste form dissolves in the molten rock. The mixture continues to melt the surrounding rock with the radioactive waste mixing with the molten material. Melting continues until a heat balance is reached and the mixture cools, solidifies, and incorporates the raw radioactive material into the rock matrix.

Rock melting is also applicable if the waste material is enclosed in a container that conducts the waste's decay heat. The rock would still melt and eventually cool to theoretically provide a tight and durable protective shell around the radioactive waste container. This approach is not currently being investigated, and there have been no credible demonstrations that rock melting is safe or economical. An obvious concern is that the molten rock would encounter subsurface water. The resulting steam formation could disperse the waste into subsurface water or the atmosphere. Long-term stability of rock melting is also open to challenge.

2.6.4 Subseabed Disposal

Subseabed disposal places containerized waste below the seabed and assumes an extended time for the containers to degrade. This time delay permits radioactive decay to reduce the waste's activity and overall hazard. As the container degrades, the radioactive material dissolves in the ocean and disperses. This is a dilution approach with container decay and waste form characteristics providing a time delay for radioactive material removal. Subseabed disposal was investigated in the 1980s by the Organization for Economic Co-operation and Development. The feasibility of this approach was terminated following significant international opposition.

There are also technical concerns regarding the feasibility of the subseabed approach. Although the premise of dilution and dispersion is technically valid, it fails to consider the effects on biological systems living near the waste. Biological systems have the capability to concentrate nutrients into their bodies. This capability reverses the expectations that are based solely on physical dilution and dispersal.

The bioaccumulation and biomagnification processes concentrate radioactive materials that enter the food chain, and the concentration factors can be significant. Therefore, it is difficult to predict the endpoint of a dilute and disperse approach. An international convention specifically prevents subseabed disposal.

2.6.5 Disposal at Sea

In the 1970s, a number of countries adopted the practice of dumping barrels of low-level radioactive waste directly into the ocean. Sea disposal has a number of negative features that facilitate the dispersion of radioactive material into the environment. Following the loss of container integrity, the dispersed material is concentrated in aquatic plants and fish through bioaccumulation and biomagnification processes. The concentration of radioactive materials in aquatic plants and fish that enter the food chain creates a health detriment if sufficient consumption occurs.

Nations that depended on fishing objected to this practice and an international convention outlawed the deliberate dumping of radioactive wastes into the sea. This high-level disposal option is no longer under active investigation.

2.6.6 Disposal in Ice Sheets

An additional approach to high-level waste disposal involved its placement on an ice sheet in Antarctica. The waste's decay heat leads to the radioactive materials package melting the ice and sinking deeper and deeper below the surface. Since Antarctica is lightly populated and isolated, it was assumed that ice sheet disposal was a safe approach that posed a limited environmental hazard. However, subsurface pockets of brine have been discovered below the ice surface. These saltwater pockets enhance the corrosion of waste containers, liberate radioactive materials, and increase the probability of dispersal. The waste's decay heat also generates steam that provides a mechanism for waste dispersal into the environment.

The ice sheet disposal approach also failed to account for global climate change that alleges to reduce the thickness and extent of the ice sheets. Issues associated with climate change and environmental dispersal of the radioactive materials suggest ice sheet disposal is not an optimum method for high-level waste disposal.

2.6.7 Disposal in Space

Space disposal of high-level waste has been a popular but highly questionable approach. The transport of high-level waste to a spaceport would encounter environmental, licensing, and security challenges. A detailed accident and security threat analysis is required. Space transport is resource intensive, which makes the space disposal option very costly. If these issues are overcome, the reliability of the space transport vehicle requires validation. Rockets and other space transport vehicles have high failure rates relative to conventional transport mechanisms, and a launchpad or atmospheric accident disperses the high-level waste over a significant area. The economic, political, and potential human toll from such an event precludes this option from significant consideration. No country is seriously considering a space disposal option.

2.6.8 Surface Storage and Shallow Land Burial

Waste storage on the earth's surface and in shallow structures has inherent security and environmental concerns. Surface storage requires a prolonged monitoring commitment of the disposal site by current and future generations. The monitoring requirement minimizes inadvertent intrusion, but security issues remain. Surface storage is vulnerable to a release of radioactive material following a degradation of institutional control measures over time or through sabotage. These negative factors preclude surface storage as a permanent solution for high-level waste disposal.

Shallow land burial has similar issues. As such, shallow land burial has not been an acceptable long-term high-level waste disposal option. In spite of the shortcomings of surface and shallow land burial for high-level waste, surface storage of spent fuel in dry storage casks is an acceptable short-term solution.

Dry spent fuel storage usually permits the surface storage of spent fuel that cooled in the reactor's fuel pool. The dry storage casks are typically steel cylinders that are either welded or bolted closed to provide a leak-tight confinement of the spent fuel. Each cask is surrounded by additional steel, concrete, or other material to provide radiation shielding to limit worker, public, and environmental doses.

Following the Fukushima Daiichi accident, there has been a focus on using passive cooling afforded by the dry storage fuel casks as opposed to the active cooling systems used for decay heat removal in a spent fuel pool. Although nuclear fuel has been successfully stored on the surface using dry storage casks, this approach is only an interim solution to long-term high-level waste storage.

2.6.9 Geologic Disposal

For more than 50 years, disposal in a geologic formation has been the option of choice for high-level waste disposal in the United States. It is also the preferred approach in other countries with spent fuel or high-level waste disposal programs.

In a geologic repository, high-level waste is placed in engineered arrays in conventionally mined cavities. The waste is confined in containers to accommodate its form, chemical content, and radiation intensity. A geologic repository includes multiple engineered and natural barriers to limit the pathways for radioactive materials to reach the environment.

Engineered barriers include the waste form, waste containers, air and water fillers, container overpacks, shaft and tunnel seals, and backfill materials. Natural barriers include the repository rock strata and overlying rock formations. Engineered barriers are designed to provide containment of the specific high-level waste forms. Geologic barriers are chosen for their stability, waste containment, and isolation characteristics.

In the United States, Yucca Mountain was selected as the site for the nation's high-level waste repository based on its geologic characteristics. The concept of a stable, geologic repository is intended to relieve future generations of the economic penalty associated with the management of high-level waste. However, this approach does not resolve the issues raised by stakeholder groups that have questioned the premise of geologic stability and the period for evaluating the licensing basis for a proposed repository.

Given the extended licensing periods, various risk scenarios must be considered. A key input into the geologic repository risk assessment is the container's durability and its ability to maintain its integrity when irradiated by the high-level waste for periods of $10^4 - 10^6$ years or longer. Container leakage will eventually occur, with waste migration in the presence of water intrusion or geologic action (e.g., earthquakes or tectonic movements). In addition, subsequent human activity (e.g., excavation or drilling) facilitates the release of radioactive material. Therefore, geologic disposal must evaluate random future events in the risk assessment and licensing decision process. Additional discussion regarding the regulatory aspects of high-level waste disposal and Yucca Mountain legal issues is provided in Chapter 7. Section 3.8 outlines additional concerns regarding the safety assumptions associated with geologic waste facilities. In particular, the 2014 release of plutonium and americium from the US Waste Isolation Pilot Plant is addressed.

2.6.10
High-Level Waste Disposal Resolution

None of the aforementioned approaches is a universally acceptable solution to long-term high-level waste disposal. Stakeholders have raised issues of safety, environmental impact, and the consequences of a terrorist attack, and these challenges continue.

There is an associated risk with any disposal option. A number of the approaches are eliminated by international convention. Political opposition and stakeholder concerns may be raised regarding any long-term disposal proposal.

It is the author's view that the most viable approach is a closed fuel cycle incorporating Generation IV reactors that utilize reprocessing and burn minor actinides and long-lived fission products. This option significantly limits and potentially eliminates the quantity of high-level waste generated in the fuel cycle. The remaining waste material would be classified as low-level waste that is addressed with currently accepted disposal options and techniques.

2.7
Twenty-First-Century Fuel Cycle Options

There are four basic options for operating the nuclear fuel cycle in the twenty-first century. These options were outlined in Section 2.3.1.6. This section addresses additional options incorporating possible twenty-first-century technology.

Table 2.12 provides a summary of the four basic fuel cycle options in terms of the spent fuel constituents (e.g., plutonium, minor actinides, uranium, and fission products). This table outlines an overall summary of the spent fuel constituents and the technology required to utilize these materials in the fuel cycle or mitigate their impact.

Option 4 presents a viable approach for minimizing high-level waste. This option recycles all spent fuel and minor actinides but may treat fission products as waste. Recycling some fission products (e.g., ^{99}Tc and ^{129}I) will be determined by cost and fuel performance. Option 4 requires multiple recycling and utilizes Generation IV reactor technology.

2.7.1
Advanced Fuel Reprocessing

In the full recycle fuel cycle (Option 4), the waste sent to a geologic repository is minimized when compared to the other three fuel cycle options. Full recycle technologies achieve actinide-free waste, and all actinides and potentially some fission

Table 2.12 Fuel cycle options[a)].

Option[b)]	Discharged fuel constituent				Technology required			
	Plutonium	Minor actinides	Irradiated uranium	Fission products	Recycle technology	Remote fabrication	Thermal reactors	Fast reactors
1	Waste	Waste	Waste	Waste	No	No	Yes	No
2	Fuel from UO_2 Waste from MOX	Waste from UO_2 Waste from MOX	Interim storage from UO_2 Waste from MOX	Waste from UO_2 Waste from MOX	One pass	No	Yes	No
3	Fuel	Waste	Interim storage	Waste	Multiple recycle	No	Yes	Yes
4	Fuel	Fuel	Fuel	Waste or recycled	Multiple recycle	Yes	Yes	Yes

a) USDOE Report (2001).
b) See Section 2.3.1.6 for a description of these options.

products are recycled to Generation IV reactors. Option 4 extracts the maximum energy from the fuel, because all actinides are burned and it contributes only fission products to the waste stream. Reprocessing technologies can be based on either aqueous or dry processes, and their development is a key aspect of Generation IV reactor viability.

When actinides form the feedstock for recycled fuel, remote fabrication technologies are required to minimize worker radiation doses. Option 4 fuel cycles also are optimized by colocating the recycling and fabrication facilities with the facilities that handle and ship radioactive materials.

Fuel incorporating minor actinides is compatible with fast neutron spectrum reactors. However, future nuclear fuel cycles will likely rely on both fast and thermal reactors. Initially, recycled fuel will be used in both thermal and fast reactors.

Advanced fuel cycle options also offer the potential to limit the required high-level waste repository capacity. Repository capacity is governed by the decay heat load, and this heat load is dominated by ^{137}Cs and ^{90}Sr for the first few hundred years and by minor actinides thereafter.

Option 4 also has the advantage that the recycled fuel containing PMAs creates an intrinsic radiation barrier to theft or diversion from the commercial fuel cycle. The presence of minor actinides and remnant fission products makes the recycled materials less attractive for weapons use as well as less accessible to theft or diversion. Incorporating actinides into fuel also avoids having inventories of weapons-grade materials in interim storage.

An advanced fuel cycle includes recycling as an initial reprocessing step with the removal of uranium and plutonium. An additional fuel cycle step processes the residual waste solution through partitioning and transmutation (PAT). The PAT operation is an alternative step to reduce the environmental burden of high-level waste.

2.7.2 Partitioning and Transmutation

PAT of nuclear material either completely or partially removes a specific isotope or group of isotopes. As an example, consider PAT operations that transmute actinides to less radiotoxic material. Once the actinides are eliminated as a hazard in the waste stream, the long-lived fission products ^{90}Sr and ^{137}Cs become the dominant concern. For a facility having an inventory of about 10 EBq of these fission product isotopes (i.e., the Hanford Site in the United States), the time to reduce the radiotoxicity to a manageable level is less than 1000 years. This is illustrated by considering the time for 10 EBq of ^{90}Sr and ^{137}Cs to undergo 20 half-lives (about 600 years). During the 600 years, 10 EBq decays to about 10 TBq.

Transmutation strategies include minor actinide burning and plutonium recycling. Each of these steps and their primary driving forces are examined in the subsequent discussion. Prior to reviewing these steps, it is necessary to review the neutron requirements for the actinide transmutation strategies.

The principal driving force for PMA burning is minimizing the nuclear proliferation potential. During reprocessing, uranium and other actinides are separated from the spent fuel. Following separation from uranium and fission products, the other actinides are recycled in reactors or accelerators. A number of specific strategies can be utilized for PMA burning including reactors and accelerator systems using thermal or fast neutrons.

2.7.2.1
PMA Burning in Generation IV Reactors

Plutonium recycling has both a resource management and nonproliferation motivation. Recycled plutonium extends fissile resources and eliminates material that might be diverted to nuclear weapons. The separation of uranium and plutonium from spent LWR fuel is performed using the reprocessing methods noted previously.

In the first reprocessing step, plutonium is recycled in thermal reactors. Later, recycling steps are accomplished using fast reactors with the option to utilize a limited number of thermal reactor cycles. In addition, an advanced fuel cycle includes partitioning other materials including (i) improved reprocessing of LWR uranium dioxide fuel with additional neptunium removal, (ii) separation of minor actinides from the reprocessing solutions, (iii) fabrication of minor actinide targets for irradiation in LWRs, and (iv) recycling of uranium and plutonium into MOX LWRs. Other partitioning options include the separation of long-lived fission products including ^{90}Sr, ^{99}Tc, ^{129}I, and ^{137}Cs.

PAT should play an essential role in future advanced fuel cycles which would reduce the long-term radiotoxic waste inventory and the dose to workers and the environment. By removing the minor actinides and long-lived fission products from reprocessing waste or burning them in a Generation IV reactor, the licensing basis for a high-level waste repository is simplified. Reducing the high-level waste (HLW) concern from 10^4-10^6 years to about 10^3 years would remove a significant concern associated with the expansion and continued development of fission reactors.

2.7.2.2
Accelerator Destruction of Actinides and Fission Products

There are numerous options for an accelerator-driven systems (ADSs) for waste incineration. These include a proton accelerator to transmute actinides via spallation reactions. As an example, the Japan Atomic Energy Research Institute (JAERI) is investigating a transmutation system using an accelerator-driven subcritical system. The system has a general objective to minimize the hazards associated with fuel reprocessing and the disposition of the residual high-level waste.

The JAERI approach utilizes a subcritical reactor with the thermal power of 800 MW that transmutes 250 kg of minor actinides annually. As proposed by JAERI, the reactor fuel includes MA nitride and is cooled using a Pb–Bi eutectic. A 1.5 GeV, 20–30 MW accelerator directs protons into the Pb–Bi target to produce spallation reaction products including neutrons. The neutrons transmute the MA fuel. Initial core loadings contain fissile plutonium to optimize neutron production while limiting the reactor to a subcritical configuration.

A number of technical challenges must be overcome to lead to a production-scale ADS facility. These issues include accelerator reliability demonstration, beam transport and window system development, high-power spallation target development, subcritical reactor physics performance, control verification, minor actinide transmutation performance, and fuel handling system development. In addition, the system cost and economic viability must be demonstrated.

A fuel cycle without Generation IV reactors or ADS would vitrify HLW and then dispose of this waste in a long-term geologic repository. With ADS and PAT, the fuel cycle would partition the HLW into two major streams. These are fission products and minor actinides.

The actinide fraction is fabricated into fuel. This fuel is transmuted into relatively small amounts of fission product waste using Generation IV reactors.

Fission products are separated into long-lived and short-lived fractions. The long-lived fraction is composed primarily of ^{90}Sr and ^{137}Cs. Medical isotopes such as ^{90}Y could also be extracted from this stream. After sufficient radioactive decay, the long-lived fraction is suitable for commercial land burial at a low-level radioactive waste facility. The short-lived fission product stream is separated into a valuable metal fraction (e.g., ruthenium, rhodium, palladium, and technetium) and medical isotopes. Remaining fission products are suitable for burial at a commercial low-level waste facility following sufficient decay to meet the facility's licensing restrictions.

Table 2.13 provides a summary of key parameters for the JAERI ADS facility. Variants and optimization of this design are likely.

Table 2.13 Initial design parameters for the proposed JAERI ADS facility[a].

Design parameter	Design selection/value
Beam particle	Protons
Beam energy	1.5 GeV
Beam power	20–30 MW
Spallation target	Pb–Bi
Coolant	Pb–Bi
Maximum k_{eff}	0.97
Thermal output	800 MW
Core height	1 m
Active core diameter	2.34 m
Minor actinide initial inventory	2.5 t
Fuel composition	60% minor actinides
	40% plutonium in mononitride form
Transmutation target	Minor actinides

a) Oigawa *et al.* (2004), Tsujimoto *et al.* (2004, 2008), and Sheffield (2009).

The development of ADS would have positive benefits in other fields. These benefits include energy generation, isotope production, and creation of an intense neutron source.

2.7.2.3
Gamma-Ray Free Electron Lasers

When sufficiently developed, gamma-ray free electron lasers (GRFELs) offer the potential for large fluence rate values and high photon energies. The energies are sufficient to initiate photon-induced reactions, and the high fluence rates would ensure reasonable reaction rates for the photodisintegration of minor actinides. These reactions accomplish the desired end state of eliminating the minor actinides from high-level waste.

The photodisintegration of ^{244}Cm illustrates the GRFEL approach. When irradiated by a GRFEL source, ^{244}Cm would undergo a number of nuclear reactions including (i) sequential (γ, n) reactions to nonactinide nuclides, (ii) photofission, (iii) photospallation, and (iv) sequential (γ, α) reactions to nonactinide nuclides.

Photon-induced reactions provide an alternative route for the elimination of PMAs. In order to be successful, the photon must have sufficient energy and fluence to irradiate the PMA fuel assembly throughout its volume. For a PMA atom density of n atoms/cm^3, a photoinduced reaction cross-section (σ), and fluence rate (ϕ), the reaction rate (R) is

$$R = n\sigma\phi \tag{2.16}$$

For an incident fluence rate (ϕ_0), the fluence rate after penetrating a depth x into the PMA material is

$$\phi(x) = \phi_0 B(\mu x, E, Z)\exp(-\mu x) \tag{2.17}$$

where $B(\mu x, E, Z)$ is the gamma-ray buildup factor, μ is the linear attenuation coefficient, E is the photon energy, and Z is the atomic number of the material shielding the photon radiation. The fluence must be sufficient throughout the fuel assembly to provide uniform transformation of the PMA material.

The time (t) required for the complete fission reaction of all PMA nuclei in the irradiated fuel is

$$t = \frac{n}{R} \tag{2.18}$$

2.7.3
Neutron Requirements for Various Fuel Cycle Options

From a neutron utilization perspective, not all generating devices efficiently operate to remove actinides from the fuel cycle. The production-to-absorption ratio of the actinides in the equilibrium core (η_{ec}) is a useful parameter to assess the suitability of a reactor or accelerator in terms of neutron utilization. Alternatively, the overall neutron balance for the complete fission of actinides can be measured

Table 2.14 Neutron performance of plutonium, minor actinide, and transuranic approaches[a].

Actinide feed component	Neutron generating device					
	Thermal accelerator		Fast reactor		Fast accelerator	
	η_{ec}	$-D$	η_{ec}	$-D$	η_{ec}	$-D$
Plutonium	1.15	0.40	1.64	1.18	1.80	1.34
Minor actinides	0.89[b]	−0.37[c]	1.28	0.71	1.33	0.79
Transuranics	1.11	0.30	2.00	1.52	1.75	1.29

a) EUR 19783 EN (2000).
b) Cannot maintain a chain reaction.
c) Cannot be completely fissioned.

in terms of the fuel neutron production parameter ($-D$). These two parameters provide an indication of the capability of a technology to initiate and sustain a successful actinide transmutation technology.

An η_{ec} value smaller than unity means that the fuel of the equilibrium core cannot maintain a chain reaction. A negative $-D$ value indicates that an actinide or an actinide mixture cannot be completely fissioned. The inability to sustain a fission reaction or failure to achieve complete fission indicates the technology is not a viable option for actinide transmutation. These parameters are influenced by the neutron spectrum and flux of the system. Evaluating a technology using either the η_{ec} or $-D$ approaches leads to the same conclusions regarding its viability for actinide transmutation.

The η_{ec} and $-D$ values provided in Table 2.14 are derived from realistic transmutation concepts. These values demonstrate that minor actinides cannot be completely burned in thermal systems. Fast reactors and accelerators are effective in transmuting PMAs and have the potential to significantly reduce the volume of high-level waste. However, their economic viability and engineering efficiency must be demonstrated before fast reactors and accelerators become an effective component of an advanced nuclear fuel cycle.

2.7.4
PAT Health Physics Considerations

A PAT facility processing high-level waste contains fission products and actinides in concentrations that present a health physics hazard. Many of these hazards are typical of any radiological facility and include the control of worker effective doses (e.g., internal and external) and limiting the release of radioactive material to the environment. These common health physics issues are discussed in numerous references and are not repeated in this text.

Issues to be addressed in the subsequent discussion focus on unique health physics issues associated with a PAT facility. These issues include criticality safety and unique radionuclides that result from PAT operations.

2.7.4.1 Criticality Safety

The spent fuel inventory includes fissile materials (i.e., ^{233}U, ^{235}U, ^{239}Pu, and ^{241}Pu), and these radionuclides present a criticality hazard. A criticality event is a major consideration for the storage of spent fuel, during fuel reprocessing operations, and during post reprocessing PAT activities. Any criticality produces an intense burst of neutron and photon radiation.

Criticality safety is enhanced using a subcritical PAT system with $k_{eff} < 1$. Subcritical systems are based on a set of controls that ensure the system does not achieve a critical mass or geometry. These controls must be protected to ensure a critical configuration is not achieved. For example, the introduction of unborated water into the system is controlled by isolating these water sources or providing double valve isolation of unborated sources. In addition, the boron concentration of the water required to ensure subcriticality is verified by periodic sampling.

2.7.4.2 Limiting Radionuclides

The following discussion assumes that minor actinides have been removed from the waste, and fission products are now the limiting waste disposal consideration. With current technology, the neutron capture process is the only practical reaction for transmuting fission products. Other candidate processes are in their initial stage (e.g., fusion neutron sources) or in development (e.g., Generation IV fast reactors and GRFELs). The transmutation of a fission product is only feasible if the reaction rate is greater than the natural decay rate of the nuclide. With the available or developing neutron sources and their associated fluence values, this feasibility requirement cannot be achieved for the most abundant fission products (e.g., ^{137}Cs and ^{90}Sr) which preclude their transmutation to a less significant radiological hazard. However, these fission products can be stored for a sufficient period (e.g., 20 half-lives) which is a significant time reduction when compared to the $10^4 - 10^6$ years licensing basis for a high-level waste repository.

Long-lived fission and activation products affect radiological assessments for a geologic repository, and some are not effectively removed using existing PAT techniques. A summary of selected properties of ^{14}C, ^{36}Cl, ^{79}Se, ^{93}Zr, ^{99}Tc, ^{126}Sn, ^{129}I, and ^{135}Cs and their associated health physics hazards are provided in Table 2.15. Eliminating these isotopes from the high-level waste stream has a significant benefit for reducing the licensing requirements for a PAT facility. Issues associated with the transmutation of these radionuclides are also noted in Table 2.15.

The results of Table 2.15 suggest that only a portion of the long-lived radionuclides in high-level waste will be successfully treated with currently available neutron PAT designs. However, PAT represents an approach that has yet to be optimized. If supplemented with other separation techniques, PAT could be used to enhance waste processing and minimize the number of long radionuclides in high-level waste.

Table 2.15 Fission and activation products important in geologic partitioning and transmutation assessments[a)].

Isotope	Source	Half-life (year)	Environmental behavior	PAT issues
^{14}C	$^{14}N(n, p)^{14}C$ activation reaction from nitrogen contamination in UO_2 fuel	5715	^{14}C can enter the environment though its solubility in groundwater and plant intakes via photosynthesis	Low neutron capture cross-sections suggest transmutation will not be effective
^{36}Cl	$^{35}Cl(n, \gamma)^{36}Cl$ activation reaction from chlorine impurities in zirconium alloy cladding	3.01×10^5	Due to its chemical characteristics, ^{36}Cl gradually dissolves in groundwater	Low neutron capture cross-sections suggest transmutation will not be effective Separation is a possible approach
^{79}Se	Fission product	3.5×10^5	Selenium behaves chemically like sulfur and is incorporated into vitrified waste Leaching from vitrified waste presents a potential environmental hazard	Accurate cross-sections must be determined for an assessment of the transmutation potential of this nuclide Separation is a possible approach
^{93}Zr	Fission product	1.5×10^6	Aquatic plants rapidly uptake soluble zirconium, but land plants tend not to adsorb it	Effective transmutation is not likely Separation is a more likely approach
^{99}Tc	Fission product	2.13×10^5	TcO_4 is soluble and presents a groundwater pathway	Partitioning of ^{99}Tc is difficult Given a relatively large neutron capture cross-section, transmutation is feasible
^{126}Sn	Fission product	2.3×10^5	^{126}Sn is partially soluble in groundwater	Effective transmutation is not likely Separation is a more likely approach

Table 2.15 (Continued)

Isotope	Source	Half-life (year)	Environmental behavior	PAT issues
^{129}I	Fission product	1.57×10^7	Iodine is one of the first radionuclides to emerge in the biosphere due to its high mobility	Transmutation of ^{129}I is difficult Confinement may be the best method to reduce its radiological impact
^{135}Cs	Fission product	2.3×10^6	Once it enters the environment, cesium is very mobile	Effective transmutation is not likely Separation is a more likely approach

a) EUR 19783 EN (2000).

Problems

2.1 A criticality accident has occurred at a French fuel reprocessing plant supporting a Generation IV supercritical water-cooled fast reactor. As the facility health physicist, you are at the command post providing technical support to the incident commander. The following questions relate to an unanticipated criticality event that occurred in a processing vessel.

(a) A search and rescue team has been assembled. The incident commander asks you if it is acceptable to send in a team to rescue a worker who is near the criticality location. What are your three primary considerations in developing your recommendation?

(b) List the primary exposure pathways and radiation sources for:
1. Workers in the processing vessel room at the time of the accident
2. Rescue workers after the criticality has been terminated
3. Other individuals within 0.1 – 10 km at the time of and following the incident

(c) Describe a method that could be used to quickly screen potentially irradiated individuals.

(d) Describe two medical interventions that could change the health outcome for an individual exposed to 7.5 Gy (whole body, deep dose) if administered during the first month following the incident.

(e) Why are large acute radiation doses (e.g., from a criticality accident) correctly expressed in units of Gy and not Sv?

2.2 You are the Radiation Protection Manager at the Higgs Boson Molten Salt Reactor in Orion, Oklahoma, operated by Ka-Boom Enterprises. During refueling operations, a spill of high-activity MSR coolant occurred with associated inhalation intakes of ^{131}I and ^{239}Pu. The facility is licensed by the

US Nuclear Regulatory Commission which uses ICRP 26/30 concepts and models as its regulatory basis.
Radiological data for the spill are tabulated below:

Quantity	Worker A ^{239}Pu intake	Worker B ^{131}I intake
Inhalation intake	5 ALI (nonstochastic)	5 ALI (nonstochastic)
Effective half-life	50 years	8 days
Organ that nonstochastic ALI is based upon	Bone surfaces	Thyroid
Tissue weighting factor for nonstochastic ALI organ	0.03	0.03

ALI, annual limit on intake.

(a) Calculate the committed dose equivalent (CDE) to the following organs and their respective committed effective dose equivalents (CEDEs) for:
1. Worker A's bone surface
2. Worker B's thyroid

(b) The physician treating Worker B proposes to remove the worker's thyroid to preclude the likelihood of thyroid cancer later in life. Is removal of the thyroid a prudent action?

(c) Both workers develop solid tumor cancers 1 year later and are suing Ka-Boom Enterprises, claiming the cancers were caused by the spill. In court, the respective attorneys claim that the worker received a dose that is five times the annual limit. Therefore, it is likely that the cancer was caused by the spill. Provide arguments to challenge the validity of this statement.

(d) On the day the spill occurs, Worker A's physician administers the chelating agent DTPA (diethylenetriaminepentaacidic acid).
1. Why is chelation appropriate for one of the exposures and not the other?
2. List factors that determine the effectiveness of DTPA.

(e) The ICRP 66 Human Respiratory Tract Model is more sophisticated than the ICRP 30 model. List improvements in the transuranic ICRP 66 lung model relative to the ICRP 30 model.

2.3 You are the Radiation Safety Officer (RSO) at the Flashbangum Corporation's fuel reprocessing facility. The facility includes a spherical, 25 cm radius, very thin-walled surge tank to recover low-enriched uranium. During a batch processing operation, an operator inadvertently diverts highly enriched uranium, which leads to a criticality in the surge tank. A technician is standing behind a 30 cm thick polyethylene shield and is 10.0 m from the center of the surge tank.

Data:

1. 1.0×10^{16} fissions occur during the criticality incident.
2. Each fission event produces 3 neutrons and 8 gamma rays.
3. The density of the polyethylene shield is 1.5 g/cm^3.

4. The dose conversion factor for 2.5 MeV neutrons is 0.002 mGy/h per 20 n/cm^2-s.
5. The gamma exposure rate conversion factor for 1.0 MeV photons is 6.0×10^5 γ/cm^2-s per 10 mGy/h.
6. The mean neutron and gamma energies are 2.5 and 1.0 MeV, respectively.
7. The neutron dose attenuation factor for 2.5 MeV neutrons through 30 cm of polyethylene is 0.005.
8. The mass attenuation coefficient for polyethylene for a fission gamma spectrum is 0.073 cm^2/g.

(a) What is the absorbed neutron dose received by the technician during the incident? Ignore the attenuation provided by the surge tank's structure and contents.

(b) What is the gamma absorbed dose received by the technician during the same incident? Ignore the attenuation provided by the surge tank's structure and contents.

(c) The facility criticality monitor is a gamma response instrument with an alarm set point of 5 mGy/h. If, during a short transient, the detector response corresponds to 1/3500 of the actual gamma absorbed dose rate, what is the maximum distance over which the device will be effective in signaling an unshielded 1 ms criticality accident with 1.0×10^{16} fissions? Neglect air absorption. Assume that an accident with 1.0×10^{15} fissions results in a gamma absorbed dose of 20 mGy at a distance of 2.0 m.

(d) List four factors that affect criticality safety.

2.4 You are the Radiation Protection Manager at Albert Snore Unit 1, the world's first operational lead-cooled fast reactor. Upon arrival at the plant, you are informed that a core cooling pipe started leaking 30 min ago, emitting a dense lead aerosol. A worker successfully stopped the spill after 20 min and has just exited the area. The worker was dressed in anticontamination clothing, a full-face respirator, and was wearing a lapel air sampler for the duration of the entry.

Data:

1. Fission product concentrations in the lead coolant:
   ^{137}Cs: 1.85×10^4 MBq/l
   ^{90}Sr: 740 MBq/l
2. Lapel air sampling rate: 4 l/min for 20 min
3. Total volume of spilled material = 500 l
4. Diameter of spill area = 10 m
5. Breathing rate of worker = 20 l/min
6. Protection factor of the respirator = 50
7. Analysis of lapel sample: (^{137}Cs (Type F) = 9×10^7 dpm and ^{90}Sr (Type S) = 2×10^6 dpm)
8. Gamma constant for ^{137}Cs $\Gamma = 8.1 \times 10^{-5}$ mGy-m^2/h-MBq
9. Dose conversion coefficient for Type F ^{137}Cs = 6.7×10^{-9} Sv/Bq
10. Dose conversion coefficient for Type S ^{90}Sr = 7.7×10^{-8} Sv/Bq

11. Particle terminal settling velocity: $\frac{d_1^2 \rho_1 g}{18\eta} = \frac{d_2^2 \rho_2 g}{18\eta}$ where $g = 9.8\ \text{m/s}^2$, η is the viscosity of air, ρ is the density of the particle, and d is the particle diameter

(a) Determine the effective dose from ^{137}Cs photons to the worker. Assume that the worker was standing at the center of the spill for 20 min and the dosimetric point of interest is 0.8 m above the spill. Neglect any effects of self-shielding.

(b) Calculate the airborne concentration of ^{90}Sr as measured by the lapel air sampler.

(c) Calculate the committed effective dose to the worker from the ^{90}Sr intake.

(d) What do the letters AMAD mean?

(e) Describe the dosimetric significance of the AMAD.

(f) If the spherical droplets have a specific gravity of 11.3 and a diameter of 5 μm, what is their AMAD?

2.5 You are employed as a senior radiological controls engineer at International Nuclear Disposal (IND), a facility that burns minor actinide waste using an accelerator-driven system. Your duties require performing evaluations to ensure regulatory compliance pertaining to worker safety including protection of workers from airborne radioactive materials and surface contamination.

You have been tasked to investigate an intake received by a male worker at IND's Leaking Falls Processing Facility in Smith & Wesson, Idaho. There were no radionuclides detected in the worker's previous routine urine sample. The worker was assigned to process and prepare radioactive waste for shipment. During your investigation, you determined that a container of dewatered resin contaminated with ^{137}Cs vented a portion of its contents while the worker was tightening the lid. He remained in the area for 30 min as the container vented. As part of your investigation, a urine bioassay sample was collected from the worker.

After reviewing the urine sample results, you conclude that the respiratory protection program has a number of weaknesses and programmatic improvements are warranted. As part of the determination of corrective actions, Questions (a)–(g) need to be answered in preparation for an upcoming NRC inspection.

The event occurred approximately 20 days prior to collection of the urine sample. The worker did not wear any respiratory protection during the waste handling operation.

Data:

1. The ventilation system in the waste preparation room delivers one room air change every 2 h.
2. Worker's breathing rate = 1.2 m^3/h.
3. Sample collection time = 24 h.
4. Sample volume = 1400 ml.

5. Analyzed portion of sample = 500 ml.
6. ^{137}Cs activity in analyzed portion = 15.9 kBq.
7. Male urinary tables are based on a urinary output of 1400 ml/day.
8. The ^{137}Cs half-life is 30.07 years.
9. The following is an excerpt from the ^{137}Cs Table in IND's Internal Dosimetry Manual:

^{137}Cs (Type S)
Particle size = 1 μm (AMAD)
Half-life = 1.10×10^4 days

Time after single intake (day)	Fraction of initial intake in	
	24 h urine	Accumulated urine
1	1.35×10^{-2}	1.35×10^{-2}
2	1.33×10^{-2}	2.68×10^{-2}
3	1.10×10^{-2}	3.78×10^{-2}
4	8.87×10^{-3}	4.67×10^{-2}
5	7.16×10^{-3}	5.38×10^{-2}
6	5.89×10^{-3}	5.97×10^{-2}
7	4.97×10^{-3}	6.47×10^{-2}
8	4.32×10^{-3}	6.90×10^{-2}
9	3.85×10^{-3}	7.28×10^{-2}
10	3.51×10^{-3}	7.63×10^{-2}
20	2.59×10^{-3}	1.04×10^{-1}
30	2.41×10^{-3}	1.29×10^{-1}
40	2.26×10^{-3}	1.52×10^{-1}

(a) What is the estimated intake of ^{137}Cs for the worker?

(b) Assume the ^{137}Cs intake for the worker was 55.5 MBq. What is the estimated average concentration of ^{137}Cs in the air to which the worker was exposed? Assume a uniform release rate from the container.

(c) Naturally occurring uranium consists of the isotopes ^{234}U, ^{235}U, and ^{238}U. By weight, the distribution is as follows:
^{234}U: 0.013 g/mol-total U $T_{1/2} = 2.5 \times 10^5$ years
^{235}U: 1.71 g/mol-total U $T_{1/2} = 7.0 \times 10^8$ years
^{238}U: 236.4 g/mol-total U $T_{1/2} = 4.5 \times 10^9$ years
Assuming equilibrium conditions, approximately what percentages of the total activity can be attributed to ^{234}U, ^{235}U, and ^{238}U?
 1. Negligible, 0.7, and 99.3%, respectively
 2. 65, 10, and 25%, respectively
 3. 49, 2, and 49%, respectively
 4. 33, 33, and 33%, respectively
 5. 99.3, 0.7%, and negligible, respectively

(d) Five uranium fuel compositions are proposed as fuel for the IND accelerator. You are to determine which of the five proposed compositions yields the highest activity fuel. The natural ^{234}U, ^{235}U, and ^{238}U composition is 0.0054, 0.7204, and 99.2742 wt%, respectively.

1. ^{234}U (1%), ^{235}U (5%), and ^{238}U (94%) obtained from centrifuge technology A
2. ^{234}U (2%), ^{235}U (10%), and ^{238}U (88%) obtained from centrifuge technology B
3. ^{234}U (0.1%), ^{235}U (8%), and ^{238}U (91.9%) obtained from SILEX
4. ^{234}U (0.1%), ^{235}U (50%), and ^{238}U (49.9%) obtained from AVLIS
5. ^{234}U (0.1%), ^{235}U (90%), and ^{238}U (9.9%) obtained from MLIS

(e) ANSI Z88.2, *Practices for Respiratory Protection*, provides recommendations for the use of supplied breathing air. This standard references other standards and specifications from organizations such as the Compressed Gas Association. Choose the best answer that agrees with recommendations of ANSI Z88.2.

1. Grade D breathing air specifications should be considered as the limits for compressed air of deteriorating quality.
2. The oxygen content of supplied breathing air shall be a minimum of 19.0% by volume.
3. Compressed oxygen may be used in supplied air or open-circuit self-contained breathing apparatus in which compressed air has previously been used.
4. 1 and 2.
5. 1 and 3.

(f) 10CFRPart 20 provides respiratory protection factors for standard types of approved devices as listed in items 1–4 below. Match the maximum allowable protection factors given in (a)–(d) to the given respiratory protection devices (1)–(4). Assume the airborne hazard is radioactive particulate material.

1. Full facepiece, negative pressure mode, air-purifying respirator	(a) 10
2. Full facepiece, pressure demand mode, self-contained breathing apparatus (SCBA)	(b) 100
3. Half-mask facepiece, negative pressure mode, air-purifying respirator	(c) 1 000
4. Full facepiece, powered air-purifying respirator	(d) 10 000

(g) Describe one type of handheld instrument routinely used for the detection of uranium on personnel as they leave contaminated areas. Your description should include the radiation type detected, any special constraints, and advantages or disadvantages of the instrument.

2.6 You are the Radiation Protection Manager at Point Beach Unit 7, which is a 1000 MWe Advanced Canadian Deuterium Reactor. Since the reactor produces copies quantities of tritium, workers at the facility are enrolled in a tritium bioassay program. Point Beach 7 utilizes a variety of workplace air monitors including flow-through ionization chambers to monitor the tritium air concentrations. Shaft and bearing maintenance are scheduled for a transfer pump in the liquid waste cleanup system. Given other priority tasks, returning the pump to service will not occur for several weeks. Assume the room ventilation is also out of service and is not returned to service before pump repair.

Data:

ICRP 26 is the regulatory basis for the NRC-licensed facility.
DAC (derived air concentration) (HT) = 2.0×10^4 MBq/m^3.
DAC (HTO) = 0.74 MBq/m^3.
Height of room = 4 m.
Room area = 200 m^2.
DCF (dose conversion factor) (acute intake) = 7.57×10^{-7} mSv-l/Bq in urine (first 24 h).
DCF (chronic intake) = 5.41×10^{-8} mSv-l/Bq in urine (average daily concentration).
Specific activity (HTO) = 53.7 TBq/g.
Specific activity (HT) = 215 TBq/g.

(a) Calculate the committed effective dose equivalent you would expect a maintenance worker to receive from a 1 h exposure to a pump room air concentration of 185 MBq/m^3 as measured by workplace air monitoring. The chemical form of tritium measured several weeks before the repair was 30% HT and 70% HTO.
(b) The individual involved in the incident submits a postincident bioassay sample collected during the first 24 h. The results indicate a tritium concentration in urine of 1850 Bq/l. Calculate the worker's committed effective dose equivalent.
(c) The committed effective dose equivalent calculated from the urine concentration differs from the value calculated from the room air concentration. Assume that the measurements and calculations were performed correctly. Provide two likely sources for this discrepancy.
(d) Identify two techniques that are used for tritium air monitoring. Specify one advantage and one disadvantage of each technique.

2.7 The Republic of North Confusia has established a facility to reprocess spent fuel from the Hopeless Power Plant. Following reprocessing, the Republic's President decided to extract plutonium for development of a nuclear weapon. To enhance the weapon's yield, she desires to separate the ^{240}Pu from ^{239}Pu using centrifuges and has constructed a facility for plutonium enrichment.

The chief scientist obtained uranium enrichment equipment, optimized to yield 5% ^{235}U, for the plutonium facility. In order to meet the President's production schedule, the uranium enrichment equipment was installed without modification. Given the clandestine nature of this facility, no oversight or review activities are conducted.

(a) Are there any health physics concerns associated with the chief physicist's decision to use the uranium enrichment equipment without modification?
(b) In attempting to conceal the clandestine enrichment facility, it is located adjacent to an electron accelerator used for materials research. The accelerator produces copies bremsstrahlung below 200 keV. Accordingly, the chief physicist modified the whole-body counting software to exclude any photons below 200 keV. To date, no positive whole-body counts have been observed for centrifuge facility workers. Are there any health physics concerns associated with the chief physicist's decision to modify the software?
(c) A leak has developed in a product line that contains 99.9% ^{239}Pu, and particulate material is uniformly distributed over a 10 m^2 area. The ^{239}Pu release activity is 100 Bq/m^3 and the release flow rate is 1 m^3/s. If the leak lasts for 14 days, what is the surface contamination level in Bq/m^2? Assume the only removal term is radioactive decay. The half-life of ^{239}Pu is 2.41×10^4 years.
(d) The leak of radioactive material has not been detected. Work activity in the contaminated area results in a resuspension factor of 2×10^{-5}/m. What is the resulting airborne concentration?
(e) A technician works for 8 h in the air concentration derived in the previous question. If his breathing rate is 1.2 m^3/h, what is the ^{239}Pu intake?
(f) If the ^{239}Pu is Class M with a dose conversion coefficient of 4.7×10^{-5} Sv/Bq, what is the resulting effective dose from the intake?
(g) Assume the facility follows the recommendations of ICRP 103. Did this event exceed any recommended limits?

2.8 You are the Radiation Protection Manager at a prototype gamma-ray high-level waste transmutation facility. The facility uses an advanced gamma-ray free electron laser that produces a photon having an energy of 1.25 MeV. It is capable of sustained operations at high fluence rates and will be used for transmuting plutonium and minor actinides (PMA) using photofission.
The gamma-ray beam strikes a cubical core perpendicular to and in the center of one of its faces. For simplicity, ignore scattering of the beam photons.
Data:

Facility and process characteristics

Parameter	Parameter value
Average photofission cross-section for PMA	1.0 μb/atom
Number density for PMA in the core	0.048 atoms/b-cm
Average photon fluence	1×10^{23} γ/cm^2-s
PMA gamma-ray attenuation coefficient	1.18/cm
Core side dimension	1.03 m
Core average gamma-ray attenuation coefficient	0.39/cm
Advanced gamma-ray free electron laser beam area	1 cm^2

Buildup factors appropriate for the facility

μx	2	4	6	8	10	20	30	35	40
$B(\mu x)$	2.85	5.30	8.31	11.8	15.8	41.3	74.5	93.5	114

(a) What time is required for the complete photofission reaction of all PMA nuclei in the irradiated fuel? Is the gamma-ray approach feasible for the problem conditions? Assume no gamma-ray attenuation in the fuel material.
(b) What process parameters could be altered to improve transmutation performance?
(c) Assume that many of the modifications identified in Question (b) are implemented and a new facility has been constructed. If 2.5×10^{-6} neutrons are produced for every photon, what is the neutron flux 10 m from the center of an unshielded core? Assume an isotropic production of neutrons and the incident photon beam delivers 3.0×10^{24} γ/s to the core. The new core geometry approximates a point source.
(d) How much concrete shielding is required to reduce the neutron effective dose rate at 25 m from the center of the core to 10 μSv/h? The flux to dose conversion factor is 25 μSv/h per 20 n/cm^2-s, the neutron attenuation factor for concrete is 0.0576/cm, and the appropriate buildup factor for the shield design is 60.2.

References

10CFR20 (2015) Standards for Protection Against Radiation, National Archives and Records Administration, U.S. Government Printing Office, Washington, DC.

10CFR50 (2015) Domestic Licensing of Production and Utilization Facilities, National Archives and Records Administration, U.S. Government Printing Office, Washington, DC

10CFR835 (2015) Occupational Radiation Protection, National Archives and Records Administration, U.S. Government Printing Office, Washington, DC

Alexander, R.E. (1974) Applications of Bioassay for Uranium, WASH-1251, US Atomic Energy Commission.

ANSI ANS-8.3-1997; R 2003 (1997) *Criticality Accident Alarm System*, American Nuclear Society, LaGrange Park, IL.

APS Report (2005) Nuclear Power and Proliferation Resistance: Securing Benefits, Limiting Risk, American Physical Society, Washington, DC.

APS SILEX Petition (2010) Petition for NRC Rule Change, American Physical Society, Washington, DC, http://www.aps.org/about/pressreleases/upload/nrcpetition201006.PDF (accessed 5 August 2015).

Ashley, S.F., Parks, G.T., Nuttall, W.J., Boxall, C., and Grimes, R.W. (2012) Thorium fuel has risks. *Nature*, **492**, 31.

Baetsle, L.H. (2001) Application of partitioning/transmutation of radioactive materials in radioactive waste management. Lectures given at the Workshop on Hybrid Nuclear Systems for Energy Production, Utilization of Actinides and Transmutation of Long-Lived Radioactive Waste, Trieste, Italy, September 3–7, 2001.

Baum, E.M., Ernesti, M.C., Knox, H.D., Miller, T.R., and Watson, A.M. (2009)

Nuclides and Isotopes, Chart of the Nuclides, 17th edn, Knolls Atomic Power Laboratory, Lockheed Martin, Schenectady, NY.

Bevelacqua, J.J. (2003) Production equations in health physics. *Radiat. Prot. Manage.*, **20** (6), 9.

Bevelacqua, J.J. (2004a) Skin contaminations, internal intakes, and ALARA. *Radiat. Prot. Manage.*, **21** (1), 11.

Bevelacqua, J.J. (2004b) Point source approximations in health physics. *Radiat. Prot. Manage.*, **21** (5), 9.

Bevelacqua, J.J. (2008) *Health Physics in the 21st Century*, Wiley-VCH Verlag GmbH, Weinheim.

Bevelacqua, J.J. (2009) *Contemporary Health Physics, Problems and Solutions*, 2nd edn, Wiley-VCH Verlag GmbH, Weinheim.

Bevelacqua, J.J. (2010a) *Basic Health Physics, Problems and Solutions*, 2nd edn, Wiley-VCH Verlag GmbH, Weinheim.

Bevelacqua, J.J. (2010b) Practical and effective ALARA. *Operational Radiation Safety. Health Phys.*, **98** (Suppl. 2), S39.

Bevelacqua, J.J. (2011) Conservatism in effective dose calculations for accident events involving fuel reprocessing waste tanks. *Health Phys.*, **101**, 48.

Bevelacqua, J.J. (2012a) Applicability of health physics lessons learned from the Three Mile Island Unit-2 accident to the Fukushima Daiichi accident. *J. Environ. Radioact.*, **105**, 6.

Bevelacqua, J.J. (2012b) The Fukushima Daiichi accident and its radiological impact on the environment. *Phys. Teach.*, **50**, 368.

Bevelacqua, J.J. (2013a) Nuclear regulation in the United States and a possible framework for an international regulatory approach. *Int. Nucl. Saf. J.*, **2** (1), 1, http://nuclearsafety.info/international-nuclear-safety-journal/index.php/INSJ/article/view/12/pdf_1 (accessed 6 August 2015).

Bevelacqua, J.J. (2013b) Methodology to ascertain core recriticality following a severe fission reactor accident. *Int. Nucl. Saf. J.*, **2** (1), 52, http://nuclearsafety.info/international-nuclear-safety-journal/index.php/INSJ/article/view/13/pdf_1 (accessed 6 August 2015).

Bevelacqua, J.J. (2014a) Health physics aspects of generation IV reactors. *Int. Nucl. Saf. J.*, **3** (1), 13, http://nuclearsafety.info/international-nuclear-safety-journal/index.php/INSJ/article/view/20/pdf_1 .(accessed 6 August 2015).

Bevelacqua, J.J. (2014b) Health physics aspects of a uranium laser isotope separation facility. *Int. Nucl. Saf. J.*, **3** (1), 38, http://nuclearsafety.info/international-nuclear-safety-journal/index.php/INSJ/article/view/21/pdf (accessed 6 August 2015).

Bevelacqua, J.J. (2014c) Radiation protection considerations for nuclear power expansion to nations with limited technical infrastructure. *Int. Nucl. Saf. J.*, **3** (3), 8, http://nuclearsafety.info/international-nuclear-safety-journal/index.php/INSJ/article/view/37/pdf (accessed 6 August 2015).

Blake, E.M. (2013) Generation III+ is about to start generating. *Nucl. News*, **56** (7), 60.

Blake, E.M., Kovan, D., and Michal, R. (2005) Looking to the future: the next 50 years. *Nucl. News*, **48**, 41.

Cember, H. and Johnson, T.A. (2008) *Introduction to Health Physics*, 4th edn, McGraw-Hill Medical, New York.

Cinotti L., *et al.* (2008) The ELSY project, paper 377. Proceeding of the International Conference on the Physics of Reactors (PHYSOR), Interlaken, Switzerland, September 14–19, 2008.

Doolittle, W.W., Bredvad, R.S., and Bevelacqua, J.J. (1992) An ALARA conscious hot particle control program. *Radiat. Prot. Manageme.*, **9**, 27.

EGG-2530 (1988) Health Physics Manual of Good Practices for Uranium Facilities, U.S. Department of Energy, Washington, DC.

Eisenbud, M. and Gesell, T.F. (1997) *Environmental Radioactivity from Natural, Industrial, and Military Sources*, Revision 4, Elsevier Science & Technology Books, Amsterdam.

Energy Policy Institute Report (2010) Economic and Employment Impacts of Small Modular Reactors, Idaho National Laboratory, Idaho Falls, ID.

European Commission Report: EUR 19783 EN (nea3108) (2000) Actinide and fission product partitioning and transmutation. 6th Information Exchange Meeting, Madrid, Spain, December 11–13, 2000.

FEMA (2013) Program Manual, Radiological Emergency Preparedness, Federal Emergency Management Agency, Washington, DC.

Generation IV International Forum (2009) *GIF R&D Outlook for Generation IV Nuclear Energy Systems*, US Department of Energy, Washington, DC.

Generation IV International Forum (2013) *Generation IV Technology Roadmap*, US Department of Energy, Washington, DC., http://www.gen-4.org/Technology/roadmap.htm.

Generation IV Roadmap Update (2014) *Technology Roadmap Update for Generation IV Nuclear Energy Systems*, US Department of Energy, Washington, DC.

Glaser, A. (2008) Characteristics of the gas centrifuge for uranium enrichment and their relevance for nuclear weapon proliferation. *Sci. Global Secur.*, **16**, 1.

Glasstone, S. (1982) *Energy Deskbook*, US Department of Energy, Oak Ridge, TN.

Hecht, J. (2011) Laser enrichment returns from the dead. *Laser Focus World*, **47** (10), 18.

HPS N13.3-1988. (1988) *Dosimetry for Criticality Accidents*, Health Physics Society, McLean, VA.

HPS N13.14-1994. (1994) *Internal Dosimetry Programs for Tritium Exposure – Minimum Requirements*, Health Physics Society, McLean, VA.

HPS N13.22-1995. (1996) *Bioassay Programs for Uranium*, Health Physics Society, McLean, VA

IAEA-EPR (2013) Actions to Protect the Public in an Emergency Due to Severe Conditions at a Light Water Reactor, International Atomic Energy Agency, Vienna.

IAEA-TECDOC-1536 (2007) *Status of Small Reactor Designs without On-site Refuelling*, International Atomic Energy Agency, Vienna.

ICRP (1979) *Limits for Intakes of Radionuclides by Workers*, Publication 30, Pergamon Press, Oxford.

ICRP (1986) *The Metabolism of Plutonium and Related Elements*, Publication 48, Pergamon Press, New York.

ICRP (1991) *Recommendations of the ICRP*, Publication 60, Pergamon Press, Oxford.

ICRP (1998) *Radiological Protection Policy for the Disposal of Radioactive Waste*, Publication 77, Elsevier, Amsterdam.

ICRP (2007) *The 2007 Recommendations of the International Commission on Radiological Protection*, Publication 103, Elsevier, Amsterdam.

International Atomic Energy Agency (2002) *IAEA Safeguards Glossary 2001 Edition*, International Atomic Energy Agency, Vienna.

International Atomic Energy Agency (2006) *Nuclear Forensics Support Reference Manual*, IAEA Nuclear Security Series No. 2 Technical Guidance, International Atomic Energy Agency, Vienna.

Knief, R.A. (1985) *Nuclear Criticality Safety, Theory and Practice*, American Nuclear Society, LaGrange Park, IL.

Kovan, D. (2013) The nuclear news interview – John E. Kelley: regaining GIF's momentum. *Nucl. News*, **56** (10), 40.

Kramer, D. (2012) As its renaissance recedes, US nuclear industry looks abroad. *Phys. Today*, **65** (11), 24.

Krass, A.S., Boskma, P., Elzen, B., and Smit, W.A. (1983) *Uranium Enrichment and Nuclear Weapons Proliferation*, Taylor & Francis Ltd, London.

LA-10860-MS (1987) Critical Dimensions of Systems Containing U-235, Pu-239, and U-233-1986 Revision, Los Alamos National Laboratory, Los Alamos, NM

Meyer, K. (2013) Expanded nuclear forensics. *Nature*, **503**, 461.

Murray, R.L. (1983) *Understanding Radioactive Wastes*, 2nd edn, Battelle Press, Columbus, OH.

National Academy of Sciences (2014a) Lessons Learned from the Fukushima Nuclear Accident for Improving Safety of U.S. Nuclear Plants. Report No. 18294, The National Academy Press, Washington, DC.

National Academy of Sciences (2014b) An All-of-Government Approach to Increase Resilience for International Chemical, Biological, Radiological, Nuclear, and Explosive (CBRNE) Events. Report No. 18814, The National Academy Press, Washington, DC.

National Academy of Sciences (2014c) The Science of Responding to a Nuclear Reactor Accident: Summary of a Symposium.

Report No. 19002, The National Academy Press, Washington, DC.

NCRP (1999) Biological Effects and Exposure Limits for "Hot Particles". Report No. 130, National Council on Radiation Protection and Measurements, Bethesda, MD.

NCRP (2002) Risk-Based Classification of Radioactive and Hazardous Chemical Wastes. Report 139, NCRP Publications, Bethesda, MD.

NCRP (2005) Performance Assessment of Near-Surface Facilities for Disposal of Low-Level Radioactive Waste. Report 152, NCRP Publications, Bethesda, MD.

Nuclear Energy Agency (2012) Actinide and Fission Product Partitioning and Transmutation. Report No. 6996, Nuclear Energy Agency, Issy-les-Moulineaux.

Nuclear Energy Agency (2014) Managing Environmental and Health Impacts of Uranium Mining. Report No. 7062, Organization for Economic Co-operation and Development, Issy-les-Moulineaux.

Nuclear News Special Section (2014) Advanced reactors. *Nucl. News*, **57** (13), 43.

Nuclear News Summary Report (2013) World list of nuclear power plants. *Nucl. News*, **56** (3), 45.

NUREG/CR-6918, Rev. 1 (2011) VARSKIN 4: A Computer Code for Skin Contamination Dosimetry, United States Nuclear Regulatory Commission, Washington, DC.

Oigawa, H., Tsujimoto, K., Kikuchi, K., Kurata, Y., Sasa, T., Umeno, M., Saito, S., Nishihara, K., Mizumoto, M., Takano, H., and Ouchi, N. (2004) R&D activities on accelerator-driven transmutation system in JAERI. OECD/NEA 8th Information Exchange Meeting on Actinide and Fission Product Partitioning and Transmutation, Las Vegas, NV, November 9–11, 2004.

Pigford, T.H. and Ang, K.P. (1975) The plutonium fuel cycles. *Health Phys.*, **29**, 451.

Salvatores, M., Slessarev, I., and Uematsu, M. (1994) A global physics approach to transmutation of radioactive nuclei. *Nucl. Sci. Eng.*, **116**, 1.

Sandia National Laboratory (2011) The Development of a Realistic Source Term for Sodium-Cooled Fast Reactors: Assessment of Current Status and Future Needs. Report SAND2011-3404, Sandia National Laboratories, Albuquerque, NM.

Schlosser, E. (2013) *Command and Control – Nuclear Weapons, the Damascus Accident, and the Illusion of Safety*, Penguin, New York.

Sheffield, R. (2009) *Accelerator Driven Systems (ADS) Fusion-Fission Workshop LA-UR 09-02736*, Los Alamos National Laboratory, Los Alamos, NM, http://web.mit.edu/fusion-fission/WorkshopTalks/Fusion_Fission_Workshop_Oct_2009_ADS_final.pdf (accessed 6 August 2015).

TID 7016, Rev. 1 (1961) Nuclear Safety Guide, U.S. Atomic Energy Commission, Washington.

Tsoulfanidis, N. (2013) *The Nuclear Fuel Cycle*, 2nd edn, American Nuclear Society, LaGrange Park, IL.

Tsujimoto, K., Oigawa, H., Kikuchi, K., Kurata, Y., Mizumoto, M., Sasa, T., Saito, K., Nishihara, K., Umeno, M., and Takei, H. (2008) Feasibility of lead-bismuth–cooled accelerator-driven system for minor-actinide transmutation. *Nucl. Technol.*, **161**, 315.

Tsujimoto, K., Sasa, T., Nishihara, K., Oigawa, H., and Takano, H. (2004) Neutronics design for lead-bismuth cooled accelerator-driven system for transmutation of minor actinide. *J. Nucl. Sci. Technol.*, **41** (1), 21.

Turner, J.E. (2007) *Atoms, Radiation, and Radiation Protection*, 3nd edn, Wiley-VCH Verlag GmbH, Weinheim.

U. S. Environmental Protection Agency (2013) PAG Manual, Protective Action Guides and Planning Guidance For Radiological Incidents, U.S. Environmental Protection Agency Draft for Interim Use and Public Comment, U. S. Environmental Protection Agency, Washington, DC.

Uranium Information Centre (2005) Advanced Nuclear Power Reactors, Nuclear Issues Briefing Paper 16, Melbourne.

Uranium Information Centre UIC (2005) Generation IV Nuclear Reactors, Briefing Paper #77, Melbourne.

US Department of Energy Report (2001) Generation-IV Roadmap Report of the Fuel Cycle Crosscut Group, US Department of Energy, Washington, DC.

US Department of Energy Report (2012) Reactor and Fuel Cycle Technology

Subcommittee Report to the Full Commission – Updated Report, US Department of Energy, Blue Ribbon Commission on America's Nuclear Future, Washington, DC.

US Department of Energy Report (2013a) FY 2012 Annual Report: Addressing the Issues of Today, US Department of Energy Office of Nonproliferation and International Security, Washington, DC.

US Department of Energy Report (2013b) Strategy for the Management and Disposal of Used Nuclear Fuel and High-Level Radioactive Waste, US Department of Energy, Washington, DC.

US Nuclear Regulatory Commission Regulatory Guide NUREG/CR-7176 (2014) Safety and Regulatory Issues of the Thorium Fuel Cycle, US Nuclear Regulatory Commission, Washington, DC.

US Nuclear Regulatory Commission Regulatory Guide NUREG-1938, Final Report (2012) Environmental Impact Statement for the Proposed GE- Hitachi Global Laser Enrichment, LLC Facility in Wilmington, North Carolina, US Nuclear Regulatory Commission, Washington, DC.

US Nuclear Regulatory Commission Regulatory Guide NUREG-2120 (2012) Safety Evaluation Report for the General Electric-Hitachi Global Laser Enrichment LLC Laser-Based Uranium Enrichment Plant in Wilmington, North Carolina, US Nuclear Regulatory Commission, Washington, DC.

USNRC Regulatory Guide 3.12 (1973) General Design Guide for Ventilation Systems of Plutonium Processing and Fuel Fabrication Plants, USNRC, Washington, DC.

USNRC Regulatory Guide 3.18 (1974) Confinement Barriers and Systems for Fuel Reprocessing Plants, USNRC, Washington, DC.

USNRC Regulatory Guide 3.7 (1973) Monitoring of Combustible Gases and Vapors in Plutonium Processing and Fuel Fabrication Plants, USNRC, Washington, DC.

USNRC Regulatory Guide 3.71 (2005) Nuclear Criticality Safety Standards for Fuels and Material Facilities, USNRC, Washington, DC.

USNRC Regulatory Guide 8.11 (1974) Applications of Bioassay for Uranium, U.S. Nuclear Regulatory Commission, Washington, DC.

Waldrop, M.M. (2012) Radical reactors. *Nature*, **492**, 26.

Wilson, P.D. (1996) *The Nuclear Fuel Cycle: From Ore to Waste*, Oxford University Press, New York.

Part III
Accidents and Nuclear Events

Part III addresses nuclear power accidents, radiological emergencies, high-level waste events, and terrorist events involving radioactive materials or nuclear weapons. The importance of these topical areas is illustrated by the extensive and prolonged media coverage of the 2011 Fukushima Daiichi accident and worldwide attention to terrorist events. In 2013, the Boston Marathon bombing involving conventional explosives emphasized the ongoing terrorist threat. The use of explosives at an internationally recognized event would have been even more dramatic if the blast was used to disperse radioactive materials.

Radiological incidents have the potential for a significant release of radioactive material. These releases can occur in a variety of facilities that utilize or produce radionuclides. The misuse of radioactive materials can also lead to a radiological emergency. Misuse events including terrorist attacks are addressed in Chapter 4.

Chapter 3 focuses on accidents and radiological emergencies that occur at nuclear power plants and other fuel cycle facilities. Nuclear power plant accidents are selected because they have the potential for significant radiological consequences and three major power reactor accidents have occurred. These events also affect national energy policies and international practices.

Other radiological events also have the potential for a significant release of radioactive material but are not included in the current text because their consequences are bounded by power reactor accidents. These events include fires in facilities enriching uranium or manufacturing power reactor fuel, research reactor accidents, and transportation events involving radioactive materials.

Accidents in nuclear weapons complex facilities can produce offsite consequences. Events involving nuclear weapons produce the most significant

Health Physics: Radiation-Generating Devices, Characteristics, and Hazards, First Edition. Joseph John Bevelacqua.

radiological consequences and are addressed in Chapter 4. Excluding a nuclear weapon's accident, one the most significant of the weapons complex events involves major releases of radioactive material from reprocessing waste tanks. Accordingly, a discussion of high-level waste accidents involving these tanks is included in Chapter 3.

3
Nuclear Accidents and Radiological Emergencies

3.1
Overview

Within the last 40 years, there have been three significant accidents involving nuclear power facilities. In 1979, a pressurized water reactor (PWR) (Three Mile Island Unit 2 (TMI-2)) had a small-break loss-of-coolant accident (LOCA) with associated fuel damage. TMI-2 was caused by a combination of operator errors and design weaknesses. Operator errors, an inadequately evaluated test procedure, and an unforgiving reactor design led to the 1986 accident at Chernobyl Unit 4, an RBMK design, located in Ukraine. These factors contributed to a power excursion that resulted in violent reactor disassembly and severe fuel damage. In 2011, the Fukushima Daiichi Nuclear Power Station (FDNPS) in Japan, consisting of six boiling water reactors (BWRs), was struck by a massive seismic event and subsequent tsunami that led to significant core damage and an off-site release of radioactive material.

Each of these power reactor events had unique operational aspects and presented significant health physics challenges. They were also of significant public interest. Media reports and information provided by the operating utility were not always representative of the actual health physics hazards or operational events. The quantification of the released radioactive material, environmental effects, or doses delivered to the public was not always clearly stated or presented in terms that were easily understood by the public. A need for improvements in risk communication was demonstrated during all three major power reactor accidents. Risk communications are addressed in Chapter 6.

The radioactive material releases and their mitigation are affected by the specific reactor design and its requisite safety systems. These design considerations are addressed in subsequent discussion.

3.2
Design Considerations

From a radiological perspective, nuclear reactor designs limit the release of radioactive materials following a severe accident. In its most basic form, reactor

Health Physics: Radiation-Generating Devices, Characteristics, and Hazards, First Edition. Joseph John Bevelacqua.

design is structured to protect three fission product barriers. Since each barrier inhibits the release of fission products, protection of these barriers is a design priority.

3.2.1
Fission Product Barriers

The radiological consequences of a reactor accident are minimized if the fission product barriers are preserved. These barriers and their status during an event are a key consideration in the IAEA's International Nuclear and Radiological Event Scale (INES) that is used to classify nuclear accidents. The INES classification of the Fukushima Daiichi event was a significant item of media interest during the first few weeks of the event, because media, industry, and government reports did not consistently convey the severity of this accident.

Most reactor types incorporate three fission product barriers that inhibit the movement of fission products contained within the fuel matrix into plant areas and the environment. As such preserving the integrity of the fission product barriers is crucial to maintaining control of radioactive material and implementing an effective health physics program.

Commercial reactors have three fission product barriers to limit the release of radioactive material to the environment. Since PWRs comprise about two-thirds of commercial light water reactors with the remainder dominated by BWRs, discussions of fission product barriers focus on their designs, characteristics, and terminology.

In a PWR, these barriers are the fuel/clad, reactor vessel and associated reactor coolant system (RCS) piping, and the containment structure or reactor building. BWR fission product barriers are the fuel/clad, reactor pressure vessel (RPV) and included piping, and the containment vessel (CV). The spent fuel pools (SFPs) in a PWR (BWR) are located in the auxiliary building (reactor building) which do not have the same capability to retain fission products as the three primary fission product barriers.

The first fission product barrier includes the fuel pellet or fuel material. The fuel material and its associated coatings or cladding retain solid and gaseous fission products. For pellet/clad configuration fuel, fission product activity is often classified as either gap activity or total fuel pin activity. Gap activity is that fission product activity residing in the gaps between the fuel pellets and the gap between the fuel pellets and the cladding. The total fuel pin activity is the gap activity and the activity contained within the fuel pellet. As noted in Chapter 2, the fuel fission product barrier is absent in molten salt reactors (MSRs), because the fuel forms a eutectic mixture with the coolant.

The second fission product barrier is the primary coolant system boundary including the reactor vessel and its included piping and components. Any break in primary piping permits radioactive material to be released into the containment structure.

The third fission product barrier is the containment structure that encloses the primary coolant system. Any breach of the containment structure creates a pathway for radioactive material to reach the environment. Penetration of any of the three fission product barriers facilitates the release of radioactive material in an uncontrolled manner. The breach of multiple fission product barriers is an indication of a major reactor accident.

The fuel of a commercial power reactor consists of UO_2 pellets enclosed in a zirconium alloy tube. Fission products are retained within the fuel pellet, and the zirconium alloy cladding supplements this fuel pellet barrier. Damaging the fuel/clad fission product barrier releases fission products to the RCS/RPV. The fuel/clad barrier was breached at TMI-2, Chernobyl-4, and Fukushima Daiichi. TMI-2 and Fukushima Daiichi events involved a loss of core cooling with subsequent fuel melting. The Chernobyl-4 fuel was ejected from the core following a power excursion.

The second barrier, the reactor vessel or RPV and included piping, was breached at TMI-2 and Fukushima Daiichi and destroyed at Chernobyl-4. With two fission product barriers damaged, only the containment structure prevents a release of fission products to the environment.

A breach of the containment eliminates the final barrier and allows fission products to escape to the environment. At TMI-2, the containment remained intact and survived a hydrogen detonation. The release of radioactive materials at TMI-2 occurred when highly contaminated water was transferred from the containment building sump to the auxiliary building sump. This transfer facilitated a fission product release through the waste gas system in the auxiliary building.

Chernobyl-4 had no containment building. The power excursion severely damaged the core and coolant system and released fission products and core materials directly to the environment.

At Fukushima Daiichi, multiple containment vessels were damaged and suspected to be leaking. The accident sequence involving the breaching of the three Fukushima Daiichi fission product barriers and subsequent release of fission products to the environment are addressed in subsequent discussion.

3.2.1.1
Fission Product Releases

The primary fission products released in a major reactor accident are radioiodine and noble gases. The Chernobyl-4 and Fukushima Daiichi accidents also involved the release of particulates including ^{137}Cs and $^{90}Sr/^{90}Y$. Regulatory aspects of a major accident with a breach of fission product barriers are provided in Chapter 7. These fission product releases affect the emergency response following an accident and the implementation of protective actions including evacuation and sheltering. Protective actions associated with power reactor accidents are addressed in subsequent discussion. Chapter 4 discusses protective actions from the perspective of a terrorist event.

3.2.1.2
Fission Product Deposition

Following a release of fission products to the environment, particulate material, including radioiodine, radiocesium, and radiostrontium, is deposited on the surface and contaminates ground and surface water. These radioactive materials are incorporated into various biota. The contaminated water, plants, and animals enter the food chain and are consumed by man. This contamination presents a challenge to subsequent land and water use, and acceptable contamination levels in foods and water must be established. The regulatory implications of the surface contamination and limits for the use of the contaminated land, food, and water are addressed in Chapter 7.

3.2.2
Accident Assumptions

Analyses that evaluate the impact of a reactor accident on the facility and the surrounding environment require assumptions regarding the status of the facility and its integral safety systems. The availability and reliability of reactor safety systems are crucial to accident analysis and are an important aspect in the licensing of facilities. Facility licensing defines a set of parameter values that determine and serve to quantify design and beyond design basis events (DBEs and BDBEs). Radiological consequences are calculated for a set of DBAs that categorize major event types.

For example, the magnitude of the design basis earthquake and subsequent induced events (tsunami and aftershocks) and their impact on the facility are key accident analysis parameters. The earthquake magnitude determines the height of seawalls for tsunami protection, the location and requirements of safety equipment (e.g., emergency diesel generators and direct current batteries), and the required defense-in-depth systems to ensure accident mitigation. Underestimating the hazard can have catastrophic consequences.

Issues associated with credibly selecting design and BDBEs are clearly illustrated by the Fukushima Daiichi accident. Underestimating the design basis earthquake and resulting tsunami led to events that disabled on-site and off-site electric power systems. Loss of these power systems disabled active safety systems that led to significant core damage and the release of radioactive material to the environment. The consequences of the Fukushima Daiichi accident and failure to adequately define credible design basis events DBEs present a long-term challenge to the nuclear industry.

Once DBEs are defined, their impacts on DBAs are evaluated. These DBAs include LOCAs, steam generator tube ruptures (SGTRs), fuel handing accidents (FHAs), and waste gas decay tank ruptures (WGDTRs). Therefore, the assumed bounding parameters defining the DBE (e.g., earthquake, floods, and tornado) and their resulting impact on plant systems (e.g., duration of power loss and extent of primary system break) govern the required plant redundancy (e.g., defense-in-depth requirements). These assumptions when coupled with plant response to an event (e.g., extent of core damage, core damage frequency (CDF), and off-site release magnitude) govern the capability of a design to be licensed. The licensed design must be capable of sustaining the stresses of normal, abnormal, and emergency conditions. As such, properly defining design basis assumptions has a significant impact on plant safety and protection of the health and safety of the public and the environment.

3.2.3 Design Basis Assumptions

The design basis concept and associated accident modeling assumptions rely on the premise that sufficient safety margins are present in the reactor and its safety systems. Models used in accident analysis also maintain sufficient conservatism to account for analysis uncertainties. In addition, adequate defense in depth is included in the reactor design to compensate for uncertainties in accident progression and analysis data. The defense-in-depth concept is validated if safety system redundancy, independence, and diversity are preserved with respect to the anticipated frequency and consequences of challenges to these systems. Safety system requirements are specified in terms of General Design Criteria specified by the regulatory agency (e.g., Appendix A to 10 CFR Part 50 for NRC licensees). Ultimately, the accident design yields an associated CDF that is part of the basis for its acceptance.

Associated with the CDF assessment and its associated design basis assumptions are specific dose consequences of the postulated DBAs. The dose consequences are evaluated in terms of a set of key parameters that are defined by licensing agencies. In the United States, the NRC defines these parameters in a series of guidance documents including Regulatory Guide 1.195, *Methods and Assumptions for Evaluating Radiological Consequences of Design Basis Accidents at Light-Water Nuclear Power Reactors*. These parameters include the fission product inventory available at the time of the accident, fraction of the core fission product inventory released into the containment, timing of release phases, and the radionuclide composition and its chemical form.

3.2.3.1 Fission Product Inventory

Fission product inventory is an important parameter in assessing the radiological consequences of a postulated accident. The inventory of core fission products available for release to the RCS is maximized by assuming the limiting values for fuel enrichment, fuel burnup, and core power.

Fission product inventories (see Table 2.2) vary with facility operating history. The inventories increase as the operating cycle length increases and the fissile material is consumed. Since only a portion of the core is replaced during a refueling outage, the actual inventory available for release varies with the time following a refueling outage. Core inventories also vary with the reactor type and power history.

TMI-2 only operated for a few months before its accident but had sufficient fission product inventory to lead to a General Emergency classification and the evacuation of the public within 16 km of the facility. Fukushima Daiichi and Chernobyl-4 had a longer operating history than TMI-2.

3.2.3.2
Release Fractions

Upon failure of the fuel and RCS fission product barriers, a fraction of the core fission product inventory is released into the containment. USNRC Regulatory Guide 1.195 provides guidance for these release fraction values.

When the fuel is melted and the cladding is breached, the core inventory release fractions for iodine and noble gases are assumed to be 1.0 and 0.5, respectively. This assumption was a gross overestimate at TMI-2, which released minimal radioiodine. A smaller than anticipated iodine source term occurred because of the release pathway from the reactor building to the auxiliary building waste gas system facilitated iodine removal prior to its release to the environment. As noted previously, the activity within the fuel is often partitioned into a gap activity component and total activity component.

Melting is assumed to release the gap activity and most of the activity within the fuel pellets. The TMI-2 accident demonstrated the conservatism of this assumption since some fission products remained in the RCS or containment and were not released to the environment. For example, iodine interacted with other fission products to form soluble chemical forms, adhered to reactor internal or containment structures, or remained in solution following the actuation of the reactor building spray system. The spray system used a solution of sodium hydroxide and water to reduce the containment building pressure.

For non-LOCA events, NUREG 1.195 assumes that only the cladding is breached. The release fractions of the core inventory, based on a cladding breach, only include the gap activity. These core inventory release fractions are given in Table 3.1.

3.2.3.3
Timing of Release Phases

Once the fuel and RCS fission product barriers fail, radiological consequence models usually assume that the liberated fission products immediately enter the

Table 3.1 Non-LOCA fraction of fission product inventory based on available gap activity[a].

Fission product	Release fraction
^{131}I	0.08
^{85}Kr	0.10
Other noble gases	0.05
Other iodines	0.05

a) Based on Regulatory Guide 1.195 (2003).

containment. To ensure maximum radiological consequences, the released core activity is assumed to be immediately available for release from containment for DBAs in which fuel damage is projected.

This immediate release approach is advocated in Regulatory Guide 1.195. However, the TMI-2 accident indicated that specific accident sequences could lead to release delays that permit physical and chemical processes to affect the source term. When calculating off-site releases and their radiological consequence, release timing must be considered to obtain an accurate assessment of off-site doses.

3.2.3.4 Radionuclide Composition

Design basis analyses define the elements in each radionuclide group that should be considered in radiological consequence models. In the United States, these are defined in terms of isotopes that dominantly contribute to the protective action guide dose values. Since US reactors include containment structures that act as a fission product barrier, noble gases, and radioiodine are the dominant source term components. Smaller quantities of particulates (e.g., cesium and strontium) are released, but these radionuclides are important environmental considerations. For example, at Fukushima Daiichi radiocesium affected the sale and distribution of various foods (e.g., rice, spinach, fish, and meat) for months following the accident.

3.2.3.5 Chemical Form

Regulatory Guide 1.195 specifies the chemical form of radioiodine released from the RCS to the containment in a postulated accident. The iodine chemical forms are assumed to be particulate, elemental, and organic with the percentages of 5, 91, and 4%, respectively. These chemical forms are also assumed for FHAs. However, the accident-specific transport pathways of iodine species following release from fuel may affect these assumed fractions.

3.2.4 Design Basis Accidents

A DBA is the postulated maximum credible event that a nuclear facility is designed and built to withstand. The safety systems, structures, and components necessary to assure public health and safety are assumed to survive the DBA. A variety of accident subtypes can be defined and include the design basis criticality, earthquake, explosion, fire, flood, and tornado events. For example, a design basis earthquake is that seismic event for which the reactor safety systems are designed to remain functional during and after the event. This assumes safety system functionality to ensure a safe shutdown condition. Inadequate design basis earthquake assumptions were a significant contributor to the Fukushima Daiichi accident. The Fukushima Daiichi design basis assumptions underestimated the earthquake and subsequent tsunami that led to multiple safety system failures.

A DBE is a postulated event used to establish the acceptable performance requirements of the safety structures, systems, and components used to control the reactor and ensure a safe shutdown condition. Events that exceed the design basis criteria are classified as beyond design basis. Although the Fukushima Daiichi accident can be classified in the beyond design basis category, TMI-2 cannot since it involved personnel errors and a control room design that did not facilitate recognition of the event. The TMI-2 accident is a type of LOCA that was included within the DBA category.

As noted previously, reactor accidents are broadly classified as DBEs and BDBEs. DBEs are events caused by a variety of factors including component failures such a break in primary system piping or steam generator tubes. BDBEs include multiple failures such as a loss of all power (off-site and on-site emergency power) or ruptures of tubes in multiple steam generators. These events are addressed in subsequent discussion. Plant procedures exist to address both DBEs and BDBEs.

There are four generic types of DBAs. These are LOCAs involving a loss of core coolant, SGTRs resulting from breaches in the tubes forming a boundary between the primary and secondary coolants, FHAs that result in damage to the fuel cladding, and WGDTRs involving a loss of integrity in structures containing fission gases and possibly radioiodine. These events are significant because they permit radioactive material to escape from engineered systems and enter plant areas or the environment in an uncontrolled manner.

Reactor accidents vary in severity, but the most significant radiological events involve core damage that lead to a radioactive release to the environment. Other events, including failure of waste gas decay tanks or spent fuel element breaches, are less severe but more likely scenarios.

The four reactor accident categories are defined as:

1) *Loss-of-coolant accidents (LOCAs)*: If the LOCA occurs in the containment building, the reactor's primary piping is breached and cooling flow to the core is reduced or lost. As a result, the temperature of the nuclear fuel increases. As the temperature increases, the fuel fission product barrier degrades, the cladding is breached or melts, and fission products are released into the primary coolant. The loss of the fuel fission product barrier is significant because it facilitates the uncontrolled release of radioactive material to the reactor coolant and increases the probability of an environmental release. If the LOCA is severe, the fuel eventually melts with the subsequent release of additional radioactive material to the primary coolant.

   Fuel cladding degradation can occur even without fuel melting. Breaches in the cladding, caused by impacts of foreign material or localized heating, release fission radionuclides into the primary coolant. Subsequent breaches in the primary coolant system and containment building offer a release path to the environment. With fuel damage and a breach in primary piping, only the containment barrier prevents a release of radioactive material to the environment. If the primary piping breach occurs in the auxiliary building of a

PWR (e.g., in the letdown line) or in the reactor building of a BWR, a release pathway to the environment exists.

Containment building failures facilitate a release to the environment. Examples of containment failures include malfunctions of purge valves, air supply valves, containment hatch valves, penetrations, and containment isolation valves.

2) *Steam generator tube ruptures (SGTRs)*: PWR steam generator tubes form a barrier between the primary and secondary coolants. Since the primary system pressure is about twice the secondary pressure, reactor coolant flows from the primary system into the secondary system if a tube is breached. If a tube rupture or leak occurs, a pathway is created that mixes the primary (radioactive) and secondary (nonradioactive) fluids. As a minimum, the secondary (clean) part of the plant becomes contaminated and its radiation levels exhibit a significant increase. In addition, a release of noble gases will occur through the condenser air ejector. This release pathway limits the iodine release because its length permits increased removal compared to a shorter steam line safety valve pathway.

   The failure of an atmospheric or steam generator safety value on a main steam line or other secondary system piping, valve, or component provides a direct release pathway for the primary coolant's radioactive material to reach the environment. This release type is more significant than an air ejector release and can involve both radioiodine as well as noble gases.

   The air ejector release pathway requires transit through the steam lines, past the atmospheric or steam line safety valves, and through the high-pressure and low-pressure turbines. In the low-pressure turbine, the steam is condensed and the noncondensable gases and limited radioiodine are released through the condenser air ejector. The increased path length permits scavenging that minimizes the iodine release for the air ejector pathway.

   A SGTR is a special class of LOCA with the primary system leak occurring through the steam generator tubes. In addition to the secondary system impacts, the primary system experiences the radiological consequences of a LOCA with the severity depending on the magnitude of the primary to secondary leakage. Off-site doses depend on the location of the release with more severe consequences resulting from open relief valve pathways.

3) *Fuel handling accidents (FHAs)*: FHAs involve mechanical damage to a fuel assembly. This damage occurs when a fuel assembly strikes an adjacent assembly, impacts a reactor vessel component, or drops. The nuclear fuel residing in the fuel storage pool or the reactor core is periodically moved during refueling operations or operations involving fuel inspection or control rod maintenance. Accidents during these evolutions damage the fuel fission product barrier and lead to a release of radionuclides into the radiologically controlled plant areas or the environment. Fission gases and radioiodine dominate the source term. If spent fuel is involved in the FHA and it has been out of the reactor for a year or more, the short-lived noble gas activity has

decayed and the dominant isotope in the release is ^{85}Kr that has a half-life of 10.76 years.

4) *Waste gas decay tank ruptures (WGDTRs)*: Waste gas decay tanks store fission gases and possibly radioiodine to permit their decay prior to the release of these radioactive materials to the environment. Failures of the tank structure, valves, or associated components release fission gases into the plant. Since these tanks reside outside the containment fission product barrier (e.g., in the auxiliary building of a PWR), a release to the environment is likely.

The extent to which these four DBAs lead to radiological consequences depends on the integrity of the reactor fuel. If the fuel fission product barrier remains intact, the LOCA and SGTR releases are characterized by the steady-state activity of the primary coolant. The radiological hazards increase with degradation in the fuel fission product barrier and the release of fission products into the primary coolant. Table 3.2 summarizes postulated Generation IV power reactor accident types, the types of radiological releases that could occur, plant systems that mitigate the release, and methods that are utilized to mitigate the release. Table 3.2 information is derived from Generations II and III facilities and their operating characteristics. Projection to Generation IV systems is uncertain since there is currently insufficient design information to accurately define their accident types and associated mitigation and termination approaches. However, Table 3.2 summary should be similar to the eventual Generation IV accident characterization.

Table 3.2 is generic and does not focus on a specific Generation IV design type. Since no formal licensing documentation is available for a Generation IV design, the radiological consequences of recent Generation III licensing basis information is provided. As an example of a specific Generation III reactor design, the DBEs and their radiological consequences for the AP-1000 are presented. Reactors utilizing the AP-1000 design are under construction in the United States and China.

The NRC Certification Review for the AP-1000 reactor provides an assessment of the DBEs for this Generation III PWR, and the radiological consequences of these events are summarized in Table 3.3. Radiological consequences are provided for the AP-1000 control room, the exclusion area boundary (EAB), and low population zone (LPZ).

As defined in US regulations, the EAB is the area surrounding the reactor, in which the reactor licensee has the authority to control activities. The USNRC quantifies the EAB as the perimeter of a 2760 ft (841 m) radius circle from the circumference of a 630 ft (192 m) circle encompassing the reactor containment structure. An LPZ is similarly defined to be the area immediately surrounding the EAB. The USNRC quantifies the LPZ as a 2 mile (3.2 km) radius circle from the circumference of a 630 ft (192 m) circle encompassing the reactor containment structure.

The Generation III radiological accident consequences are less severe than the corresponding Generation II events. This result is expected based on the Generation III design criteria with its enhanced safety performance and utilization of passive safety systems. Additional consequence reductions are expected for Generation IV facilities based on their proposed design basis requirements.

Table 3.2 Projected Generation IV design basis event accident mitigation and termination approaches.

Accident type	Release type	Mitigation	Termination
LOCA	Iodine[a)] Noble gas Particulate[b)]	NaOH spray Suppression pool Ice condensers Filtration ECCS[c)] Backup emergency power supplies[d)] Enhanced battery capacity[d)] Backup core cooling pumps[d)] Hard-piped vents with filtration[d)] Hard core facility[d)]	In-plant repairs Reestablish core cooling Isolate leak Solidified MSR reactor vessel plug melts and drains the fuel-salt coolant eutectic into storage tanks
SGTR	Iodine[a)] Noble gas	Filtration Release via the condenser ECCS[c)] Protect intact steam generators	Cool and depressurize the primary coolant system In-plant repairs
FHA (<1 year old fuel)	Iodine[a)] Noble gas	Filtration	Fuel assembly depressurizes
FHA (>1 year old fuel)	^{85}Kr	Filtration	Fuel assembly depressurizes
WGDTR	Iodine[a)] Noble gas	Filtration	In-plant repairs (e.g., tank isolation) Tank depressurizes

a) Depends on the extent of fuel barrier defects.
b) Particulates can be released in severe design basis and beyond design basis accidents.
c) Emergency core cooling system.
d) See Section 7.11.

3.2.4.1
Spent Fuel Pool Accidents

The 11 September 2001, terrorist attacks raised concerns regarding an aircraft strike into the SFP of a commercial nuclear reactor. Before these attacks, SFP accidents focused on fuel handling events, load drops into the pool, loss of pool water, and loss of fuel cooling capability. The consequences of these events were analyzed as part of the facility's DBAs, but were generally less severe than other baseline events (e.g., a LOCA). However, the large inventory of radioactive material residing in the SFP has the potential for a significant environmental impact if the activity were dispersed by accidental or intentional means. Accordingly, concerns for spent fuel accidents were expanded to include terrorist attacks using a ground force or airborne vehicle.

Table 3.3 Generation III AP-1000 radiological consequences of design basis accidents[a].

Postulated accident	Total effective dose equivalent (mSv)		
	EAB	LPZ	Control room
Loss-of-coolant accident	190	150	34
Main steam line break outside containment with an accident-initiated iodine spike	2	8	13
Reactor coolant pump shaft seizure without feedwater available	<1	<1	12
Rod ejection accident	15	24	11
Fuel handling accident	24	10	29
Small line break accident	10	4	14
Steam generator tube rupture with accident-initiated iodine spike	5	7	26
Spent fuel pool boiling	<0.1	<0.1	<0.1

a) NUREG-1793 (2006).

Concern for SFP events were again heightened by the Fukushima Daiichi accident. The severe core damage that occurred in Units 1, 2, and 3 led to hydrogen explosions that destroyed the upper level of the reactor buildings in Units 1, 3, and 4 and led to debris falling into their SFPs. Pool cooling was also disrupted by the loss of all power, which temporarily removed the capability to cool the fuel and preserve its fission product barrier. In addition, the seismic event may have damaged the structural integrity of the pools. This concern led to the structural reinforcement of the Unit 4 pool.

The 11 September 2001 terrorist attacks and 11 March 2011 Fukushima Daiichi events suggest that the original SFP design basis requires review. Events triggered by earthquakes, tsunami, loss of all electric power, hydrogen explosions, multiple unit accidents, terrorist attacks, and aircraft strikes present additional challenges to the SFP licensing bases that were not fully considered in the original design basis of Generations II and III reactors. These events have led to additional regulatory requirements that are addressed in Section 3.3.3 and Chapter 7.

Concerns for spent fuel events have prompted focus on removing fuel from the storage pools. In the United States, no high-level waste repository is available. Accordingly, other fuel storage options are receiving increased attention.

3.2.4.2
Dry Fuel Storage Accidents

Current US regulations permit two options for storing spent nuclear fuel at a commercial power reactor. In addition to storage in a water-filled pool, fuel can be stored in dry storage casks located in outdoor locations within the protected area of a reactor site. The cask design incorporates a sealed metal cylinder that encloses the spent fuel. This cylinder is enclosed within a metal or concrete shell that provides additional shielding and protection against external events. The casks are arranged either vertically or horizontally on an outdoor concrete pad.

Typically, older spent fuel is stored in dry cask systems. This fuel has a lower decay heat load, a smaller radioactive material inventory, and lower radiation levels than fuel immediately discharged from the reactor. In general, the fuel stored in a dry cask has a lower decay heat load that the fuel stored in a pool.

The dry cask storage system has the same objectives as SFP storage. First, the fuel is cooled to prevent breaching the fuel fission product barrier. Second, workers and the public are shielded from the radiation produced by the decay of fission and activation products. Third, releases of radioactive material to the environment are minimized. Finally, a criticality event must be precluded.

Dry casks meet the first two objectives utilizing passive safety systems. Spent fuel cooling is typically achieved using natural circulation air cooling and the inherent cask shielding meets the second objective. Concrete, lead, and steel shield the photon radiation, and neutron radiation is limited using concrete, polyethylene, or borated materials.

Releases to the environment are minimized by the cask package. However, accidents can disrupt the package and facilitate a release of radioactive material to the environment. Accidents are caused by both natural and man-made events.

Criticality is controlled using a physical lattice to maintain the orientation of the fuel assemblies within the dry cask. The lattice structure may incorporate borated materials to provide an additional criticality safety margin.

Monitoring and surveillance activities are performed to ensure the cask design objectives are achieved. These activities verify that the air inlet and outlet ports are free of debris and that air flows freely through the cask. Radiation and contamination surveys ensure that the casks are meeting their radiological design objectives. The casks are also monitored to verify there is no degradation of the shielding and structural materials.

In the United States, the dry cask storage design requirements are specified in 10CFR72. Dry casks are designed to ensure that spent fuel is safely stored during normal and abnormal conditions. Abnormal conditions include accidental drops or tip-over events that occur during cask transport operations. Although the casks provide a degree of protection against an external attack, their original design basis did not explicitly consider these events.

The design of a dry fuel storage facility evaluates severe natural events and extreme human-induced events. Severe wind, precipitation, and earthquake/tsunami hazards are natural events considered in cask design. Extreme human-induced events include terrorist attacks that breach the cask and lead to a release of radioactive materials including fission and activation products and fuel materials.

Terrorist attack scenarios for a dry storage facility are similar to those noted for SFPs including ground and air attacks using a variety of aircraft. Since the DBEs for both SFPs and dry cask storage are similar, it is logical to determine which storage approach is more secure from a radiological perspective.

Given existing technology, pool cooling appears to be the best approach for fuel newly discharged from a reactor. Active cooling must be provided to preserve the fuel fission product barrier. Passive air cooling is generally insufficient to remove the high initial decay heat load for newly discharged fuel. Dry storage is appropriate for older fuel with a lower decay heat load. Passive natural circulation air cooling is usually sufficient to remove the reduced decay heat load from older fuel.

From a risk perspective, there are a number of potential advantages to dry fuel storage compared with pool storage. Less fuel is at risk in a dry cask storage accident or terrorist event since a limited number of casks would be involved in an event. A pool accident involves the entire inventory of spent fuel. This suggests that the radiological consequences of a terrorist event or accident at a dry cask storage site will be less severe than an SFP event. In addition to the reduced radioactive materials inventory, dry casks have a lower decay heat load and less severe accident consequences. For example, loss of pool cooling caused by mechanical or electrical failures could lead to an increase in fuel temperatures and cladding degradation of the fuel. Under extreme circumstances, fuel melting and hydrogen generation with an associated explosion could result.

Dry cask storage relies on passive cooling and is not affected by a loss of power or cooling water inventory. These conditions have a significant impact on the fuel in a storage pool.

The recovery from an attack or event at a dry cask facility should be quicker than a corresponding SFP event. This is attributed to the limited number of casks involved, the less hazardous state of the fuel in a cask, and the passive nature of cask systems.

Since no major outdoor dry storage cask events have occurred, the aforementioned comments are based on analysis and regulatory perspectives. Future dry storage events will reveal the adequacy of the current regulatory approach.

3.2.5
Beyond Design Basis Accidents

BDBEs are incidents that involve failures of multiple safety barriers and are significant from a radiological perspective. Table 3.2 provides a summary of projected Generation IV DBAs. Since extrapolation to a Generation IV BDBE is extremely speculative, attention is focused on the most likely accidents. As noted in the Table 3.4, these Generation IV BDBEs involve combinations of DBEs or conditions more severe than assumed in the design basis assumptions.

Unless otherwise noted, the BDBEs summarized in Table 3.4 are generic and are not specific to any particular design. This is necessary since the designs are evolving and not defined to an extent to permit a specific BDBE assessment. As such, the Table 3.4 results represent a set of generic events that could occur in these reactors and encompass likely Generation IV systems, structures, and components.

Table 3.4 Generation IV beyond design basis event accident mitigation and termination.

Accident type	Release type	Mitigation	Termination
Loss of all on-site and off-site power	Iodine[a)] Noble gas Particulate[b)]	Utilize passive safety systems for core cooling Utilize passive safety systems to supply steam generator feedwater Station batteries Reflux cooling Steam-driven auxiliary feed pumps Reestablish on-site emergency power Protect primary piping integrity Filtration ECCS[c)] NaOH spray Suppression pool Ice condensers Backup power supplies[d)] Enhanced battery capacity[d)] Backup core cooling pumps[d)] Hard-piped vents with filtration[d)] Hard core facility[d)]	In-plant repairs to equipment damaged by the loss of power Reestablish core cooling with electric-driven pumps Restore on-site electric power Restore off-site electric power Stabilize the primary coolant system Solidified MSR reactor vessel plug melts and drains the fuel-salt coolant eutectic into storage tanks
LOCA coincident with a loss of power	Iodine[a)] Noble gas Particulate[b)]	Utilize passive safety systems for core cooling Utilize passive safety systems to supply steam generator feedwater Station batteries Reflux cooling Steam-driven auxiliary feed pumps Reestablish on-site emergency power Protect primary piping integrity NaOH spray Suppression pool Ice condensers Filtration ECCS[c)]	In-plant repairs to equipment damaged by the loss of power Reestablish core cooling with electric-driven pumps Restore off-site electric power Restore on-site electric power Isolate source of primary leakage Stabilize the primary coolant system In-plant repairs to terminate the LOCA Solidified MSR reactor vessel plug melts and drains the fuel-salt coolant eutectic into storage tanks

(*continued overleaf*)

Table 3.4 (*Continued*)

Accident type	Release type	Mitigation	Termination
LOCA coincident with a loss of power (*continued*)		Backup power supplies[d] Enhanced battery capacity[d] Backup core cooling pumps[d] Hard-piped vents with filtration[d] Hard core facility[d]	
Ruptures in multiple steam generators	Iodine[a] Noble gas	Utilize passive safety systems for core cooling Utilize passive safety systems to supply feedwater to intact steam generators Protect intact steam generators Filtration Release via the condenser ECCS[c]	Cool and depressurize the primary coolant system using intact steam generators In-plant repairs
Faults in multiple steam generators	Iodine[a] Noble gas	Protect intact steam generators Isolate fault locations Protect primary piping from overcooling Filtration ECCS[c]	Primary system pressure and temperature returned to acceptable values In-plant repairs
Combination of faulted and ruptured steam generators	Iodine[a] Noble gas	Protect intact steam generators Isolate fault locations After fault isolation, utilize passive safety systems for core cooling Utilize passive safety systems to supply feedwater to intact steam generators Protect primary piping from overcooling Filtration Release via the condenser ECCS[c]	Cool and depressurize the primary coolant system using intact steam generators Primary system pressure and temperature returned to acceptable values In-plant repairs

a) Depends on the condition of the fuel fission product barrier.
b) Particulates can be released in severe design basis and beyond design basis accidents.
c) Emergency core cooling system.
d) See Section 7.11.

The BDBEs include loss of power events in which all on-site and off-site power is lost, tube ruptures in more than one of the unit's steam generators, faults in multiple steam generators, combinations of steam generator faults and ruptures, and LOCAs coincident with a loss of power. A steam generator fault is a break in the secondary system piping or secondary system component failure such as a relief valve that opens and does not close. The fault provides a pathway for a release of secondary coolant to the environment and leads to overcooling of the RCS.

BDBE events are also caused by conditions more severe than considered in the design basis assumptions. The consequences of underestimating a DBE were illustrated by the 2011 Fukushima Daiichi accident. The initiating earthquake and subsequent tsunami significantly exceeded the design basis assumptions used to license the Fukushima Daiichi reactors. A total loss of power led to a severe LOCA with significant core damage.

Although the BDBEs are severe, appropriate response actions are established in procedures, and operators are trained in their use and effective implementation. Operating procedures are developed in multiple layers to address normal, abnormal, emergency DBE, and emergency BDBE operations.

For example, elevated facility radiation levels are initially addressed using an abnormal operating procedure. If these radiation levels were caused by a LOCA, operators would transition to an emergency operating procedure that focused on LOCA mitigation and termination. If a BDBE event (e.g., loss of all power leading to a station blackout) occurred during the LOCA, the LOCA procedure would transition to a procedure that would mitigate the effect of the loss of power and ongoing LOCA. In addition, a complete loss of power event during normal operations results in the immediate implementation of the BDBE procedure to mitigate the effects of the station blackout condition, restore on-site and off-site power, and protect the three fission product barriers. Even with the comprehensive procedure set, core damage can occur if the event is prolonged or severe conditions outside the facility's design basis occur. These unusual conditions occurred at Fukushima Daiichi and led to that accident.

The loss of off-site and on-site electrical power occurred during the Fukushima Daiichi accident. A total power loss disables active safety systems that jeopardize the ability to cool the core and mitigate the release of radioactive material. Since active safety systems require electric power to function, Generation IV reactors have an additional safety margin because they utilize passive systems (e.g., natural circulation) that do not require power to perform their intended function. These systems provide core cooling or supply feedwater to the secondary side of the steam generators to remove the core's decay heat. Some feedwater is provided by steam-driven auxiliary pumps, but their flow rates decrease as the core's decay heat decreases.

Upon losing power, primary coolant flow to the core ceases, but a process known as *reflux cooling* provides some core cooling. Reflux cooling is a passive process. Primary coolant is converted to steam by the core's decay heat. The steam condenses inside the cooler portion of intact primary piping and then flows back into the core. Maintaining the integrity of primary and secondary piping systems minimizes fuel damage and the release of radioactive material from the RCS. However, a prolonged loss of power increases core temperatures and eventually damages the fuel.

Station batteries provide a source of direct current (DC) that is converted to alternating current to power safety system pumps and valves. The lifetime and capability of the DC system is limited (typically a 4–8 h) so expeditious recovery of emergency power systems or off-site power is essential. One of the post-Fukushima Daiichi accident improvements is the staging of backup power supplies and backup core cooling pumps that can be installed prior to the loss of all power including the station batteries. Enhanced station battery capability is also a mitigation measure.

With a loss of power, the LOCAs severity is increased because core cooling is limited. Without electric-driven pumps to provide cooling water, the likelihood of fuel damage and melting are increased. Any degradation of the secondary system limits the core's heat sink that increases the probability of fuel damage. Core damage including fuel melting releases fission products to the RCS and possibly to the environment. The Generation IV passive safety systems mitigate a loss of power event by providing core cooling and feedwater to steam generators. However, passive systems must be tested under prolonged accident conditions, and it is likely that future events will reveal needed improvements to these designs. The hard-core concept, addressed in Chapter 7, provides an enhanced measure of safety that goes beyond the conventional defense-in-depth approach.

Multiple ruptured steam generators are a more severe version of a SGTR. As such, multiple ruptures have radiological consequences that are similar but more severe than an SGTR. These events also challenge the facility staff's ability to manage a complex event. Facility emergency procedures govern the response to tube rupture events that occur in multiple steam generators. Simultaneous ruptures in multiple steam generators have not occurred in either Generation II or III reactors.

With a single ruptured steam generator, the intact steam generators are used to provide long-term core cooling. With multiple ruptures, emphasis is placed on preserving the integrity of intact steam generators. The ruptured steam generators limit the capability to provide core cooling. Diminished core cooling capability enhances the likelihood of core damage.

Steam generator faults are breaks in secondary system piping. Faults lead to a rapid loss of secondary coolant that results in overcooling the primary system. Overcooling is significant because primary system pressure and temperature limits could be exceeded which increases the stress on primary piping and components. This additional stress enhances the potential for primary system damage including component rupture that would lead to a LOCA and subsequent core damage.

The overcooling condition exists as long as the faulted steam generator receives feedwater (secondary coolant). Recovery from a fault condition includes feedwater isolation and subsequent restoration of the primary system to design pressure and temperature conditions.

Rupture/fault combinations have characteristics of both types of events. If both events occur in the same steam generator, the combination of the loss of heat sink with a loss of primary coolant to the secondary system presents an energetic pathway for the release of primary coolant. The rupture/fault event can also lead to core damage. Generation IV passive core cooling systems mitigate a rupture/fault event and increase the time for core damage to occur.

The off-site consequences of a BDBE include the release of iodine and noble gases to the environment. Particulate material is released in a severe LOCA. The consequences of an environmental release depend on the extent of core damage, loss of fission product barriers, release rate of radioactive material, physical and chemical characteristics of the released material, meteorological conditions during the event, and release duration. The variation in release consequences as a function of these quantities is complex and scenario specific.

3.2.6 Other Events

The 11 September 2001, attacks in the United States caused regulators to review the design basis for nuclear power plants. Proposals to establish no-fly zones near reactors or require lattice-like barriers to protect reactors from an aircraft attack have been proposed.

The US Nuclear Regulatory Commission recently determined that making nuclear power plants crash proof to an airliner attack by terrorists is impractical. Protection against an air attack is the responsibility of the military and the Federal Aviation Agency. The NRC directed that the operators of nuclear plants focus on preventing radioactive material from escaping in the event of an air attack and to improve evacuation plans to protect the health and safety of the public.

Additional discussion regarding power reactor radiological events caused by intentional human intervention is addressed in subsequent discussion. These reactor events are caused by a number of initiators and are not limited to aircraft events. Nuclear terrorism is a growing concern and involves dispersing radioactive material using explosives or through attacks on a nuclear facility. Specific terrorist events are addressed in Chapter 4.

3.2.7 Probabilistic Risk Assessment

The US regulatory environment utilizes probabilistic risk assessments (PRAs) as an analysis tool to evaluate severe accidents. NRC PRAs utilize a goal for the CDF to be less than 1×10^{-4}/year and a large release frequency less than 1×10^{-6}/year. PRAs are also used to reveal design and operational vulnerabilities; strengthen programs and activities in the training, emergency operations, reliability assurance, and safety evaluation areas; and to evaluate maintenance and surveillance frequencies.

The certification of the AP-1000 design by the USNRC provides specific severe accident CDF values for a Generation III facility. The results of the AP-1000 analysis are compared to Generation II CDF values in Table 3.5.

Table 3.5 Comparison of AP-1000 Generations III and II PWR core damage frequencies.

Initiating event	Core damage frequency ($year^{-1}$)	
	Generation III	Generation II
	AP-1000[a)]	PWR[b)]
LOCAs (total)	2.1×10^{-7}	1×10^{-6} to 8×10^{-5}
Large	4.5×10^{-8}	
Spurious automatic depressurization	3.0×10^{-8}	
System actuation		
Safety injection line break	9.5×10^{-8}	
Medium	1.6×10^{-8}	
Small	1.8×10^{-8}	
Core makeup tank line break	4.0×10^{-9}	
Reactor coolant system leak	3.0×10^{-9}	
Steam generator tube rupture	7.0×10^{-9}	9.0×10^{-9} to 3.0×10^{-5}
Transients during power operations	8.0×10^{-9}	5.0×10^{-7} to 3.0×10^{-4}
Loss of off-site power/station blackout	1.0×10^{-9}	1.0×10^{-8} to 7.0×10^{-5}
Anticipated transient without scram	5.0×10^{-9}	1.0×10^{-8} to 4.0×10^{-5}
Interfacing system LOCA	5.0×10^{-11}	1.0×10^{-9} to 8.0×10^{-6}
Reactor vessel rupture	1.0×10^{-8}	1.0×10^{-7}
Total	2.4×10^{-7}	4.0×10^{-6} to 3.0×10^{-4}

a) NUREG-1793 (2006).
b) NUREG-1560 (1996).

The results summarized in Table 3.5 support the previous discussion regarding the improved safety performance of Generation III reactors relative to their Generation II counterparts. In the AP-1000 design, total CDF values are a factor of 17 – 1250 lower than the range for Generation II PWRs. This CDF reduction represents a significant improvement in safety performance provided by the Generation III design and its passive safety systems.

The use of PRAs in NRC licensing decisions and in evaluating DBEs is further addressed in Chapter 7. Chapter 7 discussion specifically addresses the regulatory process and its future direction and basis.

3.2.8 INES Event Classification Scale

Although reactors are designed to limit the off-site release of radioactive material, the accidents at TMI-2, Chernobyl-4, and Fukushima Daiichi proved that severe events occur. As demonstrated by the Chernobyl and Fukushima Daiichi events, accidents can affect neighboring countries. Therefore, it is important to have an internationally recognized method to characterize the severity of a reactor accident.

The International Nuclear Events Scale is a system used to classify the safety significance of nuclear and radiological accidents and to communicate their severity in a well-defined manner. This scale includes seven event levels that are (1) anomaly, (2) incident, (3) serious incident, (4) accident with local consequences, (5) accident with wider consequences, (6) serious accident, and (7) major accident. These levels consider the accident's impact on people and the environment, radiological barriers and controls, and safety system defense in depth.

The 11 March 2011, seismic event and subsequent tsunami affected the Fukushima Daiichi and Fukushima Daini Nuclear Power Stations and states of emergency declaration were announced at both sites. The Fukushima Daini Units 1, 2, and 4 were classified as Level 3 events, and these units safely achieved a stable, cold shutdown condition.

The Fukushima Daiichi and Chernobyl-4 events are classified as Level 7, and TMI-2 is a Level 5 event. Following the INES classification, a Level 5 event may include a limited release of radioactive material likely to require implementation of some planned countermeasures, several deaths from radiation, severe damage to the reactor core, and release of large quantities of radioactive material within an installation with a high probability of significant public release. Each of these items does not need to be met for the Level 5 assignment. At TMI-2, no radiation fatalities occurred, and the maximum public effective dose was <1 mSv.

A Level 7 event involves a major release of radioactive material with widespread health and environmental effects requiring the implementation of planned and extended countermeasures. Although both accidents are classified as Level 7 events, the radioactive material released during the Fukushima Daiichi accident is estimated to be about 10–20% of the Chernobyl-4 amount. The Fukushima Daiichi accident radioactive material releases are addressed in more detail in Chapter 7.

3.3 Major Reactor Accidents

The TMI-2, Chernobyl-4, and Fukushima Daiichi accident sequences illustrate the health physics challenges associated with these events. Three major accidents in less than 35 years suggest that commercial reactor performance is in need of improvement. The operational aspects of these accidents and their associated health physics issues are addressed in this chapter. Off-site releases of radioactive material and their associated environmental and regulatory aspects are addressed in Chapter 7.

The TMI-2 and Chernobyl-4 event descriptions focus on salient details of these accidents. An expanded presentation of the Fukushima Daiichi accident is provided since its regulatory impact is still unfolding. Additional details are presented to illustrate specific radiological aspects of the Fukushima Daiichi event.

3.3.1 Three Mile Island

The TMI Nuclear Generating Station included two PWRs located in central Pennsylvania near the state capital of Harrisburg. The reactors are located on an island in the Susquehanna River. TMI-2 operated for only a few months before the accident and TMI-1 was not affected by the event.

3.3.1.1 TMI-2 Accident Sequence

At approximately 4:00 a.m. on 28 March 1979, TMI-2 was operating at about 100% power when the plant automatically shut down when a pump, supplying feedwater to the secondary side of a steam generator, stopped operating or tripped. The loss of feedwater removed the steam generator's ability to cool the reactor, resulting in a temperature and pressure increase in the RCS. The pressure increase caused a power-operated relief valve (PORV) in the pressurizer to open as designed, but the valve failed to close when the RCS pressure returned to the normal operating range. With the PORV open, water and steam flowed out of the RCS, and a LOCA was initiated.

After significant water inventory was lost, extensive melting of the reactor core occurred, and large quantities of radioactive material were released into the RCS. The core achieved sufficient temperatures for the fuel cladding to oxidize and produce hydrogen that was released into the containment atmosphere. Sufficient hydrogen was generated to support a hydrogen explosion. The TMI-2 containment survived the hydrogen explosion with minimal damage, and no fission products were released to the environment at this stage of the accident.

3.3.1.2 Fission Product Releases

During the TMI-2 accident, approximately 50% of the noble gases and particulate cesium, 30% of the iodine, and other fission products were released from the damaged fuel into the reactor coolant. These radioactive materials decreased in concentration as the material flowed from the RCS.

Reactor coolant flowed from the core through the open PORV and into the reactor coolant drain tank located in the reactor building basement. After filling, the drain tank's rupture disk failed and core coolant exited the tank and collected in the reactor building sump. The sump filled and water in the containment basement reached a depth of about 2.1 m. This water, containing fission products, was pumped to the auxiliary building sump where fission products, primarily noble gases and some iodine, were collected by the waste gas system and released to the environment.

No removal of noble gases occurred within this pathway, but radioiodine was removed during the water transfer through chemical reactions and other removal mechanisms. The small quantity of iodine released forced a reanalysis of accident source terms and development of improved models of iodine interactions with the various plant systems, structures, and components.

The high-efficiency filtration system in the auxiliary building was designed to remove greater than 99% of the particulate activity. In addition to mechanical filtration, ventilated auxiliary building air was processed by multiple charcoal filters, which chemically removed 90–95% of the radioiodine. However, neither the mechanical filters nor the charcoal absorbers were designed to remove noble gases, which escaped directly into the environment. The quantities of radioactive material released into the environment during the TMI-2 accident involved about 0.38 EBq of noble gases and 1.3 TBq of radioiodine.

At the time of the TMI-2 accident, NRC safety analyses predicted the release of comparable quantities of noble gases and radioiodine. The limited quantity of iodine released is attributed to the unique nature of the TMI-2 release pathway.

3.3.1.3
Issues Associated with the Event

There are numerous issues associated with the TMI-2 accident. These include operational, human factors, control room design, and health physics-related items. The weaknesses revealed by the TMI-2 accident resulted in improvements in control room instrumentation that facilitate the operational and health physics response to future events. This chapter focuses on specific issues that affected the subsequent health physics response. Upgrades to emergency preparedness programs and their capability to provide radiological and environmental data were additional improvements that resulted from the TMI-2 event. In addition, improvements in communications systems to provide real-time information to risk counties, state governments, and regulators were a result of the lessons learned from TMI-2.

A number of these items affected the Fukushima Daiichi accident and are addressed in subsequent discussion. Issues including those related to public evacuations are presented in Chapter 6. Regulatory implications are outlined in Chapter 7.

3.3.2
Chernobyl

The Chernobyl site includes four reactors of the same type. Only Unit 4 was involved in the severe accident. Chernobyl-4 was a pressure tube reactor with the Soviet designation RBMK. The RBMK has a unique design, and it is graphite moderated and boiling water cooled. Vertical pressure tubes within the graphite contain low-enriched UO_2 fuel, control rods, or instrumentation. The reactor permits on-line refueling through selective pressure tube isolation. A negative feature of the design is a positive void reactivity coefficient under a range of operating conditions. The RBMK design includes emergency core cooling systems and steam suppression pools but did not incorporate a containment fission product barrier.

The TMI accident was a loss-of-coolant event resulting in core damage with minimal off-site effects. Chernobyl-4 was significantly more severe, and it resulted in fatalities and significant off-site doses. The Chernobyl-4 accident is characterized as a reactivity or power excursion, and the event was caused by the violation of standing safety requirements, an unforgiving reactor design, and failure to properly evaluate a planned evolution.

The proposed test was designed to verify that Chernobyl-4 could safely operate during a loss of off-site power by using the stored energy in the turbine generator to power safety-related equipment until the emergency diesel generators supplied auxiliary power. Although the test is similar to testing performed in US reactors, its execution was severely flawed.

The original test placed the Chernobyl-4 reactor into a safe configuration, which required a power reduction. This plan was disrupted when the plant was directed to increase power to meet local electrical demands. Upon direction to resume the test, power was again reduced. However, there was insufficient time to restore the plant to the configuration required by the original test.

In reestablishing the prerequisite power condition for the test, operators placed Chernobyl-4 into an unstable plant configuration. Since this configuration would result in an automatic reactor shutdown, the operators purposely bypassed several safety systems including the reactor control and emergency core cooling systems.

With safety systems bypassed, the Chernobyl-4 reactor was more vulnerable to a severe event. This vulnerability was exacerbated by the reactor's positive void coefficient that leads to an increase in reactivity as the volume of steam within the core increases.

Upon initiation of the test with the reactor in an unanalyzed condition, the core's steam volume increased with a coincident increase in reactivity, which dramatically increased the reactor's power level. With increasing reactor power, the increased void volume and associated reactivity addition resulted in a prompt criticality.

Prompt criticality led to rapid power and temperature increases. The increased temperature rapidly expanded fission gases within the fuel, which ruptured the fuel cladding, and released fragmented and possibly melted fuel into the coolant channels. The addition of hot material into the coolant produced additional steam. This combination of high temperatures and rapid steam production stressed the pressure tubes within the core, which subsequently ruptured. The resulting temperatures overpressurized and heated the cavity surrounding the graphite moderator. Cavity overpressurization led to ejection of a portion of the core and burning graphite moderator from the reactor vessel. Following the energetic ejection, the reactor building failed and facilitated a significant environmental release of radioactive material with subsequent radiation exposure of facility workers and the public.

The Chernobyl accident resulted in 31 fatalities to plant personnel and firefighters. About 1.9 EBq of krypton and xenon, 0.089 EBq of Cs, 1.8 EBq of radioiodine, and 0.11 EBq of other fission products were released. The release of these nuclides represents a substantial portion of the inventory of the damaged reactor's core and is significantly larger than the TMI-2 accident release source term.

3.3.2.1 Issues Associated with the Event

Operational issues associated with the Chernobyl-4 accident include bypassing safety systems and improperly evaluating a test procedure. Specific issues affecting the health physics response are provided in subsequent discussion. A number of these health physics issues also occurred during the Fukushima Daiichi accident.

On-site radiological response actions were not always performed in an as low as reasonably achievable (ALARA) manner. Weaknesses in the health physics response led to high worker exposures that resulted in a number of related health effects including death. Workers receiving 6–16 Gy had severe skin burns over 60–100% of their bodies. In this dose range, 21 deaths occurred in the cohort of 22 affected workers. Seven of 23 workers receiving doses in the 4–6 Gy range perished. In the 2–4 Gy dose group, one fatality occurred. In addition to these on-site radiological issues, off-site monitoring was less than optimal.

As with TMI-2, communication with outside organizations and information flow were major concerns. Acknowledgment of the event only occurred after fission products were detected by an operating reactor in another country. The closed nature of the Soviet Union was a contributing factor in the initial lack of information flow. Information flow also affected emergency response actions including evacuation of the public.

Both TMI-2 and Chernobyl-4 had weaknesses in the dissemination of radiation-related information. Communications and emergency response weaknesses have been common issues associated with major nuclear events at power reactors.

3.3.3 Fukushima Daiichi

The Fukushima Daiichi facility consists of six BWRs whose basic characteristics and accident-related damage associated with fission product barriers are noted in Table 3.6. A comparison to the previously addressed TMI-2 accident is also provided. Since Chernobyl-4 did not include a containment fission product barrier, it is not included in Table 3.6. From a health physics perspective, the condition of the fuel in the RPV and fuel pool and the integrity of the RPV and containment vessel (CV) are the primary concerns for power reactor accidents.

At Fukushima Daiichi, the three fission product barriers are the fuel/clad, the RPV and associated piping, and the CV. A secondary structure or reactor building, analogous to the auxiliary building in a PWR, encloses the SFP. In a PWR, the auxiliary building and containment are separate structures connected by a fuel transfer canal.

The fuel is contained within a steel RPV, and the containment vessel surrounds the RPV. The Fukushima Daiichi containment vessel includes a pear-shaped dry well and a wet well or suppression pool, which has the shape of a torus.

At Fukushima Daiichi, the reactor building includes a steel-framed service floor. The service floor is located above the RPV and contains the SFP, its support structures, and portions of emergency cooling systems. The containment vessel is below the service floor of the reactor building.

Table 3.6 Condition of the Fukushima Daiichi and Three Mile Island nuclear power plants' fission product barriers.

	Fukushima Daiichi Unit number						Three Mile Island Unit number	
	1	2	3	4	5[a]	6[a]	1[b]	2
Power output (MW_e)	460	784	784	784	784	1100	819	792
Status before event	Operating	Operating	Operating	Outage	Outage	Outage	Operating	Operating
Fuel assemblies in core	400	548	548	0	548	764	177	177
Fuel condition	Damaged (core melt)	Damaged (core melt)	Damaged (core melt)	Defueled	Not damaged	Not damaged	Not damaged	Damaged (core melt)
Reactor pressure vessel structural integrity	Partially damaged and leaking[c]	Partially damaged and leaking	Partially damaged and leaking	Not damaged	Not damaged	Not damaged	Not damaged	Minimal damage[d]
Containment vessel structural integrity	Damaged and leakage suspected[c]	Damaged and leakage suspected	Damaged and leakage suspected	Not damaged	Not damaged	Not damaged	Not damaged	Minimal damage[e]
Reactor building integrity	Severely damaged	Partly opened[g]	Severely damaged[f]	Severely damaged[f]	Undamaged	Undamaged	Undamaged	Undamaged[h]
Fuel condition in spent fuel pool	Most spent fuel is not damaged[i]	Most spent fuel is not damaged[i]	Most spent fuel is not damaged[i]	Most spent fuel is not damaged[i]	Not damaged	Not damaged	Not damaged	[j]

a) Units 5 and 6 were relatively unaffected by the earthquake/tsunami.
b) TMI-1 was not affected by the TMI-2 accident.
c) Damage estimate supported by initial 2015 muon tomography examinations Miyadera (2012).
d) Minor surface cracks were observed in the cladding in the lower portion of the TMI-2 RPV. This fission product barrier remained intact.
e) The TMI-2 containment building incurred minimal damage from the hydrogen explosion. This fission product barrier remained intact.
f) Hydrogen gas from Unit 3 is believed to have entered the Unit 4 reactor building and exploded.
g) A vent hole was opened on the rooftop to avoid a hydrogen explosion.
h) In a pressurized water reactor, this structure is the auxiliary building. It contains the spent fuel pool and portions of the emergency core cooling system.
i) Based on fission products detected in the spent fuel pool water.
j) The TMI-2 spent fuel pool contained no spent fuel. The reactor was in service only a few months before the accident.

The CV surrounds the RPV and its recirculation loops. It is a steel-lined pressure vessel encased over most of its surface by reinforced concrete. The suppression pool is located below the dry well, is connected to the dry well through a piping system designed to vent RPV pressure, and condenses and cools any vented steam.

At Fukushima Daiichi, the fuel pool resides above the containment vessel. This is in contrast to the PWR arrangement that has the SFP horizontally displaced and physically separated from the RPV. The arrangement at Fukushima Daiichi facilitated the venting of hydrogen to the reactor building, which led to damage to that structure and its components following the hydrogen explosion.

The Fukushima Daiichi accident is unique because three separate hydrogen explosions occurred following core damage. These explosions affected the SFPs in Units 1, 3, and 4. Unit 4 was in an outage condition. For this reason, both reactor specific as well as SFP-related discussions are provided.

3.3.3.1
Reactor Accident

At 2:46 p.m. local time on 11 March 2011, the Richter magnitude 9.0 Tohoku-Chiho Taiheiyo-Oki earthquake struck the FDNPS in Northeast Japan. Following the seismic event, the operating reactors (Units 1, 2, and 3) automatically shutdown and the control rods were inserted into the core. The Fukushima Daiichi reactors may have sustained some initial damage but survived the earthquake. A specific seismic damage assessment must await a detailed physical inspection of the facility.

The Fukushima Daiichi facility responded normally following the reactor trip. Containment isolation valves automatically closed, and valve closure provided an effective barrier to the release of radioactive material from the RPV. All off-site power was lost at 3:42 p.m., and emergency diesel generators started to power the electric pumps used to provide cooling water to the reactor cores and SFPs. At that time, the reactors were in a stable configuration with all fission product barriers intact. The reactors functioned as designed to this point in the accident sequence.

Following the earthquake, a tsunami struck the facility with a wave height of 14–15 m, which exceeded by almost a factor of 3, the design basis tsunami height of 5.7 m. Since the ground elevation at FDNPS Units 1–4 is 10 m above sea level and Units 5 and 6 are at an elevation of 13 m, Units 1–4 were flooded by 4–5 m and Units 5 and 6 by up to 1–2 m of sea water. This flooding contributed to the more severe damage at Units 1–4.

The tsunami breached the facility's protective seawall, and disabled the Units 1, 2, 3, and 4 emergency diesel generators at 3:45 p.m. when their fuel oil supply was disrupted. One diesel generator and its support systems survived the tsunami and powered a portion of the Units 5 and 6 safety systems. The operation of this diesel generator prevented core damage and the release of fission products from Units 5 and 6.

At that time, the FDNPS Units 1–4 were in a station blackout condition. A limited battery supply was available to power the emergency core cooling system pumps. Although stable, this reactor configuration is only effective as long as the suppression pool remains below 100 °C and power is available to provide core cooling capability.

In a few hours, the batteries were depleted and core cooling capability was lost in Units 1, 2, and 3. Core temperatures increased, and temperatures and pressures increased in the RPVs.

Rising pressure requires that the steam relief valves be opened to reduce RPV pressure, and steam is discharged to the CV and suppression pool. Consequently, the water level in the RPV decreases and the fuel is eventually uncovered. With diminished fuel cooling capability, cladding failures occur which releases fission products into the CV and suppression pool.

The FDNPS sequence of events following water addition to a degraded core is similar to the fuel failures that occurred during the TMI-2 accident. A specific Fukushima Daiichi fuel damage sequence will be forthcoming as the RPVs are defueled and the damaged core materials examined. The temperature thresholds for the various stages of core degradation at TMI-2 are suspected to have occurred at the FDNPS. The use of these temperature thresholds facilitates the subsequent discussion.

When core temperature exceeds about 1200 °C, the zirconium cladding alloy protecting the UO_2 fuel is oxidized by the water/steam in the RPV, and hydrogen is produced:

$$2\mathrm{Zr} + 2\mathrm{H_2O} \rightarrow 2\mathrm{ZrOH} + \mathrm{H_2} \uparrow \tag{3.1}$$

This reaction is exothermic which adds to the RPV heat load. With increasing RPV temperature and pressure, the hydrogen gas is vented to the suppression pool and then into the dry well.

With the loss of cooling capability, the fuel/clad temperatures increase, and additional degradation in the fuel/clad fission product barrier occurs. At about 1800 °C, the fuel cladding and adjacent steel structures in the RPV melt. Upon reaching 2500 °C, fuel rods fracture and a debris bed is created within the RPV. At about 2700 °C, uranium–zirconium eutectics melt.

If the FDNPS recovery activities proceed in a manner similar to the TMI-2, verification of the extent of fuel degradation will not be known for years. However, the radiation levels and isotopes released from the Fukushima Daiichi facility are suggestive of severe fuel damage/melting. The current assessment of fuel degradation is summarized in Table 3.6.

With severe fuel damage, fission products (e.g., Cs, I, Kr, and Xe) are liberated from the fuel and released, but the majority of the U and Pu remain in the core. The fission product aerosols are discharged from the RPV into the suppression pool, which reduces the quantity of radioactive material available for release to the environment. Similar activity reductions occurred during the TMI-2 accident. When the fission aerosols enter the dry well, the aerosols are further depleted by surface deposition.

At this stage of the event, the CV is the only remaining barrier between the fission products and the environment. The CV has a design pressure of 0.4–0.5 MPa, and the accident-induced pressure rises to about 0.8 MPa. This pressure increase is driven by the normal nitrogen inerting of the CV, added hydrogen from the zirconium cladding oxidation, and boiling within the suppression pool. Since the design pressure was exceeded, operators depressurize the CVs to ensure their long-term integrity.

Venting removes energy from the CVs and reduces their pressure to about 0.4 MPa. These positive aspects are offset by the release of fission aerosols, noble gases, and hydrogen to the upper levels of the reactor building.

In Units 1, 3, and 4, the released hydrogen explodes in the reactor building, which destroys their steel frame upper building structure and roof. Unit 4's explosion was caused by hydrogen that accumulated in its reactor building. The hydrogen is believed to have been released from the Unit 3 reactor.

The CV is damaged and suspected to be leaking in Units 1, 2, and 3. Destruction of portions of the reactor buildings in Units 1, 3, and 4 was dramatic and may have damaged structures needed for subsequent decontamination and decommissioning of the FDNPS.

In Unit 2, a hydrogen explosion may have occurred inside the CV, which damaged the suppression pool containing highly contaminated water. This resulted in the uncontrolled release of fission gases and fission products from the CV. The resulting high-dose rates led to evacuation of the site.

Initial fuel damage was mitigated upon restoring water to the RPV. This restoration involved the use of seawater.

In Unit 1, most of the core melted and formed a material mass composed of fuel, control rods, and RPV materials, which is often called *corium*. Initially, the corium was assumed to reside at the bottom of the RPV. However, it now appears that the corium has melted through the bottom of the RPV and eroded a portion of the 2.6 m thick dry well concrete. This erosion dissipated the corium heat content and permitted the mass to solidify. Much of the fuel in Units 2 and 3 appears to have melted but to a lesser extent than in Unit 1.

Corium breaching of the Unit 1 RPV is supported by the vessel's water level. The operating utility determined that the water level was more than 1 m below the bottom on the fuel, which suggests that water is leaking from the CV into the reactor building.

This accident sequence may have been exacerbated by operator action. In May 2011, the utility noted that operators might have manually shut down the core cooling systems in Unit 1 based on low RPV pressure. At TMI-2, an operator also secured core cooling systems believing that the pressurizer contained excess water. As in the case of TMI-2, it will take time for an accurate sequence of events to be firmly established. The accident sequence and extent of core damage will evolve as the RPVs are defueled.

3.3.3.2
Spent Fuel Pools Inventory

The SFPs are a significant consideration at the FDNPS because there is more fuel in the pools than in the Units 1, 2, and 3 reactor vessels. Their structural integrity must be maintained to ensure the fuel is covered with sufficient water inventory to cool the spent fuel. Loss of pool water inventory and associated fuel uncovery could result in fuel damage with a fission product release directly to the environment. The damaged Units 1, 3, and 4 reactor buildings provide little reduction in the source term and minimal aerosol depletion if additional fuel damage were to occur. The supplemental cover added to the damaged Unit 1 reactor building provided some reduction in the source term.

Fuel in the SFPs at Fukushima Daiichi is only protected by a single fission product barrier (i.e., the fuel/clad). The reactor building in a Mark I BWR does not provide the same degree of protection as the containment vessel.

Each of the Fukushima Daiichi reactors has an SFP, and there is an additional common pool. The fuel inventory in each pool represents a significant source term and is addressed in subsequent discussion.

3.3.3.2.1
Unit 1

The Unit 1 SFP has a capacity of 900 fuel assemblies. At the time of the event, it contained 292 irradiated fuel assemblies and 100 unirradiated fuel assemblies. The most recent additions of irradiated fuel assemblies occurred in March 2010. The March 12 explosion that destroyed the outer shell of the Unit 1 reactor building occurred near the SFP. Although hydrogen explosion debris landed in the Unit 1 SFP, most fuel in the pool is undamaged.

The original Unit 1 defueling plan projected defueling of the SFP to occur in 2017. In a December 2014 status report, the utility projected an additional 2 years delay with a new start date of 2019. RPV defueling is currently projected for 2025.

3.3.3.2.2
Unit 2

At the time of the accident, the Unit 2 SFP contained 587 irradiated and 28 unirradiated fuel assemblies. This pool has a capacity of 1240 fuel assemblies and it last received irradiated fuel in September 2010. During the initial phase of the accident, the operating utility was concerned that the pools would be depleted of water inventory because of the decay heat load. Initially seawater and subsequently fresh water was added to the pool. Most of the fuel in the Unit 2 pool is believed to be undamaged. Defueling plans for the Unit 2 SFP and RPV are in development and have yet to be finalized.

3.3.3.2.3
Unit 3

The Unit 3 SFP has a capacity of 1220 fuel assemblies and held 514 irradiated and 52 unirradiated fuel assemblies at the time of the accident. The Unit 3 pool received its most recent addition of irradiated fuel in June 2010.

The Unit 3 hydrogen explosion may have damaged a portion of its SFP. The operating utility was concerned about water inventory loss from the pool. On 17 March

2011, helicopters dropped seawater into the pool. Subsequent water additions sprayed water from fire trucks and other vehicles. Starting on March 24, seawater was injected into the Unit 3 pool using an existing cooling and purification line.

A 8 May 2011, water sample from the Unit 3 SFP contained elevated levels of fission products. The sample contained 140, 150, and 11 kBq/cm^3 of ^{134}Cs, ^{137}Cs, and ^{131}I, respectively. A video examination of the pool area showed debris scattered over the interior of the reactor building. However, most the Unit 3 spent fuel is not damaged. Defueling plans for the Unit 3 SFP and RPV are in development and have yet to be finalized.

3.3.3.2.4 Unit 4

Unit 4 shut down for routine maintenance in November 2010, and all fuel assemblies were transferred from the reactor to the SFP. With 1331 irradiated fuel assemblies in the pool, the thermal loading in the Unit 4 SFP was larger than in the other units. The Unit 4 pool has a capacity of 1590 assemblies and contained 204 unirradiated assemblies. It last received fuel in November 2010.

The hydrogen explosion may have caused a reduction in the cooling capability in the Unit 4 pool. Starting on 20 March 2011, water was added to the pool. On May 8, the operating utility concluded that some of the fuel in the Unit 4 SFP might have been damaged. SFP structural integrity improvements have been made to the walls of the reactor building supporting the pool. The 15 March 2011 hydrogen explosion and the March 11 seismic event may have damaged these structural members.

In 2012, the first of the 204 new fuel assemblies were removed from the Unit 4 SFP and transferred to the common SFP for detailed inspection. No fuel assembly deformation or corrosion was observed. The Unit 4 SFP defueling operations were initiated in 2013 and were completed in 2014. Unit 4's fuel was transferred to the common SFP.

The Unit 4 defueling operation is a significant milestone that eliminates a source of radioactive material and permits the pool to be used for other recovery tasks. At TMI-2, the SFP contained submerged demineralizer systems that removed radioactive material from the reactor coolant and led to a significant source term reduction.

3.3.3.2.5 Units 5 and 6

Although temperatures initially rose in the Units 5 and 6 pools, the restart of an emergency diesel generator provided power to cool these plant areas. The Unit 5(6) SFPs have a capacity of 1590(1770) fuel assemblies. At the time of the accident, there were 946(876) irradiated and 48(64) unirradiated fuel assemblies in Unit 5(6).

Fuel in the Units 5 and 6 pools was undamaged by the accident and subsequent hydrogen explosions. The Unit 5(6) RPVs were defueled in 2014(2013). These spent fuel assemblies currently reside in the respective unit's pool.

3.3.3.2.6
Common Spent Fuel Pool

In addition to pools at each of the six units at Fukushima Daiichi, the facility has a common use SFP. This common pool contains spent fuel from the six FDNPS reactors that has cooled for at least 18 months. The common pool has a capacity of 6840 assemblies and contained 6291 assemblies in March 2010. No issues with fuel integrity, pool integrity, and pool cooling capability have been reported.

3.3.3.3
Spent Fuel Pools Impacts

Spent fuel needs to be cooled and shielded. At Fukushima Daiichi, this is accomplished in SFPs and dry casks. The decay of fission products in the spent fuel generates heat that must be removed or fuel damage can occur. This fuel is cooled by water that is circulated by electric pumps through external heat exchangers that cool the spent fuel or by naturally circulating air that cools the dry storage casks. A reliable supply of on-site and off-site power is required to ensure the capability to cool fuel residing in the SFPs. Given the possibility that the hydrogen explosions weakened the pool's structural integrity, preserving the pool boundary and maintaining fuel cooling are high-priority recovery tasks.

In addition to hydrogen explosion damage, the earthquake may have produced structural degradation that will be investigated as the recovery proceeds. Damage could also have been caused by debris falling into the pools and striking fuel assemblies. To limit the impact of these issues, large-scale defueling operations have been conducted in Units 4, 5, and 6.

3.3.3.4
Issues Associated with the Event

The Fukushima Daiichi accident is unique in that it was caused by a natural event (e.g., massive earthquake and subsequent tsunami) that was addressed in the facility licensing basis. Unfortunately, the licensing basis underestimated the design basis earthquake/tsunami. As a result, the issues associated with the Fukushima Daiichi accident have a significant regulatory impact, and these items are addressed in Chapter 7.

3.4
Emergency Preparedness Programs

The accidents at TMI-2, Chernobyl-4, and Fukushima Daiichi clearly illustrated the importance of robust emergency preparedness programs. Emergency preparedness programs have two primary objectives. First, they develop plans and implementing procedures that provide the capability to mitigate the consequences of severe events in order to protect the health and safety of the public and site personnel. In addition, these programs ensure the operational readiness and capability of a facility's emergency response organization.

Emergency preparedness programs utilize dedicated utility organizations that manage the facility emergency and coordinate response actions with government agencies. State, county, and municipality governments have integrated roles and responsibilities and their respective emergency plans are coordinated with the operating utility. For the most serious events involving Site Area and General Emergencies, protective action recommendations, and their implementation require close coordination of the utility, government, and regulatory authorities.

3.4.1
Emergency Classification

An emergency classification is defined by a set of plant conditions that indicate a level of public risk resulting from a degraded facility state. In the United States, degraded plant conditions are defined in terms of a set of four emergency classifications. In order of increasing severity, these are the Unusual Event, Alert, Site Area Emergency, and General Emergency. Declaration of an emergency condition requires the activation of the facility's emergency response organization to respond to the degraded plant state.

The four classes are mutually exclusive groupings that are based on the spectrum of nuclear power plant emergencies. Each emergency classification has associated actions that must be performed including notification of off-site agencies and support organizations and mobilization of the applicable portions of the emergency response organizations to assess, mitigate, and eventually terminate the event. The emergency classes represent a hierarchy of events based on potential or actual hazards. Emergencies may be initially assigned a lower classification and then escalated to a higher classification if the plant condition deteriorates. De-escalation to a lower emergency classification also occurs as the situation improves.

Each of the four emergency classifications is determined by defined, plant-specific emergency action levels (EALs). These levels consist of specific sets of plant parameters (e.g., radiation monitoring system values, fission product barrier status, or cooling system flow capability) that are used to activate the emergency response organization. The emergency response actions include emergency classification designation, notification of government organizations, and mobilization of the facility's emergency response organization.

EALs facilitate rapid assessment and accident classification. The EALs are not selected to predetermine the necessity to implement protective actions but ensure sufficient time is provided to confirm initial plant parameter values by implementing additional on-site and off-site assessment actions. Upon declaration of a Site Area Emergency or General Emergency, protective action recommendations are determined utilizing radiological field team measurements, dose projections, or assessed plant conditions. Radiological information, relevant plant conditions, and the projected event duration are communicated to government officials as part of the utility's protective action recommendations.

In the United States, utilities adopt specific methodology to relate the effective dose and thyroid equivalent dose to the EAL values associated with an emergency classification. EAL radiation-related parameters could be chosen such that an individual exposed to these levels would receive a dose corresponding to a fraction of the Environmental Protection Agency's (EPA's) lower limit protective action guides (PAGs). For example, an Alert, Site Area Emergency, and General Emergency could be declared when the thresholds of 0.01, 0.05, and 1.0 times the PAG lower limit values are reached, respectively.

The lower limit Protective Action Guides, defined in subsequent discussion, are used as part of the basis for declaring radiological emergencies. The emergency classification philosophy is to promptly declare the highest class for which an EAL has been exceeded. For example, a Site Area Emergency is declared if one of its corresponding EALs is exceeded even if the lower Alert class was not previously declared. The emergency classification system facilitates timely evaluation of plant conditions based on comparison to defined EALs. These EALs are specific values or conditions determined by the plant's design. An application of EALs in an emergency is provided in the problem section of this chapter.

A number of the key decisions in emergency classification often require calculations of the projected dose as well as the extent of the deposition of radioactive materials in the environment. The projected dose calculations incorporate computer models that utilize plant and field team data.

Appendix E provides a list of computer models (e.g., RASCAL (radiological assessment system for consequence analysis) and MIDAS (meteorological information and dose assessment system)) and these codes are used for performing dose projections and determining ingestion pathway contamination levels following a power reactor accident. Many utilities develop facility specific models to perform projected doses. These codes are consistent with regulatory requirements.

3.4.1.1
Unusual Event

The lowest level (least severe) of the four emergency classifications is the Unusual Event. An Unusual Event is an event that defines plant conditions that are in process or have occurred which indicates a potential degradation in the level of safety. In an Unusual Event, no release of radioactive material requiring off-site response or monitoring has occurred or is expected unless further degradation occurs. Unusual events are low-level events that can include activation of plant safety systems, adverse plant radiological or operating conditions, explosions, chemical events, fires, flooding, loss of facility systems, loss of electrical power, security events, and weather conditions.

Unusual Events are based on the potential for the plant conditions to degrade to a more serious situation. An Unusual Event emergency declaration considers any uncertainty in the status of plant safety systems, the time the uncertainty may exist, and the expectation these uncertainties can be resolved in a reasonable time.

3.4.1.2
Alert

The next level of emergency classification is an Alert. An Alert classification is used to define an event that is in process or has occurred that involves an actual or potential substantial degradation in the level of safety of the plant. The Alert classification includes emergencies that are more severe than an Unusual Event. With the increased severity of an Alert classification, additional off-site emergency response agencies are notified and a larger portion of the facility's emergency response organization is activated.

Any radioactive material releases resulting from an Alert are expected to be limited to a small fraction of the EPA's protective action guide values. However, the Alert classification indicates a decrease in plant safety with potentially more severe consequences than the Unusual Event.

Alerts occur less frequently than Unusual Events. This classification includes the same general categories as noted in the Unusual Event class, but their consequences are more severe.

3.4.1.3
Site Area Emergency

The next emergency classification is the Site Area Emergency. A Site Area Emergency involves events that are in process or which have occurred that involve actual or likely major failures of plant systems needed for protection of the public. Releases of radioactive material are not expected to exceed the EPA PAGs except near the site boundary. Site Area Emergencies include significant events including LOCAs, SGTRs, and security breaches.

Protective actions are considered with the declaration of a Site Area Emergency. This emergency classification also activates on-site and off-site utility and government resources that are required to perform protective actions. If the declaration is based on radiological considerations, field teams are dispatched to perform air and direct radiation monitoring. These radiological data provide utility managers the relevant information to make protective action recommendations.

Unlike the Unusual Event and Alert levels, the Site Area Emergency may involve some radiation release to the environment with subsequent public exposure. Many accidents included in this class have the potential to degrade further to the General Emergency classification.

3.4.1.4
General Emergency

The most severe emergency classification is the General Emergency, and protective actions are implemented with this event type. A General Emergency involves actual or imminent substantial core damage or melting of reactor fuel with the potential for loss of containment integrity. Radioactive releases during a General Emergency are expected to exceed the EPA PAGs beyond the site boundary. Accidents having a large radioactive release potential (e.g., LOCAs and major security events or sabotage) can damage multiple fission product barriers. Only one

General Emergency has been declared in the United States. This event was the 1979 TMI-2 LOCA.

3.4.2
Protective Action Guides

The EPA uses ICRP 103 terminology to define its Protective Action Guides. As applied to emergency events, the EPA protective action guides for the early phase of a nuclear incident are defined for the effective dose and equivalent dose to the thyroid. The lower limit PAG values for the effective dose and equivalent dose to the thyroid for KI administration are 10 and 50 mSv, respectively, and are provided in Table 3.7. These dose values are based on plant conditions and may be either actual or projected doses.

The use of projected doses is an accepted practice but can lead to overly conservative actions. During the TMI-2 accident, containment radiation levels and associated projected doses warranted the declaration of a General Emergency. However, the actual off-site doses determined after the accident were less than 1 mSv. The declaration of a General Emergency was warranted by the available plant information, but it created considerable consternation for the public and added to the public's negative perception of nuclear power generation.

3.4.3
Protective Actions

Protective actions mitigate the consequences of a power reactor accident and minimize the resulting public radiation doses. After evaluating plant conditions, utility personnel recommend protective actions to the state government. The state government, typically the Governor with consultation with state radiation protection professionals, determines the actions necessary to protect the public and communicates these decisions. In reaching a decision, inputs from federal regulators and the utility are considered. The decision is influenced by current and projected plant conditions, competing events, weather conditions, time required to perform an evacuation, projected accident duration, projected release duration, nature of the release, and other considerations applicable to the specific event.

To facilitate the implementation of protective actions, the NRC divides the plume exposure pathway into 16 sectors with each including a 22.5° slice of the 16 km radius circle surrounding the plant. The sector concept permits the selective application of protective actions to specific areas affected by the plant release. The sectors selected for evacuation depend on the meteorological conditions and the severity of the release.

In the United States, protective actions include evacuation, sheltering, and the use of potassium iodide. Evacuation does not always involve the entire 16 km plume exposure pathway. Evacuations usually reflect the projected release direction, and default evacuation approaches are recommended under certain circumstances. For a General Emergency, all sectors within 3.2 km of the plant

Table 3.7 Planning guidance and protective action guides for radiological incidents.

Phase	Protective action recommendation	Protective action guide or planning guidance
Early	Sheltering-in-place or evacuation of the public[a)]	10–50 mSv projected dose/4 days[b)]
	Administration of prophylactic drugs (e.g., KI)	50 mSv projected child thyroid dose from radioactive iodine
	Limit emergency worker effective dose	50 mSv/year (or greater under exceptional circumstances)[c)]
Intermediate	Relocation of the public	20 mSv projected dose first year[b)] Subsequent years, 5 mSv/year projected dose
	Food interdiction	5 mSv/year projected dose, or 50 mSv/year to any individual organ or tissue, whichever is limiting
	Limit emergency worker effective dose	50 mSv/year[b)]
	Reentry	Operational guidelines[d)] (stay times and concentrations) for specific activities
Late	Cleanup	The planning process (including stakeholder participation) sets priorities and actions
	Waste disposal	The planning process (including stakeholder participation) sets priorities and actions

a) Should begin at 10 mSv. Sheltering may begin at lower levels if advantageous.
b) Projected dose is the sum of the effective dose from external radiation exposure (i.e., groundshine and cloudshine) and the committed effective dose from inhaled radioactive material.
c) Doses to emergency workers above 50 mSv may be approved by competent authority.
d) DOE/HS-0001; ANL/EVS/TM/09-(DOE, 2009).

are evacuated as well as individuals residing in the 8 km zone in the downwind and adjacent sectors. This default scheme is a "keyhole" pattern that accounts for potential wind shifts and fluctuations in the release direction. Evacuation beyond 8 km is evaluated as warranted by plant conditions and dose projections. In response to a General Emergency, individuals living in the remainder of the plume exposure pathway will likely be advised to remain indoors and monitor Emergency Alert System radio and television broadcasts.

A second protective action is sheltering that directs individuals to remain indoors to reduce their radiation dose. Sheltering is appropriate if the projected release duration is short, the release is controlled, or weather conditions do not permit a timely evacuation to occur.

The third protective action is the administration of potassium iodide (KI) in the plume exposure pathway if the projected thyroid equivalent dose is 50 mSv or greater. If taken within a few hours of the intake and at the appropriate dosage, KI prevents the thyroid from absorbing significant radioiodine and minimizes the thyroid equivalent dose. However, KI may have associated negative side effects including allergic reactions and gastrointestinal disturbances.

Although public medical evaluations for KI are not typically performed, utility personnel are often prescreened during their annual physical examinations. Personnel prescreening facilitates KI distribution to personnel during emergency operations and minimizes thyroid doses. Prescreening avoids medical side effects to sensitive individuals and is an ALARA measure that can save significant dose for releases involving an iodine source term.

The 2013 EPA Protective Action Guide Manual also added criteria for the interdiction of food and applied the PAGs and protective actions to various event types and industries. Commercial nuclear power reactors, fuel cycle facilities, radiopharmaceutical manufacturers and users, and weapons production complex facilities were included in the revised EPA manual. PAGs and protective actions were also applied to events at these facilities as well as accidents associated with space vehicle launch and reentry, terrorist attacks, and transportation. The EPA Planning Guidance and Protective Action Guides for Radiological Incidents are summarized in Table 3.7. Corresponding emergency worker guidelines are provided in Table 4.11.

3.4.4 Protective Action Recommendation Considerations

Protective action recommendations are evaluated at the Site Area Emergency and General Emergency classifications. As noted previously, the primary protective action recommendations are evacuation, sheltering, and administration of prophylactic agents. This section reviews specific considerations associated with their implementation.

3.4.4.1 Evacuation Criteria

The IAEA provides specific criteria for implementing the evacuation protective action. Evacuation as a protective action is commonly used when the public is threatened by man-made hazards (e.g., nuclear accidents, chemical spills, fires, and hazardous materials accidents) or natural events (e.g., floods, forest fires, hurricanes, landslides, seismic events, tsunami, and tornados). In most cases, the public returns to their homes in a few days. Return times can be extended if the homes were damaged or the area infrastructure was significantly degraded. When the events are short term, evacuation centers include simple accommodation in schools or other public buildings.

The implementation of a protective action should include a consideration of the dose that can be avoided by evacuation and would not be avoided by sheltering.

Evacuation decisions consider site-specific time estimates that are periodically updated, the event duration as projected by the utility, and dose projections based on plant conditions. Degrading plant conditions and security events also trigger a General Emergency classification with associated protective actions.

Although a General Emergency declaration is based on the lower limit protective action guide values, evacuation may be initiated at lower levels or when it is easily implemented (e.g., for small groups of people) . Under some circumstances, such as hazardous weather or the presence of a competing disaster, an evacuation may not be feasible and sheltering may be more appropriate. Evacuations are also complicated when large populations are involved or if transportation becomes difficult. The Fukushima Daiichi accident evacuation was impeded by the earthquake and subsequent tsunami that damaged communications capability, transportation services, and infrastructure.

3.4.4.2 Sheltering Criteria

As defined by the IAEA, sheltering as a protective action is occasionally used when man-made or natural events threaten the public. Sheltering is a passive action that entails remaining indoors to reduce the effective dose from direct radiation, surface deposition, and airborne contamination. It entails closing and possibly sealing doors and windows and securing ventilation systems to reduce the inhalation of radioactive material from the outside air. However, the effectiveness of sheltering decreases with time for most structures. For well-insulated homes with dense construction materials, sheltering reduces doses from airborne particulate material by a factor of 2 or 3. The effectiveness of sheltering rapidly decreases for buildings constructed with light materials and high air exchange rates.

As the dose rates increase, the benefits of sheltering become more important. The IAEA recommends a limit to the time that populations can remain indoors and defines a generic intervention level for sheltering of 10 mSv. This value is based on a maximum anticipated sheltering period of 2 days.

3.4.4.3 Administration of Prophylactic Agents

The prophylactic use of iodine is the administration of a nonradioactive compound (typically KI) in order to block the uptake of radioiodine by the thyroid. To be most effective, the administration must be timely. For optimum blocking, the administration of stable iodine should be accomplished several hours before and no later than a few hours postintake. Since time is critical, this protective measure is practical only if the stable iodine has been predistributed to the risk population.

Prophylactic iodine is usually accomplished in conjunction with evacuation or sheltering. The IAEA recommends a generic intervention level for the prophylactic use of iodine of 100 mGy, which is larger than the threshold of 50 mSv thyroid dose used in the United States. Because potential negative impacts such as allergic reactions to the administration exist, public health authorities should be involved in implementing this protective action measure.

3.4.5
Emergency Planning Zones

The NRC defines two emergency planning zones (EPZs) around each nuclear power plant. These are the plume exposure pathway and the ingestion pathway. The plume exposure pathway is a zone 16 km in radius around a plant, and its primary purpose is protection from airborne radioactive material released from the facility. Extending 80 km in radius around a plant is the ingestion pathway, and it protects the public from the consumption of food (e.g., crops, meat, milk, and water) that are contaminated by radioactive material. These EPZs are discussed in additional detail in Section 4.4.2.

Effectively utilizing these zones during an emergency requires considerable coordination between the operating nuclear utility and government organizations. Following dose projections and estimates of ground depositions of radioactive materials, government organizations must coordinate and implement evacuations, sheltering, administration of radioprotective agents, and food and water restrictions. As demonstrated by the Fukushima Daiichi accident, this coordination requires considerable effort and may involve other nations affected by the release of radioactive materials. Additional commentary regarding the implementation of these protective actions is addressed in subsequent discussion.

3.4.6
IAEA Evacuation Zone Issues

As noted in the previous sections, criteria for sheltering, using prophylactic agents, and evacuation, vary with the organization and government agency providing protective action recommendations. This is illustrated by differences between IAEA and US criteria. The use of differing criteria for protective actions and EPZs used by various nations and organizations affected by an accident leads to public confusion and distrust in government recommendations.

Evacuation issues related to the Fukushima Daiichi accident are addressed in Chapter 7. These issues are regulatory in nature and involve the inconsistency in national evacuation implementation criteria for a nuclear accident. The effects of accidents involving multiple nations are also addressed in Chapter 7.

3.4.7
Reentry Criteria

The IAEA considers three actions related to a public evacuation. These are the initiation of temporary relocation, termination of temporary relocation, and permanent resettlement. Temporary relocation is a public evacuation for an extended but limited time (e.g., several months) to avoid the dose derived from ground deposition, resuspended radioactive material, and contamination of the local food and water supply.

The IAEA intervention levels for initiating and terminating temporary relocation are 30 mSv in a month and 10 mSv in a month, respectively. As defined by the IAEA, the public should be temporarily relocated if the averted dose over the next month is expected to be greater than 30 mSv. The public may return when the averted dose falls below 10 mSv in a month. If the monthly dose is not expected to fall below 10 mSv within a period of a year or two, the IAEA recommends that the population be permanently relocated.

Permanent resettlement is public evacuation with no expectation of return for at least several years. The generic intervention levels for permanent resettlement are 1 Sv in a lifetime or a dose exceeding 10 mSv per month that persists beyond 1 or 2 years. Over a lifetime, the projected doses below the intervention levels for evacuation or for terminating temporary relocation could exceed the 1 Sv level to warrant permanent resettlement.

3.4.8
Psychological Effects of Evacuation

Evacuation planning focuses on protecting the public from the physical effects of ionizing radiation. It does not consider psychological impacts. During the period post evacuation, uncertainty, isolation, and fears about the effects of received radiation exposure can jeopardize the mental health of the evacuees. For example, the Japanese earthquake and tsunami that occurred in March 2011 forced evacuations related to the natural events as well as the nuclear accident at Fukushima Daiichi. As reported by *Nature*, the tsunami-related evacuees have a more positive attitude about the future. In contrast, the nuclear evacuees are less positive, have greater levels of mental stress, and suffer from depression.

A number of factors contribute to this reported trend. First, estimates of public radiation doses are uncertain because the radiation monitoring systems around Fukushima Daiichi were damaged by the earthquake and tsunami. However, the latest estimates suggest the public effective doses were 25 mSv or less. Second, survivors of the natural disaster have seen their homes rebuilt and lives restored. Some of the nuclear evacuees are still faced with both of those issues remaining unresolved.

The psychological aspects of a nuclear evacuation merit attention. Future emergency planning efforts should consider the psychological impacts noted during the Fukushima Daiichi accident. The negative public opinions generated following a nuclear emergency are exacerbated by failing to address psychological issues in an evacuated population. If nuclear power is to develop and prosper, resources should be devoted to address the psychological impacts and facilitate returning the evacuated population to a normal lifestyle in an expeditious manner.

3.5
Accident Phases

Accidents or nuclear events are typically described in terms of stages or phases. These phases characterize the facility status, active participants, and ongoing actions as the accident progresses from an active release of radioactive material to site and environmental remediation from the effects of the accident. Accidents are commonly described by three distinct elements, which are the early, intermediate, and late phases.

The duration of these phases varies with the particular accident sequence. Transitions between the phases represent significant milestones. The key participants in accident response and recovery vary with the accident phase including the extent of stakeholder involvement in the decision-making process.

The early or emergency phase is the period at the beginning of the incident when immediate decisions for the effective use of protective actions are required and actual field measurement data is generally limited or not available. Exposure to the radionuclides in the plume, short-term exposure to deposited materials, and inhalation of radioactive material are generally included when considering protective actions during the early phase. The response during the early phase includes initial emergency actions to protect public health and welfare. In the early phase of an emergency, lifesaving and first aid actions are given priority.

The intermediate phase of the response follows the early phase within as little as a few hours. It usually begins after the source and releases are under control, and protective action decisions are based on measurements of exposure and radioactive materials that have been deposited in the environment. The intermediate phase typically overlaps with the early and late phases and may continue for weeks to many months, until protective actions are terminated.

Recovery and cleanup actions designed to reduce radiation levels in the environment to acceptable levels are initiated in the late phase. The late phase ends when all recovery actions have been completed. With the additional time and increased understanding of the situation, there are opportunities to involve key stakeholders in the development of sound, cost-effective recommendations. Early phase decisions are made directly by elected public officials, or their designees, with limited stakeholder involvement. Long-term decisions should be made with stakeholder involvement, and include incident-specific technical working groups to provide expert advice to decision makers on impacts, costs, and alternatives.

3.6 Emergency Preparedness Effectiveness

To be effective, utility emergency preparedness programs must be coordinated with government organizations. The operating utility activates its emergency response organization to mitigate the accident. Utility emergency response personnel determine the nature of the accident and the capability of safety systems to mitigate the event. The operating utility provides accurate information regarding the current as well as projected plant status to government emergency management officials. This information includes the (i) availability of core cooling systems and emergency power supplies, (ii) status of the fission product barriers and their long-term capability to mitigate a release of fission products to the environment, (iii) nature of radioactive material releases and their projected duration, and (iv) capability of the utility to terminate the release.

Government officials utilize plant information to activate their emergency response organizations and implement protective actions. Coordination of government agencies and communication with the public are required to successfully implement protective actions.

Unfortunately, the three major power reactor accidents at TMI, Chernobyl, and Fukushima Daiichi clearly demonstrate operating utilities have not been completely successful in (i) ascertaining the nature of the accident in a timely manner, (ii) developing a strategy to minimize its effects, (iii) applying that strategy to mitigate the event, and (iv) communicating the accident characteristics, consequences, and anticipated duration to the public. In addition, government organizations have not adequately communicated the nature of the event to the public or fostered public confidence in their capability to oversee the event. These issues complicated the subsequent protective action decisions and their effective implementation.

The Fukushima Daiichi accident was further exacerbated by the earthquake/tsunami that disrupted communications and evacuation routes. The severe loss of life from the tsunami and fears of the release of radioactive material would challenge any emergency organization. Reactor accidents are difficult to manage even under ideal conditions. They become extremely difficult when a coincident natural disaster occurs.

3.6.1 Criteria for Safe Use of Food, Water, and Land Following a Release of Radioactive Materials

Following IAEA guidance, the control of food and water is considered for three specific situations. These situations occur when (i) alternative supplies are available, (ii) alternative supplies are scarce, and (iii) distribution involves international trade. Action levels when alternative supplies of food are available are summarized in Table 3.8. The values vary with the foodstuff and contaminating radionuclide. If these food restrictions result in nutritional deficiencies, case-by-case evaluations are required. For most situations, extreme restrictions result in relocation and alternative food is available. When this is not possible,

Table 3.8 Generic action levels for foodstuffs when alternative supplies are available[a)].

Radionuclides	Foods destined for general consumption (kBq/kg)	Milk, infant foods, and drinking water (kBq/kg)
^{134}Cs, ^{137}Cs, ^{103}Ru, ^{106}Ru, and ^{89}Sr	1	1
^{131}I	1	0.1
^{90}Sr	0.1	0.1
^{241}Am, ^{238}Pu, and ^{239}Pu	0.01	0.001

a) Crick (1996).

the radiation hazard is evaluated with respect to competing health hazards, and higher action levels are adopted.

Following an event that results in contaminated foodstuffs, a variety of countermeasures are implemented to ensure radioactive material concentrations in these commodities are maintained below the action levels. The generic action levels for foodstuff contamination levels also apply to the export of these items.

3.6.2 Stakeholder Involvement

Stakeholders include the public, special interest groups, citizens groups, and indigenous population groups (e.g., Indian tribes) that are affected by the accident or nuclear event. These groups are involved in various phases of nuclear licensing and participate in postaccident proceedings. Stakeholder involvement in nuclear licensing is addressed in Chapters 6 and 7. This chapter specifically addresses stakeholder involvement in the intermediate and late phases of a nuclear accident.

3.6.2.1 Stakeholder Groups

Stakeholders are individuals or groups that have a vested interest in the development and implementation of a policy or achievement of a specific goal. Stakeholders groups have various points of views regarding emergency management policies and differ in their expertise, areas of interest, and stage of an activity in which their involvement arises.

The most basic stakeholder associated with emergency management is the family unit or household. During a severe emergency, households evacuate and suffer economic and personal losses. Accordingly, all households have an interest in the emergency management policies developed and implemented in their communities. These interests vary by household because they have different incentives for disaster preparation and hazard mitigation. For example, homeowners have more wealth at risk than renters because they have equity in their homes. Households also vary in their capacity to deal with evacuations and possible loss of value to their homes following an accident. Differences arise from disposable income and financial resources, knowledge of the hazards and risks associated with living near a nuclear power reactor, and ability to apply this knowledge. The collection of households forms a community that represents a number of stakeholder groups.

Community stakeholders are divided into social, economic, and government groups. Social groups include religious, environmental, nonprofit, and community-based organizations.

As households are the basic unit of social stakeholders, businesses form the basic unit for economic stakeholders. Businesses are important stakeholders because their economic strength affects the sale and distribution of goods and services. A significant nuclear emergency interrupts business activities, which has an adverse impact on the local economy. Reentry and recovery decisions also have a significant influence on businesses and the economic recovery of an evacuated region.

There are various governmental stakeholders including the organizations established at the municipality, county, state, and national levels. Specific emergency management policy is developed and implemented at the state level. The federal government sets broad policy guidance and usually functions as a support organization for local and state efforts. In the case of nuclear power accidents, the federal government provides significant support to state officials.

3.6.2.2 Stakeholders in the Decision Process

In order to develop an effective response, emergency managers involve the relevant stakeholders in the decision-making process. For a nuclear power facility emergency, these decisions could involve reentry criteria following an evacuation, acceptable contamination levels in food and water, acceptable dose levels for homes and schools, decontamination criteria, decontamination methods, waste disposal locations, and criteria for resuming farming and other economic activity.

Stakeholder involvement requires trust in emergency managers and in an accurate and timely flow of information associated with decisions, recovery plans, and the disposition of community resources. Successful recovery efforts have stakeholders support.

Stakeholders must be informed of the issues and possible solutions as they are developed. The various stakeholder groups should be involved in the formulation of solutions and the selection of possible alternatives. Compromises in the desires of the various groups are inevitable, and emergency managers must be prepared to reach a solution that optimizes the needs of all effected groups.

A number of techniques are utilized to enhance stakeholder involvement. One technique is to hold public information forums explaining issues and seeking involvement of interested stakeholders. The success of this approach depends on the morale of the stakeholders, the magnitude of the accident, and its effect on their lives. At Fukushima Daiichi, the earthquake and subsequent tsunami strained resources and degraded trust in the government organizations associated with accident recovery.

A second technique for communicating and enhancing public involvement is the use of a decision hotline telephone number using a recorded message or a dedicated website providing accident recovery information. Citizens could be informed of these information sources using a mailed announcement, radio and television announcements, newspaper articles, electronic social media, and internet postings. All communication vehicles should be utilized to provide a mechanism for stakeholder feedback and questions. A procedure for promptly responding to questions and concerns is essential for cooperative stakeholder involvement.

A third approach for communicating with stakeholders is to establish direct contact through speakers at established community events. These forums address specific recovery-related topics and provide both face-to-face contact and an opportunity for dialog. Questions and concerns should be addressed promptly, accurately, and completely.

A fourth approach to fostering community involvement is the creation of stakeholder advisory committees for specific recovery activities. These committees provide an opportunity for stakeholders to participate in specific, well-defined areas and are most effective if stakeholder input receives serious consideration. An open dialog is important. Decisions should be based on an understanding of available resources, consensus criteria, and needs of the groups involved in the recovery effort. All decisions and their bases should be clearly communicated to stakeholder groups.

If stakeholders are not involved in the decision process or they feel the process is unfair or biased, they can still influence the outcome. Stakeholders have a variety of options including litigation, boycotts, public protests, and voter pressure that allows them to actively resist decisions. Given these negative processes, emergency managers should seek recovery decisions that meet stakeholder needs. This is not an easy process, and considerable patience and effort is required to achieve the recovery goals while meeting stakeholder needs.

3.6.3 Dose Reconstruction

As a major accident or nuclear event proceeds, the distribution of radioactive material dispersed into the environment is a critical concern. Mapping this distribution is often accomplished using aircraft or drones to characterize the plume and associated ground deposition of radioactive material. The distribution of radioactive materials affects subsequent land use and restrictions regarding the use of food, water, and animals within the fallout footprint.

A longer-term task is the reconstruction of doses delivered to the population from the event. Dose reconstruction is important in assessing the effects of the event, in resolving associated litigation, and in determining if reentry into the affected area is feasible. In addition, dose reconstruction guides specific areas that are to be decontaminated, selection of areas that are off-limits for reentry, and relocation areas for a returning population.

NCRP 163 provides a description of the principles and practices associated with the dose reconstruction process. Radiation dose reconstruction is the retrospective assessment of the dose delivered to representative individuals or populations. Dose reconstruction also includes estimates of the absorbed dose to individual tissues or organs for specified exposure situations in support of epidemiological studies or compensation programs.

3.6.3.1 Essential Steps and Foundation Elements

Dose reconstruction is a process that is initiated with a defined purpose and objectives and is performed in a logical and orderly manner. The dose reconstruction

Table 3.9 Summary of basic elements of the dose reconstruction process[a].

Basic step/element	Summary of step/element
Steps in the dose reconstruction process	
Definition of exposure scenarios	Description of relevant characteristics and activities of individuals or populations at locations and times when radiation exposure could have occurred Description of the sources of the radiation exposure
Identification of exposure pathways	Identification of relevant internal and external dose pathways associated with the defined exposure scenarios
Development and implementation of the methods for dose assessments	Development and implementation of assumptions, data, models, and methods used to assess dose from exposure pathways in assumed scenarios
Evaluation of uncertainties in dose assessments	An evaluation of the uncertainties and biases in the assumptions, data, models, and methods used to assess dose is essential to obtain an understanding of the degree of confidence in the dose estimate
Presentation and interpretation of results	Documentation of assumptions and methods for assessing dose and the discussion of results with respect to the purpose and objectives of the dose reconstruction
Foundation elements of the dose reconstruction process	
Data and other information	Collection, organization, evaluation, use, and presentation of all relevant information utilized in the dose reconstruction process
Quality management	Thorough documentation of the dose reconstruction process, the specific results, and associated uncertainties

a) NCRP 163 (2009).

process is divided into five essential steps and two foundation elements. These steps and elements are summarized in Table 3.9.

3.6.3.2
Radiation Dose Estimation

Dose reconstruction requires the explicit consideration of the routes of exposure and available measurement data. Radiation dose estimates include a consideration of external and internal sources, biodosimetry, and opportunistic dosimetry. Specific considerations in the calculation of the external and internal dose are summarized in Table 3.10.

Biodosimetry is the measurement of a biological response, which can be correlated to the radiation dose. It utilizes physiological, chemical, or biological effects resulting from the exposure of tissues to ionizing radiation to determine the dose to individuals or population groups. Frequently utilized

Table 3.10 Considerations for determining external and internal dose[a].

Consideration	External dose	Internal dose
Dose modeling	• Radiation dosimetry • Radiation transport	• Dosimetry • Biokinetics • Intake rates
Dose determining variables	• Exposure time (time indoors and time outdoors) • Distance from the radiation source • Shielding (workplace and residential) • Anthropometric characteristics	• Anthropometric characteristics
Covariates and attributes	• Ethnicity • Age • Disease status • Time of year • Geographic locations • Climate • Professional status/occupation • Lifestyle • Socioeconomic status	• Age • Gender • Gender-specific characteristics • Reproductive status • Ethnicity • Disease status • Diet • Socioeconomic status • Lifestyle • Religious affiliation • Variant behaviors • Workplace and residential conditions • Time of year • Excretion rate • Activity level and energy expenditure • Climate

a) NCRP 163 (2009).

biodosimetric approaches include (i) radiation-induced chromosome aberrations in lymphocytes, (ii) somatic mutations, (iii) spectroscopy of tooth enamel and bone, (iv) neutron-induced activity measurements, and (v) physiological effects of ionizing radiation including nausea, emesis, and lymphocyte depletion.

Opportunistic dosimetry is a term applied to the use of natural or man-made materials that respond to ionizing radiation in a manner that provides a measure of the dose delivered to that material. These methods include thermoluminescence, optically stimulated luminescence, chemiluminescence, neutron activation, and nuclear tract-etch detection. For example, the dose delivered to an individual or group is derived from a radiation-induced effect in these materials (e.g., light output following heating of a thermoluminescent material).

3.6.3.3
Evaluation of Uncertainty

Uncertainty analysis is essential to establishing the credibility of the calculated doses. An uncertainty analysis is based on an appropriate combination of rigorous statistical analyses and scientific judgment.

Numerical assessments of uncertainty are typically performed using Monte–Carlo simulation techniques. Each input parameter is generated from a distribution of values based on the best available information. This process is repeated using random sampling of values from the probability distribution of each input parameter to produce a distribution of dose values. The shape and magnitude of the calculated dose distribution characterizes its uncertainty.

3.6.3.4
Categories of Dose Reconstruction

NCRP 163 categorizes dose reconstructions with respect to exposures that are medical, occupational, environmental, and accidental. There are situations where more than one category could apply. For power reactor accidents, occupational dose reconstruction may be required to determine worker doses during the early phase of an event. For severe accidents, environmental dose reconstruction may be required to determine public doses.

Occupational dose reconstruction estimates prior radiation exposure resulting from employment. Personal dosimetry is normally available, and these data facilitate the dose reconstruction process. In a severe accident, varying and rapidly changing dose rates and airborne concentrations introduce uncertainty into dose reconstruction. For example, a number of internal intakes occurred during the Fukushima Daiichi accident when plant conditions were changing as the accident releases became more severe. These intakes were evaluated using postaccident data to support the dose reconstruction process.

Since radiation accidents and incidents occur without warning, the dose reconstruction must be performed in a timely manner. The specific dose reconstruction method considers all factors that significantly affect the dose, produces credible results, and is easy to use under stressful situations. Personal dosimetry is one of the most accurate methods for reconstructing doses. The use of clinical symptoms, neutron activation, biodosimetry, and opportunistic dosimetry is often required.

During an emergency, equipment shortages or inaccessibility could cause doses to be inadequately monitored. This situation occurred during the Fukushima Daiichi accident with its rapidly evolving events and the need for immediate action to restore core cooling.

Environmental dose reconstruction is performed for members of the public who may have been exposed during the early accident phase or following the consumption of contaminated food and water. Frequently, the exposed public individuals do not have radiation monitoring devices, and environmental monitoring data are sparse. These limitations place a greater emphasis on reconstructing environmental doses. The environmental dose study area may vary from a few kilometers around the facility to an impact upon neighboring nations.

Methods of environmental dose reconstruction vary in their accuracy. The most accurate methods utilize direct measurements of the individual. These methods include (i) whole- or partial-body counting to determine radionuclide deposition within the body, (ii) bioassay using urine or feces, (iii) measurement of tissue samples collected at autopsy, (iv) analysis of chromosome aberrations, and (v) electron paramagnetic resonance analysis of teeth.

An analysis of the individual's exposure environment is also performed. These methods include luminescent analysis of samples from the individual's residence. Known releases and models of radionuclide and radiation transport in air, water, and soil can also be performed. However, these methods are less accurate than the previously noted approaches.

3.6.4
Remediation of Contaminated Areas

A major nuclear event having the scale of the Chernobyl or Fukushima Daiichi nuclear power accidents will likely contaminate areas outside the facility boundary and lead to the evacuation of individuals residing near the facility. Following termination of the event, population reentry depends on the radiological conditions in the contaminated, evacuated areas. In many cases, the contaminated areas require remediation. Remediation of areas contaminated by a power reactor accident has similarities to sites contaminated by US weapons complex production activities. A number of Department of Energy sites (e.g., Hanford and Savannah River) are included in this category.

NCRP 146 presents a discussion of risk management issues associated with the remediation of sites contaminated with radioactive materials. The report focuses on sites licensed by the US Nuclear Regulatory Commission. A particular focus is the unresolved disagreement between the NRC and the EPA over questions of regulatory criteria that should be applied to site remediation to ensure adequate protection of the environment and the public.

In the United States, numerous federal laws govern the remediation of radioactively contaminated sites. These laws include the Atomic Energy Act (AEA); Comprehensive Environmental Response, Compensation, and Liability Act (CERCLA) or Superfund Resource Conservation and Recovery Act; and National Environmental Policy Act (NEPA). There are also applicable federal regulations including the License Termination Rule (LTR) [10CFR20, Subpart E] established by the NRC under the AEA and the National Contingency Plan (NCP) [40CFR300]. NRC regulations provide radiation requirements for risk management, specify annual individual doses of 0.25 mSv total effective dose equivalent, and require that doses be minimized using the ALARA principle. The EPA regulations establish requirements by specifying a goal for limiting lifetime cancer risk (usually a risk value of 10^{-4} at radioactively contaminated sites) with a separate goal of limiting contamination of water in accordance with the federal drinking water standards. At radioactively contaminated sites, the EPA guidance indicates that an annual effective dose equivalent of 0.15 mSv or less would comply with a risk goal of 10^{-4}.

In the event of a major US accident involving contamination of the environment, it is likely that these regulations as well as the recommended limits summarized in Tables 3.7 and 3.8 will be consulted for guidance in establishing acceptable criteria for reentry. Given the US regulatory history, establishing acceptable reentry standards will be a contentious process with numerous stakeholder groups involved. These conflicts are amplified by regulatory requirements that pose differing acceptance criteria for land use. This is illustrated by the conflicts between the EPA and NRC regulatory criteria for land-use requirements following the termination of a facility license.

The essential conflict between the NRC and EPA regulatory criteria is whether the NRC's 0.25 mSv/year would meet EPA's goal for lifetime cancer risk of about 10^{-4} and whether the lack of a separate provision in the LTR on protection of water resources would result in contamination of groundwater beyond federal drinking water standards. For many radionuclides, federal drinking water standards correspond to an annual effective dose equivalent of 0.04 mSv or less based on a 2 l/day consumption rate.

NRC and EPA regulations that are applicable to the remediation of radioactively contaminated sites have a number of significant differences. The EPA uses a lifetime cancer risk criterion to determine acceptable levels of residual soil contamination, but the NRC uses an annual dose criterion. It is not possible to directly compare these two criteria without examining the basis for the underlying regulatory framework including methods of site characterization, future land uses, analysis methods, and uncertainties in the underlying assumptions.

In contemporary US legislation established to ensure protection of the environment and the public, concurrent jurisdiction among federal and state agencies is the rule, not the exception. The use of concurrent regulations, having a different technical basis, complicates the remediation process. These differing bases enter into stakeholder discussions regarding reentry requirements and subsequent land use following a major nuclear power event with associated contamination of the environment.

Concurrent laws (e.g., CERCLA and NEPA) and their implementation in NRC and EPA regulations recognize that decision-making must involve stakeholders, especially nearby communities, directly affected by decommissioning and remediation. This process results in a negotiated remediation decision among stakeholders that protects the public and environment. However, the acceptable residual contamination levels are expected to vary with the accident location because stakeholder interests and perspectives vary. This lack of uniformity illustrates that the process could be improved by utilizing credible risk analysis as part of the criteria selection process.

Historically, decision-making at specific remediation sites has been primarily driven by the feasibility and costs of alternatives and the need to achieve negotiated agreements among regulators, site managers, and stakeholders. State governments have a vital role in determining acceptable remediation of radioactively contaminated areas, and the role of states should be considered to reconcile differences in NRC and EPA regulations.

Establishing uniform federal remediation standards for contaminated areas would facilitate stakeholder decisions in accepting regulatory criteria. Until this occurs, stakeholders will have an additional point of contention to complicate the choices involved with the remediation of contaminated areas and returning the areas to productive use.

3.6.5 MARSSIM

As noted in previous discussion, there is a need for a standardized approach for determining an acceptable end state following the remediation of contaminated areas. The Multi-Agency Radiation Survey and Site Investigation Manual (MARSSIM) can be used to implement established stakeholder criteria for land reutilization. MARSSIM is designed to provide a methodology to plan, implement, evaluate, and document building surface and surface soil final status radiological surveys to demonstrate compliance with established criteria. These criteria are defined in terms of dose or risk-based regulations or standards.

The Department of Defense, Department of Energy, Environmental Protection Agency, and Nuclear Regulatory Commission collaborated in the development of the MARSSIM approach. Since these four agencies have regulatory responsibility for the control of radioactive material, their consistent approach provides regulatory credibility to the MARSSIM approach. MARSSIM focuses on the demonstration of compliance during the final status survey after characterization, and any necessary remedial actions are completed.

3.7 Reprocessing Waste Tanks

Few facilities have a source term comparable to the radionuclide inventory at a commercial power reactor. Waste storage tanks that receive fuel reprocessing waste have the potential for a significant environmental impact following a major accident. Significant environmental impacts occurred in the Soviet Union following the Kyshtym waste storage tank accident in 1957. At the time of the accident, the tank is estimated to have contained about 0.74 EBq of fission products, activation products, and plutonium. This waste tank suffered an exothermic chemical reaction that dispersed about 90% of its contents within the production site and 10% off-site. The event has been ranked as Level 6 on the International Nuclear and Radiological Event Scale as a serious accident. For this reason, waste storage tanks and their environmental impacts are discussed.

This discussion is facilitated using the characteristics and source term of the Hanford Site underground reprocessing waste tanks. The Hanford example is selected for discussion because the information associated this site is more readily available than the historical data associated with Kyshtym.

3.7.1 Hanford Waste Tank Facility Background

The Hanford Site Tank Farm facilities are located in South Central Washington State. These facilities include 177 large underground tanks that store the radioactive and hazardous wastes generated during production of defense-related materials at the Hanford Site from the 1940s through the late 1980s. The tank farms are supported by facilities used to mobilize, transfer, and retrieve the waste.

During the defense mission, chemical processing of irradiated uranium fuels recovered uranium and plutonium and produced waste solutions and slurries. These waste forms contained heavy metals, organic solvents, inorganic compounds, uranium, mixed fission products, and low concentrations of plutonium. This waste was transferred from the processing facilities to the tank farms for storage. The process wastes were treated with various chemical agents, which produced the alkaline, heavy metal solutions, slurries, and inorganic salts that currently reside in the waste storage tanks.

Over 190 million liters of waste are stored in the Hanford Tank Farms. The tanks contain a mixture of liquid, sludge, and salt cake waste having both radioactive and hazardous constituents. Tank liquids occur in various locations and exist as supernatant (liquid above solids) and interstitial liquid (liquid filling the voids between solids). The sludge form is primarily solid precipitates of hydrous metal oxides created by the neutralization of acid wastes. Salt cake may exist between the supernatant and the sludge and consists of the salts formed by the evaporation of water from the waste. The waste types may not exist as distinct layers and may be intermingled.

3.7.2 Dominant Waste Tank Isotopes and Unit-Liter Doses

The radiological hazard of Hanford Tank Waste is defined in terms of the Unit-Liter Dose (ULD). A ULD is the inhalation dose obtained if an individual inhaled 1 l of waste. The ULDs provide a practical way to calculate radiological dose consequences for postulated accidents.

Tables 3.11 and 3.12 provide the 10 dominant isotopes for the set of Hanford Waste Tanks having a single shell. Table 3.11 (3.12) identifies the liquid (solid) waste constituent source term for single-shell tanks.

These source terms represent the oldest Hanford Tanks that have single-wall construction. Tables 3.11 and 3.12 provide the activity in Becquerel per liter of the waste and the corresponding ULD values for various residence times (i.e., 0, 10, 20, and 30 years) in the waste tank. The ULD values are derived from the ICRP 71 and ICRP 72 dose conversion coefficients. ULDs are used in acute inhalation dose calculations.

3.7.3
Dose Calculation Methods

The dominant isotopes in Hanford Tank Waste summarized in Tables 3.11 and 3.12 suggest that the dose received from an accident must consider both internal and external components. The dominant internal exposure pathway is inhalation. The direct dose occurs from spills, sprays, or plumes of waste mobilized by the various accident types.

Following the methodology of RPP-13033 *Tank Farms Documented Safety Analysis*, the airborne source term (Q) is the activity of airborne radioactive material generated by the accident that is available for transport to the receptor. The

Table 3.11 Unit-liter dose as a function of time for total inventory of single-shell tank liquids[a)].

Isotope	DCF (Sv/Bq)	Activity/liter (Bq/l)	ULD (Sv/l)	10-year ULD (Sv/l)	20-year ULD (Sv/l)	30-year ULD (Sv/l)
^{137}Cs	4.6E − 09	9.42E + 09	4.3E + 01	3.4E + 01	2.7E + 01	2.2E + 01
^{241}Am	4.2E − 05	6.03E + 05	2.5E + 01	2.5E + 01	2.5E + 01	2.4E + 01
^{239}Pu	5.0E − 05	2.73E + 05	1.4E + 01	1.4E + 01	1.4E + 01	1.4E + 01
^{90}Sr	3.6E − 08	3.74E + 08	1.3E + 01	1.1E + 01	8.4E + 00	6.6E + 00
^{240}Pu	5.0E − 05	4.15E + 04	2.1E + 00	2.1E + 00	2.1E + 00	2.1E + 00
^{151}Sm	4.0E − 09	2.24E + 08	9.0E − 01	8.3E − 01	7.7E − 01	7.1E − 01
^{237}Np	2.3E − 05	3.76E + 04	8.7E − 01	8.7E − 01	8.7E − 01	8.7E − 01
^{90}Y	1.5E − 09	3.74E + 08	5.6E − 01	4.4E − 01	3.5E − 01	2.7E − 01
^{238}Pu	4.6E − 05	1.18E + 04	5.4E − 01	5.0E − 01	4.6E − 01	4.3E − 01
^{154}Eu	5.3E − 08	5.90E + 06	3.1E − 01	1.4E − 01	6.5E − 02	2.9E − 02

a) RPP-13033 (2005).

Table 3.12 Unit-liter dose as a function of time for total inventory of single-shell tank solids[a)].

Isotope	DCF (Sv/Bq)	Activity/liter (Bq/l)	ULD (Sv/l)	10-year ULD (Sv/l)	20-year ULD (Sv/l)	30-year ULD (Sv/l)
^{239}Pu	5.0E − 05	1.95E + 07	9.8E + 02	9.7E + 02	9.7E + 02	9.7E + 02
^{241}Am	4.2E − 05	1.80E + 07	7.6E + 02	7.4E + 02	7.3E + 02	7.2E + 02
^{90}Sr	3.6E − 08	1.16E + 10	4.2E + 02	3.3E + 02	2.6E + 02	2.0E + 02
^{240}Pu	5.0E − 05	3.20E + 06	1.6E + 02	1.6E + 02	1.6E + 02	1.6E + 02
^{238}Pu	4.6E − 05	1.04E + 06	4.8E + 01	4.4E + 01	4.1E + 01	3.8E + 01
^{241}Pu	9.0E − 07	2.55E + 07	2.3E + 01	1.4E + 01	8.8E + 00	5.4E + 00
^{137}Cs	4.6E − 09	4.57E + 09	2.1E + 01	1.7E + 01	1.3E + 01	1.1E + 01
^{90}Y	1.5E − 09	1.16E + 10	1.7E + 01	1.4E + 01	1.1E + 01	8.5E + 00
^{231}Pa	1.4E − 04	9.20E + 04	1.3E + 01	1.3E + 01	1.3E + 01	1.3E + 01
^{227}Ac	2.2E − 04	4.41E + 04	9.7E + 00	7.1E + 00	5.1E + 00	3.7E + 00

a) RPP-13033 (2005).

Hanford Tank Farms calculations typically determine Q using the relationship:

$$Q = \text{MAR} \times \text{ARF} \times \text{RF} \times \text{LPF} \tag{3.2}$$

where MAR is the material at risk, ARF is the airborne release fraction, RF is the respirable fraction, and LPF is the leak path factor. The MAR is the amount of waste available for release when mobilized by a postulated accident. ARF is the fraction of the MAR that is suspended in air as an aerosol and available for transport. The RF is the fraction of airborne particles that can be inhaled by the receptor. LPF is the fraction of the MAR that is transported from the waste tank through its various confinement and filtration pathways to the receptor.

Given an airborne source term, the dose from the inhalation pathway is calculated using Eq. (3.3)

$$D_{\text{inh}} = Q\frac{\chi}{Q}R\,(\text{ULD}) \tag{3.3}$$

where D_{inh} is the inhalation dose (Sv), Q is the respirable source term (l), χ/Q is the atmospheric dispersion coefficient (s/m^3), R is the breathing rate (m^3/s), and ULD is the inhalation unit-liter dose (Sv/l).

3.7.4 Design Basis Accidents

Analyses of the Hanford Waste Tanks identify seven broad DBAs. These accidents are analogous to the DBAs identified in the commercial nuclear industry. The first five DBAs are operational events and classified in terms of the energy released during the accident.

Flammable gas accidents are high-energy events that involve the detonation of gases that accumulate in the waste tanks. These gases are produced from chemical reactions, radiolysis, corrosion, and other physical and chemical mechanisms. A severe flammable gas accident produces an atmospheric release of vapor, gas, and aerosolized tank waste.

Moderate energy events include the failure of the waste tank due to excessive loads and waste transfer leaks. Excessive loads causing tank failure can lead to the release of vapor, gas, and aerosolized waste to the environment. Waste transfer leaks release liquid and slurry waste from breaches in transfer lines.

The release of radioactive materials from the underground tanks can occur following the mixing of incompatible materials with the waste or the unplanned excavation or drilling into the tanks. These accidents are low-energy events. Atmospheric releases of vapors, gases, and aerosols result from the mixing of incompatible materials, and waste solids and sludges are released from unplanned excavation and drilling activities.

Other DBAs include natural events and external events including aircraft crashes into the waste tank. Additional discussion of specific tank farm DBAs, and their characteristics are provided in the problem section of this chapter.

3.7.4.1 Flammable Gas Accidents

The flammable gas accident is a detonation of gases that accumulate in the underground waste tank. A conservative estimate of the off-site doses for a flammable gas accident is in the range of 10–100 mSv. A detonation is the most severe flammable gas accident because it results in a faster flame speed than a deflagration. Accordingly, a detonation suspends the most tank waste.

The design basis flammable gas detonation destroys the center portion of the tank dome and results in a release of respirable tank waste material to the environment. Other release modes following a flammable gas event include pressure venting through the tank ventilation system and other dome penetrations, and dome cracking, and self-venting through the soil overburden. However, these release pathways are bound by the dome destruction event.

3.7.4.2 Tank Failure Due to Excessive Loads

The tank failure due to excessive loads accident involves a dome collapse due to an excessive concentrated load that results in less than 1 mSv off-site dose. A concentrated load is the most severe excessive load accident and leads to a larger release than a uniform load or load drop event. The concentrated load is modeled to shear the center portion of the dome resulting in a respirable release of waste material.

3.7.4.3 Mixing of Incompatible Materials

The mixing of incompatible materials accident is the inadvertent addition of sulfuric acid to a waste tank that generates less than 1 mSv off-site dose. This event assumes the reaction of sulfuric acid with carbonate compounds present in the tank waste to produce carbon dioxide. As the carbon dioxide gas is vented through tank penetrations, it carries a fraction of the aerosolized waste to the receptor.

3.7.4.4 Waste Transfer Leak

The waste transfer leak assumes a large pipe break that produces about 10 mSv off-site dose using conservative, deterministic methodology. Waste aerosol is generated by a splash and splatter mechanism as the waste volume falls onto an uncovered waste transfer structure. The pipe break is assumed to release respirable material over the accident duration.

3.7.4.5 Unplanned Excavation/Drilling

The unplanned excavation/drilling accident is caused by an industrial vacuum system failure that results in about 10 mSv off-site dose using conservative, deterministic methodology. A larger respirable release than other excavation sources (e.g., mechanical or manual digging) or drilling scenarios is produced by the vacuum system. Vacuum system operation causes a release of respirable contaminated soil. The accident assumes the complete failure of the vacuum filtration system.

3.7.4.6 Natural Events

An earthquake is the natural event with the highest potential for consequences because it can initiate multiple, concurrent accidents (e.g., flammable gas accidents, tank failures due to seismic loads, and waste transfer leaks). The consequences of these events were addressed in previous sections.

3.7.4.7 Aircraft Crashes into the Waste Tank

The total frequency of an aircraft crash from flight operations is approximately 1.0×10^{-6}/year. This event frequency includes contributions from general aviation, helicopter activities, commercial air carriers and air taxis, and from large and small military aircraft. An aircraft crash into a waste storage tank could produce a tank dome collapse with a subsequent fire from the aircraft's fuel. This event produces an aerosol of the tank contents, but the uncertainties in accident assumptions lead to significant variability in the calculated doses.

3.7.5 Beyond Design Basis Accidents

With respect to the Hanford Waste Tanks, beyond DBAs are those operational accidents that have more severe conditions or equipment failures than the corresponding DBAs. The DBAs summarized in Section 3.7.4 already assume severe conditions and equipment failures. Therefore, consideration of additional beyond DBAs is not warranted for the Hanford Waste Tanks.

3.7.6 Accident Analysis Conservatism

The aforementioned DBAs include a set of assumptions that are incorporated into the radiological consequence models. Many of these assumptions have a historical basis and are governed by the methods of defining the specific accident conditions. For specificity, two waste tank accidents and their associated assumptions are illustrated. These accidents are the flammable gas detonation and waste spray leak events. The accident assumptions are conservative and in some cases overly conservative.

Examples of the assumptions used in the formulation of flammable gas and waste transfer leak accidents and a brief discussion of the conservatism for the Hanford Waste Tank dose calculations include:

- Any flammable gas generated in the waste volume is instantaneously released and available for detonation at the time of the event. Gas generation decreases as the waste ages, and its concentration is reduced as gas is vented through tank openings. Pressure increases also vent gas as it accumulates.

- Flammable gases are always accompanied by spark sources, and any excursion beyond the lower flammability limit leads to a deflagration or detonation event. Spark generation and gas generation are independent events and should not be correlated in an absolute manner.
- Known physical phenomena associated with the tank's construction are ignored. These phenomena include the diffusion of hydrogen through concrete and steel structures and open penetrations that significantly reduce the flammable gas concentration.
- The tank structure is assumed to be tightly sealed, and the unsealed tank penetrations are ignored. Any airflow through the tank and through its interfacing systems via barometric or thermal breathing is assumed not to occur. The physical tank configuration facilitates airflow that reduces the flammable gas concentrations and limits or prevents the event.
- Exothermic, chemical reactions occur instantaneously, and all radioactive material is configured to be released in an optimum manner. This assumption fails to account for the aging of tank materials and the effect of prolonged radiation damage to the target reactants. Much of the assumed material that generates an exothermic reaction is degraded, no longer has the potential to produce the assumed reaction, or is significantly reduced in concentration, and no longer poses a hazard.
- Optimum conditions exist to generate the worst case (i.e., highest dose consequence) respirable aerosol in a line break accident. The aerosol reaches the environment without depletion by the ground or other structures. Respirable aerosols are only generated for a narrow range of line break configurations. These conditions are highly unlikely under the postulated line break event.
- Respirable aerosols are assumed to be generated by accident events. Known hygroscopic and gel formation characteristics of tank waste (e.g., sodium hydroxide and sodium aluminate content) that inherently limit the generation of respirable aerosols are ignored.
- Inherent physical aerosol removal mechanisms (e.g., soil overburden, intervening structures, and the particle momentum distribution) are ignored to maximize the quantity of released radioactive material.
- All released aerosol particles are respirable and produce the maximum effective dose consequence. The chemical constituents of the waste tend to produce large, nonrespirable particles that absorb water. The assumed modeling assumption is not credible.

The design basis calculations are performed in a deterministic manner using the accumulated conservatisms in the accident assumptions. A more realistic approach utilizes a stochastic calculation incorporating parameter distributions rather than individual conservative values to predict a range of dose consequences. The calculated range of dose values provides a more representative assessment of the accident than a conservative, deterministic dose assessment.

One of the inherent consequences of a conservative approach is focusing resources on hazards that do not exist. Each of these assumptions introduces conservatism that is beyond that justified by engineering judgment and scientific considerations. Each layer of conservatism adds additional safety system requirements that complicates the facility design, adds cost, burdens the operations and health physics organizations, and detracts attention from credible hazards (e.g., industrial safety concerns).

3.8 Waste Isolation Pilot Plant Accident

The Waste Isolation Pilot Plant (WIPP) near Carlsbad, NM, is the only US deep geologic repository for defense-related transuranic waste. On 14 February 2014, a chemical reaction or explosion occurred in a drum of nitrate waste containing americium and plutonium that was located in the underground salt formations of the WIPP facility. Thousands of drums are held in the 655 m deep facility. This event occurred just 15 years after WIPP opened. The event casts doubt on the capability of the Department of Energy to safely operate a defense waste facility and complicates the licensing of Yucca Mountain or a subsequent version of a high-level waste facility where a licensing basis of 10^4-10^6 years is the expected facility design basis.

Portions of the repository were contaminated with long-lived transuranic elements including plutonium and americium. A small amount of radioactive material reached the surface, and 13 workers received low-level intakes of radioactive material. The failure of the DOE and WIPP operating contractor to identify these radiological hazards and properly control them are significant issues. Placing unstable radioactive waste into WIPP also suggests weaknesses in the process used to establish the facility safety basis.

The event and its aftermath could have been considerably more significant. Maintenance resulting from a separate and apparently unrelated underground vehicle fire event on February 5 led to ventilation that was unfiltered from February 6 to 10. During that time continuous air monitors (CAMs) were not operating. Had the drum event occurred during that period, the release would have only been detected during routine radiation surveys. Under these conditions, workers would have been unknowingly exposed and higher levels of radioactive materials would have reached the environment.

During the accident, only one underground CAM was operable. Upon detecting elevated levels of radioactive materials, it sounded an alarm that led to switching the ventilation system to flow through high-efficiency particulate air (HEPA) filters to reduce the concentration of radioactive materials before the air effluent reached the environment. Shortly after the alarm, a shift manager started large capacity fans to vent the air effluent through the HEPA filters. Such an operation is normally accomplished automatically, but it was a manual operation at WIPP. This falls short of the expectations for safety standards at commercial nuclear facilities.

The ventilation system structure also did not meet nuclear safety standards since it had gaps that facilitated the release of radioactive material to the environment. These gaps were not sealed until 6 March 2014. These events suggest that WIPP was operated in a manner that fell far short of the standards expected in nuclear facilities.

The DOE accident report for the WIPP release documents a safety culture dominated by complacency. It outlines a lax safety and security culture, declining safety standards, and lack of rigorous operational standards. It repeats a number of failures encountered during the Fukushima Daiichi accident including dilution and lack of respect for safety standards, hubris, overconfidence in safety basis assumptions, and a weak safety and security culture. In addition, there was a lack of rigorous, independent oversight. DOE failed to provide adequate oversight or develop an independent group to perform that function. Oversight was also absent in other involved organizations including the operating contractor and the EPA.

The WIPP event is significant because if affects future geologic repository licensing. In addition, WIPP was being considered to expand its charter from low- and medium-level waste to include surplus weapons-grade plutonium and spent nuclear fuel. The WIPP accident casts doubt on these expanded applications. In addition, the WIPP event suggests that significant management and oversight improvements are required for the DOE to safely operate a high-level geologic waste repository.

Problems

3.1 You have been hired as a radiation protection consultant for a fuel reprocessing site that has an underground tank farm containing high-level waste. These tanks received waste following the removal of 99.5% of the uranium and plutonium from reprocessed fuel.

Waste stored in the underground tanks consists mostly of inorganic compounds. The radioactive components in the waste are primarily long-lived fission product radionuclides and some actinide elements. Mixed waste in underground waste storage tanks may contain heavy metals such as lead, chromium, zirconium, potassium, and cadmium. Detectable amounts of organic compounds, which were used in fuel reprocessing, are present.

The tanks now contain a mixture of liquid, sludge, and salt cake waste types with both radioactive and toxic chemical constituents. Liquids exist in supernatant and interstitial forms. Sludge consists primarily of metal oxides precipitated by the neutralization of acid wastes.

(a) The waste in the tank has aged at least 40 years. Which isotopes present the dominant external radiation hazard?
(b) Which isotopes present the dominant internal radiation hazard?
(c) List potential hazardous conditions for uncontrolled releases of radioactive or hazardous materials that are applicable to the underground

waste tank complex. A hazardous condition includes energy sources that could produce an uncontrolled release of radioactive or hazardous material.

(d) Based on the availability of radioactive material including fission products and fissile materials, toxic materials, chemical agents, and industrial energy sources, list and describe design basis accidents that are applicable to the high-level underground waste tanks.

3.2 You are a health physicist at a nuclear utility operating Grand Gulf Unit 8, a Generation III advanced pressurized water reactor. The utility has assigned you the responsibilities of the Radiological Control Manager (RCM) at the off-site Emergency Operations Facility (EOF) during a declared emergency during which a radioactive material release to the environment is possible.

Plant Data:

Applicable plant conditions prior to the event

Letdown radiation monitor reading (primary reactor coolant system activity)	1850 Bq/cm^3
Steam generator blowdown radiation monitor reading (secondary system activity)	1.85 Bq/cm^3
Steam generator "A" radiation monitor reading	$<1\ \mu Sv/h$
Atmospheric relief valve flow rate	$<100\ cm^3/s$
Containment pressure	0.1 psig

Applicable plant conditions following the event

Letdown radiation monitor reading (primary reactor coolant system activity)	27 MBq/cm^3
Steam generator blowdown radiation monitor reading (secondary system activity)	$1.25 \times 10^5\ Bq/cm^3$
Steam generator "A" radiation monitor reading	0.42 mSv/h
Iodine partitioning factor	0.015
Atmospheric relief valve flow rate	$1.4 \times 10^7\ cm^3/s$
Containment pressure	5.0 psig and increasing

Other Data:

Wind speed	25 km/h
Pasquill stability class	E
^{131}I Dose conversion coefficient	$20.9\ \frac{Sv\text{-}cm^3}{MBq\text{-}s}$

Steam generator "A" blowdown sample isotopic results

Radionuclide	Concentration (Bq/cm^3)
^{131}I	2.33×10^4
^{133}Xe	3.37×10^4
^{135}Xe	3.44×10^4
^{134}Cs	2.81×10^4
^{137}Cs	5.18×10^3

Pasquill class E atmospheric dispersion factors

Distance (km)	$\frac{\chi u}{Q}\left(\frac{1}{m^2}\right)$
1.6	1.57×10^{-2}
3.2	2.69×10^{-3}
8.0	1.56×10^{-3}
16.0	6.19×10^{-4}

(a) List three of the primary responsibilities of the RCM after activation of the EOF.
(b) List the three fission product barriers that protect the public from a release of radioactive material.
(c) List and define the three fission product barrier status categories used in determining off-site protective action recommendations.
(d) Based on plant data, what is the status of the fission product barriers?
(e) List four factors that can affect off-site dose calculations during a declared emergency.
(f) At this time, a release has not occurred, but you have been asked to provide an assessment of off-site doses. What is the projected thyroid equivalent dose rate 3.2 km downwind from the facility?
(g) Based upon your projected thyroid equivalent dose rate calculation and information from the plant indicating that the reactor has been stabilized and that a release of radioactivity is no longer likely to occur, describe the Protective Action Recommendation you would recommend to downwind sectors within 3.2 km of the plant.
(h) Atomic City is a small town with a population of 2500 located 3.5 km downwind from the plant. If a main steam line atmospheric relief value were opened for 15 min and then closed, what Protective Action Recommendations would you make?

(i) Using the dose rate from Question (f) and assuming an 8 h release, what emergency classification is warranted?

(j) Based on the dose calculated in Question (i), what protective actions are warranted?

3.3 You are a senior health physicist at the Wolf Creek Unit 10 Nuclear Power Station, a Generation IV gas-cooled fast reactor. The Wolf Creek facility uses continuous air monitors (CAMs) to measure airborne, beta-emitting particles near maintenance work that presents some potential for generating airborne activity. The monitors use a fixed-filter sample, a pancake-type Geiger–Müller (GM) detector contained inside a lead shield, and displays in counts per minute. It also uses a strip chart to record data for historical purposes.

It is 11 September 2042. Wolf Creek has received an advisory from the Ministry for State Security that a terrorist attack on a nuclear power facility is likely. Although the facility is under enhanced security measures, the plant manager decides to continue power operations.

A maintenance task is scheduled within a contaminated area. Prior to performing work in the area, the filter paper on the monitor is replaced and the monitor is moved into place and turned on at 08:00 h.

Data:

Worker's breathing rate	$1.2\ m^3/h$
Continuous air monitor flow rate	$1\ ft^3/min$
Filter collection efficiency	90%
Counting efficiency	0.3 counts/ disintegration
Detector background (with fresh filter paper)	70 cpm
^{60}Co dose conversion coefficient	7.1×10^{-9} Sv/Bq
^{131}I dose conversion coefficient	1.1×10^{-8} Sv/Bq
^{137}Cs dose conversion coefficient	6.7×10^{-9} Sv/Bq

(a) Particulate radon daughters are known to be present in the room at a beta concentration of $1.11 \times 10^{-5}\ Bq/cm^3$ with an effective half-life of approximately 27 min. At 09:00 h, what count rates should be observed on the monitor?

(b) At 09:00 h, work begins in the room where the air monitor is located. The crew is focused on repairing a technical specification required core cooling pump. They fail to notice that an intruder has entered the area, and she places a small shaped explosive charge near a second pump required for core cooling. The charge prematurely detonates at 09:45 h, kills the intruder, and damages all core cooling pumps and supporting systems. Since the blast is highly localized, the maintenance team remains in the room to ascertain the damage extent. The explosion causes the air in the room to become contaminated.

Over the next 10 min, the strip-chart recorder shows that the average count rate has increased by 40 000 cpm. Based on this information, what is the estimated average airborne concentration in the room assuming that the half-life of the measured activity is much greater than 10 min?

(c) The workers leave the room at 09:55 h and report their damage assessment to the emergency director. You suspect that the workers were exposed to a mixture of ^{137}Cs and ^{60}Co. To confirm these suspicions, you send the workers to have a whole body count. What are four advantages of a whole body count in this specific case over urine bioassay?

(d) A count of the filter on a high purity germanium (HPGe) gamma spectroscopy system shows that the airborne radioactivity is due to 25% ^{60}Co and the remainder due to ^{137}Cs. What is the effective dose assuming the workers remain in the room for 1 h? The average air concentration over the exposure time was 74 Bq/cm^3.

(e) Subsequent gamma spectroscopy screening of an air sample at the Technical Support Center (TSC), two floors above the site of the explosion, shows that ^{131}I is present. A TSC worker is exposed for 4 h to the contaminated atmosphere and has a ^{131}I intake of 500 MBq. What is the effective dose from this intake?

(f) The Emergency Director ordered all TSC personnel to take KI 1 h following the activation of this facility at 10:45 h. Was this action warranted?

3.4 A General Emergency has been declared at a Generation IV sodium-cooled fast reactor complex that includes fuel reprocessing and fuel fabrication facilities. As the Radiological Emergency Manager, you are directed to evaluate a release of radioactive material. The NRC licensed complex reprocesses spent reactor fuel and recovers plutonium and uranium for use as mixed oxide fuel. During the final separation process, an electrical failure caused a fire that breached the facility's engineered safety systems designed to preclude an environmental release.

The fire causes an airborne release of 2 kg of a Pu/Am mixture. The plume is predicted to drift off-site and pass over a nearby town. Assume that the material is released in a particulate form.

Data:

1. The dose conversion factor (DCF) for external exposure to the Pu/Am mixture is 2.8×10^{-5} Sv-m^2/h-g.
2. The effective DCF for inhalation of the Pu/Am mixture is 3.2×10^{-5} Sv/Bq.
3. The f_1 value for plutonium is 1.0×10^{-5}.
4. The breathing rate is 20 l/min.
5. The resuspension factor for ground deposition is 1.0×10^{-5}/m.
6. Radiological characteristics of the released Pu/Am mixture:

Isotope	wt%	Selected radiation emissions	
		Photon energy (MeV)	Photon yield (%)
^{238}Pu	0.04	0.017	11
^{239}Pu	93.3	0.017	5
^{240}Pu	5.99	0.017	11
$^{241,242}Pu$	0.32	—	—
^{241}Am	0.30	0.017	37
		0.060	36
Pu (mixture)	100	a)	<5

a) Pu (mixture) energy represents multiple photons with energies >0.03 MeV.

(a) Assume that the initial plume has passed. List two actions, which could most significantly reduce the dose to the downwind population during the first week following the accident.

(b) The EPA recommends relocation of the public based on a first year EPA Protective Action Guide (PAG) of >20 mSv effective dose. Assume the estimated dose received by residents who were outdoors during the initial plume passage ranges from 15 to 20 mSv effective dose and the estimated additional effective dose these residents are likely to receive during the first year after the accident is 13 mSv. Specify the meaning and intent of the intermediate phase PAGs.

(c) Using the information in the previous question, state your recommendations with respect to relocation of the population.

(d) To assess the off-site surface deposition of plutonium, field teams are equipped with portable handheld thin-crystal sodium iodide-based single-channel analyzers. These monitors are calibrated to either detect the 17 or 60 keV photons emitted from material involved in this release. Given that the emission ratio of the 17–60 keV photons is approximately 2.5, state two advantages of each energy calibration.

(e) Using the information from the previous question, state which photon energy you would recommend under the following two conditions: (i) dry paved road surfaces and (ii) an agricultural field following an extended rain.

(f) Calculate the effective dose from plutonium ground deposition for an individual who walks for 1 h on soil contaminated at a level of 3.7 MBq/m^2.

(g) Assume the 2 kg of Pu/Am mixture is uniformly spread over an area of 1000 m^2. Calculate the external dose received by an individual walking in the 1000 m^2 contaminated area for 8 h.

(h) Applying default assumptions from the facility's emergency plan, you estimate that a radiological worker who responded to the accident received a total effective dose equivalent of 48 mSv. Is further refinement of this dose necessary? If so, what action would you take to refine the dose estimate?

3.5 You are the Radiation Safety Officer (RSO) at Mega Nuclear Services' Rusty Pipe Fuel Reprocessing Plant supporting a Generation IV helium-cooled fast reactor complex. As part of annual radiation safety planning ordered by the Commissar for Nuclear Power Operations, you identified five accident scenarios that could occur at your facility. These accident scenarios are:

Accident 1: A worker's hand is severely lacerated by a grossly contaminated saw blade while cutting a drain line containing ^{239}Pu.

Accident 2: A diver is pulled from a spent fuel pool after a leak in the dive suit is detected. The pool water is heavily contaminated with tritium.

Accident 3: A worker is grossly contaminated on the face, hair, neck, and upper torso with ^{137}Cs following work in a highly contaminated area.

Accident 4: A researcher swallows a quantity of ^{35}S during a pipetting operation to determine primary system corrosion.

Accident 5: A worker's respirator fails in an area having a high airborne concentration of ^{131}I.

(a) List five actions that should be taken immediately in response to a generic radiological accident involving personal injury.

(b) For each of the five accident scenarios presented above, provide the preferred bioassay monitoring technique. Justify your answer, but assume your resources are not limited.

(c) Medical intervention techniques used to minimize the internal dose following an intake of radioactive material are divided into several general categories based on their protective actions. List four of these categories and give a brief description of the dose-savings principles of each.

(d) A physician working with an accident response team recommends the following intervention actions. Do you agree? Explain your answer:

1. Chelation therapy following the inhalation of 5 ALI (annual limits on intake) of ^{241}Am.
2. Lung lavage following the inhalation of 10 ALI of mixed fission products.

(e) Assume that the intakes associated with the five accidents scenarios are sufficiently high to warrant medical intervention. Give a specific intervention technique that is available for each accident and discuss any special concerns or necessary precautions.

3.6 The NRC News No. 12-129 (*NRC to Further Examine Solar Flare Issues Raised in Rulemaking Petition*) dated 18 December 2012, noted the potential safety consequences of a massive solar event that could potentially disable large portions of the US electrical grid for an extended time. Given such a prolonged, significant event, it is possible that nuclear power plants would lose off-site power or run out of diesel fuel to power emergency electrical systems. These power sources sustain safety systems that cool the nuclear fuel residing in the reactor vessel and spent fuel pool.

You are the Director of Health Physics at a multiunit boiling water reactor that uses an early Generation II design. All post-Fukushima Daiichi improvements have not yet been completed at your site. It is 1:13 p.m. when all on-site and off-site electrical power is lost at your facility. The apparent cause is a massive solar particle event that is 25 times larger than any previous event. All transformers are out of service, and the diesel generators cannot supply power to emergency buses. Battery power is also unavailable because a fire that occurred during the solar event disabled that system. The facility is in a station blackout condition, and the electrical engineering supervisor estimates it will be at least a week before limited emergency power is restored. An Alert has been declared due to the loss of all power, and the plant manager is concerned that a General Emergency declaration is possible unless core temperatures are reduced.

(a) It is now 11:44 p.m. and a General Emergency has been declared. Core temperatures are very high and reactor pressure vessel coolant samples indicate core damage has occurred. Operators have released steam from the reactor pressure vessel and depressurized the primary containment vessel. A large explosion has occurred and the upper floors of the Unit 3 reactor building were destroyed. What is the likely cause of this explosion?

(b) What are the health physics consequences of the explosion?

(c) What is the likely INES classification of the event?

(d) You are assigned to be the Health Physics Emergency Director and will manage the radiological response to the accident. An initial dose projection was performed with the following results:

Distance from facility (km)	Thyroid equivalent dose (mSv)	Effective dose (mSv)
1	2000	1000
2	900	700
5	700	500
10	400	200
16	200	50
20	40	7
50	10	4
80	5	2

Based on these results, what protective actions would you recommend?

(e) What radionuclides are most likely to affect the food chain? In what foods would these radionuclides be encountered? Confine your response to the early accident phase.

(f) Assume that several months have passed and the reactors are stabilized and near cold shutdown conditions. Radiological characterization of the facility is required. What methods would be used to develop an assessment of the radiological conditions within the Unit 3 reactor building?

(g) The operations supervisor is concerned that a recriticality occurred in one of the damaged reactors. What methods should be utilized to assess if a recriticality has occurred?

3.7 Big Salt Lick Unit 1 is a 1500 MWe molten salt reactor (MSR) that uses a mixed composition fuel including recycled uranium, plutonium, and minor actinides. During its fifth year of operation, a reactor vessel weld fails and all liquid fuel drains from the reactor vessel and into liquid fuel storage tanks. Containment integrity is breached when the liquid fuel storage tanks rupture. This failure results in a release of fission products to the environment.

The release occurs through an effective stack height of 25 m. Assume that the release occurs uniformly over a 72 h period and the wind is a constant 11 km/h.

Data:
Iodine decontamination factor for the MSR containment: 10^3
Noble gas decontamination factor for the MSR containment: 1
Activity available for release:
^{131}I: 1300 PBq
^{133}Xe: 3700 PBq

Applicable dispersion factors

Distance from release location (km)	σ_y (m)	σ_z (m)
1	30	10
2	50	20
5	120	35
10	300	50
50	800	70
100	2000	100

(a) List the three fission product barriers associated with the MSR design and any associated health physics-related weaknesses.

(b) Calculate the ^{131}I ground level concentration at plume centerline at a location 10 km downwind from the facility. Assume that the iodine aerosol behaves like a gas and all available ^{131}I is released.

(c) What is the ratio of ^{133}Xe/^{131}I ground level concentrations at 100 km from the facility?

(d) An individual inhales the uniform ^{131}I concentration determined in Question (b) for 5 h with a breathing rate of 1.2 m^3/h. What is the associated intake?

(e) If the effective dose conversion factor is 1.1×10^{-8} Sv/Bq, what is the dose to the individual receiving the ^{131}I intake?

(f) Compared to previous reactor accidents and the associated release of fission products, what is the severity of this event?
(g) What design features could be incorporated to minimize the thyroid dose?
(h) Are the decontamination factors reasonable? Provide a justification for your answer.

3.8 You are the Radiological Controls Director at Mt. St. Helens (MSH) Unit 1, a Generation III boiling water reactor utilizing an advanced torus design. Following the Fukushima Daiichi accident, the utility's board of directors placed emphasis on the emergency preparedness program and its implementation. In support of this effort, all documents related to emergency preparedness are being reviewed to ensure their quality and usefulness during an emergency event.

Data:
The following extract is one of the emergency action level statements from the MSH Emergency Plan Implementing Procedures. Use it to answer the questions associated with emergency event classification.
Emergency action level for actual or projected off-site integrated doses (effective dose or child thyroid equivalent dose):

Unusual Event: None

Alert: Off-site effective dose exceeds 0.1 mSv but less than 0.5 mSv or exceeds
0.5 mSv but less than 2.5 mSv child thyroid equivalent dose.

Site Area Emergency: Off-site effective dose ≥0.5 mSv but less than 10 mSv or ≥2.5 mSv but less than 50 mSv child thyroid equivalent dose.

General Emergency: Off-site effective dose ≥10 mSv or ≥50 mSv child thyroid equivalent dose.

(a) List the types of demographic information that should be included in the facility's emergency plan.
(b) List the general types of emergency action levels (EALs) that should be included in the emergency plan.
(c) Using the actual or projected dose EAL provided in the problem statement, what emergency classification is appropriate for the following projected doses: 0.15 mSv effective dose and 0.20 mSv child thyroid equivalent dose?
(d) Using the actual or projected dose EAL, what emergency classification is appropriate for the following projected doses: 0.45 mSv effective dose and 44.5 mSv child thyroid equivalent dose?
(e) Using the actual or projected dose EAL, what emergency classification is appropriate for the following projected doses: 30 mSv effective dose and 100 mSv child thyroid equivalent dose?
(f) If the Fukushima Daiichi accident had occurred in the United States, what emergency classification would be warranted? Based on your answer to Question (b), what EALs would have triggered this classification?

References

10CFR20 (2015) Standards for Protection Against Radiation, National Archives and Records Administration, U.S. Government Printing Office, Washington, DC.

10CFR50 (2015) Domestic Licensing of Production and Utilization Facilities, National Archives and Records Administration, U.S. Government Printing Office, Washington, DC.

10CFR72 (2015) Licensing Requirements for the Independent Storage of Spent Nuclear Fuel, High-Level Radioactive Waste, and Reactor-Related Greater than Class C Waste, National Archives and Records Administration, U.S. Government Printing Office, Washington, DC.

10CFR835 (2015) Occupational Radiation Protection, National Archives and Records Administration, U.S. Government Printing Office, Washington, DC.

40CFR300 (2015) Contingency Plan, National Archives and Records Administration, U.S. Government Printing Office, Washington, DC.

Barbera, J.A. and MacIntyre, A.C. (2007) *Medical Surge Capacity and Capability: A Management System for Integrating Medical and Resources During Large-Scale Emergencies*, 2nd edn, US Department of Health and Human Services, Washington, DC.

Batorshin, G.S. and Mokrov, Y.G. (2013) *Experience in Eliminating the Consequences of the 1957 Accident at the Mayak Production Association*, International Experts' Meeting on Decommissioning and Remediation After a Nuclear Accident, International Atomic Energy Agency, Vienna, http://www-pub.iaea.org/iaeameetings/IEM4/Session2/Mokrov.pdf (accessed 05 August 2015).

Behlung, U.H. and Hildebrand, J.E. (1986) *Radiation and Health Effects: A Report on the TMI-2 Accident and Related Health Studies*, GPU Nuclear Corporation, Middletown, PA.

Bevelacqua, J.J. (2004a) Skin contaminations, internal intakes, and ALARA. *Radiat. Prot. Manage.*, **21** (1), 11.

Bevelacqua, J.J. (2004b) Point source approximations in health physics. *Radiat. Prot. Manage.*, **21** (5), 9.

Bevelacqua, J.J. (2005) Internal dosimetry primer. *Radiat. Prot. Manage.*, **22** (5), 7.

Bevelacqua, J.J. (2008) *Health Physics in the 21st Century*, Wiley-VCH Verlag GmbH, Weinheim.

Bevelacqua, J.J. (2009) *Contemporary Health Physics: Problems and Solutions*, 2nd edn, Wiley-VCH Verlag GmbH, Weinheim.

Bevelacqua, J.J. (2010a) *Basic Health Physics: Problems and Solutions*, 2nd edn, Wiley-VCH Verlag GmbH, Weinheim.

Bevelacqua, J.J. (2010b) Radiological dispersion devices and basic radiation science. *Phys. Teach.*, **48**, 302.

Bevelacqua, J.J. (2010c) Practical and effective ALARA. *Operational Radiation Safety. Health Phys.*, **98** (Suppl. 2), S39.

Bevelacqua, J.J. (2011) Conservatism in effective dose calculations for accident events involving fuel reprocessing waste tanks. *Health Phys.*, **101**, 48.

Bevelacqua, J.J. (2012a) Applicability of health physics lessons learned from the Three Mile Island Unit-2 accident to the Fukushima Daiichi accident. *J. Environ. Radioact.*, **105**, 6.

Bevelacqua, J.J. (2012b) The Fukushima Daiichi accident and its radiological impact on the environment. *Phys. Teach.*, **50**, 368.

Bevelacqua, J.J. (2013a) Nuclear regulation in the United States and a possible framework for an international regulatory approach. *Int. Nucl. Saf. J.*, **2** (1), 1, http://nuclearsafety.info/international-nuclear-safety-journal/index.php/INSJ/article/view/12/pdf_1 (accessed 05 August 2015).

Bevelacqua, J.J. (2013b) Methodology to ascertain core recriticality following a severe fission reactor accident. *Int. Nucl. Saf. J.*, **2** (1), 52, http://nuclearsafety.info/international-nuclear-safety-journal/index.php/INSJ/article/view/13/pdf_1 (accessed 05 August 2015).

Bevelacqua, J.J. (2014a) Health physics aspects of generation IV reactors. *Int. Nucl. Saf. J.*, **3** (1), 13, http://nuclearsafety.info/international-nuclear-safety-journal/index.php/INSJ/article/view/20/pdf_1 (accessed 05 August 2015).

Bevelacqua, J.J. (2014b) Radiation protection considerations for nuclear power expansion to nations with limited technical infrastructure. *Int. Nucl. Saf. J.*, **3** (3), 8, http://nuclearsafety.info/international-nuclear-safety-journal/index.php/INSJ/article/view/37/pdf (accessed 05 August 2015).

Braun, M. (2011) *The Fukushima Daiichi Incident*. PEPA4-G, AREVA-NP GmbH, 31 March 2011, http://pdfsr.com/pdf/areva-fukushima-report.

Brodsky, A. (2014) *Handbook for Survival: Saving Lives During Radiation Releases and Other Disasters*, Fideli Publishing, Inc., Martinsville, IN.

Bromet, E.J. (2014) Emotional consequences of nuclear power plant disasters. *Health Phys.*, **106** (2), 206.

Brumfiel, G. (2011a) The meltdown that wasn't. *Nature*, **471**, 417.

Brumfiel, G. (2011b) Fukushima set for epic clean-up. *Nature*, **472**, 146.

Brumfiel, G. (2013) Fallout or fear. *Nature*, **493**, 290.

Brumfiel, G. and Cyranoski, D. (2011) Quake sparks nuclear crisis. *Nature*, **471**, 273.

Buddemeier, B.R., Valentine, J.E., Millage, J.E., and Brandt, L.D. (2011) National Capital Region Key Response Planning Factors for the Aftermath of Nuclear Terrorism. Performed under the auspices of the US Department of Energy by Lawrence Livermore National Laboratory for FEMA and Department of Homeland Security, November 2011.

Butler, D. (2011a) Radioactivity spreads in Japan. *Nature*, **471**, 355.

Butler, D. (2011b) Fukushima health risks scrutinized. *Nature*, **472**, 13.

Butler, D. (2014) Call for better oversight of nuclear-waste storage. *Nature*, **509**, 267.

Chernobyl accident Special Edition. *Nucl. Saf.*, **28**, 1 (1987).

Coleman, C.N., Weinstock, D.M., Cassagrande, R. *et al.* (2011) Triage and treatment tools for use in a scarce resources—crisis standards of care setting after a nuclear detonation. *Disaster Med. Public Health Prep.*, **5** (1), S111.

Committee on Guidance for Establishing Crisis Standards of Care for Use in Disaster Situations, Institute of Medicine (2012) *Crisis Standards of Care: A Systems Framework for Catastrophic Disaster Response*, The National Academies Press, Washington, DC.

Crick, M. (1996) Nuclear and radiation safety: guidance for emergency response. *IAEA Bull.*, **23**, 1.

Davis, M., Reeve, M., and Altevogt, B. (2014) *Nationwide Response Issues After an Improvised Nuclear Device Attack: Medical and Public Health Considerations for Neighboring Jurisdictions: Workshop Summary*, The National Academies Press, Washington, DC.

EG&G Report GEND-030 (1983) Quick Look Inspection: Report on the Insertion of a Camera into the TMI-2 Reactor Vessel Through a Leadscrew Opening, Vol. 1 and 2, EG&G Idaho, Inc., Idaho Falls, ID.

Egorov, N., Novikov, V., Parker, F., and Popov, V. (2000) *The Radiation Legacy of the Soviet Nuclear Complex*, Earthscan Publications, London.

Eichholz, G.G. (1985) *Environmental Aspects of Nuclear Power*, 2nd edn, CRC Press, Boca Raton, FL.

Fabrikant, J.I. (1981) Guest editorial: health effects of the nuclear accident at three mile island. *Health Phys.*, **40**, 151.

FEMA Program Manual (2013) *Radiological Emergency Preparedness*, Federal Emergency Management Agency, Washington, DC.

Glasstone, S. and Sesonske, A. (1963) *Nuclear Reactor Engineering*, D. Van Nostrand, New York.

Glasstone, S. and Walter, W.H. (1980) *Nuclear Power and Its Environmental Effects*, American Nuclear Society, La Grange Park, IL.

Gollnick, D.A. (2011) *Basic Radiation Protection Technology*, 6th edn, Pacific Radiation Corporation, Altadena, CA.

Gostin, L.O. and Hanfling, D. (2009) National preparedness for a catastrophic emergency: crisis standards of care. *J. Am. Med. Assoc.*, **302** (21), 2365.

GPU Nuclear Corporation (1988) GPU Nuclear Corporate Emergency Plan for Three Mile Island and Oyster Creek Nuclear Stations. 1000-PLN-1300.01, Revision 1, GPU Nuclear Corporation, Parsippany, NJ.

Gusey, I., Guskova, A., and Metter, F.A. (2001) *Medical Management of Radiation Accidents*, 2nd edn, CRC Press, Boca Raton, FL.

Hanfling, D. (2013) When the bells toll: engaging health care providers in catastrophic disaster response planning. *South. Med. J.*, **106** (1), 7.

Hanfling, D., Altevogt, B.M., and Gostin, L.O. (2012) A framework for catastrophic disaster response. *J. Am. Med. Assoc.*, **308** (7), 675.

Henderson, D.K., Malanoski, M.P., Corapi, G. *et al.* (2009) Bethesda hospitals' emergency preparedness partnership: a model for transinstitutional collaboration of emergency responses. *Disaster Med. Public Health Prep.*, **3** (3), 168.

Hrdina, C., Coleman, C.N., Bogucki, S. *et al.* (2009) The "RTR" medical response system for nuclear and radiological mass casualty incidents: a functional triage-treatment-transport medical response model. *Prehosp. Disaster Med*, **24** (3), 167.

Hubner, K.F. and Fry, S.A. (eds) (1980) *The Medical Basis for Radiation Accident Preparedness*, Elsevier/North-Holland, New York.

IAEA (1998) *Design Measures for Prevention and Mitigation of Severe Accidents at Advanced Water Cooled Reactors*, IAEA TECDOC-1020, International Atomic Energy Agency, Vienna.

IAEA (2004a) *Protection Against Internal Fires and Explosions in the Design of Nuclear Power Plants Safety Guide*, IAEA Safety Standards Series No. NS-G-1.7, International Atomic Energy Agency, Vienna.

IAEA (2004b) *Protection Against Internal Hazards other than Fires and Explosions in the Design of Nuclear Power Plants Safety Guide*, IAEA Safety Standards Series No. NS-G-1.11, International Atomic Energy Agency, Vienna.

IAEA (2004c) *Status of Advanced Light Water Reactor Designs*, IAEA TECDOC-1391, International Atomic Energy Agency, Vienna.

IAEA (2007) *Arrangements for Preparedness for a Nuclear or Radiological Emergency*, IAEA Safety Standards Series No. GS-G-2.1, International Atomic Energy Agency, Vienna.

IAEA (2011a) http://www.iaea.org/newscenter.

IAEA (2011b) *Criteria for Use in Preparedness and Response for a Nuclear or Radiological Emergency*, IAEA General Safety Guide GSG-2, International Atomic Energy Agency, Vienna.

IAEA (2014) *Experiences and Lessons Learned Worldwide in the Cleanup and Decommissioning of Nuclear Facilities in the Aftermath of Accidents*, STI/PUB/1644, IAEA Nuclear Energy Series No. NW-T-2.7, International Atomic Energy Agency, Vienna.

ICRP (1991) 1990 Recommendations of the International Commission on Radiological Protection. ICRP Report No. 60, Pergamon Press, Oxford.

ICRP (1993) Principles for Intervention for Protection of the Public in a Nuclear Emergency. ICRP Report No. 63, ICRP Publications, Oxford.

ICRP (1994) Dose Coefficients for Intakes of Radionuclides by Workers – Replacement of ICRP Publication 61. ICRP Report No. 68, Elsevier Science, Tarrytown, NY.

ICRP (1996a) Age Dependent Doses to Members of the Public from Intake of Radionuclides Part 4 Inhalation Dose Coefficients. ICRP Report No. 71, Elsevier Science, Tarrytown, NY.

ICRP (1996b) Age Dependent Doses to Members of the Public from Intake of Radionuclides Part 5 Compilation of Ingestion and Inhalation Dose Coefficients. ICRP Report No. 72, Elsevier Science, Tarrytown, NY.

ICRP (2007) The 2007 Recommendations of the International Commission on Radiological Protection. ICRP Report No. 103, Elsevier, Amsterdam.

INES (2009) *The International Nuclear and Radiological Event Scale User's Manual 2008*, International Atomic Energy Agency, Vienna.

Jadia, C.D. and Fulekar, M.H. (2008) Phytoremediation: the application of vermicompost to remove zinc, cadmium, copper, nickel, and lead by sunflower plant. *Environ. Eng. Manage. J.*, **7**, 547.

Knief, R.A. (1992) *Nuclear Engineering – Theory and Technology of Commercial Nuclear Power*, 2nd edn, Hemisphere Publishing Corporation, Washington, DC.

Kouji, H. and Harada, K.H. (2014) Radiation dose rates now and in the future for residents neighboring restricted areas of the Fukushima Daiichi Nuclear Power Plant.

Proc. Natl. Acad. Sci. U.S.A., **111** (10), pp. E914–E923.

Kuan, P., Hanson, D.J., and Odar F. (1991) Managing Water Addition to a Degraded Core. EGG-M-91375, EG&G Idaho, Inc., Idaho Falls, ID.

Kumano, Y. (2010) Integrity Inspection of Dry Storage Casks and Spent Fuels at Fukushima Daiichi Nuclear Power Station, November 16, 2010. ISSF 2010: Session 6, Tokyo Electric Power Company, Tokyo, http//www.nirs.org/reactorwatch/accidents/6-1_powerpoint.pdf (accessed 06 August 2015).

Matsui T. (2011) Deciphering the Measured Ratios of Iodine-131 and Cesium-137 at the Fukushima Reactors, http://arxiv.org/PS_cache/arxiv/pdf/1105/1105.0242v1.pdf (accessed 06 August 2015).

Merwin, S.E. and Balanov, M.I. (1993) *The Chernobyl Papers, Doses to the Soviet Population and Early Health Effects*, vol. **1**, Research Enterprises, Richland, WA.

Miyadera, H. (2012) Our Next Two Steps for Fukushima Daiichi Muon Tomography. Los Alamos National Laboratory Report LA-UR-12-20494, Los Alamos National Laboratory, Los Alamos, NM.

Murrain-Hill, P., Coleman, C.N., Hick, J.L. *et al.* (2011) Medical response to a nuclear detonation: creating a playbook for state and local planners and responders. *Disaster Med. Public Health Prep.*, **5** (1), S89.

National Academy of Sciences (2013) Messaging for Engineering: From Research to Action. National Academy of Sciences Report No. 13463, The National Academy Press, Washington, DC.

National Academy of Sciences (2014a) Lessons Learned from the Fukushima Nuclear Accident for Improving Safety of U.S. Nuclear Plants. National Academy of Sciences Report No. 18294, The National Academy Press, Washington, DC.

National Academy of Sciences (2014b) An All-of-Government Approach to Increase Resilience for International Chemical, Biological, Radiological, Nuclear, and Explosive (CBRNE) Events. National Academy of Sciences Report No. 18814, The National Academy Press, Washington, DC.

National Academy of Sciences (2014c) The Science of Responding to a Nuclear Reactor Accident: Summary of a Symposium. National Academy of Sciences Report No. 19002, The National Academy Press, Washington, DC.

NCRP (2001) Management of Terrorist Events Involving Radioactive Material. NCRP Report No. 138, NCRP Publications, Bethesda, MD.

NCRP (2003) Management Techniques for Laboratories and Other Small Institutional Generators to Minimize Off-Site Disposal of Low- Level Radioactive Waste. NCRP Report No. 143, NCRP Publications, Bethesda, MD.

NCRP (2004) Approaches to Risk Management in Remediation of Radioactively Contaminated Sites. NCRP Report No. 146, NCRP Publications, Bethesda, MD.

NCRP (2005) Key Elements of Preparing Emergency Responders for Nuclear and Radiological Terrorism. NCRP Commentary No. 19, NCRP Publications, Bethesda, MD.

NCRP (2009) Radiation Dose Reconstruction: Principles and Practices. NCRP Report No. 163, NCRP Publications, Bethesda, MD.

Nuclear News Special Report (2011) Fukushima Daiichi after the earthquake and tsunami. *Nucl. News*, **54** (4), 17.

Orient, J.M. (2014) Fukushima and reflections on radiation as a terror weapon. *J. Am. Physicians Surg.*, **19**, 48.

Rogovin, M. and Frampton, G.T., U. S. Nuclear Regulatory Commission (1980b) Three Mile Island, A Report to the Commissioners and to the Public, NUREG/CR-1250, Vols. 1–2. U. S. Government Printing Office, Washington, DC.

RPP-5924 (2003) Radiological Source Terms for Tank Farms Safety Analysis. Rev. 3, CH2M HILL Hanford Group, Inc., Richland, WA.

RPP-8369 (2003) Chemical Source Terms for Tank Farms Safety Analyses. Rev. 2, CH2M HILL Hanford Group, Inc., Richland, WA.

RPP-13033 (2005) Tank Farms Documented Safety Analysis. Rev. 1E, CH2M HILL Hanford Group, Inc., Richland, WA.

Schlosser, E. (2013) *Command and Control – Nuclear Weapons, the Damascus*

Accident, and the Illusion of Safety, Penguin, New York.

Soran, D.M. and Stillman, D.B. (1982) An Analysis of the Alleged Kyshtym Disaster. Los Alamos National Laboratory Report LA-9217-MS, Los Alamos National Laboratory, Los Alamos, NM.

Stacey, W.M. (2001) *Nuclear Reactor Physics*, Wiley-VCH Verlag GmbH & Co. KGaA, Weinheim.

Takahashi, S. (2014) *Radiation Monitoring and Dose Estimation of the Fukushima Nuclear Accident*, Springer, Tokyo.

The National Diet of Japan (2012) The Official Report of The Fukushima Nuclear Accident Independent Investigation Commission, The Fukushima Nuclear Accident Independent Investigation Commission, Tokyo.

Turner, J.E. (2007) *Atoms, Radiation, and Radiation Protection*, 3rd edn, Wiley-VCH Verlag GmbH, Weinheim.

U. S. Department of Energy (2009) Preliminary Report on Operational Guidelines Developed for Use in Emergency Preparedness and Response to a Radiological Dispersal Device Incident. DOE/HS-0001; ANL/EVS/TM/09-1, U.S. Department of Energy, Washington, DC.

U. S. Department of Energy (2014) Accident Investigation Report: Phase 1 – Radiological Release Event at the Waste Isolation Pilot Plant on February 14, 2014, Office of Environmental Management, Washington, DC.

U. S. Department of Energy Report (2014) International Nuclear Safeguards Engagement Program, National Nuclear Security Administration, US Department of Energy, Washington, DC.

U. S. Department of Energy (2015) Waste Isolation Pilot Plant Technical Assessment Team Report. SRNL-RP-2014-01198, Revision 0, Savannah River National Laboratory, US Department of Energy, Washington, DC.

U. S. Environmental Protection Agency (2013) PAG Manual, Protective Action Guides and Planning Guidance For Radiological Incidents, U.S. Environmental Protection Agency Draft for Interim Use and Public Comment, U. S. Environmental Protection Agency, Washington, DC.

U. S. Government Accountability Office (2014) Emergency Preparedness – Opportunities Exist to Strengthen Interagency Assessments and Accountability for Closing Capability Gaps. GAO-15-20, U. S. Government Printing Office, Washington, DC.

U. S. Nuclear Regulatory Commission (1975) Reactor Safety Study: An Assessment of Accident Risks in U.S. Commercial Nuclear Power Plants. U. S. Nuclear Regulatory Commission Report WASH-1400, U.S. Nuclear Regulatory Commission, Washington, DC.

U. S. Nuclear Regulatory Commission (1979a) Population Dose and Health Impact of the Accident at the Three Mile Island Nuclear Station. NUREG-0558, U. S. Government Printing Office, Washington, DC.

U. S. Nuclear Regulatory Commission (1979b) Investigation into the March 28, 1979 Three Mile Island Accident by the Office of Inspection and Enforcement. NUREG-0600, U. S. Government Printing Office, Washington, DC.

U. S. Nuclear Regulatory Commission (1979c) NRC Views and Analysis of the Recommendations of the President's Commission on the Accident at Three Mile Island. NUREG-0632, U. S. Government Printing Office, Washington, DC.

U. S. Nuclear Regulatory Commission (1979d) NRC 1979 Annual Report. NUREG-0690, U. S. Government Printing Office, Washington, DC.

U. S. Nuclear Regulatory Commission (1980a) Environmental Assessment of Radiological Effluents from Data Gathering and Maintenance Operation on Three Mile Island Unit 2. NUREG-0681, U. S. Government Printing Office, Washington, DC.

U. S. Nuclear Regulatory Commission (1981) Final Programmatic Environmental Impact Statement Related to Decontamination and Disposal of Radioactive Wastes Resulting from March 28, 1979, Accident – Three Mile Island Nuclear Station, Unit 2. NUREG-0683, Vol. 1 and 2, U.S. Nuclear Regulatory Commission, Office of Nuclear Reactor Regulation, Washington, DC.

U. S. Nuclear Regulatory Commission (1983) Answers to Questions About Updated Estimates of Occupational Radiation Doses at Three Mile Island, Unit 2. NUREG-1060, U. S. Government Printing Office, Washington, DC.

U. S. Nuclear Regulatory Commission (1984) Answers to Frequently Asked Questions About Cleanup Activities at Three Mile Island, Unit 2. NUREG-0732, Rev. 1, U. S. Government Printing Office, Washington, DC.

U. S. Nuclear Regulatory Commission (1987a) Report on the Accident at the Chernobyl Nuclear Power Station. NUREG-1250, Rev. 1, U. S. Government Printing Office, Washington, DC.

U. S. Nuclear Regulatory Commission (1987b) Interpretation of Bioassay Measurements. NUREG/CR-4884, U. S. Government Printing Office, Washington, DC.

U. S. Nuclear Regulatory Commission (1989a) Implications of the Accident at Chernobyl for Safety Regulation of Commercial Nuclear Power Plants in the United States. NUREG-1251, Vol. 1 and 2, U. S. Government Printing Office, Washington, DC.

U. S. Nuclear Regulatory Commission (1989b) The Status of Recommendations of the President's Commission on the Accident at Three Mile Island (A Ten-Year Review). NUREG-1355, U. S. Government Printing Office, Washington, DC.

U. S. Nuclear Regulatory Commission (1991) Status of Safety Issues at Licensed Power Plants. NUREG-1435, Vol. 1–3, U. S. Government Printing Office, Washington, DC.

U. S. Nuclear Regulatory Commission (1992) Summary of Chernobyl Followup Research Activities. NUREG-1422, U. S. Government Printing Office, Washington, DC.

U. S. Nuclear Regulatory Commission (1994) Lessons Learned From the Three Mile Island – Unit 2 Advisory Panel. NUREG/CR-6252, U. S. Government Printing Office, Washington, DC.

U. S. Nuclear Regulatory Commission (1996) Individual Plant Examination Program: Perspectives on Reactor Safety and Plant Performance. NUREG-1560, U. S. Government Printing Office, Washington, DC.

U. S. Nuclear Regulatory Commission (2000) Multi-Agency Radiation Survey and Site Investigation Manual (MARSSIM). NUREG-1575, Revision 1, U. S. Government Printing Office, Washington, DC.

U. S. Nuclear Regulatory Commission (2003) Methods and Assumptions for Evaluating Radiological Consequences of Design Basis Accidents at Light-Water Nuclear Power Reactors. Regulatory Guide 1.195, U. S. Government Printing Office, Washington, DC.

U. S. Nuclear Regulatory Commission (2006) Final Safety Evaluation Report Related to Certification of the AP1000 Standard Design. NUREG-1793, U. S. Government Printing Office, Washington, DC.

U. S. Nuclear Regulatory Commission (2011a) NRC Provides Protective Action Recommendations Based on U.S. Nuclear Regulatory Commission. NRC News No. 11-050, March 16, 2011, U.S. Nuclear Regulatory Commission, Washington, DC.

U. S. Nuclear Regulatory Commission (2011b) NRC Bulletin 2011-01: Mitigating Strategies, U.S. Nuclear Regulatory Commission, Washington, DC.

U. S. Nuclear Regulatory Commission (2011c) NRC Exits Monitoring Mode for Japanese Nuclear Emergency; Continues to Support U.S. Government Efforts. NRC News No. 11-083, May 16, U.S. Nuclear Regulatory Commission, Washington, DC.

U. S. Nuclear Regulatory Commission (2012) NRC to Further Examine Solar Flare Issues Raised in Rulemaking Petition. NRC News No. 12-129, December 18, 2012, U.S. Nuclear Regulatory Commission, Washington, DC.

Weightman, M. (2011) *Japanese Earthquake and Tsunami: Implications for the UK Nuclear Industry – Interim Report*, Office for Nuclear Regulation, London.

Woodard, K. (1979) Assessment of Off-site Radiation Doses from the Three Mile Island Unit 2 Accident. TDR-TMI-116, Pickard, Lowe, and Garrick, Washington, DC.

4 Nuclear Terrorist Events Including INDs and RDDs

4.1 Overview

Nuclear terrorist events, including sabotage, represent threats that utilize radioactive or special nuclear materials to disrupt society. These events can be localized attacks or create widespread devastation. The spread of radioactive materials during a terrorist event creates new challenges that require protecting the public and emergency responders from the effects of the dispersed radioactive materials. To meet these challenges, guidelines and supporting emergency response programs must be developed, successfully implemented, and coordinated with all levels of government and the public. After recovering from the immediate effects of the event, a process for the cleanup and recovery of the affected area must be developed and implemented.

The National Incident Management System (NIMS), introduced in 2004 by the Department of Homeland Security (DHS), provided a framework for coordinating emergency response capabilities at all levels of the US government, the private sector, and nongovernmental organizations. A key aspect of the NIMS was the Incident Command System (ICS). The ICS, developed in the 1970s, is an emergency management concept that facilitates an integrated organizational structure. This common structure increased emergency response efficiency and promoted effective communications. Critical decision-making is facilitated by defining well-established lines of communication and responsibilities. The NIMS also facilitated the use of significant resources (e.g., airborne assets) that enhance the assessment of the extent of physical damage and radiological characterization of affected areas.

The National Response Framework (NRF) builds upon the National Incident Management and Integrated Command Systems. The NRF describes specific authorities and best practices for managing incidents ranging from serious local events to major terrorist attacks and catastrophic natural disasters. This is a dynamic process that incorporates lessons learned from the major US natural events (e.g., Hurricanes Katrina and Rita).

This chapter considers a variety of radiological events caused by terrorist activities. Their effects on the public, emergency response actions, and health physics

Health Physics: Radiation-Generating Devices, Characteristics, and Hazards, First Edition. Joseph John Bevelacqua.

response activities are addressed. The relative hazards of the various event types are also discussed.

The discussion is initiated with the most serious terrorist threat that is the deployment of a nuclear device. Nuclear weapons pose a significant threat because their detonation produces massive destructive power with the subsequent dispersion of large quantities of radioactive materials.

4.2 Nuclear Weapons Types

Nuclear weapons are broadly classified as either fission or fusion devices. Fission weapons derive their energy from splitting a heavy nuclear system such as ^{233}U, ^{235}U, or ^{239}Pu using neutrons that results in a net energy release. Fusion weapons combine light elements (e.g., ^{2}H and ^{3}H) to produce a heavier nuclear system (e.g., ^{4}He) with a net release of energy. The energy releases from nuclear weapons are typically categorized in terms of kilotons or megatons of trinitrotoluene (TNT).

Fission and fusion produce a variety of radiation types from the initial nuclear reaction and from the radioactive decay of associated reaction products. From a health physics perspective, the initial nuclear reaction creates neutron and photon radiations that produce a detriment in the irradiated population. Subsequent alpha, beta, and photon radiations from dispersed reaction products affect populations far removed from the detonation site.

The description of nuclear weapons types is simplified and generalized to convey basic concepts. The subsequent discussion describes design considerations and does not attempt to represent an actual or conceptual weapon.

4.2.1 Fission Weapons

Fission nuclear weapons primarily use weapons-grade plutonium (WGPu) or weapons-grade uranium (WGU). The composition of WGPu is often taken to be ^{239}Pu (93.5%), ^{240}Pu (6.0%), ^{241}Pu (0.44%), ^{242}Pu (0.015%), ^{238}Pu (0.005%), and other isotopes. The isotopic composition of WGU is usually assumed to be ^{235}U (93.5%), ^{238}U (5.5%), and ^{234}U (1%). These isotopic compositions are typical of fission weapons produced during the cold war. Other compositions or radionuclides could be used for terrorist or clandestine devices. The WGU composition is derived from gaseous diffusion and gas centrifuge enrichment technologies. Advanced laser enrichment techniques lead to smaller quantities of ^{234}U in the WGU material. The reason for this difference is addressed in Chapter 2.

The fission design used in the Hiroshima weapon is one of the most basic designs and is known as a *gun-type device*. Assembling a critical mass of WGU by firing one piece of fissionable material into another was the first approach developed for a functioning nuclear device. Two subcritical masses are merged by firing a cylindrical mass through a gun barrel into the center of a second WGU mass having a machined cavity. The primary advantage of gun-type device is its simplicity, but it has a number of drawbacks. These negative features include (i) low efficiency, (ii) limited material options due to the slow insertion speed, and (iii) large size since the weight and length of the gun barrel make the weapon heavy and long.

A more complex design was incorporated into the Nagasaki weapon that utilized an implosion approach. The implosion concept involves the precise compression of subcritical masses using explosives. An implosion device is more complex than a gun weapon because it requires detonation of explosives on the outer surface of a ^{239}Pu mass, so that the detonation wave moves inward in a smooth, symmetrical manner. This shock wave impinges upon the fissionable core and compresses it to a supercritical state.

Modern weapons incorporate the basic concepts used in the Hiroshima and Nagasaki bombs. However, these weapons have larger destructive power or nuclear yield and are more efficient. Clandestine devices could involve an independent design or the theft or purchase of an existing device. The type, design, material composition, and efficiency of a clandestine device are highly uncertain and depend on the level of technology and scientific knowledge associated with group acquiring or developing the weapon.

4.2.2
Boosted Fission Weapons

In a boosted fission weapon, the extremely high temperature attained during a fission detonation is used to initiate a secondary fusion reaction. Since the rate of thermonuclear energy production is an increasing function of density and temperature, a boosted fission weapon is achieved by venting a small amount of deuterium or deuterium and tritium into the plutonium core of an implosion device. The addition of these light materials and the achievement of a thermonuclear reaction increase the explosive yield of the device.

4.2.3
Fusion Weapons

A fusion device is considerably more complex than a fission weapon. Some nations with years of experience with fission devices have not yet constructed a fusion weapon. Nuclear weapons development continues to follow the historical sequence of the first US program.

A fusion bomb is similar to a boosted fission device since the fission energy initiates a fusion reaction. Fusion weapons design varies and the subsequent discussion is limited to a basic two-stage device. A two-stage weapon consists of a fission device physically separated from light material. The fission device is called the *primary*, and light elements that will fuse following the primary detonation is the *secondary*.

A fission bomb at one end of the weapon detonates and the primary's energy compresses and heats the light elements causing these materials to implode. As the secondary implodes, fissile material at its center fissions and provides additional energy that enables the fusion fuel to ignite. The fission and fusion chain reactions exchange neutrons and boost the efficiency of each reaction. The net result is a significantly greater yield than could be achieved with a fission device.

4.2.4 Clandestine Devices

Given the previous discussion, it is likely that clandestine devices constructed by a rogue state or terrorist group will be either a gun-type weapon or an implosion device. It is also likely that a clandestine device will utilize material (e.g., WGU or WGPu) obtained by theft or diversion or created in a clandestine enrichment facility.

The projected yield of a clandestine device could vary considerably. Early generation fission weapons easily exceed the 10 kT limit that is often assumed in radiological terrorism planning, and fission devices in the hundreds of kilotons range have been constructed. These discussions become irrelevant if an existing device is purchased or stolen by the clandestine group. Theft of a thermonuclear device could lead to a weapon in the MT range. Weapons' yields of greater than tens of kilotons would significantly challenge much of the radiological planning associated with the deployment of a clandestine nuclear weapon.

It is the author's view that the 10 kT assumption is too optimistic. This view is based on the advancements that have occurred since nuclear weapons were initially deployed. In addition, there is a large pool of highly educated physicists and engineers throughout the world, with the knowledge and skills to construct nuclear weapons. Moreover, some of the technical data needed to construct a weapon and basic weapons design information are available in the open literature.

Given the proliferation or uranium enrichment technology and advancements in centrifuge and laser techniques, obtaining weapons-grade uranium is becoming more likely. The expansion of light water reactor technology to additional nations also enhances the possibility of reprocessing reactor fuel to recover plutonium with its diversion to weapons applications or terrorist groups.

The increased likelihood of obtaining weapons-grade uranium and plutonium enhances the probability of illicit weapons production and development of weapons with yields in excess of 10 kT. In addition, the expansion in the numbers of states acquiring nuclear weapons enhances the possibility of a weapon being stolen, diverted, or sold for terrorist purposes. These conditions suggest that limiting radiological planning efforts to weapons with yields of 10 kT or less should be reexamined to acknowledge the growing threat from these devices.

4.3 Nuclear Event Types

Radiological terrorism includes the use of radioactive material to contaminate a targeted area or detonation of a nuclear device to destroy a population center.

Terrorists could utilize radiological dispersal devices (RDDs); radiation exposure devices (REDs); the deliberate contamination of food, water, or other consumables with radioactive materials; the dispersal of radioactive materials from fixed radiological or nuclear facilities or materials in transit; and nuclear weapons and improvised nuclear devices (INDs) to accomplish their goals. Each of these methods is described in the subsequent discussion.

4.3.1
Nuclear Weapons and Improvised Nuclear Devices

A variety of nuclear weapons types are described in Section 4.2. Nuclear weapons are designed for mass destruction. INDs are a subtype of nuclear weapons.

An IND refers to any device designed to cause a nuclear yield using conventional explosives to create a supercritical mass of special nuclear material (usually enriched ^{235}U or ^{239}Pu). INDs are generally assumed to be less sophisticated and have a lower yield than produced by a nuclear weapon fabricated by a nation state. A malfunctioning IND could result in consequences similar to an RDD, which would primarily disperse radioactive material without a significant nuclear yield.

According to NCRP 165, the most likely nuclear weapon terrorism scenario involves the use of a single fission device with a low yield (<10 kT). The detonation would result in significant consequences to public health and safety. IND effects near the detonation site are catastrophic, and emergency support organizations located in this area are severely limited in their ability to respond. Response units in areas of elevated fallout within 16–32 km may be required to shelter in place for several hours before rendering assistance to the area near the detonation site. Unlike the response to an RDD where local infrastructure is generally intact, emergency personnel responding to a nuclear detonation originate from outside the area immediately impacted by the event. An IND detonation severely disrupts organizations located near the detonation site or ground zero.

Blast effects from an IND include imploded windows and doors; overturned vehicles; collapsed buildings; ruptured surface and subsurface utilities including gas, water, and electric power; collapsed tunnels; and loss of communications infrastructure.

Thermal effects are induced by infrared, visible, and ultraviolet radiation from the detonation and lead to widespread fires. In humans, the thermal effects include skin burns and blindness.

Early health effects are caused by prompt X-rays, gamma rays, and neutrons emitted from the detonation and from radioactive material in the fallout resulting from the nuclear explosion. The acute radiation effects depend on the total absorbed dose and the dose rate at which it is delivered.

In describing nuclear weapons and their general effects, the author follows the description and terminology of Glasstone and Dolan in their classical, open-source work *The Effects of Nuclear Weapons.* This book is a well-known and widely available reference that serves as a standard for describing nuclear weapons characteristics and effects. Accordingly, fission weapons are assumed to utilize ^{235}U and ^{239}Pu. The blast, thermal, and radiation effects following the detonation of these weapons are described in the subsequent discussion. ^{233}U weapons are not considered in this text.

4.3.1.1 General Properties of Nuclear Explosions

An explosion results from the rapid release of a large quantity of energy within a limited volume of material. The liberation of energy increases the temperature and pressure of a material, and that material is converted into hot, compressed gases that rapidly expand and create a pressure (shock) wave in the surrounding media.

A nuclear explosion's destructive action is mainly due to blast or shock effects. However, there are significant differences between chemical and nuclear explosions. First, a nuclear detonation is orders of magnitude larger than a conventional explosion. Second, the mass of nuclear material required to produce a given amount of energy is significantly less than that for a conventional explosive. Third, the temperatures reached in a nuclear detonation are much higher than those achieved in a conventional explosion. This temperature increase results in the emission of thermal radiation (e.g., heat and light) that is capable of igniting fires and causing skin burns at significant distances from the detonation site. Finally, a nuclear explosion produces neutron and gamma radiation that affects the detonation area and disperses radioactive materials that emit energy over an extended distance.

Nuclear explosions utilize fission and fusion to generate an energy release. Complete fission of 1 kg of uranium or plutonium releases as much explosive energy as about 20 kT of TNT. Fusion of 1 kg of ^{2}H would release as much explosive energy as about 65 kT of TNT.

The energy delivered from a nuclear detonation is usually divided into three types. These types include blast and shock, thermal, and nuclear energy involving initial and delayed components. The fraction of the energy from each of these types depends on the distance from the detonation site, type and yield of the device, and the environment near ground zero. For a fission weapon detonated in air at an altitude of less than 12 km, about 35% of the explosive energy is thermal radiation and 50% produces air blast and shock. The remaining 15% of the energy is released as nuclear radiation. About one-third of the nuclear energy is released within about a minute of the explosion. For a chemical explosive in air, essentially all the energy produces blast and shock effects.

The radiation emitted within the first minute, usually defined as initial radiation, is derived from the weapon's detonation and the decay of short-lived fission products. Residual radiation is created by the decay of longer-lived fission products.

The initial nuclear radiation is primarily gamma rays and neutrons, and the residual radiation is primarily gamma rays and beta particles that arise from the decay of fission products. Smaller amounts of alpha particles are emitted from residual fissile material that did not fission during the nuclear detonation. The residual radiation comprises particulate material that is deposited following the detonation (i.e., fallout) which affects the environment at distances removed from the site of the detonation.

There are five main classifications of nuclear explosions. These are the (i) air burst, (ii) high-altitude burst, (iii) underwater burst, (iv) underground burst, and (v) surface burst. These explosion types affect the resulting radiation available to contaminate the environment.

4.3.1.1.1 Air Burst

An air burst is defined as an atmospheric detonation that occurs below about 30 km. At such a detonation height, the fireball does not touch the surface of the earth. For example, a 1 MT weapon's fireball has a diameter of about 1.7 km at maximum brilliance. For it to be considered an air burst, a 1 MT device must be detonated about 0.85 km above the earth's surface. Nearly all the shock energy from an air burst appears as a blast wave. The thermal radiation from the 1 MT weapon travels over a significant distance and has sufficient intensity to cause moderately severe burns to exposed skin at a distance of about 20 km on a clear day.

A 1 MT air burst also produces significant photon and neutron radiation. A thickness of about 1.2 m of ordinary concrete provides protection from the effects of the initial radiation at a distance of about 1.6 km from the 1 MT air burst. However, the concrete shielding structure must be specifically designed to survive the blast effects at this distance.

For an air burst, the longer-lived fission products are dispersed into the atmosphere. Some of these fission products fuse with particles of soil and explosion debris. These particles remain suspended in the atmosphere and present a fallout hazard at large distances from the detonation site.

All nuclear detonations pose a direct radiation hazard, but most of the fallout from an air burst is derived from bomb components. Limited surface materials are activated or contaminated with fission products. The fallout of radioactive and contaminated materials is more significant for a surface burst.

4.3.1.1.2 High-Altitude Burst

A high-altitude burst is defined as a nuclear detonation event occurring above 30 km. Since the air density decreases with altitude, the fraction of the weapon's energy converted into blast and shock decreases with increasing height above the earth's surface. The thermal energy radiated from the high-altitude burst is affected by two factors. First, the shock wave does not readily form in the less dense air, and this permits the fireball to radiate thermal energy that would have been used to produce the air blast. This effect dominates between about 30 and 43 km, and a larger fraction of the detonation energy is released as thermal energy than at lower altitudes. Second, the decreased air density facilitates energy from the detonation to travel farther than at lower altitudes. Some of this energy warms the air at a distance from the fireball and does not produce additional damaging effects. The second effect dominates above 43 km.

The fraction of the explosion energy emitted as nuclear radiation is independent of the detonation height. The quantity of photon and neutron radiation as a function of distance varies since a significant fraction of photons result from neutron interactions with atoms comprising the atmosphere. In addition, the atmospheric density affects the attenuation of neutrons and photons. This means that for a given weapon's yield and slant path distance from the detonation altitude to the earth's surface, more initial radiation is received as the detonation height increases.

Initial and residual radiations ionize the atmosphere to create large densities of free electrons and ions at high altitudes that interfere with communications systems. The free electrons also interact with the earth's magnetic field to generate strong electromagnetic fields that have the potential to damage unprotected electrical and electronic equipment located in an extended area below the high-altitude burst. This phenomenon, known as the electromagnetic pulse (EMP), occurs over a range of frequencies and has serious implications for electronic systems. The EMP is also produced in surface and low-altitude air bursts, but these detonation types only affect a small area near the detonation location.

For a nuclear detonation of high yield and sufficient altitude, the area affected by the high-frequency EMP component extends isotropically from the burst point to the horizon. The lower-frequency EMP component yields a significant effect even beyond the horizon. For a detonation at 80 km (160 km), the EMP-affected area on the ground would have a radius of 960 km (1400 km). A detonation at 320 km above the center of the continental United States would affect the entire country as well as parts of Canada and Mexico.

EMP effects cause functional damage or operational upsets in electronic systems. Functional damage is a permanent failure of the device or component such as a fuse, transistor, semiconductor diode, or rectifier. An operational upset is a temporary impairment, and the system normally recovers from the EMP effect within a few hours.

The effects of the EMP could significantly influence the health physics response to an IND or nuclear weapons event. An EMP could render radiation instruments and supporting systems (e.g., survey meters, counting systems, electronic personnel dosimeters, computers, whole-body counting systems, and installed radiation monitoring systems) inoperable. The EMP would also affect the health physics support to medical cases since electronic medical support systems could also suffer electromagnetic-induced failures. Since the EMP effects could render expected computational tools inoperable, the subsequent discussion provides analytical tools for estimating the effects of a nuclear detonation. This appears to be a prudent choice since this book will not be affected by EMP effects.

A high-altitude burst also leads to a wide distribution of fission products. Their distribution depends on a number of factors including the meteorological conditions and detonation height.

4.3.1.1.3
Underground and Underwater Bursts

If the detonation occurs such that its center is beneath the ground (under the surface of the water), it is defined to be an underground (underwater) burst. These terms are combined in some literature as a subsurface burst since their effects are similar. Following a subsurface burst, most of the detonation energy appears as an underground or underwater shock wave. A portion of the detonation energy, which decreases as the depth of subsurface detonation increases, escapes to the surface and produces an air blast. Most of the thermal radiation and initial nuclear radiation are absorbed within a short distance from the detonation location. The residual nuclear radiation is significant because it contaminates a large volume of earth or water with fission products.

From a health physics perspective, underground and underwater bursts lead to a smaller radiation effect than the other detonation types. The shielding provided by the ground and water limits the quantity of direct radiation received by the public. Fallout is also limited because the earth and water scavenges radioactive materials before they reach the surface environment. Therefore, underground and underwater bursts produce the least detriment when compared to the other detonation types. There will be localized fission product contamination, but the effect of these radioactive materials on the environment depends on their ability to be mobilized and transported beyond the detonation site.

Many of the activation products of air (e.g., ^{11}C, ^{13}N, and ^{15}O), water (e.g., ^{3}H, ^{7}Be, ^{11}C, ^{13}N, and ^{15}O), and soil (e.g., ^{3}H and ^{22}Na) tend to be shorter lived than many fission products (e.g., ^{90}Sr and ^{137}Cs). These radionuclides and their characteristics are summarized in Appendix A.

4.3.1.1.4
Surface Burst

The final detonation type, surface burst, occurs at or slightly above the surface of the land or water. In a surface burst, the air blast and ground or water shock are produced to an extent that depends on the detonation energy and height of the burst. Surface bursts mobilize significant quantities of earth and debris and have the potential for generating large quantities of fallout.

The size of the fireball increases as the detonation yield increases. Air and surface bursts of the same yield have a different fireball radius. This arises because the fireball radius is governed by hydrodynamic effects that depend on the energy reflected from a surface. Although it is difficult to simplify nuclear detonation physics, it is possible to characterize the fireball radius R in terms of equations having the form

$$R \approx aW^b \tag{4.1}$$

where R is measured in meters, W is the detonation yield in kilotons of TNT, $b = 0.4$, and $a = 34$ m for an air burst and 44 m for a contact surface burst. More detailed weapons effects are obtained using numerical approaches such as the HOTSPOT code summarized in Appendix E.

Related to the fireball size is the detonation height (H) above which local fallout could be considered small enough to be tolerable under emergency conditions. As a rough guide, Glasstone and Dolan provide the approximate relationship:

$$H \approx 55W^{0.4} \tag{4.2}$$

For heights below the Eq. (4.2) value, local fallout should be considered as significant. Equation (4.2) provides the detonation height in meters.

Surface bursts produce the greatest volume of fallout because materials near ground zero are mobilized by the detonation. These materials, including earth and debris produced by the detonation, are contaminated by the weapon's fission products and activated by the weapon's neutron fluence.

4.3.1.2
Initial Nuclear Radiation

Most of the neutrons and a portion of the photons are emitted simultaneously with the nuclear detonation. The remainder of the photons and associated beta particles result from the decay of fission products. Delayed radiation is addressed in the next section.

The gamma rays produced during the fission process and from neutron interactions with weapons components are generated within the extremely short detonation time. Additional photons result from the decay of short-lived fission products, deexcitation of nuclear isomers, inelastic neutron scattering, and neutron capture with nuclei of elements comprising the air and their subsequent decay.

Prompt photons are generated before the weapons components are obliterated by the nuclear detonation and are strongly attenuated by these components. Delayed photons are emitted at a later stage of the detonation, after the weapons components have vaporized. The delayed photons encounter minimal attenuation before they are emitted into the atmosphere. As a result, the photon source term at a distance from the detonation point is dominated by the delayed photons and photons produced from the neutron capture by nitrogen nuclei in the atmosphere. These sources provide about 100 times more energy than the prompt photons to the total nuclear radiation produced during the first minute after the detonation.

Another source of photons exists for detonations that occur near the earth's surface. This source arises from neutron activation of materials in the earth, water, and soil. These activation photons contribute a relatively small source term except near the detonation point since their strength is determined by the activating neutron flux.

The biological effects of the nuclear detonation are addressed in the subsequent discussion. However, the gamma-ray doses producing a portion of these effects are addressed in this section. Calculations of the photon dose from a nuclear detonation depend on a variety of factors. These factors include the type of device, weapon's yield, type of nuclear detonation, distance from the detonation point,

Table 4.1 Approximate slant range in air (m) to achieve the specified gamma absorbed dose for the defined weapon's yield[a].

Absorbed dose (Gy)	Weapon's yield (kT)			
	1[b]	10[b]	100[b]	1000[c]
0.3	1400	1900	2500	3300
1	1100	1600	2300	3000
10	700	1100	1650	2400
100	400	700	1200	1800

a) Glasstone and Dolan (1977).
b) Fission weapon.
c) Fusion weapon.

and air density. The results obtained for various weapons' yields and a 0.9 sea level air density are provided in Table 4.1. Table 4.1 provides the slant distance at which a specified absorbed dose occurs for a given weapon's yield. The burst height (H) in meters used in Table 4.1 is dependent on the weapon's yield and is determined from the relationship

$$H \approx 61 W^{0.4} \tag{4.3}$$

Larger slant range values occur with increasing weapon's yield. This result reflects the fact that a larger yield produces higher doses. A given dose is produced at a greater distance as the weapon's yield (source term) increases.

In shielding photons from a nuclear detonation, a range of photon energies is considered. The required shields are thick and must be designed to withstand the thermal, blast, and shock effects from the nuclear detonation. Table 4.2 provides effective 10th-value layer thicknesses for fission products and nitrogen capture photons for common construction materials. The thickness of any material to attenuate nitrogen capture photons is about 50% larger than for fission products because the capture gammas have a higher energy.

Table 4.2 Approximate effective 10th-value thickness for fission products and nitrogen capture photons[a].

Material	Tenth-value thickness (cm)	
	Fission products	Nitrogen captures
Steel (iron)	8.4	11
Concrete	28	41
Earth	41	61
Water	61	99
Wood	97	160

a) Glasstone and Dolan (1977).

Table 4.3 Approximate slant range in air (m) to achieve the specified neutron absorbed dose for the defined weapon's yield[a].

Absorbed dose (Gy)	Weapon's yield (kT)			
	1[b]	10[b]	100[b]	1000[c]
0.3	1300	1650	2050	2500
1	1100	1450	1900	2350
10	650	1100	1450	1900
100	450	750	1100	1450

a) Glasstone and Dolan (1977).
b) Fission weapon.
c) Fusion weapon.

Neutrons only represent about 1% of the detonation energy, and they penetrate a significant distance from the burst location. A small neutron component is produced from (γ, n) reactions. The neutrons produced from fission and fusion are fast neutrons. Within the weapon, energy is lost through inelastic collisions with heavy nuclei and elastic collisions with light nuclei. Accordingly, the neutrons leaving the weapon have a broad energy spectrum.

The neutron dose from a nuclear detonation depends on the weapons design. Neutron capture reactions affect the flux and energy distribution of the emitted neutrons and depend on the materials comprising the weapons components.

Calculations of the neutron dose from a nuclear detonation depend on the factors noted previously for the photon dose. The results obtained for a various weapon's yields and 0.9 sea level air density are provided in Table 4.3. Table 4.3 provides the slant distance at which a specified neutron absorbed dose occurs for a given weapon's yield. The burst height for the weapons noted in Table 4.3 is based on Eq. (4.3).

Shielding for nuclear weapons radiation is often characterized in terms of dose transmission factors. These factors are strongly dependent of the energy and nuclides interacting with the incident radiation types. Transmission factors for a variety of construction materials are provided in Table 4.4 for the weapon's prompt gamma-ray and neutron spectra.

4.3.1.3
Delayed Nuclear Radiation and Fallout

Delayed or residual radiation is defined as any radiation type emitted after 1 min following the nuclear detonation. The sources and characteristics of residual weapon's radiation depend on the yield and type of nuclear device, the burst height, environment of the detonation site, and time of interest following the detonation. For a fission weapon detonated in air, various radiation types are produced including those derived from the decay of fission products, uranium and plutonium that did not fission, and neutron activation and reaction products

Table 4.4 Dose transmission factors for various structural material configurations[a].

Configuration	Radiation type	
	Prompt gamma rays	Neutrons
Frame house	0.8–1.0	0.3–0.8
Basement	0.1–0.6	0.1–0.8
0.91 m underground	0.002–0.004	0.002–0.01
Apartment building (upper floors)	0.8–0.9	0.9–1.0
Apartment building (lower floors)	0.3–0.6	0.3–0.8
Concrete blockhouse shelter: 22.9 cm walls	0.1–0.2	0.3–0.5
Concrete blockhouse shelter: 30.5 cm walls	0.05–0.1	0.2–0.4
Concrete blockhouse shelter: 61.0 cm walls	0.007–0.02	0.1–0.2
Shelter (partly above ground) with 0.61 m earth cover	0.03–0.07	0.02–0.08
Shelter (partly above ground) with 0.91 m earth cover	0.007–0.02	0.01–0.05

a) Glasstone and Dolan (1977).

of materials comprising the weapon's components. Surface bursts also include residual radiation derived from neutron and activation products of air, water, soil, and other materials at the detonation site. Fission products are the dominant residual radiation component.

The delayed radiation from a fusion weapon includes the fission products produced by the initiating device. Since copious fast neutrons are produced in the fusion process, the residual radiation is derived from neutron reactions with weapons components and their environment. This assumes that the fission weapon that triggers the fusion device has a low yield relative to the total device yield.

The radioactive materials that produce delayed radiation also create a significant hazard at large distances from the detonation site. This hazard is created by fallout particles that combine weapons-induced radioactive material and induced activity in air, water, soil, and other materials near the detonation site with earth and detonation debris. The fallout particles are dispersed over large areas. The weapon's yield, detonation type, and meteorological conditions govern the dispersion of fallout particles.

A localized hazard arises from the neutron activation of the materials near the detonation site. High dose rates are produced by the activated earth and weapons debris remaining after the detonation. These dose rates limit entry into areas near ground zero.

Fallout effects and characteristics are often addressed by considering early and delayed components. Early or local fallout reaches the ground in the first 24 h following the nuclear detonation. The early fallout from surface, shallow subsurface, and low-altitude air bursts disperses radioactive materials over large areas and represents a significant radiological hazard. Early fallout particles tend to be larger, heavier particles.

Table 4.5 Mean dose equivalent commitments to the year 2000 in the United States from nuclear testing through 1970[a].

Source	Dose equivalent commitment (μSv)
External	
Whole body	750
Internal (organ at risk)	
^{3}H (whole body)	20
^{14}C (whole body)	80[b]
^{90}Sr (marrow)	450
^{90}Sr (endosteal)	650
^{137}Cs (whole body)	150
^{239,240}Pu (bone)	20
^{239,240}Pu (lung)	400

a) NCRP 93 (1987).
b) Dose commitment to the year 2000. The total dose commitment delivered over many generations is 1.4 mSv.

Delayed or long-range fallout reaches the earth's surface after the first day post-detonation. It consists of fine, light particles that settle with low concentrations over a large portion of the earth's surface.

Long-term fallout resulted from atmospheric nuclear weapon's testing that was routinely performed prior to 1980. The radiation dose from atmospheric tests is delivered over years. This dose depends on the time following the detonation, the type and yield of the weapon, and its detonation environment.

The mean dose commitment from nuclear weapons tests through 1970 delivered to the US territory is summarized in Table 4.5. Most of the dose from these weapons tests has been received. The exception is the dose from ^{14}C. An estimate of the dose still to be delivered from these weapons tests is about 0.1 mSv to the whole body from external radiation. With the moratorium on atmospheric testing, fallout is no longer a significant source of exposure to the public. However, any new atmospheric nuclear detonations could significantly change this condition.

At about 1 min postdetonation, a 1 kT fission device has produced about 1000 EBq of fission products. This activity level is about 3 orders of magnitude larger than the releases from the Chernobyl accident (Section 3.2.2). Fortunately, this activity rapidly decreases with a subsequent decline in dose rates.

Fresh fission products have a dose rate that varies with time (t) postdetonation following the relationship

$$\dot{D}(t) = \dot{D}(1)t^{-1.2} \tag{4.4}$$

where $\dot{D}(t)$ is the dose rate at time t after the fission products are created, $\dot{D}(1)$ is the dose rate at unit time (e.g., 1 h or 1 day), and t is the time after the formation of the fission products. Equation (4.4) is a reasonable approximation for fission products with ages between 1 min and about 200 days. The content of Eq. (4.4) is often captured as a rule of thumb, which can be stated in a manner that notes that the dose rate decreases by a factor of 10 for any sevenfold increase in time following the detonation. An application of this rule is illustrated by the following statements. *If the dose rate at 1 h postdetonation is known, then at 7 h after the explosion the dose rate will have decreased by a factor of 10. At 49 (7 × 7) h it will have decreased by a factor of 100, and at 343 (7 × 7 × 7) h the dose rate will be one-thousandth of that at 1 h post detonation.* This rule is useful for initial planning purposes and would be valuable knowledge if limited instrumentation were available.

The total dose delivered between two times t_1 and t_2 is obtained by integrating Eq. (4.4):

$$D_{\text{total}} = \int_{t_1}^{t_2} \dot{D}(1)t^{-1.2}\mathrm{d}t = 5\dot{D}(1)(t_1^{-0.2} - t_2^{-0.2}) \tag{4.5}$$

This total dose relationship assists mission planning if entries into fallout areas are required.

The delayed fallout forms a specific geometric footprint, and an isodose contour has an elongated shape. At the detonation location, the isodose profile has a width defined as the ground zero width. The isodose profiles are also characterized by a maximum width and a maximum extent or downwind distance. The downwind distance, maximum width, and ground zero width values for selected absorbed dose rates for fallout from a surface burst are provided in Table 4.6.

These fallout patterns are highly simplified and assume a uniform wind speed and no topographical or meteorological effects. Actual fallout dose distributions are more complex since wind speeds vary, topographical characteristics perturb the fallout distribution, and meteorological effects significantly modify the simplified contours illustrated in Table 4.6.

Table 4.6 Scaling relationships for absorbed dose contours for a contact surface burst with a yield of W (kT)[a), b)].

Absorbed dose rate (Gy/h)	Downwind distance (km)	Maximum width (km)	Ground zero width (km)
30	$1.5W^{0.45}$	$0.012W^{0.86}$	$0.042W^{0.58}$
10	$2.9W^{0.45}$	$0.058W^{0.76}$	$0.096W^{0.57}$
3	$7.2W^{0.45}$	$0.21W^{0.66}$	$0.32W^{0.48}$
1	$14W^{0.45}$	$0.61W^{0.60}$	$0.62W^{0.42}$
0.3	$26W^{0.45}$	$1.2W^{0.56}$	$0.85W^{0.41}$
0.1	$38W^{0.45}$	$2.2W^{0.53}$	$1.1W^{0.41}$
0.03	$48W^{0.45}$	$3.5W^{0.50}$	$1.4W^{0.41}$
0.01	$64W^{0.45}$	$5.3W^{0.48}$	$2.4W^{0.41}$

a) Glasstone and Dolan (1977).
b) Reference wind speed of 24 km/h.

The isodose contours of Table 4.6 assume a 24 km/h wind speed. For effective wind speeds (v) in units of km/h that are greater than 24 km/h, the Table 4.6 downwind distances are multiplied by the factor F

$$F = 1 + \frac{v - 24}{96} \tag{4.6}$$

For wind speeds less than 24 km/h, the factor F is

$$F = 1 + \frac{v - 24}{48} \tag{4.7}$$

The results of Eqs. (4.6) and (4.7) and Table 4.6 are reasonably accurate for simple wind patterns (i.e., winds having minimal directional sheer and wind speeds between 13 and 72 km/h). For more complex wind profiles, the fallout patterns are not reproduced by the idealized dose rate contours assumed in Table 4.6.

4.3.1.4
Implications of IND Size

Most organizations including the National Council on Radiation Protection and Measurements (NCRP) assume that weapons with yields of 10 kT or less are the most credible IND threat and limit their consideration to that range. As noted previously, technology advances suggest that a larger weapon's yield is feasible and should be considered for emergency planning purposes. Table 4.6 provides a series of relationships that characterize the fallout footprint in terms of simplified meteorological considerations. These assumptions permit the calculation of absorbed dose contours that are approximated as an ellipse with major and minor axes equated to the downwind distance (a) and maximum width (b), respectively. Using this approximation, the area (A) enclosed by a specific isodose contour is

$$A = \frac{\pi ab}{4} \tag{4.8}$$

The effects of a 10 kT detonation assumed by the NCRP and other weapons yields are summarized in Table 4.7. In particular, weapons yields of 10, 50, 100, 500, and 1000 kT are considered, and their associated 30 Gy/h isodose curve parameters are provided. The values in Table 4.7 are derived from the information in Table 4.6 and Eq. (4.8).

Table 4.7 30 Gy/h isodose curve for a surface burst.

Weapon's yield (kT)	Downwind distance (km)	Maximum width (km)	Area enclosed by isodose curve (km^2)
10	4.23	0.0869	0.289
50	8.72	0.347	2.38
100	11.9	0.630	5.89
500	24.6	2.51	48.5
1000	33.6	4.56	120

The 30 Gy/h curve represents a significant radiological hazard. In the 10 kT NCRP assumption, the area encompassed by this curve is a small area (0.289 km^2). This area rapidly increases and illustrates that emergency response activities become significantly more complex and require greater resources as the weapons yield increases. When compared to the area affected by a 10 kT detonation, the area increase is a factor of 8.24, 20.4, 168, and 415 for 50, 100, 500, and 1000 kT weapon's yields, respectively.

Emergency response activities are also complicated by the increased blast, shock, and thermal damage associated with larger weapons' yields. These increased yields destroy infrastructure and resources that would be available following a smaller 10 kT burst. The loss in critical infrastructure and the need for resources significantly removed from the detonation site should be considered in emergency planning scenarios involving larger IND yields. Although the 10 kT yield represents an initial assumption, careful emergency planning efforts must consider larger weapons yields to account for bounding scenarios. One of the lessons from the Fukushima Daiichi accident is the need to ensure that accident assumptions are credible and bound historical occurrences. Weapons significantly larger than 10 kT exist and can be acquired or produced by a dedicated group. Therefore, it is prudent to consider more severe events and weapons yields beyond the usual 10 kT assumption.

4.3.1.5
Medical Response Activities

The biological effects from a nuclear weapons detonation are well established by the Hiroshima and Nagasaki epidemiology. Radiological effects in the exposed population depend on the delivered doses. However, addressing the progression of acute radiation effects is only part of the challenges faced by health physicists as they provide support to medical personnel.

Medical personnel must contend with radiation-induced skin burns, blindness from the weapon's light output, thermal radiation burns, and physical injuries caused by blast effects. Many of these injured individuals will also be contaminated. Health physicists will be required to participate in triage activities to ascertain the extent of internal and external contamination. These determinations govern subsequent decorporation efforts. A discussion of addressing mass casualties including contaminated individuals is provided in the subsequent discussion.

4.3.1.6
HOTSPOT

Detailed investigations of nuclear weapons effects can be assessed using the HOTSPOT computer code described in Appendix E. One of the HOTSPOT modules permits the user to model nuclear weapons effects for various yields, locations, and meteorological conditions.

Much of the previous discussion could have been addressed using HOTSPOT or other computer models. The author has chosen to use a more analytical approach.

There is value in understanding the physical phenomena associated with IND detonations and to characterize their radiological effects and affected areas in terms of simple analytical relationships. In addition, these analytical relationships will survive the EMP effects following a weapons detonation and provide a health physicist with the capability for first-order analysis if electronic capability is rendered inoperable.

4.3.2 Radiological Dispersal Devices

An RDD is a device that is intended to spread radioactive material from the detonation of conventional explosives or other means. The distributed radioactive material presents an internal radiation hazard, but external radiation effects must also be considered.

Radioactive material for an RDD could be obtained from industrial or medical facilities. Power reactor spent fuel and low-level waste also present a significant hazard if dispersed. Table 4.8 lists radionuclides that are commonly encountered in these environments or which present unique challenges if used in an RDD.

The availability of high activity sources and their potential use in an RDD has led to enhanced security measures. These measures include high dose rate source protection and security, enhanced controls on medical irradiators using large sealed sources, reliability and trustworthiness requirements for licensees of large sealed sources, additional controls of large cesium chloride sources, and replacement of large sealed sources with machine-produced radiation. The importance of the physical protection of large activity sources is promulgated in the United States through 10CFR37. Table 4.8 illustrated a portion of the possible sources of concern.

Table 4.8 also provides the gamma constant if the nuclide presents an external photon hazard, the nuclide's half-life, its common medical or industrial use, or its presence in various waste forms. As indicated in Table 4.8, devices utilizing these materials are widespread and often used in applications that do not have a level of security that is equivalent to the standards associated with power reactors or defense-related facilities.

^{60}Co, ^{137}Cs (^{137m}Ba), ^{131}I, ^{192}Ir, and ^{201}Tl emit photons and could present a significant external radiation hazard during clandestine transport if their protective shielding was removed. As noted in Table 4.8, these nuclides have widespread use and many industrial sources have significant activities that pose a major threat if utilized by a terrorist group. ^{131}I is often excluded as a possible terrorist target since it has a short half-life but is listed in Table 4.8 because it is available in large activity sources and used in multiple applications.

Other nuclides pose a more significant hazard. For example, ^{137}Cs is often used in a cesium chloride form, which is soluble in water. The 1987 Radiological Accident in Goiania, Brazil, illustrated its effects and potential impact.

In 1985, a hospital in Goiania moved to a new location, but a radiation therapy unit containing about 50 TBq of ^{137}Cs in cesium chloride form was not moved

Table 4.8 Common radionuclides utilized in medical and industrial facilities that could be incorporated into a radiological dispersal device[a),b)].

Nuclide	Gamma constant (Gy-m^2/h-MBq)	Half-life	Typical application or waste form
^{60}Co[c),d)]	3.1×10^{-7}	5.271 yr	Commercial irradiators Food irradiation Gamma knife Industrial gauges (e.g., material thickness, level, and flow) Industrial radiography Sterilization applications Stereotactic radiosurgery
^{90}Sr[c)]	0.0	28.8 yr	Biokinetic studies Medical treatment Radioisotope thermoelectric generators Remote power source
^{137}Cs[c),d)]	8.1×10^{-8}	30.07 yr	Brachytherapy Industrial gauges (e.g., density, moisture, material thickness, and flow) Food irradiation Gamma knife Industrial radiography Self-shielded irradiators Sterilization applications Radiography Well logging
^{131}I[c),d)]	5.2×10^{-8}	8.023 d	Medical diagnosis and treatment Thyroid Therapy Procedures
^{192}Ir[c),d)]	1.1×10^{-7}	73.83 d	Brachytherapy Industrial radiography Sterilization applications Food irradiation
^{201}Tl[d)]	1.1×10^{-8}	3.043 d	Heart diagnostic studies
^{210}Po[e)]	0.0	138.38 d	Heat source Clandestine poison
^{226}Ra[d),e)]	9.3×10^{-10}	1599 yr	Legacy medical isotope Self-luminous products
^{235}U[d)–f)]	3.2×10^{-8}	7.04×10^{8} yr	Nuclear reactor fuel Nuclear weapons Instrumentation (fission chambers)
^{238}U[d)–f)]	3.6×10^{-9}	4.468×10^{9} yr	Antitank weapons Projectiles Radiation shielding

(*continued overleaf*)

Table 4.8 (*Continued*)

Nuclide	Gamma constant (Gy-m²/h-MBq)	Half-life	Typical application or waste form
$^{238}Pu^{d)-f)}$	4.6×10^{-9}	87.7 yr	Neutron generators Calibration sources Pacemakers Radioisotope thermoelectric generators Remote power sources
$^{239}Pu^{d)-f)}$	1.9×10^{-9}	2.41×10^4 yr	Nuclear reactor fuel Nuclear weapons Instrumentation (fission chambers)
$^{241}Am^{d)-f)}$	1.8×10^{-8}	432.7 d	Detectors (e.g., soil moisture, hydrocarbon content, and smoke) Industrial gauges (e.g., density, moisture, material thickness, and flow) Neutron source when mixed with Be
$^{252}Cf^{d)-f)}$	8.5×10^{-8}	2.646 yr	Neutron research
Activation products[c),d)]	g)	g)	Low-level nuclear waste from reactors
Fission products[c),d)]	g),h)	g),h)	Low-level nuclear waste from reactors
Actinides and fission products[c)–f)]	h)	h)	Spent nuclear fuel

a) Bevelacqua (2010b).
b) NCRP 166 (2010).
c) Beta emitter.
d) Photon emitter.
e) Alpha emitter.
f) Spontaneous fission.
g) Varies with waste radionuclide composition.
h) Varies with fuel composition, enrichment, and burnup.

to the new facility. The therapy unit remained at the abandoned hospital location until 1987 when it was found by scrap metal hunters and dismantled. During dismantling, the ^{137}Cs source was damaged. A number of people subsequently handled the damaged source capsule. This resulted in the contamination of 250 people including 28 that sustained skin burns and 50 that ingested ^{137}Cs. In addition, four individuals died from acute radiation exposure from the breached source.

Contamination was spread over 40 city blocks, and 85 homes were significantly contaminated. Homes outside the immediate area of the event were cross-contaminated through the normal travel of affected individuals. Cleanup efforts generated 3500 m^3 of radioactive waste at an estimated cost of about \$20 million.

The Goiania event provides a preview of the potential results of an RDD event if it is not recognized as a radiological hazard. Cleanup costs and social disruption are expected to be significant, and good contamination control practices are essential to minimize the impact of an RDD attack. Clear communication with the public is needed to minimize concerns and to maintain an orderly recovery effort. The importance of clear public communication in emergency events was previously noted as a key element in the response to a power reactor accident.

Another isotope of concern is ^{90}Sr. ^{90}Sr emits beta particles and presents an internal hazard if inhaled or ingested. A major use of ^{90}Sr is in radioisotope thermoelectric generators used in the former Soviet Union to produce electricity in remote locations. These sources pose a significant threat because numerous devices are utilized, their accountability is poor, and they contain up to 50 TBq of ^{90}Sr.

^{241}Am, ^{252}Cf, ^{210}Po, ^{238}Pu, and ^{226}Ra are primarily alpha emitters. These nuclei present an ingestion and inhalation hazard. ^{252}Cf emits spontaneous fission neutrons and photons.

A dirty bomb is the popular term for an RDD that uses explosives for dispersing the radioactive material. The consequences of an outdoor RDD explosion affect only a relatively small area, but care must be exercised to limit the spread of radioactive material from the initially contaminated zone.

Following detonation, the RDD disperses the radioactive material. Upon dispersal, the radioactive material could be inhaled, ingested, or absorbed through the skin. Most radiation exposures would be too small to cause early health effects. However, internal and external contamination and psychological effects are likely.

Although deployment of an RDD has yet to occur, considerable effort has been expended in evaluating the effects of these devices and their possible deployment scenarios. Selection of characteristic RDD events is a complex process because the scenarios depend on a variety of considerations including the (i) isotope released and its radiological characteristics, (ii) physical and chemical form of the radioactive material, (iii) volatility and dispersibility characteristics of the material, (iv) dispersal environment (e.g., city or rural), (v) population density and demographics of the affected area, (vi) meteorological conditions during the release event, (vii) land use of the contaminated area, and (viii) coordination of the RDD event with other attack scenarios.

A consideration of these factors leads to a number of possible scenarios. These scenarios are included in Homeland Security recommendations and reports of the NCRP. Examples of the variety of scenarios under consideration or evaluation are included in Table 4.9. These examples illustrate the diversity in radioactive material, physical size of the source material, and hazards that are created by an RDD event.

Table 4.9 Summary of candidate RDD events, source sizes, and associated hazards[a].

Source/type of radioactive material	Physical size of the source	Hazards
RDD using portable radioactive material devices including nuclear medicine and brachytherapy sources, industrial gauges, and calibration sources	<500 cm^3 to >1 m^3 depending on the source configuration and associated shielding	• External radiation • Internal radiation • Blast from explosive detonation
RDD with large radioactive sources (e.g., ^{60}Co, ^{137}Cs, and ^{192}Ir for radiography and ^{60}Co for teletherapy or irradiation)	<500 cm^3 to >1 m^3 depending on the source configuration and associated shielding	• External radiation • Internal radiation • Blast from explosive detonation
RDD with spent nuclear fuel or reprocessing by-products	>1 m^3 depending on the source configuration and associated shielding	• External radiation • Internal radiation • Blast from explosive detonation
Terrorist attack on a commercial nuclear power reactor	Fuel assembly about 5 m in length	• External radiation • Internal radiation if fission product barriers are breached • Blast from explosive detonation
Terrorist attack on a university or national laboratory research reactor	Similar to nuclear power reactor but much smaller quantity of fuel and associated fission product material	• External radiation • Internal radiation if fission product barriers are breached • Blast from explosive detonation
Terrorist attack on radioactive material during land and rail transport	Cardboard packages, 30–55 gallon drums, low specific activity containers, and cargo containers	• External radiation • Internal radiation if the source and shipping packaging are breached • Blast from explosive detonation
IND that fails to achieve a nuclear yield	Backpack size to larger packages	• Blast from explosive detonation • High levels of alpha contamination • External radiation • Internal radiation from the dispersed radioactive material

a) NCRP 138 (2001) and Bevelacqua (2010b).

4.3.3 Radiation Exposure Devices

A RED consists of a large quantity of radioactive material clandestinely located to irradiate people. For substantial harm to occur, the RED would need to be located near the exposed individuals. Unless announced by terrorists or the device is detected, the only evidence of the RED is people seeking medical care for the symptoms of acute radiation exposure.

Serious health effects are not likely to be observed quickly unless delivered absorbed doses exceed 1.5–2.0 Gy to a large portion of the whole body. At these absorbed doses, the irradiated individuals would exhibit disorientation, dizziness, fatigue, immune system suppression, nausea, reduced blood cell production, vomiting, and weakness. There would be an increased susceptibility to infection, and at higher absorbed doses, more serious effects including death could result. The onset of symptoms depends on the exposure duration, proximity to the source, source strength, and radiation types and associated energies emitted by the source. In addition to immediate effects, long-term health effects of a RED include increased cancer risk and hereditary effects.

The effects of a RED would be localized to its immediate area. Upon detection, the RED would be removed to eliminate the hazard. However, finding the device could present a challenge unless the exposed individuals were readily associated with its clandestine location. Radiation scanning of candidate areas by manned aircraft or drones would facilitate detection of the RED.

4.3.4 Deliberate Contamination of Food, Water, or Other Consumables

The deliberate contamination of food, water, and other consumables with radioactive materials is another possible avenue for a terrorist attack. The effects depend on the radionuclides used in the intentional contamination, the level of contamination in the consumable, and the quantity of consumable ingested.

Deliberate contamination events would cause economic disruption to the sale and distribution of the affected items. Recent incidents (e.g., the Fukushima Daiichi accident) illustrate the likely chain of events and disruption patterns for specific contaminated food items. Following detection, the products are removed from sale and the public discontinues their use until the cause is determined and corrected. Although disruptions would occur, contamination events of a single food item would not be as severe or disruptive as an IND or RDD event.

The contamination of a city's water supply would have more serious consequences than the contamination of an individual food product. There is no previous accident experience to assess the effects of the contamination of a city's water supply with radioactive materials. Unless massive quantities of radioactive materials were utilized, the radiological effects would likely be relatively minor. However, the psychological consequences could be significant.

4.3.5
Dispersal of Radioactive Materials from Fixed Radiological or Nuclear Facilities or Materials in Transit

The dispersal of radioactive materials from radiological or nuclear facilities or materials in transit is similar to the effects of an RDD. Nuclear facilities and their host communities, counties, and states have emergency plans to address accident releases. The emergency response actions to a direct facility attack or act of sabotage is either covered or encompassed by existing emergency plans and emergency response actions.

The response to attacks or sabotage to a radioactive materials shipment in transit is similar to actions taken for hazardous materials (HAZMAT) events. HAZMAT initial response actions are usually adequate to isolate the affected area until a dedicated radiological response team arrives.

4.3.6
Health Physics Response and Medical Consequences of a Terrorist Event

The various radiological terrorism events lead to the irradiation of individuals through external sources, internal or external contamination, and radioactive fragments that enter the body. Each of these irradiation modes requires a specific health physics response to provide radiological information to the physician treating the patient.

External contamination of the body and clothing would be common occurrences that accompany a nuclear terrorism event. Radiation survey techniques determine the contamination levels, the areas of the body and clothing that are contaminated, and the radiation levels associated with the radioactive material. Positive nasal and mouth swabs could be used as a quick screening method to assess the presence of internal contamination. The radiation levels associated with external contamination do not normally present a radiation hazard to emergency response personnel.

Basic health physics principles dictate the removal of contaminated clothing and skin decontamination. Health physics personnel perform simple decontamination methods (e.g., using soap and water) to remove skin contamination. Medical personnel should direct more aggressive decontamination methods.

Patients are not likely to exhibit acute radiation syndrome symptoms related to their contamination. An externally contaminated individual should be checked for an internal deposition resulting from the intake of radioactive materials in dust, dirt, explosion debris, and air.

Health physics personnel implement basic contamination control methods to minimize the spread of radioactive materials to emergency response personnel. The treatment of severe injuries should not be delayed if external or internal contamination is present. Injured individuals should be handled in a manner that minimizes the spread of contamination to other individuals. However, lifesaving medical treatment should not be delayed for the implementation of contamination control measures.

The explosion of an RDD produces contaminated material and shrapnel. These fragments are propelled at significant velocities and can be imbedded in the patient. If the fragments have high activity, they can pose an external exposure hazard. Health physics surveys and localized shielding minimize the doses received by medical personnel when high activity debris is removed from the patient.

A person receiving a significant dose from an external source is at risk from the effects of the acute radiation syndrome. Specific symptoms depend on the severity of the dose and include nausea, reddening of the skin, and fatigue. Higher doses lead to more severe effects including death. Symptoms may not appear immediately and may not be observed for several days or weeks.

Dose estimates are beneficial in guiding the treatment of the irradiated individual. These dose estimates can be based on observed symptoms and their onset as well as information regarding the type of terrorist event, its severity, radioactive materials involved, their activity levels, exposure time, and position of the injured individual relative to the event location.

Internal contamination from inhalation and/or ingestion of radioactive material presents an initial challenge because the presence of this material may not be apparent. Initially, individuals are not likely to exhibit any symptoms related to radiological contamination. Internal contamination needs to be assessed using whole-body counting or other bioassay techniques. Treatment should be based on the degree of internal deposition and the projected dose.

4.4 Accident Assumptions

In order to facilitate the response to terrorist events, national and international organizations (e.g., the NCRP and International Commission on Radiological Protection (ICRP)) and government organizations (e.g., the US DHS and Environmental Protection Agency (EPA)) address terrorist events in terms of defining specific accident types and phases. Other defined assumptions and criteria include specification of emergency planning zones (EPZs), protective action recommendations, emergency response actions, reentry and recovery considerations, and utilization of volunteers in emergency response.

4.4.1 Accident Phases

The response to a radiological or nuclear terrorism incident is commonly divided into three phases. In the United States, these are the early, intermediate, and late phases. The ICRP refers to these as the rescue, recovery, and restoration phases. These phases do not have well-defined times and transitions and are not likely to be distinct. Although these phases are similar to the power reactor accident discussion in Section 3.5, there are unique aspects of the phases associated with terrorist events.

The early phase is the period at the beginning of an incident when immediate decisions for protective actions are required. Decisions and direct operations are made with limited information. The early phase following an RDD or IND can last from hours to days. However, this phase can be extended because the radiological nature of the event may not be immediately known. This phase will last longer for an IND incident.

The intermediate phase follows the early phase within as little as a few hours or in a few days. During this stage, the radiological nature of the event is well known. The main dose contributors are the direct irradiation from deposited radionuclides, inhalation of resuspended material, and ingestion of contaminated food and water. Intermediate phase actions include detailed surveys to characterize radionuclide depositions, food interdiction, and relocation of some members of the public. Activities in this phase typically overlap with early- and late-phase activities and may continue for weeks to many months, until protective actions are terminated.

The late phase begins with the initiation of restoration and cleanup actions to reduce radiation levels in the contaminated environment to acceptable levels. It ends when all remediation actions have been completed. With the additional time and increased understanding of the situation, there will be opportunities to involve key stakeholders in providing sound, cost-effective recovery recommendations.

In the United States, the NRF, Radiological/Nuclear Response Annex presents the three accident phases as the methodology for all types of radiological accidents (e.g., nuclear power plant, RDD, IND, and nuclear weapons). It describes the different decisions and actions that emergency organizations must address and how the preferred decision or action might change with the type of radiological incident. The response phases for the various types of radiological events are not different, and similar considerations apply to the various radiological incident types. However, the preferred decisions or actions will vary with the specific radiological event type.

4.4.2 Emergency Planning Zones

In the United States, power reactors utilize a 16 km radius EPZ and an 80 km ingestion pathway zone (IPZ). The emergency response actions and protective actions are dependent on the specified zone. These zones are addressed from a power reactor perspective in Section 3.4.5.

Three protective actions are implemented within the EPZ. These actions include evacuation, sheltering, and administration of thyroid blocking agents to minimize the dose to risk populations and protect their health and safety. The IPZ is designed to protect the public from consumption of food and water contaminated with radioactive materials. Protective actions within the IPZ include restrictions on milk and produce and feeding animals stored feed. Other types of planning zones have also been proposed.

For example, NCRP 165 defines zones for guiding emergency response actions to radiological or nuclear terrorism. These zones are based on radiological control criteria or structural damage considerations.

Defined zones partition the event area into different levels of risk where specific radiological controls are applied. The absorbed dose delivered to an emergency responder restricts response actions and associated stay times. NCRP 165 defines the cold, hot, and dangerous radiation zones. Zone boundaries are not determined precisely. For example, a boundary approximating 0.1 mGy/h is established by instrument measurements in the range of 0.05 – 0.2 mGy/h. In addition, zone boundaries should follow existing physical structures such as streets and fence lines that are close to the defined air kerma rate values.

The cold zone is defined by an air kerma rate $\leq$0.1 mGy/h. Establishing the hot zone is appropriate if any of the following air kerma rate or surface contamination levels are exceeded: 0.1 mGy/h, 1 kBq/cm^2 for beta and gamma surface contamination, or 0.1 kBq/cm^2 for alpha surface contamination. The dangerous radiation zone is defined for air kerma rates $\geq$0.1 Gy/h. In addition to these radiological zones, there are three additional hazard zones for a nuclear terrorist incident, and these are based on damage severity.

For RDDs and INDs, blast damage extends radially outward to distances on the order of kilometers for a nuclear detonation. Damage decreases as the distance from the initial blast increases. The region closest to the blast is defined as the severe damage zone. In the severe damage zone, buildings are destroyed and the likelihood of survivors is small. Considering the extensive property damage, minimal probability of survivors, and additional hazards, entry into the severe damage zone is not warranted until the area's radiological characteristics are known.

Beyond the severe damage zone is the moderate damage zone. In this zone, buildings are significantly damaged and rubble is dispersed throughout the area. A large number of survivors, some with severe injuries, are present, and their survival is enhanced if they receive prompt medical attention. Beyond the moderate damage zone is the light damage zone. In this zone, the primary damage is broken windows, and this zone extends for kilometers beyond the immediate blast area.

4.4.3
Protective Action Recommendations

The process for issuing a protective action recommendation following a nuclear terrorism event involves close coordination between government and civilian organizations. This process is similar to the sequence involving a power reactor accident, but its initiators are security related. This is illustrated by considering a terrorist attack on an operating nuclear power plant. The plant has an emergency preparedness program that includes a facility-specific emergency plan supported by implementing procedures. These procedures provide guidance for the activation of the utility's emergency response organization and its coordination with the local, state, and federal response agencies.

The extent of the terrorist threat, potential for a radiological release, and plant conditions determine the emergency classification (e.g., Unusual Event, Alert, Site Area Emergency, and General Emergency). As noted in Chapter 3, the declaration of a Site Area Emergency or General Emergency by the operating utility is usually accompanied by a protective action recommendation (e.g., sheltering, evacuation, or issuance prophylactic agents) to the state government. The state government in consultation with the Nuclear Regulatory Commission (NRC) and the operating utility makes the final decision regarding the protective actions that are implemented and coordinates their implementation with county and local governments. As outlined in Chapter 3, the implementation utilizes state and local police, state medical officials, communications organizations, hospitals, and medical facilities.

Other nuclear event types that do not involve a nuclear reactor (e.g., an IND or RDD terrorist events) have specific protective actions that are related to the consequences of those events. The EPA in conjunction with other federal agencies provides planning guidance and protective action guides (PAGs) for a variety of radiological events. This guidance is derived from other agency-specific recommendations. For example, the transitional DHS protective actions and associated PAGs for RDD and IND events are summarized in Table 4.10 for the early and intermediate accident phases. This direction is now integrated with the EPA guidance summarized in Table 3.7. Table 4.11 summarizes emergency worker dose guidelines for a variety of anticipated recovery actions following an IND or RDD event in the early accident phase.

Other organizations offer protective action recommendations for specific emergency conditions. For comparison, NCRP 138 dose limits and protective action recommendations for terrorist events are provided in Tables 4.12 and 4.13, respectively. These tables further illustrate the various conceptual approaches to acceptable dose limits and protective measures. Although there are similarities to the EPA recommendations, differences exist that illustrate the types of decisions and challenges faced by emergency managers in addressing IND and RDD terrorist events.

Table 4.12 summarizes the NCRP 138 recommendations for dose limitation and guidance during a terrorist event involving radiological weapons. These limits are provided for a variety of actions and exposed groups.

Ensuring public protection usually requires intervention measures to regain control of radioactive material during or after a radiological emergency. Table 4.13 summarizes countermeasures addressed in NCRP 138. Since countermeasures have an associated risk, their implementation must consider the risk/benefit of the proposed action. Table 4.13 lists available countermeasures for a variety of exposure pathways.

An additional set of possible protective measures is provided by a consideration of protective action levels (PALs) for food provided by the Province of Ontario,

Table 4.10 Transitional DHS (2008) protective action guides for RDD or IND incidents.

Accident phase	Protective action	Protective action guide
Early	Sheltering in place or evacuation of the public[a)]	10–50 mSv projected total effective dose equivalent (TEDE)
	Administration of prophylactic drugs including potassium iodide[b),c)]	50 mSv projected committed dose equivalent to child thyroid[b),c)]
	Administration of other prophylactic or decorporation agents[d)]	
Intermediate	Relocation of the public	20 mSv projected dose (TEDE) in the first year Subsequent years: 5 mSv/year projected dose (TEDE)
	Food interdiction	5 mSv/year projected dose or 50 mSv to any individual organ or tissue in the first year (whichever is limiting)
	Drinking water interdiction	5 mSv projected dose in the first year

a) Should normally begin at 10 mSv and take whichever action (or combination of actions) that results in the lowest dose for the majority of the population. Sheltering may begin at lower levels if advantageous.
b) Provides thyroid protection from radioactive iodine only.
c) KI should be administered to both children and adults at the lowest intervention threshold (i.e., >50 mSv) projected committed dose equivalent (thyroid).
d) For information on other radiological prophylactics and medical countermeasures, refer to http://www.fda.gov/cder/drugprepare/default.htm, http:/www.bt.cdc.gov/radiation, or http://www.orau.gov/reacts.

Canada. The Ontario philosophy is to express the PALs in terms of the highest projected dose likely to be received by the most exposed individual in the relevant critical group over the duration of significant releases of radioactive material.

The ingestion PALs, summarized in Table 4.14, should be utilized for food prepared for consumption and are to be applied to the sum of the activity levels for each radionuclide within a specified group. However, they are applied independently to each group. The Ontario application approach is illustrated by considering the "foods for general consumption" category. If the food item has ^{137}Cs at a level of 50% of the permitted concentration while the quantity of ^{160}Ru (which is in the same group as ^{137}Cs) is 60% of the permitted concentration, the item should not be sold or consumed. However, an item containing 50% of the permitted concentration of ^{137}Cs and 60% of the permitted concentration of an element in a different group (e.g., ^{90}Sr) would be acceptable for consumption. ^{131}I

Table 4.11 EPA response worker guidelines[a].

Effective dose guideline (mSv)	Activity	Condition
50	All occupational exposures	All reasonably achievable actions have been taken to minimize dose
100 [b]	Protecting valuable property necessary for public welfare (e.g., a power plant)	Exceeding 50 mSv is unavoidable and all appropriate actions taken to reduce dose Monitoring is available to project or measure dose
250 [c]	Lifesaving or protection of large populations	Exceeding 50 mSv is unavoidable and all appropriate actions taken to reduce dose Monitoring is available to project or measure dose

a) EPA (2013).
b) For potential doses >50 mSv, medical monitoring programs should be considered.
c) In the case of a very large incident, such as an IND, incident commanders may need to consider raising the property and lifesaving response worker guidelines to prevent further loss of life and massive spread of destruction.

Table 4.12 Dose limitation and guidance during a terrorist event involving radiological weapons[a].

Classification or action	Applicability	Limit or guidance[b]
Full mitigation	General public dose limitation	[c]
Sheltering	Avert dose to general public	5–50 mSv (effective dose)
Evacuation	Avert dose to general public	50–500 mSv (effective dose)
Administer stable iodine	Avert dose to children and pregnant women	50–500 mSv (equivalent dose)
Any single food category[d]	Avert dose to general public	10 mSv/yr (effective dose)
Relocation	Avert dose to general public	10 mSv/mo, 1000 mSv (effective dose)
Annual limit	Recovery workers (nonemergency work)	50 mSv/yr (effective dose)
Guidance for emergency action	Recovery workers (emergency work)	500 mSv (effective dose)

a) NCRP 138 (2001).
b) When two values are given, the lower value represents the lowest effective dose at which the countermeasure is likely to be justified. The larger value represents the effective dose at which the countermeasure is almost always justified.
c) Full mitigation dose limitation values are obtained through the process of justification and ALARA, and the results may be higher or lower than the NCRP limit of 1 mSv/year for individual members of the public.
d) The FDA has provided guidance for intervention in the ingestion pathway based on the total diet (FDA, 1998).

Table 4.13 Available countermeasures for various of exposure pathways[a].

Exposure pathway	Available countermeasure
External radiation exposure from nuclides in the plume	Sheltering, evacuation, and control of access to affected areas
Internal contamination due to nuclides in the plume	Sheltering, *ad hoc* respiratory protection[b], administration of stable iodine, evacuation, and control of access
External contamination from surface-deposited radioactive material	Sheltering, evacuation, control of access, and decontamination
External radiation from surface-deposited radioactive material	Sheltering, evacuation, relocation, control of access, and decontamination
Internal contamination due to resuspension	Evacuation, relocation, control of access, and decontamination
Internal deposition due to personnel contamination	Control of access and decontamination
Internal exposure due to ingestion of contaminated water and foodstuffs	Control of food, water, and use of stored animal feeds

a) NCRP 138 (2001).
b) *Ad hoc* respiratory protection includes actions such as covering the nose and mouth with a dry or wet handkerchief or washcloth.

Table 4.14 Protective action levels for food ingestion[a].

Concentration in specified form	Radionuclide/groups			
	^{89}Sr, ^{103}Ru, ^{106}Ru, ^{134}Cs, and ^{137}Cs	^{131}I	^{90}Sr	^{238}Pu, ^{239}Pu, ^{240}Pu, ^{242}Pu, and ^{241}Am (Bq/kg)
Foods for general consumption	1 kBq/kg		100 Bq/kg	10
Drinking water, infant foods, and milk	1 kBq/kg	100 Bq/kg		1

a) Province of Ontario's Nuclear Emergency Response Plan (2009).

is grouped with radiocesium for general consumption foods but is grouped with ^{90}Sr for infant food and water.

These various methods illustrate the diversity of approaches that can be utilized to address events that disperse radioactive material. Each approach has merit. Government organizations specify the approach used in response to a radiological emergency. However, confusion is minimized in an event involving multiple nations if all governments adopted a consistent set of standards. For example, inconsistent US and Japanese evacuation recommendations led to considerable confusion during the Fukushima Daiichi accident. This issue is addressed in Chapter 7.

4.4.4 Emergency Response Actions

The initial responders to a radiological terrorist event must implement a variety of actions to ensure a successful recovery. NCRP 138 outlines seven specific areas associated with radiological terrorist events that are necessary for a credible response. These areas include recognition capability, command and control, communications, psychosocial aspects, medical response, exposure guidance, and late-phase decision-making:

1) *Recognition capability*: Unless health physics support is provided or a nuclear facility is the target, first responders to the scene of a terrorist event may not recognize its radiological aspects. It is unlikely that all the first responders are trained to perform radiological measurements. Therefore, it is necessary to ensure these individuals do not receive an unacceptable radiation exposure while permitting their initial response activities to proceed. For this reason, NCRP 138 recommends personnel or response vehicles carry radiation detection equipment with preset alarm levels to alert the responders when entering an elevated dose rate area and when an unacceptable ambient dose rate or ambient dose has been reached. The radiation detection systems should be rugged, reliable, and designed with simplicity and capability to function in an environment littered with dust, rubble, and collapsed buildings. Suggested alarm levels for initial response instrumentation are summarized in Table 4.15.
2) *Command and control*: Throughout the terrorist event, coordination and control are required for an effective emergency response. From a health physics perspective, radiological data must be clearly communicated throughout the emergency response organization and with government officials. As noted in Chapter 3, difficulties with communicating radiological information have been encountered during the three major reactor accidents. With the exception of Three Mile Island Unit 2, these events did not involve a significant environmental release of radioactive material. An RDD or IND event will involve significant public confusion and apprehension, and having

Table 4.15 Initial responder radiation instrumentation alarm set points[a)].

Alarm level	Alarm set point	Comments
Initial alarm	0.1 mSv/h	Establishes an initial control location to restrict access for radiological purposes
Second alarm	0.1 Sv/h or 0.1 Sv	Turnaround level permits the initial responders to perform additional time-sensitive, critical missions beyond the point where it is recognized that there is a radiological component to the event

a) NCRP 138 (2001).

accurate radiological information is crucial for implementing effective emergency response actions.

3) *Communications*: The public should be informed of the incident as well as its radiological impact in a timely manner. Periodic updates are also essential to avoid rumors and speculation regarding the event. Communicating the radiological hazard must be performed using clear language that avoids technical terms but conveys the radiological conditions in common terms (e.g., chest X-ray dose or background radiation level). It is important to include in these communications clear statements of the uncertainties associated with these dose projections and measures that are utilized to protect the public.
4) *Psychosocial aspects*: The detonation of a nuclear device or RDD will have significant physical as well as social and psychological impacts. An IND is a weapon of mass destruction, and its detonation presents a clear radiological hazard that is well understood based on numerous atmospheric tests. However, an RDD is not a weapon of mass destruction.

   RDDs are weapons of social disruption that have minimal impact outside the initial blast and associated dispersal area. Therefore, it is essential to clearly communicate the actual hazard and distinguish an RDD from an IND. Public information programs facilitate this distinction, but responders must clearly recognize the psychosocial impact of these events. These considerations apply not only to the immediate impacts but also to the range of longer-term effects that could be expected following a terrorist event involving radioactive materials. As noted in Chapter 7, psychosocial effects have also been observed in public evacuees associated with the Fukushima Daiichi accident.
5) *Medical response*: The potential for a large number of casualties and fear of radiation, radioactive materials, and contamination may hamper the medical response to a radiological terrorist event involving an IND or an RDD. First responder radiological training and public information programs must clearly address radiological hazards and their associated health effects. For example, training and information programs must characterize internal and external contamination and ensure the first responder and public understand that these conditions are not immediately life threatening. Other considerations (e.g., physical injury and lifesaving activities) take precedence over decontaminating survivors.
6) *Exposure guidance*: Emergency responder doses should be limited, if possible, to the occupational exposure limits. Doses beyond these levels are authorized during a severe disaster if prompt, well-considered actions save lives and avert significant harm to the public. Even in these situations, as low as reasonably achievable (ALARA) principles must be applied to emergency response personnel.
7) *Late-phase decision-making*: The late-phase response includes radiological remediation and restoration of contaminated areas to their original condition. Contaminated areas could be quite large and significant cost and effort could be required for their restoration to acceptable levels. The public and affected

stakeholder groups must be full participants in establishing the radiological acceptance values. Total cost, time to accomplish the tasks, and the risks associated with established radiological criteria are important considerations in the decision-making process. Establishing acceptable radiological criteria for the recovery effort is a challenging process, and a successful outcome depends on trust and mutual respect.

Emergency responder actions vary with the defined NCRP 138 zone (i.e., light damage zone, moderate damage zone, and severe damage zone). Establishing these zones and assessing their radiological hazards are priority response tasks. A discussion of each of these zones and the associated responder actions are provided in the subsequent discussion.

4.4.4.1 Light Damage Zone

The light damage zone is characterized by broken windows as the primary damage condition. In the light damage zone, emergency responders focus medical attention on life-threatening injuries and medical conditions. This requires bypassing victims with minor injuries that would normally receive prompt medical attention. Response personnel must direct attention to the highest need.

This prioritization requires focus and dedication. Light damage zone victims may exhibit hostility when emergency response personnel bypass them. Information programs must educate the public in this reality and the need to prioritize scarce response resources during the early accident phase.

4.4.4.2 Moderate Damage Zone

In the moderate damage zone, significant building damage occurs. A priority action for this zone is to allocate sufficient resources to address emergency medical situations. These resources focus on lifesaving activities. Response actions in the severe damage zone are initially warranted if they support lifesaving activities for a large number of individuals.

4.4.4.3 Severe Damage Zone

In the severe damage zone, most buildings are destroyed and building rubble may hamper travel and emergency response. Emergency responders within the severe damage zone should shelter until air kerma rates fall below 0.1 Gy/h. To support this action, dosimeters should be available and radiation detection instrumentation periodically verified to be operable at all emergency response facilities. In the case of an IND event, instruments may be affected by EMP-induced failures. The effects of nuclear detonations, including an EMP, were previously addressed in Section 4.3.1.1.2.

While waiting for radiation levels to decrease, emergency communications systems should be activated. Emergency radios and cellular phones are activated, and communications links to other emergency response organizations and government organizations are established.

Once the air kerma rates decrease to below 0.1 Gy/h, radiological survey teams map air kerma rates within a few blocks of their initial location. The radiation profile and physical hazards are communicated to other emergency responders and government organizations to plan future response actions. If radio and phone communication are impaired, this information may be transmitted using volunteers to reach neighboring response centers.

The survey teams also record the location and extent of physical hazards (e.g., fires, chemical leaks, downed and energized power lines, and ruptured gas lines) that could affect subsequent emergency response actions. The location of the nuclear event should be determined. Airborne assets would greatly simplify the task of determining damage extent, the ground zero location, and the radiation profile in the three damage zones. If available, drones provide an effective, ALARA approach to obtain the data to support the initial response and recovery actions.

The survey teams also identify intact structures and resources that could aid emergency response actions. Staging areas for mass casualty triage are established. Citizen volunteers are solicited to assist emergency responders at triage sites, function as litter bearers, and assist in clearing evacuation routes.

4.4.4.4
Public Information Programs and Initial Actions

Emergency response activities should include active public information programs. These programs improve the initial response to a nuclear terrorist event. An effective public information program provides basic information regarding the various types of terrorist events and their characteristics. Radiological information provided to the public includes the type of radiation, nuclides of interest, and radiological hazards associated with nuclear terrorism. Basic radiation terminology, units, and health effects should be addressed in a nontechnical manner. Emphasis is placed on actions that the public can initiate to maximize their survival. The characteristics and physical behavior of fallout and the importance of sheltering are emphasized.

A key aspect of these information programs is preparation of the public to respond properly to situations that arise before emergency response personnel arrive. Informative education programs support a positive public response that is crucial to minimize injuries and enhance survivability following the initial blast from an RDD or IND.

Upon recognizing the initial indications of a large explosion resulting from either the RDD or IND, individuals should retreat from windows and utilize available cover to limit the initial blast effects. This action minimizes injuries from flying glass and other blast-induced missiles. In many situations, the time interval between the flash and audible detonation sound will be sufficient for many individuals to seek cover inside buildings or other structures. The effectiveness of these responses depends on public awareness and their confidence in the recommended actions. Practice exercises, public service announcements, and educational programs are required for these responses to be viable emergency response actions. Retreating and seeking cover will not prevent all injuries, but their implementation reduces the strain on the initial medical responders.

Following the initial blast, the most effective approach to minimize subsequent radiation effects is to limit the absorbed dose from fallout. Immediate sheltering is a simple action that can be taken by the public to minimize their radiation risk while awaiting the arrival of emergency response personnel. The effective use of sheltering improves public safety, reduces the demand for emergency response resources, and minimizes traffic congestion. For these actions to be successful, the public must understand the radiation hazards created by a terrorist event. Upon arrival, the emergency response personnel can implement an informed public evacuation using appropriate egress routes to minimize population doses. The first responders also determine absorbed dose rates and the extent and magnitude of contamination in affected areas to facilitate the evacuation in an ALARA manner.

To facilitate the sheltering protective action, local governments identify public shelters and communicate their location to the public. Signage should be prominently posted to advise the public of the presence of a shelter and its level of protection. These actions are reminiscent of cold war US civil defense programs. Their effectiveness depends on public education, awareness, and confidence in the government to adequately protect citizens in the event of a terrorist attack.

4.4.4.5
Subsequent Emergency Response Actions

Following initial radiological characterization of affected areas, appropriate routes are identified to facilitate the evacuation of individuals from the severe damage and dangerous radiation zones. In preparation for subsequent evacuation, individuals utilize stations established by the initial responders and perform self-decontamination. Self-decontamination is a necessary action since the large number of potentially contaminated individuals will exhaust resources for decontamination performed by emergency responders.

At this stage of the event, the public is either in a shelter or in a lower radiation area. The goal is to evacuate populations from elevated dose rate areas and areas with high levels of contamination from fallout to a less hazardous area with better support facilities.

Evacuation should occur only if the population is not directly exposed to fallout during their exit from the damaged areas. Sheltering is usually the best protective action for the first few hours postexplosion until the fallout hazard is characterized and mapped and a safe evacuation route is determined. To account for meteorological uncertainties, protective actions (e.g., sheltering and evacuation) should also be applied to areas adjacent to the projected fallout path. The fallout pattern broadens with time and will likely affect a large area.

Plume meander and dispersion influence the evacuation route. Changing fallout patterns lead to longer travel times and distances than initially planned. Meteorology changes also affect evacuation routes.

For an assumed 10 kT IND, early health effects including lower dose acute radiation syndrome effects are not anticipated beyond 16–32 km. The protective actions summarized in Table 3.7 should be followed and sheltering considered to reduce the potential public absorbed dose. Areas hundreds of kilometers downwind could receive effective doses of 0.01 Sv from internal and external contributions.

The response actions are affected by the IND characteristics and the detonation location. Detonation height is particularly important since it governs the amount of fallout produced. The detonation location and extent of infrastructure damage also affect the response actions and evacuation route options.

4.4.5
Preplanned Evacuation Zones

Public evacuation is a serious undertaking with inherent risks. In order to be successful, the emergency response organization must have plans to mobilize emergency responders, activate emergency response organizations, and implement protective actions. One of the key protective actions is evacuation of the public from the area affected by the terrorist event. A successful evacuation requires that sufficient resources are allocated to ensure the health and safety of the evacuating population.

Following an IND or RDD event, the inherent evacuation risks are balanced against the risks posed by fallout and associated radiation levels. The population involved in a public evacuation includes a variety of age groups with various health conditions and injuries. An evacuation decision, route, and destination zone for population relocation necessarily considers the variations in the evacuated group. Special attention is required for the elderly, hospital and nursing home patients, and those injured by the terrorist event.

The destination evacuation zone must have sufficient facilities to address injuries and medical conditions of the evacuated population. In the case of an RDD, the damage area is relatively small and neighboring resources accommodate the evacuees. This is not true for an IND event that includes mass casualties, significant damage over a larger area, destruction of infrastructure, and disruption of support facilities. The destination evacuation zones for an IND event should be located 20 or more kilometers from the initial detonation site.

The IND situation is somewhat analogous to the 2011 Fukushima Daiichi accident that involved an earthquake and tsunami that preceded the nuclear event. The natural disaster disrupted transportation routes and needed evacuation infrastructure and serves as a crude indication of the difficulties to be encountered following an IND event.

Evacuation decisions must be well defined and account for uncertainties and events that were not anticipated in the initial formulation of emergency response actions. Emergency response plans assume the existence of a degree of infrastructure and resources including emergency response facilities, response staff, transportation resources, supplies, staging areas, and access to planned evacuation zones. The RDD or IND blast may affect these assumptions, and the evacuation must account for these unanticipated disruptions. Disruptions could include damage or loss of emergency response facilities; unavailability of emergency response personnel; loss of buses, rail, or vehicular transportation resources; destruction of planned evacuation bridges and roads; and loss of medical response resources. Each of these perturbations must be overcome for the evacuation to proceed without major disruption or further loss of life. Cross-training of personnel in multiple emergency response disciplines, redundancy in critical resources, and a well-informed public can significantly improve evacuation success if disruptions beyond the initial nuclear terrorism event are encountered.

4.4.6 Reentry and Recovery Considerations

After the sources and releases of radioactive material are under control, reentry into previously evacuated areas is considered based on meeting predetermined radiological criteria. The reentry and recovery process represents a significant challenge in balancing the calculated radiation risk from low levels of ionizing radiation with the economic and social costs of restricting access to contaminated areas. Desires of citizens to return to their homes and jobs must be considered in determining realistic decontamination criteria and reentry and resettlement policies. Overly conservative criteria and decision guidelines should be avoided. Proposed guidelines and criteria are likely to be derived from risk-based models and experience. The final guidelines must consider full, open, and honest dialog between the government and stakeholders and balance the assumed risks with actual stakeholder priorities and human needs. These discussions should be held before and after a major radiological event. Candid discussions before a radiological event provide a sound basis for postevent dialog that is likely to be more stressful and emotional.

National resources including plume and dose profile mapping aircraft contribute to the early-phase response to a radiological terrorist event. However, state and local officials control reentry and recovery efforts. While national leadership and support are needed throughout a nuclear emergency, accident response history in the United States suggests that local and state authorities have decision-making responsibility during disaster recovery and resettlement. This approach is appropriate and provides for stakeholder input to influence decisions that have a significant impact of their lives.

The DHS notes that late-phase recovery after an RDD or IND incident should be achieved through a site-specific optimization process that includes a variety of considerations in addition to radiological risks. These considerations include future land uses, cleanup options and approaches, technical feasibility, cost-effectiveness, infrastructure status, local economic conditions, and public acceptance. Property decontamination is addressed by late-phase PAGs, but this activity occurs months to years after a nuclear terrorist event.

In the United States, federal recovery responsibilities are assigned to a number of agencies, including the Centers for Disease Control, DHS, EPA, Federal Emergency Management Agency, and NRC. The National Resource Framework is responsible for effectively coordinating the efforts of these organizations.

The NRC regulates nuclear plant safety and emergency planning and has the lead role for a terrorist attack on a power reactor. The DHS and FEMA have the lead on response coordination with state and local authorities. The DHS also utilizes Department of Energy (DOE) resources for radiological monitoring through the Federal Radiological Monitoring and Assessment Center. For a terrorist event involving radioactive materials, the radiological monitoring effort is coordinated with the EPA, the Department of Defense, Health and Human Services, and the Federal Bureau of Investigation. The lead role in coordinating environmental cleanup and recovery activities with state and local governments usually resides with the EPA. The CDC provides population monitoring for radiation exposure, assessments of the associated health risks, and laboratory support.

In the United States, the national response efforts have not always been timely or effective. Response failures during significant nonnuclear events such as Hurricanes Katrina and Sandy illustrate difficulties in meeting stakeholder needs. There have been delays in restoring electric power and returning the affected areas to a normal condition. In Japan, the Fukushima Daiichi accident response also illustrated the difficulties associated with postaccident response actions and in meeting stakeholder needs. Emergency planning and response have improved following each major disaster, but additional enhancements are needed to minimize human impacts following significant events. The radiological aspect of a significant nuclear disaster adds additional psychological stress in responding to these events. Additional discussion related to the regulatory and stakeholder aspects of accident response activities is provided in Chapter 7.

4.4.7 Volunteers

One positive aspect of the Fukushima Daiichi accident was the significant contribution of public volunteers assisting emergency response personnel. These individuals were involved with decontamination of homes, cleaning contaminated streets and schools, and providing emergency services.

There is also historical evidence that suggests that volunteers would assist the recovery from a major US disaster. For example, residents of the US Gulf states volunteered for recovery activities following the Deepwater Horizon oil spill. However, liability concerns precluded their full utilization.

For a severe event, volunteer participation is essential to the recovery effort. The significant role of community volunteer activities represents a paradigm shift in radiological event response. Japanese experience demonstrated that emergency response personnel alone could not sufficiently manage and implement recovery activities in a timely manner. The significant role of community volunteers is becoming apparent, and recovery planning and policy formulation should include their contribution to the pool of available resources. Utilizing trained radiation protection professionals from throughout a country as well as untrained volunteers from the affected area could have significant benefit to the successful recovery from a major radiological terrorist event. From a regulatory perspective, there is a need for dose guidelines that can be clearly applied to members of the public acting in this volunteer radiological protection capacity. The volunteers are performing an activity associated with radiation workers, but they are members of the public. Volunteer workers should be identified and trained, and their respective dose guidelines defined prior to their assignment during an emergency event.

The US Medical Reserve Corps (MRC) formed after the 11 September 2001 attacks illustrates an example of the use of volunteers in radiological emergencies. MRC units are community based and utilize volunteer resources to prepare for and respond to emergencies. MRC volunteers supplement existing emergency and public health resources but are not considered to be first responders. The MRC concept has considerable merit, and it should be utilized to ensure that public health and safety are maintained during a radiological emergency.

4.5 Radiation Protection Considerations

Radiological decisions during a terrorist event must account for all relevant, situation-specific information that affects the external and internal dose pathways to emergency response personnel and the public. The sources of the various exposure pathways for an IND are summarized in Table 4.16. These pathways include routes for the delivery of internal and external dose. The contributions

Table 4.16 Pathways and sources for an IND event[a].

Exposure pathway	Source
External exposure	Detonation of the weapon
	Radioactive plume
	Surface contamination and activation products
	Personal contamination of skin and clothing
Internal contamination	Plume inhalation
	Inhalation of resuspended radioactive material
	Inhalation or ingestion of radioactive material contaminating the body surface
	Ingestion of contaminated foodstuffs
	Skin and wound absorption of radioactive material

a) NCRP 138 (2001).

to internal and external dose are event specific and dependent of the terrorist device and the specific attack parameters. Similar considerations exist for RDDs.

Early radiation effects from a terrorist event involving radioactive material most likely arise from external radiation exposure from large particles (nonrespirable) that deposit onto surfaces. These particles present an external radiation hazard during initial first responder actions and public evacuations.

Internal contamination is unlikely to pose a significant hazard to the public during sheltering. However, internal contamination presents a hazard to first responders following the initiating event and during a public evacuation.

Following an explosion of an RDD or an IND, the most serious medical injuries are likely to occur in people close to the explosion point, and this group is also most likely to be internally contaminated. It is important to screen these individuals to ensure that medical resources are properly allocated to the most seriously injured.

Screening is the rapid assessment and measurement of external or internal contamination. It is intended to enable intervention and management of persons exposed to radioactive material from an RDD, IND, or another nuclear incident. Potentially contaminated individuals are surveyed for external contamination. Externally contaminated individuals are decontaminated prior to screening for internal contamination. The emphasis is on rapid screening to identify individuals requiring medical treatment to decorporate large activities of internally deposited radionuclides. However, in most cases physical injuries and lifesaving activities have priority over contamination removal. The management of contaminated individuals is addressed in Section 4.5.3.1.

4.5.1
Nuclides of Interest

A radiological event could involve a variety of isotopes. These isotopes can be obtained from numerous sources, but theft of domestic licensed material, theft of foreign radioactive materials, or diversion of radioactive materials streams are possible sources for an RDD. However, a smaller set of radioactive materials are likely to be utilized in an RDD.

This limited set is summarized in Table 4.8. One of the primary tasks for health physics emergency response personnel following an environmental release of radioactive material is to determine if individuals are contaminated and the radionuclides of concern.

After external decontamination of each casualty, screening for internal contamination is initiated. Emergency patients are stabilized before external decontamination is attempted beyond clothing removal.

The goal of screening potentially contaminated individuals is to determine if the activity of the internal contamination is sufficiently large to justify medical treatment via decorporation therapy to expedite biological removal. Clinical decisions for decorporation therapy are guided by the critical decision guide (CDG) value for each radionuclide. The CDG values are discussed in Section 4.5.3.1.

4.5.2 External Dose Considerations

External dose is reduced by following the NCRP 165 recommendations and good health physics practices. NCRP 165 recommends that the initial public protective action for both radionuclide dispersal incidents and nuclear detonations be sheltering followed by informed evacuation from the affected area. Until the level and extent of contamination is determined, efforts should be made to avoid being outdoors in potentially contaminated areas.

Dose limits are not recommended for emergency responders performing mission-critical tasks including lifesaving. Instead, the emergency response incident commander should establish decision points based on the evolving operational situation and mission priorities. When the emergency responder's cumulative absorbed dose reaches 0.5 Gy, withdrawal of the worker from the hot zone should be considered. The NCRP considers the 0.5 Gy cumulative absorbed dose to be a decision dose and not a dose limit.

External dose following an RDD or IND event can be estimated using the relationships summarized in Appendix C. For an event contaminating a large area as a result of an RDD detonation or fallout from an IND, the external dose from contaminated ground can be approximated by using a thin disk source. This relationship is particularly useful for an evacuation traversing contaminated ground. The isotopes contaminating the ground, the concentration per unit area of those isotopes, and the exposure time determine the external dose. However, measured dose rates are preferable to calculated results.

The exposure time is related to the evacuation time. During an evacuation traversing contaminated ground, the surface contamination could become airborne as a result of the evacuation. The airborne concentration (C_{air}) inhaled by the evacuating public is related to the surface contamination (C_{s}) by the relationship

$$C_{\text{air}} = rC_{\text{s}} \tag{4.9}$$

where r is a resuspension factor.

4.5.3 Internal Dose Considerations

The primary objective in managing persons contaminated with radioactive materials is to reduce the risk of deterministic and stochastic effects. Risk reduction is achieved by reducing the internal and external contamination levels. However, bodily injury and its immediate treatment take precedence over contamination removal.

4.5.3.1 Management of Contaminated Individuals

NCRP 161 provides quick reference information valuable to individuals responding to a contamination event. This guidance includes a decision protocol for management of persons contaminated with radionuclides. The essential elements of this guidance are summarized in Table 4.17, which includes the specific response action and physical location for taking that action. Screening of contaminated individuals by emergency response personnel is discussed in Section 4.5.4.

To assist physicians in reaching treatment decisions, NCRP 161 defines the CDG values. The numerical values for different radionuclides, excluding isotopes of iodine, in adults are based on a 50-year effective dose and deterministic effects to the bone marrow and lungs.

In formulating a CDG value, 0.25 Sv (50 year effective dose) is used for the consideration of stochastic effects. Based on ICRP 103, this effective dose represents about a 1.3% lifetime risk of a fatal cancer attributed to radiation dose. A 30-day RBE-weighted absorbed dose value of 0.25 Gy-Eq is utilized for assessing deterministic effects to the bone marrow. Finally, a 30-day RBE-weighted absorbed dose value of 1 Gy-Eq is used for evaluating deterministic effects to the lungs.

For radionuclides other than isotopes of iodine, the CDG values for children (age 0–18 year) and pregnant women are defined as one-fifth the adult values. CDG values for ^{131}I suggest that KI be administered to adults >40 years of age if the projected thyroid dose is ≥5 Gy, to adults 18–40 years of age if the projected dose is ≥0.1 Gy, and to pregnant or lactating women or individuals <18 years of age if the projected dose is ≥0.05 Gy. Table 4.18 provides a selected set of CDG values.

Internal intakes exceeding 1 CDG value suggest the need for action to remove the deposited nuclides. Decorporation therapy recommendations for selected radionuclides are summarized in Table 4.19. The drugs listed in Table 4.19 are US Food and Drug Administration approved for chelation or decorporation. All drugs listed in Table 4.19, except KI, which is sold commercially, are prescription drugs. In the United States, diethylenetriaminepentaacetic acid (DTPA), KI, and Prussian blue are available from the Centers for Disease Control through the Strategic National Stockpile. Trained medical personnel accomplish administration of these controlled agents.

Table 4.20 lists other decorporation therapy recommendations. Unless noted in Table 4.19, these recommendations are not FDA approved but are based on current research.

4.5.3.2
Internal Dose Determination

Following a terrorist radiological event, the determination of internal dose is obtained from indirect measurements such as bioassay data or environmental measurements and models that describe the transfer and bioaccumulation of radionuclides in the body. Internal dose determination has been historically performed for a variety of situations including epidemiological evaluations, retrospective dose evaluations, and applied radiation protection dose assessments.

Table 4.17 Medical management decision protocol for personnel contamination events[a].

Stage no	Response action	Location
1	*Medical assessment of radiation exposure, contamination, or injury* Providing emergency medical care to seriously injured individuals is the highest priority. The next priority is to identify exposed and contaminated individuals and those showing psychological distress. If the nuclides are known, internal intakes may be reduced using mitigation agents such as KI for radioiodine or DTPA for transuranic elements	On-site triage area
2	*External contamination assessment including survey and examination for burns, wounds, shrapnel, and hot particles* Radiological survey data assists in the development of treatment and decontamination recommendations	On-site triage area
3	*External decontamination including wounds, body orifices, and intact skin* The objective is to avoid internal intakes through skin absorption, inhalation, and ingestion. The second objective of removing contamination is to reduce skin doses and to decrease the quantity of radionuclides in wounds and their possible absorption into the blood	On-site decontamination area
4	*Evaluation and emergency care for medical and surgical procedures* Medical concerns are further assessed and treated. Any remaining external and internal contamination is diagnosed. External contamination is treated. Unless administered on-site, mitigation agents should be considered to limit the internal dose	Hospital
5	*Internal contamination assessment including the evaluation of air samples, nasal swabs, and in vitro and in vivo bioassay* Radionuclides and routes of entry are identified, and doses are estimated	Hospital
6	*The clinical decision guide (CDG) is used to minimize risks* Analysis of the internal dose is used with radionuclide-specific CDG values to evaluate the need for decorporation therapy	Hospital
7	*Medical management includes decorporation therapy if the estimated intake exceeds 1 CDG* The treatment is evaluated with periodic bioassay. Clinical follow-up includes evaluation of absolute lymphocyte decrease and evaluation of other medical issues	Hospital
8	*Follow-up medical care including long-term monitoring and treatment of late deterministic effects, latent acute radiation syndrome, psychosocial effects, internal contamination, and cancer* Accurate patient records and epidemiology studies are also important actions	Hospital and home
9	*Contaminated decedents represent a potential radiological hazard* To protect medical examiners and mortuary personnel from radiation exposure, contamination control and proper disposition of decedents are important actions	Hospital and mortuary

a) NCRP 161 (2008).

Table 4.18 Adult critical decision guide values for selected radionuclides[a].

Nuclide	Intake mode	Form	CDG intake activity (Bq)
^{3}H	Inhalation	HTO	1.4×10^{10}
^{3}H	Ingestion	HTO	1.4×10^{10}
^{32}P	Inhalation	Type M	7.1×10^{7}
^{32}P	Ingestion	Soluble	3.9×10^{7}
^{60}Co	Inhalation	Type M	3.5×10^{7}
^{60}Co	Inhalation	Type S	1.5×10^{7}
^{90}Sr	Inhalation	Type F	8.3×10^{6}
^{90}Sr	Ingestion	Soluble	8.9×10^{6}
^{90}Y	Inhalation	Type M	1.6×10^{8}
^{137}Cs	Inhalation	Type F	5.8×10^{7}
^{137}Cs	Ingestion	Soluble	2.8×10^{7}
^{226}Ra	Inhalation	Type M	1.1×10^{5}
^{239}Pu	Inhalation	Type M	7.6×10^{3}
^{239}Pu	Inhalation	Type S	3.0×10^{4}
^{241}Am	Inhalation	Type M	9.3×10^{3}

a) NCRP 161 (2008).

Table 4.19 Drugs approved for radionuclide chelation or decorporation by the FDA[a].

Drug	FDA indication for radionuclide treatment
Diethylenetriaminepentaacetic acid (DTPA) – injection	Americium Curium Plutonium
Potassium iodide (KI) – oral administration	Iodine
Prussian blue (insoluble ferric hexacyanoferrate(II)) – oral administration	Cesium Thallium

a) NCRP 161 (2008).

Ambiguities in the calculated internal dose arise from modeling assumptions and uncertainties and measurement uncertainties. These uncertainties are important considerations in research applications including epidemiological studies. In epidemiological studies, risk estimates depend on knowledge of the uncertainties in the internal dose calculation. These dose estimates rely on the survey data that characterizes a contaminated site.

In retrospective dose reconstructions (e.g., weapons fallout events, accidents, and occupational situations), there are government compensation programs (e.g., atomic veterans and nuclear weapons complex workers) that require a dose assessment and evaluation of its uncertainty. These calculations are used to

Table 4.20 Decorporation therapy recommendations for selected radionuclides[a),b)].

Radionuclide	Possible treatments	Preferred treatment
Antimony	British anti-Lewisite (BAL) Penicillamine	BAL
Cerium	DTPA	DTPA
Cobalt	Dimercaptosuccinic acid (DMSA) DTPA Ethylenediaminetetraacetic acid (EDTA) *N*-Acetyl-l-cysteine (NAC)	DTPA
Mixed fission products	Management depends on predominant radionuclides present at the time of interest (e.g., early: iodine; late: strontium, cesium, and others)	Depends on time of interest
Manganese	Deferoxamine (DFOA) DTPA EDTA	DTPA
Potassium	Diuretics	Diuretics
Technetium	Potassium perchlorate	Potassium perchlorate
Tritium	Force fluids	Water diuresis
Zinc	DTPA EDTA Zinc sulfate as a diluting agent	DTPA
Zirconium	DTPA EDTA	DTPA

a) NCRP 161 (2008).
b) See Table 4.19 for FDA-approved treatments.

determine the probability of causation of a radiation-induced disease. Radiation therapy or diagnostic procedures often require an assessment of the administered activity to minimize unnecessary patient dose and ensure that the administered activity achieves the desired medical objective.

In applied radiation protection, the need for the evaluation of uncertainties in internal dose estimates is more limited. Dose limit recommendations of the ICRP are based on values of dose per unit intake that are usually applied without any consideration of uncertainty. This approach meets regulatory requirements.

In many cases, the evaluation of uncertainties requires analysis and interpretation of incomplete data and other supporting information. It necessarily relies on professional judgment, which is inherently subjective.

For a terrorist event, a variety of models and methods will be employed to reconstruct public doses. These calculations are initially utilized to guide medical treatment. Following the initial accident phase, internal dose calculations and their associated uncertainty are utilized in personnel compensation programs and litigation directed at resolving claims of affected members of the public.

4.5.3.2.1
Internal Dose Methodology

An overview of the general methods of performing internal dose calculations is provided in Appendix D. These methods are used to define models that form the basis for calculations determining the internal dose following an RDD or IND event. Bioassay measurements provide a key input to these models.

Internal dose calculations often utilize computational models. A summary of selected computer codes is provided in Appendix E. These models address a number of aspects of the internal dose calculation that include:

1) Determination of the intake or the amount of radioactive material entering the body
2) Assessment of the uptake or the fraction of the intake that is absorbed into the body fluids
3) Assessment of the fraction of the uptake transferred to particular organs/tissues and the subsequent behavior of the radioactive material in these structures. This assessment aspect involves biokinetic models that facilitate the calculation of the time-dependent activity in each source organ or transfer compartment per unit activity absorbed into the body fluids.
4) Determination of the absorbed dose per unit decay in each source organ or transfer compartment.

When measurements are available, some of the aspects listed above need not be included in the calculation. The immediate availability of a measurement technique depends on the terrorist event type and the extent of damage caused by its deployment. For an RDD, much of the area infrastructure near the blast site will remain intact and a variety of bioassay tools will be available to estimate the internal dose. If an IND is detonated, the local resources available for internal dose assessment could be severely limited.

4.5.3.2.2
Types and Categories of Uncertainties

Internal dose calculations have an inherent uncertainty. The sources of uncertainty are broadly divided into two categories. Uncertainties introduced by the bioassay or environmental measurements used to determine the activity of a radionuclide in the human body or in an environmental media compromise the first category. The second category includes uncertainties in parameter values and the biokinetic and dosimetric models used in the internal dose calculation.

Uncertainties in measurements used in the internal dose assessment arise principally from calibration methods in which the response of the detector is determined in a well-characterized radiation field, and then a measurement is made with the detector in an unknown field. For example, *in vivo* measurement uncertainty arises from variations in (i) counting statistics, (ii) detector positioning, (iii) background count rate, (iv) body dimensions, (v) overlaying structures, and (vi) activity distribution. The detector calibration and spectrum evaluation introduce additional uncertainty in the determination of internal dose.

Intakes are chronic or acute. In a radiological terrorist event, they can occur via inhalation, ingestion, skin absorption, and through wounds contaminated with radioactive material. The particle size of debris containing the radioactive material depends on the type of event and the magnitude of its explosive component.

Immediately following the detonation of an RDD or IND, individuals could receive an airborne intake of radioactive material. Material is also resuspended during an evacuation. Each of these pathways has inherent uncertainty and associated differences in particle sizes, chemical forms, and retention characteristics.

Internal intakes also occur following evacuation from the affected area. These intakes result following the consumption of contaminated food and water from the affected area. The extent and magnitude of the contamination depends on the severity of the event and the availability of food and water supplies. An indication of the complexity of the internal intake issue is illustrated by examining intakes in a normal operating environment.

The inhalation pathway occurs predominantly in an occupational setting and to a lesser extent in an environmental venue. An important source of uncertainty is the physical and chemical form of the radionuclide. If the radionuclide is attached to an aerosol particle, the size distribution, shape, and density are additional sources of uncertainties. In environmental exposures, particle sizes are usually smaller and soluble forms of radionuclides are more common than in an occupational setting. Other uncertainties include knowledge of the air concentration and characterization of the individual's breathing rate during the intake period.

Ingestion intakes are more common in an environmental setting than in occupational situations. In most environmental situations, bioassay measurements are not available and intakes are derived from the radionuclide concentrations in the consumed foodstuffs and their estimated consumption rates. The consumption rates vary with respect to the time of the year, region of the country, and degree of urbanization in the affected area. Only rough estimates of the radionuclide concentrations in foodstuffs and the corresponding intakes are possible because (i) there is a large variety of foodstuffs, (ii) the consumption rates of individual foodstuffs vary widely, (iii) radionuclide concentrations vary between foodstuffs, and (iv) radionuclide concentrations in foodstuffs vary with time.

Additional factors tend to increase the uncertainties in the calculated intake from food consumption. These factors include the (i) delay between harvesting, production, and consumption, (ii) activity loss during culinary preparation and cooking, and (iii) fraction of food consumed that is contaminated.

Intake uncertainties will also exist following a terrorist event. These uncertainties become more significant because greater activities of radioactive material are dispersed and normal control measures may be less rigorously enforced. Uncertainties in the intake are reflected in the calculated internal dose.

Doses from internal sources depend on the intake pathway, the activity transferred to blood, bioaccumulation of the radionuclide in the organs of the human body, and energy deposited in the organ of interest. Based upon available data, models are developed to predict the transfer and bioaccumulation of a radionuclide in a source organ and to estimate the energy deposited in a target organ. The current baseline internal dosimetry models are the ICRP 66 Human Respiratory Tract Model (HRTM) and the ICRP 100 Human Alimentary Tract Model (HATM). Uncertainties exist in the model structure and parameter values used in the HRTM and HATM. These models are described in Appendix D.

The main sources of uncertainty in the HRTM include the selection and modeling of the respiratory tract (e.g., splitting the alveolar interstitial region into three subdivisions), selection of model parameter values, intake parameters, and physiological parameters. Physiological parameters include the assumed breathing rates and transfer rates between the various model regions. Intake parameters include the size, shape, and solubility of the inhaled particle and its activity.

The selection of model parameters include the ventilation rate, fractional deposition in the various lung regions, absorption rates for the various types (F, M, and S), and rate of particle transport from the defined lung regions. The use of default absorption types instead of site- and material-specific values, selection of absorption and mechanical clearance rates, and fractional deposition values for the assumed lung regions are an additional source of HRTM uncertainty.

Similar uncertainties are associated with the HATM. These uncertainties include (i) modeling of the divisions of the alimentary tract, (ii) use of first-order linear differential equation kinetic models, (iii) selection of transit times between compartments, (iv) selection and modeling of anatomical features, (v) location of target regions for cancer induction including target depths and configurations, and (vi) selection of absorption coefficients.

4.5.3.2.3
Methods Used to Evaluate Bioassay Data and Associated Uncertainty

Least squares methods are usually used to obtain intake values from measurements of activity in bioassay samples. The methods typically assume a single intake, the biokinetic model and its parameters are credible, and all measurements are independent and representative of the model assumptions.

The intake relationships and their associated uncertainty are provided for three specific approaches: (i) uniform absolute error (unweighted least squares), (ii) ratio of the means, and (iii) average of the slopes. In the subsequent relationships, the following notation is used: I is the intake (maximum likelihood estimate), x_i is the set of bioassay measurements, i is the ith measurement, σ_i is the standard deviation of the set of measurements x_i, n is the number of measurements, R_i is the excretion or retention function at the time of measurement i, and ΔI is the intake variance.

The uniform absolute error method assumes that all measurements have the same variance ($\sigma_i^2 = \sigma^2$). With this assumption, the intake and its uncertainty are

$$I = \frac{\sum_{i=1}^{n} R_i x_i}{\sum_{i=1}^{n} R_i^2} \tag{4.10}$$

$$\Delta I = \left[\sigma^2 \left(\sum_{i=1}^{n} R_i^2\right)^{-1}\right]^{\frac{1}{2}} \tag{4.11}$$

The ratio of the mean error method assumes that the variance of the measurement is proportional to the magnitude of the expected value

$$\sigma_i^2 = kIR_i \tag{4.12}$$

where k is a constant. With this assumption, I and ΔI are given by

$$I = \frac{\sum_{i=1}^{n} x_i}{\sum_{i=1}^{n} R_i} \tag{4.13}$$

$$\Delta I = \left[kI \left(\sum_{i=1}^{n} R_i\right)^{-1}\right]^{\frac{1}{2}} \tag{4.14}$$

The average of the slope error method assumes that the variance of the measurement is proportional to the square of the expected value:

$$\sigma_i^2 = kI^2 R_i^2 \tag{4.15}$$

With this assumption, I and ΔI are

$$I = \frac{\sum_{i=1}^{n} \frac{x_i}{R_i}}{n} \tag{4.16}$$

$$\Delta I = \frac{I}{(kn)^{\frac{1}{2}}} \tag{4.17}$$

Although each of these methods can be used to determine the intake and its uncertainty, their listed order reflects their general usefulness and applicability. For example, the NRC in NUREG/CR-4884 *Interpretation of Bioassay Measurements* utilizes the uniform absolute error approach.

4.5.4 Emergency Screening of Contaminated Individuals

The methods and approaches used to screen contaminated individuals depend on their number and the contamination levels. In a nuclear facility, individual contamination events are managed using whole-body counting and hand-held survey instrumentation. For contaminated individuals associated with a terrorist event, other techniques must be utilized. The large number of contaminated individuals and the event environment govern these approaches. This section addresses screening aspects specifically related to initial response actions following a terrorist event involving the dispersal of radioactive materials.

4.5.4.1 External Contamination

In the event of a terrorist attack that involves a release of radioactive material, members of the public are screened for external contamination. If the release is associated with an explosion, individuals closest to the blast are most likely to be contaminated. These individuals are also the most likely to be injured.

Injured individuals may be taken to hospitals before the use of radioactive materials has been identified. If radioactive material is known or suspected to be associated with the explosion, individuals who are not injured will congregate at hospitals for decontamination. The flood of injured and individuals concerned with being contaminated could overwhelm the medical facility and delay treatment to the critically injured.

To prevent this onslaught, an emergency response plan for screening uninjured individuals must be available. This plan should be developed with public input and acceptance and be clearly communicated. Periodic public information sessions and continuous outreach are necessary to avoid public panic in the event of a terrorist event involving radioactive materials. A key element of an emergency response plan is direction of uninjured persons to preidentified screening sites. Stadiums, arenas, and locations with ample parking are reasonable candidates for contamination screening centers.

The screening centers should be stocked with the requisite decontamination equipment and supplies, and the emergency plan must identify staffing resources to implement timely public contamination monitoring. Screening equipment includes a variety of instrumentation (e.g., Geiger–Muller (GM) detectors, ion chambers, and portal monitors). Personnel trained to use this equipment must be available to staff the screening centers.

If emergency plans have not been developed or are not clearly communicated to the public, hospitals should establish a screening center. The center should be located outside the hospital, near a large parking area, to prevent uninjured individuals from overwhelming the facility.

Hospitals with a nuclear medicine department have portable radiation detection instrumentation and personnel qualified to operate the devices. These resources can be used for initial screening operations until additional emergency response personnel are mobilized and arrive at the designated screening locations.

Although GM detectors are the simplest and the most commonly available devices for contamination screening, they are not appropriate for all isotopes. GM detectors are not the optimum detector for low-energy beta emitters and low-energy photon emitters. Therefore, GM detectors should be used with caution. The choice of instrumentation and applicable radiation types and energies associated with potential RDD candidate radionuclides should be addressed in emergency response training. Without an emergency plan and appropriately trained personnel, a radiological terrorist event has the potential for significant societal disruption.

If screening facilities are unavailable or overcrowded, public address announcements should direct uninjured persons in proximity to the radioactive material release event to return to their homes. Upon returning home, individuals should remove and wash all clothing, shower and wash their hair, clean their shoes with a wet paper towel, and then report to an available screening facility. Individuals should bring their washed clothes and cleaned shoes to the screening facility for contamination monitoring.

These actions are important intermediate steps. Removal of clothing typically eliminates about 90% of the external contamination. Public cooperation is important because the treatment of injuries takes precedence over decontamination of individuals contaminated with low levels of radioactive material.

For those individuals entering the hospital, it is unlikely that any residual patient contamination constitutes a radiological hazard to medical personnel. The initial patient decontamination involves removal of clothing. If contamination is still detected, soap and water should be used to further decontaminate the patient. When soap and water does not remove all the radioactive material, there is the possibility that the contamination is internal. Since most internal contamination enters the body through inhalation and ingestion, the main areas of the body containing radioactive material are the mouth, nasal passages, chest, and abdomen. Internal contamination screening is addressed in the next section.

4.5.4.2
Internal Contamination

The radiation types that are most often encountered in internal depositions of radioactive materials include photons, beta particles, and alpha particles. The screening method for an internal deposition depends on the radiation types emitted by the materials deposited in the body. During the immediate aftermath of a terrorist event, the presence of radioactive materials or specific radionuclides may not be known.

Photon-emitting radionuclides are identified when portable spectrometers analyze on-site samples. Surface barrier detectors readily identify alpha-emitting radionuclides. Unless portable spectrometers are available, the identification of the specific low-energy beta- and alpha-emitting radionuclides normally awaits analysis by a laboratory located outside the area affected by the event.

Under normal circumstances, radioactive material enters the body through inhalation, ingestion, skin absorption, and puncture wounds. In a terrorist event contaminated shrapnel resulting from the explosion may be embedded in the body.

As with any projectile, radioactive shrapnel should be surgically removed as quickly as possible. The exact location of the material within the wound can be determined using computed tomography, gamma cameras, or a diagnostic X-ray examination. Following removal, the shrapnel should be placed in a shielded container to minimize irradiation of hospital personnel.

Internal radionuclides emitting photons and high-energy beta particles are readily detected with *in vivo* whole- or partial-body counting techniques. Qualitative indications of an internal deposition of radioactive materials are derived from nasal swabs, and quantitative results are obtained through bioassay sample results (e.g., urine sampling and fecal sampling). The location of the radioactive material is determined from whole- or partial-body counting techniques if the internal radioactive material also emits photons. Decorporation is warranted if the CDG values are reached (see Section 4.5.3.1).

A number of alpha emitters (e.g., ^{241}Am, ^{239}Pu, ^{226}Ra, ^{234}U, ^{235}U, and ^{238}U) could be incorporated into a terrorist device. The energies of the emitted alpha particles are similar and travel <1 mm in tissue. Therefore, alpha particles that are deposited internally cannot be externally detected. These alpha emitters also emit low-energy photons that can be detected with sufficient preparation and care using spectrometers or *in vivo* counting, but these will not likely be available for an initial screening following the detonation of a terrorist device. Initial contamination screening will most likely use nasal swabs.

Pure beta emitters that could be utilized in a terrorist device include ^{3}H, ^{14}C, ^{32}P, ^{35}S, ^{90}Sr, and ^{90}Y. Low-energy beta emitters used in research or medical applications (e.g., ^{3}H, ^{14}C, and ^{35}S) would not be detected by most handheld survey instrumentation. Ion chambers that could detect low-energy beta-emitting radionuclides may not be initially available.

The detection of low-energy beta-emitting radionuclides typically requires bioassay with subsequent analysis by liquid scintillation counting. During the initial screening phase, these techniques may not be available. Ion chambers, spectrometers, and whole-body counting techniques detect the bremsstrahlung radiation that is emitted by high-energy beta emitters (e.g., ^{32}P and ^{90}Y). Handheld survey instruments can be utilized to detect high-energy bremsstrahlung radiation.

A terrorist device could utilize photon-emitting radionuclides. ^{60}Co, ^{125}I, ^{131}I, ^{137}Cs, and ^{192}Ir are candidate RDD photon-emitting radionuclides that may be identified by their characteristic spectrum and are readily detected using portable spectrometers (e.g., high purity germanium (HPGe)). ^{125}I is more challenging to detect when deposited internally since it has a low-energy photon, which is poorly detected except when it is localized in the thyroid.

Screening for an internal deposition of radioactive material is an important first step in patient treatment because it identifies at-risk individuals. Confirming the presence of radioactive material and identifying the radionuclides present is the second step. The third step is determining the activity of internal radionuclide contamination. Quantification of the activity deposited within the body is used by the physician to determine if the use of decorporation agents is advisable.

Some decorporation materials have limited side effects, but others have definite medical risks. Therefore, the consequences of using these materials are evaluated with respect to the radiation risk. It is not necessarily good medical practice to treat low levels of contamination since more harm could be associated with decorporation than the radiation dose. However, the physician determines the patient's treatment. The health physicist provides a radiological advisory function.

4.6
Mass Casualty Considerations

Mass casualty situations following a terrorist event involving radioactive materials result in a spectrum of injured individuals some of whom are contaminated. Life-threatening medical injuries are given priority because they represent the most serious risk. In most cases, the radioactive contamination is less serious than the physical injuries. The medical triage principle applies in assigning medical resources to contaminated, injured patients. The management decision protocol for addressing mass casualties resulting from radiological terrorism is outlined in Table 4.17.

Individuals that have less severe injuries should be screened for contamination while they are waiting for medical treatment. The procedures for screening individuals were previously addressed.

Medical personnel assigned to manage patients who have been contaminated with radioactive material, especially in a mass casualty situation, may not have had prior education, training, or experience in managing this situation. A mass casualty situation is further complicated because individuals may reach a medical facility before radioactive material has been detected or associated with the terrorist event. The screening approaches for the detection of external and internal contamination are an initial tool to sort contaminated from noncontaminated individuals.

4.7
Stakeholder Involvement

Stakeholders have an important role in the development of emergency response programs that address an RDD or IND event. Stakeholder input and participation strengthens the program, improves the acceptance of the response actions, and contributes to the effectiveness of associated public information programs. The success of these programs and improved stakeholder knowledge of the response to a terrorist event enhances protection of the public through facilitating appropriate initial actions during an IND or RDD emergency. Emergency response actions, including sheltering, have a positive radiological benefit, protect the public from external exposure, and minimize the internal dose following the initial event.

During any subsequent evacuation, public understanding of basic radiological terminology and the characteristics of radioactive material minimizes the likelihood of panic and facilitates an orderly evacuation from the area of the initial RDD or IND detonation. Public information programs address aspects of an evacuation, and this information would partially ameliorate the stress and confusion encountered during the implementation of an evacuation protective action.

Stakeholders have significant involvement in the recovery phase. Important decisions regarding reentry criteria and acceptable contamination and radiation levels should have public input and acceptance. The experience and lessons gained during the recovery phase of the Fukushima Daiichi accident form the basis for a portion of public information programs.

Stakeholder involvement is not easily managed, and emergency response personnel could face significant public opposition and anger in the aftermath of a terrorist event. This is particularly important when stakeholder neighborhoods are significantly affected by nuclear terrorism. The probability of a negative reaction is enhanced if the public is not knowledgeable of radiological issues and has only an emotional response to the terrorist event. These conditions can be partially ameliorated through effective public information programs and stakeholder participation that address the various possible events, their likely conditions, and associated response activities.

4.8 Contamination Remediation

Contamination of large areas could result from an RDD or IND detonation. At least a portion of these areas will be reentered and require remediation to return them to unrestricted access. Although numerous isotopes could be present, one of the limiting isotopes is ^{137}Cs. The importance of ^{137}Cs was demonstrated following the Fukushima Daiichi accident with the associated contamination of soil, water, fish, vegetables, and animal products.

NCRP 154 provides a comprehensive summary of radiocesium, its physical and chemical properties, and important parameters that affect its influence on the environment. Much of the subsequent discussion is derived from that publication.

Operation of a number of nuclear facilities also demonstrates that ^{137}Cs often is the limiting long-term contributor to the environmental radiation dose received by humans and other organisms. Over the past few decades, ^{137}Cs has been one of the most important residual radionuclides in a number of DOE weapon's complex facilities (e.g., Savannah River Site, Oak Ridge Reservation, and Hanford Site), at fuel reprocessing facilities, at many waste disposal facilities, in soils due to weapons test fallout, and in areas affected by the Chernobyl and Fukushima Daiichi accident. Recent concerns also involve the use of ^{137}Cs in a terrorist weapon such as an RDD or dirty bomb.

Table 4.21 Estimates of ^{137}Cs released to the environment from human activities.

Source	Release activity (PBq)
Global atmospheric nuclear weapons testing	948[a)]
Chernobyl accident	70[a)]
Nuclear power production (fuel reprocessing)	40[a)]
Fukushima Daiichi accident	10–20[b)]
Goiania accident	0.05[a)]
Kyshtym accident	0.04[a)]
Windscale accident	0.02[a)]
Cosmos 954 accident	0.003[a)]

a) NCRP 154 (2007).
b) IAEA (2012a).

Radiocesium is produced primarily as a fission product in nuclear weapons' detonations and in nuclear reactors. ^{134}Cs, ^{135}Cs, and ^{137}Cs are the dominant radiocesium isotopes and could be incorporated into an RDD. Table 4.21 provides a summary of the major contributions to ^{137}Cs in the global radiation environment.

^{137}Cs has several characteristics (see Appendix A) that enhance its importance as a dominant contributor to radiation dose. In particular, ^{137}Cs has a 30.07-year half-life and a 0.514 MeV maximum energy beta particle, and its 2.552 min daughter ^{137m}Ba emits a 661.7 keV photon. Cesium is readily transported through the environment and food chain. When in solution, it is effectively absorbed into plants and assimilated by animals. It also has an affinity to attach to common clay minerals found in soils and sediments.

The environmental transport of cesium depends on a number of parameters. Cesium accumulation varies by orders of magnitude between different biological systems within a single environment and among different ecosystems. Much of the observed behavior follows from the chemical properties of cesium and its interaction with soil and sediment. An influential factor in these interactions is the clay mineral abundance. Other chemical characteristics that affect the transport of cesium are the soil or sediment cation exchange capacity and pH and the soluble potassium levels in soil.

Unlike most other radionuclides, the passage of radiocesium through animal food chains often increases from one species to the next food chain member. For example, predatory animals concentrate ^{137}Cs in their soft tissues to a higher degree than their prey.

A number of strategies for the remediation of areas contaminated with ^{137}Cs can be formulated, but the selected approach should incorporate stakeholder input. These strategies range from no action to engineered cleanup and restoration and depend on the levels of contamination and size of the contaminated area. Optimizing a strategy often involves a cost–benefit analysis that includes risk assessment methodology. The risk analysis includes an evaluation of the severity of human health and ecological impacts.

Numerous techniques are available to mitigate the consequences of an environmental radiocesium release. These approaches utilize microorganisms, plants, chemicals, and various soil constituents. Microorganisms alter the mobility of radiocesium, but these organisms do not remove cesium.

Phytoremediation includes both phytoextraction and phytostabilization. Phytoextraction is the concentration of contaminants into harvestable portions of plants. Phytostabilization is the use of plants to minimize off-site losses of contaminants through erosion and leaching. The defined cleanup goals, level of radionuclide contamination, depth of contamination penetration into soil, presence of other toxic materials, and site-specific climatic conditions influence the selection and successful utilization of these techniques.

Chemical remediation techniques are also available to mitigate the cesium contamination. Methods including fluid extraction, oxidation, and peroxide treatments are generally appropriate for volatile and semivolatile organic compounds but do not facilitate the remediation of radiocesium. Chemical immobilization or photodegradation methods have been applied with success for some contaminants. Heavy metal mobility is reduced by up to 80% with chemical treatments.

These techniques may not be necessary for cesium because it binds strongly with lattice-type clay materials. However, the use of fertilizer has been shown to have a significant effect on the accumulation of radiocesium in plants. Nitrogen fertilization enhances cesium uptake by increasing the plant growth rate. Fertilization with large quantities of potassium results in a reduction in cesium uptake. Ammonium sulfate fertilizers are used for growth stimulation and to displace the exchangeable fraction of cesium in the soil with the ammonium ion, which increases the uptake availability of cesium.

Another physical approach to cesium remediation is the application of illite-type 2 : 1 clays. This material effectively immobilizes a large fraction of biologically available radiocesium when time is allowed for equilibration.

In situ vitrification is a physical method for immobilizing radiocesium in soil. An electrical current between electrodes initiates soil melting at temperatures approaching 2000 °C. Upon cooling, a volume reduction is achieved and the contaminants are immobilized in the vitrified soil mass. However, vitrification is expensive, can cause ecological damage, and is impractical for large-scale operations.

Plowing reduces concentrations by mixing the upper 20–30 cm of soil. This method also reduces long-term resuspension, which limits the inhalation potential, and lowers the external ^{137}Cs exposure. Plowing must be carefully implemented, because it can cause resuspension and the further spread of contamination.

Other countermeasures for environmental releases of radiocesium include relocation of the population and limiting the ingestion of contaminated food. These interim measures are not universally acceptable to an evacuated population. The Fukushima Daiichi evacuation clearly illustrates the limited patience of an evacuated population and the strong desire to return to their homes and farms and resume normal lives.

Agricultural countermeasures can be effective but are generally appropriate for a limited time frame. These approaches include (i) removing contaminated lands from production, (ii) importing uncontaminated feed for livestock, (iii) deep plowing fields to reduce radionuclide concentrations in the soil layer in contact with plant roots, (iv) addition of potassium fertilizer to reduce the cesium uptake by crops, (v) choosing crops that take up less cesium, and (vi) adding Prussian blue to the diets of livestock which reduces the uptake of radiocesium from the digestive tract. These short-term countermeasures require implementation funding. Once again, stakeholder participation will significantly enhance the acceptance and implementation of these techniques.

Contaminated water and measures to protect water supplies affect people living in cities, small towns, and farms. Any measures to protect water supplies should incorporate shareholder input. These countermeasures include (i) regulating water flow through reservoirs, (ii) decreasing the use of surface water, (iii) increasing the use of groundwater supplies, (iv) adding purification steps to the treatment of drinking water, (v) eliminating the use of contaminated water sources, (vi) restricting the use of water from lakes and reservoirs, (vii) restricting fishing, and (viii) limiting the use of water for other traditional purposes such as irrigation. Each of these approaches affects various stakeholder groups and should address their concerns prior to implementation.

If an urban environment is contaminated, then additional decontamination methods are available. Decontamination methods include washing buildings, cleaning residential areas, removing contaminated soil, and washing roads. The efficiency of these various methods depends on the building design and construction, type of radionuclide deposition (wet or dry), physicochemical composition of the fallout, and time postevent. Following the Fukushima Daiichi accident, the public was instrumental in participating in the decontamination of their personal residences and neighborhoods. Volunteer efforts will likely be needed for the implementation of any long-term remediation effort to be successful.

Decisions on methods and priorities for addressing contaminated areas are complex and governed by a numerous considerations. The extent of the contamination, levels of contamination, future use of the contaminated area, cleanup costs, impact on human health, and ecological risk reduction are a few of the considerations that are part of formulating a cleanup strategy. However, the technical approach is only a portion of the solution. Stakeholder involvement is an important consideration that should not be overlooked.

Problems

4.1 A radiological dispersal device utilizing ^{239}Pu is being transported to Seattle by a terrorist group. Near the Hanford Site, its transport vehicle collides with a gasoline tanker. The resulting accident causes device detonation, contamination of the surrounding land, and a massive brush fire.

You work for the United Nuclear Environmental Services, a Hanford contractor, and have responded to the fire. The following questions are related to this event and associated health physics actions.

Data:

Filter alpha self-absorption = 0.4 (i.e., 60% of the alphas are absorbed in the filter)

Filter collection efficiency = 0.8

Detector active area = 60 cm^2

Background count: 180 counts in 60 min

First sample count: 500 counts in 10 min

Second sample count (1 h later): 360 counts in 10 min

Detector efficiency for alpha particles assuming a uniform distribution over the detector area = 0.3 cpm/dpm

Active filter area = 500 cm^2

^{239}Pu Type M effective dose conversion factor per unit inhalation intake = 4.7×10^{-5} Sv/Bq

Breathing rate = 1.2 m^3/h

Effective half-life for radon (^{222}Rn) progeny = 30 min

(a) You take a 1 m^3 air sample at the downwind location. Calculate the ^{239}Pu airborne activity in Bq/m^3 correcting for the contribution from radon (^{222}Rn) progeny. Assume no thoron progeny are present and neglect decay correction during counting.

(b) Calculate the lower limit of detection (LLD) for this counting system in cpm.

(c) Calculate the committed effective dose to a person standing at the sampler location. Assume the release occurs over a period of 4 h and the average ^{239}Pu activity concentration is 20 Bq/m^3.

(d) List five ways of improving the dose estimate for off-site individuals.

(e) List five possible methods to reduce the effective dose to individuals from brush fires or other high resuspension events.

4.2 An RDD was detected in a vehicle and you are assigned to provide radiological oversight of the team that is diffusing the device. You are using an open-air ionization chamber to perform surveys of the vehicle. This instrument is calibrated in conventional units of exposure rate (R/h). The ionization chamber was calibrated to give the correct response at a temperature of 0 °C and at an atmospheric pressure of 760 mm of Hg. A portable HPGe detector indicates the device incorporates ^{137}Cs and you estimate the total activity in the RDD is 3.7×10^7 MBq.

Data:

Detector active volume = 235.5 cm^3 (5 cm radius and 3 cm long).

Density of air = 1.29 kg/m^3.

At the time of the event, the temperature was 20 °C and the atmospheric pressure was 720 mm Hg.

Assume negligible humidity.

(a) What detector current is generated by an exposure rate of 1 R/h? Assume the detector is uniformly irradiated by the radiation field.

(b) A measurement of 12.6 R/h is obtained on a hot day of 35 °C and 740 mm Hg. Calculate the exposure rate at standard conditions. Assume the same current is produced in each case.

(c) The RDD detonates and personnel shelter behind a very thick wall that has a long 1 cm wide vertical crack extending through the wall thickness. You position the ionization chamber and use the crack to obtain a detector response of 20 mR/h when the active volume of the ion chamber is centered over the crack. Assuming that the crack length exceeds the dimensions of the ion chamber, provide an estimate of the true exposure rate. Assume electronic equilibrium and the following measurement conditions: 0 °C and 760 mm Hg. The detector is centered over the crack with the 3 cm length perpendicular to the plane of the crack.

4.3 You are the senior radiological controls technical advisor to the Director of Disaster Management in the Department of Homeland Security. The Director informs you that four separate dirty bombs were detonated within the last hour in Chicago, New York, San Francisco, and Washington, DC. The Chicago blast dispersed ^{32}P using a large truck bomb that detonated in the business district and contaminated several city blocks. Approximately 200 people are injured and another 150 presumed dead in the highly damaged area within 200 m of the detonation site.

In New York, a ^{60}Co RDD was prematurely detonated within the Holland Tunnel that has collapsed. The magnitude of the RDD is unknown but thefts of ^{60}Co sources having a cumulative activity of at least 500 TBq were reported in the New York area over the last 4 weeks. Contamination is limited to the interior of the tunnel and to the areas immediately outside the tunnel entrances. At the time of the blast, traffic was light. Based on tunnel cameras, 12 passenger cars, 3 light trucks, and 3 vans are buried in the collapsed area. The vehicles and people exiting the tunnel are contaminated, but the levels are well below the critical decision guide values.

The San Francisco detonation dispersed mixed fission products. The detonation was massive and occurred at the top of a tall parking structure in the center of the city. At least 30% of the city is contaminated.

Washington DC's detonation appears to be a failed IND. The device detonated near the White House, and ^{239}Pu contamination has been detected in numerous government buildings within a radius of 1 mile around the White House.

The Director requests your initial action plan in order to allocate resources. She has asked the following questions.

(a) If an average ^{32}P contamination level of 250 dpm/100 cm^2 has been measured, what initial actions should be taken in Chicago? The highest measured absorbed dose rate is 0.05 mGy/h.

(b) Contamination levels outside the Holland Tunnel are below 100 dpm/100 cm^2. Initial tunnel entries find radiation levels below 0.03 mGy/h and contamination levels below 1000 dpm/100 cm^2. It appears that 250 ft of the tunnel are collapsed and that the highest radiation and contamination levels will be encountered within the collapsed section. Initial sound readings detect no movement, and very few people are expected to have survived. The debris is not easily removed and consists of rock, concrete, steel structural members, and reinforcing bars. What initial actions do you recommend?

(c) The radiation and contamination levels from the San Francisco blast indicate that a large quantity of fuel reprocessing waste including ^{90}Sr and ^{137}Cs is the source of the fission products in the RDD. Initial surveys suggest that radiation and contamination levels vary and about 10% of absorbed dose rates are in the 1–2 Sv/h range. The remaining dose rates are in the 5–200 mSv/h range. Since the RDD detonated in an elevated structure, few blast injuries occurred. No damage to structures other than the parking facility was reported. Most of the injuries resulted from falling debris, and all injured individuals and fatalities were recovered during the initial response actions of emergency personnel. What are the next actions that emergency response personnel should perform?

(d) The plutonium contamination levels are in the range of 1000–1 000 000 dpm/100 cm^2 with the highest levels near the White House. Government buildings have contamination levels in the range of 10–10 000 dpm/100 cm^2. Initial screening suggests that several hundred people have positive nasal smears and about 25% of people evacuated from the affected areas have contamination on skin or clothing. Approximately 250 serious injuries are reported within and outside the White House. What immediate emergency response actions are required?

4.4 A ^{137}Cs radiological dispersal device was activated on a hill overlooking Pittsburgh. The wind was blowing toward the city and remained in that direction for the next 24 h. The device was silent and slowly dispersed radioactive material that reached the city. This material was not detected because the city's limited radiation monitoring system was in a scheduled outage to upgrade its capability. The subsequent investigation revealed that the dispersal device functioned over an 8 h period and produced a constant release rate.

Measurable concentrations of deposited ^{137}Cs were detected at an outdoor concert area located 2 km from the RDD location. The concert started at 4 p.m. and lasted until 3 a.m. the following day. Most people attended the entire event.

Data:

The measured ^{137}Cs deposition on the soil in the concert area = 518 kBq/m^2.

Wind speed = 5 m/s in the direction of Pittsburgh.

During the entire event, the meteorology was Pasquill stability class C characterized by the following dispersion coefficients at the concert site: σ_y = 205 m and σ_z = 120 m.

Nominal ^{137}Cs deposition velocity = 0.002 m/s.

Effective release height = 40 m.

Breathing rate = 0.8 m^3/h.

^{137}Cs (Type F) effective dose conversion factor for the released particle size = 4.6×10^{-9} Sv/Bq.

The ^{137}Cs half-life is 30.07 year.

(a) What is the ^{137}Cs inhalation intake to a person present at the concert during the 8 h time that the air was contaminated?

(b) What effective dose was delivered to the individuals attending the entire concert?

(c) Assume that the calculated 8-h average air concentration at the park site is 11.1 kBq/m^3. On that basis, how much ^{137}Cs was released from the terrorist device? Assume the park is on the plume centerline and the air concentration is the ground-level value.

(d) What additional information would confirm the release scenario and reduce the uncertainty in the release estimate?

(e) A system incorporating a high-purity germanium detector is procured to count environmental samples taken as a result of this incident. List five tasks that should be routinely performed to ensure the quality of the counting system measurements.

4.5 You are the Radiation Protection Manager at the Point Gravel Nuclear Power Plant (PGNPP), which includes two Generation II boiling water reactors. The PGNPP site is on the shore of Lake Huron in Michigan. Both units are operating at full power and all plant conditions are nominal.

Lake Huron has experienced a significant increase in traffic since the discovery of large natural gas reserves. Each day, numerous liquefied natural gas tankers pass within a mile of the PGNPP. Earlier in the day, a group of terrorists seized several of these tankers and their course has been diverted to the plant. These tankers have now collided with plant structures that house PGNPP safety equipment.

These collisions damage the tankers and cause the liquefied gas to vaporize. Boiling liquid expanding vapor explosions (BLEVEs) occur which severely damage reactor buildings, service buildings, turbine buildings, and the electrical switchyard. The explosion of the first tanker causes all off-site power to be lost.

With the loss of all off-site power, both reactors trip. The station emergency diesel generators start and power safety equipment and the cores are being cooled in a normal manner. Upon the explosion of the second tanker, the emergency diesel generators and DC battery systems are disabled and cannot be repaired, and the site's normal and emergency electrical distribution systems are severely damaged. The second explosion also destroys the facility structure that houses the backup power supplies, core cooling pumps, and supporting equipment mandated by the response to the Fukushima Daiichi accident. A station blackout condition now exists at the facility.

Explosions of the third and fourth tanker disables all systems designed to provide emergency core cooling. Following the fourth explosion, all core cooling has been lost and core temperatures rapidly increase. Operators have no capability to cool the reactor cores, and off-site power cannot be restored for days. In addition, these explosions killed about 60% of station personnel and another 20% are severely injured. The ability of facility personnel to respond to the emergency has been severely diminished.

Station operators are forced to vent hydrogen to the reactor buildings, and both units experience Fukushima Daiichi-type hydrogen explosions that damage both spent fuel pools, which are now leaking. At the present time, both cores have melted and all fission product barriers have failed. The fuel in the spent fuel pool has been damaged by hydrogen explosion debris. A local fire department is attempting to add water to the spent fuel pools. Efforts to cool the reactor cores have been unsuccessful due to damage caused by the BLEVEs. A number of facility structures are burning, and the fires are not under control.

Data:

Weather forecast: Winds are 1 – 10 km/h in variable directions.

Initial dose assessment summary:

Distance from the PGNPP (km)	Projected effective dose (mSv)	Projected child thyroid equivalent dose (mSv)
1	500	2000
3	400	1600
5	350	1400
8	250	1000
11	150	600
14	100	400
16	80	300
32	10	50
80	2	8

(a) The PGNPP Radiation Protection Manager has been assigned to the off-site Emergency Operations Facility. The station's Emergency Director will brief the Michigan governor and wants an assessment of what radionuclides are being released to the environment. What radionuclides will likely be released?

(b) A preliminary dose projection has been performed. The wind is blowing in the direction of a population center located 8 km from the facility. Based on the initial dose assessment summary, what emergency classification should be declared?

(c) What protective action recommendations should be provided to the state of Michigan?

(d) The Engineering Director has developed a plan to restore core cooling but will require emergency team members to receive 1000 mGy absorbed dose and 6000 mGy thyroid equivalent dose. These doses are based on estimated dose rates and air concentrations. Volunteers for the team include a declared pregnant worker in her 10th week of gestation, a 30-year-old female, a 30-year-old male, a 50-year-old female, a 55-year-old male, and a 60-year-old male. All are qualified to perform the task. The Maintenance Manager asks for your advice in forming the four-person team. What is your advice?

(e) You are asked to provide a radiological briefing to the repair team. What topics do you address? What mitigating actions do you take?

(f) The repair team has been successful and completed its tasks. Their dosimetry indicates an average of 800 mGy absorbed dose. Whole-body counting suggests an average thyroid dose of 400 mGy. Why are these doses inconsistent with the initial dose estimates?

(g) A physician asks you to assist him in addressing the emergency team to discuss any anticipated health effects. What information do you provide to the physician?

(h) With the termination of the release, the state is concerned about minimizing the economic impact of the accident. Should crop use be restricted?

(i) What recommendations should be made regarding farm animals?

(j) What criteria should be used for future land use?

4.6 A nuclear weapon has been detonated in Los Angeles near the center of the city. The yield of the weapon has not yet been determined. Based on the damaged area, it appears to significantly exceed the 10 kT planning basis incorporated into the Los Angeles emergency response planning documents. You have been designated by the Homeland Security Secretary (HSS) to be the Radiological Control Director for all radiological aspects of the Los Angeles rescue and recovery effort.

(a) A group of city employees are in an underground shelter, and their remote radiation detection system measured an absorbed dose rate of 1 Gy/h 1 day postdetonation on the surface above their location. Based on HPGe spectra, the radiation is predominantly attributed to fallout from the detonation. All radiation detection instrumentation failed shortly after these measurements. The group has food and water reserves for 4 months, and the dose rates within the shielded shelter are 0.3 μGy/h. The HSS has directed that no rescue operations be initiated until emergency response team members receive no more than 0.1 Gy. The rescue operation will take 24 h to complete and requires excavation and clearing bomb debris to reach the trapped city workers. How long must the team be delayed before initiating emergency recovery operations? Assume that fallout is the only significant radiological source term.

(b) Dr Elmo Ka-Boom, the HHS's Chief Nuclear Physicist assigned to your group, determined that the detonation was a surface burst with a yield of 100 kT. At what downwind distance from the detonation site will the postdetonation absorbed dose rate fall below 0.1 Gy/h? Assume a 24 km/h wind speed and constant meteorological conditions.

(c) How does the downwind distance calculated in the previous question change for a wind speed of 50 km/h?

(d) The emergency planning basis was derived from the assumed 10 kT yield. How does the actual yield of 100 kT detonation affect the emergency planning basis?

4.7 You are a senior health physicist with Special Operations Associates, Inc., a security firm that provides consultation services associated with radiological terrorism. The mayor of Paris requested an assessment of the city's vulnerability to a terrorist attack, and you have been assigned to provide general information to her radiological safety committee. The committee raised the following questions and operational situations, and you are to address these issues during an upcoming meeting.

(a) What radionuclides could be utilized in a terrorist attack? What are the typical source activities? What are the conventional uses for these sources?

(b) The committee is concerned with the city's monitoring capability to ensure detection of radionuclide intakes. What methods are utilized to detect photon emitters incorporated in the body, and what detectors are utilized to optimize detection?

(c) You have been asked to develop a set of drill and exercise scenarios. Part of this task is the development of spectra including various radionuclides. What radionuclides correspond to the gamma-ray peaks noted in the following Drill-02A Spectrum: 0.0136, 0.0711, 0.140, 0.186, 0.316, 0.364, 0.662, 1.17, and 1.33 MeV?

(d) For each of the radionuclides noted in the previous question, list the recommended bioassay sample type and analysis method for assessing an internal deposition of that radioactive material.

(e) During response to a suspected dirty bomb, a number of injured individuals have positive nasal smears. The first responders suspect the weapon contained ^{241}Am. What are the radiological characteristics of ^{241}Am and what are the options for detecting its intake?

(f) What records including types of information should be maintained for all medical and emergency response personnel involved in a radiological terrorist event?

4.8 A radiological dispersal device was detonated during the Rose Bowl Parade in Pasadena, California. The radioactive material was prepared in a form that maximized the production of respirable particles. ^{60}Co has been detected in large quantities at the detonation site and hundreds of individuals have positive nasal smears including a number of victims with positive alpha counts. The Department of Homeland Security expedited the sample analysis and determined that the alpha-emitting radioactive material is ^{210}Po. The only available bioassay technique is urine sampling. Related attacks destroyed city facilities that housed its whole-body counters.

You are the Radiation Safety Officer at the University of Sunny California and have been requested to provide health physics consultation and perform an initial dose assessment for a severely contaminated, injured victim who was near the explosion. The individual is a 20-year-old pregnant female who is in labor.

Data:

^{60}Co urine bioassay results:

Time post intake (d)	Intake retention fraction	Activity (MBq)
1	2.59×10^{-2}	1.05
5	2.26×10^{-3}	0.12
10	1.27×10^{-3}	0.065

^{210}Po urine bioassay results:

Time post intake (d)	Intake retention fraction	Activity (kBq)
1	4.67×10^{-4}	0.15
5	6.34×10^{-4}	0.22
10	5.78×10^{-4}	0.20

Adult clinical decision guide values:
  ^{60}Co: 1.5×10^{7} Bq
  ^{210}Po: 1.1×10^{5} Bq

Child clinical decision guide values:
  ^{60}Co: 3.0×10^{6} Bq
  ^{210}Po: 2.2×10^{4} Bq

Applicable inhalation effective dose conversion factors for ^{60}Co:
  Adult: 1.0×10^{-8} Sv/Bq
  Infant: 4.2×10^{-8} Sv/Bq

Applicable inhalation effective dose conversion factors for ^{210}Po:
  Adult: 3.3×10^{-6} Sv/Bq
  Infant: 1.5×10^{-5} Sv/Bq

Half-lives for ^{60}Co:
  Physical: 5.27 year
  Biological: 40 days (based on an initial estimate for the victim)

Half-lives for ^{210}Po:
  Physical: 138 days
  Biological: 10 days (based on an initial estimate for the victim)

(a) What is the likely source of the ^{60}Co and ^{210}Po?
(b) What is the injured woman's intake of these radioactive materials? Base your results on the ^{60}Co and ^{210}Po urine sample results at 1, 5, and 10 days post exposure.
(c) What is the effective dose to the adult from the internal intakes?
(d) The physician is concerned that the fetus will receive radioactive material from the mother. He has decided to accelerate the delivery. Will this action limit the fetal dose?
(e) What effective dose is likely to be delivered to the victim during the first 10 days post intake? The physician will use your answer as part of his decision to implement decorporation therapy.
(f) What effective dose is likely to be delivered to the infant from the intake? A whole-body count of the premature infant is performed shortly after his birth. The results suggest ^{60}Co and ^{210}Po intakes of 5.0 MBq and 38 kBq, respectively.
(g) The physician requests that you provide specific decorporation agents that are appropriate for the adult's intake. What compounds are appropriate for this situation?

References

10CFR37 (2015) Physical Protection of Category 1 and Category 2 Quantities of Radioactive Material, National Archives and Records Administration, U. S. Government Printing Office, Washington, DC.

Ahearne, J.F. (2010) Radiological terrorism first responders and communicating risk. *Bridge: Natl. Acad. Eng.*, **20**, 45.

American College of Radiology Report (2006) Disaster preparedness for radiology professionals response to radiological terrorism, in *A Primer for Radiologists, Radiation Oncologists and Medical Physicists*, Government Version 3.0, American College of Radiology, Reston, VA.

Armed Forces Radiobiology Research Institute (2003) *Medical Management of Radiological Casualties*, 2nd edn, Armed Forces Radiobiology Research Institute, Bethesda, MD.

Baldini, E. (2010) Preventing the theft of dangerous radiological materials. *Police Chief*, 77, 30, http://www.policechiefmagazine.org/magazine/index.cfm?fuseaction=display_arch&article_id=2199&issue_id=92010.

Baum, E.M., Ernesti, M.C., Knox, H.D., Miller, T.R., and Watson, A.M. (2009) *Nuclides and Isotopes, Chart of the Nuclides*, 17th edn, Knolls Atomic Power Laboratory, Lockheed Martin, Schenectady, NY.

Becker, S.M. (2004) Emergency communication and information issues in terrorism events involving radioactive materials. *Biosecur. Bioterror.*, **2** (3), 195.

Becker, S.M. (2005) Addressing the psychosocial and communication challenges posed by radiological/nuclear terrorism: key developments since NCRP Report No. 138. *Health Phys.*, **89** (5), 521.

Benjamin, G.C. (2010) Medical preparedness and response to nuclear terrorism. *Bridge: Natl. Acad. Eng.*, **20**, 39.

Bevelacqua, J.J. (2003) Production equations in health physics. *Radiat. Prot. Manage.*, **20** (6), 9.

Bevelacqua, J.J. (2004a) Skin contaminations, internal intakes, and ALARA. *Radiat. Prot. Manage.*, **21** (1), 11.

Bevelacqua, J.J. (2004b) Point source approximations in health physics. *Radiat. Prot. Manage.*, **21** (5), 9.

Bevelacqua, J.J. (2005) Internal dosimetry primer. *Radiat. Prot. Manage.*, **22** (5), 7.

Bevelacqua, J.J. (2008) *Health Physics in the 21st Century*, Wiley-VCH Verlag GmbH, Weinheim.

Bevelacqua, J.J. (2009) *Contemporary Health Physics, Problems and Solutions*, 2nd edn, Wiley-VCH Verlag GmbH, Weinheim.

Bevelacqua, J.J. (2010a) *Basic Health Physics, Problems and Solutions*, 2nd edn, Wiley-VCH Verlag GmbH, Weinheim.

Bevelacqua, J.J. (2010b) Radiological dispersion devices and basic radiation science. *Phys. Teach.*, **48**, 302.

Bevelacqua, J.J. (2010c) Practical and effective ALARA. *Health Phys.*, **98** (Suppl. 2), S39.

Bevelacqua, J.J. (2011) Conservatism in effective dose calculations for accident events involving fuel reprocessing waste tanks. *Health Phys.*, **101**, 48.

Bevelacqua, J.J. (2013a) Nuclear regulation in the United States and a possible framework for an international regulatory approach. *Int. Nucl. Saf. J.*, **2** (1), 1, http://nuclearsafety.info/international-nuclear-safety-journal/index.php/INSJ/article/view/12/pdf_1 (accessed 04 August 2015).

Bevelacqua, J.J. (2013b) Methodology to ascertain core recriticality following a severe fission reactor accident. *Int. Nucl. Saf. J.*, **2** (1), 52, http://nuclearsafety.info/international-nuclear-safety-journal/index.php/INSJ/article/view/13/pdf_1 (accessed 04 August 2015).

Bevelacqua, J.J. (2014a) Health physics aspects of generation IV reactors. *Int. Nucl. Saf. J.*, **3** (1), 13, http://nuclearsafety.info/international-nuclear-safety-journal/index.php/INSJ/article/view/20/pdf_1 (accessed 04 August 2015).

Bevelacqua, J.J. (2014b) Radiation protection considerations for nuclear power expansion to nations with limited technical infrastructure. *Int. Nucl. Saf. J.*, **3** (3), 8, http://nuclearsafety.info/international-nuclear-safety-journal/index.php/INSJ/article/view/37/pdf (accessed 04 August 2015).

Bromet, E.J. (2014) Emotional consequences of nuclear power plant disasters. *Health Phys.*, **106** (2), 206.

Brandt, L.D. (2009) Mitigation of Nuclear Fallout Risks Through Sheltering and Evacuation. Report SAND2009-7367C, November 18, 2009, Sandia National Laboratories, Albuquerque, NM.

Brandt, L.D. and Yoshimura, A.S. (2009) Analysis of Sheltering and Evacuation Strategies for an Urban Nuclear Detonation Scenario. Report SAND2009-3299, June 2009, Sandia National Laboratories, Albuquerque, NM.

Brodsky, A. (2014) *Handbook for Survival: Saving Lives During Radiation Releases and Other Disasters*, Fideli Publishing, Inc., Martinsville, IN.

Brodsky, A., Johnson, R.H. Jr., and Goans, R.E. (eds) (2004) *Public Protection from Nuclear, Chemical and Biological Terrorism*, Medical Physics Publishing, Madison, WI.

Buddemeier, B. (2010a) Reducing the consequences of a nuclear detonation recent research. *Bridge: Natl. Acad. Eng.*, **20**, 28.

Buddemeier, B.R. (2010b) Reducing the Consequences of a Nuclear Detonation: Recent Research. LLNL-JRNL-427932, Lawrence Livermore National Laboratory, Berkeley, CA.

Buddemeier, B.R. and Dillon, M.B. (2009) Key Response Planning Factors for the Aftermath of Nuclear Terrorism. LLNL-TR-410067, August 2009, Lawrence Livermore National Laboratory, Berkeley, CA.

Burchfield, L.A. (2009) *Radiation Safety Protection and Management for Homeland Security and Emergency Response*, John Wiley & Sons, Inc., Hoboken, NJ.

Department of Homeland Security (2006) Preparedness directorate; protective action guides for radiological dispersal device (RDD) and improvised nuclear device (IND) incidents. *Fed. Regist.*, **71** (1), 174 Tuesday, January 3, 2006 Notices.

Department of Homeland Security (2008) National Response Framework. Friday, August 1, 2008 Notices. January 2008, http://www.fema.gov/pdf/emergency/nrf/nrf-core.pdf (accessed 04 August 2015).

Department of Homeland Security, Federal Emergency Management Agency (2008) Planning guidance for protection and recovery following radiological dispersal device (RDD) and improvised nuclear device (IND) incidents. *Fed. Regist.*, **73** (149), 45029 (Friday, August 1, 2008 Notices).

Egorov, N.N., Novikov, V.M., Parker, F.L., and Popov, V.K. (2000) *The Radiation Legacy of the Soviet Nuclear Complex*, Earthscan, London.

FDA (1998) Guidance on accidental radioactive contamination of human food and animal feeds: recommendations for state and local agencies; availability. *Fed. Regist.*, **63** (156), 43402–43403.

FEMA Program Manual (2013) *Radiological Emergency Preparedness*, Federal Emergency Management Agency, Washington, DC.

Fetter, S., Frolov, V.A., Miller, M., Mozely, R., Prilutskii, O.F., Rodionov, S.N., and Sagdeev, R.Z. (1990) Detecting nuclear warheads. *Sci. Global Secur.*, **1** (3-4), 225.

Garwin, R.L. (2010) A nuclear explosion in a city or an attack on a nuclear reactor. *Bridge: Natl. Acad. Eng.*, **20**, 20.

Glaser, A., Barak, B., and Goldston, R.J. (2014) A zero-knowledge protocol for nuclear warhead verification. *Nature*, **510**, 497.

Glasstone, S. and Dolan, P.J. (1977) *The Effects of Nuclear Weapons*, 3rd edn, US Department of Defense and US Department of Energy, Washington, DC.

Glasstone, S. and Lovberg, R.H. (1960) *Controlled Thermonuclear Reactions: An Introduction to Theory and Experiment*, D. Van Nostrand Company, Inc., Princeton, NJ.

Goans, R.E. and Wasalenko, J.K. (2005) Medical management of radiological casualties. *Health Phys.*, **89**, 505.

Gollnick, D.A. (2011) *Basic Radiation Protection Technology*, 6th edn, Pacific Radiation Corporation, Altadena, CA.

Gusey, I., Guskova, A., and Metter, F.A. (2001) *Medical Management of Radiation Accidents*, 2nd edn, CRC Press, Boca Raton, FL.

Haphuriwat, N., Bier, V.M., and Willis, H.H. (2011) Deterring the smuggling of nuclear weapons in container freight through detection and retaliation. *Decis. Anal.*, **8** (2), 88.

Hecker, S.S., Lee, S.C., and Braun, C. (2010) North Korea's choice-bombs over electricity. *Bridge: Natl. Acad. Eng.*, **20**, 5.

Homeland Security Council (2009) *Planning Guidance for Response to a Nuclear Detonation*. Developed by the Homeland Security Council Interagency Policy Coordination Subcommittee for Preparedness & Response to Radiological and Nuclear Threats, Department of Homeland Security, Washington, DC.

Horton, D.K., Berkowitz, Z., and Kaye, W.E. (2003) Secondary contamination of ED personnel from hazardous materials events, 1995–2001. *Am. J. Emergency Med.*, **21**, 199.

Hubner, K.F. and Fry, S.A. (eds) (1980) *The Medical Basis for Radiation Accident Preparedness*, Elsevier/North-Holland, New York.

IAEA (1988) *The Radiological Accident in Goiania*, IAEA Publication STI/PUB/815, 1988, International Atomic Energy Agency, Vienna.

IAEA (2002) *Preparedness and Response for a Nuclear or Radiological Emergency*, IAEA Safety Standards Series No. GS-R-2, International Atomic Energy Agency, Vienna.

IAEA (2004a) *Protection Against Internal Fires and Explosions in the Design of Nuclear Power Plants Safety Guide*, IAEA Safety Standards Series No. NS-G-1.7, International Atomic Energy Agency, Vienna.

IAEA (2004b) *Protection Against Internal Hazards other than Fires and Explosions in the Design of Nuclear Power Plants Safety Guide*, IAEA Safety Standards Series No. NS-G-1.11, International Atomic Energy Agency, Vienna.

IAEA (2007) *Arrangements for Preparedness for a Nuclear or Radiological Emergency*, IAEA Safety Standards Series No. GS-G-2.1, International Atomic Energy Agency, Vienna.

IAEA (2011) *Criteria for Use in Preparedness and Response for a Nuclear or Radiological Emergency*, IAEA General Safety Guide GSG-2, International Atomic Energy Agency, Vienna.

IAEA (2012a) *Emergency Preparedness and Response Operations Manual for Incident and Emergency Communication*, IAEA Publication EPR–IEComm, International Atomic Energy Agency, Vienna.

IAEA Publication (2012b) *Communication with the Public in a Nuclear or Radiological Emergency*, International Atomic Energy Agency, Vienna.

ICRP (1977) *Recommendations of the International Commission on Radiological Protection*, ICRP Publication 26, Pergamon Press, Oxford.

ICRP (1991) *1990 Recommendations of the International Commission on Radiological Protection*, ICRP Publication 60, Pergamon Press, Oxford.

ICRP (1994) *Human Respiratory Tract Model for Radiological Protection*, ICRP Publication 66, Pergamon Press, Oxford.

ICRP (2006) *Human Alimentary Tract Model for Radiological Protection*, ICRP Publication 100, Elsevier, Amsterdam.

ICRP (2007) *The 2007 Recommendations of the International Commission on Radiological Protection*, ICRP Publication 103, Elsevier, Amsterdam.

Levanon, I. and Pernick, A. (1988) The inhalation hazard of radioactive fallout. *Health Phys.*, **54** (6), 645.

Maiello, M.L. and Groves, K.L. (2006) Resources for nuclear and radiation disaster response. *Nucl. News*, **49** (9), 29.

Mettler, F.A. and Upton, A.C. (1995) *Medical Effects of Ionizing Radiation*, 2nd edn, W.B. Saunders, Philadelphia, PA.

Minister of Community Safety and Correctional Services of Ontario (2009) The Province of Ontario's Nuclear Emergency Response Plan, Minister of Community Safety and Correctional Services of Ontario, Toronto.

Morral, A.R., Price, C.C., Ortiz, D.S., Wilson, B., LaTourrette, T., Mobley, B.W., McKay, S., and Willis, H.H. (2012) *Modeling Terrorist Risk for the Air Transportation System*, RAND Corporation, Santa Monica, CA, http://www.rand.org/content/dam/rand/pubs/monographs/2012/RAND_MG1241.pdf. (accessed 1 August 2013).

Musolino, S.V. and Harper, F.T. (2006) Emergency response guidance for the first 48 hours after the outdoor detonation of an explosive radiological dispersal device. *Health Phys.*, **90** (4), 377.

Musolino, S.V., Harper, F.T., Buddemeier, B., Brown, M., and Schlueck, R. (2013) Updated emergency response guidance for the first 48 h after the outdoor detonation of an explosive radiological dispersal device. *Health Phys.*, **105** (1), 65.

National Academy of Sciences (2014a) An All-of-Government Approach to Increase Resilience for International Chemical, Biological, Radiological, Nuclear, and Explosive (CBRNE) Events. National Academy of Sciences Report No. 18814, The National Academy Press, Washington, DC.

National Academy of Sciences (2014b) The Science of Responding to a Nuclear Reactor Accident: Summary of a Symposium. National Academy of Sciences Report No. 19002, The National Academy Press, Washington, DC.

National Research Council (2012a) Crisis Standards of Care: A Systems Framework for Catastrophic Disaster Response. National Research Council Report, Institute of Medicine, The National Academies Press, Washington, DC.

National Research Council (2012b) Disaster Resilience: A National Imperative. National Research Council Report, The National Academies Press, Washington, DC.

National Research Council (2013a) Performance Metrics for the Global Nuclear Detection Architecture. National Research Council Report, The National Academies Press, Washington, DC.

National Research Council (2013b) Public Response to Alerts and Warnings Using Social Media: Report of a Workshop on Current Knowledge and Research Gaps. National Research Council Report, The National Academies Press, Washington, DC.

NCRP (1987) Ionizing Radiation Exposure of the Population of the United States. NCRP Report 93, NCRP Publications, Bethesda, MD.

NCRP (1993) Limitation of Exposure to Ionizing Radiation. NCRP Report 116, NCRP Publications, Bethesda, MD.

NCRP (2001) Management of Terrorist Events Involving Radioactive Material. NCRP Report 138, NCRP Publications, Bethesda, MD.

NCRP (2004) Approaches to Risk Management in Remediation of Radioactively Contaminated Sites. NCRP Report 146, NCRP Publications, Bethesda, MD.

NCRP (2005) Key Elements of Preparing Emergency Responders for Nuclear and Radiological Terrorism. NCRP Commentary No. 19, NCRP Publications, Bethesda, MD.

NCRP (2007a) Cesium-137 in the Environment: Radioecology and Approaches to Assessment and Management. NCRP Report 154, NCRP Publications, Bethesda, MD.

NCRP (2007b) Uncertainties in the Measurement and Dosimetry of External Radiation. NCRP Report 158, NCRP Publications, Bethesda, MD.

NCRP (2008) Management of Persons Contaminated with Radionuclides. NCRP Report 161, NCRP Publications, Bethesda, MD.

NCRP (2009) Uncertainties in Internal Radiation Dose Assessment. NCRP Report 164, NCRP Publications, Bethesda, MD.

NCRP (2010a) Responding to a Radiological or Nuclear Terrorism Incident: A Guide for Decision Makers. NCRP Report 165, NCRP Publications, Bethesda, MD.

NCRP (2010b) Population Monitoring and Radionuclide Decorporation Following a Radiological or Nuclear Incident. NCRP Report 166, NCRP Publications, Bethesda, MD.

NCRP (2013) Preconceptual and Prenatal Radiation Exposure: Health Effects and Protective Guidance. NCRP Report 174, NCRP Publications, Bethesda, MD.

NCRP (2014) Decision Making for Late-Phase Recovery from Major Nuclear or Radiological Incidents. NCRP Report 175, NCRP Publications, Bethesda, MD.

NTI Report (2014) *2014 Nuclear Threat Initiative (NTI) Nuclear Materials Security Index, Building a Framework for Assurance, Accountability, and Action*, 2nd edn, Nuclear Threat Initiative, Washington, DC.

Orient, J.M. (2014) Fukushima and reflections on radiation as a terror weapon. *J. Am. Physicians Surg.*, **19**, 48.

Radzinsky, R. and Perkovich, G. (2010) Iran's nuclear program status, risks, and consequences. *Bridge: Natl. Acad. Eng.*, **20**, 13.

Ricks, R.C. and Fry, S.A. (eds) (1989) *The Medical Basis for Radiation Accident Preparedness II*, Elsevier/North-Holland, New York.

Rosoff, H. and von Winterfeldt, D. (2007) A risk and economic analysis of dirty bomb attacks on the Ports of Los Angeles and Long Beach. *Risk Anal.*, **27** (3), 533.

Schlosser, E. (2013) *Command and Control – Nuclear Weapons, the Damascus Accident, and the Illusion of Safety*, Penguin, New York.

Smith, D.S. and Stabin, M.G. (2012) Exposure rate constants and lead shielding values for over 1,100 radionuclides. *Health Phys.*, **102**, 271.

Takahashi, S. (2014) *Radiation Monitoring and Dose Estimation of the Fukushima Nuclear Accident*, Springer, Tokyo.

Tenforde, T.S., Schauer, D.A., Goans, R.E., Mettler, F.A. Jr., Pellmar, T.C., Poston, J.W. Sr., and Taylor, T.P. (2010) Health aspects of a nuclear or radiological attack. *Bridge: Natl. Acad. Eng.*, **20**, 50.

United States Government Accountability Office (2013) Nuclear Terrorism Response Plans-Major Cities Could Benefit from Federal Guidance on Responding to Nuclear and Radiological Attacks. GAO-13-736, United States Government Accountability Office, Washington, DC.

U. S. Department of Energy (2014) International Nuclear Safeguards Engagement Program. U. S. Department of Energy Report, National Nuclear Security Administration, US Department of Energy, Washington, DC.

U. S. Environmental Protection Agency (2013) *PAG Manual, Protective Action Guides and Planning Guidance For Radiological Incidents*, U.S. Environmental Protection Agency Draft for Interim Use and Public Comment, U. S. Environmental Protection Agency, Washington, DC.

U. S. Government Accountability Office (2014) Emergency Preparedness – Opportunities Exist to Strengthen Interagency Assessments and Accountability for Closing Capability Gaps. U. S. Government Accountability Office GAO-15-20, U. S. Government Printing Office, Washington, DC.

U. S. Nuclear Regulatory Commission (1979) Emergency Planning for Fuel Cycle Facilities and Plants Licensed Under 10 CFR Parts 50 and 70. U. S. Nuclear Regulatory

Commission Regulatory Guide 3.42, Rev. 1, U.S. Nuclear Regulatory Commission, Washington, DC.

U. S. Nuclear Regulatory Commission (1983) Emergency Planning for Research and Test Reactors. U. S. Nuclear Regulatory Commission Regulatory Guide 2.6, Rev. 1, U.S. Nuclear Regulatory Commission, Washington, DC.

U. S. Nuclear Regulatory Commission (1987) Interpretation of Bioassay Measurements. U. S. Nuclear Regulatory Commission Regulatory Guide NUREG/CR-4884 (BNL-NUREG-52063), U.S. Nuclear Regulatory Commission, Washington, DC.

U. S. Nuclear Regulatory Commission (2005) Emergency Planning and Preparedness for Nuclear Power Reactors. U. S. Nuclear Regulatory Commission Regulatory Guide 1.101, Rev. 5, U.S. Nuclear Regulatory Commission, Washington, DC.

Yoshimura, A.S. and Brandt, L.D. (2009) *Analysis of Sheltering and Evacuation Strategies for an Urban Nuclear Detonation Scenario*, SAND2009-3299, Sandia National Laboratories, Albuquerque, NM.

Part IV
Nuclear Medicine and Public Health

The public receives most of its radiation exposure from natural background sources and nuclear medicine procedures. However, the public often does not recognize that it receives this exposure. When recognized by the public, these radiation doses are more acceptable than radiation received from other sources (e.g., nuclear power plants). Occasionally, public interest is aroused by new technologies. The use of airport security scanners raised public concerns for doses that are well below natural background levels. The public also receives unanticipated exposure from the inadvertent entry of radioactive materials into consumer products typically through scrap metal recycling.

Part IV reviews existing, evolving, and emerging medical procedures that utilize radioactive materials and various radiation types. Public exposures resulting from consumer products, security procedures, emerging commercial space tourism, and nuclear accidents are also addressed. The topics addressed in Part IV are becoming more important as illustrated by the impacts of the Fukushima Daiichi accident and its influence on evacuation planning. Public doses are an important issue as nations wrestle with stakeholder concerns associated with new radiation sources and the expanding use of radioactive materials.

Part IV begins with a discussion of nuclear medicine and the associated radiation doses. Although traditional techniques are noted, the focus is directed toward new technologies and evolving treatment protocols.

Health Physics: Radiation-Generating Devices, Characteristics, and Hazards, First Edition. Joseph John Bevelacqua.

5
Nuclear Medicine

5.1
Overview

This chapter addresses diagnostic and therapeutic methods for cancer and other disease detection and treatment that use ionizing radiation. Selected nonionizing techniques are also outlined. The focus is on emerging or proposed technologies.

Nuclear medicine is broadly classified in terms of imaging or diagnostic techniques and therapy protocols. Imaging includes radiography, fluoroscopy, computed tomography (CT), tracer studies, and radiopharmaceutical administration whose purpose is to obtain information without damaging healthy tissue. These imaging approaches supplement other techniques including ultrasound (US), optical, and magnetic resonance imaging (MRI) to gather information for subsequent evaluation and treatment planning.

Nuclear medicine therapeutic techniques are alternatives to invasive surgery and chemotherapy. Their purpose is the destruction of diseased tissue by depositing absorbed dose preferentially in the tumor site. Therapeutic nuclear medicine techniques include radionuclide administration, external beam therapy, and brachytherapy.

Radiation therapy includes the use of X-rays, gamma rays, neutrons, protons, heavy ions, and other radiation types to kill cancer cells and eradicate tumors. Radiation is generated from external beams or radioactive materials placed within the body near cancer cells. Systemic radiation therapy uses a radioactive source (e.g., radiolabeled monoclonal antibodies) that traverses blood to reach a targeted tissue.

Cancer detection and treatment are facilitated using a variety of approaches including imaging/diagnostic nuclear medicine techniques; comprehensive, personalized treatment of tumors; external beam radiation therapy; and the delivery of radioactive materials to the tumor site. Radiation therapy techniques can be accomplished in conjunction with other treatment methods to enhance the integrated treatment protocol.

This chapter provides a description of the various techniques that use radioactive materials or various radiation types to treat tumors or other diseases. Before addressing specific nuclear medicine methods, a general discussion of imaging and therapy methods is outlined.

Health Physics: Radiation-Generating Devices, Characteristics, and Hazards, First Edition. Joseph John Bevelacqua.

5.2
General Nuclear Medicine Categories

General nuclear medicine categories are broad descriptions of the methods used to image the body or preferentially deliver an absorbed dose to the tumor site. These nuclear medicine procedures include diagnostic X-ray imaging, diagnostic radionuclide administration, computed tomography, therapeutic radionuclide administration, external beam therapy, and brachytherapy.

These general approaches are enhanced by specialized techniques that optimize the delivery of absorbed dose to a specific location. Methods used to enhance the delivery of absorbed dose to specific body locations include image-guided radiation therapy (IGRT), intensity-modulated radiation therapy (IMRT), stereotactic radiosurgery (SRS) and stereotactic body radiotherapy (SBRT), and radioimmunotherapy (RIT). Specific absorbed dose delivery methods (e.g., heavy ion therapy) are addressed in subsequent discussion.

Table 5.1 provides a list of diagnostic radionuclides currently utilized in imaging applications. A list of therapeutic and brachytherapy radionuclides is provided in Table 5.2. New applications for radioactive materials are addressed in subsequent discussion.

5.2.1
Established Imaging Approaches

Ionizing radiation is effective in providing noninvasive imaging methods to obtain information about body functions and diseased or damaged tissue. Low-energy X-rays are a workhorse diagnostic tool. Photon-emitting radionuclides are useful imaging tools, and a variety of diagnostic radionuclides has been developed to provide additional imaging capability. Computers permit the administration of radionuclides to be utilized in conjunction with numerical techniques to enhance imaging methods and their capabilities.

5.2.1.1
X-Ray Imaging Radiography and Fluoroscopy

Diagnostic X-ray procedures include radiography and fluoroscopy. Radiography typically involves the exposure of selected tissue to a broad beam of X-rays. Attenuation of the X-ray beam is influenced by intervening tissue that leads to variable contrast of the imaging medium (typically photographic film). These contrast differences are the basis for image formation that reveals details of the tissue structures under investigation.

Fluoroscopy procedures capture the X-ray beam in an imaging system that produces a video output after it traverses the patient. The video output permits real-time imaging that facilitates diagnosis or enhances an ongoing procedure (e.g., placement of a medical device into the patient). The fluoroscopy output can be recorded for subsequent viewing and evaluation. Fluoroscopy procedures provide greater flexibility and expand the information obtained from snapshot X-ray

Table 5.1 Diagnostic radionuclides used in nuclear medicine.

Radionuclide	Half-life[a)]	Gamma-ray energy (keV)[a)]	Organ imaged/ intended purpose
^{11}C	20.36 min	511	Heart
^{13}N	9.97 min	511	Heart
^{15}O	2.037 min	511	Brain
^{18}F	1.8295 h	511	Bone Brain
^{35}S	87.2 days	[b)]	Extracellular fluid volume
^{42}K	12.36 h	1524.6	Brain Renal blood flow Tumor detection
^{43}K	22.3 h	372.8 and 617.5	Heart
^{47}Ca	4.536 days	1297.1	Calcium metabolism studies
^{51}Cr	27.702 days	320.1	Red blood cell volume and mass determination Spleen
^{64}Cu	12.701 h	1345.8	Brain Wilson's disease
^{67}Ga	3.2611 days	93.3, 184.6, and 300.2	General tumor agent
^{75}Se	119.78 days	136.0, 264.7, and 279.5	Parathyroid glands Pancreas
^{81m}Kr	13.1 s	190.4	Lung
^{85}Sr	64.84 days	514.0	Bone
^{87m}Sr	2.805 h	388.5	Bone
^{99m}Tc	6.01 h	140.5	Bladder Bone Bone marrow Brain Gastric function Heart Kidney Liver Lungs Salivary gland Spleen Thyroid Multiple other organs
^{111m}In	2.8049 days	171.3 and 245.4	Bone marrow Gastric function Labeled blood products Prostate

(continued overleaf)

Table 5.1 (*continued*)

Radionuclide	Half-life[a]	Gamma-ray energy (keV)[a]	Organ imaged/ intended purpose
^{113m}In	1.658 h	391.7	Brain Liver Spleen
^{123}I	13.221 h	159.0	Brain Renal function Thyroid
^{125}I	59.4 days	35.49	Liver function Pancreatic function Thyroid
^{131}I	8.023 days	364.5	Brain Cardiac function Liver function Pulmonary function Renal function Thyroid Pancreatic function
^{127}Xe	36.4 days	172.1 and 202.9	Brain Lung
^{133}Xe	5.243 days	80.99	Brain Lung
^{169}Yb	32.03 days	63.1 and 198.0	Brain Cisternography
^{197}Hg	2.672 days	77.3	Brain Renal function Renal system
^{198}Au	2.6949 days	411.8021	Liver
^{201}Tl	3.043 days	135.3 and 167.4	Heart
^{203}Hg	46.61 days	279.2	Brain Renal system

a) Baum *et al.* (2010).
b) Beta emitter.

images. However, patient doses from fluoroscopy generally exceed those from X-ray imaging.

5.2.1.2
Diagnostic Radionuclide Administration

Most diagnostic radionuclides are photon emitters with short half-lives. The goal in nuclear imaging is to obtain information while delivering minimal absorbed dose to the patient.

Radionuclide administrations are used to detect the presence and location of infections, blood clots, myocardial infarctions, pulmonary emboli, occult bone fractures, and tumors. These techniques also provide information to assess organ

Table 5.2 Therapeutic and brachytherapy radionuclides used in nuclear medicine.

Radionuclide	Half-life[a)]	Radiation type and energy[a),b)] (MeV)	Disease treated/ intended purpose
^{32}P	14.28 days	β: 1.709	Leukemia Bone cancer pain Pancreatic cancer Head–neck tumors Ovarian cancer Rheumatoid arthritis
^{60}Co	5.27 years	γ: 1.1732 and 1.3325	Teletherapy
^{67}Cu	2.58 days	β: 0.39, 0.48, and 0.58 γ: 0.0933 and 0.1846	Breast cancer Colorectal cancer Lymphoma Rheumatoid arthritis
^{90}Sr	28.8 years	β: 0.546	Treatment of benign eye conditions
^{90}Y	2.669 days	β: 2.281	General cancer agent Liver cancer Treatment of benign eye conditions
^{103}Pd	16.99 days	γ: 0.3575 and 0.0398	Brachytherapy
^{125}I	59.4 days	γ: 0.03549	Brachytherapy
^{131}I	8.023 days	β: 0.606 γ: 0.3645	Hyperthyroidism Thyroid cancer
^{137}Cs	30.07 years	β: 0.514 γ: 0.6617	Brachytherapy
^{145}Sm	340.0 days	γ: 0.0612	Brachytherapy
^{153}Sm	1.928 days	β: 0.64 and 0.69 γ: 0.0697 and 0.1032	Bone cancer pain Leukemia Spinal cord tumors
^{177}Lu	6.65 days	β: 0.497 γ: 0.1129 and 0.2084	Bone cancer pain Heart disease
^{186}Re	3.718 days	β: 0.933 and 1.071 γ: 0.1372	Bone cancer pain
^{188}Re	17.004 h	β: 1.962 and 2.118 γ: 0.155041	Bone cancer pain

(continued overleaf)

Table 5.2 (*continued*)

Radionuclide	Half-life[a)]	Radiation type and energy[a), b)](MeV)	Disease treated
^{192}Ir	73.83 days	β: 0.535 and 0.672 γ: 0.3165 and 0.4681	Brachytherapy
^{198}Au	2.6949 days	β: 0.961 γ: 0.4118021	Brachytherapy Limit spread of ovarian cancer Lymphoma
^{210}Bi	5.01 days	α: 4.687	Treatment of benign eye conditions
^{210}Pb	22.3 years	β: 0.017 and 0.061 γ: 0.0465	Treatment of benign eye conditions
^{213}Bi	45.6 min	α: 5.87	Leukemia
^{223}Ra	11.435 days	α: 5.607 and 5.7164	Bone cancer pain
^{225}Ac	10.0 days	α: 5.829, 5.731, and 5.793	Solid tumors
^{241}Am	432.7 years	α: 5.4430 and 5.4857	Brachytherapy

a) Baum *et al.* (2010).
b) Maximum beta energies are provided.

function without disturbing the tissue under observation. This is accomplished by attaching a photon-emitting radionuclide to a biologically active molecule to form a radiopharmaceutical. The molecule facilitates deposition of the radiopharmaceutical into the location of interest. For example, radioiodine compounds are preferentially absorbed into the thyroid that facilitates imaging that organ. Following administration, the radiopharmaceutical is monitored using a suitable measurement system (e.g., gamma camera) to detect photons escaping from the organ being investigated.

5.2.1.3
Computed Tomography

Diagnostic procedures include imaging the body's cross-section using tomography. Tomography has an advantage over planar imaging because it is designed to view underlying organs using computational techniques. The technique obtains multiple axial slices through a region of interest.

There are two basic forms of computed tomography: single-photon emission computed tomography (SPECT) and positron emission tomography (PET). The SPECT method uses a rotating gamma camera and radionuclides utilized in conventional planar imaging.

In contrast to the use of a rotating gamma camera, PET cameras are a stationary array of instruments that detect the two 511 keV photons produced by the annihilation of a positron–electron pair. PET scanners rely on coincidence detectors located 180° apart or time-of-flight circuitry to determine the spatial coordinates of the annihilation event. The coincidence events are assembled to produce an image of the scanned organs.

PET techniques utilize positron-emitting radionuclides including ^{11}C, ^{13}N, ^{15}O, and ^{18}F. The emitted positrons interact with electrons in the tissue of interest to produce the annihilation photons. The short-lived positron-emitting radionuclides (e.g., ^{11}C, ^{13}N, and ^{15}O) are necessarily produced in the medical facility (e.g., using a cyclotron), while ^{18}F is typically shipped to the facility from a vendor.

5.2.2
Established Therapy Applications

Radiation therapy utilizes radioactive material or beams of various radiation types to deposit energy in the body to destroy diseased tissue. These therapy approaches utilize beta- and alpha-emitting radionuclides; external electron, proton, and heavy ion beams; brachytherapy; and RIT to deliver absorbed dose to the tumor or diseased tissue location.

5.2.2.1
Therapeutic Radionuclide Administration

Therapeutic radionuclides are designed to deposit significant energy in a target tissue. As such, they incorporate beta- and alpha-emitting radionuclides into the radiopharmaceutical. A summary of radiotherapy nuclides is provided in Table 5.2.

Therapeutic radionuclides have a variety of uses including the suppression of hyperthyroidism, reduction in pain levels associated with metastatic cancer, and treatment of cancers and tumors. The beta-emitting radiopharmaceutical destroys a tumor mass when deposited locally, but the radiation can also affect healthy tissue if its range extends beyond the tumor boundary. High-energy beta radiation and associated bremsstrahlung from nuclides including ^{32}P and ^{90}Y have a range that often extends beyond the target tissue. The irradiation of healthy tissue has a negative impact on patient recovery and can affect their subsequent quality of life.

The irradiation of healthy tissue is minimized if alpha-emitting radionuclides are used in the radiopharmaceutical. Alpha particles have a much shorter range in tissue than the high-energy beta particles, and their use has positive advantages in terms of the preferential delivery of absorbed dose to cancer cells. However, the use of alpha-emitting radionuclides in therapy applications is currently limited by their availability and cost-effective production.

5.2.2.2
External Beam Therapy

Large therapy doses are delivered using accelerators and teletherapy units including ^{60}Co systems. External beams deliver various radiation types that either directly attack the tumor using electrons, photons, neutrons, protons, and heavy ions or interact with a material that is incorporated into the tumor volume (e.g., boron neutron capture therapy (BNCT) utilizing ^{10}B). Beam therapy should also minimize the absorbed dose in tissue between the skin and tumor mass.

There is wide diversity in penetration depths and energy deposition characteristics associated with the beam's radiation type and its energy. For photon beams, divergence and attenuation reduce the photon fluence as a function of the depth in tissue. However, the electron density builds to an equilibrium value while traversing tissue. The combination of these effects produces an absorbed dose curve that rises to a maximum and then decreases with increasing depth into tissue. Electron backscatter increases the surface absorbed dose to a value between 15 and 100% of the maximum dose. The depth of the maximum absorbed dose increases with increasing beam energy. For example, the absorbed dose curve for 15 MeV photons peaks at about 2.7 cm depth, and clinically useful radiation is available to about 8 cm tissue depth.

With electron beams, the primary electrons lose energy in tissue and produce high ionization densities per unit length as they reach their maximum range. For tissue depths beyond the maximum range, the electron absorbed dose decreases rapidly to a value of only a small percentage of the maximum absorbed dose.

For electron energies below 1 MeV, the maximum absorbed dose occurs near the skin surface. Since most lesions are below the surface of the skin, it is advantageous to use higher-energy electron beams to reach the desired tissue depth. By properly selecting the beam energy, the tumor is attacked while the underlying healthy tissue is spared. For example, a chest wall tumor should be treated without damaging the underlying lung tissue. As the electron beam energy increases from 4 to 20 MeV, the shape of the absorbed dose curve shifts from a surface peak to a broader plateau extending into tissue. Beyond 20 MeV, the plateau expands, and the advantage of sparing healthy tissue at depth is lost. In general, the useful electron energy range is between 4 and 20 MeV.

Proton beams produce a relatively low constant absorbed dose that terminates in a narrow peak at the end of the absorbed dose curve. The absorbed dose is highly localized which produces high tumor energy deposition and lower healthy tissue doses. However, the ability to track a conventional beam is limited. Tracking is important to ensure the entire tumor volume is irradiated. Beam tracking and the use of additional radiation types in external beam therapy are addressed in subsequent discussion.

5.2.2.3
Brachytherapy

Brachytherapy involves the placement of radioactive material in direct contact with the tumor volume. It is primarily used for tumors that are accessible through natural body cavities or near body surfaces. An ideal brachytherapy radionuclide has a range that confines its energy deposition profile to the tumor volume. Traditional brachytherapy sources involve radionuclides that emit photons, beta particles, and conversion electrons. Brachytherapy can be used exclusively or in combination with other therapies including surgery, external beam radiotherapy, and chemotherapy. Common brachytherapy radionuclides, summarized in Table 5.2, include ^{125}I, ^{137}Cs, ^{192}Ir, and ^{198}Au.

The brachytherapy source material is enclosed in a protective capsule or wire, which allows the ionizing radiation to escape and irradiate the surrounding tissue but prevents the radioisotope from dissolving in body fluids and being translocated to other tissues. Brachytherapy sources are removed following treatment or (with some short-lived radionuclides) allowed to remain in place. In principle, the brachytherapy source only irradiates a localized tissue volume, and the absorbed dose delivered to healthy tissues is minimized. As such, brachytherapy offers the potential for more localized dose delivery than is often possible with external beams.

A brachytherapy treatment protocol is usually completed in less time than other radiotherapy techniques and may be performed on an outpatient basis. Treatment results suggest that the brachytherapy success rates are comparable to surgery and external beams. The results tend to improve when multimodal approaches are utilized. Brachytherapy also has a relatively low incidence of significant adverse side effects. However, these side effects affect a patient's quality of life.

Brachytherapy is commonly used to treat cancers of the cervix, prostate, breast, and skin. It is also utilized to treat tumors in other body tissues including the anus, brain, digestive tract, esophagus, eye, female reproductive tract, gall bladder, head, neck, lip, male reproductive tract, mouth, nasopharynx, oropharynx, rectum, respiratory tract, soft tissue, tongue, and urinary tract.

Brachytherapy sources are often defined in terms of their delivered absorbed dose rates. Low-dose rate (LDR) brachytherapy sources produce an absorbed dose rate up to 2 Gy/h. LDR brachytherapy is commonly used for cancers of the oral cavity, oropharynx, and prostate as well as other sarcomas. Medium-dose rate (MDR) brachytherapy is characterized by delivered absorbed dose rates in the range of 2 – 12 Gy/h. High-dose rate (HDR) brachytherapy delivers an absorbed dose rate in excess of 12 Gy/h. The most common applications of HDR brachytherapy are in tumors of the breast, cervix, esophagus, lungs, and prostate.

Brachytherapy can also involve pulsed sources. Pulsed dose rate (PDR) brachytherapy utilizes short pulses of radiation to simulate the overall rate and effectiveness of an LDR treatment. The PDR source is typically an ^{192}Ir source (on the order of 4×10^4 MBq) that is assembled and driven the same way as an HDR system. PDR brachytherapy is used for gynecological and head and neck tumors.

The general PDR approach simulates a continuous LDR interstitial treatment lasting several days with a series of short (e.g., about 10 min) HDR irradiations on a predetermined frequency (e.g., about an hour). PDR is designed to exploit computer-controlled remote afterloader technology using the catheters of a brachytherapy implant. The residence times in each position are adjusted to obtain the required absorbed dose. When the source is not stepping through the implant, it is retracted. The PDR approach has a number of advantages: (i) the patient is only irradiated a portion of the time, (ii) the medical facility needs a smaller source inventory, and (iii) the delivered absorbed dose to the tumor is optimized using computer control.

NCRP 155 makes an arbitrary division between sealed and unsealed radioactive sources. All unsealed sources are considered as radiopharmaceutical therapy sources. It should be noted that some unsealed source applications fit the category of a device as defined by the US Food and Drug Administration. Many sealed sources are considered brachytherapy sources.

Table 5.3 lists radionuclide sources historically utilized in therapy implantation applications. Also listed are current and future brachytherapy source materials. The physical characteristics of brachytherapy radionuclides are summarized in Table 5.4. Table 5.5 provides the specific bremsstrahlung constants in soft tissue and bone for selected photon and beta–gamma therapeutic radiopharmaceuticals.

Most of these brachytherapy sources utilize photon, beta, and conversion electron sources. The use of low-energy photons, low-energy beta particles, and low-energy conversion electrons enhances energy deposition into the tumor volume and minimizes the dose to healthy tissue.

Tumor dose is also maximized using alpha-emitting radioactive material. The short range of the alpha particles maximizes the absorbed dose delivered to the tumor. Candidate nuclides are provided in Table 5.2. In addition to alpha particles, the use of neutron radiation in brachytherapy applications is under development.

^{252}Cf has a half-life of 2.646 years, and it decays primarily through alpha emission (96.9%) with spontaneous fission occurring with a yield of about 3.1%. Currently, it is the only neutron-emitting radioisotope that has been used in

Table 5.3 Radionuclides used for implantation[a].

Technique	Traditional	Current	Future
Intracavitary and intraluminal applications			
Low-dose rate	^{226}Ra	^{137}Cs	^{241}Am, ^{192}Ir, and ^{169}Yb
High-dose rate	^{60}Co	^{60}Co and ^{192}Ir	^{192}Ir and ^{169}Yb
Interstitial implants			
Preloaded	^{226}Ra	^{137}Cs	—
Afterloaded	—	^{192}Ir	^{125}I, ^{103}Pd, and ^{169}Yb
High-dose rate	—	^{192}Ir	^{192}Ir and ^{169}Yb
Permanent implants			
Conventional dose rate	^{222}Rn	^{198}Au	^{198}Au and ^{131}Cs
Ultralow dose rate	—	^{125}I and ^{103}Pd	^{125}I and ^{103}Pd

a) NCRP (2007).

Table 5.4 Physical characteristics of brachytherapy radionuclides[a].

Radionuclide	Half-value layer (mm Pb)	Physical form
^{226}Ra[b]	12	Tubes and needles
^{222}Rn[b]	12	Seeds
^{60}Co	12	Plaques and needles
^{137}Cs	6.5	Tubes and needles
^{192}Ir	3	Seeds in ribbons, wires, and source on cable
^{125}I	0.025	Seeds
^{103}Pd	0.008	Seeds
^{198}Au	3.3	Seeds
$^{90}Sr/^{90}Y$	c)	Plaques
^{241}Am	0.12	Tubes
^{169}Yb	0.48	Seeds
^{131}Cs	0.030	Seeds
^{145}Sm	0.060	Seeds

a) NCRP (2007).
b) Legacy source.
c) Not provided.

Table 5.5 Specific bremsstrahlung constants for selected radionuclides commonly used in therapeutic radiopharmaceuticals.[a]

	Specific bremsstrahlung constant (pGy-m^2/MBq-h)	
Radionuclide	Soft tissue $Z_{eff} = 7.9$	Bone (calcium) $Z_{eff} = 21$
Photon and beta emitters		
^{131}I	18.2	48.3
^{177}Lu	9.12	24.2
^{153}Sm	14.1	37.7
^{186}Re	28.7	76.3
^{188}Re	36.5	96.9
Beta emitters		
^{32}P	96.0	256
^{33}P	15.6	41.5
^{89}Sr[b]	77.4	200
^{90}Y	134	356

a) NCRP (2007).
b) Although ^{89}Sr emits a gamma ray, it is grouped with the beta emitters because the gamma-ray yield is negligibly low (<0.01%).

radiotherapy applications. In addition to spontaneous fission neutrons and alpha particles, ^{252}Cf decay products emit photons and beta particles.

When applied to a brachytherapy source, a ^{252}Cf irradiation of a tumor primarily includes neutrons and photons. The ^{252}Cf alpha particles and beta particles have very short ranges (<1 mm) in tissue and usually do not penetrate the capsule wall of the source. Approximately, one-third of the absorbed radiation dose in tissue surrounding a ^{252}Cf brachytherapy source is due to gamma rays, and two-thirds of the dose is due to neutrons.

^{252}Cf brachytherapy has been used in clinical trials involving a number of cancer types including cervical, esophageal, lip, mouth, oral cavity, rectal, skin, tongue, and soft tissue. It has also been used in combination with BNCT.

Boron-enhanced neutron brachytherapy (BENBT) is a combination of neutron brachytherapy and BNCT. In BNCT, a molecule containing boron is administered and localizes in the tumor site. The tumor is then irradiated with thermal neutrons, which produce heavy ions via the $^{10}B(n, \alpha)^{7}Li$ reaction. The heavy ions from the reaction have a short range and are effective in depositing absorbed dose within the tumor volume. A portion of the fast neutrons produced from the spontaneous fission of ^{252}Cf are reduced to thermal energies as they penetrate the tumor. With BENBT, absorbed dose is deposited in the tumor from direct ^{252}Cf neutron radiation, and additional heavy ion absorbed dose is delivered from the $^{10}B(n, \alpha)^{7}Li$ reaction.

Other brachytherapy techniques involve the permanent implantation of radioactive material imbedded into glass and resin microspheres and low-energy applications. A discussion of the radionuclides and the characteristics of the microsphere approach is provided in Section 5.5.7. Low-energy brachytherapy is outlined in Section 5.2.3.1.

5.2.2.4
Radioimmunotherapy

RIT is a combination of radiotherapy and immunotherapy. Immunotherapy uses an engineered molecule (monoclonal antibody) to bind to cancer cells. Monoclonal antibodies function analogously to natural antibodies that attack bacteria and viruses.

In RIT, a monoclonal antibody is combined with radioactive material. When injected, the monoclonal antibody binds to the cancer cell and delivers large absorbed doses directly to the tumor. Radionuclides including ^{90}Y and ^{131}I are being used in RIT.

RIT is currently used to treat various forms of lymphoma. This approach has the potential to treat a variety of disease types including brain tumors, colorectal cancer, leukemia, melanoma, ovarian cancer, and prostate cancer. Imaging supporting RIT is often performed using a gamma camera.

The various radiation therapy techniques deposit some energy into healthy tissue that results in a biological detriment and associated side effects. Side effects from radiation therapy are addressed in Section 5.3.

5.2.3
Targeted Delivery of Dose

Radiation therapy techniques have a common goal of preferentially depositing energy within the tumor volume. These techniques include low-energy brachytherapy, IGRT, IMRT, and SRS and stereotactic body radiotherapy (SBRT). The next four sections address these techniques.

5.2.3.1 Low-Energy Brachytherapy

Low-energy photon sources offer the potential to treat cancers using catheters. A low-energy photon emitter has a shorter range than a higher-energy photon and facilitates longer implantation times. For example, ^{169}Yb with a half-life of 32.03 days emits a series of photons with the highest yields occurring below 200 keV. It also emits electrons with the dominant yields having energies below 6 keV. With these properties, nuclides such as ^{169}Yb offer additional cancer therapy options (e.g., breast tumor therapy) with extended treatment times without the need for heavily shielded rooms.

Low-energy brachytherapy also offers the potential for localized energy deposition for unique therapy applications. These applications include the disruption of a tumor's blood supply by destroying its vascular structure. Tumor vascular disruption has been accomplished using chemotherapy and radiotherapy. Section 5.5.7 provides additional discussion of vascular disruption therapy.

Low-energy photons for use in therapy applications are also delivered using miniature X-ray sources. Therapy applications include interstitial radiosurgery and intravascular irradiation to prevent restenosis. Typical generating voltages are in the 20–40 kV range. The technique has additional therapy applications since the biological effectiveness of these low-energy photons is large when compared to higher-energy gamma rays. This occurs because the photoelectric absorption is very significant at low energies.

5.2.3.2 Image-Guided Radiation Therapy

IGRT is the use of imaging techniques during radiation therapy to direct the delivery of absorbed dose to the tumor site. In IGRT, the devices delivering the absorbed dose include imaging technology to permit viewing the tumor during the therapy procedure. The imaging device facilitates necessary adjustments to more precisely target the tumor and avoid irradiating adjacent healthy tissue.

Computed tomography (CT), magnetic resonance imaging (MRI), positron emission tomography (PET), ultrasound (US), and X-ray imaging have been traditionally used in IGRT. Recent IGRT methods incorporate light-emitting diode (LED) output directed on the patient's body or implanted magnetic transponders.

IGRT is often used to treat tumors in tissues that can move (e.g., lungs, liver, and prostate) during the nuclear medicine procedure. The IGRT approach is also used in conjunction with IMRT, external beam therapy, SRS, and SBRT.

5.2.3.3 Intensity-Modulated Radiation Therapy

IMRT is an external beam therapy method that uses a linear accelerator to generate photons having energies on the order of 10 MeV. This technique precisely delivers absorbed dose to a tumor or specific areas within the tumor.

Computed tomography or MRI is often used in conjunction with calculations to determine the absorbed dose profile that conforms to the tumor shape. Intravenous contrast agents may be injected during imaging to better define the tumor geometry. Typically, combinations of multiple intensity-modulated fields arising from different beam orientations generate an optimized absorbed dose profile that maximizes tumor dose.

Since the absorbed dose to healthy tissue is minimized with the IMRT approach, higher absorbed doses are delivered to tumors with fewer side effects when compared to conventional radiotherapy techniques. Given its complexity, IMRT requires slightly longer treatment times and additional planning than conventional radiotherapy.

Currently, IMRT is being used most extensively to treat central nervous system disease, head and neck tumors, and prostate cancer. IMRT has also been used to treat breast, lung, gastrointestinal, gynecologic, and thyroid tumors. It may also be beneficial for treating pediatric tumors.

IMRT often requires fractionated treatment sessions. The total number of IMRT sessions and associated absorbed dose depend on the type, location, and size of the tumor; doses to adjacent tissues; and the patient's overall health. Patients are usually scheduled for 10 and 30 min IMRT sessions, 5 days a week for 5–8 weeks.

5.2.3.4
Stereotactic Radiosurgery and Stereotactic Body Radiotherapy

SRS is a precise form of radiotherapy that was initially used to treat tumors and abnormalities of the brain. The techniques developed with SRS are also being applied to treat other cancers using a procedure called SBRT. SRS and SBRT offer an alternative to invasive surgery.

SRS is a nonsurgical technique that precisely delivers absorbed dose at much higher levels than traditional radiation therapy. This precision minimizes the dose to adjacent healthy tissue.

SRS and SBRT are based on three-dimensional imaging and localization techniques to deliver highly focused gamma-ray or X-ray beams that converge on a tumor or abnormality. This approach is also be used in conjunction with IGRT to further improve the precision of the treatment.

Three-dimensional imaging including CT, MRI, and PET/CT is used to locate the tumor or abnormality within the body and define its exact geometry. These imaging results assist in defining the orientation of the photon beams that converge on the tumor. Although SRS is typically completed in 1 day, two to five fractionated treatments are often used for tumors larger than 1 in. in diameter.

SBRT treats a variety of tumors that occur within the abdomen, head, liver, lung, neck, prostate, and spine. Three machine types are utilized to perform SBRT. A gamma knife incorporates beams of highly focused gamma rays aimed at the target region and is typically used to treat intracranial tumors. Linear accelerators deliver high-energy photons for the treatment of larger tumors. A third approach uses proton or heavy ions to attack the tumor site.

5.2.4 Source Security

Therapy facilities utilize large activity sources and have significant inventories of radioactive materials. As noted in Chapter 4, many of these sources could be stolen and used in the construction of a radiological dispersal device or a radiological exposure device. As such, security measures are warranted to ensure therapy sources are utilized for their intended purpose.

Following the 11 September 2001 attacks, the NRC issued security orders to protect selected types of radioactive materials from theft or diversion. 10CFR37 was established to replace these orders and establish regulations for physical protection measures, fingerprinting, and background checks for any licensee that possesses an aggregated Category 1 or Category 2 quantity of radioactive material. Requirements for access to radioactive material and the use, transfer, and transport of this material are included in 10CFR37.

Medical facilities utilizing radioactive materials are affected by this rule. In a September 2012 report, the Government Accountability Office (GAO) found significant security issues at some medical facilities. At one hospital having a 56 TBq ^{137}Cs irradiator at its blood bank, the security door utilized a combination lock. The lock combination was written on the doorframe in an open-access hallway. At another hospital, two ^{137}Cs research irradiators containing 74 and 222 TBq were housed in a building open to the public. A security camera located in a hallway outside the irradiator room was not directed to view the room entrance, and one of the irradiators was on a wheeled structure that facilitated its movement. In addition, there were no cameras or security measures inside the room. The room is located near a loading dock, and monitoring of corridors leading to the loading dock did not clearly indicate ongoing activity. These types of security lapses facilitate the theft of radioactive materials for a terrorist device.

Category 1 and 2 quantities are defined in 10CFR37 for a limited set of radionuclides. The activity limits (provided in parenthesis) in TBq for Category 1 radionuclides are ^{241}Am (60), $^{241}Am/Be$ (60), ^{252}Cf (20), ^{60}Co (30), ^{244}Cm (50), ^{137}Cs (100), ^{153}Gd (1000), ^{192}Ir (80), ^{238}Pu (60), $^{239}Pu/Be$ (60), ^{147}Pm (40 000), ^{226}Ra (40), ^{75}Se (200), ^{90}Sr (1000), ^{170}Tm (20 000), and ^{169}Yb (300). The Category 2 limits are a factor of 100 lower than the aforementioned Category 1 values.

5.3 Side Effects from Radiation Therapy

Although radiation therapy has positive medical benefits, it can also result in negative side effects. Side effects associated with radiation therapy include issues directly related to the delivery of absorbed dose to the tumor and radiation damage to healthy tissue adjacent to the target.

5.3.1 General Description

Early side effects occur during or shortly after the radiation administration and typically last for a few weeks. These side effects include fatigue and skin conditions that result in blistering, dryness, itching, irritation, peeling, redness, sensitivity, and swelling. Other effects include diarrhea, difficulty swallowing and eating, digestion problems, hair loss, headaches, nausea and vomiting, soreness and swelling in the treatment area, and urinary and bladder ailments.

Late side effects, occurring months or years following treatment, appear less frequently than early side effects and are often permanent conditions. Tissue changes in the brain, colon, lung, kidney, joints, mouth, rectum, and spinal cord are common late effects. Other late effects include cardiovascular disease (CVD), infertility, lymphedema, and secondary cancer.

5.3.2 Second Primary Cancers and Cardiovascular Disease

Radiation remains a key component in successful cancer treatment with 50% of all patients estimated to have received radiation therapy. For many patients, radiation-induced late effects follow cancer survival. Secondary primary cancer (SPC) and CVD are two of the most frequent and important side effects associated with radiation therapy.

The risk of SPC following radiation therapy is supported by considerable data. For example, increases in the risk of secondary cancers have been documented for bone, brain, breast, lung, thyroid, leukemia, and soft tissue. As noted in NCRP 170, quantitative risk estimates for CVD are not yet firmly established, but are under investigation.

Basic radiobiology provides a foundation for understanding the fundamental cell damage mechanisms that lead to the incidence of SPC and CVD. Genetic conditions are known to predispose the development of multiple primary cancers. These genetic conditions also provide insight into the interaction between high-dose radiation and genetic susceptibilities in causing SPCs.

As noted previously, conventional radiation therapy delivery methods include beam therapy using a variety of radiation types, ^{60}Co teletherapy units, and brachytherapy. NCRP 160 notes that modern radiation therapy approaches have significantly changed over the last few decades and patient doses have increased. These increased doses and SPC and CVD effects suggest that the benefits from radiation therapy must be weighed in terms of the associated risks. The data in NCRP 160 and 170 should assist physicians in determining the optimum treatment approach.

5.3.3 Future Considerations

The use of radiation therapy and development of advanced radiation treatment modalities to treat cancer contribute to the long-term improvement in patient survival. This improvement is accompanied with the negative side effects associated with SPC and CVD. The exact nature of the molecular and genetic basis for these side effects has yet to be determined. Optimal cancer screening and interventional techniques require additional research including an improved understanding of methods to limit SPC and CVD through epidemiological, laboratory, and clinical studies.

The number of patients undergoing radiation therapy is less than 1% of the number receiving diagnostic procedures. However, the absorbed dose delivered to the target volume is $5 \times 10^3 - 5 \times 10^4$ times as large as the diagnostic organ doses. The magnitude of the therapy doses and increasing use of these techniques suggest the importance of understanding late-term effects. Understanding these effects will improve therapy approaches and enhance the quality of life of patients following radiation treatments.

5.4 Emerging Therapy Approaches

The twenty-first century will witness the development of new therapy approaches and the optimization of existing techniques. For example, enhancements in proton therapy and development and further implementation of heavy ion therapy are in progress. Additional approaches using pions, muons, and antimatter have appeared in the literature. The expanded use of alpha radiopharmaceuticals, neutron therapy, tumor vascular disruption techniques, and nanotechnology has also been proposed.

In order to develop or optimize a therapy procedure, a number of general items must be considered. These items include fundamental physics considerations, dose delivery methods, incorporation of multimodal therapy, and tracking the absorbed dose profile. Each of these considerations affects the selection and optimization of the therapy approach and its subsequent viability as a successful treatment option.

5.4.1 Fundamental Physics Considerations

The purpose of a nuclear therapy procedure, whether the energy is delivered by a radiopharmaceutical, brachytherapy implant, or beam of a particular radiation type, is to deliver absorbed dose selectively to a tumor site. The tumor location varies with the type of cancer and its stage of development. In general, the tumor extent is defined by the volume $\xi(x, y, z)$ where the growth is assumed to lie within a three-dimensional manifold residing within the region specified by

$$\begin{aligned} x_1 &\leq x \leq x_2 \\ y_1 &\leq y \leq y_2 \\ z_1 &\leq z \leq z_2 \end{aligned} \tag{5.1}$$

An ideal treatment protocol preferentially deposits energy within the volume of the tumor. The deposited energy must be sufficient to irradiate the tumor's central mass as well as the various fibers and extensions that are produced during its growth cycle.

This requires that the delivered dose profile $D(x, y, z)$ matches the tumor geometry:

$$V_{\mathrm{D}}(x, y, z) \approx \xi(x, y, z) \tag{5.2}$$

where $V_{\mathrm{D}}(x, y, z)$ is the volume in which the absorbed dose is deposited. For energy (E), the absorbed dose (D_{R}) at a specific location for radiation type R is given by the relationship

$$D_{\mathrm{R}}(E, x, y, z) = \frac{1}{\rho(x, y, z)} \left| -\frac{\mathrm{d}E_R\,(x, y, z)}{\mathrm{d}\eta(x, y, z)} \right| \phi_{\mathrm{0R}}(x_0, y_0, z_0) e^{-\mu_{\mathrm{R}}(E)\eta(x,y,z)} \tag{5.3}$$

where $E_{\mathrm{R}}(x, y, z)$ is the energy of the radiation type R deposited at location (x, y, z), $\rho(x, y, z)$ is the density at location (x, y, z), $\frac{1}{\rho}\left|-\frac{\mathrm{d}E_{\mathrm{R}}}{\mathrm{d}\eta}\right|$ is the mass stopping power, $\eta(x, y, z)$ is the path length of the radiation associated with location (x, y, z), $\mu_{\mathrm{R}}(E)$ is the energy-dependent attenuation coefficient for radiation type R, and $\phi_{\mathrm{0R}}(x_0, y_0, z_0)$ is the entrance flux into the body or tumor depending on the mode of delivery of the radiation type R. The various approaches used to preferentially deliver dose to the tumor are addressed in subsequent discussion.

In order to destroy the tumor, both the central mass and its extensions must be irradiated. A highly localized beam could destroy the central tumor mass but leave the extensions relatively intact and capable of further growth. Therefore, some spreading in the absorbed dose profile is desirable. An optimum therapy approach has the capability to localize and track the beam to ensure it irradiates the entire tumor volume.

5.4.2 Dose Delivery Methods

A variety of modalities are utilized for diagnostic imaging and radiation treatment. Table 5.6 summarizes various radiation therapy and imaging modalities and their associated dose levels. Radiation doses from multiple procedures are often recorded separately, and most dosimetry studies only focus on a single modality or procedure.

Table 5.6 *Radiation therapy* modalities, energies, and associated dose levels[a)].

Procedure	Radiation type	Energy	Dose level		
			Primary target dose (Gy)	Dose (Gy) outside the treatment volume	Effective dose (Sv)
Radiation therapy					
External beam	Photons, electrons, protons, neutrons, and heavy ions	^{60}Co:1.17 and 1.33 MeV 4–250 MeV from X-ray and proton beams 90–330 MeV/nucleon for heavy ions	≤100 (Gy × RBE)	Dose level: Low: <5 Intermediate: 5–50 High: >50	[b)]
Brachytherapy	Photons, electrons, neutrons, and electronic X-ray sources	<2 MeV	~60	~1	[b)]

Table 5.6 *(continued)*

Procedure	Radiation type	Energy	Dose level		
			Primary target dose (Gy)	Dose (Gy) outside the treatment volume	Effective dose (Sv)
Radiation therapy (continued)					
Radioimmunotherapy	Photons, electrons, and alphas	<5 MeV	~100	~10	b)
Diagnostic imaging					
Radiography	Photons	<150 keV	b)	b)	<0.001 per scan
Multidetector CT	Photons	<140 keV	b)	b)	~0.01 per scan
Interventional fluoroscopy	Photons	<140 keV	b)	b)	~0.2 per scan
Nuclear stress test	Photons	<2 MeV	b)	b)	~0.01 per test
Hybrid PET/CT	Photons Positrons	0.511 MeV	b)	b)	~0.04 per scan
Cone-beam CT procedures for IGRT	Photons	keV to MeV	b)	b)	0.5–1 (for a total of 30 fractions)

a) Bevelacqua (2005b) and NCRP (2011).
b) Not reported.

5.4.3 RBE Considerations

For deep-seated tumors, an ideal therapy protocol has a relative biological effectiveness (RBE) distribution that has its maximum values within the tumor volume and minimal RBE outside the tumor site. Heavy ions such as ^{12}C exhibit a low RBE near the body surface with a good likelihood for healthy tissue repair. In comparison, a strongly elevated RBE occurs near the end of the heavy ion's range, and the damaged deoxyribonucleic acid (DNA) in this region is more difficult to repair. This is a reasonable therapy condition because the healthy tissue receives a smaller absorbed dose than the tumor, and it will exhibit a greater likelihood for repair. However, it is not an optimum condition since healthy tissue is receiving absorbed dose, which has the potential for negative side effects.

As the heavy ion mass increases beyond carbon, higher RBEs (relative to carbon) result in healthy tissue receiving increased absorbed dose in the entrance area. The entrance RBE increases with increasing atomic number of the heavy ion beam. Since the RBE affects the effectiveness of the repair of damaged DNA, the repair capability of the irradiated tissue becomes relevant. Healthy tissue repair is an important consideration for slowly growing tumors that have a significant repair capacity and are usually radioresistant. The effects of absorbed dose deposition are addressed in more detail in subsequent discussion.

5.4.4 Multimodal Therapy Options

Multimodal therapy combines two or more distinct options to treat a disease. These treatment methods commonly include surgery, chemotherapy, and radiation therapy. Physicians often utilize a combination of treatment options to attack a disease such as cancer.

Chemotherapy involves the systematic administration of chemical agents to attack a tumor or disease. The approach is similar to radiation therapy in that chemical agents are used to target and kill cancerous cells. As with radiotherapy, chemotherapy also affects healthy tissue, and negative effects such as temporary hair loss and nausea can occur.

In addition to standard approaches, experimental treatments that have demonstrated success in clinical trials can also be utilized as part of the therapy approach. Other treatments including optical and acoustic techniques, gene therapy, and immunotherapy are utilized as part of the multimodal approach.

The selection of a particular option or combination of approaches depends on the patient's health, physician's experience and success with a particular protocol, and nature of the disease being treated. Emerging treatment techniques presented in subsequent discussion offer additional options for disease destruction.

5.4.5
Tracking Dose Profiles

The capability to track the absorbed dose delivered to a tumor site is a desirable consideration for ensuring that a treatment method is successful in depositing energy at the desired location. Having absorbed dose delivered to healthy tissue complicates patient recovery and creates undesirable side effects. For example, the side effects of the treatment of prostate cancer include incontinence, erectile dysfunction, and prolonged recovery time.

Tracking methods depend on the dose delivery method. For a radiopharmaceutical, tracking is a secondary consideration if the molecular species is preferentially deposited within the tumor boundary. Dose localization within the tumor boundary necessitates the use of a short-range radiation type (e.g., low-energy photons, low-energy conversion electrons, low-energy beta particles, or alpha particles). These radiation types facilitate the selective deposition of absorbed dose within the tumor volume.

Tracking the absorbed dose location becomes more important if external beam therapy is utilized. The beam can be tracked under certain circumstances. For example, a heavy ion is stripped of neutrons as it penetrates tissue. Stripping neutrons from the nucleus increases the proton to neutron ratio, which moves the nucleus further from the line of stability. Having excess protons residing within the nucleus favors the positron emission and electron capture processes. Positron emission results in the creation of two annihilation photons, and these photons are detected using PET. The detection of these photons provides their annihilation location, which yields information regarding the location of the external beam of heavy ions. An example of neutron stripping from a heavy ion penetrating tissue is illustrated by considering an external beam of ^{12}C ions:

$$\begin{aligned} &^{12}\mathrm{C} \rightarrow {}^{11}\mathrm{C} + \mathrm{n} \rightarrow {}^{10}\mathrm{C} + \mathrm{n} + \mathrm{n} \\ &^{11}\mathrm{C} \rightarrow {}^{11}\mathrm{B} + \beta^+ + \nu_\mathrm{e} \\ &^{10}\mathrm{C} \rightarrow {}^{10}\mathrm{B} + \beta^+ + \nu_\mathrm{e} \end{aligned} \tag{5.4}$$

The electron neutrinos produced in Eq. (5.4) produce no biological detriment.

Photons can also be used to track the radiation type used in the therapy application. For example, initial antiproton therapy applications have been published. The irradiation of tissue with antiprotons yields a variety of reactions including

$$\mathrm{p} + \overline{\mathrm{p}} \rightarrow 3\pi^+ + 3\pi^- + \gamma'\mathrm{s} \tag{5.5}$$

Photons can be tracked using computed tomography to determine the location of the $\mathrm{p} + \overline{\mathrm{p}}$ event.

5.5
Evolving, Emerging, and New Therapy Approaches

As noted in the introduction to this chapter, the focus is on emerging, developing, and proposed treatment methodologies. These approaches include external beam therapy using protons, heavy ions, pions, muons, neutrons, and antiprotons; microsphere brachytherapy; and a variety of applications to deliver radioactive material to a specific tumor site. Contemporary techniques are not addressed in detail, and the reader is referred to the references to this chapter for their description.

External particle beam therapy applications are based on the localized deposition of energy into tumors utilizing the characteristics of the Bragg curve. While photons lose intensity in an essentially exponential manner as they penetrate tissue, charged particles deposit most of their energy in the Bragg peak near the end of their range. The Bragg peak exhibits a sharp spike for protons and heavy ions but is somewhat broader for charged pions. Since antiprotons interact with protons in tissue, the antiproton energy deposition curve represents a combination of the Bragg peaks for these radiation types and their annihilation products. These approaches are addressed in subsequent discussion.

Expanded presentations are provided for two theoretical treatment approaches. The first is tumor vascular disruption using alpha-emitting and low-energy beta–gamma-emitting radionuclides, which is presented in Section 5.5.7.4. A second technique utilizing nanotechnology is the internal radiation-generating device concept that is addressed in Section 5.6.3.2.

5.5.1
External Beam Proton Therapy

In the mid-1940s, Wilson theorized that proton beams could be utilized to deliver a highly localized absorbed dose to tumors while minimizing the dose to adjacent healthy tissue. This theory was confirmed at the Lawrence Berkeley Laboratory.

The depth dose distribution of proton beams differs significantly from that of photon or neutron beams. Protons have an increasing energy deposition with penetration distance that culminates in a Bragg peak near the end of its range. At depths beyond the Bragg peak, the proton absorbed dose decreases rapidly. By selecting appropriate proton beam energies, the individual Bragg peaks can be utilized to irradiate the entire tumor volume. Therefore, improved dose localization compared to the use of photons or neutrons is achieved. Proton therapy is used to treat a variety of disease types including pediatric intracranial tumors, ocular tumors, head and neck tumors, prostate cancer, lung cancer, liver cancer, and sinonasal malignancies.

As an illustration of the output spectrum from an external proton beam, the normalized dose ($D_{\text{norm}}(r)$) from 20, 40, and 60 MeV protons as a function of penetration depth in water is provided in Figure 5.1. In Figure 5.1, the normalization is relative to the peak dose (D_{peak}) for each energy:

$$D_{\text{norm}}(r) = \frac{D(r)}{D_{\text{peak}}} \tag{5.6}$$

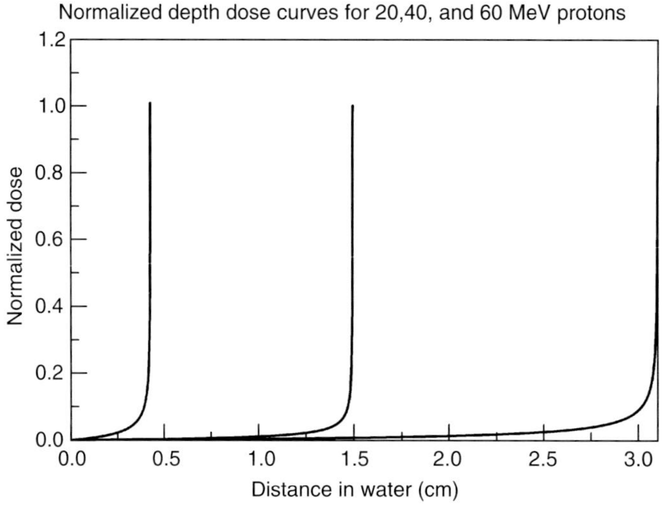

Figure 5.1 Depth dose curves for 20 (left curve), 40 (middle curve), and 60 (right curve) MeV protons in water. The peak dose is normalized to unity. This figure was initially published in Bevelacqua (2010c).

where r is the penetration depth. The dose is normalized such that the peak dose has a unit value. The dose distribution has the expected Bragg shape. Straggling effects are not included in Figure 5.1.

Since the proton dose distributions exhibit a sharp Bragg peak, larger tumors require a range of proton energies to ensure its entire volume is irradiated. A range of proton energies provides a series of Bragg peaks that deliver significant absorbed dose to the tumor site.

5.5.2
External Beam Heavy Ion Therapy

A heavy ion is usually defined to be a nucleus having a mass heavier than a proton. This definition is consistent with most therapy literature and reflects terminology utilized to describe ion beam development as it progressed from proton, deuteron, and alpha particle beams to beams of heavier nuclides.

Conventional beam radiotherapy with photons and electrons is limited because healthy tissue is irradiated during treatment. In order to overcome the physical and biological limitations of conventional radiotherapy, the use of heavy ions was proposed by Wilson. Heavy ions facilitate the deposition of a higher absorbed dose to a deep-seated tumor than achieved with proton therapy. This is accomplished because heavy ions exhibit an inverse absorbed dose profile with dose increasing with penetration depth. The maximum absorbed dose occurs at the Bragg peak with its elevated stopping power.

For energies relevant to therapy applications, the stopping power ($-\mathrm{d}E/\mathrm{d}x$) of an ion is dominated by electronic collisions. Using relativistic quantum mechanics,

Bethe derived the following equation for the stopping power of a charged particle in a uniform medium:

$$-\frac{dE}{dx}=\frac{4\pi k^2 z^2 e^4 n}{mc^2\beta^2}\left[\ln\frac{2mc^2\beta^2}{I(1-\beta^2)}-\beta^2\right] \quad (5.7)$$

where k is the electric constant, z is the atomic number of the ion, e is the magnitude of the electric charge, n is the number of electrons per unit volume in the medium interacting with the ion, m is the electron rest mass, c is the velocity of light in a vacuum, β is the velocity of the particle relative to the speed of light (v/c), v is the velocity of the ion, and I is the mean excitation energy of the medium interacting with the ion.

Using relativistic mechanics, β is determined from the total energy (W) and rest energy (E_o) of the ion:

$$W = E + E_o \quad (5.8)$$

$$E_o = m_o c^2 \quad (5.9)$$

$$W=\frac{m_o c^2}{\sqrt{1-\beta^2}} \quad (5.10)$$

where E is the ion's kinetic energy and m_o is the ion's rest mass. Equations (5.8–5.10) lead to an expression for the ion's kinetic energy:

$$E = W - E_o = m_o c^2\left(\frac{1}{\sqrt{1-\beta^2}}-1\right) \quad (5.11)$$

Equation (5.11) is solved for β:

$$\beta=\left[1-\left(\frac{m_o c^2}{E+m_o c^2}\right)^2\right]^{1/2}=\left[1-\left(\frac{E_o}{W}\right)^2\right]^{1/2} \quad (5.12)$$

The mean excitation energy I is often represented by the following empirical formulas for an element with atomic number Z:

$$I \cong 19.0\,\text{eV},\ Z = 1 \quad (5.13)$$

$$I \cong (11.2 + 11.72Z)\,\text{eV},\ 2 \le Z \le 13 \quad (5.14)$$

$$I \cong (52.8 + 8.71Z)\,\text{eV},\ Z > 13 \quad (5.15)$$

Once the stopping power is known, it is used to calculate the ion's range, which is the distance it travels before coming to rest. The reciprocal of the stopping power is the distance traveled per unit energy loss. Therefore, the range $R(E)$ of a charged particle is the integral of the reciprocal of the negative stopping power from the initial kinetic energy E_i to the final kinetic energy of a stopped particle ($E = 0$):

$$R(E) = \int_{E_i}^{0} \left(\frac{dE}{dx} \right)^{-1} dE \tag{5.16}$$

Equation 5.16 is often written in terms of the stopping power:

$$R(E) = \int_{0}^{E_i} \left(-\frac{dE}{dx} \right)^{-1} dE \tag{5.17}$$

As an ion beam loses energy, it broadens in energy, position, and angle. For example, the Bragg peak spreads in energy and has a distinctive width. Each of these spreading mechanisms affects the delivered absorbed dose at the tumor site. Accordingly, energy straggling, range straggling, and angle straggling are briefly addressed.

For a beam of ions, the width of the Bragg peak is obtained by the summation of multiple scattering events that yield a Gaussian energy loss distribution often referred to as *energy straggling*:

$$\frac{N(E)dE}{N} = \frac{1}{\alpha\pi^{1/2}} \exp\left[-\frac{\left(E - \overline{E}\right)^2}{\alpha^2} \right] \tag{5.18}$$

Energy straggling represents the specific number $N(E)$ of particles having energies in the range E to $E + dE$ divided by the number of particles N, with mean energy $\overline{E}$ after traversing a thickness x_o of absorber. The distribution parameter or straggling parameter (α) expresses the half-width at the $(1/e)$th height and is given by the expression

$$\alpha^2 = 4\pi z^2 e^4 nZx_o \left[1 + \frac{KI}{mv^2} \ln\left(\frac{2mv^2}{I} \right) \right] \tag{5.19}$$

where K is a constant depending on the electron shell structure of the absorber and has a value between 2/3 and 4/3 and Z is the atomic number of the absorber. It is also possible to recast Eq. (5.19) to represent the full width at half maximum (FWHM) height.

In an analogous manner, the *range straggling*, expressed as the number of particles $N(R)$ with ranges R to $R + dR$ divided by the total number of particles of the same initial energy, is given by the equation

$$\frac{N(R)dR}{N} = \frac{1}{\alpha\pi^{1/2}} \exp\left[-\frac{\left(R - \overline{R}\right)^2}{\alpha^2} \right] \tag{5.20}$$

where $\overline{R}$ is the mean range.

Upon entering a medium of thickness x_o, a collimated beam experiences multiple collisions that broaden the beam and cause it to diverge. This phenomenon is called *angle straggling*, and the mean divergence angle ($\overline{\theta}$) is given by the relationship

$$\overline{\theta^2} = \frac{2\pi z^2 e^4}{\overline{E}^2} nZ^2 x_o \ln\left(\frac{\overline{E} a_o}{zZ^{4/3} e^2} \right) \tag{5.21}$$

Table 5.7 ^{12}C ion range and straggling widths in water.

Energy (MeV/nucleon)	Range @ peak dose location (cm)[a]	Straggling FWHM (cm)[a]	Range (cm)[b]
90	2.13	0.07	2.14[c] (2.12)[d]
198	8.28	0.23	8.54[c] (8.45)[d]
270	14.43	0.5	14.5[c] (14.3)[d]
330	20.05	0.7	20.2[c] (19.9)[d]

a) Weber (1996).
b) Equation 5.16.
c) The mean excitation energy is based on Eqs. (5.13–5.15).
d) Values in parenthesis are based on the SPAR Code (see Appendix E).

where a_o is the Bohr radius:

$$a_o = \frac{\hbar^2}{kme^2} \tag{5.22}$$

The position of the Bragg peak and straggling FWHM are summarized in Table 5.7. The values are provided for ^{12}C ions with energies between 90 and 330 MeV/nucleon. Table 5.7 indicates that the particle range is a reasonable approximation to the Bragg peak for heavy ions.

Table 5.8 provides the range in water for a number of heavy ions including ^{4}He, ^{12}C, ^{16}O, ^{20}Ne, ^{40}Ca, ^{63}Cu, ^{92}Mo, ^{107}Ag, ^{142}Nd, ^{172}Hf, ^{184}Os, ^{197}Au, ^{209}Bi, ^{238}U, and ^{236}Np. The ion ranges are evaluated for typical therapy energies between 90 and 330 MeV/nucleon. Water is selected as a surrogate for tissue composition. The use of water simplifies the calculation without significantly introducing a large error.

The results of Table 5.8 illustrate that desired irradiation locations are achieved by selecting specific ion and energy combinations. This characteristic is highly desirable in providing an effective treatment protocol. The ability to target a specific location by selecting the ion and its energy makes heavy ions an attractive tool for future therapy applications.

The ion range estimate is a preliminary step in formulating a therapy protocol. Therapy planning also involves an estimate of delivered dose and the distribution of that dose within the tumor volume.

For a volume irradiated by a parallel beam of particles of a particular radiation type, the absorbed dose (D) as a function of penetration distance x into this volume is

$$D(x) = \frac{1}{\rho}\left(-\frac{dE}{dx}\right)\Phi(x) \tag{5.23}$$

where ρ is the density of the material (e.g., bone, tissue, and tumor) attenuating the heavy ion, $-dE/dx$ is the stopping power, and Φ is the heavy ion fluence. The particle fluence varies with penetration distance according to the relationship

$$\Phi(x) = \Phi(0)\exp(-\mu x) \tag{5.24}$$

Table 5.8 Heavy ion ranges in water (cm) for selected energies[a)].

	Ion energy (MeV/nucleon)			
Ion	**90**	**198**	**270**	**330**
^{4}He	6.42	25.6	43.4	60.5
^{12}C	2.14	8.54	14.5	20.2
^{16}O	1.60	6.40	10.8	15.1
^{20}Ne	1.28	5.12	8.67	12.1
^{40}Ca	0.64	2.56	4.34	6.05
^{63}Cu	0.48	1.92	3.25	4.53
^{92}Mo	0.34	1.34	2.26	3.15
^{107}Ag	0.31	1.24	2.10	2.93
^{142}Nd	0.25	1.01	1.71	2.38
^{172}Hf	0.21	0.85	1.44	2.01
^{184}Os	0.20	0.82	1.38	1.93
^{197}Au	0.20	0.81	1.37	1.91
^{209}Bi	0.19	0.78	1.32	1.83
^{238}U	0.18	0.72	1.22	1.70
^{236}Np	0.18	0.70	1.18	1.65

a) Bevelacqua (2005b).

where $\Phi(0)$ is the entrance fluence and μ is the macroscopic reaction cross-section (linear attenuation coefficient). The linear attenuation coefficient is defined as

$$\mu = n\sigma \tag{5.25}$$

where n is the number of atoms of absorbing material per unit volume and σ is the total microscopic reaction cross-section for the heavy ion–tissue interaction.

For practical radiotherapy applications, the energy of the primary beam is varied to alter the Bragg peak position. By varying the energy, the tumor volume is irradiated with a series of overlapping Bragg peaks.

The distribution from each energy is summed to obtain the total dose profile. When performing the summation, the absorbed dose is modified by an energy-dependent weighting factor. Kraft notes that for ^{12}C the RBE initially increases by factors of 2–4 when the heavy ion slows. From a practical standpoint, the complex heavy ion interaction sequence must be known when variations of the RBE are included in dose specification and optimization.

When calculating the absorbed dose to a complex medium such as tissue, the methodology is modified. In particular, modifications to the linear attenuation coefficient and stopping power are required.

For a medium, such as tissue composed of hydrogen (5.98×10^{22} atoms/cm^3), oxygen (2.45×10^{22} atoms/cm^3), carbon (9.03×10^{21} atoms/cm^3), and nitrogen (1.29×10^{21} atoms/cm^3), the attenuation coefficient is the summation over the product of each component element attenuation coefficients times their number density:

$$\mu = \sum_i \mu_i n_i \tag{5.26}$$

In a similar fashion, the stopping power for a medium, composed of a number of elements i having charge Z_i, number density n_i, and mean excitation I_i, is obtained through a modification of Eq. (5.7). For a complex medium, the following substitution is made in Eq. (5.7):

$$\frac{n}{\ln I} \rightarrow \sum_i \frac{n_i Z_i}{\ln I_i} \tag{5.27}$$

When a heavy ion beam interacts with tissue, the interactions leave nuclei in an excited state, and these excited nuclei decay by a variety of processes including particle and photon emission. These radiation types must be considered in therapy dose planning because they broaden the Bragg peak.

As an example, consider a therapy protocol in which primary ^{16}O ions impinge on tissue and a portion of the beam produces ^{15}O and neutron fragments. The delivered absorbed dose (D) is composed of three components:

$$D = D(^{16}\text{O}) + D(^{15}\text{O}) + D(\text{n}) \tag{5.28}$$

where the first term accounts for the ^{16}O absorbed dose from the primary beam. ^{16}O fluence decreases as it penetrates tissue:

$$\Phi(^{16}\text{O}, x) = \Phi(^{16}\text{O}, 0)\exp(-\mu x) \tag{5.29}$$

where μ is the ^{16}O total macroscopic reaction cross-section in tissue. The second and third terms in Eq. (5.28) are the absorbed dose contributions that arise from the fragmentation of $^{16}\text{O} \rightarrow {}^{15}\text{O} + \text{n}$. Neutron and ^{15}O fluence depend on a number of factors including the ^{16}O fragmentation cross-section as a function of energy and angle, the ^{15}O and neutron reaction cross-sections in tissue, and their associated energy-dependent tissue interactions.

The dose distribution depends on the number of primary beam particles, fragments, and reaction products. For each component, the energy, interaction angular distribution, and RBE are key aspects of therapy dose planning. In particular, the increased entrance RBE to healthy tissue that occurs with increasing ion atomic number must be balanced against the larger RBE values that occur at the tumor site. The optimum ion and energy combination depends on the specific tumor location and its characteristics. In addition, higher Z ions should be evaluated to optimize therapy dose delivery. As future therapy beams are developed, the option of selecting a variety of beams becomes more likely. ^{12}C beams have been developed and utilized in practical applications, but limited beams are available for $A \geq 20$ ions.

One of the key aspects of the delivered dose is the beam's spatial distribution. The spatial distribution can be measured by monitoring the positron–electron annihilation photons resulting from primary beam fragmentation products. For the Eq. (5.28) example, ^{15}O would provide a measure of the beam profile by monitoring the annihilation of the emitted positrons using PET.

Table 5.9 lists candidate heavy ion therapy beams and their possible PET fragmentation products. For simplicity, Table 5.9 considers the most likely positron-emitting fragments that occur by neutron removal from the primary beam ion. The results of Table 5.9 suggest that heavy ion beams from elements spanning the entire periodic table can be monitored using PET techniques to ensure the beam's effectiveness in tumor irradiation.

An examination of Table 5.9 suggests that *in situ* beam monitoring via PET is possible throughout most of the periodic table. However, it becomes more difficult as the mass increases beyond about $A = 200$.

Another key aspect of Eqs. (5.24) and (5.25) is the use of the total microscopic reaction cross-section to obtain the total macroscopic reaction cross-section. The total microscopic reaction cross-section is obtained from data parameterizations or the use of optical model codes such as Distorted Wave University of Colorado Kunz (DWUCK) or MERCURY. These computer codes are summarized in Appendix E.

Parametric fits to available cross-section data use established relationships including trends in nuclear radii, reaction kinematics, and energy dependence. The optical model codes require parameterization of the entrance and exit channel reactions, nuclear structure information for the transferred particles, spectroscopic information, and specification of kinematic information related to the reaction under investigation. Each of these approaches has inherent shortcomings, and careful evaluation is required. The best practice is to use measured data. However, the use of models is often required because a complete set of cross-sections are often not available.

5.5.3
External Pion and Muon Beams

Pions and muons are additional candidates for external beam therapy. Table 5.10 summarizes the ranges of these radiation types and energies that would be of potential interest in therapy applications using external beams. For comparison, the ranges of protons and alpha particles are provided. The ranges in Table 5.10 suggest that a variety of radiation types could be utilized in therapy applications.

Pion absorbed dose deposition in the Bragg peak tends to be broader than proton and heavy ion peaks. In particular, three features of the negative pion depth dose curve are relevant to therapy applications. First, as the penetration depth increases with increasing beam energy, the width of the Bragg peak is broadened. Second, a higher-energy beam penetrates deeper into tissue and has a smaller Bragg peak to entrance plateau absorbed dose ratio. Finally, the absorbed dose in the region beyond the Bragg peak (about 10% of the peak value) is due mostly to the high-energy muons and electrons that arise from pion decay (e.g., $\pi^- \rightarrow \mu^- + \bar{\upsilon}_\mu$) and muon decay (e.g., $\mu^- \rightarrow e^- + \upsilon_\mu + \bar{\upsilon}_e$). There is also some contribution from protons and neutrons. For example, the range (Bragg peak width) at negative pion energies of 150, 170, 190, and 207 MeV are approximately 10 (2 cm), 15 (3 cm), 21 (4 cm), and 27 cm (5 cm), respectively.

Table 5.9 Candidate heavy ion fragmentation products that can be monitored using PET[a].

Primary ion	Positron-emitting fragmentation products[b]	Positron-emitting fragmentation product half-life[c]
^{4}He	—	—
^{12}C	^{9}C	127 ms
	^{10}C	19.308 s
	^{11}C	20.36 m
^{16}O	^{13}O	8.6 ms
	^{14}O	70.62 s
	^{15}O	2.037 min
^{20}Ne	^{17}Ne	109 ms
	^{18}Ne	1.667 s
	^{19}Ne	17.22 s
^{40}Ca	^{35}Ca	25.7 ms
	^{36}Ca	101 ms
	^{37}Ca	181 ms
	^{38}Ca	0.44 s
	^{39}Ca	861 ms
^{63}Cu	^{62}Cu	9.74 min
	^{61}Cu	3.35 h
	^{60}Cu	23.7 min
	^{59}Cu	1.36 min
	^{58}Cu	3.21 s
^{92}Mo	^{91}Mo	15.5 min
	^{91m}Mo	64 s
	^{90}Mo	5.7 h
	^{89}Mo	2.0 min
	^{88}Mo	8 min
	^{87}Mo	14 s
^{107}Ag	^{106}Ag	24.0 min
	^{105}Ag	41.3 days
	^{104}Ag	1.15 h
	^{104m}Ag	33 min
	^{103}Ag	1.1 h
	^{102}Ag	13.0 min
	^{102m}Ag	7.8 min
^{142}Nd	^{141}Nd	2.49 h
	^{139}Nd	30 min
	^{139m}Nd	5.5 h
	^{137}Nd	38 min
^{172}Hf	^{171}Hf	12.2 h
	^{169}Hf	3.25 min
	^{168}Hf	25.9 min
	^{167}Hf	2.0 min

(continued overleaf)

Table 5.9 (*continued*)

Primary ion	Positron-emitting fragmentation products[b)]	Positron-emitting fragmentation product half-life[c)]
^{184}Os	^{183}Os	13 h
	^{181}Os	2.7 min
	^{181m}Os	1.75 h
	^{179}Os	6.5 min
^{197}Au	^{196}Au	6.167 days
	^{194}Au	1.58 d
	^{192}Au	4.9 h
^{209}Bi	^{207}Bi	32.0 years
	^{206}Bi	6.243 days
	^{205}Bi	15.31 days
^{238}U	—	—
^{236}Np	^{234}Np	4.4 days

a) Bevelacqua (2005b).
b) The fragmentation products are limited to the loss of five neutrons from the primary ion to display the most likely nuclides.
c) Baum *et al.* (2010).

Table 5.10 Pion, muon, proton, and alpha particle ranges in water (cm) for selected energies[a)].

	Energy (MeV)			
Ion	10	20	30	40
π^+ or π^-	0.56	1.92	3.89	6.33
μ^+ or μ^-	0.70	2.36	4.71	7.58
p	0.12	0.42	0.88	1.48
^{4}He	0.011	0.036	0.074	0.12

a) Bevelacqua (2010c).

Some pions are lost from the external beam by decay into a muon and a neutrino. This is important in pion therapy because muons are contaminants of pion beams. Although about 211 times heavier than the electron, muon interactions are similar to electron interactions. As such, they could offer a mechanism for the selective energy deposition of absorbed dose at a tumor site. For example, the Bragg peak width for a 30 MeV negative muon is about a centimeter.

Both muons and charge pions could be utilized in therapy applications if the beams could be matched to the tumor depths and sizes. Although absorbed dose localization is not as good as exhibited by the proton and heavy ion Bragg peaks, charged pions and muons offer another tool for therapy applications. These applications could be enhanced if internal radiation-generating devices described in Section 5.6.3.2 produced these beams.

5.5.4 External Beam Antimatter Therapy

The deposition of energy from a photon, electron, muon, pion, proton, or heavy ion at a tumor site is typically 100 MeV or less. Although this energy deposition has value in therapy applications, higher-energy depositions are possible through annihilation events involving hadrons. An initial application of this concept is antiproton therapy incorporating the annihilation of a proton and antiproton.

The conversion of the mass of a proton–antiproton pair during an annihilation event provides a source of high-energy deposition that could have potential applications in cancer therapy. In a manner similar to the energy deposition of protons and heavy ions, antiprotons deposit most of their kinetic energy at the end of their range in the Bragg peak. The antiproton annihilates when it strikes a proton and deposits additional energy. Upon annihilation, new particles and photons are created and deposit a portion of their energy within the tumor site. However, most of the available annihilation energy of 1.88 GeV is removed from the tumor site when charged pions, neutrons, and photons escape from the target volume.

Initial investigations at the *Centre (Organisation) Européen pour la Recherche Nucléaire* (CERN) estimate that the dose deposition from antiprotons is similar to that reported for a proton but the Bragg antiprotons deposit an additional 30 MeV within a few millimeters of the annihilation event. Although the additional local energy deposited is small compared to the total proton–antiproton annihilation energy, it doubles the absorbed dose deposited per particle in the Bragg peak compared to protons. In addition, the RBE of the particles contributing to the additional dose is higher than that for protons because it includes recoiling heavy fragments produced in the annihilation event. A portion of the remaining annihilation energy leaving the tumor site and exiting the body could potentially be used for real-time imaging of the dose distribution.

The resulting antiproton depth dose curve can be compared with that of protons measured under similar conditions. These curves indicate that the ratio of the dose at the antiproton stopping power peak to that in the plateau region is about twice that found for protons. Proton–antiproton annihilation events result in the production of an average of 4–5 charged and neutral pions with a mean energy about 400 MeV plus 3 high-energy gamma rays. Since the ranges of 100, 300, 500, and 1000 MeV charged pions in water are 27, 120, 217, and 453 cm, respectively, much of the charged pion energy does not deposit at the tumor site. However, the photons offer the potential to track the beam and optimize energy deposition using computed tomography techniques.

Although the dose to healthy tissue appears to be sizeable in these initial investigations, the antimatter approach has yet to be fully optimized. As such, it is another potential tool for twenty-first-century applications.

A theoretical therapy approach involves complex antimatter such as $^{12}\overline{\mathrm{C}}$ or $^{16}\overline{\mathrm{O}}$. The absorbed dose profile as a function of depth into tissue would resemble the conventional heavy ion curve, but there would be an enhancement of dose deposition in the Bragg peak and at the annihilation location. The available energy from a $^{16}\overline{\mathrm{O}} + {}^{16}\mathrm{O}$ annihilation event is substantial and includes a contribution from charged pions and their decay products.

Although antinuclei, particularly those that could annihilate with the major constituents of tissue, could deliver significant dose, their production, storage, and administration are problematic. No antinuclear systems as heavy as $^{12}\overline{\mathrm{C}}$ or $^{16}\overline{\mathrm{O}}$ have yet to be produced, and their production is beyond current technology. In fact, antinuclei heavier than $A = 4$ have yet to be experimentally observed.

5.5.5
Alpha Radiopharmaceuticals

Beta-emitting radionuclides are commonly used in medical therapy applications. The beta radiation destroys tumors, but their range often extends beyond the tumor site. An alternative to beta emitters is the use of alpha-emitting radionuclides. Targeted alpha therapy offers a significant benefit for patients as well as hospital staff. However, the use of targeted alpha therapy is limited by the availability of appropriate radionuclides and the ability to produce them in a cost-effective manner.

The use of alpha-emitting radionuclides introduces the potential for internal intakes of these materials. Accordingly, appropriate contamination control practices are warranted to minimize surface contamination and preclude internal intakes. The production of these materials by accelerator techniques introduces the potential for external exposures either from accelerator beams or their associated reaction products.

Table 5.2 summarizes the physical properties of alpha-emitting radionuclides currently in use or under development. In some cases, these nuclides are members of a decay chain and can be derived from a supply of the parent nuclide. For example, ^{213}Bi and ^{225}Ac are daughters of ^{233}U and have been used in leukemia and solid tumor treatment, respectively.

^{223}Ra has therapy applications to relieve bone cancer pain. This radionuclide is produced in an accelerator and can be extracted from the target material using chemical separation techniques. A common production mode of ^{223}Ra is

$$^{226}\mathrm{Ra} + \mathrm{n} \rightarrow {}^{227}\mathrm{Ra} \xrightarrow{\beta^-} {}^{227}\mathrm{Ac} \xrightarrow{\beta^-} {}^{227}\mathrm{Th} \xrightarrow{\alpha} {}^{223}\mathrm{Ra} \tag{5.30}$$

Other alpha emitters under investigation for therapy applications include ^{149}Tb, ^{211}At, ^{212}Bi, and ^{224}Ra.

Accelerators, reactors, and chemical separation from decay chains offer the most practical methods of producing these alpha emitters. Ideally, generators analogous to the ^{99}Mo generators used to produce ^{99m}Tc will be developed to facilitate the use and application of alpha-emitting radionuclides.

One possible candidate is a ^{225}Ac^{213}Bi generator with the ^{213}Bi obtained following elution from a column containing ^{225}Ac and its daughter products. This generator is based on the chemical properties of the nuclides comprising the ^{233}U

Table 5.11 Comparison of ^{131}I, ^{90}Y, and ^{213}Bi dosimetry for acute myeloid leukemia[a].

	Mean absorbed dose per unit administration (mSv/MBq)			Absorbed dose ratios	
Isotope	Red marrow	Liver	Whole body	Red marrow/liver	Red marrow/whole body
^{131}I	2.7	0.8	0.16	3.4	14.4
^{90}Y	6.8	4.0	0.49	1.9	13.9
^{213}Bi	9.8	5.8	0.0004	1.7	27 300

a) Jurcic *et al.* (2002).

decay chain. A possible production/decay chain for an ^{225}Ac ^{213}Bi generator is

$$^{232}\text{Th} + \text{n} \rightarrow {}^{233}\text{Th} \xrightarrow{\beta^-} {}^{233}\text{Pa} \xrightarrow{\beta^-} {}^{233}U \xrightarrow{\alpha} {}^{229}\text{Th} \xrightarrow{\alpha}$$
$$^{225}\text{Ra} \xrightarrow{\beta^-} {}^{225}\text{Ac} \xrightarrow{\alpha} {}^{221}\text{Fr} \xrightarrow{\alpha} {}^{217}\text{At} \xrightarrow{\alpha} {}^{213}\text{Bi} \tag{5.31}$$

The generator could also be produced from an available supply of ^{233}U.

As an illustration of the advantage of targeted alpha therapy, ^{213}Bi trial results are compared to therapeutic administrations of ^{131}I and ^{90}Y to treat myeloid leukemia. This leukemia type arises from abnormal growth in the blood-forming tissue of the bone marrow or spinal cord.

A number of differences are noted when comparing conventional ^{131}I and ^{90}Y myeloablation therapy to ^{213}Bi therapy. These include the longer half-lives, use of multiple infusions to deliver ablative doses, and need for hospitalization and radiation isolation of the patient for the ^{90}Y and ^{131}I approaches. The higher-energy ^{90}Y beta particles and ^{131}I photons have a longer range when compared to the ^{213}Bi alpha particles. Irradiation of healthy tissue is significantly minimized when utilizing ^{213}Bi as noted in Table 5.11.

Table 5.11 illustrates the capability to localize dose in the tissue of interest using alpha therapy. In treating myeloid leukemia, the target tissue is the bone marrow. As noted in Table 5.11, limited dose is provided to the whole body from the targeted alpha therapy. Liver doses from ^{213}Bi are comparable to ^{90}Y administrations, but larger than the ^{131}I doses. However, the standard deviations in the absorbed doses estimates are often large.

The other nuclides included in Eq. (5.31) also offer possible alpha radionuclide therapy options. ^{221}Fr (4.79 min) and ^{217}At (32 ms) emit alpha particles with energies of 6.127 and 6.341 and 7.067 MeV, respectively. These energies are higher than the currently utilized alpha-emitting radionuclides noted in Table 5.2.

5.5.6
Neutron Therapy

Neutrons have also been used in therapy applications because they have a larger RBE value than protons, pions, muons, photons, and electrons. Although neutron therapy produces a significant dose deposition, this positive feature must be balanced with side effects due to the poor dose localization.

Neutron therapy is performed employing a range of energies including the thermal, epithermal, and fast regimes. Neutron capture therapy uses the thermal and epithermal energy regions, and fast neutron therapy (FNT) incorporates the highest energy regime.

In neutron capture therapy, an isotope with a large absorption cross-section for thermal or epithermal neutrons (e.g., ^{10}B) is introduced into the body usually through injection into the blood. The isotope is part of a molecule that is designed to accumulate in a particular tissue. For example, neutron capture therapy is used to activate ^{10}B-tagged molecules that accumulate in brain tumors (glioblastoma). BNCT selectively irradiates these tumor cells. Selectively delivering ^{10}B to the tumor site and then irradiating the tumor with thermal/epithermal neutrons achieves dose localization. The short range of the $^{10}\text{B}(\text{n}, \alpha)^{7}\text{Li}$ reaction products limits most of the dose to the boron-loaded tumor cells.

FNT uses neutron beams with energies of about 15 MeV. The therapeutic effect is achieved from recoiling protons and heavier fragments resulting from fast neutron interactions. A variety of tumors, including those of the head and neck, salivary glands, and soft tissue, are treated with fast neutrons. For BNCT or FNT, neutrons can be generated through a variety of reactions including $^{2}\text{H}(\text{d, n})^{3}\text{He}$ and $^{3}\text{H}(\text{d, n})^{4}\text{He}$.

Neutron therapy can be highly effective in treating inoperable, radioresistant tumors occurring anywhere in the body. Fast neutrons have the capability to control very large tumors because FNT does not depend on the presence of oxygen to kill the cancer cells. In addition, the FNT approach is not affected by the life cycle stage of the cancer cells. Given their large RBE, the neutron absorbed dose required to kill cancer cells is about one-third the dose required with photons, electrons, or protons. Moreover, a full course of neutron therapy is delivered in only 10–12 treatments compared to 30–40 treatments needed for low linear energy transfer (LET) radiation.

While neutron beam effectiveness depends on the patient, cancer stage, and histology, a number of tumors have been treated using this approach. These applications include tumors residing in the abdomen and pelvis (e.g., prostate, kidney, and uterus), chest (e.g., lung), extremities (e.g., bone, cartilage, and soft tissue), head and neck (e.g., nasopharynx, oral cavity, pharynx, salivary glands, and tongue), skin (e.g., melanoma), and trunk (e.g., bone, cartilage, and soft tissue).

One of the limitations for standardization of neutron therapy approaches is the fact that neutron facilities have pronounced differences in beam characteristics, energy spectra, and collimation techniques. This therapy approach is also limited to facilities having a source of neutron radiation. The capability of producing neutrons is usually limited to facilities having a research reactor or accelerator.

Although neutron therapy has been extensively utilized, it offers a number of emerging applications. These applications offer the potential to address cancers that currently have limited options. Possible applications require additional research and include inoperable pancreatic and rectal cancers. Additional applications include treatment of bladder, cervix, esophagus, and uterine tumors.

5.5.7 Radionuclide Vascular Therapy

Conventional radiotherapy often involves the deposition of the radionuclide within a tumor mass. The radiation from the decay of the radionuclide is used to deposit absorbed dose within the tumor mass and reduce its size. With sufficient absorbed dose, the tumor can be eradicated.

An alternative technique adopts an approach used in chemotherapy. Instead of attacking the tumor mass, a radionuclide is utilized to destroy the tumor's blood supply and deprive it of nutrients. Vascular disruption agents have been incorporated into chemotherapy and radiotherapy procedures. These approaches are also known as *antiangiogenic* or *radioembolization therapies.*

Radiotherapy approaches to vascular disruption have been extensively applied to liver cancers utilizing ^{90}Y microspheres. Other radionuclides (e.g., ^{32}P) have been less thoroughly investigated, and radiation types other than high-energy beta particles have not been systematically investigated.

5.5.7.1 Tumor Vasculature

One of the most striking characteristics of solid tumors is their vascular configuration. In normal tissues, the vasculature structure is arranged to provide optimum nourishment conditions. However, growing tumors have a chaotic vasculature that is not fully developed or adequate to optimally nourish the tumor cells. Given this condition, the tumor's vasculature can be disrupted with an appropriate agent.

Common defects in a tumor's vascular structure include vessels that are dilated and have elongated shapes, blind ends, bulges, leaky sprouts, and abrupt changes in diameters. Accordingly, blood flow in these vessels is sluggish and irregular. This flow pattern furnishes less nourishment than delivered to normal cells and results in hypoxic areas that are characteristic of solid tumors. These hypoxic conditions limit the effectiveness of both chemotherapy and radiotherapy. The lack of oxygen provides a degree of radioresistance to tumor cells when compared with oxygenated cells. Since a tumor's growth is dependent on sufficient nourishment, its viability is affected by disrupting the blood supply. In principle, eliminating a tumor's blood supply offers an alternative means to facilitate or supplement its destruction.

5.5.7.2 Current Radiological Efforts

Radiological efforts at tumor vascular disruption have focused on ^{90}Y. ^{90}Y was a logical choice for antiangiogenic therapy since the dose to destroy a tumor is $\geq$70 Gy. This absorbed dose is easily achieved using ^{90}Y. However, the ^{90}Y beta particles have significant range and extend well beyond the vascular target.

Table 5.12 Properties of resin and glass ^{90}Y microspheres[a)].

	Microsphere type	
Parameter	**Resin**	**Glass**
Diameter (μm)	20–60	20–30
Density (g/cm^3)	1.6	3.6
Activity per microsphere (Bq)	50	2500
Number of microspheres per 3 GBq vial ($\times 10^6$)	40–80	1.2
^{90}Y form	Yttrium bound to resin	Yttrium in glass matrix

a) Kennedy *et al.* (2007).

The maximum beta energy of 2.27 MeV has a range in tissue of about 1.1 cm, which delivers considerable absorbed dose beyond the target vascular structure. The properties of ^{90}Y microspheres used in therapy applications are summarized in Table 5.12.

Medical reviews suggest the ^{90}Y approach is a safe and effective therapy method for selected patients. However, a number of negative features are associated with ^{90}Y microsphere therapy. These negative aspects include the following:

1) ^{90}Y bremsstrahlung affects healthy tissue well beyond the vasculature.
2) Resin microspheres may have trace ^{90}Y on their surface, which can be excreted via urine and deliver absorbed dose to healthy tissues in the excretion pathway. This radiological concern does not exist for glass microspheres.
3) For glass microspheres, 150 Gy is the dose delivered to the target tissue. The recommended cumulative lung dose should be maintained below 30 Gy to prevent radiation pneumonitis.
4) ^{90}Y microsphere patients can experience mild postembolization syndrome for up to 3 days after treatment. Symptoms associated with this condition include abdominal pain, fatigue, and nausea.
5) Radioembolization to nontarget tissues can cause other acute damage that results in gastrointestinal ulceration, pancreatitis, and radiation pneumonitis. Late effects can include radiation-induced liver disease.

5.5.7.3
Theoretical Approach

Destroying a tumor's vascular structure using radiotherapy could be performed as a stand-alone protocol, as the first therapy step followed by a chemical vascular disrupting agent (VDA), as a final step in a therapy procedure initiated by a VDA, or as part of an alternating sequence involving both radiotherapy and chemical VDAs. This section addresses candidate radionuclides and radiation types that could be used to disrupt a tumor's vasculature.

Table 5.13 Characteristics of various blood vessel types[a].

Blood vessel type	Wall thickness	Lumen diameter
Aorta	2 mm	25 mm
Artery	1 mm	4 mm
Arteriole	20 μm	30 μm
Capillary	1 μm	8 μm
Venule	2 μm	20 μm
Vein	0.5 mm	5 mm
Vena cava	1.5 mm	30 mm

a) Barrett *et al.* (2012).

For an effective radiotherapy approach, the absorbed dose delivered to the blood vessels should be maximized while minimizing the dose to healthy tissue. Delivering stray dose to the tumor would also be acceptable, but minimizing dose to healthy tissue is strongly desired.

The blood supply to a tumor could be disrupted by damaging the vessel wall, causing it to become restricted or increase its leakage. Delivering absorbed dose preferentially to the blood vessel wall facilitates disruption. Table 5.13 summarizes wall thickness for a variety of human blood vessel types.

Table 5.13 suggests there are a variety of blood vessel types that could service a developing tumor. A review of the literature indicates that developing tumor vessel wall sizes are typically less than 100 μm. This wall size includes arterioles, which is the assumed base case for the discussion presented in this chapter.

5.5.7.4
Other Candidate Microspheres

An alternative to the use of ^{90}Y is provided by radionuclides that emit low-energy photons, low-energy beta particles, or alpha particles. These radionuclides would replace ^{90}Y as the radioactive material loading the microspheres.

For simplicity, a single microsphere is assumed to deliver the requisite disruption dose to the arteriole wall. The actual microsphere activity is distributed over a large number of microspheres (see Table 5.12) that are collectively designed to disrupt the tumor's vasculature.

Candidate microspheres are assumed to be loaded with either alpha, low-energy beta, or low-energy beta–gamma-emitting radionuclides. Desirable characteristics for the radionuclide and candidate microsphere to facilitate tumor blood vessel disruption include the following:

1) The nuclide should have a short half-life. In this chapter, an arbitrary half-life limit of 100 days is chosen.
2) The range of the radiation types emitted by the nuclide should be shorter than the range of the ^{90}Y beta particles and associated bremsstrahlung in tissue.
3) The absorbed dose to the arteriole wall should be at least 100 Gy.

4) The dose delivered to healthy tissue should be minimized.
5) The microsphere has the capability to preferentially attach to the wall of an arteriole supplying blood to the tumor.
6) The candidate radionuclide is chemically compatible with a polymer microsphere that can be absorbed into the body. Absorption is assumed to occur after nearly all of the radioactive material decays.

Although these characteristics provide a basis for the calculations presented in this chapter, they have not been optimized to produce a viable alternative to the ^{90}Y microsphere approach. However, they provide an initial set of reasonable parameters to determine the characteristics of a replacement microsphere.

5.5.7.4.1
Microspheres Using Alpha-Emitting Radionuclides

The dose delivery capability of microspheres loaded with alpha-emitting radionuclides is determined if simplifying assumptions are made. First, the microsphere is assumed to be sufficiently small such that it provides minimal attenuation of the emitted alpha particles. Second, the alpha microsphere is assumed to attach preferentially and remain attached to the arteriole wall. Given these limitations, the absorbed dose rate delivered by the microsphere as a function of time t, $\dot{D}(t)$, is

$$\dot{D}(t) = \dot{D}(0)e^{-\lambda t} \tag{5.32}$$

where λ is the physical disintegration constant for the alpha-emitting radionuclide and $t = 0$ corresponds to the time that the microsphere attaches to the arteriole wall. Assuming that the microsphere remains attached to the arteriole wall until all the radioactive material decays, the total absorbed dose (D_T) delivered to the tumor vasculature is

$$D_\mathrm{T} = \int_0^\infty \dot{D}(0)e^{-\lambda t}\mathrm{d}t = \frac{\dot{D}(0)}{\lambda} \tag{5.33}$$

The initial absorbed dose rate is written in terms of a dose conversion factor K:

$$\dot{D}(0) = \frac{A(0)}{4\pi r^2}K \tag{5.34}$$

where r is the distance to the point of interest in the arteriole wall. Substituting the value of the initial dose rate from Eq. (5.33) into Eq. (5.34) yields the initial attached microsphere activity:

$$A(0) = \frac{4\pi r^2 \lambda D_\mathrm{T}}{K} \tag{5.35}$$

where D_T is the total dose delivered to the arteriole wall to facilitate vascular disruption (100 Gy), r is chosen to be 15 μm as a representative depth into the arteriole wall, and K is the ICRP 116 dose conversion coefficient corresponding to the emitted alpha particle energy. If the radionuclide emits multiple alpha particles,

Table 5.14 Candidate alpha-emitting nuclides for loading microspheres[a].

Nuclide	Half-life	Activity to deliver 100 Gy at 15 μm depth in an arteriole wall (kBq)
^{149}Tb	4.12 h[b]	0.152
^{206}Pb	8.80 days[c]	0.00226
^{211}At	7.214 h[c]	0.0588
^{212}Bi	1.009 h[c]	0.407
^{222}Rn	3.823 days[c]	0.00494
^{223}Ra	11.43 days[c]	0.0016
^{224}Ra	3.66 days[c]	0.00499
^{225}Ac	10.00 days[c]	0.00179
^{227}Th	18.718 days[c]	0.00094
^{230}U	20.8 days[c]	0.00085
^{240}Cm	27.0 days[c]	0.00061
^{246}Cf	35.7 h[c]	0.0103
^{253}Es	20.47 days[c]	0.00076

a) Derived from Bevelacqua (2014).
b) Baum *et al.* (2010).
c) Shleien *et al.* (1988).

the dose conversion factor is evaluated at an averaged energy ($\overline{E}$) given by

$$\overline{E} = \frac{\sum_{i=1}^{N} E_i Y_i}{\sum_{i=1}^{N} Y_i} \tag{5.36}$$

where E_i (Y_i) is the energy (yield) of the ith alpha particle and N is the number of alpha particles emitted by the radionuclide.

Table 5.14 provides the activities of candidate alpha microspheres that deliver 100 Gy at a depth of 15 μm into an arteriole wall supplying a tumor. Only the parent nuclide decays are used in these initial calculations, and no daughter contributions are considered.

The radionuclides listed in Table 5.14 are somewhat arbitrary since they were limited to alpha energies in the 3–8 MeV range. The requisite alpha activities to accomplish arteriole disruption are in general smaller than the 0.05–2.5 kBq values used in ^{90}Y microspheres. This is expected since the calculated activities are values corresponding to the time of the microsphere's attachment to the arteriole wall. The manufactured activity depends on the nature of the fabrication process, the time between fabrication and injection, and the time for the microsphere to attach to the arteriole wall following injection. However, the requisite activity levels appear to be achievable with current technology and suggest that alpha particle microspheres can be designed to implement tumor vascular disruption.

The calculations of Table 5.14 are based on a first-order approximation intended to provide an order of magnitude estimate of the requisite activity to initiate vascular disruption. A more sophisticated design effort is required to develop a viable alpha radionuclide microsphere therapy approach, but the results of Table 5.14 suggest that the alpha microsphere approach is viable in the near term. Vascular disruption could also be achieved using internal radiation-generating devices discussed in Section 5.6.3.2.

One of the challenges of using alpha-emitting radionuclides is their difficulty of production and associated availability. Therefore, the radionuclide selected to load the alpha microsphere should be readily available. As such, ^{222}Rn is an attractive possibility if a microsphere can be designed to retain the gas. Retention is certainly achievable as evidenced by the retention of fission gasses in the ceramic uranium dioxide fuel pellet used in commercial power reactors. The ^{222}Rn daughters yield additional dose to the tumor vasculature, which also enhances the approach.

Some of the nuclides presented in Table 5.14 (e.g., ^{240}Cm, ^{246}Cf, and ^{253}Es) are difficult to produce and would not represent likely microsphere candidates. These nuclides are presented to illustrate the impact of higher-energy alpha particles in the range of 6–7 MeV.

The selected alpha-emitting radionuclide will have superior dose localization characteristics when compared to ^{90}Y. Designing an appropriate microsphere is dependent on the characteristics of the selected radionuclide and its physical and chemical characteristics.

5.5.7.4.2
Microspheres Using Low-Energy Beta–Gamma-Emitting Radionuclides

The beta–gamma microsphere concept was investigated using a modification of the ISO-PC code (see Appendix E). In illustrating this option, the beta–gamma source activity was uniformly embedded in a 15 μm radius carbon microsphere with a density of 2 g/cm^3. The model density and size of the microsphere lie within the range of the ^{90}Y microspheres defined in Table 5.12. ISO-PC defined tissue is selected as the composition of the blood vessel material. Table 5.15 summarizes the results of the ISO-PC calculations.

The beta–gamma activity to deliver 100 Gy is obtained using the alpha microsphere calculation as a guide. The total absorbed dose (D_T) is delivered by the complete decay of the activity initially attaching to the blood vessel wall $A(0)$. To obtain the activity A' that delivers 100 Gy to the arteriole wall, a relationship analogous to Eq. (5.35) is utilized:

$$A' = A(0)\frac{100\,\mathrm{Gy}}{D_\mathrm{T}} = A(0)\frac{100\,\mathrm{Gy}}{\dot{D}(0)}\lambda \tag{5.37}$$

For consistency with the alpha microsphere calculation, $t = 0$ corresponds to the time the microsphere attaches to the arteriole wall.

The half-life, ISO-PC nuclide library reference number, and activity to deliver 100 Gy to a depth of 15 μm into the arteriole wall are provided for

Table 5.15 Candidate low-energy beta–gamma-emitting nuclides for loading microspheres[a].

Nuclide	Half-life	ISO-PC nuclide library number[b]	Activity to deliver 100 Gy at 15 μm depth in an arteriole wall (kBq)
^{32}P	14.28 days	459	81.7
^{33}P	25.3 days	056	762
^{35}S	87.2 days	460	440
^{47}Sc	3.349 days	463	47.6
^{72}Se	8.5 days	409	1.95
^{82}Sr	25.36 days	540	1.40
^{83m}Kr	1.86 h	045	1060
^{90}Y	2.669 days	084	311
^{99m}Tc	6.008 h	140	457
^{103}Pd	16.99 days	570	3.64
^{125m}Te	58.0 days	270	1.35
^{125}I	59.4 days	595	1.09
^{169}Er	9.39 days	630	1420
^{189}Ir	13.2 days	665	0.475
^{193m}Pt	14.33 days	677	384

a) Derived from Bevelacqua (2014).
b) ISO-PC library reference numbers define the energy structure of source nuclides. Specific details are provided by Rittmann (2004).

each nuclide listed in Table 5.15. Calculations are also provided for ^{32}P and ^{90}Y to facilitate a comparison with nuclides currently used in microsphere applications.

The model predicts that ^{32}P and ^{90}Y microspheres require 81.7 and 311 kBq, respectively, to deliver 100 Gy to the selected arteriole wall location. As noted in Eq. (5.37), the activity to deliver 100 Gy depends on the radiation type and energy of the radionuclide embedded in the carbon microsphere as well as its half-life. Therefore, nuclides with similar energies and half-lives (e.g., ^{125m}Te and ^{125}I) require similar activity levels to deliver 100 Gy to the arteriole wall. For ^{125m}Te and ^{125}I, the requisite activity is 1–2 kBq. As noted in the alpha particle discussion, the fabricated microsphere activity is larger than $A(0)$ and depends on the specific fabrication process, administration protocol, and characteristics of the selected radionuclide.

The predicted activity levels to achieve 100 Gy are in a range that can be readily incorporated into the microsphere, and there are numerous nuclide options for incorporation into beta–gamma microspheres. In addition, the various nuclides listed in Table 5.15 offer considerable flexibility in developing a beta–gamma microsphere approach.

The choice of a beta–gamma radionuclide for microsphere clinical trials depends on its availability and compatibility with the final microsphere design.

However, the diversity of isotopes summarized in Table 5.15 suggests that a number of options are available for beta–gamma microspheres to provide improved dose localization in comparison with ^{90}Y.

5.6 Nanotechnology

Nanotechnology offers a number of options for radiation therapy applications. These devices have the potential to operate at the cellular level. Nanotechnology applications include the delivery of localized radiation dose to a tumor cell or collection of cells and the possibility of physically repairing damaged cellular structures.

Nanomaterials and nanodevices are available, in development, or planned to facilitate cancer detection, *in vivo* molecular or cellular imaging, and the delivery of radioisotopes to a tumor site. In addition to the delivery of radioactive material, postulated nanomachines (e.g., internal radiation-generating devices) could deliver a radiation beam to the tumor site.

5.6.1 Principles of Nanomedicine

Cancer is a disease in which cell replication fails to be regulated by inherent physiological processes. Cancerous cells form a tumor but in their earliest stages lack their own blood vessels. In this early stage, cancer cells rely on surrounding tissue for their nutrients (e.g., oxygen and glucose). Interior tumor cells release proteins to signal their nutrient deficiency, and these proteins diffuse outward until they reach the blood vessels. This process stimulates the growth of new blood vessels that supply the tumor with nutrients to sustain its rapid growth.

As noted previously, a tumor's rapid growth produces blood vessels that are irregular and have larger gaps than healthy vessels. These gaps vary and are typically in the range of a few hundred nanometers to a few micrometers that are larger than the 2–6 nm pore size in normal blood vessels. Nanoparticles sized in the 10–300 nm range pass through the tumor's blood vessels, but do not penetrate into healthy tissue. Loading the nanoparticles with radioactive material or chemotherapy drugs facilitates the destruction of cancerous tissue without affecting normal tissue. Nanoparticles accumulate in the tumor mass, but competing biological processes can limit their incorporation into the tumor. The use of nanoparticles to deposit radioactive material is similar to the microsphere vascular disruption approach previously addressed in Section 5.5.7. However, the microsphere sizes are on the order of 20–60 μm, while the nanoparticles are on the order of 10–100 nm.

The first consideration limiting nanoparticle incorporation into a tumor mass is the human immune system that treats nanoparticles and viruses in a similar manner. Both nanoparticles and viruses are attacked by phagocytes in a manner analogous to the body's defense system response to bacteria. The attacking phagocytes adhere to the nanoparticles and are transported to the liver and spleen. This action prevents their effectiveness in attacking the tumor and delivers absorbed dose to these structures.

A second effect is the high fluid pressure that develops in tumor cores. In healthy tissue, fluid leaks from blood vessels into surrounding tissue, and it is collected by the lymphatic system, which returns it to the bloodstream. Solid tumors lack an effective lymphatic drainage mechanism, and the leaking fluid increases pressure within the tumor volume. This increase in pressure relative to healthy tissue limits the penetration of the nanoparticle into the tumor volume.

Penetrating into the tumor mass represents a challenge that must be overcome. Overcoming the effects noted previously must be achieved for nanotechnology to reach its full potential. Fortunately, there are approaches to overcome these barriers. For example, antiangiogenic drugs lower the pressure at the tumor core, which facilitates nanoparticle penetration. Multistage nanoparticles are also under investigation. These particles combine the larger particles' ability to accumulate in the tumor mass with smaller particles' ability to penetrate tumor tissue.

Designing nanoparticles to leak out of the tumor's blood vessels into the tumor's mass is only an initial step in tumor destruction. The effectiveness in positioning the nanoparticle within a tumor mass is determined by the physical properties of these particles, their immediate environment, and interaction effects between the particle and its environment. These effects include fluid dynamics within the tumor blood vessels, sheer forces on blood vessel walls, velocity of the nanoparticle, and permeability of local tissues.

There are numerous clinical trials involving the use of nanoparticles to treat cancer. Many involve the use of nanoparticles to transport proven chemotherapy agents. Others utilize new chemical agents and multimodal therapy with radioactive material. Much research remains to develop and optimize nanoparticles into a proven cancer therapy technique.

5.6.2 General Nanotechnology-Based Therapy Techniques

As noted previously, nanotechnology-based therapy techniques include the delivery of radioactive materials or chemotherapy drugs to the tumor site. Other techniques include the use of nanotechnology radiation therapy (NTRT), nanomachines, and magnetic fluid hyperthermia (MFH) technology. Each of these general nanotechnology approaches are addressed in subsequent discussion.

5.6.2.1 Nanotechnology Radiation Therapy

NTRT offers the potential to deliver localized radiation directly to a tumor site while minimizing radiation doses to adjacent healthy tissue. NTRT is a logical extension of brachytherapy. The NTRT device can include a single or combination of radionuclides.

NTRT can also be considered to be a second-generation microsphere approach. The microspheres discussed in Section 5.5.7 are designed to attach to a blood vessel wall. NTRT provides this capability as well as other methods to facilitate tumor destruction.

This technology offers a number of potential advantages over conventional brachytherapy and microsphere devices. The NTRT particles are much smaller, and a large number of devices are administered to reach the desired location and deliver the requisite absorbed dose to the tumor. Nanotechnology devices incorporate biodegradable materials that permit repeated treatments. This is in contrast with ceramic and glass microspheres and other brachytherapy materials such as titanium that are not biodegradable.

Nanotechnology has the potential to administer both chemotherapy and radiotherapy agents, and it can be applied to a range of tumors. In addition, NTRT is delivered directly into the tumor that minimizes collateral dose while maximizing delivered tumor dose. The nanoparticle is designed to target specific types of tumors as well as tumor structures. Nanoparticles offer considerable potential, and their use should increase with the application of innovative materials and radionuclides.

5.6.2.2 Nanomachines

Advancing twenty-first-century technology offers the potential for the development of nanomachines. These machines could repair cellular structures or damage induced by the direct or indirect effects of ionizing radiation. Direct cellular repair by these devices is possible because they have the capability of extended circulation in the bloodstream and are small enough to obtain access to target cells and tissues. Nanomachines are engineered to escape the body's defense mechanisms that remove foreign materials (e.g., the endosome–lysosome process), which is necessary for them to enter the cell. Escaping this defense mechanism is enhanced if their construction uses biocompatible materials.

5.6.2.3 Magnetic Fluid Hyperthermia Therapy

The nonradiological use of nanoparticles to facilitate tumor destruction is being implemented. For example, MFH has been successful in animal studies. Laboratory rats were implanted with a glioblastoma and then given MFH treatment. The treated animals lived an average of 35 days, which is about four times the life expectancy of an untreated rat with brain cancer. Human trials are pending further development to ensure localization of the thermal effect.

The MFH technique uses nanoparticles made of iron oxide and coated with a material such as glucose to facilitate absorption into a tumor mass. Using the MFH approach, nanoparticles are injected into the tumor. With a higher metabolism than normal cells, the tumor cells preferentially absorb the glucose-coated nanoparticles. Following absorption into the tumor cells, a magnetic field is applied. The interaction of the nanoparticles with the magnetic field produces heat, and local temperatures can increase to about 45 °C. This temperature increase causes significant damage to tumor cells. The success of the MFH approach depends on the ability to preferentially localize the nanoparticles within the tumor mass.

5.6.3 Specific Nanoparticle Applications

The previous discussion outlined general nanoparticle techniques that have potential therapy applications. Subsequent discussion addresses techniques that are more completely defined, are currently being implemented, or are under investigation. These techniques include nanoparticles loaded with radioactive materials. Other techniques are theoretical and beyond current technology but have been sufficiently defined to suggest their future development and implementation. These emerging techniques include internal radiation-generating devices.

A variety of other techniques are available to treat cancer either by advanced imaging or therapy techniques. These techniques incorporate both ionizing and nonionizing radiation. Ionizing radiation techniques include hybrid medical imaging and Cherenkov luminescence imaging (CLI).

In addition to the aforementioned ionizing radiation techniques, emerging nonionizing techniques are available to diagnose disease and repair its detrimental effects. These nonionizing techniques include photodynamic therapy (PDT), low-coherence interferometry (LCI), nonlinear interferometric vibrational imaging (NIVI), and optical coherence tomography (OCT). Each of these techniques is addressed in subsequent discussion.

5.6.3.1 Nanoparticles Loaded with Radioactive Materials

Nanoparticles are designed with a specific composition and morphology. These particles also provide sufficient surface area and structure to facilitate the attachment of radioactive or chemical agents to a tumor site. For example, the transport and delivery of therapeutic radionuclides are facilitated by coating polymers with reactive functional groups of the radioactive material.

The polymer coatings are designed to reduce the uptake of radioactive materials in healthy tissue and to extend their time of circulation within the blood. Surface morphology, particle size, and surface charge are important characteristics that determine the biodistribution of nanoparticles. These characteristics are adjusted to facilitate the delivery of the nanoparticles to the target site.

Active and passive targeting techniques are used to enhance the transport of nanoparticles to tumors. With passive targeting, the nanoparticles reach the tumor site through its permeable vascular structure, accumulate, and remain due

Table 5.16 Characteristics of selected therapeutic radionuclides used in conjunction with nanoparticles[a].

Radionuclide	Radiation type emitted	Maximum particle range in tissue (mm)
^{221}Ac	α	0.08
^{225}Ac	α, β	0.1
^{212}Bi	α, β	0.09
^{213}Bi	α, β	0.09
^{223}Ra	α, β	< 0.1
^{212}Pb	α, β	< 0.1
^{149}Tb	α	< 0.1
^{131}I	β, γ	2.0
^{90}Y	β	12.0
^{67}Cu	β, γ	1.8
^{186}Re	β, γ	5.0
^{177}Lu	β, γ	1.5
^{64}Cu	β	2.0
^{67}Ga	e, γ	< 0.1
^{123}I	e, γ	< 0.1
^{125}I	e, γ	< 0.1

a) Zhang *et al.* (2010).

to its limited lymphatic drainage pathways. In an active mode, nanoparticles are designed to target specific tumor molecules. The differences in the characteristics of tumor specific and healthy tissue molecules represent a significant opportunity to selectively deposit nanoparticles in the tumor cells.

To facilitate selective deposition within a tumor mass, nanoparticles incorporate therapeutic radionuclides emitting short-range radiation. These include radionuclides emitting alpha particles, low-energy beta particles, and Auger electrons. Selected radionuclides that could be used in conjunction with nanoparticles are provided in Table 5.16.

5.6.3.2
Internal Radiation-Generating Devices

As noted in the previous discussion, heavy ions, neutrons, protons, and other radiation types have numerous applications for treating a variety of cancers. These techniques currently focus on beams originating outside the body. External beams selectively irradiate the tumor mass but still deliver some dose to healthy tissue. This stray dose affects the patient's recovery and subsequent quality of life.

As an alternative to the previous therapy approaches, a theoretical approach would use radiation-generating devices or antimatter implanted in the body to preferentially irradiate a tumor. With an internal placement, these devices deliver various radiation types that selectively irradiate the tumor mass.

Internal radiation-generating or antimatter devices do not yet exist, and significant development is required before they could be utilized for effectively treating cancer. However, this text presumes the existence of these devices and develops their requisite characteristics and possible arrangement configurations within a tumor. These characteristics include the selection of the internal beam energy and radiation type.

Although internal radiation-generating devices are conceptual, the requisite technology to construct these devices is moving closer to fruition. Laser-driven electron accelerators about the size of the eye of a needle have been developed. These accelerators are optical cavities that optimally have a size on the scale of the light's wavelength. Using shorter-wavelength radiation would bring the scale of these devices to the size envisioned for internal radiation-generating devices.

Since one of the purposes of this chapter is to determine the characteristics of internal irradiating devices, Table 5.8 provides the results of calculations of the range in water for a number of heavy ions. These results demonstrate that specific target irradiation locations are achieved by selecting appropriate ion and energy combinations. The capability to target a specific location by selecting the ion and its energy is a positive feature that makes heavy ions an attractive tool for external beam therapy applications and supports their potential use in an internal beam device. For example, a ^{197}Au beam has a range ranging between 0.20 and 1.91 cm for energies between 90 and 330 MeV/nucleon. For this same energy interval, longer ranges are achieved using lighter ions, and shorter ranges are obtained with heavier ions. Therefore, the internal device concept has the flexibility to irradiate the entire tumor mass by adjusting the beam energy and radiation type.

Absorbed dose is only part of the therapy protocol. A successful therapy approach delivers the beam to the desired target volume. Ideally, the beam is monitored and redirected as warranted. An external heavy ion beam can be tracked using PET. If the range of the particles is much less than the tumor size, then internal beam tracking becomes a secondary consideration. As envisioned, an internal radiation-generating device would have sufficient beam control to ensure its delivery preferentially to the tumor mass.

5.6.3.2.1
Candidate Radiation Types

An internal device could incorporate additional radiation types beyond those considered in Table 5.8. For example, pions, muons, and protons could be utilized as well as lower-energy heavy ions. Tables 5.8 and 5.10 summarize the ranges and energies of these radiation types that would be of potential interest in therapy applications using internal devices. An initial internal radiation-generating device will likely be of limited capability and its output limited to radiation types such as protons, electrons, or alpha particles.

The ranges in Tables 5.8 and 5.10 suggest that a variety of radiation types could be utilized in an internal beam device. Ranges on the order of a centimeter are achieved using 10–20 MeV pions and muons, 30–40 MeV protons, 100–200 MeV alpha particles, and energies on the order of 90 MeV/nucleon for ^{12}C, ^{16}O, ^{20}Ne ions, and heavier ions.

As noted previously, antiprotons have therapy applications. For an antiproton device, the reaction of the antimatter with a tumor's protons produces a variety of radiation types including charged pions, muons, and photons. Using charged particles, enhanced absorbed dose deposition occurs in the Bragg peak. If the technological issues associated with antimatter production and containment are overcome, antiprotons would be an option for a production-scale internal radiation-generating device.

5.6.3.2.2
General Characteristics and Arrangement

In order to illustrate the feasibility of using internal radiation-generating devices for therapy applications, a simple cubic Cartesian configuration is assumed to irradiate a unit tumor volume. The device/unit tumor volume configuration can be replicated to irradiate tumors of various sizes. Using a unit cell concept is somewhat arbitrary but simplifies the discussion of delivering absorbed dose to the tumor site.

The postulated Cartesian configuration consists of 27 devices within a cube arranged in three layers with nine devices in each layer ($3 \times 3 \times 3$ configuration). The number of devices is arbitrary, but selecting a specific configuration facilitates an initial presentation of the internal machine concept. The coordinates of the accelerators are written in terms of a scaled dimension ξ:

$$\xi = \frac{R}{d} \tag{5.38}$$

where d is the internal device grid spacing and R is the maximum ion range used to irradiate the unit cell. A scaled approach permits a more generalized discussion and avoids the need to adjust these dimensions to account for specific ion–energy selections.

The 27 devices are assumed to reside at the following Cartesian locations (x, y, z): $(0, 0, z)$, $(\xi, 0, z)$, $(\xi, -\xi, z)$, $(0, -\xi, z)$, $(-\xi, -\xi, z)$, $(-\xi, 0, z)$, $(-\xi, \xi, z)$, $(0, \xi, z)$, and (ξ, ξ, z) for $z = -\xi$, 0, and ξ. Additional devices would enhance the ability to deliver dose in a more uniform manner across the tumor, but the use of a limited set of devices illustrates the flexibility of the internal radiation-generating device approach. Various ion–energy combinations permit considerable flexibility in irradiating a target volume.

Uniform irradiation of a tumor mass is achieved using three basic options: (i) a sparse array of fixed devices with long-range capability, (ii) a dense array of fixed devices with limited-range capability, and (iii) mobile devices. Using a small number of machines is attractive because it limits the number of devices that must be implanted and controlled. Removal of the devices and negative impacts of their residence in the body are minimized with option (iii).

Using a dense device array minimizes the need to have machines with tunable energy and ion capability. This option requires care in placing and tracking the devices. Their eventual removal from the body must also be addressed.

Mobile devices limit the need for large numbers of machines. The ability to impart mobility to the device is a requirement that is avoided with fixed machines.

Given these three options, a list of desired internal device characteristics can be tabulated. In general, the initial internal radiation-generating devices should have the capability to (i) rotate to irradiate 4π steradians, (ii) produce a range of ion–energy combinations, (iii) be controlled in real time, (iv) rapidly change their output (radiation type, energy, and fluence), (v) produce a variable fluence such that the delivered dose is uniform over the range of the device, (vi) position itself at a desired location, and (vii) monitor the delivered dose profile using PET or other techniques to verify that it is preferentially irradiating the tumor volume.

Delivering a uniform absorbed dose (D) requires careful control of the fluence, ion type, and energy. These parameters are varied during the irradiation time (T) to deliver a uniform dose within the unit cell:

$$D = \sum_{i=1}^{N} \int_{-\xi}^{+\xi} \int_{-\xi}^{+\xi} \int_{-\xi}^{+\xi} \int_{0}^{T} \frac{1}{\rho(x_i, y_i, z_i)} \left(-\frac{\mathrm{d}E(x_i, y_i, z_i, t)}{\mathrm{d}r(x_i, y_i, z_i)} \right) \times \dot{\Phi}(x_i, y_i, z_i, t)\, \mathrm{d}x_i \mathrm{d}y_i \mathrm{d}z_i \mathrm{d}t \quad (5.39)$$

where $r(x_i, y_i, z_i)$ is the distance measured from each device, $\dot{\Phi}(x_i, y_i, z_i, t)$ is the time-dependent fluence rate, N is the number of implanted devices, and i labels the individual device.

A mobile accelerator concept requires optimization of the number of devices and their associated range requirements. The devices would be implanted and then follow a programmed trajectory in the tumor. The devices could follow a raster pattern with a pitch that is less than the radiation type's range. As the device's trajectory reaches the tumor boundary, the output range would be decreased to minimize the absorbed dose to healthy tissue. A limited number of mobile devices per x–y plane would be utilized.

For the aforementioned unit cell configuration with a volume of $8\xi^3$, mobile devices could be deployed at predetermined locations within this volume. Their trajectories are optimized to minimize the number of devices. However, in subsequent discussion a simple raster pattern trajectory in the x–y plane at a specified slice height is assumed.

Table 5.17 summarizes the z-location of mobile devices restricted to x–y planes as a function of the number of implanted devices (N). Assuming a range (R) of useful dose output for each device, the N devices would be located at the following z-positions to permit complete irradiation of the tumor volume:

$$Z = 0, \pm 2R, \pm 4R, \ldots, \pm(N-1)R \text{ for odd } N \quad (5.40)$$

$$Z = \pm R, \pm 3R, \pm 5R, \ldots, \pm(N-1)R \text{ for even } N \quad (5.41)$$

Table 5.17 Number and location of six or fewer mobile accelerators.

Number of devices	z-Location (scaled dimension[a])	Range (scaled dimension[a])
1	0	1
2	$\pm\frac{1}{2}$	$\frac{1}{2}$
3	$0,\ \pm\frac{2}{3}$	$\frac{1}{3}$
4	$\pm\frac{1}{4},\ \pm\frac{3}{4}$	$\frac{1}{4}$
5	$0,\ \pm\frac{2}{5},\ \pm\frac{4}{5}$	$\frac{1}{5}$
6	$\pm\frac{1}{6},\ \pm\frac{1}{2},\ \pm\frac{5}{6}$	$\frac{1}{6}$

a) The scaled dimension is defined in Eq. (5.38).

with

$$R = \frac{1}{N} \tag{5.42}$$

The mobile device concept permits the use of the smallest number of machines. For example, a single mobile device placed in the $z = 0$ plane with a useful dose range of ξ would be sufficient to cover the entire $8\xi^3$ volume of the previously defined unit cell. The slice location and number of devices will ultimately depend on the machine's output capability and reliability.

Considering the current state of technology, it is likely that the initial internal radiation-generating device will be a relatively basic unit having limited rotational and translational capability. This prototypical device will likely be implanted and have limited radiation output capability. Therefore, it is reasonable to select a fixed device having the capability to generate limited radiation types as a base case. Given these considerations, a fixed, proton-generating device is selected for subsequent review in this chapter. Although the device is limited to isotropic proton generation, it is assumed to have the capability to provide a range of energies.

5.6.3.2.3
Absorbed Dose Calculations

Equations (5.23) and (5.24) are used to calculate the absorbed dose from the $3 \times 3 \times 3$ array of fixed internal radiation-generating devices within a Cartesian lattice. Stopping powers are determined using Eq. (5.7), and energy-dependent cross-sections are obtained from Shen's parameterization.

For an initial evaluation of the efficacy of an internal device irradiating a tumor, 27 irradiators are utilized within a cubical volume. These devices are distributed as noted previously in three parallel $x-y$ planes with each plane containing nine devices. The planes are located on the cube surfaces and the cube midplane. Within a given plane, the devices are located at each corner, at the midpoint of each edge, and in the center of the plane.

Based on previous discussion, a variety of radiation types could be used to irradiate a tumor volume. From a technological perspective, it is assumed that the initial output of the device will be relatively uncomplicated. Accordingly, a spectrum of protons is selected to be the output of the device. To illustrate the concept, the spectrum consists of 10, 20, 30, 40, 50, 60, 70, and 80 MeV protons.

As an illustration of the output spectrum from an internal radiation-generating device, the normalized dose ($D_{\text{norm}}(r)$) from 20, 40, and 60 MeV protons is provided in Figure 5.1. The dose distribution has the expected Bragg shape, but straggling effects are not included. This is a reasonable first approximation since only the general characteristics of internal irradiating devices are being illustrated. A spectrum of proton energies facilitates the irradiation of the entire tumor volume. A uniform distribution of proton dose requires a continuous proton energy distribution.

For the initial calculations, the 27 proton-generating devices are distributed in a 10 cm × 10 cm × 10 cm volume of water. Each device is assumed to radiate isotropically. The results of irradiating a 10 cm × 10 cm × 10 cm water volume with 27 internal devices generating an output of 10, 20, 30, 40, 50, 60, 70, and 80 MeV protons are illustrated in Figure 5.2. The fluence of at each proton energy is selected to be the same.

Figure 5.2 represents the superposition of isodose surfaces to yield the total absorbed dose distribution within the three-dimensional volume. The three-dimensional absorbed dose distribution is plotted as viewed by an observer looking down at the absorbed dose profile. Since the total absorbed dose of Figure 5.2 is the superposition of a number of manifolds (e.g., the 1% isodose

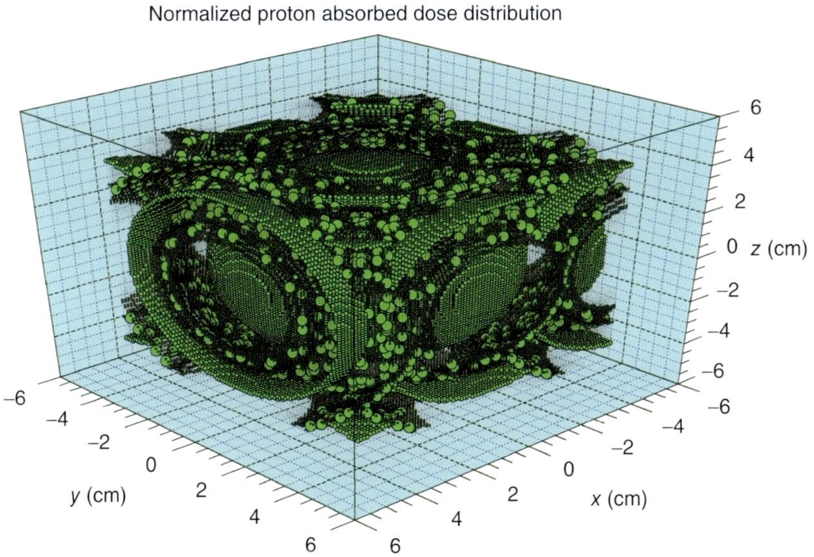

Figure 5.2 Normalized absorbed dose distribution from 27 internal devices generating a spectrum of 10, 20, 30, 40, 50, 60, 70, and 80 MeV protons in water. The absorbed dose is proportional to the plotted circle radius. This figure was initially published in Bevelacqua (2010c).

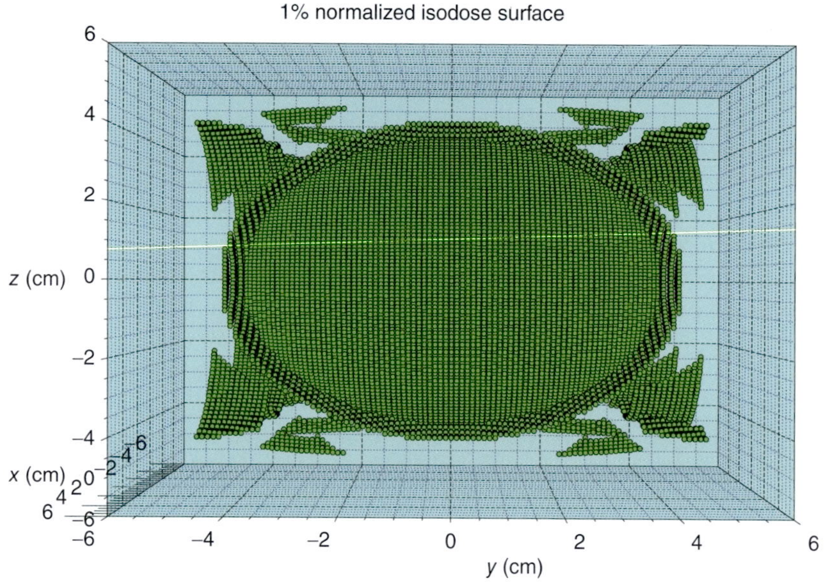

Figure 5.3 The 1% isodose distribution (relative to the peak absorbed dose) from 27 internal devices generating a spectrum of 10, 20, 30, 40, 50, 60, 70, and 80 MeV protons in water. The dose is proportional to the plotted circle radius. The 1% absorbed dose distribution is viewed from the *x*-axis. This figure was initially published in Bevelacqua (2010c).

surface of Figures 5.3 and 5.4), the structure of the surface is governed by the proton output spectrum, fluence, attenuating medium characteristics, ion stopping power, and reaction cross-section as noted in Eqs. (5.23) and (5.24). The plots of Figures 5.2–5.4 present a symbol size (i.e., circle) with the absorbed dose being proportional to the radius of the plotted circle.

The dose distribution of Figure 5.2 is complex and illustrates the symmetry associated with the 27 internal irradiators. These devices do not yield a uniform dose distribution, but do effectively irradiate the tumor volume. The average dose delivered to the 10 cm × 10 cm × 10 cm volume is dependent on the proton device output. For example, a device output using energy groups of 10 MeV; 10 and 20 MeV; 10, 20, 30, and 40 MeV; and 10, 20, 30, 40, 50, 60, 70, and 80 MeV leads to an average dose over the 10 cm × 10 cm × 10 cm volume of 5.89×10^{-6}, 5.75×10^{-4}, 3.00×10^{-2}, and 9.79×10^{-2} relative to the peak dose, respectively.

The increase in the absorbed dose delivered to the tumor is apparent from consideration of the aforementioned dose delivered from the 1, 2, 4, and 8 proton energy group calculations. Increasing the number of output energy groups spanning the 10–80 MeV range from 8 to 16, 32, 64, or more groups will continue to increase the ratio of average to peak absorbed dose delivered to the tumor site.

These results suggest that by increasing the number of energy output groups of the device, uniform tumor irradiation becomes more probable. In addition,

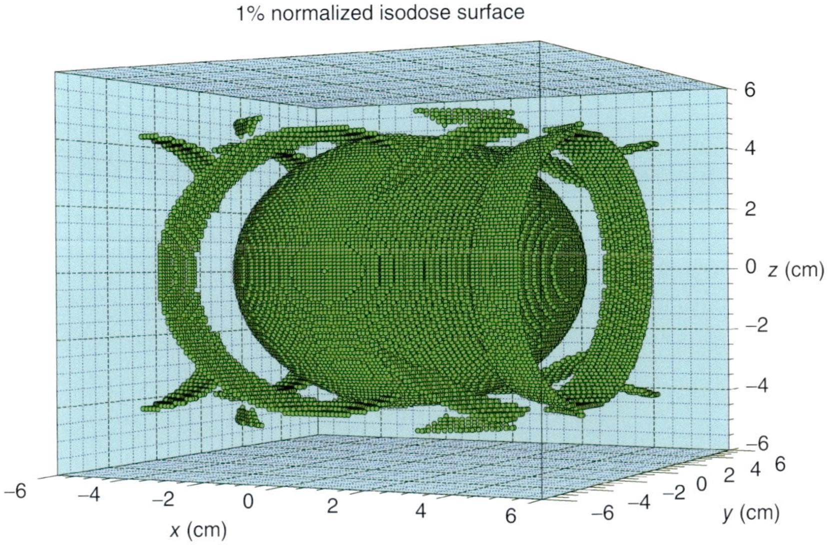

Figure 5.4 The 1% isodose distribution (relative to the peak absorbed dose) from 27 internal devices generating a spectrum of 10, 20, 30, 40, 50, 60, 70, and 80 MeV protons in water. The dose is proportional to the plotted circle radius. The 1% absorbed dose distribution is viewed from the x–y plane. This figure was initially published in Bevelacqua (2010c).

inspection of Figures 5.1 and 5.2 demonstrates that dose localization is achievable from the characteristics of the individual proton dose distributions.

To further illustrate the complexity of the dose field, the 1% isodose surface is presented in two orientations in Figures 5.3 and 5.4. The inherent complexity and symmetry of the isodose surfaces are illustrated in these figures.

The 1% isodose surface represents the set of points having an absorbed dose that is 0.01 times the peak dose value. An isodose surface is a three-dimensional structure formed from the output of all proton-generating devices. Since the isodose surface is a superposition of the output of each internal device, two views of the surface are provided to clearly illustrate the manifold's shape.

Figure 5.3 illustrates the 1% isodose surface with the observer on the x-axis looking at the y–z plane. The inherent symmetry of the center device output is apparent as well as the influence of the irradiators in the midplane and cube's surface.

Figure 5.4 illustrates the same 1% isodose curve as Figure 5.3 with the observer in the x–y plane. The influence of each proton-generating device is more apparent. This figure also illustrates one of the surfaces that combine to form the complex absorbed dose structure of Figure 5.2.

The discussion of the characteristics of the detailed three-dimensional absorbed dose profile illustrates the complexity of therapy planning when implementing a new technology. Internal radiation-generating devices are a logical twenty-first-century extrapolation of existing external beam therapy and emerging laser-powered electron accelerators advanced by Travish and Yoder.

5.6.3.2.4 Angular Absorbed Dose Dependence

The radial absorbed dose characteristics of an internal radiation-generating device were addressed in the previous section. In this section, the angular absorbed dose output of these devices is reviewed. For a tissue volume irradiated by a beam of ions of a given energy (E), the absorbed dose ($D(r, \theta)$) as a function of penetration distance r (i.e., the distance from the internal radiation-generating device) into tissue at an angle θ relative to the beam direction is obtained from the relationship

$$D(r, \theta) = \frac{1}{\rho}\left(-\frac{\mathrm{d}E}{\mathrm{d}r}\right)\Phi(r, \theta) \tag{5.43}$$

where standard spherical coordinates (r, θ, ϕ) are used. For specificity, the ranges of the spherical coordinates are $0 \leq r \leq \infty$, $0 \leq \theta \leq \pi$, and $0 \leq \phi \leq 2\pi$.

The particle fluence varies with tissue penetration depth and angle according to the relationship

$$\Phi(r, \theta) = \Phi(0, 0)\exp(-\Sigma(\theta)r) \tag{5.44}$$

where $\Phi(0, 0)$ is the entrance fluence into tissue as it leaves the internal radiation-generating device at 0° relative to the beam direction and $\Sigma(\theta)$ is the angular macroscopic reaction cross-section for a given nuclear reaction at energy E defined as

$$\Sigma(\theta) = n\frac{\mathrm{d}\sigma(\theta)}{\mathrm{d}\Omega} \tag{5.45}$$

In Eq. (5.45), n is the number of target atoms per cm^3, $(\mathrm{d}\sigma(\theta)/\mathrm{d}\Omega)$ is the microscopic differential cross-section (cm^2/atom-sr) for the reaction of interest, and $\mathrm{d}\Omega$ is the spherical coordinate area element ($r^2 \sin(\theta)\mathrm{d}\theta\,\mathrm{d}\phi$). As defined in standard models and their associated codes (see Appendices E–G), the differential cross-section is isotropic in the ϕ coordinate. Given these cross-section model definitions, $\Sigma(\theta)$ has units of 1/cm-sr. Additional commentary regarding these models is provided in Appendices E–G.

The results of the cross-section calculations incorporated into Eq. (5.45) are provided in Figure 5.5 for a proton energy of 19.8 MeV as a function of the center-of-mass angle (θ_{cm}). Figure 5.5 shows the anticipated angular dependence with the peak cross-section occurring in the beam direction and the cross-section generally decreasing with increasing angle.

The shape and magnitude of the angular macroscopic cross-section suggest that for a given distance from the internal radiation-generating device, the magnitude of the absorbed dose will be a minimum in the beam direction and generally increase as θ_{cm} increases. This expected pattern is a consequence of Eqs. (5.44) and (5.45) and the shape and magnitude of the angular macroscopic cross-section.

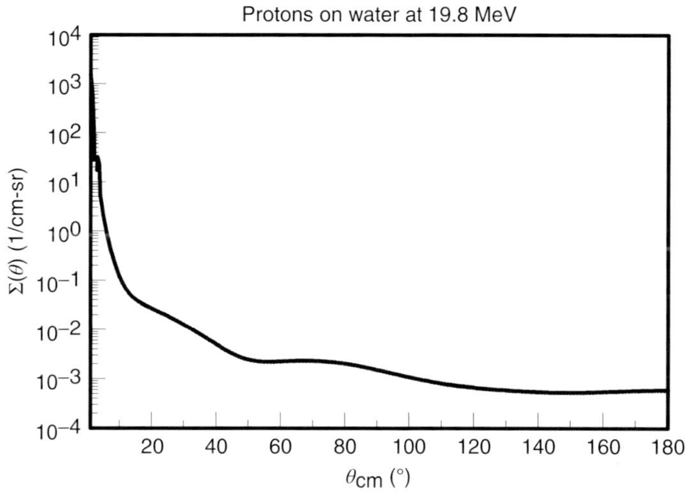

Figure 5.5 Angular macroscopic cross-section (1/cm-sr) as a function of center-of-mass angle for protons on water at 19.8 MeV. This figure was initially published in Bevelacqua (2012).

The absorbed dose as a function of distance and angle is determined by using the cross-section data of Figure 5.5 and Eq. (5.43). The resulting absorbed dose calculations are summarized in Figure 5.6 and consider three generated proton energies.

Figure 5.6 presents the normalized absorbed dose as a function of θ_{cm} for 40 MeV (top curve), 60 MeV (middle curve), and 80 MeV (bottom curve) for internal radiation-generating device proton energies. The curves represent the absorbed dose angular distributions at a distance into water where the proton energy has degraded to 19.8 MeV and the cross-section curve of Figure 5.5 is applicable. This occurs at 1.1, 2.7, and 4.8 cm penetration depths for 40, 60, and 80 MeV protons, respectively. Beyond about $\theta_{cm} = 30°$, the absorbed dose results of Figure 5.6 have a relatively flat shape.

This forward angle absorbed dose depression requires that the device have the capability to rotate through a critical angle (θ_c). Having the internal radiation-generating device rotate though an angle of θ_c fills the angular absorbed dose depression. Therefore, it appears that the initial discussion of Section 5.6.3.2.3 that the internal radiation-generating devices rotate to traverse 4π sr can be relaxed if the device has the capability to rotate through a sweep angle θ_c. As an illustration, the results of Figure 5.6 suggest that the sweep angle would be about 30°.

The critical angle varies with energy and depth and will ultimately be based on a specific internal radiation-generating device design. Although the results of Figure 5.6 are encouraging and suggest that the internal radiation-generating device concept is sound from a basic physics perspective, the evolving design remains an academic exercise until sufficient technology becomes available to permit device fabrication and testing.

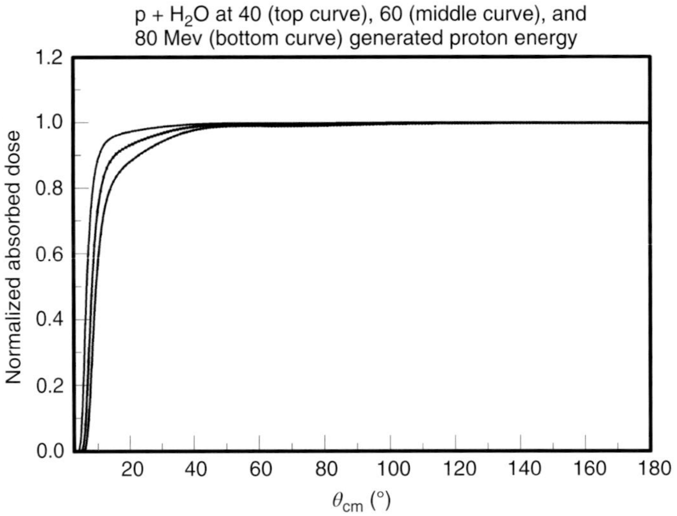

Figure 5.6 Normalized absorbed dose distribution as a function of center-of-mass angle for 40 MeV (top curve), 60 MeV (middle curve), and 80 MeV (bottom curve) protons on water. The curves correspond to the distance from the device where the proton energy degrades to 19.8 MeV (i.e., 1.1, 2.7, and 4.8 cm for 40, 60, and 80 MeV, respectively). This figure was initially published in Bevelacqua (2012).

5.6.3.2.5
Comparison with Brachytherapy

Brachytherapy devices are implanted and irradiate the tumor from an interior body location. As such, brachytherapy and internal radiation-generating devices perform a similar function, and their comparison is warranted.

Brachytherapy seeds are macroscopic devices that are roughly the size of a grain of rice for prostate applications. As such, their insertion into the body is accomplished in an invasive manner. The size of an internal radiation-generating device is microscopic, and these devices will be injected directly into the blood. The actual device size is governed by future design constraints, but optimized devices with a size on the order of a cell (about 10 μm) or smaller are envisioned. The size differential between brachytherapy seeds and internal radiation-generating devices is another factor motivating their development and use.

Given this anticipated size, an injection is effectively used to administer radiation-generating devices directly into the blood vessels, which provides a pathway to their desired location. It is presumed that the devices have the capability to be directed to the desired location and are sufficiently small to not interfere with normal body functions. As such, internal radiation-generating devices are not restricted to body cavities or areas accessible by a brachytherapy implant, but can reach any desired tumor location. Therefore, the internal radiation-generating device has the potential for greater flexibility than a brachytherapy implant. The microscopic size of an internal radiation-generating device will also minimize the negative impacts associated with conventional brachytherapy.

Brachytherapy sources including resin and glass microspheres (see Section 5.5.7) also have the capability to be injected and are smaller than traditional seeds. However, the microspheres become permanent implants, typically irradiate healthy tissue (e.g., liver), and are limited by their radionuclide loading. As such, brachytherapy microspheres do not have the range of capabilities available in an internal radiation-generating device.

Short-term side effects of brachytherapy normally last a few days and include bruising, swelling, bleeding, and discomfort within the implanted region. In a small subset of patients, brachytherapy causes longer-term side effects. These effects are usually attributed to damage or disruption of adjacent tissues or organs. These longer-term side effects are usually mild or moderate in nature. For example, urinary and digestive problems may occur for prostate brachytherapy treatments. Breast or skin brachytherapy can produce scar tissue around the treatment area. In the case of breast brachytherapy, the breast tissue may swell and become tender.

Both short-term and long-term side effects of brachytherapy are eliminated by the internal radiation-generating device approach. Although conceptual, an internal radiation-generating device has significant potential for preferentially depositing energy at the tumor site without damaging healthy tissue or producing negative side effects.

In addition to development and implementation uncertainties, the costs associated with the use of internal devices are highly uncertain. However, as has been the case with most technological advances (e.g., electronic calculators, computers, and consumer electronic products), the initial costs are usually high but rapidly decline as the devices are optimized and mass-produced. Therefore, internal radiation-generating devices have the potential to eventually reduce health-care costs associated with radiation therapy. This contention must be demonstrated and is strongly linked to the technological feasibility of developing internal radiation-generating devices.

5.6.3.2.6
Application to Vascular Disruption Using Various Radiation Types

In Section 5.5.7.4, radionuclides that could affect vascular disruption were investigated. Similar results can be achieved using internal radiation-generating devices. These devices meet the desired characteristics to maximize dose to the tumor's vascular walls while minimizing the dose to healthy tissue. As such, they also avoid a portion of the negative aspects of the ^{90}Y microsphere therapy noted previously. The design characteristics and capabilities of internal radiation-generating devices are used to investigate the vascular disruption characteristics of protons, heavy ions, and low-energy photons.

For a tissue volume irradiated by a beam of ions of a given energy, the absorbed dose (D) as a function of penetration distance into tissue is given by

Eqs. (5.23–5.25). Calculations were performed for proton, ^{4}He, ^{12}C, ^{20}Ne, and ^{40}Ca beams delivered by internal radiation-generating devices. All beams were assumed to be fully ionized (e.g., ^{40}Ca ions have a +20e charge).

The photon absorbed dose is derived from a relationship, which assumes that the internal radiation-generating device is located at the arteriole wall:

$$D = \frac{S}{4\pi r^2} \frac{\mu_{\text{en}}}{\rho} EB(\mu x)e^{-\mu x} \tag{5.46}$$

where S is the total number of photons that irradiates the arteriole wall, r is the distance from the radiation-generating device, μ_{en}/ρ is the mass energy absorption coefficient, E is the photon energy, B is a buildup factor, x is the material thickness between the device and the target tissue, and μ is the attenuation coefficient. Since higher-energy photons have poor dose localization, low-energy photons are investigated as a possible vascular disruption agent.

Since the base case considered in this chapter is the 20 μm thickness of an arteriole wall, the focus is delivering a requisite dose to this tissue region and for blood vessel wall thicknesses ≤100 μm that likely service tumors. The target dose delivered to this tissue is assumed to be sufficient to disrupt the vessel wall, which is on the order of 100 Gy. No attempt to optimize dose delivery has been made, and ion fluences to reach the 100 Gy dose level are 5×10^9, 5×10^8, 1×10^8, 5×10^7, and 1×10^7 ions/cm^2 for protons, alpha particles, ^{12}C, ^{20}Ne, and ^{40}Ca, respectively. 1×10^{10} photons are utilized in the calculations using Eq. (5.46).

In subsequent absorbed dose calculations, the internal radiation-generating device is assumed to reside at the inner arteriole wall. The results of Figures 5.7–5.12 provide absorbed dose profiles for blood vessel wall depths ≤100 μm. Water is assumed to be the medium comprising the vessel wall.

Figure 5.7 provides absorbed dose profiles for protons with energies between 0.5 and 2.3 MeV. Dose localization within an arteriole wall could be achieved using a 1.0–1.5 MeV proton beams. The results of Figure 5.7 suggest that the 100 Gy target dose for blood vessel destruction can be achieved using low-energy protons. These results also suggest that delivering the target absorbed dose to the 20–100 μm depth is readily achieved using protons with energies between 1.0 and 2.3 MeV.

Figure 5.8 summarizes ^{4}He absorbed dose curves for 3–8 MeV alpha particles. These energies correspond to the values achieved by alpha particles emitted by many radionuclides. The results summarized in Figure 5.8 suggest that sufficient absorbed dose at the requisite depths can be delivered by alpha energies below 8 MeV.

Alpha particles with energies below 3 MeV will not penetrate the arteriole wall. The arteriole wall is disrupted, with minimal dose to surrounding tissue, by alpha particles in the 4–5 MeV energy range. This energy range is obtained by numerous alpha-emitting radionuclides.

The results of Figure 5.8 confirm that alpha-emitting radionuclides are an alternative to the use of ^{90}Y in microspheres. This alternative could be implemented in the near term and would not require the advanced technology utilized in an internal radiation-generating device. The use of an alpha-emitting radionuclide in a microsphere to affect tumor vascular disruption is addressed in Section 5.5.7.4.1.

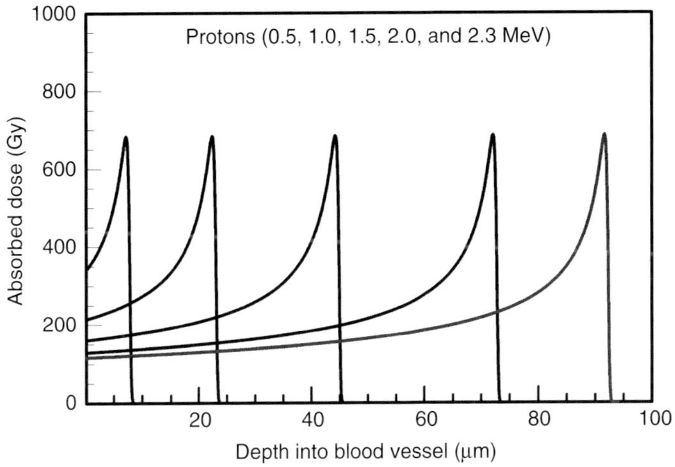

Figure 5.7 Absorbed dose profiles for 0.5 (far left curve), 1.0, 1.5, 2.0, and 2.3 (far right curve) MeV protons in water. The absorbed dose curves peak at a greater depth with increasing proton energy. For all energies, the total proton fluence is 5.0×10^9 p/cm^2. The protons are delivered by an internal radiation-generating device. This figure was initially published in Bevelacqua (2014).

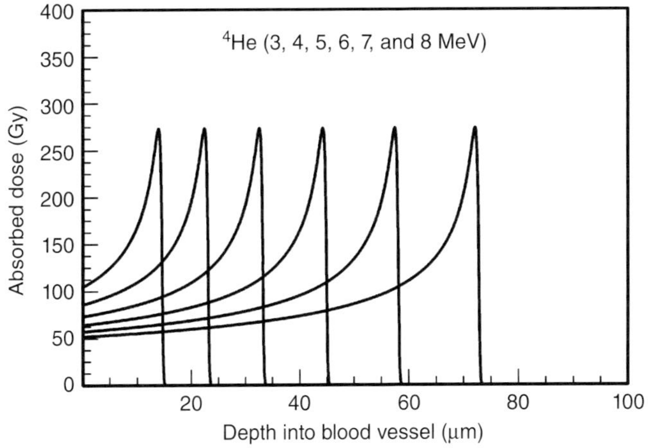

Figure 5.8 Absorbed dose profiles for 3.0 (far left curve), 4.0, 5.0, 6.0, 7.0, and 8.0 (far right curve) MeV ^{4}He ions in water. The absorbed dose curves peak at a greater depth with increasing ^{4}He ion energy. For all energies, the total ion fluence is 5.0×10^8 ^{4}He ions/cm^2. The ions are delivered by an internal radiation-generating device. This figure was initially published in Bevelacqua (2014).

Heavy ion beams of ^{12}C, ^{20}Ne, and ^{40}Ca ions and their penetration through the 20–100 μm range are summarized in Figures 5.9–5.11, respectively. ^{12}C ions below about 20 MeV will not penetrate the arteriole wall, and 20–50 MeV ions will deposit sufficient energy into a range of vessel wall thicknesses in the

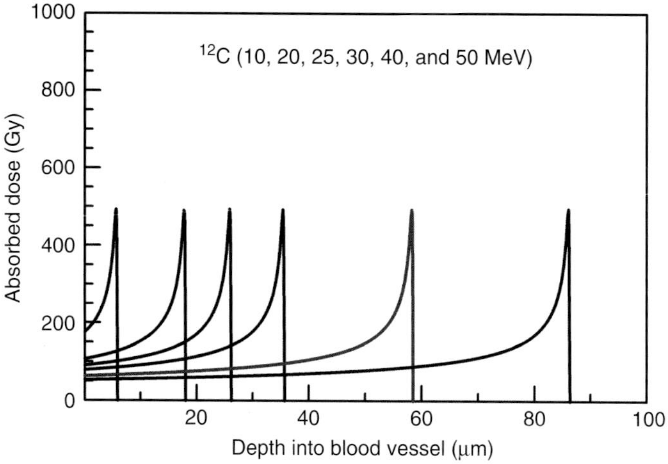

Figure 5.9 Absorbed dose profiles for 10.0 (far left curve), 20.0, 25.0, 30.0, 40.0, and 50.0 (far right curve) MeV ^{12}C ions in water. The absorbed dose curves peak at a greater depth with increasing ^{12}C ion energy. For all energies, the total ion fluence is 1.0×10^8 ^{12}C ions/cm^2. The ions are delivered by an internal radiation-generating device. This figure was initially published in Bevelacqua (2014).

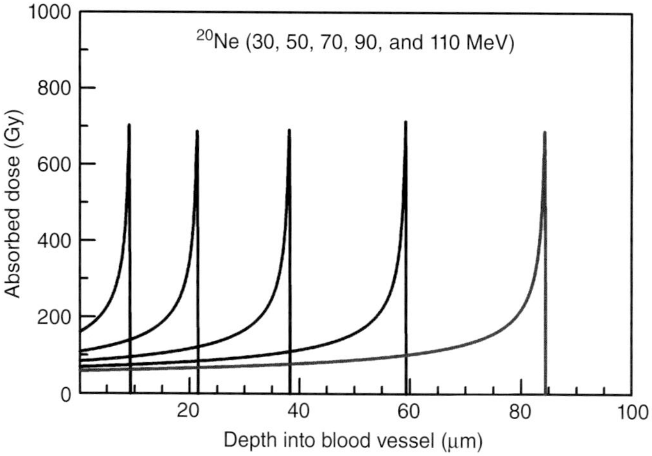

Figure 5.10 Absorbed dose profiles for 30.0 (far left curve), 50.0, 70.0, 90.0, and 110.0 (far right curve) MeV ^{20}Ne ions in water. The absorbed dose curves peak at a greater depth with increasing ^{20}Ne ion energy. For all energies, the total ion fluence is 5.0×10^7 ^{20}Ne ions/cm^2. The ions are delivered by an internal radiation-generating device. This figure was initially published in Bevelacqua (2014).

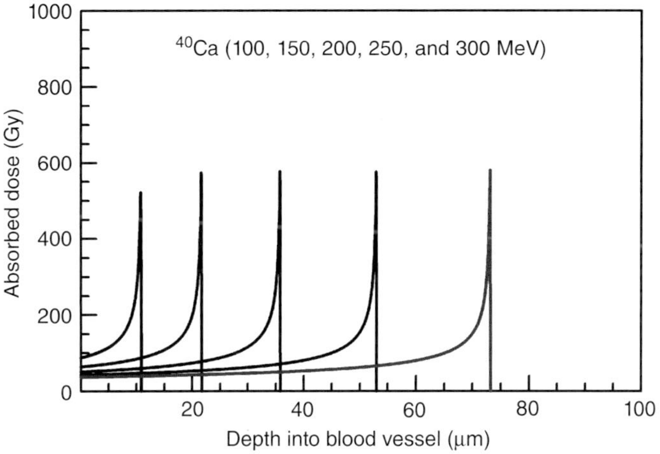

Figure 5.11 Absorbed dose profiles for 100.0 (far left curve), 150.0, 200.0, 250.0, and 300.0 (far right curve) MeV ^{40}Ca ions in water. The absorbed dose curves peak at a greater depth with increasing ^{40}Ca ion energy. For all energies, the total ion fluence is 1.0×10^7 ^{40}Ca ions/cm^2. The ions are delivered by an internal radiation-generating device. This figure was initially published in Bevelacqua (2014).

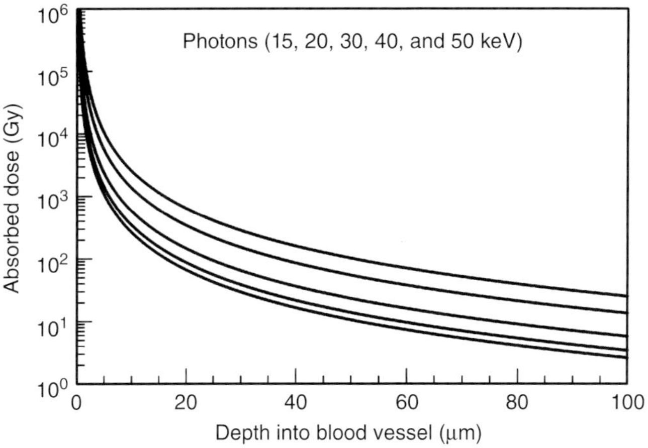

Figure 5.12 Absorbed dose profiles for 15.0 (top curve), 20.0, 30.0, 40.0, and 50.0 (bottom curve) keV photons in water. The absorbed dose curves decrease in magnitude with increasing photon energy. 1.0×10^{10} γ are delivered by the internal radiation-generating device. This figure was initially published in Bevelacqua (2014).

20–100 μm range to produce vascular disruption. Arteriole wall disruption can be achieved using 20–30 MeV ^{12}C ions. However, generation of ^{12}C, ^{20}Ne, and ^{40}Ca ions presents a greater challenge than producing lighter ions in a first-generation internal radiation-generating device.

^{20}Ne ions below 30 MeV will not penetrate the arteriole wall. As illustrated in Figure 5.10, ^{20}Ne ions in the range of 50–110 MeV will be sufficient to reach the range of vessel wall thicknesses addressed in this chapter. Arteriole wall disruption is achieved using 50–70 MeV ^{20}Ne ions.

In a similar manner, ^{40}Ca ions require 150–200 MeV to disrupt the arteriole wall. Figure 5.11 illustrates the penetration of 100–300 MeV ^{40}Ca ions through vessel wall thicknesses below 100 μm.

Figure 5.12 illustrates that photon energies in the range of 15–50 keV can deposit the requisite absorbed dose to disrupt an arteriole wall. Significant dose is also deposited in the 20–100 μm range by the 15–50 keV photons summarized in Figure 5.12. However, protons and ^{4}He, ^{12}C, ^{20}Ne, and ^{40}Ca ions achieve better dose localization. As noted in Figure 5.12 appreciable photon dose is deposited outside the 100 μm target range.

5.6.3.3
Hybrid Medical Imaging

Hybrid medical imaging is the combined use of anatomical and functional imaging to improve the individual diagnostic procedure. These techniques include X-ray, ultrasound, optical, CT, MRI, and PET imaging approaches. Although the individual techniques are well established, their combined use offers the potential for improved imaging capability.

Previous discussion noted a variety of imaging approaches that focused on the use of photon-emitting radiation. There are other methods to perform imaging studies including ultrasound, optical, and MRI techniques. Table 5.18 compares the characteristics of these methods to the previously discussed computed tomography and PET methods. In Table 5.18, this set of diagnostic techniques is compared in terms of selected imaging parameters including anatomical detail, spatial resolution, clinical penetration, sensitivity, and molecular resolution. The comparisons are made on a qualitative basis (i.e., poor, satisfactory, good, and excellent).

Table 5.18 Characteristics of selected imaging techniques[a].

	Imaging techniques				
Imaging parameter	**Ultrasound**	**Optical**	**CT**	**MRI**	**PET**
Anatomical detail	Satisfactory	Good	Good	Excellent	Poor
Spatial resolution	Satisfactory	Good	Good	Excellent	Satisfactory
Clinical penetration	Satisfactory	Poor	Excellent	Excellent	Poor
Sensitivity	Poor	Poor	Poor	Poor	Excellent
Molecular resolution	Poor	Poor	Poor	Satisfactory	Excellent

a) Lewis and Kalemis (2011).

5.6.3.3.1 PET/CT

An optimized PET/CT approach is superior to the independent use of CT or PET. PET/CT provides good or excellent capability for the imaging parameters summarized in Table 5.18. An initial application of hybrid imaging used PET in conjunction with computed tomography. Although PET/CT has been useful in cancer investigation, CT has some limitations associated with soft tissue features that can be improved using contrast agents. The absorbed doses from CT imaging should be carefully monitored when this technique is utilized in pediatric applications or when repeated scanning is required.

5.6.3.3.2 PET/MRI

An alternative approach combines PET with MRI. A review of Table 5.18 suggests the potential advantage of the PET/MRI imaging approach. If fully optimized, PET/MRI provides excellent capability in terms of the imaging parameters summarized in Table 5.18.

In comparison with CT, MRI provides good contrast in soft tissue through the alignment of the hydrogen nucleus (proton) magnetic moment by a strong magnetic field. A transient radiofrequency field flips the spin of a portion of the protons. When these hydrogen nuclei return to their initial state, they radiate at the same radiofrequency. An image is produced by applying a magnetic field gradient such that the resonant frequency is a function of position within the body.

In contrast, PET techniques detect the distribution of positron-emitting radionuclides within the body. This is accomplished by detecting the pair of 511 keV annihilation photons using coincidence counting or time-of-flight techniques. Detection is facilitated by using scintillation detectors connected to photomultiplier tubes.

PET/MRI techniques are available from a number of manufacturers, but the designs are evolving. This evolution is needed to resolve concerns associated with photon attenuation in combined PET/MRI systems. Photons traversing the patient's body are absorbed or attenuated and not counted. Methods to compensate for photon attenuation are under development.

Another issue is the electronic coupling between the PET and MRI systems. The radiofrequency pulses from the MRI system may cause the PET electronics to lose counts during transmission of the radiofrequency pulses. Techniques used in particle physics research (e.g., silicon photomultipliers) will eventually be incorporated into future generations of PET/MRI scanners to eliminate the aforementioned issues. Full optimization of this technique has not yet been achieved.

5.6.3.3.3 X-ray/MRI

Another hybrid system combines X-ray and MRI techniques. A potential application of such a system is reduction of blood pressure caused by advancing cirrhosis in the liver. As cirrhosis advances, fibrous scars develop in the liver. These scars impede the flow of blood as it enters the liver from the intestines and raises blood pressure. Without treatment, the weakest blood vessels in the stomach and esophagus could rupture and result in severe, potentially fatal, internal bleeding.

Conventional treatment methods reduce blood pressure through the use of shunts. The shunts lower blood pressure by bypassing the blockage in the liver. Shunts are placed between the portal vein that supplies the liver and blood vessels inside the liver.

Shunt insertion is an inefficient process. Using X-rays, surgeons detect the hollow needle used to open a flow path in the blood vessels. Detecting the blood vessels involves blindly poking until blood flow through the needle indicates that a vessel has been found. To understand the current limitations, it is necessary to address the strengths and limitations of conventional detection techniques.

X-rays readily detect materials or tissues having higher Z values (e.g., a metal instrument or bone) than soft tissue. The discrimination of subtle features in soft tissues requires a method based on principles that differ from the X-ray approach. In contrast to X-ray techniques, MRI reveals subtle tissue features and facilitates insertion of the shunt at the desired location. The combined X-ray/MRI system has the capability to monitor the needle location via X-rays and the blood vessels using MRI.

The X-ray and magnetic resonance features noted previously form an ideal solution to facilitating shunt placement. A combined X-ray/MRI approach has not yet been fully optimized because the interaction of the X-ray generation and MRI magnetic field systems degrades image quality. Upon resolution, the X-ray/MRI technique has significant potential as an imaging technique.

5.6.3.4
Cherenkov Luminescence Imaging

Cherenkov radiation occurs when a charged particle traverses an optically transparent medium with a velocity greater than the speed of light in that medium. As the particle moves, it excites the medium's electrons. When the electrons return to their ground state, electromagnetic radiation is emitted. Cherenkov radiation appears as a weak bluish glow that is often seen in the spent fuel pool of a nuclear reactor.

Cherenkov radiation also occurs when radioactive material decays in the human body, and this radiation can be used for imaging. This new technique is often called *Cherenkov luminescence imaging*.

Although similar imaging is obtained using established techniques (e.g., PET), CLI has the potential to be more cost effective. CLI also provides a link between PET and optical imaging. A possible application includes surgery verification to ensure that all cancerous tissue has been removed. This is an ambitious goal since the CLI signal is relatively weak and has limited capability to penetrate tissue.

5.6.3.5 Photodynamic Therapy

PDT is a nonionizing radiation technique that has the potential to be a viable option for cancer treatment. It is more selective than other treatments (e.g., radiation therapy and chemotherapy) and causes less damage to healthy tissue. It selectively occurs in the preferential intake of photoreactive material or photosensitization agents by tumor cells.

Optimizing the combination of photosensitization and conjugate agents is required to advance the technique. PDT currently introduces the photosensitization agent into the patient, and this material preferentially accumulates within the tumor. Light of the desired wavelength is provided by a diode laser or LED. The relevant wavelength is directed into the tumor region and activates the photosensitization agent without damaging healthy tissue. Following activation, the photosensitization agent transforms a portion of its energy to molecular oxygen to create excited oxygen, which destroys the cancer cells by oxidation. As designed, limited adjacent healthy tissue is damaged.

Effectively transmitting light to the tumor site is an open issue. Initial PDT studies targeted various types of melanoma because the near-infrared wavelengths were readily transmitted a few millimeters through the skin's surface to the tumor site. Recent advances in light transport systems facilitate reaching deeper tumors. These delivery systems also expand the range of useful wavelengths and associated photosensitization agents.

Optimizing a photosensitization agent to attach itself to a tumor site is a challenge, because the human immune system attacks some agents, which reduces the overall effectiveness of the PDT technique. Current research efforts are utilizing lipoprotein shells and gold nanoparticle delivery systems to enhance deposition within the tumor.

5.6.3.6 Low-Coherence Interferometry

LCI is a diagnostic imaging technique that combines the advantages of OCT (Section 5.6.3.8) and light scattering techniques. Angle-resolved LCI utilizes the capability of this technique to isolate scattering from subsurface tissue layers. Light scattering spectroscopy is used to derive structural information associated with these tissue layers by analyzing the angular scattering data. This structural information differentiates healthy and diseased tissue and reveals their spatial extent.

5.6.3.7 Nonlinear Interferometric Vibrational Imaging

NIVI combines the good resolution of Raman spectroscopy with the high count rates associated with coherent anti-Stokes Raman scattering microscopy. Current techniques yield the accuracy associated with Raman spectroscopy with speeds 200–500 times faster. This offers the potential for rapid three-dimensional tissue imaging which enhances the diagnosis and characterization of cancer.

NIVI has been applied to mammary tissue to investigate the molecular mechanisms associated with breast cancer formation and detection. It offers the potential for detection with a greater sensitivity than current imaging methods.

5.6.3.8 Optical Coherence Tomography

OCT is the optical analogue of ultrasound and assembles images using computational techniques. The OCT technique scans a sample and combines backscattered photons with a reference beam using an interferometer. A light source with limited coherence maintains the scattered photons out of phase with the reference beam when collected at the detector. This permits the selective measurement of scattered photons. Image contrast is derived from changes in the refractive index within tissue.

In addition to the information derived from changes in the refractive index, OCT has the capability to detect structures that alter the phase, amplitude, or polarization of light within the imaged tissue. For example, polarization shifts indicate changes in collagen, the most common protein in mammals.

Tissue is translucent at red and near-infrared wavelengths. At 800–1300 nm, tissue penetration depths are in the range of 1–3 mm. Although OCT is a useful noninvasive imaging technique, its short range has been a limiting factor in expanding its usefulness. This technique has application to diagnostic studies of the eye, pulmonary and circulatory systems, and skin.

5.6.3.8.1 Eye Imaging

To date, OCT has been most successful in ophthalmology where the eye's transparency facilitates light penetration throughout its volume. As such, OCT permits high-resolution examination of the retina. OCT is also used to examine arterial plaque with superior clarity. Other applications include techniques that combine OCT with other diagnostic techniques including fluorescence and Raman techniques.

Most OCT eye scans focus on the retina with about 5 μm resolution for three-dimensional imaging. This resolution permits monitoring of blood vessels and the observation of age-related defects including macular degeneration, which facilitates the determination of treatments to limit further damage. High-resolution retinal imaging also enables detection of the early onset of glaucoma. Cornea imaging is also facilitated since OCT can measure the eye's topology and internal structure. These are important considerations in optimizing refractive surgery.

5.6.3.8.2 Circulatory and Pulmonary Applications

OCT has been recently applied to cardiology by inserting fibers into arteries to scan their internal structure and assess plaque stability. Plaque detachment is a potential contributor to arterial blockage and subsequent heart damage. OCT techniques are also being developed to examine the embryonic heart as a means to treat congenital heart defects. Other investigations include applications in the gastrointestinal and pulmonary systems and breast cancer detection. However, the 1 – 3 mm depth for current OCT imaging is a limiting factor in extending this technique to other imaging applications.

5.6.3.8.3
Skin Applications

Another area for OCT application is a noninvasive method to map the network of blood vessels in the epidermis in three dimensions to facilitate the monitoring and treatment of skin cancer. The approach is sufficient to distinguish the structure of the skin's vasculature, which provides a clear indication of cancer development. Characteristics of a tumor's vascular structure are provided in Section 5.5.7.1.

5.6.3.9
Personal Genomics

One of the most promising medical approaches is the use of personal genomics as a guide to disease treatment and prevention. Only one in five cancer drugs is effective in treating a specific patient. With a progressive disease, time is critical. Having the capability to select a treatment approach based on genetic information would enhance the probability of patient survival.

With improved knowledge of the human genome, the potential exists for genetic-based therapies. These therapies are optimized using specific genetic data and increase the probability of successfully treating the patient.

5.6.3.10
Second-Generation Nanotechnology

Second-generation nanotechnology is a hunter-killer device that would detect malignant tissue and destroy it regardless of its location. A second-generation device would include a radionuclide that emits a short-range radiation type or internal radiation-generating device coupled with antibodies. The antibodies would find the malignant tissue or ideally its progenitor cells and then destroy it with the localized deposition of absorbed dose. Radioimmunoconjugates represent a proactive approach to cancer treatment that would eradicate the disease prior to tumor formation.

An advanced technique combining personal genomics and second-generation nanotechnology would represent a significant advance for treating cancer and all other diseases. The combination of therapy techniques offers an approach that would significantly improve cancer survival by destroying nascent cells before their growth and subsequent development.

5.7 Other Considerations

As an alternative to beam therapy, Italian physicians removed a patient's liver, irradiated it using BNCT, and then reimplanted it. The technique has been dubbed TAOrMINA after the Italian for "advanced treatment of organs by means of neutron irradiation and autotransplant."

The organ removal technique allows physicians to deliver high doses directly to diseased targets without irradiating healthy tissue. This novel technique eliminates issues of beam localization and selective energy deposition that are importantl considerations in external beam therapy.

TAOrMINA was used to treat a 48-year-old man with multiple liver tumors. The operation took 21 h, and the last report indicated that the man was alive and well. At 1-year postsurgery, his liver was functioning normally, and the scans did not revealed any signs of tumors. Even if additional trials indicate that the method is effective against liver and other cancers, this drastic technique would be reserved for patients with limited treatment options. It would likely be used only if the patient was strong enough to survive the organ removal and subsequent transplant procedure. However, it does provide another therapy option.

Problems

5.1 You are the Radiation Safety Officer at the University of Pittsburgh Medical Center. The senior radiologist has developed a therapy procedure that utilizes gold colloid nanoparticles to deposit ^{131}I into thyroid tumor blood vessels to facilitate tumor reduction. She plans to treat a series of patients and requests a dose assessment for the proposed therapy technique. Patients will be treated in a separate area of the outpatient department.

Data:

Administered activity = 7400 MBq of $^{131}\mathrm{I} = Q_0$

Extrathyroid uptake fraction = 93% = F_1

Thyroidal uptake fraction = 7% = F_2

Occupancy factor for the first 8 h = 5% = E_1

Occupancy factor from the first 8 h to total decay = 10% = E_2

Maximum absorbed dose at 1 m from a patient treated with gold colloid nanoparticles containing ^{131}I is given by

$$D_\infty = \left(\frac{34.6\,\Gamma Q_0}{(100\,cm)^2}\right)\left(\begin{array}{l} E_1 T_p\,(0.8)\,(1 - e^{-\ln 2(0.33d)/Tp}) + E_2 F_1 T_{1\mathrm{eff}} e^{-\ln 2(0.33d)/Tp} \\ \qquad + E_2 F_2 T_{2\mathrm{eff}} e^{-\ln 2(0.33d)/Tp} \end{array}\right)$$

where

Γ = absorbed dose rate constant for $^{131}\mathrm{I} = 5.2 \times 10^{-8}$ Gy-m^2/MBq-h

T_p = physical half-life of $^{131}\mathrm{I} = 8.04$ days

For the proposed thyroid cancer approach, the following effective half-life values are applicable:

$$T_{1\text{eff}} = 0.32\,\text{days}$$
$$T_{2\text{eff}} = 7.3\,\text{days}$$

(a) Based on the problem data, calculate the absorbed dose to an individual that is positioned at a distance of 1.0 m from the patient receiving the ^{131}I administration. Are written safety instructions required for this patient?

(b) Assume the individual cannot be treated as an outpatient. State two restrictions that would allow you to release her from the hospital.

(c) Calculations show the individual can be released as an outpatient. What three general requirements could you apply to minimize dose to members of her family?

(d) What additional instructions, if any, would you provide if the patient had a 15-month-old child at home?

(e) What additional instructions, if any, would you provide if the patient had a 15-year-old child at home?

5.2 You are the staff medical physicist at the University of Beijing. A senior researcher has developed a new external beam cancer therapy approach using negative K mesons. The charged K meson has a mass of 494 MeV and a mean lifetime of 1.24×10^{-8} s. K mesons are produced by an accelerator when protons with energies of a few gigaelectronvolts strike a metal target. A beam of negative K mesons is extracted from the accelerators and used to irradiate cancer patients.

Like all charged particles, K mesons or kaons that enter the body slow down. By properly selecting the incident energy, a beam of negative K mesons comes to rest at the site of a tumor. When it stops in matter, a negative K meson is captured by a positively charged atomic nucleus. The negative K meson interacts with the nucleus, releasing a variety of radiation types including energetic neutrons, protons, and heavier fragments. Based on the researcher's initial calculations, the average distribution of emitted particles and energies for capture by a ^{16}O nucleus is provided in the following table. Similar data are predicted to describe capture by carbon and nitrogen nuclei.

Reaction products and average energies from capture of a stopped negative K meson by a ^{16}O nucleus

Emitted particle	Average kinetic energy per capture (MeV)
Fast neutrons	215
Protons	60.0
Heavy fragments[a)]	70.6
Gamma rays	21.2
Total	366.8

a) Includes spallation products.

(a) What would be the chief potential advantage of using a negative K meson beam versus using a ^{60}Co beam for treating a tumor?

(b) Which of the following has the greatest effect in causing an initially parallel beam of charged kaons to broaden as it penetrates tissue?
1. Multiple Coulomb scattering of the particles by atomic nuclei
2. The generation of delta rays along the particle paths
3. Energy loss straggling
4. Range straggling
5. Collisions of the particles with atomic electrons

(c) What types of radiation are the most important considerations in the shielding design for a negative K meson beam?
1. Neutrons, prompt photons, muons, and protons
2. Neutrons, prompt photons, and pions
3. Neutrons, prompt and residual photons, and muons
4. Neutrons and prompt and residual photons
5. Neutrons and residual photons

(d) When the accelerator is not operating, which of the radiation types would most likely contribute the greatest effective dose to a technician working in the treatment area where the patients are exposed to a negative K meson beam?
1. Gamma rays and beta particles
2. Gamma rays and neutrons
3. Neutrons, gamma rays, and beta particles
4. Gamma rays, muons, and pions
5. Muons, beta particles, and gamma rays

(e) The difference between the negative K meson rest energy (494 MeV) and the average total kinetic energy released per capture (367 MeV) shown in the table is:
1. Carried away by undetected neutrinos
2. Spent in overcoming nuclear binding energies
3. Not zero, because the table gives only average values
4. Lost by the kaon when captured
5. Emitted as bremsstrahlung during rearrangement of the atomic electrons about the produced nuclear fragments

(f) A negative K meson beam from an accelerator will also likely contain:
1. Pions, negative muons, and electrons as well as some neutrons
2. Negative muons only
3. Neutrons only
4. Electrons and photons
5. Electrons only

(g) The range of a K meson in material of low atomic number is determined to be $11.5\,\mathrm{g/cm^2}$. What is the range in centimeters, in soft tissue, having a density of $0.95\,\mathrm{g/cm^3}$?

(h) From the data given in the table, estimate the average absorbed dose in a 4.0 cm radius sphere of water surrounding the capture site of a stopped negative K meson by a ^{16}O nucleus. State the assumptions used in the estimate.

(i) In addition to the data in the table, what other information would you need in order to make a more accurate calculation of the average absorbed dose in the last problem?

(j) If beams of negative pions, negative K mesons, and ^{12}C ions are available, which of these beams has the greatest potential for selectively depositing dose within the tumor volume?

5.3 You are a medical physicist at the Higgs Medical Research Institute assigned to the team developing an iridium–iron–iodine colloid that preferentially deposits radioiodine in thyroid cancer cells. The current project assignment involves treatment of thyroid disease using various radioiodine isotopes. In the current treatment series involving Graves' disease, a nuclear medicine physician administers a small amount of ^{123}I to determine the uptake and to perform initial imaging. The physician plans to treat the patients having Graves' disease with ^{131}I.

Data:

Thyroid uptake in the patient = 60%
Thyroid mass in the patient = 100 g
Thyroid mass in reference man = 20 g
Absorbed dose per unit cumulated activity (S factor) for thyroid as source and target organ = 1.57×10^{-3} mGy/MBq-s
Assume the contribution from all other source organs to the thyroid (target organ) is negligible
Effective half-life of radioiodine in the patient's thyroid = 5 days
Physical half-life of ^{131}I = 8.04 days

(a) The physician decides to deliver an absorbed dose of 70 Gy to the thyroid. Calculate the ^{131}I activity administered to the patient to deliver the prescribed dose.

(b) If the patient was administered 1480 MBq, calculate the cumulative external effective dose to his spouse under the following conditions:
- Sleeping arrangements: distance is 1 m.
- The thyroid is the only source of exposure.
- Time spent in the vicinity (1 m) of the spouse over a period of 24 h = 8 h.
- Specific gamma-ray dose constant at 1 m = 5.2×10^{-5} mSv/h-MBq.

(c) Assume the dose equivalent to the patient's spouse is 2.5 mSv. Is the licensee in compliance with the radiation protection limits of 10CFR35 if the patient is released from the hospital immediately after administration? For this question, assume the patient is a mother nursing 3-month-old twins and was given no release instructions by the medical facility. The effective dose to each infant is estimated to be 1.25 mSv.

(d) Give four general precautionary measures that you would suggest to a patient treated for the condition of Graves' disease upon release from the hospital.

5.4 As the newly appointed Radiation Safety Officer at the University of Eastern Alaska, you are reviewing the medical school's proposal for advanced brachytherapy research using injectable, remotely controlled sources of ^{192}Ir. The principal investigator wants to use remotely controlled brachytherapy (RCB) in the middle of a treatment room that already houses a superficial therapy system.

Three walls adjoin unoccupied areas. The remaining wall adjoins an area, which will contain the control panel for the RCB and the associated X-ray systems. The RCB system is supplied quarterly with fresh ^{192}Ir sources having a total administered activity of 0.37 TBq.

Data:

1. HVL (half-value layer) for $^{192}\text{Ir} = 4$ cm concrete.
2. HVL for 125 kVp X-rays = 2 cm concrete.
3. Kerma rate constant for $^{192}\text{Ir} = 1.1 \times 10^{-4}$ mGy-m^2/MBq-h.
4. The room size is 4.9 m × 4.9 m.
5. The control panel wall is 15 cm thick.
6. The output of the X-ray system is 180 mGy at 30 cm source to survey distance for 125 kVp and 5 mA-min system settings.
7. The kerma in the control panel area due to the X-ray beam is 0.0125 mGy with machine settings of 125 kVp and 10 mA-min. These values correspond to a 1 min exposure.
8. The weekly workload of the superficial therapy system is 750 mA-min.
9. The weekly design kerma limit for the new RCB research area is 1.0 mGy.
10. Apply NCRP dose recommendations using 1 Gy = 1 Sv for photon radiation.
    (a) You do not know the composition of the control panel wall. The X-ray system is 3.7 m from the wall, points toward it, and runs at 125 kVp, 10 mA (the maximum possible values) for 8 min. For these parameters, the kerma outside the wall due to the X-ray beam is 0.1 mGy. Calculate the equivalent thickness of the wall in centimeters of concrete. Ignore buildup and assume that the HVL does not change with penetration depth.
    (b) The principal investigator intends to treat five to eight patients per week with the RCB system. The average treatment time per patient is 4 min (at maximum activity). Assuming the maximum activity loading, calculate the workload for the RCB system.
    (c) For the conditions described in the previous problem, how much shielding should you add to the control panel wall before using the RCB system?
    (d) The space above the treatment room ceiling is normally unoccupied. However, technicians occasionally go on the roof during treatment hours to conduct radiation surveys. When calculating the shielding requirements for the ceiling of the RCB treatment area, what

would be the appropriate values for the use factor and the occupancy factor?

(e) In a recent accident involving an RCB unit, a patient went home with the source inside her body. She allegedly died from the resulting exposure. Describe two precautions that could have prevented this type of event from occurring.

5.5 You are a duty shift health physicist employed by a large metropolitan hospital with an excellent nuclear medicine department. A dirty bomb has been detonated, and a contaminated survivor is brought to the emergency room for treatment. The attending physician managing all health-care actions has requested your assistance with radiological issues associated with patient treatment.

Data:
Area of uniformly contaminated skin = 50 cm^2
Dose rate factor = 1.35×10^{-3} mGy-cm^2/Bq-h
Sloughing fractional removal rate = 5% per day
Pancake probe efficiency (counts/disintegration) = 0.1
Pancake probe area = 15 cm^2

(a) Assume that the radiological hazards to medical personnel while treating the injury are minimal. List four radiological items regarding the incident and individual that you as the health physicist should provide to the physician.

(b) After all routine decontamination techniques have been attempted, the residual contamination level is still of concern to the physician. The physician is contemplating a radical technique for removing stubborn skin contamination. What four considerations concerning risk to the patient would you provide as advice to the physician while she is considering using the radical technique?

(c) The physician conditionally decides that no treatment be given if the dose from the stubborn contamination is limited to 1 Gy. What persistent level of long-lived contamination in counts per minute could be left on the skin such that the limiting dose is not exceeded? State all assumptions.

5.6 You are assigned as the lead health physicist to a research team charged with developing and constructing a prototype internal radiation-generating device incorporating nanotechnology. The team leader is experienced in nanotechnology, but does not have an extensive knowledge of health physics or ionizing radiation. Accordingly, he has submitted a list of questions for your action. Your answers will assist the team in refining the desired operating characteristics of the prototype.

The preliminary design specifications for the device are:

1. The size will permit injection into a body.
2. The device can be directed to a specific body location.

3. The device preferentially irradiates the tumor with minimal dose to healthy tissue.

(a) What are the relative hazards of the following radiation types that are candidates for use in the device: photons, electrons, muons, protons, charged pions, alpha particles, ^{12}C nuclei, ^{16}O nuclei, ^{56}Fe nuclei, thermal neutrons, antiprotons, and anti ^{12}C nuclei?

(b) What are the interaction products of low-energy antiprotons in tissue?

(c) For each of the radiation types created in question (b), what is their applicability to meeting the design characteristics of the device?

(d) Given the radiation types noted in part (a), which offer the best potential for localization of dose within the tumor?

(e) Based on your input and the engineering staff's assessment, the team leader decides to base the prototype on low-energy protons. What information is needed to determine the absorbed dose as a function of penetration depth in tissue?

5.7 You are supporting a nanotechnology project whose primary objective is to develop a technique to destroy a tumor's blood supply. The senior physician has constructed nanoparticles containing radioactive material that preferentially attach to an arteriole wall supplying blood to a tumor. His approach is based on the observation that a tumor's rapid growth produces blood vessels that are irregular and have larger gaps in their walls than healthy vessels. These characteristics lead to a vascular structure that is not sufficient to optimally nourish the tumor. Doses in the 100 Gy range damage the blood vessels, prevent tumor growth, and eventually lead to starvation of the cancer cells.

Data:

1. Arterioles have a lumen diameter of 30 μm and a mean wall thickness of 20 μm.
2. ^{113}I has a half-life of 5.9 s and emits a 2.61 MeV alpha particle with an associated dose conversion coefficient of 364 pGy-cm^2/alpha. It also emits photons with energies 0.463 and 0.622 MeV. Assume all yields are 100%.

(a) Characterize the appropriateness of the following radionuclides for the tumor vascular disruption research: ^{3}H, ^{32}P, ^{60}Co, ^{125}I, ^{201}Tl, and ^{252}Cf. The research goal is to preferentially deposit energy into the arteriole wall and significantly limit the dose to the healthy tissue.

(b) Based on the results of the previous question, list desirable characteristics for the radionuclide incorporated into the nanoparticles to facilitate tumor blood vessel destruction.

(c) Assume that an ^{113}I generator has been procured and this isotope is incorporated into the nanoparticle. What activity of ^{113}I is required to produce an integrated absorbed dose of 100 Gy at the outer arteriole wall? Assume that it takes 1 min for the material to attach to the arteriole wall following injection into the body and there is a 1 min delay between milking the generator, forming the nanoparticle, and injecting the material into the patient. Nanoparticle production is accomplished through a muon-catalyzed reaction with ^{113}I and a gold colloid. Base the activity estimate on the dose delivered by the alpha particles. Treat the collection of nanoparticles as an unattenuated point source and ignore any attenuation by the arteriole wall.

(d) Upon milking, estimate the unshielded gamma absorbed dose rate 1 cm from the ^{113}I source determined in the previous question. Assume a point isotropic source for the capsule containing the nanoparticles, and the nanoparticles do not attenuate the ^{113}I photons.

(e) Based on the absorbed dose rate calculated in the previous question, what ALARA (as low as reasonably achievable) provisions should be implemented for milking the ^{113}I generator and for administering the isotope to the patient?

(f) The methodology used in the initial calculations is not sufficiently rigorous to finalize a microsphere design. What calculations should be performed to verify that the ^{113}I alpha particles have sufficient range to disrupt the vascular wall?

5.8 A 25 MeV pulsed electron linear accelerator has been installed at the London Memorial Veterans Hospital. The accelerator is part of a multimodal cancer therapy project sponsored by the European Union. In view of a compressed installation and operational acceptance schedule, accelerator testing begins before the interior sides of the concrete walls of the target room have been painted. The proposed schedule has the accelerator in operational testing from 7:00 a.m. to 3:00 p.m. followed by wall painting from 4:00 p.m. to 11:00 p.m.

Table 1 summarizes the three elements of interest in this problem, their atomic weights, and their concentrations in the concrete walls. The density of concrete is 2.37 g/cm^3.

Table 1 Target elements in concrete.

Target element	Atomic weight	Concentration (g/cm^3)
Na	22.99	0.012
K	39.10	0.008
Fe	55.85	0.018

Table 2 provides the thermal neutron macroscopic activation cross-section for the production of ^{24}Na, ^{42}K, and ^{59}Fe. In addition, photonuclear reactions produce ^{22}Na [^{23}Na(γ, n)^{22}Na] with a 2.60 year half-life, ^{38}K [^{39}K(γ, n)^{38}K] with a 7.63 min half-life, and ^{55}Fe [^{56}Fe(γ, n)^{55}Fe] with a 2.75 year half-life. The accelerator has a thin tungsten target with a neutron yield of 0.001 neutrons per incident electron. The average beam current is 200 μA. The electron beam travels from South to North. The North, East, and West walls are all 3 m from the accelerator target, which is unshielded.

Table 2 Neutron activation reactions of interest.

Target isotope	Abundance (%)	Product	Half-life	Cross-section (cm^2/g)
^{23}Na	100	^{24}Na	15.0 h	1.39×10^{-2}
^{41}K	6.77	^{42}K	12.4 h	1.22×10^{-3}
^{58}Fe	0.31	^{59}Fe	45.6 days	3.01×10^{-5}

(a) Calculate the direct neutron fluence rate (flux) at a distance 3 m North of the target. Assume an isotropic emission of neutrons.
(b) For a thermal neutron fluence rate of 2.0×10^7 n/cm^2-s at one of the concrete walls, calculate the activity (Bq/cm^3) of ^{24}Na in 1 cm^3 of concrete at saturation. Assume the accelerator can run for an extended period.
(c) For a thermal neutron fluence rate of 2.0×10^7 n/cm^2-s at one of the concrete walls, calculate the activity (Bq/cm^3) of ^{24}Na in 1 cm^3 of concrete after 8 h of beam operations at 3:00 p.m.
(d) For a thermal neutron fluence rate of 2.0×10^7 n/cm^2-s at one of the concrete walls, calculate the activity (Bq/cm^3) of ^{24}Na in 1 cm^3 of concrete 8 h later after the painters go home at 11:00 p.m.
(e) Calculate the ratio of the ^{42}K and ^{24}Na saturation activities.
(f) Give one reason why it would be inappropriate to use a bare, unmodified BF_3 proportional counter to measure the neutron flux at the North wall inside the accelerator room.
(g) Before the painters enter the accelerator room, you survey the walls with a shielded pancake GM (Geiger–Müller) probe, and find that the readings on the North, East, and West walls are essentially the same. Explain why the readings for the East, West, and North walls are the same after the initial run.
(h) Five years later, accelerator operations are suspended to permit an upgrade that increases the beam energy and current. Two weeks after the last use of the accelerator, you conduct a similar survey and find that the East and West walls are close to background but that the North wall is still showing significant activation. Explain why the North wall shows activation on your last survey but the East and West walls do not.

References

10CFR37 (2015) Physical Protection of Category 1 and Category 2 Quantities of Radioactive Material, National Archives and Records Administration, US Government Printing Office, Washington, DC.

Adriani, G., de Tullio, M.D., Ferrari, M., Hussain, F., Pascazio, G., Liu, X., and Decuzzi, P. (2012) The preferential targeting of the diseased microvasculature by disk-like particles. *Biomaterials*, **33** (22), 5504.

Akabani, G., Kennel, S.J., and Zalutsky, M.R. (2003) Microdosimetric analysis of alpha – particle-emitting targeted radiotherapeutics using histological images. *J. Nucl. Med.*, **44**, 792.

Amor-Coarasa, A., Milera, A., Carvajal, D., Gulec, S., and McGoron, A.J. (2014) ^{90}Y- DOTA-CHS microspheres for live radiomicrosphere therapy: preliminary *in vivo* lung radiochemical stability studies. *J. Radiother.*, **2014**, 1.

Barrett, K.E., Barman, S.M., Boitano, S., and Brooks, H. (2012) *Ganong's Review of Medical Physiology*, 24th edn, McGraw-Hill Medical, New York.

Baum, E.M., Ernesti, M.C., Knox, H.D., Miller, T.R., and Watson, A.M. (2010) *Nuclides and Isotopes – Chart of the Nuclides*, 17th edn, Knolls Atomic Power Laboratory, Bechtel Marine Propulsion Corporation.

Benalcazar, W.A. (2010) High-speed nonlinear interferometric vibrational imaging of biological tissue with comparison to Raman microscopy. *IEEE J. Sel. Top. Quantum Electron.*, **16**, 824.

Berry, C.C. and Curtis, A.S.G. (2003) Functionalisation of magnetic nanoparticles for applications in biomedicine. *J. Phys. D: Appl. Phys.*, **36**, R167.

Bethe, H. (1930) Zur theorie des durchgangs schneller korpuskularstrahlung durch materie. *Ann. Phys. (Leipzig)*, **5**, 325.

Bevelacqua, J.J. (2003) Production equations in health physics. *Radiat. Prot. Manage.*, **20** (6), 9.

Bevelacqua, J.J. (2004a) Skin contaminations, internal intakes, and ALARA. *Radiat. Prot. Manage.*, **21** (1), 11.

Bevelacqua, J.J. (2004b) Point source approximations in health physics. *Radiat. Prot. Manage.*, **21** (5), 9.

Bevelacqua, J.J. (2005a) Internal dosimetry primer. *Radiat. Prot. Manage.*, **22** (5), 7.

Bevelacqua, J.J. (2005b) Systematics of heavy ion radiotherapy. *Radiat. Prot. Manage.*, **22** (6), 4.

Bevelacqua, J.J. (2009) *Contemporary Health Physics: Problems and Solutions*, 2nd edn, Wiley-VCH Verlag GmbH, Weinheim.

Bevelacqua, J.J. (2010a) *Basic Health Physics: Problems and Solutions*, 2nd edn, Wiley-VCH Verlag GmbH, Weinheim.

Bevelacqua, J.J. (2010b) Practical and effective ALARA. *Operational Radiation Safety. Health Phys.*, **98** (Suppl. 2), S39.

Bevelacqua, J.J. (2010c) Feasibility of using internal radiation-generating devices in radiotherapy. *Health Phys.*, **98**, 614.

Bevelacqua, J.J. (2012) Angular absorbed dose dependence of internal radiation-generating devices in radiotherapy. *Health Phys.*, **102**, 2.

Bevelacqua, J.J. (2014) Tumor vascular disruption using various radiation types. *PeerJ*, **2**, e320. doi: 10.7717/peerj.320, https://peerj.com/articles/320/ (accessed 03 August 2015).

Bevelacqua, J.J. (2015) ^{210}Po Microsphere Radiological Design for Tumor Vascular Disruption, *PeerJ.*, **3**, e1143; doi: 10.7717/peerj. Available at https://peerj.com/articles/1143/.

Bilbao, R.I. and Reiser, M.F. (2008) *Liver Radioembolization with ^{90}Y Microspheres*, Springer, Berlin.

Bittner, M., Grosu, A., Wiedenmann, N., and Wilkens, J.J. (2013) A systematic review of antiproton radiotherapy. *Front. Phys.*, **1**, 37. doi: 10.3389/fphy.2013.00037

Blatter, C., Weingast, J., Alex, A., Grajciar, B., Wieser, W., Drexler, W., Huber, R., and Leitgeb, R.A. (2012) In situ structural and microangiographic assessment of human skin lesions with high-speed OCT. *Biomed. Opt. Express*, **3** (10), 2636.

Boas, G. (2012) Moving noninvasive cancer imaging into the clinic. *Biophotonics*, **19** (3), 28.

Boyer, A.L., Goitein, M., Lomax, A.J., and Pedroni, E.S. (2002) Radiation in the treatment of cancer. *Phys. Today*, **55** (9), 38.

Brenner, D.J. (1997) Radiation biology in brachytherapy. *J. Surg. Oncol.*, **65**, 66.

Brenner, D.J., Leu, C.-S., Beatty, J.F., and Sheferk, R.E. (1999) Clinical relative biological effectiveness of low-energy X-rays emitted by miniature X-ray devices. *Phys. Med. Biol.*, **44**, 323.

Brentjens, R.J. *et al.* (2013) CD19-targeted T cells rapidly induce molecular remissions in adults with chemotherapy-refractory acute lymphoblastic leukemia. *Sci. Transl. Med.*, **5** (177), 177ra38.

Burke, P.A. and DeNardo, S.J. (2001) Antiangiogenic agents and their promising potential in combined therapy. *Crit. Rev. Oncol. Hematol.*, **39**, 155.

Carmeliet, P. and Jain, R.K. (2000) Angiogenesis in cancer and other diseases. *Nature*, **407**, 249.

Carr, B.I. (2004) Hepatic arterial ^{90}Y glass microspheres (Therasphere) for unresectable hepatocellular carcinoma: interim safety and survival data on 65 patients. *Liver Transpl.*, **10**, S107.

Cember, H. and Johnson, T.A. (2008) *Introduction to Health Physics*, 4th edn, McGraw-Hill Medical, New York.

Charlton, L.A. and Robson, D. (1973) MERCURY. Florida State University Technical Report No. 5, Florida State University, Tallahassee, FL.

Chaudhury, P.K., Hassanain, M., Bouteaud, J.M., Alcindor, T., Nudo, C.G., Valenti, D., Cabrera, T., Kavan, P., Feteih, I., and Metrakos, P. (2010) Complete response of hepatocellular carcinoma with sorafenib and ^{90}Y radioembolization. *Curr. Oncol.*, **17** (3), 67.

Cherry, S.R., Sorenson, J.A., and Phelps, M.E. (2003) *Physics in Nuclear Medicine*, 3rd edn, Elsevier Health Sciences, Amsterdam.

Christian, P.E. and Waterstram-Rich, K. (2007) *Nuclear Medicine and PET/CT Technology and Techniques*, 6th edn, Elsevier Health Sciences, Amsterdam.

Chu, W.T., Ludewigt, B.A., and Renner, T.R. (1993) Instrumentation for treatment of cancer using proton and light-ion beams. *Rev. Sci. Instrum.*, **64**, 2055.

Dadachova, E., Burns, T., Bryan, R.A., Apostolidis, C., Brechbiel, M.W., Nosanchuk, J.D., Casadevall, A., and Pirofski, L. (2004a) Feasibility of radioimmunotherapy of experimental pneumococcal infection. *Antimicrob. Agents Chemother.*, **48**, 1624.

Dadachova, E., Howell, R.W., Bryan, R.A., Frenkel, A., Nosanchuk, J.D., and Casadevall, A. (2004b) Susceptibility of the human pathogenic fungi *Cryptococcus neoformans* and *Histoplasma capsulatum* to gamma- radiation versus radioimmunotherapy with alpha- and beta-emitting radioisotopes. *J. Nucl. Med.*, **45**, 313.

Dadachova, E. and Casadevall, A. (2005) Antibodies as delivery vehicles for radioimmunotherapy of infectious diseases. *Expert Opin. Drug Deliv.*, **2**, 1075.

Dadachova, E., Nakouzi, A., Bryan, R.A., and Casadevall, A. (2003) Ionizing radiation delivered by specific antibody is therapeutic against a fungal infection. *Proc. Natl. Acad. Sci. U.S.A.*, **100**, 10942.

Day, C. (2005) Hybrid imaging system combines X rays and magnetic resonance to improve surgical procedures. *Phys. Today*, **58** (6), 22.

Dezarn, W.A., Cessna, J.T., DeWerd, L.A., Feng, W., Gates, V.L., Halama, J., Kennedy, A.S., Nag, S., Sarfaraz, M., Sehgal, V., Selwyn, R., Stabin, M.G., Thomadsen, B.R., Williams, L.E., and Salem, R. (2011) Recommendations of the American Association of Physicists in Medicine on dosimetry, imaging, and quality assurance procedures for ^{90}Y microsphere brachytherapy in the treatment of hepatic malignancies. *Med. Phys.*, **38** (8), 4824.

Folkman, J. (1971) Tumor angiogenesis: therapeutic implications. *N. Engl. J. Med.*, **285**, 1182.

Friedman, A.M., Seevers, R.H. Jr.,, and Spencer, R.P. (1987) *Radionuclides in Therapy*, CRC Press, Boca Raton, FL.

Grossman, J. (2012) Nanotechnology in cancer medicine. *Phys. Today*, **65** (8), 38.

Hamid, O. *et al.* (2013) Safety and tumor responses with lambrolizumab (anti–PD-1) in melanoma. *N. Engl. J. Med.*, **369**, 134.

Hendee, R.R., Ibbott, G.S., and Hendee, E.G. (2004) *Radiation Therapy Physics*, 3rd edn, Wiley-Liss, New York.

Hendee, W.R. and Ritenour, E.R. (2002) *Medical Imaging Physics*, 4th edn, Wiley-Liss, New York.

Heo, S.H., Kim, H.J., Ha, J.M., and Cho, S.O. (2012) A vacuum-sealed miniature X-ray tube based on carbon nanotube field emitters. *Nanoscale Res. Lett.*, **7**, 258.

Hirao, Y., Ogawa, H., Yamada, S., Sato, Y., Yamada, T., Sato, K., Itano, A., Kanazawa, M., Noda, K., Kawachi, K., Endo, M., Kanai, T., Kohno, T., Sudou, M., Minohara, S., Kitagawa, A., Soga, F., Takada, E., Watanabe, S., Endo, K., Kumada, M., and Matsumoto, S. (1992) Heavy ion synchrotron for medical use—HIMAC project at NIRS-Japan. *Nucl. Phys.*, **A538**, 541c.

Holzscheiter, M., Bassler, N., Agazaryan, N., Beyer, G., Blackmore, E., DeMarco, J., Doserm, M., Durand, R., Hartley, O., and Iwamoto, K. (2006) The biological effectiveness of antiproton irradiation. *Radiother. Oncol. J.*, **81**, 233.

IAEA (1997) *Nuclear Data for Neutron Therapy: Status and Future Needs.* IAEA TECDOC 992, International Atomic Energy Agency, Vienna.

ICRP (1996) *Radiological Protection and Safety in Medicine*, ICRP Publication 73, Elsevier, Amsterdam.

ICRP (2000a) *Radiation Dose to Patients from Radiopharmaceuticals*, ICRP Publication 80, Elsevier, Amsterdam.

ICRP (2000b) *Pregnancy and Medical Radiation*, ICRP Publication 84, Elsevier, Amsterdam.

ICRP (2001a) *Avoidance of Radiation Injuries from Medical Interventional Procedures*, ICRP Publication 85, Elsevier, Amsterdam.

ICRP (2001b) *Prevention of Accidents to Patients Undergoing Radiation Therapy*, ICRP Publication 86, Elsevier, Amsterdam.

ICRP (2001c) *Managing Patient Dose in Computed Tomography*, ICRP Publication 87, Elsevier, Amsterdam.

ICRP (2005a) *Release of Patients after Therapy with Unsealed Radionuclides*, ICRP Publication 94, Elsevier, Amsterdam.

ICRP (2005b) *Prevention of High-Dose Rate Brachytherapy Accidents*, ICRP Publication 97, Elsevier, Amsterdam.

ICRP (2006) *Radiation Aspects of Brachytherapy for Prostate Cancer*, ICRP Publication 98, Elsevier, Amsterdam.

ICRP (2007) *Managing Patient Dose in Multi-Detector Computed Tomography (MDCT)*, ICRP Publication 102, Elsevier, Amsterdam.

ICRP (2010) *Conversion Coefficients for Radiological Protection Quantities for External Radiation Exposures*, ICRP Publication 116, Elsevier, Amsterdam.

ICRP (2014) *Radiological Protection in Ion Beam Radiotherapy*, ICRP Publication 127, Sage, London.

ICRU (1976) Determination of Absorbed Dose in a Patient Irradiated by Beams of X- or Gamma-Rays in Radiotherapy Procedures. ICRU Report 24, ICRU Publications, Bethesda, MD.

ICRU (1977) Neutron Dosimetry for Biology and Medicine. ICRU Report 26, ICRU Publications, Bethesda, MD.

ICRU (1985) Dose and Volume Specification for Reporting Intracavitary Therapy in Gynecology. ICRU Report 38, ICRU Publications, Bethesda, MD.

ICRU (1986) Modulation Transfer Function of Screen-Film Systems. ICRU Report 41, ICRU Publications, Bethesda, MD.

ICRU (1987) Use of Computers in External Beam Radiotherapy Procedures with High-Energy Photons and Electrons. ICRU Report 42, ICRU Publications, Bethesda, MD.

ICRU (1989) Clinical Neutron Dosimetry-Part 1: Determination of Absorbed Dose in a Patient Treated by External Beams of Fast Neutrons. ICRU Report 45, ICRU Publications, Bethesda, MD.

ICRU (1992) Photon, Electron, and Neutron Interaction Data for Body Tissues. ICRU Report 46, ICRU Publications, Bethesda, MD.

ICRU (1993a) Phantoms and Computational Models in Therapy, Diagnosis and Protection. ICRU Report 48, ICRU Publications, Bethesda, MD.

ICRU (1993b) Prescribing, Recording, and Reporting Photon Beam Therapy. ICRU Report 50, ICRU Publications, Bethesda, MD.

ICRU (1996) Medical Imaging – The Assessment of Image Quality. ICRU Report 54, ICRU Publications, Bethesda, MD.

Jain, R.K. and Stylianopoulos, T. (2010) Delivering nanomedicine to solid tumors. *Nat. Rev. Clin. Oncol.*, **7** (11), 653.

Johns, H.E. and Cunningham, J.R. (1983) *The Physics of Radiology*, 4th edn, Charles C. Thomas Publisher, Springfield, IL.

Jurcic, J.G., Larson, S.M., Sgouros, G. *et al.* (2002) Targeted alpha particle immunotherapy for myeloid leukemia. *Blood*, **100**, 1233.

Kennedy, A., Nag, S., Salem, R., Murthy, R., McEwan, A.J., Nutting, C., Benson, A., Espat, J., Bilbao, J.I., Sharma, R.A., Thomas, J.P., and Caldwell, D. (2007) Recommendations for radioembolization of hepatic malignancies using yttrium-90 microsphere brachytherapy: a consensus panel report from the radioembolization brachytherapy oncology consortium. *Int. J. Radiat. Oncol. Biol. Phys.*, **68** (1), 13.

Kennedy, A.S. and Salem, R. (2003) Comparison of two ^{90}Yttrium microsphere agents for hepatic artery brachytherapy. Proceedings of the 14th International Congress on Anti-Cancer Treatment, Vol. 1.

Kereiakes, J.G. and Rosenstein, M. (1980) *Handbook of Radiation Doses in Nuclear Medicine and Diagnostic X-Rays*, CRC Press, Boca Raton, FL.

Khan, F.M. (2003) *Physics of Radiation Therapy*, 3rd edn, Lippincott Williams and Wilkins, Philadelphia, PA.

Khandhar, A.P., Ferguson, R.M., Simon, J.A., and Krishnan, K.M. (2012) Enhancing cancer therapeutics using size-optimized magnetic fluid hyperthermia. *J. Appl. Phys.*, **111**, 07B306.

Kraft, G. (1990) The radiobiological and physical basis for radiotherapy with protons and heavier ions. *Strahlenther. Onkol.*, **166**, 10.

Kraft, G. (2000) Tumor therapy with heavy charged particles. *Prog. Part. Nucl. Phys.*, **45**, S473.

Kramer, D. (2013) Tighter security ahead for nuclear materials in health care. *Phys. Today*, **66** (3), 30.

Kunz, P.D. (1986) Computer Code DWUCK, University of Colorado, Bolder, CO; available as part of Radiation Safety Information Computational Center Computer Code Collection: Code Package PSR-235 DWUCK, Nuclear Model Computer Codes for Distorted Wave Born Approximation and Coupled Channel Calculations, Oak Ridge National Laboratory, Oak Ridge, TN, http://rsicc.ornl.gov (accessed 23 August 2005).

Kwok, R. (2011) A physicist in the cancer lab. *Symmetry*, **8** (3), 16.

Lewis, D.M. and Kalemis, A. (2011) PET and MRI: providing the full picture. *CERN Cour.*, **51** (9), 39.

Li, Q., Furusawa, Y., Kanazawa, M., Kanai, T., Kitagawa, A., Aoki, M., Urakabe, E., Sate, S., and Wei, Z. (2006) Enhanced biological effect induced by a radioactive ^{9}C-ion beam at the depths around its Bragg peak. *Nucl. Instrum. Methods*, **B245**, 302.

Löbenberg, R. (2003) Smart materials: applications of nanotechnology in drug delivery and drug targeting. IEEE Proceedings of the International Conference on MEMS, NANO and Smart Systems, p. 82.

Marks, D.L. and Boppart, S.A. (2004) Nonlinear interferometric vibrational imaging. *Phys. Rev. Lett.*, **92**, 123905-1.

Marmier, P. and Sheldon, E. (1969) *Physics of Nuclei and Particles*, vol. **1**, Academic Press, New York.

Martinez, A.A., Gonzalez, J.A., Chung, A.K., Kestin, L.L., Balasubramaniam, M., Diokno, A.C., Ziaja, E.L., Brabbins, D.S., and Vicini, F.A. (2000) A comparison of external beam radiation therapy versus radical prostatectomy for patients with low risk prostate carcinoma diagnosed, staged and treated at a single institution. *Cancer*, **88**, 425.

Mettler, F.A. and Guiberteau, M.J. (2005) *Essentials of Nuclear Medicine Imaging*, 5th edn, Elsevier Health Sciences, Amsterdam.

NCRP (2007) Management of Radionuclide Therapy Patients. NCRP Report No. 155, NCRP Publications, Bethesda, MD.

NCRP (2009) Ionizing Radiation Exposure of the Population of the United States. NCRP Report No. 160, NCRP Publications, Bethesda, MD.

NCRP (2011) Second Primary Cancers and Cardiovascular Disease after Radiation Therapy. NCRP Report No. 170, NCRP Publications, Bethesda, MD.

Olsena, D.R., Brulanda, Ø.S., Frykholmc, G., and Norderhaugd, I.N. (2007) Proton therapy – A systematic review of clinical effectiveness. *Radiother. Oncol.*, **83**, 123.

Pilat, M.J. and LoRusso, P.M. (2006) Vascular disrupting agents. *J. Cell. Biochem.*, **99**, 1021.

Pinelli, T., Zonta, A., Altieri, S., Barni, S., Braghieir, A., Pedroni, P. *et al.* (2002) TAOrMINA: from the first idea to the application to the human liver, in *Research and Development in Neutron Capture Therapy* (eds M.W. Sauerwein, R. Moss, and A. Wittig), Monduzzi Editore, International Proceedings Division, Bologna, p. 1065.

Powsner, R.A. (2006) *Essential Nuclear Medicine Physics*, 2nd edn, Blackwell Publishing Limited, Oxford.

Radiation Safety Information Computational Center (1985) *SPAR: Calculation of Stopping Powers and Ranges for Muons, Charged Pions, Protons, and Heavy Ions,*

Radiation Safety Information Computational Center Computer Code Collection: Code Package CCC-228, Oak Ridge National Laboratory, Oak Ridge, TN, http://rsicc.ornl.gov (accessed 06 August 2015).

Rittmann PD. (2004) *ISO-PC Version 2.2-User's Guide*. Richland, WA: Fluor Government Group.

Rivard, M. (1999) Dosimetry for ^{252}Cf neutron emitting brachytherapy sources: protocol, measurements, and calculations. *Med. Phys.*, **26**, 1503.

Rosenberg, S.A. (2011) Cell transfer immunotherapy for metastatic solid cancer- what clinicians need to know. *Nat. Rev. Clin. Oncol.*, **8** (10), 577.

Savage, L. (2012) PDT for cancer depends on improved photosensitizers. *Biophotonics*, **19** (3), 24.

Scheidenberger, C. and Geissel, H. (1998) Penetration of relativistic heavy ions through matter. *Nucl. Instrum. Methods Phys. Res., Sect. B*, **135**, 25.

Selman, J. (1976) *The Basic Physics of Radiation Therapy*, 2nd edn, Charles C. Thomas, Springfield, IL.

Shapiro, J. (2002) *Radiation Protection: A Guide for Scientists, Regulators, and Physicians*, 4th edn, Harvard University Press, Cambridge, MA.

Shen, W.Q., Wang, B., Feng, J., Zhan, W.L., Zhu, Y.T., and Feng, E.P. (1989) Total reaction cross section for heavy-ion collisions and its relation to the neutron excess degree of freedom. *Nucl. Phys. A*, **491**, 130.

Shleien, B., Slaback, Jr L.A., and Birky, B.K. (1998) *Handbook of health physics and radiological health*, 3rd edition. Baltimore: Lippincott, Williams, and Wilkins.

Sihver, L., Schardt, D., and Kanai, T. (1998) Depth-dose distributions of high-energy carbon, oxygen and neon beams in water. *Jpn. J. Med. Phys.*, **18**, 1.

Skarsgard, L.D. (1998) Radiobiology with heavy charged particles: a historical review. *Phys. Med.*, **14**, 1.

Slater, J.D. (2006) Clinical applications of proton radiation treatment at Loma Linda University: review of fifteen-year experience. *Technol. Cancer Res. Treat.*, **5** (2), 81.

Smith, A.R. *et al.* (1977) Dosimetry of pion therapy beams. *Med. Phys.*, **4**, 408.

Sullivan, A.H. (1985) A measurement of the local energy deposition by antiprotons coming to rest in tissue-like material. *Phys. Med. Biol.*, **30**, 1297.

Travish, G. and Yoder, R.B. (2011) Laser-powered dielectric-structures for the production of high-brightness electron and X-ray beams. *Proc. SPIE*, **80790**, 80790K.

Turner, J.E. (2007) *Atoms, Radiation, and Radiation Protection*, 3rd edn, Wiley-VCH Verlag GmbH, Weinheim.

U. S. Environmental Protection Agency (2013) PAG Manual, Protective Action Guides and Planning Guidance For Radiological Incidents. U.S. Environmental Protection Agency Draft for Interim Use and Public Comment, U. S. Environmental Protection Agency, Washington, DC.

United States Government Accountability Office (2012) Additional Actions Needed to Improve Security of Radiological Sources at U.S. Medical Facilities. Government Accountability Office Report GAO-12-925, United States Government Accountability Office, Washington, DC.

Wang, C.-K. (2012) Progress in californium-252 neutron brachytherapy, in *Brachytherapy* (ed K. Kishi), InTech, ISBN: 978-953-51-0602-9, http://www.intechopen.com/books/brachytherapy/progress-in-cf-252-based-neutron-brachytherapy (accessed 04 August 2015).

Washington, C.M. and Leaver, D.T. (2003) *Principles and Practice of Radiation Therapy*, 2nd edn, Elsevier Health Sciences, Amsterdam.

Weber, U. (1996) Volumenkonforme Bestrahlung mit Kohlenstoffionen. PhD thesis. Universität Gh Kassel, Germany.

Wijesinghe, P.I. (2007) Energy Deposition Study of Low-Energy Cosmic Radiation at Sea Level. Physics and Astronomy Dissertation. Paper 14, http://scholarworks.gsu.edu/cgi/viewcontent.cgi?article=1013&context=phy_astr_diss. (accessed 12 October 2015)

Wilson, R.R. (1946) Radiological use of fast protons. *Radiology*, **47**, 487.

Wolbarst, A.B. (2005) *Physics of Radiology*, 2nd edn, Medical Physics Publishing, Madison, WI.

Wolchuk, J.D. *et al.* (2013) Nivolumab plus ipilimumab in advanced melanoma. *N. Engl. J. Med.*, **369**, 122.

Zhang, L., Chen, H., Wang, L., Liu, T., Yeh, J., Lu, G., Yang, L., and Mao, H. (2010) Delivery of therapeutic radioisotopes using nanoparticle platforms: potential benefit in systemic radiation therapy. *Nanotechnol. Sci. Appl.*, **3**, 159.

6
Public Radiation Exposures and Associated Issues

6.1
Overview

Public attitudes regarding radiation and its associated biological effects are influenced by numerous events. Major power reactor accidents significantly affect public perceptions and attitudes toward technologies utilizing radiation or radioactive materials. These events trigger emotional reactions as do discussions of terrorist attacks utilizing dirty bombs or improvised nuclear devices.

More subtle situations involving radiation or radioactive materials and their associated doses are also encountered. Airport scanners used for security evaluations also influence public perception. In a similar fashion, the increased use of nuclear medicine procedures raises public awareness of the expanding use of radioactive materials and their associated doses.

The radiation exposure incurred by aircrews and space tourists and the associated hazard of solar flares present a potential public health concern. These and other sources of radiation exposure affecting the public and their associated issues are addressed in this chapter.

6.2
Public Radiation Exposures and Associated Effects

The local geophysical environment, level of technology, and cultural habits influence public radiation exposures. Accurately documenting public doses is a challenging task that requires a careful compilation, categorization, and quantification of the radiation sources. Fortunately, tabulations of public radiation exposures have been performed for many areas of the world including the United States.

NCRP 160 (National Council on Radiation Protection and Measurements) documents contributions from ionizing radiation sources to the population of the United States based on 2006 data. This report updates the information presented in NCRP 93 published in 1987. The radiation exposure to the US population is defined in terms of five broad categories including (i) exposure to

Health Physics: Radiation-Generating Devices, Characteristics, and Hazards, First Edition. Joseph John Bevelacqua.

ubiquitous background radiation including radon; (ii) exposure to patients from medical procedures; (iii) exposure from consumer products or activities involving radiation sources; (iv) exposure from industrial, security, medical, educational, and research radiation sources; and (v) occupational exposure.

The results are presented as annual values for the average effective dose to an individual in a group exposed to a specific source, collective effective dose (S), and the average effective dose per individual in the US population (E_{US}). S and E_{US} are useful quantities to compare different radiation sources.

6.2.1 Ubiquitous Background Radiation

Ubiquitous background radiation includes naturally occurring sources of ionizing radiation. NCRP 160 divides ubiquitous background radiation into four specific subcategories. These subcategories are (i) internal exposure from inhalation of radon (^{222}Rn) and thoron (^{220}Rn) and their progeny, (ii) external exposure from space radiation derived from solar particle events (SPEs) and galactic cosmic radiation (GCR), (iii) internal exposure from radionuclides deposited in the body, and (iv) external exposure from terrestrial radiation primarily ^{40}K and the ^{238}U and ^{232}Th natural decay series. The ubiquitous background exposure category is summarized in Table 6.1.

Ubiquitous radiation contributes 3.11 mSv to an average member of the US population. In terms of the four subcategories defined NCRP 160, 73% of the effective dose is attributed to radon and thoron, 11% to space radiation, 9% to radionuclides in the body, and 7% to terrestrial radiation.

Table 6.1 Ionizing radiation exposure to the US population from the ubiquitous background in 2006[a].

Ubiquitous background component	Effective collective dose (person-Sv)	Average effective dose to US population (mSv)	Average effective dose for the exposed group (mSv)
Internal radiation (inhalation of radon and thoron)	684 000	2.28	2.28
External radiation (space)	99 000	0.33	0.33
Internal radiation (ingestion)	87 000	0.29	0.29
External radiation (terrestrial)	63 000	0.21	0.21
Total	933 000	3.11	3.11

a) NCRP 160 (2009).

6.2.1.1
Cosmogenic Radionuclides

Naturally occurring radionuclides include cosmogenic and primordial radionuclides. Most cosmogenic radionuclides decay by beta, gamma-ray, or X-ray emission; have low to intermediate atomic numbers; and are created by the interactions of cosmic radiation with target atoms in the atmosphere and earth. The most significant cosmogenic nuclides include ^{3}H, ^{7}Be, ^{14}C, and ^{22}Na. ^{14}C is the major cosmogenic radionuclide that contributes to internal exposure.

6.2.1.2
Primordial Radionuclides

Most of the primordial radionuclides are members of the ^{232}Th, ^{235}U, and ^{238}U natural decay series. The radiation emission from the ^{238}U and ^{232}Th decay series is a significant contributor to the average effective dose to the US public.

The dominant primordial radionuclides are ^{40}K and ^{87}Rb. ^{40}K contributes about one-third of both the external terrestrial and internal effective doses derived from natural sources. Since ^{87}Rb is a pure beta emitter, it contributes minimal external dose, and its contribution to internal dose is about 2 orders of magnitude less than ^{40}K. In terms of internal radiation exposure, the most significant contributors are isotopes of uranium, thorium, radium, radon, polonium, bismuth, and lead derived from the ^{238}U and ^{232}Th decay series and ^{40}K.

6.2.1.3
Radon

Radon is produced from the ^{238}U series and thoron is a member of the ^{232}Th decay chain. Due to its unique nature and significant effective dose contribution, radon is included as a separate category in NCRP 160. The average radon concentration in US homes is 43.3 Bq/m^3, and the mean outdoor concentration is 15.1 Bq/m^3. For radon, an effective dose conversion factor (DCF) of 10 mSv/WLM (working level month) is adopted by NCRP 160. The thoron effective DCF is within the range of 3.3–3.8 mSv/WLM.

6.2.1.4
Anthropogenic Radionuclides

Human activities create radionuclides and release radioactive material into the environment. These activities include atmospheric nuclear weapons testing; fuel cycle activities supporting weapons production and nuclear power operations; radionuclides used in medicine, research, and industry; and nuclear reactor accidents. Radiation from weapons testing fallout has been significantly reduced since major atmospheric testing ceased around 1980. The radioactive material released during the Chernobyl accident did not significantly contribute to radiation exposure in the United States. However, it was a significant contributor in Europe and Asia. Most of the radioactive materials released from the Fukushima Daiichi accident were deposited in Japan, but residual activity has been detected in neighboring areas.

6.2.1.5
Radiation from Space

Protons and alpha particles with smaller admixtures of light nuclei are the dominant space radiation types. There are four dominant sources of space radiation incident on the earth and its atmosphere. These sources are (i) energetic particles associated with SPEs (e.g., solar flares); (ii) cosmic rays coming from interstellar space at the edge of the heliopause; (iii) GCR originating outside the Solar System, but within the Milky Way galaxy; and (iv) extragalactic cosmic radiation.

The incident particulate radiation has sufficient energy to induce nuclear reactions and produce high-energy muons, electrons, photons, and neutrons. Neutrons are attenuated by the atmosphere, contribute relatively low effective doses at sea level, and yield significant doses at higher altitudes. Aircrew members are dominantly exposed at these higher altitudes, and their doses are addressed in subsequent discussion.

6.2.1.6
Solar-Induced Disruptions and Radiation Effects

Solar-induced disruptions most frequently gather public attention when they interrupt communications and electrical power systems. Interruptions of television and radio signals are annoying but relatively minor events. However, more significant disruptions have occurred and affected millions of people. For example, solar storms are known to disrupt cellular phone service, global positioning systems (GPS), electrical power grids, and television and radio signals.

In March 1989, a solar storm much less severe than the 1859 Carrington event, addressed in Section 6.2.1.6.1, disrupted the US, Canadian, and Swedish power grids for several hours. The resulting damages and loss in revenue were estimated to be in the hundreds of millions of dollars. A 1994 solar event caused communications satellite malfunctions that disrupted television and radio service throughout Canada. The loss of communications capability during a nuclear event could have a significant impact on emergency management and supporting actions.

The Northeast blackout of 2003 was another massive widespread power outage caused by a solar event. This event affected parts of the Northeastern and Midwestern United States and Ontario, Canada. The blackout affected an estimated 10 million people in Ontario and 45 million people in eight US states.

Given the number, magnitude, and influence of these solar events, their characteristics and severity are addressed in subsequent discussion.

Solar flare-induced power surges have been cited as causing electrical transformer component melting at a nuclear plant in New Jersey that led to a partial loss of station power. The loss of power at a nuclear generating station has significant consequences. This was clearly illustrated by the Fukushima Daiichi accident.

In 2005, X-rays from another solar storm disrupted satellite-to-ground communications and GPS navigation signals for about 10 min. Since aircraft and ships utilize GPS information for landing and docking, disruptions have serious consequences.

6.2.1.6.1
Solar Event Characteristics

Solar flare radiation or SPEs are ejections of matter from the Sun. Their composition reflects the mass constituent characteristics of solar plasmas. Therefore, they are composed predominantly of protons with admixtures of alpha particles and heavier nuclei. The intensity and composition of solar flare radiation vary with the specific event. Carbon, nitrogen, and oxygen dominate the $Z > 2$ particles and constitute about 1% of the solar flare fluence rate.

Typical flare events last from 1 to 4 days although somewhat longer durations have been observed. On an annual basis, 8–11 significant solar flares occur. Solar physics models are not sufficiently advanced to predict the timing, duration, and intensity of a flare event. This uncertainty and the magnitude of these SPEs present a significant radiation hazard to astronauts in low earth orbit (LEO), during moon missions, and on planned missions to other planets. Solar flare radiation is also a significant consideration in the emerging space tourism industry that may provide orbital and suborbital travel within the next decade.

Electronic technologies and their associated components are vulnerable to SPEs. The effects of a massive solar flare are similar to the electromagnetic pulse effects noted in Chapter 4. For example, cell phone communications, GPS, and radar transmissions are vulnerable to large SPEs. In addition, commercial satellites are at risk from large-scale events such as the Carrington flare.

Humans working in space would also be at risk. Astronauts and space tourists would have limited time from the initial indication of a major SPE to find shelter from energetic solar particles and photons. Although it takes hours to days for a SPE to reach the earth, the time from detection to required action is often much shorter. Accordingly, spacecraft should have adequate shielding to attenuate the various SPE radiation types.

In space, SPE doses can be quite large. An August 1972 SPE was one of the largest dose events of the space era, and it occurred between two Apollo missions. However, ice core data from Antarctica indicate that the largest SPE in the past 500 years was probably the Carrington flare of 1859. A comparison of the Carrington flare to other large SPEs is summarized in Table 6.2. These data are further evaluated and their implications explored in subsequent discussion.

6.2.1.6.2
Low Earth Orbit Radiation Environment

Manned low earth orbit activities are influenced by the various components of the space radiation environment. The relative importance of each of the components depends on the specific LEO parameters including the spacecraft trajectory (e.g., altitude, orientation, and orbital characteristics), mission timing relative to periodic solar activity, mission duration, and spacecraft shielding characteristics. Space tourism activities are also affected by the LEO radiation environment. Tourism will soon begin with suborbital flights and will likely be extended to orbital environments with increasing duration.

Table 6.2 Large solar energetic proton events during 1859–2000[a].

Date	>30 MeV proton fluence (10^9 protons/cm^2)
August–September 1859	18.8
1895	11.1
November 1960	9.7
1896	8.0
1894	7.7
1864	7.0
July 2000	6.3
1878	5.0
August 1972	~5

a) Cliver and Svalgaard (2004).

LEO environments are normally dominated by energetic charged particles including electrons, protons, and heavy ions. The environment is also significantly influenced by large emissions of solar flares and the temporal and spatial fluctuations of the particles trapped by the earth's magnetic field.

Nuclear interactions of neutrons, protons, and heavy ions with the spacecraft, earth's atmosphere, and the human body produce secondary particles that contribute to a space tourist's or an astronaut's effective dose. In contrast, most of the electrons do not penetrate the wall of a spacecraft, but could penetrate suits worn during an extravehicular activity (EVA).

Table 6.3 summarizes the LEO radiation environment by particle type, source of the particle, particle energy, and ability to penetrate an EVA suit and the spacecraft. The unrestricted linear energy transfer (L_∞) in water is also provided.

The proton fluence for energies greater than 30 MeV is typically in the range of 10^6-10^{10} protons/cm^2. Table 6.4 provides a summary of SPEs from solar cycles 19 to 22 that are likely to exceed the NCRP 132 dose recommendations for LEO activities. These fluence values illustrate the variation that can be encountered during a solar cycle. The variations have a significant impact on the doses delivered to space tourists in LEO. Specific NCRP 132 dose recommendations are provided in subsequent discussion.

A comparison of Tables 6.2 and 6.4 illustrates uncertainties in the evaluation of SPE fluence data. For example, the August 1972 proton ($E > 30$ MeV) data vary by a factor of about 2 (~5 to 8×10^9 protons/cm^2).

6.2.1.6.3 Low Earth Orbit Dose Limits

In the twenty-first century, public access to LEO will significantly increase. There are a number of firms developing LEO vehicles with the intent of providing public transportation services. Although currently cost prohibitive to most members of

Table 6.3 Characterization of the LEO radiation environment[a].

Particle type	Source	Energy (MeV)	L_∞ (keV/μm)	Ability to penetrate		Comments
				EVA suit	Spacecraft	
Electrons	Trapped particles[b]	0.5–6	~0.2	Yes	Yes	Electrons dominate the effective dose for aluminum shields with a density thickness <0.15 g/cm^2
Electrons	Decay products from interactions with trapped GCR ions[b] Atmospheric scattering	1 to >1000	0.2 to >3	Yes	Yes	The effective dose contribution is about 10 times greater than the trapped electron effective dose
Protons	Trapped particles[b]	<10	>5	No	No	At low energies, protons pose a limited effective dose concern
Protons	Trapped particles[b] SPEs	10–400	0.3–5	Yes	Yes	As proton energies increase, the effective dose increases
Light ions	SPEs	10–400	0.3–5	Yes	Yes	The effective dose depends on the nature of the SPE and the specific ions and their energies
Ions ($Z > 1$) and charged secondary fragments	GCRs	>50 MeV/nucleon	1–1000	Yes	Yes	Pion production occurs, but the pion contribution to the effective dose is not well characterized
Charged target fragments	Nuclear interactions from all sources	<10 MeV/nucleon	2–1200	Yes	Yes	Large effective doses are possible
Neutrons	Nuclear interactions	0.1–500	[c]	Yes	Yes	Large effective doses are possible

a) NCRP 142 (2002b).
b) Particles are trapped in the earth's electromagnetic field.
c) Neutrons interact with atomic nuclei to produce highly ionizing charged particles.

Table 6.4 Proton fluence levels of significant solar events of cycles 19–22 likely to exceed the NCRP 132 recommendations[a)].

Date	Fluence (protons/cm^2)	
	$E > 10$ MeV	$E > 30$ MeV
23 February 1956	2×10^9	1×10^9
10–11 July 1959	5×10^9	1×10^9
14–15 July 1959	8×10^9	1×10^9
16–17 July 1959	3×10^9	9×10^8
12–13 November 1960	8×10^9	2×10^9
15 November 1960	3×10^9	7×10^8
18 July 1961	1×10^9	3×10^8
18 November 1968	1×10^9	2×10^8
11–13 April 1969	2×10^9	2×10^8
24–25 January 1971	2×10^9	4×10^8
4–9 August 1972	2×10^{10}	8×10^9
13–14 February 1978	2×10^9	1×10^8
30 April 1978	2×10^9	3×10^8
23–24 September 1978	3×10^9	4×10^8
16 May 1981	1×10^9	1×10^8
9–12 October 1981	2×10^9	4×10^8
1–2 February 1982	1×10^9	2×10^8
25–26 April 1984	1×10^9	4×10^8
12 August 1989[b)]	8×10^9	2×10^8
29 September 1989[b)]	4×10^9	1×10^9
19 October 1989[b)]	2×10^{10}	4×10^9
26 November 1989[b)]	2×10^9	1×10^8

a) Wilson *et al.* (1999).
b) The listed 1989 SPEs had an extended duration.

the public, it is likely that prices will decrease and access to LEO will expand. With expanded access, space tourist LEO radiation protection limits will be established. The radiation environment description and dose limit recommendations such as those published in NCRP 132 are considerations for establishing dose limits and regulatory standards for space tourist activities in LEO. In order to determine the direction of these standards, current astronaut limits are reviewed.

The NCRP 132 LEO recommendations are established for short-term exposure, limiting health effects, and career doses. Included in the NCRP 132 recommendations are career whole-body exposure limits for lifetime excess risk of total cancer of 3% (Table 6.5), 10-year career limits based on 3% excess lifetime risk of cancer mortality (Table 6.6), and dose limits for all ages and both genders (Table 6.7).

The NCRP 132 risk estimates are subject to large uncertainties. Part of this uncertainty is inherent in the nature of SPEs. These uncertainties include limits of scientific knowledge, risk model limitations, and lack of data to adequately characterize the risk. In addition, these uncertainties lead to shielding requirements that place significant limitations on space vehicle design and flight duration. Given

Table 6.5 Career whole-body exposure limits for a lifetime excess risk of total cancer of 3% as a function of age at exposure[a].

Age (years)	Female (Sv)	Male (Sv)
25	1.0	1.5
35	1.75	2.5
45	2.5	3.25
55	3.0	4.0

a) NCRP 132 (2000).

Table 6.6 Ten-year career limits based on three percent excess lifetime risk of cancer mortality[a].

Age at exposure (years)	Effective dose (Sv)	
	Female	Male
25	0.4	0.7
35	0.6	1.0
45	0.9	1.5
55	1.7	3.0

a) NCRP 132 (2000).

Table 6.7 Recommended dose limits for all ages and both genders[a].

Time frame	Blood-forming organs (Gy-Eq)	Eye (Gy-Eq)	Skin (Gy-Eq)
Career	[b]	4.0	6.0
1 years	0.50	2.0	3.0
30 days	0.25	1.0	1.5

a) NCRP 132 (2000).
b) The career stochastic limits in Table 6.6 are adequate for protection against deterministic effects.

these uncertainties, risk estimates suggest that for each week in space outside the earth's magnetosphere, there is a 1 in 500 chance that unshielded space tourists will receive a lethal dose from solar flare radiation.

The recommendations of Tables 6.5–6.7 greatly exceed current US public dose limits for nuclear facility operations, which are 1 mSv/year from licensed activities. This limit is unrealistic given the radiation levels associated with the LEO environment. LEO space tourism limits must consider the anticipated radiation environment, the volunteer nature of public space tourist participation, and possible health effects. Given these conditions, regulatory limits for space tourists will likely exceed the occupational limits for radiation workers as embodied in US Federal Regulations (10CFR20 and 10CFR835).

Public space tourist participation must involved informed consent. This consent is based on radiation protection training including a review of the LEO radiation environment, the SPE hazard, and possible biological effects from this environment. Following this training, a hazards acceptance statement should be signed to eliminate future legal action related to the space tourist's radiation exposure.

Space tourism regulatory dose limits should incorporate a number of considerations including the following:

1) Given their susceptibility to the biological effects of ionizing radiation (BEIR), no minors are permitted to utilize a LEO space tourist service.
2) Pregnant individuals are excluded from LEO space tourism to protect the developing embryo/fetus.
3) Given the voluntary nature of public space tourist participation, passenger radiation dose limits are based on the NCRP 132 limits specified in Tables 6.5–6.7. The recommendations summarized in Table 6.6 suggest that no individual younger than age 25 participate in the voluntary space tourism activity.
4) Space tourists are physically and mentally capable of meeting the challenges of the LEO environment.

To bound the radiation hazards of the LEO environment, a worst-case SPE is selected. The 1859 Carrington flare is defined as this bounding event. This selection represents a 500-year frequency flare event. Before the Fukushima Daiichi event, the author would have selected a 50–100-year frequency flare event (e.g., about 10 times the 29 September 1989 flare). However, the Fukushima Daiichi event suggests that improbable but historically viable events be selected as a credible design basis assumption.

Absorbed doses from Carrington-type SPEs as a function of aluminum shield thickness are summarized in Table 6.8. For the Carrington flare, bone marrow doses of 1–3 Gy are possible inside a spacecraft. A shielded room with about 18 cm of aluminum is needed to reduce the Carrington flare absorbed doses to the applicable NCRP 132 recommended deterministic doses (30-day blood-forming organ (BFO) dose limit of 0.25 Gy-Eq).

Table 6.8 Carrington flare absorbed dose estimates[a].

Shielding (g/cm^2 Al)	Skin (Gy)	Eye (Gy)	BFO (Gy)
1	35.4	23.4	2.81
2	6.65	6.02	1.71
5	2.82	2.73	1.09

a) Derived from Townsend (2004).

The results of Table 6.8 suggest that a space tourist vehicle in LEO will not be adequately shielded to accommodate a Carrington-type flare. A practical solution to minimize passenger doses is to utilize satellite radiation warnings or onboard radiation instrumentation to indicate elevated radiation levels. These indications would signal the pilot to abort the LEO flight trajectory and reenter the atmosphere for immediate landing. The combination of reduced altitude and timely landing significantly reduces passenger and crew radiation exposures.

6.2.2
Medical Exposure

In NCRP 160, the medical exposure of patients was separated into five subcategories including (i) computed tomography (CT), (ii) conventional radiography and fluoroscopy, (iii) interventional fluoroscopy, (iv) nuclear medicine, and (v) external beam radiotherapy. The radiation impacts from each of these subcategories are provided in Table 6.9.

NCRP 160 performed a dose assessment for external beam radiotherapy, but the results are not included in Table 6.9 because there are unique circumstances associated with this treatment subcategory. The treated individuals received average effective doses of 0.4 Sv, but less than 3% of the US population was exposed to this medical procedure. Effective doses to tissues near the treatment volume could exceed 1 Sv.

The results for medical exposure of patients excluding radiotherapy documented in NCRP 160 are much higher than the NCRP 93 values. The NCRP 160 results show a significant increase in the collective dose (a factor of 7.3) and effective dose per member of the US population (a factor of 5.7) compared to the NCRP 93 values. The increase in delivered dose is primarily attributed to the increased utilization of computed tomography, interventional fluoroscopy, and nuclear medicine procedures.

Table 6.9 Ionizing radiation exposure to the US population from the medical exposure of patients in 2006[a)].

Medical exposure component	Effective collective dose (person-Sv)	Average effective dose to US population (mSv)	Average effective dose for the exposed group (mSv)
Computed tomography	440 000	1.47	[b)]
Nuclear medicine	231 000	0.77	[b)]
Interventional fluoroscopy	128 000	0.43	[b)]
Conventional radiography and fluoroscopy	100 000	0.33	[b)]
Total	899 000	3.00	[b)]

a) NCRP 160 (2009).
b) Not determined because the number of exposed patients is not known. The number of medical procedures is noted in NCRP 160.

Medical exposure of patients accounts for an average dose of 3 mSv to an average member of the US population. The percent contribution from each of the medical subcategories to the 3 mSv value is computed tomorgraphy (49%), nuclear medicine (26%), interventional fluoroscopy (14%), and conventional radiography and fluoroscopy (11%).

6.2.2.1
Trends in CT Medical Exposure

The greatest growth in the number of computed tomography procedures occurred in the late 1990s and early 2000s. Use of CT technology is likely to continue to increase because additional clinical applications are being developed, the technique is quick and easy to perform, and it provides high-quality diagnostic information. With the expected expansion in CT usage, additional exposures are expected, and the collective effective dose from this imaging technique is likely to increase.

6.2.2.2
Mammography Doses

Mammography is the only medical X-ray imaging procedure that is regulated by the Food and Drug Administration. The effective dose from mammography is typically about 0.18 mSv for two views of each breast. The mean glandular dose for the total breast tissue is about 1.8 mGy per view.

6.2.2.3
Dose from Emerging Techniques

Chapter 5 outlines emerging diagnostic and therapeutic technologies that could reduce both patient and staff radiation doses. These technologies include internal radiation-generating devices, microspheres loaded with short-range radiation types to induce vascular disruption, and optical methods.

6.2.2.3.1
Internal Radiation-Generating Devices

Internal radiation-generating devices are conceptual nanoaccelerators that are injected into the body and deliver a beam of a selected radiation type and energy to a specific body location. The devices only emit radiation when operating. These devices emit no radiation until they are directed to the tumor site and are remotely controlled. Therefore, the dose to medical personnel can be eliminated, and the patient dose is limited to the tumor area. Proper control of the beam particle and its energy minimizes absorbed dose to healthy tissue. Information regarding beam placement can be obtained from the internal radiation-generating device telemetry or from computed tomography as noted in Chapter 5. Therefore, internal radiation-generating devices will significantly reduce the radiation doses delivered to medical staff.

6.2.2.3.2
Microsphere Disruption of Tumor Vasculature

The use of microspheres loaded with radioactive materials emitting short-range radiation types could reduce the effective dose to medical personnel particularly technicians involved with administration of the material. The external radiation should be significantly less than the doses received from ^{32}P- and ^{90}Y-loaded microspheres. Patient doses should also be limited to the vascular region, and the dose to healthy tissue should be significantly reduced.

6.2.2.3.3
Optical Methods

The various optical methods outlined in Chapter 5 do not involve ionizing radiation, which eliminates the hazard. However, the nonionizing hazards of these and emerging radiation types may cause secondary effects to healthy tissues.

6.2.3
Consumer Products and Activities

NCRP 160 defines the Consumer Products and Activities category in terms of seven subcategories. These subcategories are (i) building materials, (ii) commercial air travel, (iii) cigarette smoking, (iv) mining and agriculture, (v) combustion of fossil fuels, (vi) highway and road construction materials, and (vii) glass and ceramics. The contributions from these sources are summarized in Table 6.10.

The consumer products and activities collective dose and the average dose to a member of the US population are 39 000 person-Sv and 0.13 mSv, respectively. The percent contribution from each of the seven subcategories and other sources is cigarette smoking (35%), building materials (27%), commercial air travel (26%), mining and agriculture (6%), other sources (3%), combustion of fossil fuels (2%), highway and road construction materials (0.6%), and glass and ceramics ($<0.03\%$). Consumer products and activities are not a major source of effective dose in the United States.

In the United States, current energy costs and environmental policy suggest a de-emphasis of coal and oil and expanded utilization of natural gas. Table 6.10 suggests that an increased utilization of natural gas will lead to an associated increase in its dose contribution. However, this contribution is quite small in comparison to other sources of public dose (i.e., radon).

Numerous commercial products use radioactive materials or produce radiation during their operation. These products include building materials, cigarettes, dental prostheses, glassware, electronic tubes, fossil fuels, natural gas, aerosol smoke detectors, luminous watches and clocks, ophthalmic glass, thorium products including gas mantles and welding rods, and tobacco products. Specific commentary is provided for selected items. These are cigarettes, smoke detectors, and thorium tungsten welding electrodes.

Table 6.10 Summary of the number of people exposed, average annual effective dose, and annual collective effective dose from consumer products and activities[a].

Source	Number of people exposed (millions)	Average effective dose for the exposed group (μSv)	Annual collective effective dose (person-Sv)
Cigarette smoking	45	300	13 500
Building materials	150	70	10 500
Commercial air travel	b)	b)	10 300
Mining and agriculture	250	10	2 500
Other sources[c]	b)	b)	1 000
Natural gas cooking	155	4	620
Coal	300	1	300
Highway and road construction materials	6	40	240
Glass and ceramics	b)	b)	<10
Total	b)	b)	38 970

a) NCRP 160 (2009).
b) Not reported in NCRP 160.
c) Includes dental prostheses, ophthalmic glass, luminous watches and clocks, gas and aerosol (smoke) detectors, electron tubes, and thorium products including gas mantles and welding rods.

Cigarettes are a popular tobacco product that contains a number of naturally occurring radioactive materials including ^{226}Ra, ^{228}Ra, ^{210}Pb, and ^{210}Po. As noted in NCRP 160, the average effective dose from smoking one cigarette per day is 18 μSv. In addition, these radioactive materials deposit their alpha energy in the bronchial epithelium and, with the chemical constituents in cigarette smoke, are a major contributor to lung cancer.

Ionization smoke or aerosol detectors contain 20–50 kBq of an alpha-emitting radionuclide (e.g., ^{241}Am). The emitted alpha radiation ionizes the air between two electrodes and facilitates the flow of electric current across the air gap, which is subjected to a small potential difference. Smoke particles interrupt the penetration of the alpha particles through the air and trigger an alarm when the current is interrupted.

Thorium tungsten welding electrodes are used in electric arc welding. These electrodes are used in the aircraft, construction, food-processing equipment, nuclear power plant construction, and petrochemical industries. The content of natural thorium in these electrodes is usually in the range of 1–4 wt% in the form of thorium dioxide (ThO_2). ^{228}Th and ^{232}Th are the dominant isotopes present in the welding electrodes.

Given the widespread use of radioactive materials, inadvertent entry into consumer products is likely. Entry often occurs during scrap metal recycling.

The inadvertent or intentional diversion of radioactive materials into commercial products is also associated with the loss of control of radioactive sources.

6.2.3.1
Source Control

In the United States, the Nuclear Regulatory Commission (NRC) maintains a Nuclear Material Events Database (NMED) that includes source control issues reported by NRC licensees, Agreement States, and nonlicensees. Lost or abandoned radioactive sources, commonly denoted as orphan sources, represent a radiation protection concern because there is the potential for uncontrolled irradiation and contamination of the public.

Incidents involving orphan sources are increasing. Sources have been found in uncontrolled locations, which amplify concerns regarding their misuse and subsequent irradiation of the public. According to the International Atomic Energy Agency (IAEA), there are more than 2 million sealed sources in the United States, and about 375 are lost, stolen, or abandoned annually. Since 1986, only 40% of lost and stolen sources have been recovered. Orphan radioactive sources are a global concern, and their use as a terrorist weapon is widely recognized. The impact of the intentional dispersal of radioactive material is addressed in Chapter 4.

The loss of control of orphan sources or the theft of sources currently in service presents a significant radiological concern. Events involving sources occur when these devices are no longer in service or are forgotten, misplaced, abandoned, or lost. Therefore, it is beneficial from a safety and security viewpoint for orphan sources to be identified, tracked, controlled, and properly disposed. Staff turnover, poor inventory control, lack of acceptable disposal options, and high disposal costs contribute to the proliferation of orphan sources.

Lost or misplaced sources are likely for small and mobile devices (e.g., brachytherapy sources). These sources can be orphaned if not properly controlled. As a control measure, radiation detectors should be installed at exit points from the facilities where mobile sources are used. If not properly controlled, fixed sources, such as teletherapy units, present an additional risk because their shielding material has scrap value.

Scrap metal merits special attention since orphan sources have been inadvertently incorporated into various scrap metal recycling activities. The recycling and reuse of materials and equipment have increased because of their salvage values and demand for scrap metals. Scrap metal recycling is addressed in more detail in Section 6.2.3.1.1. Mobile sources used in industrial radiography also have the potential to become orphaned.

Given the competitive nature of the industrial radiography sector, there is a risk for sources to be abandoned, lost, or stolen. Radiation sources are also incorporated into industrial gauges used to measure the thickness, density, or moisture content of materials. Although the loss or theft of a radioactive source can occur at any time, the risk for a source to become orphaned increases at the end of its useful life. There are numerous examples where sources have been orphaned after being removed from equipment and placed in storage or left in the equipment in a facility that is no longer in active service. Each of these circumstances increases the likelihood for sources to be orphaned and for the subsequent loss of control of radioactive material.

One solution to the orphan source issue is the return of out of service sources to the supplier or manufacturer for reuse or recycling. Financial provisions for returning the source to the supplier or manufacturer are an effective motivator for this approach. Some countries (e.g., France) require the import of a source to be conditional on its export at the end of its useful lifetime or when its import use conditions or associated work scope are completed. In support of the import–export requirement, the concept of recommended working life (RWL) is established as a control measure. The RWL of a sealed source would be defined by the source supplier and specified in the purchase agreement. To ensure proper returns, regulatory organizations would establish a fund with the disposal costs provided by a specified funding source (e.g., the source suppliers or shared between supplier and purchaser).

Another option to minimizing future orphan sources is to find substitute technologies that do not require radioactive materials. For example, ^{241}Am in ionization smoke detectors can be replaced with optical or electronic devices. Other substitutions include the use of ultrasonic methods in density and level gauges to replace photon sources such as ^{137}Cs. In addition, linear accelerators could replace ^{60}Co teletherapy units.

When radioactive material substitution is not feasible, radionuclides with shorter half-lives should be utilized. Although the substitute sources require more frequent replacement, their use minimizes the hazards associated with the disposal of long-lived radionuclides.

If administrative and regulatory controls fail, the final line of defense is source detection before loss of control occurs. As noted by the IAEA, instrumentation used to detect orphan sources includes a variety of types. High-sensitivity, pocket-sized instruments used for personnel monitoring and protection provide a broad detection approach, particularly if the devices have an alarm function to indicate the presence of a source. Handheld survey instruments are used to detect and measure source dose rates. Fixed systems or portal monitors provide an automated alarm at strategic locations including the entrance to scrap yards, exit points from facilities using radioactive sources, and exit points from nuclear facilities.

Monitoring systems should also be located where the flow of goods, vehicles, and people is concentrated. This suggests that monitors be placed at border crossing points, ports of entry (e.g., airports and seaports), and highway/railway locations. Regulatory authorities must foster the installation and use of monitoring instruments at these and other locations. Installing monitoring systems to detect orphan sources and the illicit trafficking of radioactive materials is a complex task. Coordination is needed between government organizations controlling customs and border enforcement, national, and local law enforcement and radiation protection experts.

Table 6.11 Search region parameters [a].

Monitor type	Distance		Speed (km/s)
	Vertical (m)	Horizontal (m)[b]	
Pedestrian	0–1.8	0–1.5	<0.0018
Car	0–2	<4	<8
Truck and bus	0.7–4	3–6	<8

a) IAEA STI/PUB/1262 (2006a).
b) Parallel to the direction of movement.

The IAEA provides recommendations for systems used to monitor and detect orphan sources. The first IAEA recommendation involves the requirement that a fixed installed monitoring system is sensitive to gamma radiation. At a mean effective dose rate of 0.2 μSv/h, an alarm should be triggered when the dose rate is increased by 0.1 μSv/h for a period of 1 s. The second IAEA requirement involves the region monitored by the detector. Table 6.11 summarizes the search region in which the alarm levels should be applicable. The final IAEA recommendation involves the false alarm rate. For operational conditions, the false alarm rate should be less than one alarm per day for background effective dose rates of up to 0.2 μSv/h.

6.2.3.1.1
Contaminated Scrap Metal

As noted in Section 6.2.3.1, contaminated metal scrap represents a potential public radiation concern. Millions of tons of scrap metal from domestic and international sources are recycled each year. Recycled metal sources include appliances, automobiles, construction materials, steel containers, and miscellaneous steel products. This material must be verified to be contamination-free. Any radioactive material in the scrap has the potential to threaten the health and safety of the public, expose metal processing workers, and could be incorporated into consumer products.

Metal scrap becomes contaminated when abandoned, lost, or stolen radioactive sources enter the recycling process. Lost or abandoned sources can be inadvertently included with clean scrap during demolition activities involving facilities that previously contained radioactive sources. This is likely when demolition contracts do not identify the presence of these sources.

Contaminated scrap is also produced when uncontrolled radioactive material is recycled. For example, fertilizer production, oil and gas drilling, and petrochemical production produce pipe scale that contains naturally occurring radioactive material. When the piping or equipment containing this scale is recycled, radioactive material enters the scrap stream. Radioactive material also enters the scrap stream when material that is below regulatory limits is recycled.

Other devices utilizing radioactive material are also contributors to contaminated scrap. These devices include industrial gauges and instrumentation (e.g., aircraft cockpit displays). The devices may not be recognized as radioactive material since the gauges may be missing labels or be masked by a layer of paint, dirt, or covers. With their true character obscured, these devices are often assumed to be a common component such as a gauge or indicating device.

Specific examples of contaminated consumer products include brushed steel tissue box holders, radioactive elevator buttons, cheese graters, and recliners with radioactive metal brackets. The radiation exposure to consumers from these products is generally low. However, contaminated consumer products present a concern since there was no control over the radioactive material content incorporated into the manufactured item.

The brushed steel tissue box holders received considerable media attention in 2012. The highest radiation dose rate reported was about 0.2 mSv/h on the surface of a box and about 1 μSv/h at a distance of a meter. These dose rates correspond to about 2 MBq of ^{60}Co in the most contaminated boxes. Given the likely occupancy factors, these effective dose rates are not hazardous, but illustrate the potential concerns associated with the loss of control of the radioactive material.

The tissue box holders were sold as a commercial product. The original radioactive material was undetected, became part of the recycled metal, and was incorporated into the final steel product. This steel product was sold to a manufacturer that fabricated the boxes. Again, there were no controls to detect the radioactive material as it progressed through the manufacturing supply chain. Controls were also not present to detect the contaminated steel as it was shipped to the manufacturer, fabricated into a consumer product, packaged, shipped, received by the distributor, and shipped to the retail outlets. The lack of controls to detect the radioactive material within the supply chain is a concern that has the potential for significant public exposures. Moreover, as noted in Chapter 4, the inability to detect radioactive material has national security implications.

The issues of source control and security are important topics that have regulatory and public safety implications. Source security is addressed in Section 6.2.3.1.2.

Although there are no regulatory limits that specifically apply to scrap metal, ANSI/HPS N13.12-1999 recommends a criterion 10 μSv/year total effective dose equivalent above background for clearance of materials from regulatory control. The NCRP recommends that 10 μSv be accepted as a negligible annual individual dose, and the IAEA includes a 10 μSv/year value in its Basic Safety Standards clearance values.

6.2.3.1.2
Security Groups and Source Categories

The IAEA categorizes radioactive sources to facilitate their assignment to security groups. The security levels, source categories, activity ranges, and source examples

Table 6.12 IAEA categories for radioactive sources for the purpose of assigning them to security levels[a)].

Security level	Source category[b)]	Source activity/*D*	Source examples
A	1	$\geq 10^3$	Radioisotope thermoelectric generators Irradiators Teletherapy sources Fixed multibeam teletherapy (e.g., gamma knife) sources
B	2	≥ 10 to $<10^3$	Industrial gamma radiography sources High-/medium-dose rate brachytherapy sources
C	3	≥ 1 to <10	Fixed industrial gauges that incorporate high-activity sources Well logging gauges
c)	4	$\geq 10^{-2}$ to <1	Low-dose-rate brachytherapy (except eye plaques and permanent implants) Industrial gauges that do not incorporate high-activity sources Bone densitometers Static eliminators
c)	5	>exempt to $<10^{-2}$	Low-dose-rate brachytherapy eye plaques and permanent implant sources X-ray fluorescence devices Electron capture devices Mossbauer spectrometry sources Positron emission tomography check sources

a) IAEA Nuclear Security Series No. 11 (2009).
b) Category 1 – Extremely dangerous if not safely managed or secured.
Category 2 – Very dangerous if not safely managed or secured.
Category 3 – Dangerous if not safely managed or secured.
Category 4 – Unlikely to be dangerous.
Category 5 – Not dangerous.
c) Apply basic security measures. Measures should be established to ensure the safe use of the source, adequately protect it as an asset, and verify its presence at set intervals.

are summarized in Table 6.12. The activity limits are specified in terms of *D* values that are provided for selected radionuclides in Table 6.13.

Since human health is of paramount importance, the categorization system *D* value is based on the potential for a radioactive source to cause a deterministic health effect. The *D* value for a given radionuclide is the activity of a source,

Table 6.13 *D* values for selected radionuclides[a].

Radionuclide	*D* value (TBq)
^{241}Am	0.06
^{252}Cf	0.02
^{244}Cm	0.05
^{60}Co	0.03
^{137}Cs	0.1
^{153}Gd	1.0
^{192}Ir	0.08
^{147}Pm	40
^{238}Pu	0.06
^{239}Pu	0.06
^{226}Ra	0.04
^{75}Se	0.2
^{90}Sr/Y	1.0
^{170}Tm	20
^{169}Yb	0.3

a) IAEA Nuclear Security Series No. 11 (2009).

which, if not under control, could cause severe deterministic effects. Loss of control involves a range of events that include external exposure from an unshielded source and internal exposure following dispersal of the source material. Dispersal could result from fire and explosions that occur inadvertently or through a terrorist act.

The recommended security measures for a source of radioactive material are specified for each security level. For Security Level A, measures should be established to deter unauthorized access and to detect unauthorized access and acquisition of the source in a timely manner. These measures should be sufficient to delay access until a security response is activated. Loss of Security Level A sources represent a serious concern since they would involve high effective dose rates if the protective shielding were lost through recycling or intentional removal by a terrorist action.

For Security Levels B and C, measures should be established to deter unauthorized access. Security Level B measures should have the capability to detect unauthorized access and acquisition of the source in a timely manner. For Security Level C, the capability should exist to verify the presence of the source at set intervals.

The nuclides of Table 6.13 represent the judgment of the IAEA regarding the sources that are most likely to be placed in Category 1, 2, and 3. As noted by the IAEA, other commonly used radionuclides (e.g., ^{198}Au, ^{109}Cd, ^{57}Co, ^{55}Fe, ^{68}Ge, ^{63}Ni, ^{103}Pd, ^{210}Po, ^{106}Ru/Rh, and ^{204}Tl) are unlikely to be used in individual radioactive sources with activity levels that would place them within these categories. Accordingly, these nuclides would not normally be subject to national registries or to import and export controls.

6.2.4 Industrial, Security, Medical, Educational, and Research Activities

The NCRP 160 Industrial, Security, Medical, Educational, and Research Activities (ISMERA) category is divided into six subcategories. This category contributes 1000 person-Sv and 0.003 mSv average effective dose to the population of the United States. The subcategories and their percentage contribution to the ISMERA category are caregiving or other contact with nuclear medicine patients (72%); nuclear power generation (15%); industrial, medical, educational, and research activities (13%); Department of Energy (DOE) installations ($\ll$1%); decommissioning and radioactive waste ($\ll$1%); and security inspection systems ($\ll$1%).

In comparison to other sources of public dose, this category is not a significant contributor. However, two areas merit additional commentary. Nuclear power operations and its impact on the public are a constant area of media and public interest. The use of airport radiation scanning of passengers has been another area of active media discussion. These topics are addressed in the next two sections.

6.2.4.1 Nuclear Power Operations

It is interesting to note that only a minimal contribution to public doses results from nuclear power operations. These doses do not justify the public's concerns regarding nuclear power and provide another example of failing to balance the benefits of a technology against its risks. In the case of US nuclear power operations, the safety record has been good in spite of the regulatory issues noted in Chapter 7. The Three Mile Island Unit 2 (TMI-2) accident permanently removed an operating reactor from service but resulted in minimal off-site doses and no environmental impact. Fukushima Daiichi and Chernobyl-4 caused serious societal disruption and resulted in larger releases of radioactive material. The regulatory implications of these accidents and their impact on the public are addressed in Chapter 7.

6.2.4.2 Airport Scanners

In the United States, the Transportation Security Administration (TSA) is responsible for ensuring security in all modes of transportation. As part of this charter, the TSA uses backscatter X-ray, millimeter wave, and terahertz device technology to scan passengers for prohibited items including explosives, hazardous materials, metal objects, narcotics, weapons, and other potentially dangerous items.

Scanning devices are classified as either active or passive systems. Active systems irradiate objects with radiation to facilitate detection of potentially dangerous materials and objects. Passive systems detect the electromagnetic radiation emitted from the body.

Privacy concerns have been raised regarding the display and storage of the images derived from the full-body scanning devices. These concerns are addressed through software that modifies the image to reveal less personal detail.

Concerns for individual privacy are a subset of a more general issue regarding the ethics of overt and covert radiation exposure for national security purposes. The delivery of a radiation dose must have a well-defined purpose and be evaluated in terms of the risk, benefit, privacy, and legal basis. Alternatives to the use of radiation exposure must also be evaluated.

National security should not be used as the sole justification for the use of radiation exposure without a thorough and comprehensive evaluation. This evaluation should address the concerns of stakeholder groups as well as government officials and the national interest.

6.2.4.2.1
X-Ray Backscatter Technology

General-use backscatter technology utilizes X-ray beams that scan the body's surface. The body and other objects placed or carried on the body reflect the X-rays, and the reflected radiation is converted into an image using computational techniques.

Backscatter technology delivers low effective doses and meets applicable standards including ANSI/HPS N43.17-2009. These standards contain a requirement that the effective dose is less than 0.1 μSv per scan. The facility performing the scan is operated to ensure that no scanned individual receives in excess of 0.25 mSv in any 12-month period.

The first full-body scanners were used in Amsterdam in 2009. By the end of 2011, radiation concerns led to the European Union banning these devices at its airports.

In the United States, the ionizing and nonionizing output of full-body scanners also raised public concern. Apprehensions typically arise from a lack of scientific knowledge and the public's negative perception of radiation.

6.2.4.2.2
Millimeter Wave Technology

Millimeter wave technology uses radiofrequency radiation to scan the body. The millimeter wave imaging devices utilize radiation in the 1 – 10 mm wavelength range, and submillimeter systems operate at 0.1 – 1.0 mm wavelengths.

Energy reflected from an individual is used to construct a computer-generated image of the body and objects placed or carried on the body. The radiation emitted by millimeter wave devices meets the exposure standards for nonionizing radiation. As a comparison, the energy projected by millimeter wave technology is about 3 orders of magnitude less than the radiation emitted in cell phone transmissions.

6.2.4.2.3
Terahertz Scanners

Terahertz systems include both passive and active techniques. The passive devices detect the electromagnetic radiation emitted from the body. These systems are essentially passive differential radiometers that detect concealed objects by analyzing the heat signature emitted from an individual. An individual's blackbody radiation provides sufficient energy to facilitate the detection of concealed items and yields an image that meets privacy concerns.

6.2.4.2.4
Future Trends

Security services appear to favor millimeter wave and terahertz systems to provide sufficient detection capability while minimizing public concerns. The public's fear of ionizing radiation is a major consideration in this technology selection. However, active systems emit electromagnetic radiation, and the biological effects resulting from prolonged operation of these systems have yet to be characterized in a manner that is consistent with the level of rigor applied to systems using ionizing radiation.

6.2.5
Occupational Exposure

The NCRP 160 Occupational Exposure category is subdivided into six subcategories. This category contributes 1400 person-Sv and an average US population dose of 0.005 mSv. The subcategories and their percentage contribution to the Occupational Exposure category are medical (39%), aviation (38%), commercial nuclear power (8%), industry and commerce (8%), education and research (4%), and government, DOE, and the military (3%). Table 6.14 summarizes the occupational effective dose by subcategory.

In 2006, the percentage of workers exceeding 50 mSv was less than 0.1%. The average effective worker dose in these subcategories is about 1 mSv.

The highest average effective dose occurs in aviation and is larger than the occupational dose received in the commercial nuclear power industry. Collective medical and aviation worker doses are about a factor of 5 higher than the collective doses received in the commercial nuclear power and industry and commerce subcategories. The reduction in nuclear power doses from the NCRP 93 values is significant and represents the successful implementation of as low as reasonably achievable (ALARA) programs at commercial nuclear power reactors.

The data of Table 6.14 represent excellent dosimetry data and should be utilized in future assessments involving the biological effects of ionizing radiation. Although this data was not directly utilized in the BEIR VII report, it should be utilized in subsequent reports to ensure an accurate assessment of radiation risks and to ascertain the validity of the linear-nonthreshold (LNT) model. The reader is referred to Appendix H for additional discussion of the basis for this model in radiation protection regulations.

Table 6.14 Ionizing radiation exposure to the US population from occupational exposure in 2006[a].

Occupational exposure component	Collective effective dose (person-Sv)	Average effective dose to US population (mSv)	Average effective dose for the exposed group (mSv)
Medical	550	b)	0.8
Aviation	530	b)	3.1
Commercial nuclear power	110	b)	1.9
Industry and commerce	110	b)	0.8
Education and research	60	b)	0.7
Government, DOE, and the military	40	b)	0.6
Total	1400	0.005	1.1

a) NCRP 160 (2009).
b) Not provided in NCRP 160.

6.2.5.1
Definition of Radiation Workers

A radiation worker is an individual whose work activities involve exposure to radiation, radioactive material, or other sources of ionizing radiation. Occupationally exposed workers at NRC licensed facilities (e.g., power reactors, universities, and hospitals) and DOE facilities (e.g., national laboratories and weapons complex sites) are radiation workers because the source of their exposure is controlled by the licensee.

Other work activities involve exposure to ionizing radiation, but these employees are not considered radiation workers because they are exposed to radiation that is not controlled by a licensee. Two unique groups (e.g., aircrew member and morticians/medical examiners) are included in this categorization. Each of these groups is addressed in the next two sections.

6.2.5.2
Aircrew Radiation Exposures

In a Section 6.2.1, cosmic radiation was shown to be an important component of the natural ubiquitous radiation source. Because the earth's atmosphere shields solar and cosmic radiation, the associated doses increase by a factor of 2 for every 1500–2000 m increase in altitude. These radiation levels are substantially increased by solar flares and cosmic radiation events. Cosmic radiation and SPEs provide the dominant source of aircrew effective dose. The transportation of radioactive materials is an additional source of aircrew exposure.

In 1994, the Federal Aviation Administration (FAA) formally recognized that commercial aircrews are exposed to ionizing radiation. The FAA recommended that aircrews be informed about their radiation exposure and associated health risks. In addition, aircrews should be assisted in reaching informed decisions with regard to their work environment.

The FAA applied the ICRP 60 (International Commission on Radiological Protection) recommendations to commercial aircrews. However, these recommendations were not formalized as regulatory requirements.

The FAA recommended a 5-year average effective dose of 20 mSv/year, with no more than 50 mSv in a single year. For a declared pregnant aircrew member, the recommended fetal limit is an equivalent dose of 1 mSv. US airlines have not adopted radiation protection or dose monitoring programs for aircrew members. This is in contrast with European flight attendants that began radiation monitoring in 2000.

The effective dose received by commercial pilots and flight attendants is often greater than those received by traditional radiation workers regulated through DOE and NRC licenses. However, there is no US regulatory requirement to inform aircrew members of the possible health risks associated with their radiation exposure or monitor and record these effective doses. This nonrequirement is curious because solar and cosmic radiation can generate occupational level doses. In addition, high photon doses are produced by atmospheric phenomena.

The earth's atmosphere can produce regions with high-strength electric fields associated with thunderclouds and lightening. During these conditions, high-energy electrons are created, and their bremsstrahlung yields a spectrum of photon radiation. This phenomenon is known as a *terrestrial gamma-ray flash* (TGF). The TGF duration is on the order of a few milliseconds, but photons with energies of up to tens of megaelectronvolts are created and produce a significant radiation dose. If an aircraft is near the TGF event, a radiation dose exceeding occupational limits could occur. Dwyer *et al.* estimate that the radiation dose received by passengers and crew members inside an aircraft near the TGF could potentially approach 0.1 Sv in less than 1 ms. The US FAA considers the worst-case TGF to be about 30 mSv. Both of these estimates exceed the ICRP 103 occupational dose recommendation of 20 mSv/year.

6.2.5.3
Morticians and Medical Examiners

Medical examiners, coroners, and morticians could be exposed to radiation as part of their normal duties. For example, these duties involve processing a decedent following discharge from a hospital after a radiopharmaceutical treatment or implantation of a brachytherapy source. These medical professionals could also be subjected to ionizing radiation while processing radioactively contaminated victims following the detonation of a nuclear weapon, activation of a radiological dispersal device, reactor accident, or transportation accident involving radioactive material.

An explosion could also embed small fragments of debris containing radioactive material in tissue. These fragments could emit enough radiation to cause medical examiners to exceed regulatory limits for radiation workers. Although imbedded radioactive material can be surgically removed, care must be taken to minimize dose to medical examiners, coroners, and morticians. Other radiological considerations associated with emergency medical operations are addressed in Chapters 3 and 4.

6.3 Summary of Doses to the US Population

NCRP 160 noted that the collective dose from all sources to the US population is 1 870 000 person-Sv. An average effective dose to a member of the US population of about 300 million is 6.2 mSv. Most of this dose is derived from ubiquitous background (50%) and medical exposure of patients (48%). Consumer products and activities contribute about 2%. The ISMERA and Occupational Exposure categories contribute about 0.05% each. Table 6.15 provides the effective doses from each of these categories.

Details of each of these categories were summarized previously. Based on these tables, the dominant sources of US radiation exposure are radon and thoron (37%), computed tomography (24%), and nuclear medicine (12%). The external plus internal ubiquitous background sources contribute 13%, and the medical subcategories (interventional fluoroscopy, conventional radiography, and conventional fluoroscopy) yield 12% of the average population dose.

6.4 Public Dose Limits

Following 10CFR20, public dose results from exposure to radiation or radioactive material released by a licensee. Public dose also includes exposure from any other source of radiation under the control of a licensee. It does not include dose derived from an occupation, background radiation, medical administrations, or voluntary participation in medical research programs. An NRC licensee is required to conduct operations such that the total effective dose equivalent to individual members of the public from licensed activities does not exceed 1 mSv in a year.

The current public dose recommendations of the NCRP are similar to those of ICRP 103. NCRP 116's primary guidance for the general population is that the annual effective dose will not exceed 1.0 mSv. Both NCRP and ICRP provide for larger public doses under certain circumstances.

The public receives higher doses if they are family members of a patient receiving radionuclides as part of radiotherapy. Excluding children and pregnant women, NCRP 155 recommends a limit of 5 mSv to members of a radiotherapy patient's family.

Dose limits are important considerations in communications with the public, in assessing nuclear emergencies, and in licensing procedures for new commercial power reactors. Each of these areas is addressed in subsequent discussion.

Table 6.15 Ionizing radiation exposure to the US population in 2006[a].

Exposure category	Collective effective dose (person-Sv)	Average effective dose to US population (mSv)	Average effective dose for the exposed group (mSv)
Ubiquitous Background	933 000	3.11	3.11
Medical Exposure	899 000	3.00	[b]
Consumer Products and Activities	39 000	0.13	0.001 – 0.3[c]
Industrial, Security, Medical, Educational, and Research Activities	1 000	0.003	0.001 – 0.01[c]
Occupational Exposure	1 400	0.005	1.1
Total	1 870 000[d]	6.2[d]	[e]

a) NCRP 160 (2009).
b) Not determined because the number of exposed patients is not known. The number of medical procedures is noted in NCRP 160.
c) The range of values applies to the subcategories in this exposure category.
d) Rounded value.
e) Not provided in NCRP 160.

6.5 Risk Communication

Risk communication is a complex process based on trust, clear and accurate information, and honesty. Challenges range from communicating the hazards of airport security scanners to the consequences of a major reactor event such as the Three Mile Island (TMI) or Fukushima Daiichi accidents. Individuals responsible for communicating risk information to the public face two key challenges. First, risk must be communicated in a manner that acknowledges the emotional aspect of the event and provides information to alleviate public concerns. Second, communication must also be done in a manner that engages the public to become an effective partner in addressing and understanding the event's risks.

Risk communication is complicated because the public does not have a complete understanding of radiation and radioactive materials and their associated biological effects. Radioactive materials and radiation also tend to be regarded negatively by the public. The public is more accepting of radiation if it is received in a voluntary medical procedure. Public reaction to radiation following a power reactor accident is much less acceptable since it is a nonvoluntary or imposed exposure situation. Voluntary and imposed situations are particular risk attribute types. Other risk attributes and associated risk types are illustrated in Table 6.16.

Table 6.16 Risk preference types[a].

Risk attribute	Risk type	
	Preferred	**Undesired**
Situation	Voluntary	Imposed
Required Action	Controlled by the individual	Controlled by others
Benefit	Clearly positive	Little or none
Consequence	Distributed uniformly	Distributed unfairly
Event type	Natural	Man-made
Nature	Statistical	Catastrophic
Origin	Caused by a trusted source	Caused by a source that is not trusted
Hazard	Familiar	Exotic
Group	Adults	Children
Impact	Affects the individual	Affects others

a) Adler and Kranowitz (2005).

In general, radiation risks are characterized by a set of undesirable risk types. For example, the radiation dose received following a power reactor accident is imposed, controlled by others, perceived to have little benefit, affects individuals in the vicinity of the nuclear facility, is man-made, is not well understood, and affects children more severely than adults. As summarized in Table 6.16, all of these factors contribute to the difficulty associated with communicating risks associated with ionizing radiation. In addition, the public is more suspicious of communications coming from a representative of the nuclear facility (not a trusted source) than a physician or university professor (trusted source). The individual delivering the message is often as important as the message itself.

With radiation, the associated benefit relative to the risk strongly influences public perception. The public views radioactive material or radiation used in medical applications as having a high benefit and low risk and an acceptable practice. The high degree of trust in physicians also influences the acceptability of the medical use of radiation. However, the public regards industrial uses of radiation as less desirable. This perception is influenced by the public's general belief that the government and nuclear industry management are not completely trustworthy. These perceptions have been reinforced by government and industry risk communications following the TMI and Fukushima Daiichi accidents.

The evaluation of risk is always personal. As such, the communication of risk to the public requires an understanding of how risks affect people individually and as part of stakeholder groups.

Risk communication is interactive and dynamic. It involves information exchange between individuals, groups, and institutions. Risk communication can be defined as the approach used to inform the public of the potential issues and benefits of specific projects, programs, or events. It includes all communication with the government, media, stakeholders, and public regarding programs or events. Risk communication is most important when it involves topics that are controversial or not fully accepted.

Effective risk communication between a facility operator and the public should include a number of elements. These include acceptance of the public as a full partner having an interest in the facility and its operations. Public input is relevant and dialogue is important. Interactions with the public and other stakeholders must be open, frank, and honest. Communications must be clear and unambiguous.

Risk communications should be carefully planned and reviewed by facility management. The effectiveness of these exchanges is routinely evaluated to improve the risk communications process. Communications must also meet the needs of media and involve credible individuals who facilitate dialogue between the operating organization and stakeholder groups.

These general guidelines have not always been followed. The TMI-2 and Fukushima Daiichi accidents provided recent examples of issues associated with risk communication. This chapter focuses on the communications aspects of these accidents. The regulatory issues are addressed in Chapter 7.

The author also notes that understanding public norms and acceptable practices is also important in communicating radiation risk to individuals. During a public tour of TMI-2 a few years following the accident, the author was asked about a radiation posting encountered along the tour route. The purpose of the posting and the radiation levels in the area were noted and compared to the dose received from cigarette smoking. An elderly female individual, who was a smoker, informed the author in an emotional, expletive-laced outburst that it was her choice to smoke, but the radiation from the accident was imposed on her without consent. Since that event, the author has not used the smoking example in discussing radiation doses and their associated risk.

6.5.1
TMI-2 Accident

One of the most challenging aspects of the TMI-2 accident was communicating timely, accurate, and complete information to news outlets, regulators, and the public. This was difficult to achieve in the complex, evolving situation that occurred during the TMI-2 accident. As a result, some of the plant information provided during the first few days of the accident was conflicting and confusing. With changing plant conditions, various utility spokespersons offered opinions that reflected personal insights rather than a coordinated, accurate response. As a result, the utility's credibility as a reliable source of information quickly eroded. In a rapidly unfolding event, it is not surprising that some initial statements would later prove to be inaccurate as more information was obtained. Other organizations including the NRC encountered similar problems in their communications with the public.

The utility was accused of not acknowledging the severity of the accident during the height of the event. Part of the blame for this confusion was the initial judgment that there was no immediate public danger and that the accident was not serious. Since events were rapidly unfolding, it was difficult to have timely communications while addressing plant conditions during the first few days of the accident. The utility's emphasis was maintaining the plant in a safe operating condition and minimizing the radioactive releases and off-site exposures to the public. In pursuing these objectives, the utility sometimes had to place communications with the public in a secondary role. In addition, individuals responsible for accident management also were briefing the media. These individuals were well versed in plant operations, but not in risk communications.

These conditions also contributed to inconsistent media reports. At times, the utility and government provided differing accident descriptions. When this occurred, media representatives gave conflicting reports, which further added to public confusion. Nuclear jargon, competing nuclear experts, and the communications differences between the various government agencies also contributed to media inconsistencies.

Inaccuracies and erroneous information have serious consequences. The discovery of hydrogen gas in the reactor coolant system and the possibility of additional releases of radioactive material triggered another series of missteps. Unfounded concern of a hydrogen explosion in the reactor vessel caused the Pennsylvania's Governor to issue an advisory that pregnant women and preschool children living within an 8 km radius of the plant be evacuated. The Governor also closed all schools in the area and suggested that people living within 16 km of TMI-2 remain indoors. Subsequent information proved there was never a danger of an explosion within the reactor vessel. These actions were a direct result of inadequate communication and the lack and availability of accurate and timely information. For example, the fact that pressurized water reactors use hydrogen gas for reactor coolant system oxygen control was not effectively communicated to stakeholders and would have assisted in placing the perceived hazard into a proper risk context.

Throughout the TMI-2 accident, the public, press, government, and the utility had different impressions of the event because information conveyed by the utility and regulator did not always meet the recipient's expectations. These impressions affected subsequent actions and led to the impression that the utility was not providing a complete, accurate description of events, off-site releases, and associated plant conditions. In hindsight, the utility did not have sufficient communications staff to satisfy the media's demands, which reinforced the impression that information was being withheld from the public.

From a practical perspective, the operating utility is in the best position to manage the event. External group assistance may be beneficial to provide technical assistance, logistical support, and infrastructure support after the initial event is under control, but operating utility personnel hold the specific plant knowledge needed to manage the early phase of an event. Government officials do not have an equivalent level of plant specific knowledge and should not attempt to micromanage the utility's accident response. The government licenses and regulates a facility to operate safely and manage off-normal events. Micromanagement suggests that its licensing and regulatory process was less than adequate.

6.5.1.1
Enhanced Communications Organization

Given the communications issues that emerged during the accident, an expanded communications organization was subsequently developed at TMI to facilitate the dissemination of information to the public, press, government organizations, and other stakeholders. This aggressive action was implemented to ensure that the public was informed of recovery events in a complete, timely, and accurate manner.

TMI supplemented its expanded communications organization with an active Speakers Bureau that facilitated staff and manager presentations to stakeholders and public tours to ensure the recovery process was viewed as open and transparent. Numerous public meetings were held to ensure recovery activities were understood. Although challenging to implement with a skeptical public, continued communications efforts have a positive, lasting benefit to a long-term recovery plan.

6.5.1.2
Off-Site Monitoring Network

Following the TMI-2 accident, public distrust of utility radiological information prompted the installation of an array of pressurized ion chambers around the site. These ion chambers included the capability for the real-time display of the local radiation level. Remote displays were located at TMI-2, at the emergency operations centers (EOCs) of the five risk counties, and at the State of Pennsylvania EOC in Harrisburg, PA. In addition, the radiation levels were published on a daily basis in local newspapers. These actions restored a measure of credibility and improved the public's understanding of the radiation effects of TMI-2 recovery activities.

The public received the availability of these data in a very positive manner. Many members of the public monitored the daily radiation levels as a means to measure the impact of recovery activities. When the radiation levels failed to increase and continued to represent background levels, public confidence in the recovery process increased.

6.5.2
Chernobyl

The 26 April 1986 Chernobyl accident occurred in the former Soviet Union in the Ukraine. The USSR was a closed society and did not operate under the same communications rules as the United States and Japan. However, even in a closed society, a nuclear accident cannot be hidden from the public.

Public notification was intentionally delayed because the accident occurred at night and the initial protective action was to shelter in place. When public notification was made in the morning, it occurred by door-to-door visits. Public evacuation was permitted by private auto at noon on April 27 and notification to evacuate by bus occurred a few hours later. The government did not use a siren system or television for public communications.

The NRC in NUREG-1250 comments that the extent of the Soviet Union's public education and information program on radiological emergency response was uncertain. For example, peasants refused to evacuate unless their farm animals were evacuated, and the refusal of some peasants to destroy contaminated milk may have resulted from a lack of public education regarding radiation risk. However, peasants were reported to have taken potassium iodide. Public education programs directed at the 45 000 residents of the local village of Pripyat were likely given the fact that the village was evacuated in a 3 h period.

The Ukrainian Health Minister provided protective action information to the Kiev residents through television broadcasts. These broadcasts were used to advise city residents of wind shifts that redirected the radioactive plume toward the city. As with the TMI-2 accident, many rumors were spread, and government officials and newspapers were used to refute those rumors. In addition, there were considerable public uncertainty and concern during the first few days of the Chernobyl accident because information from the site was not disseminated to the public.

In terms of risk communication, the TMI-2 and Chernobyl accidents have similar themes related to providing accurate and timely information. This commonality is remarkable in view of the different systems of government that regulated these facilities.

6.5.3
Fukushima Daiichi

The Fukushima Daiichi accident further illustrates the consequences of poor risk communications. Although many of the communications issues encountered during the TMI-2 accident were repeated, the Fukushima Daiichi event was significantly exacerbated by a massive earthquake and subsequent tsunami that devastated the facility and the surrounding geographical area.

The communications errors exhibited at TMI-2 and Fukushima Daiichi caused the public to question the capability and trustworthiness of utilities and governments to safely manage and regulate nuclear power facilities. This is important since the history of nuclear power operations has fostered public suspicions created by previous accidents and events. A brief history of these events is summarized in Chapter 7.

At Fukushima Daiichi, these suspicions were raised by the response of the government and facility to the accident and their communications with the public. In particular, the utility and government failed to inform the public in a timely and consistent manner regarding the release of radioactive material and associated radiation levels as the accident unfolded. There was also a lack of coordination since radiation maps were available, but not disseminated. The availability of these radiation maps would have assisted in directing the public to the optimum evacuation routes.

The consistent message from the three major reactor accidents is that nuclear power facilities do not operate in a vacuum and their effects on the environment have significant consequences for the individual facility as well as the nuclear industry. Public confidence is essential if nuclear power is to have a sustainable future. One of the keys to achieving public confidence is a well-supported risk communication organization based on honesty, candor, and the flow of accurate, timely information.

Reviews of the Fukushima Daiichi accident suggest that a number of options exist for improving the relationship with the public. In addition to the activities taken following the TMI-2 accident, communications enhancements could include improving systems to disseminate plant and radiological data in a timely and consistent manner. Data quality must be improved to avoid corrections and revisions of earlier reports. Such corrections and revisions do not inspire public confidence and are often interpreted in a negative manner. Needed improvements in the Fukushima Daiichi communications area are also noted in the 2012 Japanese Diet Report summarized in Chapter 7.

The availability of consistent radiological data (e.g., dose rates, facility contamination levels, and radioactive material content in milk, fish, water, plants, and other foodstuffs) and its timely communication to the public should be improved. A possible improvement item was utilized as a response action at TMI-2.

As noted in Section 6.5.1.2, an array of pressurized ion chambers installed around the TMI site provided the capability for the real-time display of the local radiation level. An extension of the TMI-2 remote display approach could be a useful addition to the Fukushima Daiichi risk communications effort. In addition to providing direct radiation levels, the display of plant parameters including effluent monitor readings could be provided on a real-time basis. The availability of these data would foster informed public debate and improve understanding of the environmental impact of recovery activities. Improved information dissemination is a necessary step in public debates associated with the risks and benefits of nuclear power facilities.

Real-time reporting of contaminated groundwater activity levels and the extent of underground liquid plume migration at the Fukushima Daiichi site would also improve confidence in the operating utility. This is particularly important in view of the worldwide interest in liquid releases from the Fukushima Daiichi facility and its radioactive liquid waste storage tanks.

6.6 Public Involvement in Nuclear Licensing

The licensing of nuclear facilities should provide a well-defined process that holds the operator and the regulator accountable for safe facility operation. Stakeholders must have confidence that the facility will operate safely and not have a negative impact on the environment. This is a reasonable expectation that is often met with suspicion because the accidents at TMI-2, Chernobyl-4, and Fukushima Daiichi reinforce concerns that facilities are not always operated in a safe manner. The process of licensing and operating nuclear facilities must permit all stakeholders, especially host communities, to participate and ensure that their interests and concerns have been completely addressed. Public involvement in nuclear licensing is important for all facilities including power plants, industrial facilities using radioactive materials, and high-level waste storage facilities.

6.6.1 Emergency Response

Following the 11 September 2001 attacks, emergency preparedness programs were expanded to more completely respond to terrorist-based events. With these improvements, emergency preparedness programs included increased security measures. Although a terrorist event could alter the initial event response, the final consequences of that event are similar to the design basis accidents used in licensing the facility.

The evacuations during the Fukushima Daiichi accident renewed the public's interest and focused concern on the emergency preparedness activities associated with nuclear facilities including power plants. Evacuations are common events resulting from floods, hurricanes, tornados, fires, industrial accidents, and transportation accidents. However, evacuations associated with a nuclear emergency engender a more significant public reaction.

Although emergency response actions are an integral aspect of nuclear power reactor licensing, the NRC reexamined the role of emergency preparedness programs following the accidents at the TMI, Chernobyl, and Fukushima Daiichi. The TMI-2 accident showed the need for improved planning, response, and communication by federal, state, and local governments to deal with reactor accidents. Emergency preparedness revisions related to Fukushima Daiichi include expanded response organizations, improved availability of equipment and supplies, enhanced emergency response procedures, increased availability of backup core cooling systems and power supplies, and improved spent fuel pool level instrumentation. As such, emergency preparedness is a dynamic aspect of nuclear facility licensing and is required to adapt to changing natural and man-made threats to the facility.

6.6.2 Stakeholder Involvement

Diverse stakeholder views and interests must be considered in the licensing process. Including stakeholders in this process is more stressful but results in a facility license that has a broader range of support. Although it may not be possible to satisfy the concerns of all groups, the licensing process benefits from the participation of a wide range of stakeholders. The task of communicating information and engaging different interest groups should be a significant consideration in facility licensing. Stakeholders and the public should understand how decisions were reached, different options and opinions were considered, and issues and concerns were resolved. As discussed in Chapter 7, this was an early failure of the US nuclear power facility licensing process. This failure is one of the factors that led to stakeholder suspicions and public mistrust in the nuclear power reactor licensing process and in the oversight of operating facilities.

Stakeholder and public involvement in the licensing process has a number of potential advantages. One possible benefit is that stakeholder participation could increase confidence and acceptance of the licensing decision. A second possible benefit is that well-qualified public intervenors introduce valid concerns that require agency staffs to be more thorough and articulate in their analyses and justifications for decisions. The third possible benefit is the introduction of substantive recommendations, new information, or additional viewpoints not previously considered in the regulatory process. In general, the presentation of new information and viewpoints constitutes the public's primary contribution. Stakeholders also have the potential to provide technical insight and guidance. In hindsight, underestimating the design basis earthquake/tsunami assumptions was a major contributor to the Fukushima Daiichi accident. These assumptions could have been challenged by either the public or stakeholder groups and that action could have precluded this accident. Any delay in the licensing of the Fukushima Daiichi facility to accommodate more credible design basis assumptions would have been well justified based on the accident severity and its aftermath.

In subsequent discussion a number of examples of opportunities for stakeholder participation are provided. Following this generic discussion, the specific involvement of stakeholders in the licensing process in Canada and the United States is addressed.

6.6.2.1
Incorporation of Nuclear Energy in a National Energy Plan

Governments usually formulate their national energy plans utilizing discussions between the regulatory agencies and the affected industries. Stakeholders are typically informed of the regulatory outcomes, but are not involved in the initial policy decision discussions. The TMI, Chernobyl, and Fukushima Daiichi accidents led some countries to initiate a comprehensive debate on energy policy that enhances the likelihood of public input into the decision-making process. This process may not be efficient or orderly, but it permits the formation of a broader perspective and an improved chance for a national consensus.

6.6.2.2
Development of Nuclear Regulation Legislation

Legislation that provides the statutory authority for nuclear regulators or ministries follows an established process in most nations. The legislation is the responsibility of elected governments that represent stakeholders. Development of second-level nuclear legislation (e.g., regulations governing the licensing process and radiation protection regulations) is typically delegated to a ministerial or regulatory body. The participation of stakeholders in the second level is established in many countries. Participation at the second level is difficult because stakeholders may not have the technical knowledge and skills to understand details of the proposed regulation. For example, the adoption of the ICRP 103 methodology for radiation protection has a number of aspects in which stakeholder involvement is possible. These include protection of the environment, identification of important local species, and participation in emergency response decisions. The implication of new dose limits and their impact on the workforce could also be an appropriate topic for stakeholder involvement. For example, the adoption of a lower fetal dose limit would enhance protection of the embryo fetus but could impact the worker's professional development and advancement.

6.6.2.3
Decisions to Build a Major Nuclear Facility

Construction of a major nuclear facility such as a new nuclear power plant, fuel cycle installation, or high-level waste repository is a significant decision that affects a number of stakeholder groups. In many nations, stakeholders' participation is incorporated into the regulatory process, but the process is not always well defined. In democratic societies, the construction of a major nuclear facility is not likely without a regulatory process involving the affected population. The IAEA notes that successful stakeholder involvement occurred in Finland for the site selection for a final spent fuel repository. Participation by the parliament, local authorities, and the public was time consuming, but the decision on the type and site of the repository was successfully approved with a large consensus.

6.6.2.4
Emergency Planning Development and Implementation

Stakeholders residing near a nuclear installation would obtain a greater understanding of the facility and its inherent risks if they participated in the development of its emergency plan. Since local technical experts (e.g., police, fire fighters, and emergency medical personnel) and elected officials must deal with any future emergency, their participation enhances the likelihood of the development of a plan that can be successfully implemented. These stakeholders should participate in drafting and commenting on the emergency plan as it is being developed. Stakeholders must also verify that the assumed equipment and services are available and that the plan is capable of being successfully implemented. They should also be encouraged to participate in drills and exercises to test the efficacy of the emergency plan and in the development of corrective actions to improve aspects of the plan that did not function as intended.

The emergency plan addresses releases of radioactive material to the environment. These releases are characterized in terms of their magnitude, duration, and effect on the public. Emergency drills and exercises evaluate the response to a radioactive material release and associated stakeholder actions.

6.6.2.5
Facility Releases of Radioactive Material

Responding to the uncontrolled releases of radioactive material is an integral aspect of a facility's emergency plan. However, normal operations require the periodic, controlled release of radioactive material from a facility. For example, a nuclear power plant releases minimal amounts of radioactive materials during containment purges that ensure a negative pressure is maintained, waste gas decay tank and liquid releases following verification that regulatory limits are met, and air ejector noble gas releases that ensure vacuum conditions are maintained in the condenser. These controlled radioactive releases can be misunderstood if the public does not have an understanding of the facility and its routine operations. For example, environmental monitoring programs periodically assess the impact of facility operations on the environment. These programs collect samples of food, fish, water, air, soil, and sediment as well as monitor dose rates to determine the impact of facility operations. The samples are analyzed to determine radionuclides present and their origin (e.g., naturally occurring or due to facility operations). Communicating these results to the public provides stakeholders with a detailed tabulation of the radiological impact of facility operations on their community. If openly provided in a timely manner, these communications should alleviate public fears of facility operations.

Typically, regulators inform the affected population after a release of radioactive material has occurred, but there are established methods to ensure the participation of the affected stakeholders before a major planned release (e.g., the postaccident TMI-2 containment venting). In the power reactor examples noted previously, these methods could be further developed to ensure stakeholder participation in the routine planned release process. Stakeholder participation could be part of the release planning process that validates regulatory compliance. Although initial planned releases would likely be contentious, issues would rapidly disappear as the public had a better understanding of the facility, its regulatory requirements, the consequence of the release, and the processes involved in a planned release.

6.6.2.6
Environmental Restoration of Legacy Sites

Legacy environmental restoration areas include the Hanford and Savannah River Sites that were initially operated by the US Department of Energy (USDOE) for weapons production activities. Production and fuel reprocessing activities involved planned and unplanned releases of radioactive materials to the environment that created areas of contaminated soil. These areas offer an increased potential to further contaminate water and biota and have received increased public attention. For the USDOE sites, significant public participation has occurred in the remediation process, but the approach is slow and often tedious. This is an artifact of the involvement of diverse groups with differing viewpoints and goals. For example, Hanford stakeholders include the States of Oregon and Washington, local communities, tribal organizations, environmental and cultural groups, nonprofit groups, student organizations, individual citizens, and business organizations.

The adoption of the USDOE process by other nuclear sectors and nations should consider their specific licensee requirements, public interaction history, and national goals and objectives. A more efficient process can be developed if stakeholder involvement occurs prior to facility operations. At Hanford, public involvement only occurred following years of operations and releases of radioactive material to the environment. The releases associated with these operations occurred during the Cold War era when facility operations and operating practices were closely guarded secrets. This secrecy and the releases of radioactive material created concerns and fears that have not been completely resolved and continue to affect stakeholder attitudes.

One of the more challenging environmental restoration activities involves the establishment of cleanup criteria for facilities, neighboring areas, and water sources. Establishing these criteria is challenging because they affect the subsequent availability and use of the land for public or commercial purposes.

6.6.2.7
Nuclear Facility Decontamination and Decommissioning

Nuclear facility decontamination and decommissioning (D&D) is a stakeholder concern because the long-term disposition of the site including decisions regarding acceptable concentrations of radioactive materials following site remediation activities are involved. Experiences in the United States (e.g., Shippingport Atomic Power Station D&D) and Europe suggest that these activities are technically feasible. Stakeholder participation should be an integral aspect of future D&D efforts since the activities can be scheduled with public input without significantly affecting the project mission and goals.

6.6.2.8
Radioactive Waste Management

Radioactive waste management is another contentious area that is difficult to solve without stakeholder participation. The challenges of the Yucca Mountain High-Level Waste Repository are a prime example of the legal and societal difficulties that are encountered when a waste management process fails to account for stakeholder input and concerns. These concerns also involve political considerations that significantly complicate the licensing process.

Stakeholder interest is particularly intense for high-level waste facilities that remain in place for prolonged periods. The control of this waste is an obvious concern because accidental releases or terrorist attacks could disperse the waste and have a significant environmental and economic impact.

6.6.2.9
Transportation of Radioactive Materials

The transportation of radioactive materials from a nuclear installation is typically accomplished with minimal public input. This is an appropriate response for low-level radioactive materials, but public concerns increase when high-level waste (e.g., spent or irradiated fuel assemblies) is involved in the shipment. Since waste shipments are normal plant activities, general shipment information should be shared with the public. However, the threat of nuclear terrorism requires that specific shipment details be withheld to minimize the terrorist threat.

Generic information associated with spent fuel shipments can be shared to improve public understanding of the process and the measures that exist to ensure its safety. This information includes the regulatory requirements associated with shipping spent fuel, considerations in selecting transportation routes, emergency planning provisions, and radiological requirements. For example, the public should be informed of the rigorous methodology associated with developing a Safety Analysis Report for Packaging (SARP). The safety analysis is comprehensive and evaluates the normal and accident conditions of transport. Portions of this analysis and its careful review by the public should engender confidence that the shipment can be performed safely without the loss of control of radioactive material.

Spent fuel is shipped in a Type B Cask rigorously designed to contain its radioactive material contents under normal and accident conditions that are specified in 10CFR71. The radioactive material package, consisting of the cask and spent fuel, is certified by the NRC after a rigorous analysis and approval process.

The certification is based on detailed analyses of the package to perform under normal and accident conditions. These conditions include surviving accidents involving fire, drops onto unyielding pins, drops onto hard surfaces from various heights, and submersion in water. In addition to containing the radioactive materials, the package must remain subcritical after a series of hypothetical accident conditions. This is achieved by maintaining sufficient structural integrity to maintain a critically safe geometry supplemented by nuclear poisons to ensure that under both normal and accident conditions, the effective neutron multiplication factor (k_{eff}) of the package does not exceed 0.95.

The package contains sufficient shielding to meet the 49 Code of Federal regulations (CFR) transportation dose rate requirements. If these dose rates are met for the normal and accident conditions, the package meets the requirements specified 10CFR71, and the SARP is reviewed. If the regulatory review accepts the SARP, the NRC issues a certificate of compliance (COC). The COC formally approves the package and may specify additional requirements and restrictions for the shipment.

This rigorous analysis and review process can be used to inspire public confidence in a particular aspect of facility operations. However, these generic details must be conveyed to the public in an understandable manner.

6.6.2.10
Security of Nuclear Sites and Special Nuclear Material

The public has a vested interest in a nuclear facility being maintained in a safe, secure status. Nuclear security information is a sensitive area that requires limited access and a degree of secrecy. In addition, a nuclear site may utilize special nuclear materials that have direct use in nuclear weapons and are a prime terrorist target. Therefore, the information that can be shared with stakeholders is limited, but the topic should be discussed to the extent practicable to minimize public concerns. This must be accomplished without divulging any safeguards information that would place the facility in jeopardy. The public should be invited to witness aspects of the security training that are not proprietary to gain a measure of confidence in the organization and its capability to perform its intended function. For example, selected training of security staff including small arms qualification could be open to the public.

6.6.3
Canadian Process for Licensing a High-Level Waste Facility

It is well known that the United States has failed to develop a high-level waste repository. The Yucca Mountain facility is the most recent project failure, and the current licensing status is addressed in Chapter 7. In a similar manner, Canada failed to license a high-level waste facility.

Canada's Nuclear Waste Management Organization (NWMO) was established in 2002 after the failure of a technically oriented approach to establish a high-level waste repository. A commission reviewed the previous approach and concluded that while the program successfully evaluated the repository's scientific and technical aspects, it failed to engender public confidence.

The NWMO adopted lessons learned from worldwide nuclear waste management experience and initiated innovative steps in formulating its approach. One of its innovations was to understand the values and concerns of citizens and then to evaluate the available repository options in view of citizen input. After extensive interactions with Canadian citizens and stakeholder organizations, the NWMO adopted a sequential, adaptive approach called Adaptive Phased Management (APM).

The APM plan incorporates integrity, excellence, engagement, accountability, and transparency as fundamental values for its technical approach and management system. The technical method envisions disposal in a geologic formation with an option of shallow underground storage. It includes the potential for retrievability, continuous monitoring, flexible design, and ongoing technical and social research. The management system incorporates collaborative and phased decision making; continuous learning; open, inclusive, and transparent engagement; and pursuit of a willing and informed host community.

APM is a deliberate, transparent, and highly engaged process that has led communities to volunteer to pursue discussions with the NWMO. Community discussions are held to gather information before considering if site surveys should be conducted. Canada's progress provides insight into a success path for the establishment of high-level waste repositories. It also contains insight for licensing other nuclear facilities including nuclear power plants.

6.6.4
US Reactor Licensing Process

US NRC licensing proceedings are an important venue for public debate regarding nuclear power facilities. Licensing proceedings provide a mechanism for avoiding litigation regarding the construction and operation of commercial nuclear facilities. In a licensing hearing, government agencies, intervenor groups, and the public debate the worth of the proposed nuclear facility. In the United States, the hearing process has historically failed to provide closure, and lengthy delays have been encountered during the subsequent litigation process.

Numerous objectives are served by intervention in licensing proceedings before the NRC. These objectives vary from violent opposition to any nuclear power plant to concerns about specific aspects of the facility. Other organizations, such as state and local government agencies, may have no specific issue with the proposed plant but want to ensure that the licensing process and technical review are performed in a comprehensive and transparent manner.

The NRC's reactor licensing process for new power reactors is defined in 10CFR52. This process allows input from intervenors to address a variety of issues including potential safety and environmental problems. If the 10CFR52 process is unsuccessful, the litigation process could lead to extended construction and operation delays.

The licensing process incorporates well-defined phases. For example, a utility can apply for a combined Construction and Operating License (COL) for a new reactor. In utilizing the COL option, a nuclear utility references a preapproved standardized reactor design that achieved certification through a specified regulatory process. In principle, the COL approach should provide a regulatory framework that avoids facility concerns, accelerates reactor construction and operation, and satisfies public interest and intervenor groups.

Unfortunately, reactor licensing is not a simple sequence of events as outlined in 10CFR52. Reactor licensing is a complex process that includes groups with widely diverging views, goals, and priorities. The process becomes even more complex when external events introduce new aspects into the intended regulatory sequence. The Fukushima Daiichi accident introduced a significant perturbation into the US regulatory process.

Intervenors are concerned that the licensing process is being followed in a manner that defeats its intended purpose. For example, reactor manufacturers are revising designs after their certification has been completed, and the NRC is accepting the applications for hearings for reactor designs that have not achieved regulatory certification. In some cases, the NRC did not have a schedule for completion of its design review to complete the certification process.

When a COL application is scheduled for a hearing, intervenors have a limited time (typically 60 days) to review the application and file any contentions. When key safety and environmental information is incomplete because the designs themselves are incomplete or being revised, it is difficult to evaluate the specific issues and how to characterize their severity during the hearing process. As argued by the intervenor groups, this leads to contentions that are filed beyond the allowed review period. These contentions are typically rejected. However, they may eventually be litigated which could significantly extend the licensing process. Court challenges to reactor licenses are possible if a late contention is rejected or if safety and environmental information has not been finalized or is under revision at the time a license is issued. The US regulatory process is addressed in detail in Chapter 7.

6.7
Litigation

Legal actions occur when conflicts within the regulatory process exist or regulatory decisions fail to adequately resolve the concerns of stakeholder groups. Litigation is a key aspect of licensing a nuclear facility and procedures are codified in national laws. For example, US nuclear facilities are addressed in Title 10 of the Code of Federal Regulations with general requirements outlined in 10CFR2.338 (*Settlement of issues; alternative dispute resolution*). Litigation aspects and procedures for specific license types are scattered throughout Title 10. Although the litigation requirements are intended for resolving good-faith differences, groups opposed to the licensing of nuclear facilities have used these provisions as a delay tactic. These delays increase the facility cost, generate negative perceptions of the facility's merit, focus on perceived facility risks, confuse the public, and discourage the operator from pursuing further development.

These negative outcomes do not meet the intent of the regulations, which were designed to enhance safety, involve stakeholders, and streamline the licensing process. Nuclear facilities and the full use of nuclear power and nuclear materials for humanity's betterment must be evaluated on their merits and societal needs. It is hoped that significant stakeholder involvement earlier in the process will restore certainty to the regulatory process and limit litigation to technically valid dispute resolution. The revisions to nuclear licensing in the United States for a single COL have streamlined the process. However, unanticipated events, such as the Fukushima Daiichi accident and its associated impact on licensing requirements, will likely complicate the existing process.

6.8
Environmental Protection

Environmental protection is of significant interest to the public, and this interest is heightened when radiation and radioactive materials are involved. A nation's approach to the protection of the environment depends on the ethical and cultural basis of a society, the methodology used to provide protection, identification of desired environmental endpoints, and specific environmental elements to be protected. The level of technology and economic strength of the responsible government significantly influence these factors.

Environmental protection approaches consider national as well as international priorities. Activities within a nation should not damage the environment of other states. The protection activities should maintain ecosystems and biodiversity by preserving processes that are important for proper functioning of the biosphere. Environmental media including air, water, soil, and sediment must be preserved to ensure ecosystems are maintained. To assess impacts, reference animals and plants (RAPs) are defined, and impacts are quantified in terms of doses delivered to these species and any associated level of detriment induced by these doses.

A system of environmental protection has been introduced and developed by the ICRP. The ICRP system for environmental protection is similar to the approach used to protect humans. Environmental protection is based on biological effects as the basis for determining doses for protection purposes.

6.8.1
Reference Animals and Plants

The ICRP environmental system is defined in terms of RAPs that are broadly representative of significant and ubiquitous wildlife and are provided in Table 6.17. The ICRP assumes that radiation effects must be manifest in individual organisms in order for any population or ecosystem effect to develop. Therefore, if doses are maintained below RAP detriment levels, minimal radiological environmental impact is expected.

One of the difficulties encountered by any environmental approach is the limited set of detriment data as a function of dose for the considered species.

Table 6.17 Wildlife groups, corresponding reference animals and plants, and their environment[a].

Wildlife group	Reference animals and plants	Environment
Amphibians	Frog	Freshwater Terrestrial
Aquatic birds	Duck	Freshwater Marine
Pelagic fish	Trout	Freshwater
Large mammals	Deer	Terrestrial
Large plants	Pine tree	Terrestrial
Crustaceans	Crab	Marine
Fish	Flatfish	Marine
Seaweeds	Brown seaweed	Marine
Small mammals	Rat	Terrestrial
Small plants	Wild grass	Terrestrial
Annelids	Earthworm	Terrestrial
Insects	Bee	Terrestrial

a) Larsson (2012).

This limitation is minimized within the RAP approach that allows for collecting biological effects data into a generic database and evaluating these data in terms of protective measures. The detrimental effects are grouped into four categories: mortality, morbidity, reduced reproductive capability, and frequency of harmful mutations.

6.8.2
Derivation of Protective Measures

Once the RAP databases that relate exposure and effects are developed, these data are utilized to derive benchmark doses to guide protective actions. These benchmark doses become the basis for subsequent environmental protection of the individual RAPs through the development of derived concentration reference levels (DCRLs). The DCRLs are dose rate bands that cover 1 order of magnitude within which a biological detriment might exist.

Table 6.18 outlines the rationale for establishing a DCRL for a hypothetical RAP. Adopted RAP DCRL values are based on specific data. Table 6.18 values are illustrative and do not apply to a specific RAP.

For the 12 RAPs currently adopted by the ICRP (see Table 6.17), the DCRLs fall within three dose ranges. For deer, duck, pine tree, and rat, the DCRL range is 0.1 – 1 mGy/day. A DCRL range of 1 – 10 mGy/day is applicable for flatfish, frog, trout, and wild grass. For bee, brown seaweed, crab, and earthworm, the DCRL range is 10 – 100 mGy/day.

Once the DCRL values are identified, appropriate environmental levels can be established for air, soil, and water concentrations and surface deposition values. These environmental levels provide guidance for appropriate cleanup levels following a release of radioactive materials. The guidelines supply important parameters for decisions involving recovery activities. In addition, the DCRL values are a useful input for discussions with stakeholder groups in addressing

Table 6.18 Approach to develop DCRL values for a hypothetical reference animal or plant[a].

Dose rate range (mGy/days)	Hypothetical observation	Relative biological concern
$10^3 - 10^4$	Significant mortality	High to very high
$10^2 - 10^3$	Population disturbance from prolonged exposure	High for prolonged exposure
$10^1 - 10^2$	Morbidity effects	Increases with dose
$10^0 - 10^1$	Reproductive effects	Based on the DCRL range
10^{-1} to 10^0	No effects observed	Very low to low
10^{-2} to 10^{-1}	No effects observed (on the order of natural background)	None to very low

a) ICRP 108 (2008) and Larsson (2012).

allowed contamination levels and subsequent land use. These values could also be used as a guide to required cleanup activities following releases of radioactive material into the environment. DCRL values are applicable for a variety of release events including reactor accidents and terrorist events.

6.9 Unresolved Issues Associated with Major Reactor Accidents

Since most operating reactors are pressurized water reactors and boiling water reactors, the radiological issues from these accidents are most appropriate for assessing the event response in terms of its effect on the public. When considered collectively, the TMI-2, Chernobyl-4, and Fukushima Daiichi accidents suggest that radiation protection of the public could have been improved. Although the affected populations did not receive doses that caused severe acute effects, radiation protection improvements are warranted.

6.9.1 Radiation Risk

During these accidents and following stabilization of the reactors, antinuclear groups and the media argued that the health risk from the public's radiation exposure was higher than suggested by the government or facility estimates. These claims were based on assertions that the risk coefficients underestimated the radiation effects and that the models used to predict detriment underestimated the risk. Although such claims are not based on science, they tend to excite the public and cause undue stress during periods in which cooperation and trust are needed. Substantial scientific evidence supports the conservative ICRP risk coefficients used for bounding radiological protection purposes.

Public debate during an accident is not an optimum forum for fostering understanding of the science associated with radiation protection. The public must be educated prior to a significant event, and this process should be initiated during facility licensing and throughout facility operations. This instruction should be continuous and include educational institutions, public service groups, speaker's bureaus, and public forums.

6.9.2 Radiation Units Used to Quantify Radiation Dose

The units used for quantifying the radioactive material release and the associated radiation dose of individuals contribute to the communications problems. For example, the public does not understand the various quantities used in radiation protection. These include the various dosimetric and radiometric quantities such as activity and activity concentration. In addition, the use of the same unit for the equivalent dose to an organ and effective dose without always specifying which quantity is used has fostered confusion in interpreting the effects of these doses. This is particularly true when assessing thyroid doses (equivalent dose) from radioiodine and the dose to the whole body (effective dose) which are quantified in terms of Sv. Although the ICRP system of quantities and units is well suited for operational radiation protection, it is less suited for communication with the public especially during emergency conditions. Radiation units and associated terminology are issues that merit public education that is best accomplished over time through educational and outreach programs.

6.9.3 Internal Dose Assessment

Internal exposures are perceived by the public to be of greater hazard than the same exposure from external sources. The public's concern is that a dangerous material is residing within the body and continually delivering a dose to an individual. There is also a concern that an internally deposited radioactive material could be transferred to a family member or friend.

A well-established tenant of the ICRP methodology is that radiation risk depends on the effective dose and not on whether that dose is delivered from outside or inside the body. The media and the public either ignore this fact or are unaware of its existence. For a given effective dose, the same risk is expected, whether irradiation is from an internal or external source. In addition, the ICRP radiation protection methodology tends to be more conservative for internal exposures because its methodology determines the committed dose over a specified time rather than the dose immediately delivered to the individual. These distinctions should be addressed through public education and information programs.

6.9.4 Emergency Management Guidance

International as well as national emergency management guidance is available for a major accident involving a large release of radioactive materials into the environment. However, a number of issues associated with emergency management have been raised by the ICRP in their analysis of the Fukushima Daiichi accident. The Fukushima Daiichi accident was unique because it involved a prolonged release of radioactive materials from multiple units. Guidance had previously assumed the accident would occur as an acute release from single unit.

At TMI-2 the established emergency planning zones were utilized during the evacuation. However, during the Fukushima Daiichi accident, the emergency planning zones were extended to account for significant releases following severe core damage and hydrogen explosions that dispersed fission products into the environment.

Emergency response personnel are challenged in applying international guidance during the management of an emergency exposure situation. This was particularly true at Fukushima Daiichi since there were challenges associated with the protracted release period and with extending existing emergency planning zones. Prioritizing emergency protective measures, planning for lifting these measures, and transitioning from an emergency exposure situation to an existing exposure situation were additional issues of concern that occurred during the Fukushima Daiichi accident. The absence of quantitative recommendations for lifting emergency protective measures continues to create problems for emergency managers.

The aforementioned international emergency management issues should be addressed. Resolution of these issues through associated guidance or regulation would have credibility if proposed by a regulatory organization (e.g., the US NRC), scientific organization (e.g., the ICRP), or international body (e.g., IAEA). As noted by various authors, the issues that appear to require the development of additional guidance include the (i) management of prolonged emergency exposure situations, (ii) expansion of the emergency planning zones, (iii) coordinated management of multiunit site accidents and accidents involving multiple sites, (iv) prioritization of emergency protective measures, (v) lifting of emergency protective measures, (vi) coordinating consistent emergency response actions when an accident impacts multiple nations, and (vii) transition from the emergency exposure situation to an existing exposure situation. International emergency management actions are discussed from a regulatory perspective in Chapter 7.

6.9.5
Emergency Medical Response

The Fukushima Daiichi accident illustrated common weaknesses in emergency medical response. This response was significantly complicated because the accident occurred in conjunction with a severe earthquake and resulting tsunami and involved medical professionals with limited radiological knowledge and experience.

The radiological knowledge of medical personnel should be enhanced. Medical professional training will ideally include a basic knowledge of radioactive materials and radiation as well as the biological effects of ionizing radiation. This training defines the hazards to individuals contaminated with radioactive materials and response to personal contamination situations. Medical professionals responding to a severe reactor accident including physicians, nurses, and radiation technologists should be trained in emergency radiological measures. Basic radiation protection science must be integrated into the core curriculum in medical and nursing schools and in emergency medical response training. Responding to contaminated injured individuals should also be emphasized in drills and exercises that involve medical personnel.

The roles and responsibilities of volunteer health physics experts for medical and radiation safety support during emergencies also require clarification. For example, the Medical Reserve Corps (MRC) concept being developed and implemented in the United States offers an initial framework for the utilization of health physics resources during a radiological emergency. However, utilization of these resources for a major long-term event (e.g., improvised nuclear device detonation or multiunit reactor accident with severe core damage) has yet to be formally established. The MRC concept is addressed in Section 4.4.7.

6.9.6 International Emergency Criteria

There is emergency guidance for occupational radiation workers, and existing ICRP guidance can be utilized by radiation protection professionals to address public evacuations and associated doses to the evacuees and emergency response personnel. However, the utilization of this guidance during the Fukushima Daiichi accident by government officials and the associated public confusion suggest that additional clarification is needed. Based on the Fukushima Daiichi accident, additional clarification or development of new criteria is suggested for the radiological protection of the following groups: rescuers and volunteers, public evacuees, and the public reentering evacuated areas. Dose criteria for the public including pregnant women and children need to be better quantified and communicated.

The Fukushima Daiichi accident also suggested that other emergency criteria needed to be better defined. In particular, international clarification is needed in the following areas: (i) de-escalating from an emergency classification, (ii) providing public dosimetry and dose monitoring, (iii) addressing environmental contamination, (iv) addressing public psychological issues caused by the accident, and (v) communicating effectively with the public. As noted in Section 6.9.6.3, de-escalating from an emergency classification is well defined in US regulations.

6.9.6.1 Protecting Rescuers and Volunteers

A severe nuclear accident taxes the capability of facility emergency workers, and additional resources might be required to stabilize the reactor and return the unit to a cold shutdown condition. These resource requirements include fire fighters, military organizations, police departments, medical personnel, and volunteers assisting in rescue and recovery operations. These groups are not normally trained as radiation workers, and national regulations or international standards groups do not address their dose restrictions.

During the Fukushima Daiichi accident, there was some confusion regarding the appropriate dose limits applicable to rescuers and volunteers. Utility emergency personnel had their dose limits increased to the 250 mSv emergency dose limit, which created an issue for the press and some stakeholders. Radiation protection systems provide limited general guidance, but do not specifically address volunteers who are willing to take high risks for saving lives or perform other beneficial activities. This general guidance does not provide specific criteria and guidance that could be applied during a severe emergency. Examples of these general criteria are provided in Chapters 3, 4, and 7.

National and international radiation protection systems do not specifically address emergency dose limits or recommendations for volunteer workers or response personnel who are not classified as radiation workers. Establishing dose guidelines for these workers is important because they may be highly exposed to radiation during accident mitigation activities.

6.9.6.2
Public Evacuations

The TMI-2 and Fukushima Daiichi accidents involved public evacuations based on projected radiological conditions. An evacuation decision has serious consequences because people are removed from the safety of their homes and placed in a transient condition that is extremely stressful. As such, the evacuation process can cause public harm because stress aggravates existing medical conditions. In addition, accidents occur during the movement of the public to locations unaffected by the radiological emergency.

It is appropriate to question the merit of an evacuation order to ensure that the benefits outweigh the risks. The merit of an evacuation is addressed through the ICRP principle of justification that is usually applied to the introduction of new sources of radiation, which are expected to increase the dose delivered to an individual. Justification is applicable to the introduction of disruptive protective actions such as evacuation, which are expected to decrease public dose. Therefore, an evacuation is only justified in terms of the benefit achieved from the protective action.

Applying justification in an emergency situation such a severe reactor accident is difficult because the risks and benefits are not definitively known. For example, decisions to evacuate the public from areas of elevated doses can present a dilemma if the projected doses do not represent a well-defined risk. Without the evacuation some incremental radiation dose is incurred that theoretically increases the possibility of a future radiation detriment. If the evacuation is ordered, the possible radiological detriment is eliminated, but the public is subjected to the actual detriments associated with the evacuation. Additional guidance regarding the application of the justification principle to evacuations should be provided by national and international organizations.

Improved guidance should better quantify the evacuation risk and provide a credible assessment of the radiological risk. This will be a difficult task as long as radiological organizations continue to utilize the LNT model for assessing the biological effects of ionizing radiation.

It is well known that the use of the LNT model has significant implications for nuclear regulations affecting routine operations. It is less obvious that these linear models affect emergencies by setting the criteria for implementation of protective actions including evacuation of the public during a severe reactor accident. By adopting the overly restrictive LNT hypothesis, optimum decisions may not be realized during emergencies. LNT usage increases costs during routine operations. It can also lead to a poor evacuation decision that affects the lives of the public directly impacted by the protective action. As such, the LNT hypothesis needs to be reviewed in terms of the harm it could potentially cause during an evacuation.

6.9.6.3
De-escalating from an Emergency Classification

During the Fukushima Daiichi accident, there were issues associated with the transition from an emergency exposure situation created by the accident to an existing exposure situation that involves long-term accident recovery. A key issue involves defining when the emergency exposure situation is terminated and the existing exposure situation starts. The Japanese perceive that it would be easier to judge when the emergency exposure situation shifts to an existing exposure situation if the ICRP recommendations were more quantitative.

This issue is not a major issue for a US reactor. Emergency planning involves a clear entrance into an emergency classification (i.e., Unusual Event, Alert, Site Area Emergency, and General Emergency) as well as de-escalation based on well-defined emergency action levels that are quantified in terms of specific criteria. As noted in Chapter 3, when these criteria are met, the emergency classification is entered, and when the condition is no longer met, de-escalation is achieved. Senior managers, well versed in facility operations and its emergency planning bases, perform escalation and de-escalation based on specific emergency action level criteria. Emergency action levels are addressed in more detail in Chapter 3. Changes in protective actions following de-escalation also have well-defined guidance (e.g., the 2013 *EPA Protective Action Guides and Planning Guidance for Radiological Incidents*).

6.9.6.4
Reentry into Evacuated Areas

Reentry into evacuated areas following the TMI-2 accident presented minimal difficulty since the accident involved an insignificant release of long-lived fission products to the environment. Moreover, reentry was not complicated by a natural disaster that occurred coincidently with the Fukushima Daiichi accident. The 2005 Hurricane Katrina and 2012 Hurricane Sandy events that occurred in the United States illustrated the difficulties associated with reoccupying areas following natural disasters.

The reentry of populations following the Chernobyl-4 and Fukushima Daiichi accidents was more difficult because there were significant releases of long-lived fission products into the environment. For an accident involving a major fission product release, it is a challenge to reoccupy a contaminated area that has been evacuated. This was one of the lessons learned from the Chernobyl-4 accident. A similar situation currently exists in a portion of the areas in the Fukushima Prefecture that were evacuated.

Returning the public to their homes in areas that are contaminated or have elevated radiation levels is challenging. The reentry situation in Japan was further complicated because the earthquake and subsequent tsunami damaged critical infrastructure. In addition, the large loss of life and property damage resulting from the natural disaster required significant resources. Without the natural disaster, these resources could have been applied to Fukushima Daiichi accident recovery efforts.

During the Fukushima Daiichi Nuclear Power Station (FDNPS) accident, questions were raised regarding the appropriate dose limit for a returning population. In the United States, the EPA provides specific reentry guidance (Table 3.7). The ICRP also provides guidance for emergency public exposure situations that are summarized in Table 7.10. While the ICRP recommendations are not explicit on how to handle de-escalating protective actions, it is reasonable to consider that the return from a temporary evacuation leads to an existing exposure situation. International standards organizations should better define reentry conditions with an emphasis on criteria that the public can readily understand.

6.9.6.5
Emergency Public Dose Limits

During the Fukushima Daiichi accident, there was confusion in the media and within various stakeholder groups regarding the appropriate dose limit for protection of the public. Japanese regulatory authorities adopted a public reference level of 20 mSv/year for the emergency at Fukushima Daiichi. However, individuals in the affected areas were confused with the basis and justification for the various public dose limits that were perceived to be a combination of normal operations, emergency operations, and postemergency protection criteria.

Based on press reports and reviews, it appears that the public has concerns regarding dose limits above the ICRP planned exposure situation limit of 1 mSv/year. This dose criterion is a challenge when addressing a significant radiological event. The criteria used for restricting public doses can be controversial because the limits involve judgments regarding the acceptability of radiological risks. This is a concern if the bases for the criteria are not clearly defined. Such radiological criteria are difficult for the public to understand and for the government to explain during an evolving accident situation.

Although the current ICRP recommendations account for public dose restrictions, they may not clearly present their bases in a manner than can be easily understood by the public and government officials responding to a protracted nuclear emergency. For example, the reference levels recommended for dealing with a nuclear emergency are larger than the limits used for planned situations. Although they still provide sufficient protection to members of the public, it is essential that the bases for dose criteria are clearly understood by the affected individuals. In spite of the confusion, the adopted dose limits, public evacuation, and food restrictions directed by the Japanese government reduced the effective dose received by the public.

6.9.6.6
Emergency Dose Limits for Children and Fetuses

Japanese parents expressed concern regarding the protection of their children from the effects of radiation and radioactive materials released during the Fukushima Daiichi accident. The parents were concerned that the dose levels established to protect the general population were also being applied to the more radiosensitive children. Parents expressed concern that the 20 mSv/year reference level is too high for children in view of the 1 mSv/year dose recommendation for the public for normal operating situations.

In a similar fashion, pregnant women expressed concerns regarding the effects of radiation on the health of their unborn child. The intake of radioactive material is also of concern to these pregnant women.

A consensus standard for radiological emergencies that specifically addresses the protection of children, infants, and unborn children has not been published. In ICRP 103, there is about a 30% difference between the detriment-adjusted nominal risk coefficient for the total population including children and the risk coefficient for the adult population. However, given the concerns expressed during the Fukushima Daiichi accident and recent data regarding the radiation risk for children, the emergency dose limits for children merit additional consideration.

6.9.6.7
Public Dosimetry and Dose Monitoring

The Fukushima Daiichi accident raised issues related to dose monitoring of the public as a means to further their protection. The Japanese public questioned their safety because they were not monitored while receiving accident-related exposure, but workers at the FDNPS were monitored. To the Japanese public this seemed questionable because the same radioactive materials were irradiating both groups. A second issue involves specification of the policy for environmental monitoring following an accident. These issues contributed to public anxiety.

Recommendations for public radiation monitoring during residence in long-term contaminated areas resulting from a nuclear accident are available. However, there is a lack of international guidance for radiation monitoring of the public during the immediate aftermath of an accident. This deficiency should be addressed in a manner that can be applied consistently from an international perspective. Uniform international monitoring is important because large-scale accidents can involve multiple nations. Both the Chernobyl-4 and Fukushima Daiichi accidents affected multiple nations as well as their economies.

6.9.6.8
Environmental Contamination

The TMI-2 accident had minimal environmental impact because the release of long-lived fission products was inconsequential. Both Chernobyl-4 and the Fukushima Daiichi accidents contaminated the environment, and some of the radioactive material was deposited in neighboring areas. At Fukushima Daiichi, the released radioactive materials contaminated water, soil, food, and surface areas occupied by the public. This contamination was of public concern and forced the government to restrict land use and the consumption of crops, milk, water, and other consumables.

The contamination of the Fukushima and neighboring prefectures required the Japanese government to address the degree of contamination and the radiological conditions required to reenter evacuated areas. The public wanted to resume a normal life following the evacuation, but the government struggled with the radiological levels to permit reentry and with characterizing the areas in terms of their contamination, required remediation, and habitability. Government delays in reentry decisions frustrated the public, and the government failed to resolve a basic public issue: *When can I return to my home?*

Contaminated soil and to a lesser extent rubble from the earthquake/tsunami present a legacy recovery issue. The contaminated media vary greatly in terms of their radioactive material content. Only a fraction of the contaminated media contains significant quantities of radioactive materials. However, the public perception is that all contaminated material is hazardous and should be treated as radioactive waste.

In addressing contaminated food and liquids, the Japanese government utilized guidelines that differed from published recommendations. The adopted values differed from the United Nations' Codex Guideline Levels for radionuclide levels in internationally traded food, and these values were periodically modified. Changing guideline values contributed to confusion regarding the acceptable levels of radioactive materials in consumer goods including food and milk. Accordingly, the international regulation of consumer products containing radioactive materials is in need of clarification and standardization to minimize the confusion that occurred during the Fukushima Daiichi accident.

The Fukushima Daiichi accident illustrated the need for clear international guidance regarding the disposition of goods originating in an environment contaminated with radioactive materials. Guidance associated with contaminated environments has been controversial and does not always have a well-defined basis. Control of consumer products from these areas is an additional unresolved radiological protection issue associated with the Fukushima Daiichi accident. Successfully meeting these radiological issues will require guidance that should address the remediation of contaminated land areas, disposing of contaminated materials including soil, and controlling contaminated consumer products.

The remediation of contaminated land is a significant undertaking. For example, the Japanese Environment Ministry estimates that about 30×10^6 m^3 of contaminated soil and vegetation from the Fukushima Prefecture may require disposal. This is a complex task and involves a number of activities. For a private residence these tasks include (i) removing contaminated plants and weeds, (ii) decontaminating roofs and outdoor building surfaces, (iii) cleaning gutters and drainpipes, (iv) decontaminating yards and driveways, and (v) decontaminating indoor surfaces from contaminated air infiltration and dust. Houses in elevated radiation areas had their roofs decontaminated using high-pressure equipment.

Yard areas were decontaminated through the removal of surface soil. In addition, roads were rinsed, mud in roadside ditches removed, tree branches trimmed, and dead leaves disposed. In agricultural areas, contaminated topsoil was removed.

In addition to questions regarding the effectiveness of these decontamination measures, there are associated issues related to their implementation. These issues include the impact on farmland topsoil and long-term fertility of the agricultural land. Decontamination methods also have the potential to further disperse radionuclides following the washing of radioactive materials into surface waters, rivers, and sewage systems.

6.9.6.9
Psychological Consequences

A natural consequence of most major disasters is physiological stress and a subsequent psychological impact. This occurred during the TMI-2, Chernobyl-4, and Fukushima Daiichi accidents. The Fukushima Daiichi accident produced significant psychological consequences because it combined the stress created by a major power reactor accident with the aftermath of a catastrophic earthquake and tsunami. Psychological consequences include chronic anxiety, depression, despair, grieving, intense anger, posttraumatic stress disorder, severe headaches, sleep disturbances, and increased smoking and alcohol use. For some individuals, these consequences lead to additional physical health consequences.

International safety standards and recommendations do not currently account for these psychological consequences and their impact on the detriment caused by a major reactor accident. Emergency planning should recognize the necessity for addressing psychological consequences and their long-term health impact. Responding to public mental health needs following a nuclear accident or terrorist event raises numerous challenges that are difficult to quantify and properly address.

6.9.6.10
Public and Media Communications

The Fukushima Daiichi accident again demonstrated that communications between the operating utility and government and between the government and public are in significant need of improvement. Poor radiation risk and dose communications affected the public's perception of the accidents at TMI, Chernobyl, and Fukushima Daiichi.

These issues were magnified during the Fukushima Daiichi accident since it involved international media. The demands of the international media during a major accident and the continuous sharing of accurate information are significant challenges when an accident involves multiple nations. International guidance is needed to ensure the coordination of information dissemination during a nuclear accident affecting multiple nations. This guidance should address the dissemination of accurate and timely radiological data in an understandable format.

The use of social networks and smart communications devices should be incorporated into international communications and emergency plans as timely methods of distributing information. In addition, groups having public credibility and trust (e.g., university professors and physicians) should be included in international communications plans as a method to enhance information exchange with the public and media. Accurate and timely information would have partially alleviated the stress and some of the challenges associated with the Fukushima Daiichi accident particularly if it was consistently delivered by a trusted source.

Problems

6.1 Six months ago, terrorists transported a dirty bomb incorporating natural uranium to Philadelphia. During transport, the bomb prematurely detonated near a rural farming community. Radiological surveys indicate that uranium has begun to leach into individual groundwater wells serving this community. You have been hired by the Citizens Roundtable against Zero Yield, a Pennsylvania stakeholder group, to perform a dose assessment for a residential scenario.

Data:

1. ^{238}U concentration in groundwater (C_W) = 1.85 Bq/l used for irrigation. For simplicity, only consider the contribution from ^{238}U and ignore any daughters.
2. Assume all irrigation is overhead irrigation. Irrigation rate (I_R) = 2.5 l/m^2-day.
3. Effective weathering and decay constant of uranium on plant surfaces, $\lambda = 0.12$/day.
4. Translocation factor, transfer of radionuclides from plant surfaces to edible parts for nonleafy vegetables, $T_v = 0.1$.
5. Fraction of deposited activity retained on plant surfaces, $r_v = 0.25$.
6. Plant yield (nonleafy vegetables), $Y_v = 4$ kg plant wet weight/m^2.
7. Crop growing period = 90 days.
8. Soil-to-plant concentration factor (nonleafy vegetables dry weight basis), $B = 0.012 \frac{\text{kg(soil)}}{\text{kg(plant dry)}}$.
9. Mass-loading factor for resuspension to edible proportions (also termed the *crop external contamination factor*), $M_L = 0.1 \frac{\text{kg(plant wet)}}{\text{kg(soil)}}$.
10. Wet-to-dry weight conversion factor, $W_{W-d} = 0.25 \frac{\text{kg(plant wet)}}{\text{kg(plant dry)}}$.
11. Consumption rate of produce, $Q = 50$ kg (wet weight)/year.
12. ^{238}U (type F) ingestion dose conversion factor, DCF = 4.4×10^{-8} Sv/Bq.
13. The area is encountering a severe drought and all watering is by irrigation.

(a) Calculate the deposition rate r_d in $\frac{\text{Bq}}{\text{kg(plant wet weight)-day}}$ to edible parts of plants from direct application of overhead irrigation for ^{238}U.

(b) Although plant samples are obtained and directly analyzed when possible, some plant concentrations must be modeled. Using a daily direct irrigation deposition rate to the edible parts of plants of 0.037 Bq/kg-day for ^{238}U, calculate the ^{238}U concentration in the plants at the end of the growing season (from direct deposition only).

(c) Assuming an equilibrium concentration of ^{238}U in soil of 7.77 Bq/kg, calculate the plant concentration at the end of the growing season as a result of root uptake and resuspension.

(d) Assuming a uranium concentration of 0.0296 Bq/kg plant wet weight from direct deposition and a concentration of 0.0592 Bq/kg plant wet weight from root uptake and resuspension, calculate the effective dose to an individual from 1 year of produce consumption.

(e) List four factors that may influence the plant uptake of uranium.

(f) List five other exposure pathways from the terrorist device that are not considered above.

(g) Your organization is offering bioassay monitoring to residents concerned about their internal exposure. List two methods of determining the concentration of uranium in the body and an advantage and disadvantage of each method.

6.2 You have been retained by Century 22 Realtors to perform risk estimates for a large proposed residential development in an area of higher than normal radon levels. Measured radon emanation at the soil surface and the radon flux in the first floor of a slab foundation home without any mitigation are available.

Data:

J_i = radon flux into home = 0.074 Bq/m^2-s
J_o = radon flux at the soil surface = 0.185 Bq/m^2-s
F_{eq} = equilibrium factor = 0.4
S = building area = 200 m^2
H = building room height = 2.5 m
k_v = ventilation removal rate constant = ventilation flow rate (F)/room volume (V) = 0.5/h
R = lifetime excess cancer mortality risk per WLM = 5.5×10^{-4}/WLM
F = occupancy factor = 0.7
L = life expectancy = 70 years

Radon and its short-lived daughters

Nuclide	Alpha energy (MeV)	Half-life
^{222}Rn	5.49	3.82 days
^{218}Po	6.00	3.1 min
^{214}Pb	—	27 min
^{214}Bi	—	19.9 min
^{214}Po	7.68	164 μs

(a) Calculate the steady-state indoor radon concentration in the first floor living space.
(b) Assume the answer to question (a) was 0.518 Bq/l. What is the exposure to the short-lived radon progeny in working level months (WLMs) per year?
(c) List four sources of uncertainty in the application of the results from epidemiological studies of populations of underground miners to health effects in the general population.
(d) The current radon risk model is based on empirical studies (i.e., developed from epidemiological studies of underground uranium miners). Another type of model could develop risk estimates based on radon's effects on the respiratory tract. List four sources of uncertainty in this dosimetry model for the respiratory tract as applied to risk estimates for radon exposures.
(e) List four methods to reduce the radon entry into a home or building.
(f) Another potential concern is the radon in the water supply to the home. Which of the following statements represents the best estimate of the water to air transfer factor for the reduction in concentration of radon in water (in Bq/l) to the indoor air concentration (in Bq/l)?
   1. 10 to 1 reduction (i.e., a 10 Bq/l water concentration to a 1 pCi/l air concentration)
   2. 100 to 1 reduction
   3. 1000 to 1 reduction
   4. 10 000 to 1 reduction
   5. 100 000 to 1 reduction

6.3 You are the duty health physicist at a small regional medical center. A private aircraft crashed near the medical center, and there was a single survivor, a 25-year-old female. The unconscious woman was brought into the emergency room, severely bleeding and with several broken bones. When admitted, the Wham-o-Dyne computed tomography (CT) scanner was out of service, and a "trauma" series of diagnostic X-rays were taken. Following the X-ray series, fluoroscopy was required as part of an effort to investigate the possibility of internal injuries. When the woman regained consciousness, she informed her physician of the pregnancy with a date of conception about 1 month prior to the accident. The pregnancy was not recorded on the medical charts and was not considered during her initial examination and imaging.

Data:

1. A trauma series of diagnostic X-ray examinations was performed including a single exposure of the head/neck, chest, abdomen, pelvis, and lumbar, thoracic, and cervical spine. All X-ray projections are anterior–posterior (AP).

2. Four minutes of fluoroscopy time at 2 mA was logged for the abdominal procedures.
3. For the machines used in examining the woman, the fetal dose is determined to be 45% of the entrance skin exposure to the mother.
4. Radiological information for the diagnostic AP series is summarized in the following table:

Procedure	Entrance skin exposure (ESE) ($\times 10^{-4}$ C/kg)
Chest	0.07
Pelvic	0.79
Head/neck	1.14
Abdomen	1.08
Lumbar spine	1.40
Thoracic spine	1.33
Cervical spine	0.39

5. Fluoroscopy entrance skin exposure = 14.9 mGy/mA-min

(a) Calculate the absorbed dose to the fetus. You need not calculate radiation doses from every procedure if you can justify omitting the calculations.

(b) Assume that the radiation dose calculated in question (a) was 35 mGy. What three pieces of advice would you give the woman's physician regarding terminating the pregnancy or letting it proceed?

(c) What are the three pieces of information necessary to determine the risk of injury to the fetus in this incident?

(d) During subsequent communications with the woman's physician, you are informed that she appeared to have skin burns. List five reasonable explanations why this could occur.

(e) List five machine parameters that will affect fetal radiation exposure from CT, X-ray, or fluoroscopy.

6.4 A van transporting stolen radioactive materials was involved in an accident on a major interstate highway. The news media has surrounded the affected area because initial reports suggested the detonation of a dirty bomb. Since you are the Radiation Safety Officer at a nearby university, the State Police requests your assistance in source recovery, in radiological assessment, and in responding to media radiological inquiries. Upon arriving at the crash site, you use a portable high-purity germanium detector and ionization chamber and determine that the ^{60}Co and ^{137}Cs isotopes are present at the accident site and that elevated dose rates exist.

Data:

^{60}Co radiological data

Gamma emissions	1.17 MeV @ 100%	
	1.33 MeV @ 100%	
Gamma constant	3.1×10^{-4} mGy-m^2/h-MBq	
Attenuation coefficients	μ(^{60}Co for Pb)	0.679/cm
	μ(^{60}Co for water)	0.0707/cm
	μ (^{60}Co for air)	7.75×10^{-5}/cm

^{137}Cs radiological data

Gamma emissions	0.662 MeV @ 85.1%
Attenuation coefficients	
μ(^{137}Cs for water)	0.0894/cm
μ_{en}(^{137}Cs for water)	0.0327/cm
μ(^{137}Cs for air)	0.0001/cm

Additional information

Lead blanket specifications
31 cm × 62 cm × 2.5 cm (equivalent lead),
10.4 kg with a polyvinyl chloride cover

Density of lead 11.4 g/cm^3

Dose buildup factors for a point source at 1.25 MeV

μx	Water	Air	Lead
0.5	1.47	1.47	1.20
1	2.08	2.08	1.38
2	3.62	3.60	1.68
3	5.50	5.46	1.95
4	7.68	7.60	2.19
5	10.1	10.0	2.43
6	12.8	12.7	2.66
7	15.8	15.6	2.89
8	19.0	18.8	3.10
10	26.1	25.8	3.51
15	47.7	47.0	4.45
20	74.0	72.8	5.27

(a) A collection of small sources are within a 200 cm^2 area in a stream having a depth of 1 m. The absorbed dose rate from the sources is 30 mGy/h at the water surface. A portable high purity germanium (HPGe) detector is used to determine that all activity is ^{60}Co. A mobile crane will be used to remove the sources. Calculate the absorbed dose rate in the crane cab 10 m above the water surface directly above the sources when the sources are submerged in 1.0 m of water. State any assumptions used in the calculations.

(b) Calculate the absorbed dose rate in the crane cab 10 m above the sources when the sources are lifted just above the water surface.

(c) A small ^{137}Cs rod is lying on the side of the highway. The absorbed dose rate in air from the source is 100 mGy/h at 30 cm. What is the activity of the source?

(d) A 3.0 m long, thin-walled, 1.0 cm diameter pipe is found in a field near the crash site. The dose rate at the midlength of the pipe is 9 mGy/h at 1 m from the pipe in air. Calculate the activity per unit length of the pipe. Portable HPGe scans suggest all activity in the pipe is ^{60}Co. State any assumptions used in the calculation.

(e) The dose equivalent rate at 1 m from an additional small ^{60}Co source is 1.5 mGy/h. Calculate the minimum number of layers of lead-wool blankets (polyvinyl chloride (PVC) covered lead-wool used for shielding) needed to reduce the area around the source to 1 mGy/h at a distance of 30 cm.

(f) A local stakeholder group questions the adequacy of the cleanup and the final radiological survey. The group is concerned that some radioactive sources remain in the areas beyond the paved highway. How do you resolve their concerns?

6.5 It is a clear spring day with a temperature of 20 °C. The sun is shining and there is a variable 5 km/h wind. No precipitation is forecast.

You are the Director of the Bureau of Radiation Protection for the State of Pennsylvania. A reactor accident involving a loss of core cooling is in progress at a pressurized water reactor in your state. The Governor has requested that you participate in discussions with the operating utility. An Alert was declared based on a diminished capability to cool the core.

(a) The Governor asks that you compile a list of state and local government responses for the current Alert. She also requests a list of actions if the accident escalates to a Site Area Emergency or General Emergency. Develop the requested list.

(b) The utility has escalated to a Site Area Emergency after failure of the fuel fission product barrier. However, no release is in progress, and the utility projects no release will occur. The utility recommends sheltering in place as the protective action recommendation. Do you concur with this recommendation?

(c) An airborne release is now in progress. What radionuclides are of concern?

(d) The following dose projection has been received from the utility:

Distance (km)	Effective dose (mSv)	Thyroid equivalent dose (mSv)
1.6	0.08	0.01
3.2	0.05	0.001
4.8	0.02	<0.001
8.0	0.005	<0.001
16	0.001	<0.001

What protective actions are warranted?

(e) The reactor coolant system has been breached, and fuel has been severely damaged. Containment pressure has decreased, indicating that all three fission product barriers have been breached. In addition, containment radiation monitors are off scale high, and utility field teams report a direct radiation measurement of 175 mSv at 3.2 km and 10 mSv at 16 km from the facility. No thyroid doses were provided. The utility recommends sheltering in place as a protective action. A release duration of 24 h is projected by the utility. Do you concur with this recommendation?

6.6 The year is 2044 and Utopian Air has established routine commercial air travel between Los Angeles and LEO-1, a low earth orbit hotel and casino. You are Utopian's space health physicist, and part of your duties are to forecast solar events, determine their dose consequences, and advise the Flight Operations Director of radiation hazards.

Utopian's major lift vehicle has a shell that provides 3 cm of equivalent aluminum shielding. There is no emergency shelter on the lift vehicle. Passengers and crew are exposed for a maximum of 2 h to solar radiation between launch and docking with LEO-1. The hotel provides 4 cm of aluminum shielding and has an emergency shelter having 15 cm of equivalent aluminum. In view of the unpredictability of solar events, passengers sign a claims waiver regarding radiation exposure incurred during a major radiation event.

The average radiological conditions encountered during a flight to LEO-1 include proton and heavy ion radiation. For the current solar cycle, typical spectrometer output by particle type for a 2 h flight is provided in the following table:

Particle type	Two hour integrated fluence (particles/cm^2)	Dose conversion factor (pGy-cm^2/particle)
Protons	3×10^5	3.0×10^3
Heavy ions	4×10^4	7.0×10^3

Attenuation coefficients applicable during the current solar cycle and a massive solar particle event are:

Particle type	Attenuation coefficient during normal solar conditions (1/cm)	Attenuation coefficient during a massive solar particle event (1/cm)
Protons	0.20	0.15
Heavy ions	0.35	0.30

(a) What is the absorbed dose received during a typical Utopian Air flight from Los Angeles to LEO-1?
(b) What is the absorbed dose during a Utopian Air flight from Los Angeles to LEO-1 during a massive solar particle event? The event's proton and heavy ion properties are noted in the following table. Assume the flight crew is unaware of the event and take no action to mitigate its effects.

Particle type	Two hour integrated fluence (particles/cm^2)	Dose conversion factor (pGy-cm^2/particle)
Protons	5×10^9	4.5×10^3
Heavy ions	6×10^8	9.5×10^3

(c) How could the absorbed dose values of question (b) be reduced?
(d) What doses would be received if the passengers and crew were within the shielded LEO-1 hotel emergency shelter during the massive solar event? Assume the event duration is 5 h.
(e) What is the absorbed dose on the earth's surface at sea level resulting from this event? Assume the atmosphere has an effective thickness of 25 km. For the massive solar event, the atmosphere decreases the absorbed dose by a factor of 2 for every 2000 m for protons and 1500 m for heavy ions.
(f) If the solar event of question (b) continued for a 4-week period, what is the effective dose on the earth's surface? Assume the radiation weighting factors for protons and heavy ions are 2 and 20, respectively.
(g) For the effective dose calculated in the previous question, what recommendations would you make to the Utopian Air Flight Operations Director to minimize the radiation hazards?

6.7 An object, apparently a meteorite, impacted the earth's surface near the Barrow Nuclear Power Plant (BNPP). The town council requested BNPP assistance since the Alaska State Police reports that the object is radioactive. As the BNPP's Radiation Protection Manager, you have been directed to assist the state government in assessing the radiological hazard of the object.

Upon responding to the area, the State Police provide you with a survey map. The map documents a 10 mGy/h absorbed dose rate at 1 km from the impact crater. Following your direction, a senior radiological controls technician confirms this absorbed dose rate. Subsequently, a state police helicopter flew over the impact site, and the pilot estimates that the object appears to be a meteorite about 1 m in diameter.

(a) The radionuclide composition of the meteorite is unknown. How would you determine the radionuclides present in the meteorite?
(b) The State Police want to approach the crater. Calculate the dose rate at 100 m. Based on the calculation, would you recommend that the State Police relocate their command post to a distance of 100 m from the crater? Ignore any shielding provided by the earth.
(c) What techniques could be employed to permit access to the vicinity of the crater?
(d) A portable HPGe detector scan from a drone 100 m above the crater yields the following peak energies and associated count rates following a 2 s scan:

Gamma energy (MeV)	Detector efficiency (counts/disintegration)[a]	Detector counts @ 100 m
0.0496	0.035	1.3×10^6
0.0516	0.037	1.5×10^6
0.66	0.022	1.6×10^8
1.17	0.0093	5.5×10^8
1.33	0.0092	5.5×10^8

a) Efficiency is based on a standard counting geometry with a source-detector distance of 10 cm.

Based on the 2 s scan, what isotopes are present in the meteorite? From a consideration of these isotopes, is the meteorite composed of naturally occurring radioactive material?
(e) What is the estimated activity of ^{60}Co and ^{137}Cs in the meteorite? The yield for the ^{60}Co photopeaks is 1.0 and the yield for the ^{137}Cs/^{137m}Ba photon is 0.851.
(f) Given the activity calculated in the previous question, what absorbed dose rate is expected to be measured at a distance of 1 km from the impact location? The gamma constants for ^{60}Co and ^{137}Cs are 3.1×10^{-4} and 0.81×10^{-4} mGy-m^2/h-MBq, respectively. Ignore attenuation and the contribution from other radionuclides.

6.8 You are employed by the Washington State Department of Ecology as a senior health physicist. A Spokane Chiefs Trucking Company manager calls to request state radiological assistance because a large, shielded package appears to contain radioactive materials. The package was dropped while being lifted by a warehouse crane. After impacting the floor, the package ruptured, spilled its contents, and triggered a radiation alarm. The trucking manager reports that a long metal rod and small objects are observed on the warehouse floor. There is also a spill of power in the shape of a large circular disk.

The manager transmitted a portion of the bill of lading, which suggests that the package contains radioactive materials and should have been labeled Radioactive Yellow III. Print and electronic media have arrived at the warehouse, and you are requested to provide technical assistance. None of the warehouse personnel are trained as radiation workers, and they have a very limited knowledge of radiation and radioactive materials.

Data:

1. All sources are ^{60}Co. The shipping papers and supporting documentation describe four separate radioactive sources.
2. The half-life of ^{60}Co is 5.27 year.
3. The ^{59}Co(n, γ)^{60}Co cross-section is 37 b.
4. Source 1 is a 2 cm × 3 cm × 0.2 cm irradiated metal scrap. The scrap was initially 100% ^{59}Co and had a mass of 10 g.
5. Source 2 is a 2.5 mm spherical particle with an activity of 1.1 TBq
6. Source 3 is a 10.0 m length of 0.5 cm diameter tubing with a total activity of 3.0 TBq.
7. Source 4 is a 20.0 m diameter powder spill with a total activity of 15.1 TBq. The spill is in the shape of a thin disk (0.2 cm thick).
8. The ^{60}Co gamma constant is 3.1×10^{-7} Gy-m^2/MBq-h
9. Neglect all self-shielding in the sources in answering the following questions.

(a) What are your instructions to the trucking company manager to minimize the warehouse workers' doses?

(b) What instrumentation do you bring to the warehouse to determine the isotopes involved in the warehouse incident?

(c) Upon arrival at the site, you find the shipping papers in the cab of the transport vehicle. The shipping papers indicate that Source 1 was produced in an activation reaction. Six months ago, a small scrap of material was removed from the reactor vessel of a power reactor. Records attached to the shipping papers indicate that it had been irradiated for 10 years. The scrap material is natural cobalt and was subjected to an average thermal neutron fluence rate (flux) of 1.0×10^{10} n/cm^2-s and a fast neutron fluence rate of 5.0×10^{10} n/cm^2-s. Based on the problem data, what is the activity of Source 1?

(d) What absorbed dose rate in air is expected at a point that lies 2.0 m from Source 1?

(e) Calculate the absorbed dose rate at a point 0.3 m from Source 2.

(f) Calculate the absorbed dose rate at a distance of 2.0 m from the end of the tubing (Source 3) containing a uniform distribution of Co-60.

(g) Calculate the absorbed dose rate at a point 10.0 m above the centerline of the spill (Source 4) which contains a uniform distribution of Co-60.

(h) Your boss directs you to erect a rope barrier at an absorbed dose rate of 10 μGy/h around the warehouse. Use Source 2 to establish this zone. At what distance from Source 2 should the rope barrier be established?

References

10CFR2 (2015) Rules of Practice for Domestic Licensing Proceedings and Issuance of Orders, National Archives and Records Administration, U.S. Government Printing Office, Washington, DC.

10CFR20 (2015) Standards for Protection Against Radiation, National Archives and Records Administration, U.S. Government Printing Office, Washington, DC.

10CFR52 (2015) Licenses, Certifications, and Approvals for Nuclear Power Plants, National Archives and Records Administration, U.S. Government Printing Office, Washington, DC.

10CFR71 (2015) Packaging and Transportation of Radioactive Material, National Archives and Records Administration, U.S. Government Printing Office, Washington, DC.

10CFR835 (2015) Occupational Radiation Protection, National Archives and Records Administration, U.S. Government Printing Office, Washington, DC.

49CFR (2015) Transportation, National Archives and Records Administration, U.S. Government Printing Office, Washington, DC.

Adler, P.S. and Kranowitz, J.L. (2005) *A Primer on Perceptions of Risk, Risk Communications and Building Trust*, The Keystone Center, http://www.eyeofthestormleadership.com/pg21.cfm (accessed 07 August 2015).

ANSI ANSI/HPS N13.12-1999. (1999) *Surface and Volumetric Radioactivity Standards for Clearance*, Health Physics Society, McLean, VA.

Bailey, S. (2000) Aircrew radiation exposure-an overview. *Nucl. News*, **43** (1), 32.

Barish, R.J. (2004) Radiation risk from airline travel. *J. Am. Coll. Radiol.*, **1**, 784.

Barish, R.J. (2008) *The Invisible Passenger: Radiation Risks for People Who Fly*, 2nd edn, Advanced Medical Publishing, Madison, WI.

Barish, R.J. (2009) Health physics and aviation: solar cycle 23 (1996–2008). *Health Phys.*, **96**, 456.

Barish, R.J. and Dilchert, S. (2010) Human resource responsibilities: frequent flier radiation exposure. *Respon. Rights J.*, **22**, 361.

Benton, E.R. and Benton, E.V. (2001) Space radiation dosimetry in low-earth orbit and beyond. *Nucl. Instrum. Methods Phys. Res.*, **B184**, 255.

Bevelacqua, J.J. (2003) Production equations in health physics. *Radiat. Prot. Manage.*, **20** (6), 9.

Bevelacqua, J.J. (2004a) Skin contaminations, internal intakes, and ALARA. *Radiat. Prot. Manage.*, **21** (1), 11.

Bevelacqua, J.J. (2004b) Muon colliders and neutrino dose equivalents: ALARA challenges for the 21st century. *Radiat. Prot. Manage.*, **21** (4), 8.

Bevelacqua, J.J. (2004c) Point source approximations in health physics. *Radiat. Prot. Manage.*, **21** (5), 9.

Bevelacqua, J.J. (2005a) An overview of the health physics considerations at a 21st century fusion power facility. *Radiat. Prot. Manage.*, **22** (2), 10.

Bevelacqua, J.J. (2005b) Internal dosimetry primer. *Radiat. Prot. Manage.*, **22** (5), 7.

Bevelacqua, J.J. (2008) *Health Physics in the 21st Century*, Wiley-VCH Verlag GmbH, Weinheim.

Bevelacqua, J.J. (2009) *Contemporary Health Physics, Problems and Solutions*, 2nd edn, Wiley-VCH Verlag GmbH, Weinheim.

Bevelacqua, J.J. (2010a) *Basic Health Physics, Problems and Solutions*, 2nd edn, Wiley-VCH Verlag GmbH, Weinheim.

Bevelacqua, J.J. (2010b) Radiological dispersion devices and basic radiation science. *Phys. Teach.*, **48**, 302.

Bevelacqua, J.J. (2010c) Practical and effective ALARA. Operational Radiation Safety. *Health Phys.*, **98** (Suppl. 2), S39.

Bevelacqua, J.J. (2012a) Applicability of health physics lessons learned from the Three Mile Island Unit-2 accident to the Fukushima Daiichi accident. *J. Environ. Radioact.*, **105**, 6.

Bevelacqua, J.J. (2012b) in *Particle Physics*, Chapter 5 (ed E. Kennedy), InTech, pp. 91–114, ISBN: 978-953- 51-0481-0, http://www.intechopen.com/books/particle-physics/muon-colliders-and-neutrino-effective-doses (accessed 07 August 2015).

Bevelacqua, J.J. (2012c) Environmental radiation effects from muon and tau colliders and their impact on facility licensing. *J. Environ. Radioact.*, **113**, 123.

Bevelacqua, J.J. (2012d) The Fukushima Daiichi accident and its radiological impact on the environment. *Phys. Teach.*, **50**, 368.

Bevelacqua, J.J. (2013a) Nuclear regulation in the United States and a possible framework for an international regulatory approach. *Int. Nucl. Saf. J.*, **2** (1), 1, http://nuclearsafety.info/international-nuclear-safety-journal/index.php/INSJ/article/view/12/pdf_1 (accessed 07 August 2015).

Bevelacqua, J.J. (2013b) Methodology to ascertain core recriticality following a severe fission reactor accident. *Int. Nucl. Saf. J.*, **2** (1), 52, http://nuclearsafety.info/international-nuclear-safety-journal/index.php/INSJ/article/view/13/pdf_1 (accessed 07 August 2015).

Bevelacqua, J.J. (2014a) Health physics aspects of generation IV reactors. *Int. Nucl. Saf. J.*, **3** (1), 13, http://nuclearsafety.info/international-nuclear-safety-journal/index.php/INSJ/article/view/20/pdf_1 (accessed 07 August 2015).

Bevelacqua, J.J. (2014b) Radiation protection considerations for nuclear power expansion to nations with limited technical infrastructure. *Int. Nucl. Saf. J.*, **3** (3), 8, http://nuclearsafety.info/international-nuclear-safety-journal/index.php/INSJ/article/view/37/pdf (accessed 07 August 2015).

Blue Ribbon Commission on America's Nuclear Future (2012) Report to the Secretary of Energy, US Department of Energy, Washington, DC.

Bozkurt, A. and Xu, X.G. (2004) Fluence-to-dose conversion coefficients for monoenergetic proton beams based on the VIP-man anatomical model. *Radiat. Prot. Dosim.*, **112**, 219.

Bramlitt, E.F. and Shonka, J.J. (2015) Radiation exposure of aviation crewmembers and cancer. *Health Phys.*, **108**, 76.

Bromet, E.J. (2014) Emotional consequences of nuclear power plant disasters. *Health Phys.*, **106** (2), 206.

Burrell, M.O., Wright, J.J., and Watts, J.W. (1968) An Analysis of Space Radiation and Dose Rates. NASA-TN-D-4404, National Aeronautics and Space Administration, Washington, DC.

Cliver, E.W. and Svalgaard, L. (2004) The 1859 solar–terrestrial disturbance and the current limits of extreme space weather activity. *Sol. Phys.*, **224**, 407.

CODEX Standard 193-1995. (1995) *Codex General Standard for Contaminants and Toxins in Food and Feed*, Food and Agriculture Organization of the United Nations, New York.

Covello, V.T., Sandman, P.M., and Slovic, P. (1988) *Risk Communication, Risk Statistics, and Risk Comparisons: A Manual for Plant Managers*, Chemical Manufacturers Association, Washington, DC.

Cucinotta, F.A., Schimmerling, W., Wilson, J.W., Peterson, L.E., Badwhar, G.D., Saganti, P.B., and Dicello, J.F. (2001) Space radiation cancer risks and uncertainties for mars missions. *Radiat. Res.*, **156**, 682.

Cucinotta, F.A., Wu, H., Shavers, M.R., and George, K. (2003) Radiation dosimetry and biophysical models of space radiation effects. *Gravit. Space Biol. Bull.*, **16**, 11.

Dwyer, J.R., Smith, D.M., Uman, M.A., Saleh, Z., Grefenstette, B., Hazelton, B., and Rassoul, H.K. (2010) Estimation of the fluence of high-energy electron bursts produced by thunderclouds and the resulting radiation doses received in aircraft. *J. Geophys. Res.*, **115**, D09206.

Elias, B. (2012) Airport Body Scanners: The Role of Advanced Imaging Technology in Airline Passenger Screening. R42750 CRS Report. Congress Congressional Research Service, Washington, DC.

EPA (1992) Manual of Protective Action Guides and Protective Actions for Nuclear Incidents. EPA-400-R-92-001, U. S. Environmental Protection Agency, Washington, DC.

GPUNC (1988) GPU Nuclear Corporate Emergency Plan for Three Mile Island and Oyster Creek Nuclear Stations. GPUNC Report 1000-PLN-1300.01, Revision 1, General Public Utilities Nuclear Corporation, Parsippany, NJ.

IAEA (2006a) *Safety and Security of Radioactive Sources: Towards a Global System for the Continuous Control of Sources throughout Their Life Cycle*, IAEA STI/PUB/1262, International Atomic Energy Agency, Vienna.

IAEA (2006b) *Stakeholder Involvement in Nuclear Issues: INSAG-20*, IAEA

STI/PUB/1276, International Atomic Energy Agency, Vienna.
IAEA (2009) *Security of Radioactive Sources*, IAEA Nuclear Security Series No. 11, International Atomic Energy Agency, Vienna.
IAEA (2014) *Experiences and Lessons Learned Worldwide in the Cleanup and Decommissioning of Nuclear Facilities in the Aftermath of Accidents* (STI/PUB/1644). IAEA Nuclear Energy Series No. NW-T-2.7, International Atomic Energy Agency, Vienna.
ICRP (2003) A framework for assessing the impact of ionising radiation on non-human species, ICRP Report No. 91. *Ann. ICRP*, **33** (3), 207.
ICRP (2007) The 2007 recommendations of the international commission on radiological protection, ICRP Report No. 103. *Ann. ICRP*, **37** (2-4).
ICRP (2008) Environmental protection: the concept and use of reference animals and plants, ICRP Report No. 108. *Ann. ICRP*, **38** (4-6), 3.
ICRP (2009) Application of the commission's recommendations to the protection of people living in long-term contaminated areas after a nuclear accident or a radiation emergency, ICRP Report No. 111. *Ann. ICRP*, **39** (3).
ICRP (2010) Environmental protection: transfer parameters for reference animals and plants, ICRP Report No. 114. *Ann. ICRP*, **39** (6).
ICRP (2012) Task Group 84 Report, Initial Lessons Learned from the Nuclear Power Plant Accident in Japan vis-à-vis the ICRP System of Radiological Protection. ICRP ref 4832-8604-9553, November 22, 2012.
ICRP (2013) Assessment of radiation exposure of astronauts in space, ICRP Report No. 123. *Ann. ICRP*, **42** (4).
ICRP (2014a) Protection of the environment under different exposure situations, ICRP Report No. 124. *Ann. ICRP*, **43** (1).
ICRP (2014b) Radiological protection in security screening, ICRP Report No. 125. *Ann. ICRP*, **43** (2).
Kennedy, A.R. and Todd, P. (2003) Biological countermeasures. *Gravit. Space Biol. Bull.*, **16** (2), 37.
Larsson, C.-M. (2012) Biological basis for protection of the environment. *Ann. ICRP*, **41** (3-4), 208.
Miroshnichenko, L.I. (2003) *Radiation Hazard in Space*, Springer, New York.
National Academy of Sciences (2012) Analysis of Cancer Risks in Populations Near Nuclear Facilities: Phase I. National Academy of Sciences Report No. 13388, The National Academy Press, Washington, DC.
National Academy of Sciences (2013) Messaging for Engineering: From Research to Action. National Academy of Sciences Report No. 13463, The National Academy Press, Washington, DC.
National Academy of Sciences (2014a) Best Practices for Risk- Informed Decision Making Regarding Contaminated Sites: Summary of a Workshop Series. National Academy of Sciences Report No. 18747, The National Academies Press, Washington, DC.
National Academy of Sciences (2014b) Review of NASA's Evidence Reports on Human Health Risks: 2013 Letter Report, National Academy of Sciences, Washington, DC, http://www.nap.edu/catalog.php?record_id=18575 (accessed 08 August 2015).
National Academy of Sciences (2014c) Health Standards for Long Duration and Exploration Spaceflight: Ethics Principles, Responsibilities, and Decision Framework, National Academy of Sciences, Washington, DC, http://www.nap.edu/catalog.php?record_id=18576 (accessed 08 August 2015).
National Academy of Sciences (2014d) Analysis of Cancer Risks in Populations Near Nuclear Facilities: Phase 2 Pilot Planning. National Academy of Sciences Report No. 18968, The National Academy Press, Washington, DC.
National Academy of Sciences (2014e) The Science of Responding to a Nuclear Reactor Accident: Summary of a Symposium. National Academy of Sciences Report No. 19002, The National Academy Press, Washington, DC.
National Research Council, Committee on the Biological Effects of Ionizing Radiation (1980) *The Effects on Populations of*

Exposures to Low Levels of Ionizing Radiation (BEIR III), National Academy Press, Washington, DC.

National Research Council (1990) *The Health Risks from Exposure to Low Levels of Ionizing Radiation, BEIR V*, National Academy Press, Washington, DC.

National Research Council (2006) *Health Risks from Exposure to Low Levels of Ionizing Radiation, BEIR VII Phase 2*, The National Academies Press, Washington, DC.

National Research Council (2011) *Recapturing a Future for Space Exploration: Life and Physical Sciences Research for a New Era*, The National Academies Press, Washington, DC.

National Research Council (2012) *Research for a Future in Space: The Role of Life and Physical Sciences*, The National Academies Press, Washington, DC.

National Research Council (2013) *Solar and Space Physics: A Science for a Technological Society*, The National Academies Press, Washington, DC.

NCRP (1987) Ionizing Radiation Exposure of the Population of the United States. NCRP Report No. 93, NCRP Publications, Bethesda, MD.

NCRP (1993) Limitation of Exposure to Ionizing Radiation. NCRP Report No. 116, NCRP Publications, Bethesda MD.

NCRP (2000) Radiation Protection Guidance for Activities in Low-Earth Orbit. NCRP Report No. 132, NCRP Publications, Bethesda, MD.

NCRP (2002a) Managing Potentially Radioactive Scrap Metal. NCRP Report No. 141, NCRP Publications, Bethesda, MD.

NCRP (2002b) Operational Radiation Safety Program for Astronauts in Low-Earth Orbit: A Basic Framework. NCRP Report No. 142, NCRP Publications, Bethesda, MD.

NCRP (2006a) Information Needed to Make Radiation Protection Recommendations for Space Missions Beyond Low-Earth Orbit. NCRP Report No. 153, NCRP Publications, Bethesda, MD.

NCRP (2006b) Management of Radionuclide Therapy Patients. NCRP Report No. 155, NCRP Publications, Bethesda, MD.

NCRP (2007) Radiation Protection in Educational Institutions. NCRP Report No. 157, NCRP Publications, Bethesda, MD.

NCRP (2009) Ionizing Radiation Exposure of the Population of the United States. NCRP Report No. 160, NCRP Publications, Bethesda, MD.

Ness, W.N. (ed.) (1965) *Introduction to Space Science*, Gordon and Breach Publishers, New York.

Ness, W.N. (1968) *The Radiation Belt and Magnetosphere*, Blaisdell Publishing Company, Waltham, MA.

Ney, E.P. and Stein, W.A. (1962) Solar protons, alpha particles, and heavy nuclei in November 1960. *J. Geophys. Res.*, **67**, 2087, http://nssdc.gsfc.nasa.gov/planetary/mars/marsprof.html (accessed 14 August 2015).

Nuclear Waste Management Organization (2005) *Choosing a Way Forward: The Future Management of Canada's Used Nuclear Fuel (Final Study – Summary)*, Nuclear Waste Management Organization, Toronto.

Overton, G. (2013) Will full-body scanners keep you safe and secure? *Laser Focus World*, **49**, 45.

The National Diet of Japan (2012) The Official Report of The Fukushima Nuclear Accident Independent Investigation Commission, The Fukushima Nuclear Accident Independent Investigation Commission, Tokyo.

Townsend, L.W. (2004) Implications of the space radiation environment for human exploration in deep space. Joint Organization of the Tenth International Conference on Radiation Shielding (ICRS-10) and Thirteenth Topical Meeting on Radiation Protection and Shielding (RPS-2004), Funchal, Madeira Island (Portugal), May 9–14.

Townsend, L.W., Shinn, J.L., and Wilson, J.W. (1991) Interplanetary crew exposure estimates for the August 1972 and October 1989 solar particle events. *Radiat. Res.*, **126**, 108.

Tribble, A.C. (2003) *The Space Environment: Implications for Spacecraft Design*, Princeton University Press, Princeton, NJ.

U. S. Environmental Protection Agency (2013) *PAG Manual, Protective Action Guides and Planning Guidance for Radiological Incidents*, U.S. Environmental

Protection Agency Draft for Interim Use and Public Comment, U. S. Environmental Protection Agency, Washington, DC.
U. S. Federal Aviation Administration (2013) Occupational Exposure to Ionizing Radiation for Crews of Suborbital Spacecraft: Questions & Answers. DOT/FAA/AM-13/23, U. S. Federal Aviation Administration, Washington, DC.
U. S. Government Accountability Office (2014) Emergency Preparedness – Opportunities Exist to Strengthen Interagency Assessments and Accountability for Closing Capability Gaps. Report GAO-15-20, U. S. Government Printing Office, Washington, DC.
U. S. Nuclear Regulatory Commission (1974) Termination of Operating Licenses for Nuclear Reactors. Regulatory Guide 1.86, U. S. Government Printing Office, Washington, DC.
U. S. Nuclear Regulatory Commission (1979a) Population Dose and Health Impact of the Accident at the Three Mile Island Nuclear Station. NUREG-0558, U. S. Government Printing Office, Washington, DC.
U. S. Nuclear Regulatory Commission (1979b) Investigation into the March 28, 1979 Three Mile Island Accident by the Office of Inspection and Enforcement. NUREG-0600, U. S. Government Printing Office, Washington, DC.
U. S. Nuclear Regulatory Commission (1979c) NRC Views and Analysis of the Recommendations of the President's Commission on the Accident at Three Mile Island. NUREG-0632, U. S. Government Printing Office, Washington, DC.
U. S. Nuclear Regulatory Commission (1979d) NRC 1979 Annual Report. NUREG-0690, U. S. Government Printing Office, Washington, DC.
U. S. Nuclear Regulatory Commission (1980a) Environmental Assessment of Radiological Effluents from Data Gathering and Maintenance Operation on Three Mile Island Unit 2. NUREG-0681, U. S. Government Printing Office, Washington, DC.
U. S. Nuclear Regulatory Commission (1980b) Criteria for Preparation and Evaluation of Radiological Emergency Response Plans and Preparedness in Support of Nuclear Power Plants. NUREG-0654, FEMA-REP-1, Revision 1, U. S. Government Printing Office, Washington, DC.
U. S. Nuclear Regulatory Commission Rogovin, M. and Frampton, G. T. (1980c) Three Mile Island, A Report to the Commissioners and to the Public, NUREG/CR-1250, Vols. 1–2, U. S. Government Printing Office, Washington, DC.
U. S. Nuclear Regulatory Commission (1981) Final Programmatic Environmental Impact Statement related to decontamination and Disposal of Radioactive Wastes Resulting from March 28, 1979, Accident – Three Mile Island Nuclear Station, Unit 2. NUREG-0683, Vol. 1 and 2, U. S. Government Printing Office, Washington, DC.
U. S. Nuclear Regulatory Commission (1983) Answers to Questions About Updated Estimates of Occupational Radiation Doses at Three Mile Island, Unit 2. NUREG-1060, U. S. Government Printing Office, Washington, DC.
U. S. Nuclear Regulatory Commission (1984) Answers to Frequently Asked Questions About Cleanup Activities at Three Mile Island, Unit 2. NUREG-0732, Rev. 1, U. S. Government Printing Office, Washington, DC.
U. S. Nuclear Regulatory Commission (1987a) Report on the Accident at the Chernobyl Nuclear Power Station. NUREG-1250, Rev. 1, U. S. Government Printing Office, Washington, DC.
U. S. Nuclear Regulatory Commission (1987b) Interpretation of Bioassay Measurements. NUREG/CR-4884, U. S. Government Printing Office, Washington, DC.
U. S. Nuclear Regulatory Commission (1989a) Implications of the Accident at Chernobyl for Safety Regulation of Commercial Nuclear Power Plants in the United States. NUREG-1251, Vol. 1 and 2, U. S. Government Printing Office, Washington, DC.
U. S. Nuclear Regulatory Commission (1989b) The Status of Recommendations of the President's Commission on the Accident at Three Mile Island (A Ten-Year

Review). NUREG-1355, U. S. Government Printing Office, Washington, DC.

U. S. Nuclear Regulatory Commission (1991) Status of Safety Issues at Licensed Power Plants. NUREG-1435, Vol. 1–3, U. S. Government Printing Office, Washington, DC.

U. S. Nuclear Regulatory Commission (1992) Summary of Chernobyl Followup Research Activities. NUREG-1422, U. S. Government Printing Office, Washington, DC.

U. S. Nuclear Regulatory Commission (1994) Lessons learned From the Three Mile Island – Unit 2 Advisory Panel. NUREG/CR-6252, U. S. Government Printing Office, Washington, DC.

U. S. Nuclear Regulatory Commission (2003) Methods and Assumptions for Evaluating Radiological Consequences of Design Basis Accidents at Light-Water Nuclear Power Reactors. Regulatory Guide 1.195, U. S. Government Printing Office, Washington, DC.

U. S. Nuclear Regulatory Commission (2011a) NRC Provides Protective Action Recommendations Based on U.S. Nuclear Regulatory Commission. NRC News No. 11-050, March 16, U.S. Nuclear Regulatory Commission, Washington, DC.

U. S. Nuclear Regulatory Commission (2011b) NRC Bulletin 2011-01: Mitigating Strategies, U.S. Nuclear Regulatory Commission, Washington, DC.

U. S. Nuclear Regulatory Commission (2011c) NRC Exits Monitoring Mode for Japanese Nuclear Emergency; Continues to Support U.S. Government Efforts. NRC News No. 11-083, May 16, U.S. Nuclear Regulatory Commission, Washington, DC.

U.S. Public Health Service (1995) Risk Communication: Working with Individuals and Communities to Weigh the Odds. Prevention Report, February/March 1995, http://odphp.osophs.dhhs.gov/pubs/prevrpt/Archives/95fm1.htm.

Weightman, M. (2011) Japanese Earthquake and Tsunami: Implications for the UK Nuclear Industry – Interim Report, Office for Nuclear Regulation, London.

Wilson, J.W., Simonsen, L.C., Shinn, J.L., Kim, M.-K., Cucinotta, F.A., Badava, F.F., and Atwell, A. (1999) Astronaut Exposures to Ionizing Radiation in a Lightly-Shielded Spacesuit, Paper Number 1999-01-2173, http://papers.sae.org/1999-01-2173.

Woodard, K. (1979) Assessment of Off-Site Radiation Doses from the Three Mile Island Unit 2 Accident. TDR-TMI-116, Pickard, Lowe, and Garrick, Washington, DC.

Part V
Regulatory Issues, Limitations, and Challenges

Government agencies and the nuclear industry are facing significant challenges, which require revised operating strategies and regulatory approaches. These challenges were illustrated by the aftermath of the 11 March 2011, Fukushima Daiichi accident in Japan. The effects on neighboring countries associated with contaminated food were expected, but the nature of the accident, failure of regulators to properly assess natural hazards in the licensing process, and degradation of plant safety systems were unexpected and challenged existing paradigms. In addition, inconsistencies in US and Japanese evacuation recommendations highlighted the consequences of inconsistent national regulations.

Additional challenges result from the increasing global interest in the use of nuclear energy for power generation and its expansion into nations that do not have the technological capabilities of nations having significant nuclear operating experience. This expansion and the growing concern of terrorist attacks increase the threat to the security of nuclear installations and place greater emphasis on the relationships between security, safety, and emergency response capabilities.

Additional issues are associated with existing facilities, including effects of aging equipment and infrastructure, extensions of operating licenses, and reevaluations of design and beyond design basis events. These and other issues place additional stress on an industry that will find little relief from challenges and stakeholder concerns regarding their safe operation and environmental impact.

Health Physics: Radiation-Generating Devices, Characteristics, and Hazards, First Edition. Joseph John Bevelacqua.

7 Regulatory Considerations

7.1 Overview

An ideal regulatory framework is proactive, is accepted internationally, is supported by stakeholders, anticipates accident events, constantly challenges accepted operating practices, and prevents major accidents that result in the release of fission products to the environment. The 1979 Three Mile Island Unit 2 (TMI-2), 1986 Chernobyl Unit 4, and 2011 Fukushima Daiichi Unit 1–4 accidents suggest that the conventional regulatory framework has not been completely successful in achieving these goals. Three major reactor accidents in a span of 32 years offer a sobering reminder that the current regulatory approach has not produced the desired results and that change is warranted.

A number of authors have reviewed the cultural and societal aspects of various nuclear regulation models following the Fukushima Daiichi Nuclear Power Station (FDNPS) accident. Their work focused on regulatory performance with an emphasis on Chinese, Japanese, and US regulations and examined the impacts of the accident on European regulations. Most of this work reviewed the relationship between government and the nuclear industry but did not examine the failure of regulatory agencies to anticipate or preclude specific events. A consensus implies that the US approach is currently the best available regulatory model, but it also requires improvement. Some authors propose that the relationship between the nuclear industry and government involves potential conflicts of interest and the revolving door between industry and government should be eliminated. These authors also suggest that reactor safety be entrusted to an independent agency that has sufficient authority to be an effective regulator and that nuclear safety be handled by international rules without border limitations.

In this chapter, a complementary approach is presented. This approach utilizes the conclusion that the US regulatory system has advantages over other systems. Given this conclusion, the focus is on the Nuclear Regulatory Commission (NRC) that licenses and monitors commercial power reactors in the United States. Historical operational events and occurrences are reviewed to determine systemic weaknesses in the US regulatory system. These weaknesses suggest

Health Physics: Radiation-Generating Devices, Characteristics, and Hazards, First Edition. Joseph John Bevelacqua.

possible regulatory improvements. This review also offers a possible set of elements for international nuclear regulation.

The discussion begins with a brief review of US nuclear operating history and significant events that shaped the direction of nuclear regulation. The current US regulatory approach, its response to the Fukushima Daiichi accident, and proposed enhancements are also presented. Emphasis is placed on the health physics aspects of nuclear regulation including the preservation of fission product barriers and emergency response to reactor accidents.

7.2 Twentieth-Century Regulatory Challenges

A number of events illustrate the challenges faced by the NRC and its predecessor organization, the Atomic Energy Commission (AEC). The AEC had dual responsibility for developing and regulating the US nuclear power industry. These responsibilities were often in conflict, and these contentions were a key factor in stimulating the opposition to nuclear power and stakeholder concerns regarding its development.

The US nuclear regulatory history presented in this chapter is derived primarily from NRC reports. US nuclear regulations were initially governed by the AEC and more recently by the NRC. These organizations controlled the commercial nuclear power regulatory environment in the United States.

An examination of significant events provides a historical perspective regarding the evolution of US nuclear regulation. It also illustrates the strengths and weaknesses in the process and suggests areas where improvements should occur. Following examination of these events, specific recommendations for future regulatory options are provided. These options encompass near-term revisions to US regulations and eventual transition to an international regulatory process that will enhance reactor safety.

Issues of reactor safety regulation have been historically linked to the protection of the three fission product barriers. These barriers are the fuel and associated cladding, the reactor vessel and included piping, and the containment structure. Fission product barriers prevent the release of radioactive material to the environment. Preserving these barriers is an important requirement for a successful regulatory system.

7.2.1 Containment Fission Product Barrier

Prior to the mid-1960s, the AEC considered that the containment building was the final line of defense against the release of fission products. However, it became apparent that under some circumstances the containment building could be breached. Given this possibility, protecting the public from a fission product release relied heavily on a reliable, properly designed, and functional emergency core cooling system (ECCS).

The ECCS includes passive, low-pressure, and active, high-pressure injection systems to ensure core cooling which minimize the release of fission products to the environment. Spray systems, filter banks, ventilation systems, and air circulation units support the ECCS function.

The containment integrity concern prompted the AEC to conduct tests of the ECCS. These tests were not conclusive and their results led to additional questions regarding ECCS effectiveness. The test results cast doubt on previous AEC assertions regarding the adequacy of reactor safety systems and had the potential to undermine public confidence in nuclear power technology. Unfortunately, the AEC attempted to prevent the test results from becoming public and withheld the information from Congress.

7.2.2 Emergency Core Cooling System Contention

As a result of the containment effectiveness issue, the Union of Concerned Scientists (UCS) and some scientists at US national laboratories questioned the reliability of the ECCS. Questions regarding ECCS reliability arose following tests of subscale systems that failed to fully support the design predictions. The regulatory staff concluded that the issues identified in ECCS testing would be corrected in subsequent designs and used engineering judgment to justify system requirements. However, the resolution of this safety issue was not universally accepted.

The ECCS controversy damaged the credibility of the AEC and strengthened its critics. Rather than acknowledging the ECCS issue and fully evaluating the system's uncertainties, the AEC acted in a manner that it hoped would not undermine public confidence in reactor safety. The AEC's actions added credibility to the reliability allegations and created an atmosphere of distrust that would influence the disposition of future reactor safety issues. Had the AEC acknowledged the potential significance of the ECCS tests, devoted additional time to evaluate the associated uncertainties, and involved stakeholders such as the UCS in resolution of the issue, the dispute might have been settled in a manner that would have avoided much of the distrust that developed and continues today.

The ECCS reliability issue continued to generate controversy into the 1970s, and extensive ECCS hearings were conducted in 1972. However, they produced contentious testimony and media reports that reflected negatively on the AEC's safety program, revealed divisions among government experts, and further damaged the agency's credibility.

7.2.3 Lyons High-Level Waste Disposal Facility

Another issue that undermined public confidence in the AEC and strengthened its opponents was the methodology used by the Commission to select a high-level radioactive waste disposal site. In 1970, following Congressional and scientific pressure regarding the need for a high-level waste disposal facility, the AEC decided to develop a permanent repository in an abandoned salt mine near Lyons, Kansas. This decision was made without conducting comprehensive geologic and hydraulic evaluations. These omissions were revealed when the state of Kansas and scientists challenged the appropriateness of the Lyons site. The resulting dispute intensified when Congress and state officials continued their challenges. Many of the issues raised during the Lyons site debate were again raised 40 years later with attempts to license the Yucca Mountain site. These dual failures to license a high-level waste repository indicate weaknesses in the licensing and political processes used to design and support the regulatory framework.

This Lyons dispute ended in 1972, when the concerns expressed by the site opponents proved to be valid. The high-level waste and ECCS issues strengthened the opponents of the AEC and added credibility to groups opposed to the development of nuclear power. These issues as well as concerns over reactor design, reactor safety, quality assurance, and the probability of a major reactor accident continued and amplified the debate over nuclear power and further weakened the credibility of the AEC as an effective regulator of the US nuclear industry. The continuing controversies required the AEC to dedicate resources that would have been more productively utilized in resolving these issues and forging a more cooperative relationship with its detractors. For a second time, the AEC placed safety and reliability in a subordinate role to maintaining public perceptions of the viability of nuclear power production.

7.2.4
Transition from AEC to NRC

As the AEC's credibility was damaged by the Lyons and ECCS disputes, increased criticism was directed at its dual responsibility for developing and regulating the nuclear power industry. The weakening of AEC credibility led Congress in 1974 to divide it into two separate organizations with the NRC assigned responsibility for nuclear regulation. Its other functions were transferred to a separate agency that is now the Department of Energy (DOE). The NRC inherited the AEC's credibility issues, antinuclear stakeholders, and growing public uncertainty regarding the safety of nuclear power.

Two events added to the NRC's challenges during its first few months of existence. These were the Browns Ferry Nuclear Plant Fire in March 1975 and publication of WASH-1400, the NRC Reactor Safety Study, also known as the *Rasmussen Report*.

7.2.5
Browns Ferry Fire

One of the first major issues for the NRC following its creation involved a reactor safety issue associated with a fire at the Tennessee Valley Authority's Browns Ferry Nuclear Plant in Alabama. This fire burned for about 7 h and nearly disabled the safety systems of one of the site's units. The fire was caused by a technician who used a lighted candle to search for air leaks in an area containing electrical cables powering portions of the plant's control room and safety systems. This event further damaged public opinion regarding the ability of the NRC to ensure reactor safety. The event also raised concerns regarding fire protection programs and their adequacy to protect reactor safety systems. In addition, the Browns Ferry event directed attention to common mode failures in which a single failure triggers a sequence of events that damages defense-in-depth (DID) or redundant safety systems.

7.2.6 Reactor Safety Study

The NRC's WASH-1400 Reactor Safety Study was commissioned by the AEC in 1972. The Massachusetts Institute of Technology developed the study's methodology with the assistance of NRC regulatory staff. This study's purpose was to estimate severe accident probabilities, since these estimates had not previously been determined in a rigorous manner.

It is somewhat surprising that a reactor could be licensed without a quantification of the risk that it posed to the public. Failure to consider risk in the licensing process represents a significant omission that was rectified in WASH-1400. However, the WASH-1400 issue is another example of a regulatory weakness.

WASH-1400 used methodologies including fault tree analysis to conclude that the severe accident risks from a nuclear power accident were very small when compared to risks from other events such as aircraft crashes, dam failures, earthquakes, explosions, fires, hurricanes, tornadoes, and toxic chemical spills. Although WASH-1400 was a cutting-edge effort that addressed the complex event sequence of a severe accident, both nuclear proponents and opponents criticized this study. One common objection was that the study failed to include additional sequences that could lead to a severe accident. Opponents asserted that the analysis and supporting data did not support the report's conclusions regarding the relative risks associated with a severe nuclear power accident. Given the controversy, the NRC withdrew its full endorsement of the report's executive summary in 1979.

Some of the critics of WASH-1400 were concerned about low-probability, high-consequence events. One of these events (Event v) involved a small-break loss-of-coolant accident that occurred at TMI-2. However, Event v was not addressed as a likely operational event in WASH-1400.

7.2.7 TMI-2 Accident

In 1979, the TMI-2 pressurized water reactor (PWR) had a small-break loss-of-coolant accident (LOCA) with associated core damage. TMI-2 was caused by a combination of operator errors and design weaknesses.

The TMI-2 accident galvanized nuclear power opponents and intensified public debate and criticism of the NRC. Following mechanical failures and human errors, the TMI-2 reactor core was uncovered and about one-half of its fuel melted. In spite of NRC oversight, a primary system's pressure relief valve failed to close following a reactor trip and created a flow path for the loss of primary coolant. Although the ECCS functioned as designed, operators significantly reduced its flow rate because available instrumentation suggested that the primary system was filling with water. Operators believed this condition presented a danger of overpressuring the reactor coolant system. When the actual condition of low core water level was discovered and ECCS flow restored, core water inventory was reestablished. However, core uncovery and fuel melting had already occurred.

Reviews of the TMI-2 accident questioned the adequacy of NRC oversight of maintenance activities (e.g., pressurizer relief valve failing to close). In addition, the design of the control room did not provide a clear indication of the event (e.g., easily observable indication of the pressure relief valve failure to close, high water level in the reactor building basement, and actual pressurizer water level) which contributed to the plant operator's failure to determine that a LOCA was in progress. Moreover, a similar event previously occurred at the Davis–Besse Power Plant in Ohio, but the NRC failed to comprehensively evaluate this event or disseminate information regarding the event to other nuclear plants. In addition to highlighting these regulatory failures, the TMI-2 accident clearly demonstrated that a serious reactor accident with severe core damage could occur.

The TMI-2 corrective actions initiated by the NRC included greater emphasis on human factors in plant performance for minimizing the type of operator errors that contributed to the accident. These actions strengthened requirements for operator training, testing, and licensing. The NRC also promoted the use of control room simulators and performed assessments of control rooms and their instrumentation. In addition, the NRC's resident inspector program was expanded to include at least two representatives at each plant site. Focus was placed on the review and dissemination of operating data from nuclear power plants. Emergency preparedness programs were expanded and rigorously evaluated with drills and exercises. These actions strengthened the regulatory program and have prevented another major US accident.

In spite of these positive improvements, the TMI-2 accident represented regulatory failures on several levels. First, the event was not identified as a likely severe accident in the WASH-1400 analysis. Second, the licensed control room design and requisite instrumentation did not facilitate operator response or contribute to a timely recognition of the accident conditions. Third, operational experience from other nuclear power plants, which could have prevented the TMI-2 event, was not communicated to the industry. When coupled with the failure to address the high-level waste issue, containment building and ECCS safety issues, fire protection effectiveness, and other reactor safety issues, the AEC/NRC record of success and effectiveness of nuclear regulation at the time of the TMI-2 accident certainly could be questioned.

The TMI-2 accident also led to the creation of industry groups to promote improved operational performance. These groups included national as well as international organizations. Two of the more significant groups were the Institute of Nuclear Power Operations (INPO) and the World Association of Nuclear Operators (WANO).

7.2.8
Salem ATWS Events

Unfortunately, additional reactor events with safety significance continued to occur. These events involved similarities to TMI-2 in that safety-related equipment failed to function and maintenance issues contributed to the problem.

In February 1983, inadequate surveillance and testing of reactor shutdown circuitry at the Salem Nuclear Power Plant in New Jersey led to the failure of a reactor to trip when plant conditions warranted an automatic shutdown. Similar events occurred on two separate occasions. This condition, known as an Anticipated Transient without Scram (ATWS) event, placed the reactor in a condition that was outside its intended design basis since it did not shut down as warranted by plant conditions.

The first ATWS event occurred on February 22. The licensee failed to perform a thorough and systematic review of the February 22 event. The post trip review was inadequate because plant management did not aggressively investigate the causes of the event. In addition, there was a lack of questioning attitude, diligence, and attention to detail in the response to the reactor trip. In spite of the safety significance of the ATWS event, the NRC failed to investigate the failure and prevent plant start-up.

A second ATWS event occurred on February 25. After consultation with the NRC, the licensee agreed to defer plant start-up until a more comprehensive review of the event could be conducted.

The Salem ATWS events are important because maintenance and testing did not reveal a significant safety issue associated with the reactor trip circuitry. In addition, the INPO previously identified a deficiency in Salem's preventive maintenance program, but the NRC failed to investigate this issue in a comprehensive manner. The NRC systematic appraisal of licensee performance for the period September 1981 to August 1982 did not identify these maintenance problems. However, an NRC Resident Inspector Report in January 1983 noted the need for the licensee to develop a formal preventive maintenance program for reactor trip breakers. Both the NRC and the INPO identified a significant deficiency that went uncorrected and contributed to the Salem ATWS events. The Salem events provide another example where the NRC's regulatory approach failed to detect a significant safety issue or direct the development of timely corrective actions.

It is the author's view that both national organizations (e.g., INPO) and international groups including the International Atomic Energy Agency (IAEA) and WANO have a role in improving nuclear regulation. This role should be apparent in national as well as fully integrated international regulatory structures and will be addressed in the subsequent discussion.

7.2.9 Chernobyl

Chernobyl Unit 4 was an RBMK design that utilized a graphite-moderated core. Operator errors, an inadequately evaluated test procedure, and an unforgiving reactor design led to the 1986 accident at Chernobyl Unit 4. These factors contributed to a power excursion that resulted in violent reactor disassembly and severe fuel damage. The event culminated in the ejection of a portion of the core and graphite moderator from the reactor pressure vessel and released fission products directly into the environment. Chernobyl-4 released more fission products than the TMI-2 and Fukushima Daiichi accidents and was exacerbated because no containment fission product barrier was included in the RBMK design.

Since the RBMK design was considerably different than PWR and BWR (boiling water reactor) designs, the NRC emphasized that a Chernobyl-type accident could not occur in commercial US plants. The NRC also noted that US reactors have redundant safety systems and a containment fission product barrier that mitigate the release of fission products into the environment.

Nuclear critics used the Chernobyl-4 accident as a prime example of the hazards associated with nuclear power and the need for a more demanding regulatory approach. The Chernobyl-4 accident was another setback for nuclear power advocates and their desire to garner public support. Chernobyl's environmental impact provided sobering evidence that a major accident could occur, have severe environmental impacts, and lead to evacuated areas that would remain restricted for an extended time.

The Chernobyl accident emphasized the need for management oversight of reactor operations that occur infrequently. US reactors placed an increased emphasis on monitoring infrequently performed tests and evolutions. These infrequent operations required additional oversight by senior plant management. In addition, any proposed evolution was rigorously evaluated for its impact on plant safety. Thorough briefings were performed to ensure all personnel were aware of the nature, limitations, and restrictions of the infrequently performed evolution.

7.2.10 Towers Perrin Report

In 1997, the Towers Perrin consulting firm prepared a report for the Nuclear Energy Institute that expressed concern that NRC policies and practices distracted plant management, undermined public trust, and increased operating costs. The report noted that the NRC did not make a significant effort to distinguish safety from nonsafety issues and appropriately prioritize these items. It also claimed that the NRC's actions resulted in a diversion and dilution of licensee resources from the most important safety issues.

The Towers Perrin report illustrates the importance of ensuring that regulators focus on safety significant issues. To achieve this focus, a tool for evaluating hazards in terms of their safety significance is needed.

The report's conclusions were consistent with growing interest within the NRC and the nuclear industry to utilize probabilistic risk assessments (PRAs). PRAs were viewed as a more effective method to assess hazards and to prioritize resources to more effectively address and eliminate their effects. The PRA approach contrasts with the conventional NRC deterministic analysis methodology and the DID approach which was instrumental in preventing a significant radiological release during the TMI-2 accident. However, the NRC considered that the PRA approach was secondary to the DID philosophy, and it was used primarily to identify overly conservative regulatory requirements.

During its assessment of the PRA methodology, the NRC adopted a Maintenance Rule that required strong maintenance programs at commercial nuclear power plants. This rule was a positive safety development in view of previous operational events that included maintenance issues associated with the failure of reactor trip circuitry at Salem and failure of a pressure relief valve to close during the TMI-2 accident.

Although risk-informed regulation offered potential benefits, it was not designed to detect the wide spectrum of safety issues that could occur at an operating nuclear power facility. This situation was demonstrated when a series of problems occurred at the Millstone Power Station in Connecticut. The safety issues at Millstone warranted attention, but risk analysis would not necessarily identify them as priority safety issues.

7.2.11
Millstone Safety Allegations

Allegations arose in the early 1990s when several Millstone plant employees claimed they were punished for raising safety issues. The NRC investigated the employee concerns but determined these issues did not have major safety significance and were addressed by the licensee. Although a \$100 000 fine was imposed on the licensee, this action did not satisfy the critics of the NRC.

Media scrutiny intensified when new Millstone allegations were revealed. In 1993 and 1994, the NRC levied additional fines for procedural violations that were viewed as serious management issues. Millstone employees raised another issue related to outage practices involving offloading the reactor core to the spent fuel pool (SFP). This practice was in violation of NRC requirements that precluded a complete core offload for plants of the Millstone type.

The fuel offload issue involved specific plant safety requirements and the ability of the NRC to enforce those requirements. The continuing Millstone controversy was addressed in a 1996 NRC Inspector General (IG) Report that faulted the regulator for failing to recognize and impose corrective actions. The numerous Millstone issues illustrated the difficulty that the NRC had with plants that did not perform at the level required by agency standards and in correcting the associated issues in an effective and timely manner. These issues also demonstrated that once a problem was identified, it would eventually be corrected. However, the regulator had not yet mastered the ability to develop regulations, inspection practices, and management controls that would anticipate problem areas or implement timely corrective actions.

7.3
Twenty-First-Century Regulatory Challenges

The regulatory issues that developed and matured in the twentieth century continued into the twenty-first century. Additional technical and regulatory issues emerged and another severe accident occurred. The first significant event of twenty-first century focused on terrorism and the potential for a terrorist event to be directed at a nuclear power reactor.

7.3.1
11 September 2001 Attacks

The inability of the NRC to anticipate safety issues was again illustrated by the 11 September 2001 terrorist attacks on the World Trade Center in New York and the Pentagon near Washington, DC. These attacks raised two additional safety issues, which the existing licensing basis of some nuclear plants had not fully addressed.

The first issue involved the nuclear power facility's design basis as related to the effects of an aircraft impact on the integrity of the three fission product barriers. A second issue was the plant's vulnerability to a terrorist attack resulting in a release of fission products to the environment. Accordingly, the NRC ordered a series of security measures, and again Congress challenged their rigor and effectiveness. More importantly, the original design basis for operating reactors did not specifically address these challenges or ensure the full spectrum of natural and man-made events were evaluated before a plant license was issued.

The ramifications of the September 11 attacks added to the continuing theme regarding the adequacy of the facility design basis. In particular, the adequacy of the containment building and SFP designs to withstand the impact of a contemporary commercial aircraft was questioned. In September 2004, the NRC reported that an aircraft strike at a nuclear power plant could cause a radioactive material release. In addition, a 2005 National Academy of Science report concluded that a successful terrorist attack would be difficult to achieve but is a credible threat. This report argued that there was no regulatory requirement to protect the facility from this type of hazard. Once again, the NRC was involved in a design basis controversy.

7.3.2 Davis–Besse Reactor Vessel Head Erosion

During the aircraft attack controversy, a serious operational issue arose at the Davis–Besse plant, which is a PWR. In 2002, an inspection of the upper reactor vessel head discovered significant material degradation, which created an American football-sized cavity. This degradation was caused by borated water that leaked onto the reactor vessel head through cracks in a control rod drive mechanism nozzle and the weld that attached the nozzle to the reactor pressure vessel head. The erosion of the reactor vessel head structural material involved about 32 kg of steel, which only left the thin (about 1 cm) stainless steel cladding intact as the only pressure boundary preventing a LOCA.

Both the utility and NRC failed to identify the issue and take timely action to correct the conditions that initiated and continued the erosion process. The NRC's failure is also of concern since the corrosion issue was related to a previous NRC inquiry regarding the cracking of control rod drive mechanism nozzles.

In August 2001, the NRC instructed PWR owners to inspect these nozzles by December 2001. However, the inspection date could be delayed if the NRC staff judged the specific plant's risks were acceptably small. The operators of Davis–Besse requested a delay in the inspection date until a scheduled first quarter 2002 outage. The NRC staff approved the request and determined that the plant could be safely operated until that date.

The discovery of the significant reactor vessel head erosion suggested that the NRC was in error in granting an extension to perform the requisite inspection after the December 2001 due date. The NRC IG responded to a UCS charge that the NRC failed to adequately regulate the Davis–Besse plant and that a LOCA could have resulted from failure of the reactor pressure boundary. The IG strongly criticized the NRC's safety performance and found that the agency had considered the financial impact to the licensee rather than making public health and safety its highest priority. Although the NRC disputed the IG's safety conclusion, it did conclude that a break in the cladding could have led to a LOCA and that the corrosion of the reactor vessel head was an enormous failure of both the NRC and operating utility. However, the NRC denied that the cladding failure would have led to a massive release of radioactive material to the environment. DID was emphasized as an effective means to prevent the release of fission products to the environment.

The Davis–Besse event is troubling from a regulatory perspective because many indications of the reactor vessel head degradation were present but were not recognized by operating utility and regulatory personnel. These indications include radiation monitoring system filter systems being clogged by boric acid and corrosion particles, the buildup of boric acid deposits on containment air cooler fins, and boric acid deposits on the reactor vessel head. This event was not prevented because the NRC, plant personnel, and industry groups failed to adequately review and analyze relevant operating experience; plant personnel failed to ensure that safety issues received proper attention; and the NRC failed to include known or available facility information into its assessments of Davis–Besse performance.

Once again, the NRC provided sound corrective actions after the event. However, it failed to be proactive and preclude another major regulatory failure in spite of multiple indications that a significant corrosion issue existed.

7.3.3 Yucca Mountain High-Level Waste Repository

The Yucca Mountain High-Level Waste Repository Site is approximately 100 miles northwest of Las Vegas, Nevada. This site is intended to store high-level waste, including spent fuel from commercial power reactors, in an underground facility that has stable geologic characteristics. The NRC has the responsibility to license the facility, and the site was selected and designed by the DOE.

Selection of the Yucca Mountain site by the DOE was reminiscent of the previous Lyons site selection process since significant opposition was expressed by the host state, and litigation followed site selection. In spite of this opposition, the NRC received an application from the DOE in 2008 for a license to construct and operate the first US geologic repository for high-level nuclear waste at Yucca Mountain. This submittal was a significant milestone, because it transferred focus from DOE's efforts to select a repository site to the NRC's review of the repository design to determine its suitability as a high-level nuclear waste storage facility.

The NRC's regulatory process involves technical design reviews and hearings that are conducted concurrently. Technical licensing reviews assess the merits of the repository design. Adjudicatory proceedings assess challenges by the public and other stakeholders regarding the technical and legal aspects of the DOE license application. Based on the results of the licensing review and the hearings, the Commission determines the appropriateness to authorize construction of the Yucca Mountain repository.

In 2011, the NRC regulatory process was interrupted by a variety of factors that included similarities to the previously raised Lyons site objections. Licensing actions resumed in 2014, but funding issues and further litigation have the potential to disrupt the Yucca Mountain effort. It appears that the DOE and NRC have repeated the errors of the AEC in its attempt to license the Yucca Mountain repository.

The examples of this section summarized events that reveal weaknesses in the US regulatory process. In the next section, these weaknesses are reviewed within the context of the Fukushima Daiichi accident.

7.3.4 Fukushima Daiichi Accident

In March 2011, the FDNPS in Japan, consisting of six BWRs, was struck by a significant seismic event and subsequent tsunami. These events culminated in severe core damage in three reactors.

The Fukushima Daiichi accident occurred when the facility encountered an earthquake and resulting tsunami that exceeded the design basis assumptions.

As a result of this accident sequence, the DID safety systems failed to provide the intended margin of safety. Their failure rapidly led to the loss of all fission product barriers with releases of radioactive materials from multiple units into the environment. From a regulatory perspective, failures of both the design basis foundation and DID philosophy suggest that a reevaluation of the basis for licensing nuclear plants is warranted. A review is also warranted because the recommended corrective actions for US plants that were derived from analyses of the Fukushima Daiichi event continue to utilize the DID philosophy to preclude major events.

A review of design basis assumptions is also warranted because additional natural events have occurred outside conventional weather patterns. In October 2011, Hurricane Sandy struck the northeast coast of the United States and led to significant flooding of New York City and the coasts of New Jersey and New York. The storm surge exceeded the assumed maximum flood levels, disrupted power and transportation systems, and destroyed hundreds of homes. The severity of Hurricane Sandy and the failure of government agencies to anticipate and control the resulting storm surge, when combined with the failure to adequately plan for the Fukushima Daiichi earthquake and subsequent tsunami, suggest that the ability of regulators to establish design requirements to anticipate extreme natural events is questionable and requires significant improvement.

There were a number of national and international reviews of the Fukushima Daiichi accident. Although each has a unique and valuable perspective, this chapter focuses on the NRC review and review by the Japanese government. A brief review of industry initiatives is also provided.

7.3.4.1
NRC Review

The NRC's review of the Fukushima Daiichi accident is important because its reactor and Mark I BWR containment designs are also utilized in the United States. This review focused on DID actions, training, procedures, and programs. In its accident assessment, the NRC continued to follow its basic regulatory philosophy and did not introduce any new regulatory approaches.

Following its review of the FDNPS event, the NRC recommended that licensees take a number of actions including:

1) *Design Basis Seismic and Flooding Systems, Structures, and Components (SSCs)*: Seismic and flood protection SSCs are to be reevaluated and upgraded as necessary.
2) *Station Blackout (SBO) Mitigation Capability*: The capability of SBO systems to mitigate design basis and beyond design basis events needs to be reevaluated and strengthened as necessary. Emergency preparedness programs and equipment must be capable of addressing multiunit and prolonged SBO situations.
3) *Mark I and Mark II BWR Containments*: A reliable hardened vent must be provided for these reactor types.

4) *Spent Fuel Pools (SFPs)*: An installed seismically qualified means to spray water into the SFP must be provided. The enhanced SFP water addition capability must include associated instrumentation and safety-related power. Safety-related instrumentation must be capable of withstanding design basis natural phenomena to monitor spent fuel parameters including water level, temperature, and radiological conditions.
5) *On-site Emergency Response Capabilities*: Emergency operating procedures, severe accident management guidelines, and emergency damage mitigation guidelines must be strengthened and integrated. More realistic training and exercises must be provided for all staff expected to implement these guidelines during an emergency.

These actions again reflect the NRC's history of providing credible corrective actions after an event occurs. In an ideal regulatory approach, these recommendations should have been included in the original licensing basis of nuclear power plants. Consistently failing to have an inclusive design basis and reacting to events by only issuing corrective actions following the event is not a successful, long-term regulatory approach. It further suggests that a change in regulatory approach is warranted and that previous assumptions and practices require significant revision.

7.3.4.2
Japanese Diet Commission Review

The results of the Japanese Diet Commission review of the Fukushima Daiichi accident provides a somewhat different perspective and focuses on the regulatory process. This is in contrast to the NRC recommendations that focused on plant systems, programs, and staffing. The major conclusions of the Japanese Diet Commission review include:

1) The accident's root causes were the organizational and regulatory systems that supported faulty rationales for decisions and actions.
2) The operating utility was too quick to cite the tsunami as the cause of the nuclear accident and deny that the earthquake caused any damage.
3) Organizational problems within the utility (e.g., level of knowledge, training, and equipment inspection) limited accident response.
4) Emergency response issues existed because roles and responsibilities were not well defined.
5) Regulators failed to implement adequate evacuation plans, and an inadequate crisis management system contributed to public confusion during the evacuation.
6) The government and regulators are not fully committed to protecting public health, safety, and welfare of the evacuees.
7) The safety of nuclear energy in Japan cannot be assured unless the regulatory process is changed by eliminating its insular attitude of ignoring international safety standards.

8) The operating utility did not fulfill its responsibilities as a private corporation, and its relationship with the regulators was used to weaken proposed safety regulations.
9) The latest technological findings from international sources should be reflected in existing nuclear energy laws and regulations.
10) Root causes must be addressed and preventive measures implemented to preclude future accidents.

The Diet Commission conclusions are consistent with the analysis of other reports and publications. It also has a number of regulatory items that are appropriate for consideration of future US and international regulations. In particular, recommendations 1, 7, and 9 will be addressed in the subsequent discussion.

7.3.4.3 FLEX Strategy

The US nuclear industry responded to the Fukushima Daiichi event by endorsing the Nuclear Energy Institute's FLEX strategy. This strategy is based on enhancing DID systems against power loss and subsequent failure of core and SFP cooling systems. The FLEX concept places backup safety equipment (e.g., air compressors, battery chargers, battery packs, electrical generators, and pumps) at each power reactor site and at satellite locations to create additional DID capability. Since communications systems were disrupted during the Fukushima Daiichi accident, the FLEX strategy also enhances communications capability. These communications systems include satellite phones that operate during severe natural events that could disrupt conventional methods of communications.

The backup power systems emphasize portability to provide maximum flexibility in supporting safety systems. These systems have variable size, typically between 0.15 and 3.0 MW, to meet the individual backup system power requirements.

The FLEX strategy also includes additional equipment to monitor SFPs to ensure that safe temperatures and water levels are maintained. These additional systems require that personnel be trained in their use and maintenance.

The FLEX strategy acknowledges that older Generation II reactors have safety vulnerabilities and backup systems that are not as robust as required to mitigate credible natural events. NEI's FLEX strategy provides a near-term solution to minimize these inherent vulnerabilities. The subsequent discussion provides an alternative regulatory philosophy that offers an avenue to enhance reactor safety beyond that offered by the FLEX approach.

7.3.5 Waste Disposal

In August 2012, the NRC stopped issuing operating licenses until it addressed issues associated with nuclear waste policy that were raised by a federal appeals court ruling. In June 2012, the court ruled that the NRC's approach to managing nuclear waste was inconsistent with federal environmental standards.

Until this ruling, the NRC based its waste management approach on its Waste Confidence Decision (WCD) when it issued licenses for proposed plants or extended the licenses of existing plants. Under the WCD approach, the NRC asserted that it could issue licenses because it had confidence that a permanent fuel repository would be licensed. However, termination of Yucca Mountain funding damaged the credibility of the WCD approach.

In addition to striking down the WCD, the court also rejected the NRC's assertion that spent fuel could be stored in a facility pool and dry casks for up to 60 years beyond a plant's licensed life. The court ruled that the NRC must assess accidents associated with these fuel storage configurations. This ruling further complicated the storage of high-level waste in the United States and was an additional obstacle for licensing new US reactors.

The WCD again raised issues associated with the adequacy of the NRC licensing basis assumptions and associated analyses. Although this issue was resolved in 2014, failure to comprehensively analyze spent fuel events is another example of the NRC's failure to perform its licensing responsibilities in a complete and comprehensive manner.

7.4
Proactive Vice Reactive Philosophy

Reacting to nuclear events and proposing regulations to prevent these events from recurring is an approach that has not been successful. The three major accidents that have occurred must be the last events if the nuclear industry is to prosper and gain sustained public acceptance. Reactor safety is a requirement for the sustainable development and deployment of nuclear power plants. To ensure safety, the industry needs to improve performance. The open question is to determine the best approach to improve nuclear safety and security and to optimize the regulatory framework to enhance these goals.

What are the characteristics of a regulatory organization that enhances reactor safety? This organization should be proactive and constantly challenge assumptions that were previously accepted or assumed. For example, challenges and subsequent changes in the design basis earthquake and tsunami assumptions at the Fukushima Daiichi facility could have mitigated the events that led to the March 2011 accident.

The regulator should be independent of the nuclear industry and political and economic considerations. Regulators must establish and maintain effective communication with stakeholders and have the confidence and respect of the public and the operating utility. Credibility is essential and the regulatory staff must have the requisite skills and experience to understand the technology being regulated. The regulator must remove the perception that it functions as a traffic officer writing citations for minor violations to justify its existence and focus on major safety issues. Inspections must go beyond matching plant programs with regulatory requirements. Plant performance, equipment material condition, staff expertise and capability, management focus, safety system performance, and validity of design basis assumptions must be constantly evaluated with a perspective beyond the traditional audit mentality.

International consistency is required to avoid public confusion regarding the credibility of the approach used to monitor and regulate reactor operations and respond to emergency events. The confusion generated by conflicting Fukushima Daiichi accident evacuation distance decisions by US and Japanese officials did not foster credibility and public confidence in the regulatory process. In addition, regulatory performance should receive the same level of scrutiny and review as the industry's operational record.

7.5 Accident Analysis and Risk Assessment

Risk analysis identifies the major accident contributors and their possible impacts on event sequences and their underlying causes. An effective risk analysis focuses on the major contributors to a specific outcome (e.g., core damage). These contributors include accident sequence assumptions, assumed design basis event parameters including loss of power duration and earthquake magnitude, component degradation, human errors, malfunctions, mechanical failures, and system failures.

Risk assessments have inherent uncertainties. If the uncertainties are reasonably low, risk assessment studies place the facility risk (e.g., core damage frequency with a subsequent release of radioactive material to the environment) into perspective with respect to other well-defined risks (e.g., natural disasters and industrial accidents). If the uncertainties are large, the analysis improves understanding of the interaction of system components and insight into the systems that have the most safety significance. Even if these risk assessment conclusions are based on design uncertainties, the results guide improvements that enhance safety including the addition of safety system pumps or backup power systems to enhance core cooling capability.

7.5.1 Design Basis Accidents

The NRC defines a design basis accident as a postulated event that a nuclear facility must be designed and built to withstand without loss of the SSCs necessary to ensure public health and safety. Details of these accidents were enumerated in Chapter 2, and this chapter reviews regulatory issues associated with their definition and characterization.

The definition of a design basis accident requires an assessment of licensing basis assumptions that were formulated to develop credible accidents. This includes an assessment of natural events (e.g., earthquakes, floods, hurricanes, tsunami, and precipitation) and man-made events (e.g., transportation accidents and security events). Determining the appropriate design basis assumptions is crucial. If the assumptions are not bounding, the design basis accidents will not be sufficient to ensure the fission product barriers are protected and releases of radioactive material are minimized.

The consequences of selecting appropriately limiting design basis assumptions were illustrated by the Fukushima Daiichi accident that failed to account for a severe Richter magnitude 9 earthquake. The design basis earthquake underestimate led to a resulting design basis tsunami that was not bounding. These decisions generated a facility design that was inadequate to mitigate the 11 March 2011 earthquake/tsunami. Although the extent of direct earthquake damage to the Fukushima Daiichi facility is uncertain, it was sufficient to disrupt off-site power. The loss of off-site power was initially mitigated by actuation of the facility emergency diesel generators, but these systems were disabled by flooding that occurred when the tsunami struck the Fukushima Daiichi facility. The lifetime of the station batteries was insufficient to provide sufficient power to core cooling water systems to prevent core damage.

The Fukushima Daiichi accident clearly emphasized the risk of understating design basis assumptions. In this case, design deficiencies included tsunami protection against flooding, the location of emergency power supplies to eliminate their flooding potential, and capacity of station batteries to prevent core damage during the loss of off-site power and on-site emergency generators.

Underestimating design basis assumptions is a significant regulatory failure because it undermines confidence in the process used to define plant requirements. The US regulatory response to the Fukushima Daiichi accident requires that licensees evaluate their design basis assumptions in terms of the associated accident sequence and identify any weaknesses.

These weaknesses are identified by design reviews concerning the facility's vulnerability to seismic and flooding events. The design basis assumptions regarding flooding and seismic events and their original basis must be challenged and verified to be adequate. This requires reanalysis of historical events and their impact on the facility and its capability to protect the three fission product barriers. The obvious question is why the NRC did not satisfactorily address these seismic and flooding issues before US plants were initially licensed.

7.5.2
Beyond Design Basis Accidents

The NRC defines a beyond design basis accident as an event sequence that is possible but not fully considered in the design process. These events were judged by the regulator to be too unlikely or beyond the scope of design basis accidents. Within the US regulatory approach, beyond design basis accidents are analyzed to understand the capability of a reactor design. Specific beyond design basis accidents are addressed in Chapter 2.

From a regulatory perspective, these events are judged in the licensing process to not pose a credible threat to the public. As demonstrated by the Fukushima

Daiichi accident, underestimating design basis events led to an underestimate of the design basis earthquake and resultant tsunami. The 11 March 2011 earthquake and subsequent tsunami clearly illustrated an important weakness in the regulatory process that failed to consider a natural event that had a historical basis.

Another emerging aspect of beyond design basis accidents is their assumed initiator and its frequency and magnitude. These are inherent assumptions that must be considered in evaluating a reactor design. For example, accidents are often predicated based on 100- or 500- year events, which suggest these event patterns are reasonably well predicted by a normal sequence of events. However, recent research suggests that there is a new normal, because historical event patterns could have been altered.

For example, the 2012 Hurricane Sandy storm surge in New York and the surrounding area was identified as a once-every-500-year event. This frequency is based on the current climate and its historical variation. However, the real risks may be higher if climate models are utilized in the projection. In addition, the 1-in-500-year estimate did not fully account for the unique nature of Sandy. Sandy was a combination of a tropical cyclone and a severe winter snowstorm. It was a hybrid storm and recent literature suggests assessing the risk from a hybrid storm needs to be improved.

These natural events and their analysis are also complicated because climate change advocates suggest that 100- and 500-year events will have a more frequent return. For example, historical 100-year events could occur every 3–20 years and historical 500-year events could return every 25–240 years. In addition, the severity of the events could increase. These results taken with the Fukushima Daiichi earthquake/tsunami and Hurricane Sandy storm surge suggest that beyond design basis events require renewed scrutiny as do their underlying assumptions.

7.6 Licensing Process and Technical Basis

As noted previously, the licensing process evaluates the technical merits of the power reactor application as well as stakeholder and intervenor concerns associated with the proposed facility. Resolving the concerns of stakeholder and intervenor groups has been a long-standing issue in the development of public support for nuclear power. Possible approaches to strengthening the regulatory process and effectively addressing stakeholder concerns are proposed in the subsequent discussion.

7.6.1 Stakeholder Involvement

The US regulatory approach attempts to foster an open relationship with stakeholders. Historical evidence summarized previously suggests that the current approach has not been completely successful. In the United States, issues arose during resolution of stakeholder concerns regarding the containment building and ECCS adequacy and additional concerns have been raised in the licensing of new plants. In view of the ramifications of the Fukushima Daiichi accident, stakeholder issues have increased and assumed a renewed focus and urgency. As noted previously, the satisfactory resolution of stakeholder issues remains a major impediment to the advancement and acceptance of nuclear power. Approaches to improve stakeholder involvement and the resolution of associated issues are addressed in the subsequent discussion.

7.7 National and International Standards

National nuclear safety standards are defined by individual sovereign states to govern the design, construction, and operation of nuclear power plants within their borders. As such, they meet national needs and incorporate cultural and technological values unique to that state. National regulators use individual standards to evaluate performance of their reactors. There is international interest in assessing the feasibility of aligning these individual national standards.

It has been argued that when the essential regulatory elements are fully aligned and harmonization of safety standards is achieved, overall operating performance, design consistency, and safety improvements will be realized. There are numerous arguments to support or negate this international harmonization proposition.

Operating experience suggests that a standard reactor design improves compliance with regulatory requirements and safety performance. What happens if standard designs are utilized in several countries and these nations combine their regulations into a single set of harmonized requirements? If differing national safety standards are harmonized, a reactor concept could enter into the licensing process without major changes and become an internationally standardized design. This argument suggests that harmonization of safety requirements leads to standardization of reactor designs. However, the argument could be challenged by examining the less difficult issue of harmonizing radiation protection regulations.

Many national regulatory agencies follow the recommendations of the International Commission on Radiological Protection (ICRP), but a number of countries do not have harmonized regulations. In addition, radiation protection regulations are not standardized among the various US government agencies. For example, the Occupational Safety and Health Administration follows the 1959 recommendations of ICRP 2. NRC licensees utilize the 1977 recommendations of ICRP 26. The DOE adopts ICRP 26 for its dose limits but uses the 1991 recommendations of ICRP 60 for its radiation protection infrastructure. Most of the world currently follows ICRP 60. The most recent recommendations are published in ICRP 103 and its implementation will proceed in most of the world following the publication of supporting reports.

How will the world's reactor licensing regulations become harmonized when radiation protection regulations are not yet unified either in the United States or internationally? The harmonization of nuclear reactor licensing will be a significant and difficult undertaking. Harmonization must not degrade safety or reduce standards in countries with a high technological base when the standards also apply to less developed nations.

In order to illustrate the benefits of harmonization, a brief summary of possible regulatory enhancements is presented. The World Nuclear Association (WNA) and the IAEA have proposed arguments similar to the subsequent discussion.

7.7.1 Benefits of Standardization for Nuclear Safety

The standardization of reactor designs should lead to higher levels of safety and performance. Improved reactor safety performance is based on the capability to utilize design and operating experience during plant construction, commissioning, operation, and decommissioning. The WNA argues that collective experience and reliability databases provide the underpinning for enhanced safety.

During the design phase, new plants incorporate current technology and lessons learned from the current operating fleet. In the construction phase, subsequent plants benefit from accumulated construction experience. During operations, a global fleet of standardized nuclear plants offers a common operating and maintenance experience base that should improve capacity factors and overall safety performance and provide a basis for continuous improvements. Improvements in maintenance and enhancing safety system operability and reliability are logical endpoints of the WNA proposals.

Standardization also has the potential for detrimental effects on safety. If there were a limited number of standardized designs, an unknown defect in a particular design would affect all reactors of that type. However, utilizing a large number of reactors of a given design could increase the likelihood of discovering a flaw if operating and maintenance experience were available to all fleet members and regulatory agencies.

The WNA argues that design changes and modifications could be organized and implemented in an efficient manner. This assumes that utilities, reactor vendors, and regulators cooperate based on internationally harmonized regulations, voluntary initiatives, and reporting requirements. For example, the civil aviation industry issues airworthiness directives that are utilized by affected countries to correct design or operating issues.

The author questions this assumption given the likelihood of litigation to determine the assignment of fault and compensation for lost power production. In addition, reactors are more complex than aircraft and require greater time for issue resolution than other industries. The effect of public involvement and antinuclear intervention also poses a risk to continued operation if a generic design issue is discovered.

7.7.2
Benefits of Standardization for Regulators

The WNA argues that the harmonization of national standards would facilitate increased international regulatory cooperation. By sharing safety evaluations and supporting methodology, regulatory reviews of reactor designs could be improved. The transfer of regulatory methodologies could facilitate the development of nuclear energy in emerging nuclear countries. Effective collaboration is enhanced if rules and standards are harmonized internationally.

Collaboration based on harmonized safety requirements would improve quality inspections in construction and component manufacturing. Given the diversity of contractors and subcontractors, regulatory collaboration could enhance manufacturing oversight. The WNA assumption is contingent on maintaining quality, unbiased inspections. This will be difficult to achieve within the framework of international political considerations. The difficulty the United Nations encounters in issue resolution and oversight suggests that although international regulation has certain advantages, it will be difficult to achieve. In addition, the likelihood of maintaining a high quality level is an open issue.

Issues associated with standardization have also arisen in the United States in implementing its combined construction and operating license. The new generation of US reactor construction is based on the concept of standardization. Under this principle, the reactor type used at one site is essentially a replica of the same model used at another site. This approach permits nuclear safety issues to be resolved in advance with no need to address these issues in licensing procedures for individual reactor plants. However, the new construction process is also being affected by the amendment process, with individual licensees requesting numerous modifications of the certified design to suit their own reactor plans. Whether the amendment process undermines the standardization goal will be initially determined by litigation that has been the arbiter of last resort in US nuclear licensing. Ultimately, the long-term safety performance of the standard designs determines their success and viability.

The regulatory process is also perturbed by major international events including terrorist attacks, earthquakes, floods, storms, and major reactor accidents. Although the effects of the Fukushima Daiichi accident are still unfolding, the aforementioned NRC recommendations have perturbed the regulatory process in the United States and other nations. As noted by the Fukushima Daiichi regulatory response, various nations have differing views regarding the best approach to address the associated issues. Resolving issues that challenge harmonized regulatory processes will be a difficult and political venture with nations promoting a solution that is beneficial to its interests. Issue resolution will also be time consuming and costly which presents challenges to maintaining a harmonized approach if it is achieved.

7.7.3
Benefits for the Nuclear Industry

Standardized designs tend to reduce engineering and construction costs. In addition, standardization reduces risk associated with the licensing process, limits construction issues, and enhances cost predictability for new nuclear plants. In principle, a vendor could market a reactor without the need for design changes, unless justified by site-specific circumstances. Utilities would gain in the ability to choose a design without major challenges.

These arguments are credible in a world without politics and national interests. However, they fail to recognize current reality and the opposition of governments to nuclear development when it is perceived to be associated with the desire to acquire nuclear weapons or associated technology such as fuel reprocessing. The WNA arguments must be refined to accommodate export controls, licensing restrictions, and proliferation concerns before its goals can become reality.

7.7.4
Future Directions

Although costly in terms of time and effort, international regulatory relationships have merit and should be explored. There are a number of relatively simple steps to build these relationships and develop a means for further development. For example, regulatory agencies, utilities, vendors, and manufacturers should encourage staff exchange to share knowledge and experience. Building a harmonized set of regulations and standards will not be easily accomplished, but many of the arguments of the IAEA and WNA have merit.

International nuclear safety regulation must recognize both technical and political aspects. Nuclear activities and practices, including standard setting, licensing, inspection, and enforcement, require that nuclear safety and security issues receive attention at the highest political level.

As an initial focus of international harmonization, radiation protection regulations should be harmonized and based on the most recent guidance of the ICRP. The lessons learned in facilitating a transition to ICRP-based radiological regulations would provide guidance to address the more complex issue of harmonizing nuclear reactor regulations, standards, and licensing.

7.8
Accidents Affecting Multiple Nations

Major nuclear accidents (i.e., INES (International Nuclear and Radiological Event Scale) Level 7 events) such as Chernobyl-4 and Fukushima Daiichi that involve a significant release of radioactive material can affect the host as well as neighboring nations. Chernobyl-4 and Fukushima Daiichi not only deposited radioactive material outside their national borders but also affected international commerce. Sales of food from the vicinity of these reactors were necessarily restricted because they were contaminated with fission products.

The radiological effects of the three major reactor accidents on the environment are outlined in the subsequent discussion. The accident sequences were described in Chapter 3. This chapter focuses on the impact of these accidents on the environment and regulatory systems.

7.8.1
TMI-2

The TMI-2 accident had a very limited impact on the local environment and minimal impact on any area outside the immediate reactor location. This environmental effect is assessed by examining the radionuclides released from the TMI-2 accident. The TMI-2 releases to the environment are summarized in Table 7.1.

During the TMI-2 accident, the release pathway included transfer of radioactive material from the reactor coolant to the containment building, transfer from containment to the auxiliary building, and release to the environment through the auxiliary building waste gas system.

Table 7.2 provides a summary of a portion of the core inventory released during the TMI-2 and Chernobyl events. Minimal amounts of Cs, Te, and other particulate fission products were released from TMI-2. As a comparison, less than 25% of the available particulate inventory was released from the Chernobyl-4 accident.

A number of public effective dose assessments were performed to determine the radiological impact of the TMI-2 accident. There is general agreement that the effective dose was primarily derived from noble gases. Noble gases also contributed to the skin dose. Skin doses, excluding the shielding provided by clothing, were approximately four times the corresponding whole-body doses.

The maximum individual dose was calculated using the highest off-site environmental thermoluminescent dosimeter (TLD) output. This TLD was located about 0.8 km east-northeast of the plant and recorded a dose of 0.83 mSv for the period 28 March 1979 to 7 April 1979. Since no member of the public resided closed than this TLD, its output represents an upper bound of the public dose.

Individual and population doses were calculated from dosimeter and meteorological data and population distributions around TMI-2. The population dose was calculated by summing each individual dose for the 2 million people that resided within an 80 km radius of the plant. The radiation dose received by the public from the TMI-2 accident is summarized in Table 7.3.

The accident description provided in Chapter 3 and the off-site release consequences presented in this chapter clearly illustrate that TMI-2 was less severe in comparison to the Chernobyl-4 and Fukushima Daiichi accidents. Although the TMI-2 accident involved an evacuation, it was not based on actual off-site radiological doses. The evacuation was based on elevated containment radiation levels and concerns regarding conjectured hydrogen accumulation within the reactor vessel. However, there was considerable confusion during the accident

Table 7.1 Airborne radioactivity released to the environment during the Three Mile Island Unit 2 accident.[a)]

Radionuclide	Quantity (PBq)	Half-life
Noble gases		
^{133}Xe	310	5.2 days
^{133m}Xe	6.3	2.2 days
^{135}Xe	56	9.1 h
^{135m}Xe	5.2	15.3 min
^{85}Kr	1.8[b)]	10.8 years
^{88}Kr	2.3	2.8 h
Radioiodine		
^{129}I	1.1×10^{-10}	1.6×10^{7} years
^{131}I	$<1.1 \times 10^{-3}$	8.02 days
^{133}I	1.5×10^{-4}	20.8 h
Cesium and strontium		
^{134}Cs	3.7×10^{-10}	2.1 years
^{136}Cs	1.1×10^{-11}	13.1 days
^{137}Cs	1.5×10^{-9}	30.1 years
^{138}Cs	7.4×10^{-10}	32.2 min
^{89}Sr	2.2×10^{-9}	50.6 days
^{90}Sr	2.2×10^{-9}	28.8 years
Activation products		
^{3}H	5.4×10^{-3}	12.3 years
^{58}Co	1.5×10^{-8}	70.9 days
^{60}Co	3.3×10^{-9}	5.3 years
Alpha-emitting radionuclides		
Gross alpha	3.0×10^{-9}	[c)]

a) Behling and Hildebrand (1986).
b) Includes the 1980 reactor building purge.
c) Varies with radionuclide.

Table 7.2 Comparison of the Chernobyl and TMI-2 accident source terms.[a)]

Constituent	Percent of inventory released from the core	
	Chernobyl	TMI-2
Noble gases	100	<8
Iodine	40	$<2 \times 10^{-5}$
Cs	25	—
Te	>10	—
Particulates	3–6	—

a) Knief (1985).

Table 7.3 Summary of radiation doses resulting from the TMI-2 accident.[a)]

Exposed group	Dose			
	Whole body		Thyroid dose	
	Individual (mSv)	Collective (person-Sv)	Individual (mSv)	Collective (person-Sv)
Highest individual	<1	16–53[b)]	<0.2	14–28
Average dose to an individual within a 16 km radius	0.08	—	0.01	—
Average dose to an individual within an 80 km radius	0.015	—	—	—

a) Behling and Hildebrand (1986).
b) Most probable estimate is 33 person-Sv.

that created uncertainty regarding the future radiological conditions and the ability to control and mitigate the accident.

Based on advice from the NRC and concern for the public, the Pennsylvania governor advised those individuals most susceptible to the effects of radiation (e.g., pregnant women and children) to leave the area within an 8 km radius of TMI-2. In addition, schools within the 8 km radius were closed. The governor noted that he was exercising caution based on the continued presence of radioactive material in the area and the possibility of further radioactive material releases.

Public evacuation following an accident and the associated stress and confusion create a negative view of the regulatory process and the safety of nuclear power operations. These issues are addressed in the subsequent discussion.

7.8.2 Chernobyl-4

Chernobly-4 was the most serious accident in the history of the nuclear industry and the first to have major international ramifications. Although the effects from the TMI-2 accident were confined to the immediate area surrounding the facility, Chernobyl-4 affected portions of Asia and Europe. The reactivity excursion that ruptured the Chernobyl-4 reactor vessel ejected radioactive material into the environment.

Major releases of radioactive gases, condensed aerosols, and fuel particles from Chernobyl-4 continued for 10 days following the 26 April 1986 reactivity excursion. The total release of radioactive material was 5–10 EBq, which included 1.8 EBq of ^{131}I, 0.085 EBq of ^{137}Cs, 0.01 EBq of ^{90}Sr, and 0.003 EBq of plutonium radioisotopes. Noble gases comprised about 50% of the total release activity. These releases are summarized in Table 7.4.

The Table 7.4 values illustrate that the release pathway has a profound impact on the isotopes that reach the environment. At TMI-2, the release pathway involved

Table 7.4 Releases from Chernobyl-4 and TMI-2.[a)]

Isotope(s)	Activity released off-site (EBq)	
	Chernobyl-4	TMI-2
Kr and Xe	1.9[b)]	0.38
Cs	0.089	2.6×10^{-12}
Iodine	1.8	1.3×10^{-6}
Other fission products	0.11	4.4×10^{-12}

a) IAEA (2006b).
b) The noble gas release quantity was estimated to be 6.5 EBq in IAEA (2012). See Table 7.7.

a torturous path that limited the quantity of iodine released. The Chernobyl-4 accident was a direct atmospheric release since the facility had no containment fission product barrier. For a severe accident, WASH-1400 predicted a large iodine release. This prediction was validated at Chernobly-4, but TMI-2 illustrated the weaknesses in applying the assumed source term to all accident scenarios.

The released radioactive materials contaminated significant portions of land surrounding the Chernobyl site. Over 200 000 km^2 of Europe were contaminated with ^{137}Cs with levels above 37 kBq/m^2. Greater than 70% of this area was in Belarus, Russia, and Ukraine which were the areas most affected by the accident. Contamination levels varied significantly and were elevated in locations that received precipitation when the radioactive plume traversed these areas. Most of the strontium and plutonium radionuclides were depleted from the plume within 100 km of the facility since they tended to have a larger particle size.

An estimated 350 000 emergency and recovery operation workers supported mitigating and recovering from the accident during 1986–1987. About 240 000 recovery workers participated in mitigation activities at the site and within the 30 km exclusion zone surrounding the reactor. In later years, the numbers of recovery personnel or *liquidators* increased to 600 000, but only a small fraction of these were exposed to high levels of radiation.

In 2006, the IAEA reported that more than 5 million people resided in Belarus, Russia, and Ukraine in areas that are classified as contaminated with ^{137}Cs levels above 37 kBq/m^2. In addition, about 400 000 people lived in more contaminated areas with ^{137}Cs levels above 555 kBq/m^2 that required radiological control. Within this population, 116 000 people were evacuated in the spring and summer of 1986 from the 30 km exclusion zone. In addition, 220 000 people were relocated in subsequent years.

The average accumulated doses from recovery workers and affected populations are summarized in Table 7.5. Natural removal processes and recovery countermeasures have reduced affected area radiation levels by a factor of several hundred. The majority of the contaminated areas are now available for resettlement and economic activity. However, some restrictions have been retained in the Chernobyl exclusion zone and in other limited areas.

Table 7.5 Summary of average accumulated doses to affected populations from Chernobyl fallout.[a]

Population category	Number	Average dose (mSv)
Liquidators (1986–1989)	600 000	~100
Evacuees from highly contaminated zone (1986)	116 000	33
Residents of "strict-control" zones (1986–2005)	270 000	>50
Residents of other "contaminated" areas (1986–2005)	5 000 000	10–20

a) IAEA (2006b).

Table 7.6 Summary of radiation doses resulting from the Chernobyl accident to the general population.[a]

Exposed group	Dose		
	Average individual whole body (Sv)	Average individual skin dose (Sv)	Maximum child thyroid dose (Sv)
Village residents 2 km from the plant	0.014	0.1–0.2	2.0
Farmers within 3–15 km of the plant	0.43[b]	—	—
Average dose to individuals within Europe and Asia	0.002[b]	—	—

a) Gollnick (2011).
b) Fifty-year dose commitment.

Table 7.6 provides a summary of radiation levels that also includes the effects beyond the immediate Chernobyl area. This table lists whole-body, skin, and child thyroid doses for village residents within 2 km of the facility and to farmers within 3–15 km of the facility. As an illustration of the wide-ranging effects of the Chernobyl-4 accident, average doses to individuals within Europe and Asia are also provided. The 50-year committed doses to Europe and Asia represent a significant fraction of the annual background effective dose.

7.8.3 Fukushima Daiichi

The Fukushima Daiichi accident involved the loss of all power, degraded core and SFP cooling capability, core damage, high dose rates, and the release of fission products to the environment. These conditions required that protective actions be initiated to protect the public. Protective action implementation was challenging because the accident was initiated by a massive earthquake and tsunami that damaged critical infrastructure.

7.8.3.1
Protective Actions

During the course of the accident, a number of protective actions were issued including orders for evacuation, sheltering, and administration of stable iodine. These orders had a significant impact on public perceptions of the accident both within Japan and throughout the world. Other actions included the administration of stable iodine and topsoil removal. Each of these actions is addressed in the subsequent discussion.

7.8.3.1.1
Evacuation Orders

The Fukushima Daiichi and Fukushima Daini nuclear power stations were affected by the earthquake and subsequent tsunami. Fukushima Daiichi is located about 12 km north of Fukushima Daini. Although the accident consequences were more significant for Fukushima Daiichi, both power stations were involved in evacuation orders.

Nuclear emergency declarations were made at the Fukushima Daini and FDNPS. The events at Units 1, 2, and 4 at Fukushima Daini were classified as Level 3 (Serious Incident) INES events, and these units were successfully placed into a cold shutdown condition. Units 1–3 at Fukushima Daiichi were classified as INES Level 7 (Major Accident) events, and Unit 4 was classified as a Level 3 INES event.

The initial evacuation order for residents within a 3 km radius of Fukushima Daiichi was issued on March 11 at 9:23 p.m. This order also included sheltering in place for residents between 3 and 10 km of the FDNPS. Three separate orders were issued on March 12:

1) At 5:44 a.m., residents within a 10 km radius of Fukushima Daiichi were directed to evacuate.
2) At 5:39 p.m., residents within a 10 km radius of Fukushima Daini were directed to evacuate.
3) At 6:25 p.m., the evacuation radius around Fukushima Daiichi was extended to 20 km.

Sheltering in place for residents within 20–30 km of Fukushima Daiichi was ordered on March 15 at 11:06 a.m. On March 25, the government of Japan advised the residents within a 20–30 km radius of Fukushima Daiichi to voluntarily evacuate.

As a matter of comparison, the US government on March 16 advised its citizens within 80 km of the Fukushima Daiichi facility to leave the area. On April 15, the US State Department lifted its voluntary evacuation advisory for families of government employees in Tokyo and other Japanese cities. The US recommendation to avoid travel within 80 km of Fukushima Daiichi was inconsistent with the Japanese government recommendations.

On April 22, the Japanese government announced the expansion of the evacuation zone to selected areas beyond the 20 km radius. Residents of the new areas were asked to evacuate by the end of May. The decision was made since residents could be exposed to effective dose rates of 20 mSv/year if they stayed in their homes. This area included five municipalities north west of the FDNPS. Evacuation of families with babies and children up to kindergarten age and pregnant women living outside the 20 km zone from the FDNPS began on May 15.

The Japanese government also established a no entry zone within 20 km of the FDNPS and designated parts of areas within 20–30 km of the facility as areas in which residents should remain indoors and be prepared to evacuate on limited warning. This order replaced the previous 20–30 km voluntary evacuation order. In addition, the Japanese government restricted rice farming in designated areas during 2011.

The evolving accident, expanding evacuation zone, restrictions on crops, and inconsistency between US and Japanese evacuation orders fostered public uncertainty and confusion. These conflicts added to public stress and did little to engender confidence in the nuclear industry or the regulatory process.

7.8.3.1.2
Administration of Stable Iodine

On March 14, Japan distributed 230 000 units of stable iodine to evacuation centers around the Fukushima Daiichi and Fukushima Daini nuclear power plants. The Japanese government recommended that local authorities instruct evacuees leaving the 20 km zone to ingest stable iodine. This order was made on March 16 and recommended a single administration with an amount dependent on age: babies (12.5 mg), 1 month to 3 years (25 mg), 3–13 years (38 mg), and 13–40 years (76 mg). The government recommended no administration for evacuees 40 years of age or older.

7.8.3.1.3
School Topsoil Removal

Cities in Fukushima prefecture removed the top 1–2 cm of topsoil from school grounds to permit children to resume outdoor activity. Soil removal was requested by parents and teachers to limit student radiation doses.

In May 2011, the Japanese government decided that burying contaminated soil was an effective disposal approach. Burying the soil 50 cm underground reduced that radiation level by 90%. Removal of soil is a significant remediation action that requires a disposal facility. Licensing the soil disposal facility is a regulatory action requiring stakeholder input.

7.8.3.2 Releases of Radioactive Material

The Fukushima Daiichi evacuations were ordered because significant quantities of radioactive materials were released from the damaged reactors. A summary of the Fukushima Daiichi accident releases is provided in Tables 7.7 and 7.8. The fission products released during the accident provided an early indication that significant core damage had occurred. Table 7.7 summarizes airborne releases of radioactive materials. Liquid releases to the ocean are provided in Table 7.8.

The airborne releases include contributions from noble gases, ^{131}I, ^{134}Cs, and ^{137}Cs. Releases of ^{137}Cs are particularly important since its 30-year half-life ensures an extended environmental impact on food, water, and land use.

Table 7.7 illustrates the severity of the Chernobyl-4 accident relative to Fukushima Daiichi. Although the Chernobyl-4 radioactive material releases are significantly larger than the FDNPS emissions, public discussions often focus on Fukushima Daiichi. This is likely attributed to the fact that it is a more recent event. However, videos of the upper levels of three reactor buildings being destroyed by hydrogen explosions are powerful images that trigger a significant emotional reaction. When coupled with protracted media reports regarding radiation, radioactive material releases to the air, leakage of contaminated water

Table 7.7 Estimated releases into the air from the March 2011 Fukushima Daiichi accident.[a)]

Organization	Released amount (PBq)			
	Noble gas	^{131}I	^{134}Cs	^{137}Cs
IAEA	~500	~500	~10	~10
IRSN	2000	200		30[c)]
Chernobyl-4[b)]	6500	1800	Not reported	85

a) IRSN Report (2011) and IAEA Report (2012).
b) Chernobyl-4 data provided as a comparison.
c) Combined radiocesium acivity.

Table 7.8 Estimated releases into the sea from the March 2011 Fukushima Daiichi accident.[a)]

Organization	2011 period of assessment	Released activity (PBq)		
		^{131}I	^{134}Cs	^{137}Cs
IAEA	26 March to 30 September	11	3.5	3.6
IAEA	21 March to 30 April	11.4	b)	3.6
IRSN	21 March to mid-July	b)	b)	27

a) IRSN Report (2011) and IAEA Report (2012).
b) Not reported.

to the ocean, displaced populations, and contamination of food and water, nuclear power proponents have significant issues to overcome to restore public confidence in this electrical generation technology.

The Fukushima Daiichi accident releases to the sea are summarized in Table 7.8. Sea release values for ^{131}I, ^{134}Cs, and ^{137}Cs are provided. The activity released to the sea was considerable less than the corresponding air activity. However, these releases affected the international sale of food that was potentially contaminated with radioactive materials. Moreover, these releases have continued for years following the accident. In 2013 and 2014, significant public attention was directed at releases of contaminated water that leaked from above-ground storage tanks.

The air and ocean releases of radioactive material heightened the international reaction to the Fukushima Daiichi accident. For example, the World Health Organization (WHO) issued an advisory regarding travel to Japan. During the early phase of the accident, individuals were advised by the WHO to avoid travel to the areas most affected by the earthquake and tsunami. Japanese authorities prohibited travel within the evacuation and exclusion zones surrounding the Fukushima Daiichi site. The size of these zones was determined by local dose rates and contamination levels and was reduced as mitigation activities were performed.

The contaminated areas include agricultural lands and fishing grounds. Agricultural products and fish were contaminated with low levels of fission products. Although these levels do not present a significant health hazard, their presence led to restrictions on international commerce with Japan.

Restrictions on food were imposed following the accident. Many of these restrictions remained in place well beyond the termination of major releases from the Fukushima Daiichi facility. Even a year after the Fukushima Daiichi accident, foreign import restrictions were imposed on Japanese food items due to lingering radiation concerns. At that time, 16 countries and regions banned the import of Japanese-produced food. Only four countries (Canada, Chile, Mexico, and Myanmar) lifted restrictions including requirements for Japanese exporters to submit radiation screening certifications. Kuwait and Mauritius in southern Africa imposed total embargos on Japanese-made food items. Fourteen other countries and regions, including China and Taiwan, continued to suspend the imports of some Japanese food items. In addition, 57 countries and regions required Japanese exporters to submit government certificates of origin and radiation screening. At 1-year postaccident, 73 countries and regions maintained import controls on Japanese food. If the Chernobyl-4 accident is a guide, it will take an extended period for all of the restrictions to be eliminated.

7.9 Emergency Response

Discussions of US regulatory shortcomings were presented in a previous section. This section focuses on issues that affect emergency response actions following a major power reactor accident involving core damage. The TMI-2, Chernobyl-4, and Fukushima Daiichi accidents involved unique events that outpaced the capability of the operating utility to simultaneously manage the on-site accident and provide timely information to support emergency response actions by governments. This weakness is ultimately a regulatory issue because the regulator licenses a facility and certifies to stakeholders that the utility is capable of addressing all normal and emergency issues in a manner that protects the health and safety of the public.

The TMI-2 accident was the first major commercial power accident that revealed emergency preparedness weaknesses. Accident response and failure to recognize that a LOCA was in progress were major weaknesses that led to core damage and the escalation of the event to a General Emergency classification. The resulting release of noble gas fission products and uncertainty regarding the accident's severity led to conflicting communications with the public.

The regulator and operating utility presented conflicting accounts of the accident progress and its severity. These conflicts are illustrated by the confusion regarding the NRC's conjectured accumulation of hydrogen within the core region and the possibility of an explosion of this gas. Conflicting assessments of the hydrogen hazard created considerable confusion and led to a precautionary evacuation. These TMI-2 response deficiencies resulted in a number of emergency preparedness enhancements that were promulgated by regulators.

Following the TMI-2 accident, emergency preparedness programs were required to perform immediate NRC notifications for a specified set of events. Drills and response plans were upgraded. The licensee was also required to evaluate drills and exercises several times a year and implement corrective actions for identified weaknesses. A portion of these exercises include participation by state and local agencies, the Federal Emergency Management Agency, and the NRC. In order to enhance emergency performance, additional equipment and monitoring instrumentation were installed to enhance accident identification and mitigation. Additional instrumentation was also installed to monitor radiation levels. These modifications improved response capability, and their effectiveness was evaluated through drills and exercises.

As noted by the Nuclear Energy Agency, the Chernobyl-4 accident was a unique event that should not be utilized as the reference accident for future emergency planning purposes. However, the Chernobyl-4 accident did reveal a number of deficiencies in emergency preparedness and radiation protection. In addition, it was the first major reactor accident to affect multiple nations and to cause radiation-related fatalities in the workforce.

Initial response actions suggested that emergency planning organizations were unprepared for the scope and magnitude of the Chernobyl-4 accident. These plans did not contain sufficient decision criteria to effectively manage the event as the accident evolved. In addition, clear lines of authority were not defined and too many organizations were involved in the decision process. These deficiencies suggested that enhancements to the Chernobyl-4 emergency planning infrastructure were needed. In particular, enhancements were needed to improve emergency communications systems, intervention team response, worker dose control and limitation, and radiation monitoring network capability. Mobile ground monitoring teams, aerial monitoring, and plume tracking were also in need of significant improvement. The Fukushima Daiichi accident revealed additional radiological response weaknesses and the need for further improvements in a number of aforementioned areas.

International evaluations of the Chernobyl-4 accident concluded that the intervention plans were too complex and time consuming to be efficiently implemented. Intervention actions and criteria for their initiation should have an international basis to ensure emergency plans involving multiple nations are implemented in a timely, efficient, and consistent manner.

The Chernobyl-4 radioactive material release was large and energetic and dispersed fission products over multiple continents. Since radioactive material was dispersed over large areas, the Chernobyl-4 accident demonstrated the need to include the multiple nations in the emergency response plans. Chernobyl-4 also demonstrated that a nation could be affected by nuclear accidents occurring within its borders and from foreign sources. This situation was repeated at Fukushima Daiichi. The international impact of the released fission products prompted cooperation and coordination of emergency response actions and activities. The Fukushima Daiichi accident also prompted the development of international emergency exercises.

Characterizing the accident severity and communicating the hazards to the public in a clear, well-defined manner was a significant Chernobyl-4 challenge. This issue was resolved by developing the International Nuclear Event Scale. The INES facilitates communication with the public on the severity of nuclear accidents and is currently adopted by a large number of countries to characterize nuclear events. Its effectiveness was illustrated as a consistent means of characterizing the Fukushima Daiichi accident.

Chernobyl-4 provided motivation for international agreements involving food items and their import following a major reactor accident. Monitoring imported food to ensure its safety was one of the first control measures instituted following the Chernobyl-4 accident and continued to be performed during Fukushima Daiichi.

The Fukushima Daiichi accident was exacerbated by a major earthquake/tsunami that complicated emergency response actions. Shortcomings in emergency response were highlighted in the National Diet of Japan Report on the Fukushima Daiichi accident.

The Diet report noted that emergency response issues existed because roles and responsibilities were not well defined. This continues to be a common theme in major power reactor events. Resolution of role responsibilities can be addressed through national and international exercises that challenge and stress the full extent of emergency response plans and procedures and the capabilities of their emergency organizations.

At Chernobyl-4 and Fukushima Daiichi, regulators failed to implement adequate evacuation plans, and an inadequate crisis management system contributed to public confusion during the evacuation. In the case of Fukushima Daiichi, the nuclear evacuation was further complicated by earthquake and tsunami damage.

At Fukushima Daiichi, the government and regulators were not fully committed to protecting public health, safety, and welfare of the evacuees. This is an incredible admission in the Japanese Diet Report which invalidates a basic premise of emergency planning, namely, that the government acts for the benefit of its citizenry.

The Diet report concluded that the safety of nuclear energy in Japan cannot be assured unless the regulatory process is changed by eliminating its insular attitude of ignoring international safety standards. This was also a key lesson from the Chernobyl-4 accident. Emergency planning considerations must include multinational considerations when a significant event could affect other countries.

7.10 Emerging Issues

Nuclear power proponents and opponents express varied opinions regarding the proper dose limits for workers and the public. These discussions are often fueled by changing national and international recommendations that have been proposed by the National Council on Radiation Protection and Measurements and the ICRP. Recommendations are often accepted by regulators and become national requirements. Concerns regarding appropriate dose limits for workers and the public intensify in the aftermath of a power reactor accident. The most recent example of these concerns is associated with the Fukushima Daiichi accident and the allowable dose limit for workers during accident response, the appropriate dose limit for the public during an evacuation, the appropriate dose limit for the public to reoccupy their homes, and the acceptable dose for children to return to their schools.

In addition to these radiological concerns, the consequences of reactor aging as well as releases of radioactive material to the environment are continuing areas of stakeholder interest. These issues have been raised in regulatory proceedings to extend power reactor operating licenses. Associated with these stakeholder concerns are discussions involving releases of tritium from degrading underground piping systems. The tritium and reactor aging issues are addressed in the subsequent discussion. Reactor safety and management of nuclear waste remain continuing stakeholder concerns.

7.10.1 Public Dose Limit Considerations

Dose limits are traditionally established for the whole body and specific organs for an individual year or for a specified period to calculate a committed or cumulative value. These limits are established for normal conditions. Additional limits are also established for emergency conditions associated with protective action recommendations.

7.10.1.1
Normal Conditions

During normal operations, public dose limits are often based on a fraction of the occupational values. The most recent guidance for dose limits for occupational and public exposures is provided in ICRP 103. These ICRP dose recommendations are provided in Table 7.9. Occupational limits have steadily declined and the question arises regarding the appropriateness of limits when they appear to be continually reduced. This has been apparent as the ICRP has progressed in its evaluations through Reports 2 (1959), 26 (1977), 60 (1991), and 103 (2007). In the United States, public effective dose limits for NRC-licensed activities are limited to 1 mSv/year.

Until the linear-nonthreshold (LNT) hypothesis is revised, continued reductions in dose limits should be anticipated. Associated public concern and confusion regarding declining dose limits remain a stakeholder concern associated with the operation and safety of nuclear facilities.

Occupational limits are also provided for pregnant women. As with the aforementioned limits, the pregnancy limits vary with the specific ICRP report. For example, ICRP 60 limited the effective dose to 2 mSv to the surface of the mother's abdomen or 1 mSv from the intake of radionuclides. ICRP 103's pregnancy limit is based on limiting the fetal dose to 1 mSv. In contrast, the

Table 7.9 ICRP 103 recommended dose limits in planned exposure situations.[a)]

Type of limit	Occupational	Public
Effective dose	20 mSv/year, averaged over defined periods of 5 years[b)]	1 mSv in a year[c)]
Annual equivalent dose		
Lens of the eye[d)]	20 mSv/year, averaged over defined periods of 5 years[b)]	15 mSv
Skin[e),f)]	500 mSv	50 mSv
Hands and feet	500 mSv	—

a) Limits on effective dose are the sum of the relevant effective dose from external exposure and the committed effective dose from intakes of radionuclides. For adults, the committed effective dose is computed for a 50-year period after intake. For children, it is calculated for the period up to age 70.
b) With the provision that the dose does not exceed 50 mSv in any year.
c) In special circumstances, a higher value of effective dose could be allowed in a single year. However, the average over 5 years is limited to 1 mSv/year.
d) The occupational limit was revised in an April 2011 ICRP Statement.
e) The limitation on effective dose provides sufficient protection for the skin against stochastic effects.
f) Averaged over 1 cm^2 of skin regardless of the area exposed.

NRC's pregnancy limit is 5 mSv during the term of the pregnancy. These inconsistencies lead to public confusion and do not inspire confidence in the dose limits established by regulators when these various values are cited in licensing discussions.

7.10.1.2
Emergency Conditions

Emergency dose limits influence public evacuation and sheltering and the administration of thyroid blocking agents. Following these initial protective actions, limits are established to restrict the dose from intakes of food and water and to determine if evacuated individuals are permitted to return to their homes. These limits are more than academic interest. As demonstrated by the Fukushima Daiichi accident, dose limits have a profound impact on evacuated individuals and the ability to return to their homes and resume normal lives.

In the United States, the EPA establishes public limits for emergency conditions. These limits have been adopted by the NRC and applied to commercial power reactor accidents. The Department of Homeland Security also uses these limits for terrorist events. These limits are addressed in the subsequent discussion.

7.10.1.2.1
EPA Guidelines

Upon classification of a Site Area or General Emergency, protective actions are usually implemented. Protective actions are guided by actual or projected off-site doses and plant conditions. These actions include sheltering, evacuation, and the administration of radioprotective chemicals. Table 3.7 summarizes the EPA protection action guides (PAGs) for the early, intermediate, and late phases of a nuclear incident.

Projected dose is the anticipated dose to be delivered to the public given a set of existing plant conditions and the anticipated release duration. Protective actions are governed by a number of considerations including the projected dose. If plant conditions change (e.g., the release rate changes, core conditions change, equipment status is altered, or the meteorological condition changes), the projected doses are updated.

Protective Action Guides for radiological dispersal device (RDD) and improvised nuclear device (IND) incidents are included in the EPA guidelines. The purpose of this guidance is to aid federal decision-makers in protecting the public and emergency responders from the effects of radiation during a terrorist-related emergency. These guidelines are provided in Tables 3.7 and 4.11.

7.10.1.2.2
ICRP Guidelines

Emergency exposures are unexpected situations that occur during the operation of a planned situation or from a malicious act. Before describing the ICRP guidelines, it is necessary to define terminology specific to that methodology. In an emergency exposure situation, the reference level is the total residual dose an individual would not exceed during a single acute exposure or during a protracted exposure (annual basis). For example, in emergency exposure situations, the criteria in ICRP 60 are specified in terms of averted dose (intervention levels), but the ICRP 103 recommendations are defined in terms of incremental dose (reference levels).

The dose that is expected to occur from the emergency event, should no protective actions be utilized, is called the *projected dose*. Residual dose is the dose that results following the implementation of a protection strategy. Each protective measure eliminates or avoids a certain dose, which is called the *averted*

Table 7.10 ICRP emergency exposure situations.

Categories of exposure	ICRP 60 Intervention levels[a)–c)]	ICRP 103 Reference levels[a),c)]
Occupational exposure[d),e)]		
• Lifesaving (informed volunteers)	• No dose restrictions[d)]	• No dose restrictions if benefit to others outweighs the rescuer's risk[e)]
• Other urgent rescue operations	• ~500 mSv; ~5 Sv (skin)[d),f)]	• 1000 or 500 mSv[e)]
• Other rescue operations	• Not provided	• ≤100 mSv[e)]
Public exposure[e),g)]		
• Foodstuffs	• 10 mSv/yr[g)]	• —
• Distribution of stable iodine	• 50–500 mSv (thyroid)[g),f)]	• —
• Sheltering	• 5–50 mSv in 2 days[g)]	• —
• Temporary evacuation	• 50–500 mSv in 1 week[g)]	• —
• Permanent relocation	• 100 mSv first year or 1000 mSv[g)]	• —
• All countermeasures combined in an overall protection strategy	• —	• Typically between 20 and 100 mSv/year according to the situation

a) Effective dose unless otherwise specified.
b) Averted dose.
c) Intervention levels refer to averted dose for specific countermeasures and remain valuable for optimization of individual countermeasures when planning a protection strategy. As a supplement to reference levels for evaluation of protection strategies, these levels refer to residual dose.
d) ICRP 60 (1991).
e) ICRP 96 (2005). Effective doses below 1000 mSv should avoid serious deterministic effects, and effective doses below 500 mSv should avoid other deterministic effects.
f) Equivalent dose.
g) ICRP 63 (1993).

dose. As noted in Table 7.10, optimization of protective measures that comprise the overall protection strategy is a complex process that includes a number of considerations.

For emergency exposure situations, ICRP 60 and ICRP 63 recommend no response below the action levels. Recommend action level values for the averted dose are appropriate for protective actions where intervention is usually justified. These protective actions include sheltering, administration of stable iodine, evacuation, and relocation. ICRP 103 recommends optimization below the reference levels. The ICRP recommends an upper value of the projected dose or reference level received from all pathways below which optimization is applied. Specific ICRP 60, 63, 96, and 103 recommendations are summarized in Table 7.10.

The application of the ICRP 103 reference levels involves (i) characterizing the exposure situation, (ii) setting a reference level, and (iii) optimizing protection accounting for the specific circumstances. This is an iterative process and yields an improvement in the level of protection for existing and emergency situations.

The ICRP emphasizes the need for optimization and justification of protection strategies for application during an emergency. Optimization is influenced by the reference levels. Protective actions and dose evaluations are part of the optimization process.

7.10.1.2.3
NCRP Guidelines

Radioactive materials are dispersed by natural means (e.g., the diffusion of radon gas through soil), through an unlikely event (e.g., a power reactor accident) (Chapter 3), and through an act of sabotage or terrorism (Chapter 4). Following the 11 September 2001 attacks, concerns for the intentional dispersal of radioactive material have increased.

In order to protect the public, protective actions are usually required to limit dose during or after a radiological emergency. Because all protective actions have an associated risk, the use of a particular action involves an assessment of risk/benefit. For terrorist events, NCRP 138 recommends sheltering of the public at an effective dose threshold of 5 mSv. Evacuation of the public is justified at a 50 mSv effective dose threshold and administration of stable iodine at 50 mSv equivalent dose to the thyroid. These NCRP 138 values and their applicability are summarized in Table 4.12. The NCRP 138 evacuation threshold is higher than the 10 mSv EPA requirement noted in Table 3.7.

7.10.2
Radiation Worker Dose Limits

The issue of radiation worker dose limits was raised during the early accident phase at Fukushima Daiichi when a number of workers exceeded the 250 mSv effective dose limit. Worker dose limits are a topic of interest because efforts have been made to harmonize national radiation protection regulations. As noted previously, there are a variety of national and international approaches to establishing worker dose limits.

Radiation workers have established dose limits for normal operations, planned special exposures (PSEs), and emergencies. Normal operations doses are specified in national regulations (e.g., 10CFR20 and 10CFR835 in the United States) and in NCRP (e.g., NCRP 116) and ICRP (ICRP 60 and 103) publications. PSEs are unique to the US regulatory environment and are defined in the Code of Federal Regulations for US Nuclear Regulatory licensees in 10CFR20 and for USDOE licensees in 10CFR835. Emergency dose limits and recommendations are specified by national as well as international organizations. Each of these worker dose limits is addressed in the subsequent discussion.

7.10.2.1
Normal Operations

The ICRP 103 dose limit recommendations are summarized in Table 7.9. These limits provide a dose framework for normal operations and operating conditions. In the United States, regulatory limits follow ICRP recommendations, but the adoption of the most recent ICRP publications is often delayed by years. Current NRC regulations are based on ICRP 26. The USDOE utilizes a combination of ICRP 26 and ICRP 60 recommendations.

7.10.2.2
Planned Special Exposures

PSEs are a unique provision of the US regulatory environment. Under the provisions of a PSE, the licensee may authorize an adult worker to receive doses in addition to those limited by occupational exposure, if specific criteria are met. A PSE is authorized only in an exceptional situation when alternatives that might avoid the higher dose are unavailable or impractical. The licensee or employer, if the employer is not the licensee, specifically authorizes the PSE in writing before the exposure occurs.

Before the PSE is implemented, the licensee must ensure that the individuals involved meet specified conditions. The individuals must be (i) informed of the purpose of the planned operation; (ii) informed of the estimated doses, associated potential risks, and specific radiation levels or other conditions that might be involved in the task; and (iii) instructed in the measures to be taken to keep the dose as low as reasonably achievable (ALARA) considering other risks that may be present. Prior to permitting an individual to participate in a PSE, the licensee (i) ascertains prior doses during the lifetime of the individual, (ii) ensures that the doses from the PSE will not exceed applicable limits, (iii) maintains records of the PSE, and (iv) submits a report to the NRC within 30 days of the PSE.

The applicable restrictions include limits for individual and cumulative PSE exposures. Cumulative PSE exposure is the individual's total PSE exposure during their lifetime.

For PSE exposures, dose limits equal to the annual occupational limits are not to be exceeded. For cumulative PSE exposures, five times the annual dose limits during the individual's lifetime may not be exceeded. PSE exposures are not voluntary and are not considered emergency exposures.

PSE exposures are allowed by regulation but are rarely used due to the stringent regulatory requirements. In principle, PSEs could be used for tasks when skilled workers are in short supply. For example, a unique situation could arise when a single uniquely qualified welder is available, and he is near the annual dose limits. If a qualified welder was required for a time-critical task required to ensure nuclear safety, a PSE could be evoked. This action permits task completion without exceeding the annual occupational dose limits.

7.10.2.3
Emergency Operations

Three major power reactor accidents have occurred and evacuations were implemented for each event. The TMI-2 accident involved a voluntary evacuation but did not result in significant off-site doses or levels of contamination. Chernobyl-4 and Fukushima Daiichi led to mandatory evacuations and areas outside the facility boundary were contaminated with fission products. In addition, worker exposures at Chernobyl-4 and Fukushima Daiichi were significantly higher than encountered during the TMI-2 accident.

During an emergency, decisions are made that affect the health of workers and the public. In order to facilitate these decisions, guidelines are needed to quantify dose limits for lifesaving activities and for the protection of facility equipment and property. Table 7.10 summarizes emergency worker dose guidance provided by the ICRP. US Environmental Protection Agency guidance is provided in Tables 3.7 and 4.11. It is interesting to note that the NRC provides no specific emergency guidance. The NRC uses the EPA guidance that is applied to all US commercial nuclear power events.

The EPA limits emergency worker effective doses to 50 mSv/year during the early accident phase, which is assumed to last 4 days. Higher effective doses can be incurred under exceptional circumstances. The EPA notes that protecting valuable property necessary for public welfare (e.g., a power plant) can utilize worker limits of 100 mSv and lifesaving activities can be authorized up to 250 mSv. In the case of a very large incident, such as an IND attack, incident commanders may raise the property and lifesaving response worker guidelines to prevent further loss of life and massive spread of destruction. In the United States, a dose authorized by incident commanders in excess of 250 mSv would likely be voluntary.

The Fukushima Daiichi accident emergency worker dose limit was established at 250 mSv. This value is consistent with ICRP recommendations.

The possibility of worker doses in excess of 250 mSv in a severe power reactor or IND event introduces an emergency management concern. Since doses above 250 mSv are likely, emergency managers may face a lack of volunteers during a high-dose event. The US guidelines are inconsistent with the most recent ICRP 103 guidance noted in Table 7.10. For example, ICRP 103 imposes no dose restrictions if the benefit to others outweighs the rescuer's risk for lifesaving activities. Emergency management decisions and worker activities can take very different paths when using ICRP 103 or EPA guidance. Based on the doses likely to be encountered following a nuclear detonation or severe core melt event, it appears that more definitive guidance is required for emergency workers and emergency managers to avoid delays encountered in addressing voluntary exposure situations. The ICRP 103 approach offers a more realistic view of the possible consequences of an IND event or severe reactor accident. Training and emergency exercises should reflect situations encountered in events where worker effective doses exceed 250 mSv.

Inconsistency between ICRP and US emergency worker doses is a regulatory issue that should be reconciled. These discrepancies are most apparent for high-dose emergencies and have the potential to create confusion and possible litigation regarding reasonable and prudent worker dose values that could be encountered during a severe emergency.

7.10.3 Future Dose Limits

Future dose limits should not be restrained by the current regulatory restrictions and biases. These limits should not adopt overly conservative assumptions that unnecessarily restrict the beneficial use of radiation and radioactive materials. The limits must protect radiation workers and the public and reflect the best available science.

7.10.3.1 LNT Hypothesis

The regulatory basis for radiation protection recommendations and limits assumes the validity of the LNT hypothesis. LNT is based on the premise that even the smallest amount of radiation causes a biological detriment (e.g., mutations that increase the risk of cancer). The current radiation safety basis using the LNT hypothesis was introduced following the observation of linear dose dependence of leukemia in atomic bomb survivors and the observation of linear dose dependence of mutations in drosophila for high-dose radiation. Linking the two high-dose radiation data sets and extrapolating these sets linearly to low doses is an assumption that merits challenge since radiation protection regulations are based on its validity. Appendix H provides additional commentary on the basis for radiation protection regulations.

The current radiation protection and dose limit regulatory environment is deeply rooted in the LNT hypothesis. Although there is research that contradicts the LNT hypothesis, it remains and is likely to remain the basis for radiation protection regulations.

International radiation protection organizations that develop recommendations supporting future rulemaking also support the LNT hypothesis. These organizations include the ICRP, the National Council on Radiation Protection and Measurements, and the US National Academy of Sciences.

7.10.3.2
Challenges to the LNT Hypothesis

Support for the LNT hypothesis is not universal, and numerous organizations including the American Nuclear Society, French Academy of Sciences, French National Academy of Medicine, and Health Physics Society have expressed various degrees of opposition to the LNT approach. The LNT hypothesis has been thoroughly discussed by health physicists, but recent publications review its inherent assumptions from a physiological perspective.

There are shortcomings of the LNT hypothesis that have not been fully evaluated. The LNT hypothesis does not incorporate a number of processes that are present in cellular repair and damage mitigation. For example, biological mechanisms involved in cellular repair are time dependent and dose rate dependent. These mechanisms are not incorporated in the LNT approach since the hypothetical model does not consider when a DNA break occurs. The LNT hypothesis also does not include the evolutional development of a species and its adaptation to the natural radiation environment. An evolving species would minimize the low-dose radiation influence as a risk factor in its survival.

In a similar manner, the LNT hypothesis does not account for DNA repair at low doses. Ionizing radiation damage to DNA involves a double-strand break that severs the double helix. These breaks are repaired or reconnected in the cell by the aggregation of cellular proteins. At low doses, these cellular repair mechanisms are efficient. However, at high doses, the more extensive DNA damage tends to form clusters. These damage clusters facilitate improper repairs that can lead to a health detriment. Specific detriments include mutations (chromosome rearrangements) or cancer (malfunctioning cells). Since DNA repair is less effective at high doses, it is problematic to extrapolate the high-dose results to low doses when DNA repair effectiveness differs. This simple description also provides an explanation to the increased risk of cancer at high doses, but it does not validate the LNT hypothesis.

Other mechanisms, including adaptive response, suggest that a biological insult (e.g., radiation exposure) enhances the body's ability to address further insults by activating its defense mechanisms. Adaptive response suggests that a low dose of radiation preconditions the body to withstand additional radiation exposure.

There is also evidence to suggest that low doses of ionizing radiation stimulate cellular defense mechanisms that protect the individual against disease. This process is known as *hormesis* and has been observed experimentally in lower life forms. Hormesis and adaptive response present additional challenges to the LNT hypothesis.

A future system of radiation dose limits cannot ignore the specific differences in biological repair effectiveness at low and high doses. In addition, hormesis and adaptive response must be evaluated without regard to historical bias. The use of dose and dose rate effectiveness factors acknowledges the inherent difference between high- and low-dose exposures. However, a complete set of factors must be considered in establishing a valid model for radiation detriment.

For example, Doss notes that autopsy studies have shown that the presence of cancer cells is not a decisive factor in the physical manifestation of clinical cancer. However, immune system suppression in organ transplant patients more than doubles the cancer risk. This supports an important immune system role in limiting occult cancers. Doss further notes that low-dose radiation elevates immune response, and so it may reduce rather than increase the risk of cancer. The beneficial effects of low-dose radiation have been noted in numerous publications. However, the most recent Biological Effects of Ionizing Radiation (BEIR VII) report reviewed but did not accept the role of hormesis and its challenge to the LNT hypothesis. The BEIR VII report also supports the LNT hypothesis and rejects threshold effects in dose–response mechanisms.

The LNT hypothesis focuses attention on DNA damage leading to further health detriments including cancer and hereditary effects. DNA damage is only one factor in assessing detriment, and medical researchers suggest that it is not a decisive factor. By focusing on DNA damage, the LNT hypothesis ignores the response of the immune system, which is an important factor in determining the physical detriment. In addition, adaptive response appears to be a valid effect that stimulates the immune system and permits it to function at an optimum level to counter the ionizing radiation detriment.

From a physiological perspective, there are three fundamental issues in the current radiation safety basis established using the LNT hypothesis. First, the LNT hypothesis focuses its attention on DNA damage and mutations which are not the only factors affecting the onset and propagation of cancer. Second, the LNT approach ignores the effect of the immune system response which is an important factor modulating the occurrence of cancer. The effect of radiation on immune system response is not linear, since low-dose radiation stimulates the immune system and high-dose radiation suppresses it. Third, the LNT model ignores the large variability in cancer rates by specifying no threshold. Lifetime cancer risks are likely to have large errors arising from the variability in confounding factors. Moreover, cancer rates also vary from year to year.

These issues suggest a thorough review of the LNT radiation safety basis is warranted. Although it is the basis for current radiation protection regulations, there are numerous publications that suggest there is no justification for continuing the use of the current LNT radiation safety paradigm. The LNT hypothesis has contributed to an unjustified fear of low-dose radiation and has inhibited the study of potentially beneficial applications of low-dose radiation. If the LNT hypothesis is discarded, what radiation protection approach would replace it?

7.10.3.3
New Regulatory Options

It has been over 60 years since the last significant change in the basis for radiation protection regulations. Prior to the 1950s, skin erythema was a major concern with radiation use, and physicians treated common diseases and conditions with radiation. In the 1950s, the observation of the increased incidence of leukemia in atomic bomb survivors shifted the radiation protection regulatory basis. Following the observation of detrimental effects in studies of atomic bomb survivors, genetic effects became the dominant concern that led to the adoption of the current regulatory basis. With the adoption of the LNT hypothesis, advisory bodies such as NCRP and ICRP reduced the radiation dose limits. With a number of studies suggesting the validity of adaptive response and radiation hormesis, it is time to thoroughly review all data to determine if a new radiation protection basis is warranted and if the LNT hypothesis should be abandoned or supported with data that reinforces its validity.

If justified by an unbiased review, a revised radiation protection basis should recognize adaptive response, the existence of thresholds for radiation detriment, and the potential for the beneficial effects of low-dose radiation. Making this change will be challenging since it is contrary to the recommendations of most advisory bodies, current government regulations, and public perception regarding the effects of low-dose radiation. Attempts to change the current regulations will be viewed with suspicion by the public because of the widespread fear of radiation fostered by media coverage of significant radiological events (e.g., the power reactor accidents at TMI, Chernobyl, and Fukushima Daiichi). In addition, most members of the public have a limited knowledge of radiation and its associated health effects.

These challenges are significant and should be addressed in new regulations and their associated justification. New regulations should incorporate all available radiation data and not rely solely on high-dose data. Specific effects including adaptive response, hormesis, and thresholds must be thoroughly evaluated in terms of data sets that include the traditional high-dose data from the Japanese atomic bomb survivors and high-dose therapy patients as well as data sets that have not been thoroughly incorporated in the past. These data sets include occupational radiation protection dosimetry from power reactors, medical facilities, universities, fuel cycle facilities, and government employees including the military; environmental data from areas of the world having elevated background radiation levels; and low-dose medical imaging data.

In addition to the inclusion of all data, the new radiation protection rules should consider a variety of dose–response models and not solely rely on LNT models. Risk models should be expanded to include other approaches that go beyond the historical absolute and relative risk approaches. Excess risk functions should also be expanded and utilize contemporary methods to fit data and not rely on historical models utilizing step functions. Epidemiologists, health physicists, and radiation biologists must also perform rigorous evaluations to ascertain the radiation-induced effects and the associated doses leading to these effects. Additional discussion of these issues is provided in Appendix H.

The scientific community should form a diverse group of professionals to aggressively investigate the LNT hypothesis. Adopting the proposed approach will present significant difficulties since the pro- and anti-LNT groups are firmly entrenched in their respective positions. In addition, an educational outreach campaign should be conducted to correct the current misconceptions in the scientific community, public, and government regarding the pathogenesis of clinical cancer and the biological effects of low-dose radiation. Research to demonstrate the beneficial health effects of low-dose radiation is essential to reduce the fear of radiation and facilitate an understanding of the limits of the LNT hypothesis. Finally, scientific accountability is needed. Statements and pronouncements by both pro- and anti-LNT advocates must be presented in terms of a complete, peer-reviewed methods and data analysis.

Alternatives to the LNT hypothesis are numerous but must conform to experimental observations. For example, inclusion of a threshold dose would provide additional credibility to radiation protection regulations. The threshold represents a dose below which no biological effect or detriment would occur. The threshold value would require careful evaluation, but a number of options exist. For example, the lowest dose where acute radiation effects are observed (e.g., chromosome aberrations in blood at 50 mGy) would provide an upper bound for a threshold.

Another possibility would set the threshold at the level of a typical annual background or environmental dose (e.g., 3 mGy in the United States). The environmental level is supported by the large variability in the earth's background radiation level and the lack of observed radiation-related health effects in high-dose areas of the world (e.g., India and Iran).

An additional regulatory format could include a threshold and then an assumed linear extrapolation from the threshold dose to high-dose data. This approach creates a *de minimis* dose that would be exempt from regulatory control with dose limits based on values above this threshold.

If the 3 mGy exemption were adopted, it would have a significant impact on the practice of radiation protection. For example, application of this approach to power reactors would significantly reduce the radiation protection requirements. Since many workers do not exceed 3 mGy/year, radiation protection programs could focus on the more hazardous activities and not be burdened by regulatory concerns regarding minor skin contamination events, low-level intakes of radioactive material, and excessive ALARA reviews for worker doses below the *de minimis* threshold.

7.10.4
Reactor Aging

Materials aging and their associated degradation affect the SSCs that comprise a nuclear power reactor. The environment associated with a power reactor increases the likelihood for equipment degradation over time. This environment includes elevated temperatures and pressures, a variety of radiation types including neutrons, thermal and mechanical stress, and corrosive materials. Proposals to extend reactor service lifetimes beyond the initial license extensions place additional demands on the performance of reactor materials. The performance of the materials associated with the three fission product barriers is particularly important from a health physics perspective. Reliable performance of these barriers must be maintained over the reactor lifetime to ensure the rigorous control of radioactive materials and preservation of the three fission product barriers.

Since fuel is periodically replaced, this fission product barrier is not affected by long-term reactor aging. The reactor pressure vessel is part of the reactor coolant system barrier, and it is susceptible to damage from the radiation generated through the fission process. Neutron damage must be carefully evaluated since it has the potential to reduce the strength and effectiveness of the reactor vessel in providing a fission product barrier. Reactor vessel damage is particularly significant since its replacement is cost prohibitive. Significant damage to the reactor vessel effectively precludes subsequent facility operations.

The containment fission product barrier is also prone to the effects of reactor aging. The performance of concrete and other containment components (e.g., cable insulation and penetration seals) must be verified over prolonged periods when subjected to a combination of chemicals, humidity, mechanical stress, moisture, oxygen, radiation, temperature, and vibration. Data regarding the effects of the various degradation agents over time on the performance of containment materials are sparse and need to be improved. Extensions of facility operating lifetimes must fully consider the performance of the fission product barriers and their capability to retain radioactive materials.

7.10.5 Tritium Leakage

Groundwater contamination has occurred at many of the reactor sites in the United States. The NRC concludes that 10–20% of these groundwater contamination events are attributed to leaks from underground piping systems. Tritium leakage from these piping systems is receiving heightened media, public, and regulatory attention.

The NRC requires that a licensee minimize spills, leaks, and other unplanned releases of radioactive materials into the environment. Licensees are required to monitor periodically for site contamination. This monitoring includes surface soil, water, subsurface soil, and groundwater. The NRC does not currently have a mandatory requirement for licensees to conduct radiological remediation during operations.

NRC regulations permit limited levels of radioactive materials to be released into the environment. The amount of radioactive materials released from underground piping system leaks has been small relative to the discharge limits summarized in Table 7.11. In addition, the Environmental Protection Agency imposes drinking water standards for radioactive isotopes. Although these limits primarily apply to public drinking water systems, many states also utilize them as groundwater protection standards. For tritium, the EPA set a maximum contaminant level of 20 000 pCi/l. To date, none of the reported underground piping system leaks have exceeded NRC or EPA limits in off-site locations.

Underground piping system tritium leakage at nuclear power plants is expected to continue. Nuclear power plant aging and associated piping system corrosion contribute to this trend. To date, reported underground piping system leakage has not created a public health detriment. Since the release of radioactive material to

Table 7.11 US radiation protection limits associated with power reactor underground piping system leakage.[a)]

Regulatory basis	Annual limit	Basis
10CFR50, Appendix I Liquid release limit	30 μSv whole body and 100 μSv to any organ of an individual who lives in close proximity to the facility boundary	A fraction of the natural background radiation dose and an attainable objective that nuclear power plants could reasonably meet
10CFR20.1301(e) The EPA radiation standard as incorporated into the NRC regulations	250 μSv to the whole body, 750 μSv to the thyroid, and 250 μSv to any other organ of an individual member of the public	Limit is cost effective in reducing potential health risks
10CFR20.1301(a)(1) NRC dose limit	1 mSv to any individual members of the public	ICRP recommendation Roughly equivalent to background radiation from natural sources excluding radon and thoron

a) GAO-11-563 (2011).

the environment is a radiological concern, leakage must continue to be monitored and evaluated.

Any environmental release of radioactive material increases stakeholder concerns that the public's health and safety are at risk. This concern is at least partially addressed by the NRC's groundwater monitoring requirements. The NRC has concluded that licensees' groundwater monitoring programs exceed agency requirements. However, these requirements and the actual leakage quantities need to be communicated to the public to ensure the hazards associated with the tritium leakage are clearly understood.

The corrosion of underground piping, particularly in safety-related systems, is a radiological and environmental concern. However, limitations in the industry's ability to measure the wall thickness of an underground pipe without excavation prevent licensees from determining the absolute structural integrity of these systems. The inability of verifying underground piping system's structural integrity increases the risk to public health and safety. Without the ability to verify piping system integrity, nuclear power plants cannot assure that underground piping will continue to function properly between inspection intervals.

The current regulatory efforts for addressing groundwater contamination are likely to lead to additional intervenor action with a subsequent degradation of public confidence. Congressional inquiries and actions by state governments following tritium leakage events have challenged the adequacy of existing regulatory requirements and undermined public confidence. Additional regulatory efforts to characterize and define the impacts of these leaks with an associated dialog with stakeholders would limit public and political uncertainty.

The nuclear power industry has not standardized data collection and analysis methodologies used to characterize the impacts of leaks. Rigorous risk assessment approaches, at locations other than the specific sampling collection points, have not yet been developed. In order to obtain a more comprehensive view of a leak's consequences, monitoring wells should be located in areas governed by the site's hydrogeologic characteristics. In addition, licensees need credible models that can predict contamination movement as a function of time. These models would estimate the time required for contamination to migrate off-site or contaminate drinking water and assess the impacts to public health and the environment.

To enhance public confidence, monitoring data and risk assessments should be transparent and independently reviewed. This could be accomplished by including a complete set of groundwater data as part of annual environmental reports. In addition, the licensees' groundwater monitoring programs should be available for review by stakeholders. Specific program elements such as the number, location, and depth of monitoring wells should be reviewed with stakeholder groups to enhance dialog and to minimize allegations. Public confidence would also be enhanced by subjecting the licensees' dose models to independent review. The results should verify that monitoring well data are sufficient to determine the tritium contamination levels and locations and the associated public doses.

7.10.6
Reactor Safety

Given the three major power reactor accidents, reactor safety is a continuing stakeholder and regulatory issue. The previous discussion noted weaknesses in the regulatory environment that contributed to the TMI-2, Chernobyl-4, and Fukushima Daiichi accidents. The discussion in Section 7.11.2.2 provides a proposal for future nuclear regulation to mitigate power reactor accidents.

7.10.7
Nuclear Waste Management

The current US regulatory environment and stakeholder concerns have not been conducive to establishing a high-level waste repository for spent fuel and fuel reprocessing waste. There has been some progress in selected countries for establishing a high-level waste facility, but the United States has been unable to resolve this issue. With stakeholder involvement, an alternative solution to the high-level waste issue is possible.

The implementation of the Generation IV reactors (see Chapter 2) offers the possibility for eliminating high-level waste because the limiting radionuclides in these wastes are recycled and included in the fuel. A Generation IV reactor burns the minor actinides and selected long-lived fission products and the remaining waste is shorter lived. For example, if the Generation IV reactor eliminates all minor actinides and long-lived fission products, the remaining waste is dominated by ^{90}Sr, ^{90}Y, and ^{137}Cs. These isotopes are considerably easier to manage than the high-level waste created by Generation II and III reactors. Although a Generation IV reactor fuel cycle minimizes the need for a high-level waste repository, the health physics aspects of a repository are briefly considered.

The protection of humans and the environment from releases of radioactive materials from a geologic disposal facility is a key licensing consideration. These disposal facilities are designed to contain and isolate high-level waste for periods that are comparable with geologic time scales. Geologic disposal is particularly suited for high-level radioactive waste including spent fuel and fuel reprocessing waste where long-term containment and isolation are required.

The required level of oversight for the disposal facility is another important regulatory consideration and it varies over time. This oversight includes health physics surveys and monitoring to ensure that the radioactive material is properly controlled. In ICRP 122, three facility regulatory phases are considered and summarized in Table 7.12. The first regulatory phase is the time of direct oversight (e.g., operational phase) when the disposal facility is being operated and is under active health physics control. A second phase involves indirect oversight. During this phase, the disposal facility is partly or fully sealed (i.e., postclosure period). Direct regulatory or societal oversight would continue for a period. It would then be supplemented or replaced by indirect oversight including monitoring of the repository and its release pathways for radionuclide releases, verification that restrictions on land use are maintained, and ensuring that facility records are maintained. The third phase (i.e., postclosure period) involves no oversight when knowledge of the disposal facility is lost. In the second and third phases, protection relies on the passive controls incorporated as part of the facility design.

Table 7.12 Radiological exposure situations for a geologic disposal facility.[a)]

Disposal facility status	Type of oversight: Direct	Indirect	None
Design basis evolution	Planned (normal and potential) exposure situation	Planned (potential) exposure situation	Planned (potential) exposure situation
Nondesign basis evolution	Emergency exposure situation at the time of exposure, followed by an existing exposure situation	Emergency exposure situation at the time of exposure, followed by an existing exposure situation	Emergency and/or existing exposure situation, once exposure is recognized
Inadvertent human intrusion	Not relevant	Not relevant	Emergency and/or existing exposure situation, once exposure is recognized

a) ICRP 122 (2013).

In the third phase, the geologic disposal facility might release radioactive material to the environment. In application of the optimization principle, ICRP 122 recommends an annual dose constraint for the population of 0.3 mSv/year. For doses in the future, ICRP 122 recommends a radiological risk constraint for the population of 1×10^{-5} per year. However, ICRP 103 warns that estimates of the future effective dose lose a direct connection to health detriment after a time span of a few generations. This is related to the evolution of society, human habits and characteristics, and the uncertain evolution of the biosphere. These uncertainties include severe, beyond design basis events and inadvertent human intrusion.

Design basis event selection and evaluation are critical because the future is uncertain. The Fukushima Daiichi accident illustrated the results of underestimating the design basis assumptions for a facility. If severe, beyond design basis events occur during periods of direct or indirect oversight, protection measures are implemented by existing regulatory organizations. If these events occur when no oversight of the disposal facility exists, then there is no guarantee that a competent authority will be able to understand the source of the exposure and implement necessary protective measures to terminate or mitigate the release.

Inadvertent human intrusion into the geologic disposal facility is not a credible event during the direct or indirect oversight periods. In the period of no oversight, inadvertent human intrusion could occur. The consequences of intrusion are uncertain because regulatory authorities if they exist may not understand the source of the radiation exposure.

7.10.8
Climate Change

Nuclear plants offer a low-carbon source of electric power that make a significant contribution to minimizing air pollution. These facilities also mitigate many of the concerns and associated contentions advocated by proponents of global climate change. The climate change topic is similar to the LNT hypothesis in that the proponents and opponents have entrenched positions and constructive communication between the camps is poor. If the climate change advocates are correct, nuclear power will play a significant role in mitigating the effects of greenhouse gases (GHGs).

A recent NASA climate change study suggests that global nuclear power has prevented an average of 1.84 million air pollution-related deaths and eliminated 64 Gt of CO_2-equivalent GHG emissions. These results are based on comparisons to equivalent generation from the use of fossil fuels. The NASA report also notes that large-scale expansion of natural gas use would not mitigate the climate change concern. Natural gas usage causes more air pollution-related deaths than the expansion of nuclear power. The NASA report provides additional motivation for an expansion of nuclear generating capacity and the need for strong health physics programs to ensure worker and public doses are maintained at current or lower levels.

Global climate change concerns often focus on rising temperatures and parameters that influence these increases. GHG concentrations and measurements used to infer temperatures in the distant past are issues that receive considerable attention in climate change discussions. A discussion of GHGs and the inference of temperatures from oxygen isotope ratios are presented to illustrate a portion of the climate change debate that can be influenced by health physicists. The subsequent discussion is not intended to be complete but provides an overview of the climate change issue, its relationship to the nuclear fuel cycle, and a specific area of data analysis that depends on the physical properties of oxygen isotopes.

A number of important climate change topics are noted but not discussed. This limitation is necessary because the primary purpose of this book is health physics and associated radiation-generating technologies. Important considerations in the climate change arena include (i) adequacy of deterministic models; (ii) validity of atmospheric thermal transport algorithms; (iii) incorporation of historic solar variability; (iv) full consideration of ocean currents and wind patterns and their influence on thermal transport; (v) accurate modeling of water vapor, its interaction with other atmospheric gases, and its historic variability; (vi) use of parameter distributions based on expert opinion rather than experimental data; (vii) model applicability as a function of time; (viii) validity of model input values and assumptions; (ix) model benchmarking, verification, and validation; and (x) numerical stability and convergence of computational models as a function of time.

7.10.8.1
Greenhouse Gases

Climate change and GHG mitigation are increasingly important issues that are evaluated by policymakers considering future energy sources. The concentrations of GHGs and their effect on rising temperatures have been asserted and vigorously debated. Given these concerns and the need for reliable and sustainable electrical generation, policymakers are evaluating a variety of energy sources including biomass, coal, geothermal, hydroelectric, natural gas, nuclear, oil, solar, and wind technologies.

Climate change is a broad and diverse topic and a complete discussion is beyond the scope of this book. Therefore, this chapter's focus is upon a limited set of topics that are relevant to nuclear regulation in general and health physics in particular. Accordingly, a very brief review of GHGs is provided since these gases and their respective concentrations have a postulated correlation with global temperature. Temperature changes have also been correlated with the ratio of oxygen isotopes, and these ratios are briefly examined.

The quantity of GHGs emitted by various electrical generating technologies is an important parameter in licensing proceedings and regulatory actions. Table 7.13 summarizes emissions from a set of these technologies. Most GHG emissions are produced when electricity is generated at a power plant and are designated as stack emissions. Emissions are also generated by other fuel cycle

Table 7.13 The range of total greenhouse gas emissions from various electricity production technologies.[a)]

Technology	Emissions (g C_{eq}/kW h)	
	Stack	Other fuel cycle activities
Biomass	—	8.4–16.6
Coal: 1990s Technology	216–278	48–79
Coal: 2005–2020 Technology	181	25
Hydroelectric	—	1.1–64.6
Lignite: 1990s Technology	247–359	7–14
Lignite: 2005–2020 Technology	217	11
Natural gas: 1990s Technology	99–157	21–31
Natural gas: 2005–2020 Technology	90	16
Oil: 1990s Technology	195–215	24–31
Oil: 2005–2020 Technology	121	28
Nuclear	—	2.5–5.7
Solar: 1990s Technology	—	27.3–76.4
Solar: 2010–2020 Technology	—	8.2
Wind	—	2.5–13.1

a) IAEA (2000).

components that include mining, processing, transportation, waste removal, and decommissioning activities.

The IAEA data summarized in Table 7.13 suggests that nuclear power generation produces essentially no GHG stack emissions. The nuclear fuel cycle also yields one of the lowest emissions of grams of carbon per kilowatt-hour (g C_{eq}/kW h) of any generating option. In general, fossil fuel technologies have the highest carbon emissions, and nuclear, hydroelectric, modern solar, and wind technologies have the lowest GHG emissions.

All energy production technologies generate waste as part of their fuel cycles. For nuclear power, the principal concern is a small quantity of radioactive waste that is managed through confinement. In comparison, significantly larger volumes and masses of waste result from the combustion of fossil fuels. These waste materials include ash, GHGs, toxic gases, particulates, and heavy metals. Fossil fuels use a dispersion strategy that dilutes and releases wastes to the environment.

Only limited quantities of radioactive materials are released from nuclear power plants, and these amounts are a small fraction of the 10CFR50 Appendix I dose limits. Meeting the challenge of reducing GHGs must rely on low-emission sources of electricity. This reinforces the viability of the nuclear power option.

7.10.8.2
Oxygen Isotopes

There are three stable isotopes of oxygen which are ^{16}O, ^{17}O, and ^{18}O with natural atom percent abundances of 99.757, 0.038, and 0.205%, respectively. The relative amounts of these isotopes in a geologic sample (e.g., water or ice) depend on the environmental conditions and their physical behavior. The impact of the behavior of these isotopes in a specific molecule such as H_2O forms the basis for ascertaining aspects of the climate (e.g., temperature) at a particular geologic time. Climate science typically examines the ratio of ^{18}O and ^{16}O isotopes.

The change in the natural abundance values for a specific set of environmental conditions is referred to as isotope fractionation. Condensation of water vapor preferentially removes ^{18}O, and evaporation preferentially removes ^{16}O. These effects have been utilized to perform an assessment of the earth's surface temperature from water and ice samples.

As an illustration, consider evaporation of surface water. The lighter $H_2{}^{16}O$ molecules evaporate more rapidly than the heavier $H_2{}^{18}O$ molecules. Therefore, there is less $H_2{}^{18}O$ in the atmosphere and the concentration of ^{18}O in precipitation decreases. The decrease depends on the surface temperature and other environmental conditions driving the evaporative process.

These assertions follow from a consideration of energy and momentum conservation relationships when examining a body of surface water at a given temperature with a specified distribution of oxygen isotopes. Water molecules evaporate from the surface, condense in the atmosphere, and then return to the surface as precipitation in a cyclic manner.

The dynamics of the aforementioned water cycle is understood by equating molecular energy relationships that illustrate the velocity of a molecule decreases with increasing mass if the temperature and all other environmental conditions are constant:

$$\frac{1}{2}mv^2 = kT \tag{7.1}$$

or

$$v = \sqrt{\frac{2kT}{m}} \tag{7.2}$$

where m is the molecule's mass, v is the molecule's velocity, k is Boltzmann's constant, and T is the absolute temperature of the water environment.

If the water cycle is in an equilibrium condition, the momentum (p) of the individual molecules in a volume element is equal:

$$p(^{18}O) = p(^{16}O) = m(^{18}O)v(^{18}O) = m(^{16}O)v(^{16}O) \tag{7.3}$$

$$v(^{18}O) = \frac{m(^{16}O)}{m(^{18}O)}v(^{16}O) \tag{7.4}$$

Therefore, the velocity of the $H_2{}^{16}O$ molecule is greater than the velocity of $H_2{}^{18}O$ with velocity dependent on temperature following Eq. (7.2). This increased $H_2{}^{16}O$ velocity enables it to evaporate preferentially. Consequently, precipitation is depleted in ^{18}O. In a similar manner, the pool is enriched in ^{18}O. Therefore, cold areas such as Antarctica and Greenland have about 5% less ^{18}O than ocean water.

By examining the oxygen isotope ratios, details of temperature can be inferred and correlated to specific geologic periods. Analyzing ice cores and correlating the core with a geologic age permits an assessment of the temperature of that era based on the $\delta(^{18}O)$ value.

In many studies, $\delta(^{18}O)$ is defined as

$$\delta(^{18}O) = \frac{r_{\text{sample}} - r_{\text{standard}}}{r_{\text{standard}}} \tag{7.5}$$

where r is the ratio of the ^{18}O to the ^{16}O abundance. In the pool example, $\delta(^{18}O)$ is positive in the surface water and negative in atmospheric water vapor. These ratios also vary with latitude. For example, in ice sheets, $\delta(^{18}O)$ is negative and has a value of −0.030 in Greenland and −0.055 in Antarctica. Midlatitude rivers have values on the order of −0.015. Tropical oceans have $\delta(^{18}O)$ values that vary with depth. The associated surface and deepwater values are 0 to −0.002 and +0.003 to +0.004, respectively.

These concepts permit the oxygen isotope ratios to act as a surrogate for temperature. This concept is based on the following observations: (i) precipitation has less ^{18}O than the ocean, and (ii) the ^{18}O content of precipitation at a given latitude decreases with decreasing temperature. This means that less ^{18}O will be detected in glacier ice.

There are a few issues associated with this argument. First, it assumes the abundance of oxygen isotopes and their associated ratios are constant in time. These ratios can change with variations in solar activity, volcanic activity, and geologic stability. Second, the models for determining these ratios need to include a variety of hydrodynamic and thermodynamic effects that must be benchmarked. This has not yet been fully demonstrated. Third, sample representativeness must be demonstrated. This requires a large sample and rigorous statistical analysis. Fourth, removal terms from samples must be rigorously quantified and included in models. In addition, all results should be verified using $\delta(^{17}O)$ values as well as a comparison of similar ratios involving ^{17}O and ^{18}O. By correlating all available data and not just a portion embodied in $\delta(^{18}O)$, a more representative temperature and associated uncertainty are obtained.

7.11 US Regulatory Improvements

The Section 7.2 discussion sets the stage for improving the regulatory process and an eventual transition to the international regulation of nuclear power plants. This transition appears to be an inevitable consequence of major nuclear accidents that affect multiple nations. The need for consistency was illustrated by confusion caused by inconsistencies in US and Japanese evacuation recommendations during the Fukushima Daiichi accident. A uniform set of regulations would have eliminated this confusion and not created public anxiety regarding a safe evacuation distance. In order to introduce additional motivation for the need for international regulations, discussion is initiated with appropriate modifications and improvements that could be made to the US regulatory process.

7.11.1
US Regulatory Enhancements

A number of improvements in the US regulatory process are in development. These include a State-of-the-Art Reactor Consequence Analysis and proposed risk-informed regulatory management framework. Although these are improvements that offer the potential for increased safety, they do not provide a new regulatory approach or significant departure from the current regulatory philosophy.

7.11.1.1
State-of-the-Art Reactor Consequence Analyses

The US regulatory history illustrates numerous areas where improvement could have been achieved, but opportunities were missed. Fortunately, the NRC has attempted to learn from previous failures and applied these lessons to improve the US regulatory process. Part of the improvement process is the desire to enhance analysis tools for assessing events in order to focus attention on safety significant issues.

Improvements in the NRC's regulatory process are illustrated by the most recent generation of reactor phenomena and consequence models designed to develop the best estimates of the off-site impact from potential severe reactor accidents. The NRC's most recent methodology is called the State-of-the-Art Reactor Consequence Analyses (SOARCA). This project evaluates plant improvements and changes that were not reflected in earlier NRC models. SOARCA includes system improvements, improvements in training and emergency procedures, and emergency response and security upgrades. It also incorporates the effects of plant changes such as power uprates and higher core burnup.

The SOARCA methodology is intended to be comprehensive and was initiated prior to the Fukushima Daiichi accident. As such, the SOARCA and FDNPS accident offer a benchmark to its predictive capability. It is interesting to note that several classes of accidents that were integral to the FDNPS events were not considered as part of the SOARCA. These omissions include:

1) *Multiunit Accidents*: The FDNPS is a six-unit facility. Hydrogen explosions occurred in multiple units including the operating units (Units 1 and 3) and Unit 4, which was defueled at the time of the accident. This omission does not provide confidence in the adequacy or completeness of the SOARCA philosophy since it omitted the multiunit aspect of the FDNPS accident.
2) *Low Power and Shutdown Unit Accidents*: FDNPS Unit 4 had its entire fuel inventory offloaded to its SFP. The events in Units 1–3 led to a hydrogen explosion in Unit 4 that severely damaged its reactor building and possibly damaged the SFP and its included fuel. Damage to the Unit 4 SFP is possible, and it has been structurally reinforced. The potential for a defueled unit to suffer a severe hydrogen explosion with its subsequent impact on the SFP was not considered in the initial SOARCA.
3) *Extreme Seismic Event that Led Directly to Gross Containment Failure with Subsequent Core Damage*: The extent of damage at the FDNPS attributed to

the seismic event will be determined as the recovery effort proceeds. Comprehensive seismic inspections will not be possible until facility radiation and contamination levels are significantly reduced. However, the tsunami caused by the massive earthquake facilitated the SBO condition and subsequent core damage, but this sequence was not considered in the SOARCA. This omission further challenges the adequacy of the SOARCA to anticipate natural events that could damage fission product barriers and lead to a significant release of radioactive material to the environment.

4) *Spent Fuel Pool Accidents*: The hydrogen explosions in Units 1, 3, and 4 damaged their respective reactor buildings. Debris from the hydrogen explosions fell into the pools and may have mechanically damaged fuel and the associated fuel cooling safety systems. The SOARCA did not address the sequence of events associated with the FDNPS SFPs. This is another demonstrated weakness in the NRC's ability to anticipate and preclude significant events.

7.11.1.2
Risk-Informed Regulatory Framework

In 2011, the NRC reviewed the need for modifications to its regulatory framework in order to enhance safety and improve regulatory consistency across its various programs. This effort focused on a more comprehensive, risk-informed, performance-based regulatory approach for power reactors, research reactors, materials facilities, low-level waste sites, high-level waste repositories, uranium recovery facilities, fuel cycle facilities, spent fuel storage areas, and the transportation of radioactive materials. The risk-informed framework envisioned an approach that could be in existence within the next 10–15 years.

Although the risk-informed approach has merit, it focuses on a variety of programs. Would the NRC or an international regulator be more effective by solely focusing on nuclear power plants? The NRC's global nuclear regulation approach can be contrasted with the INPO and WANO that have a singular, nuclear power focus. This focus has been successful in improving nuclear power plant performance and should be considered in future regulatory models.

The risk-informed study concluded that the DID concept remains valuable, but it is not uniformly applied and more guidance is required for its optimization. Optimization would be enhanced using a risk-informed and performance-based regulatory approach. The NRC's risk management approach recognizes that adequate protection of public is not synonymous with absolute plant safety.

The report reaffirms that the concept of design basis events and design basis accidents continues to be a sound licensing approach. However, the design basis concept has not been refined to incorporate the operating history of power reactors and a variety of analysis techniques including PRA. The NRC also suggested the creation of a design-enhancement regulatory category for the treatment of beyond design basis accidents. In addition, the methodology used to assess the frequency and magnitude of external hazards should be determined using both deterministic and PRA techniques.

7.11.2 Nuclear Regulatory Options

The previous discussion provides historical examples that the NRC model has not been completely successful in precluding accidents or off-normal events. Planned NRC improvements follow the conventional regulatory model and do not offer the likelihood of sufficient improvements in predicting or preventing the next major reactor accident. These shortcomings include licensing issues associated with underestimating the design basis (e.g., protection against beyond design basis earthquake events illustrated by the Fukushima Daiichi accident and reanalysis of design basis flooding and seismic hazards), design issues (e.g., TMI-2), failing to recognize accident precursor events (e.g., the Davis–Besse reactor vessel head erosion event), and maintenance issues during operations (e.g., Salem ATWS events). The historical review also illustrates that the agency has incrementally improved its performance and continues to evolve and improve its regulatory model. However, given the nature of the TMI-2 and FDNPS accidents and the ATWS and reactor vessel head erosion events, change is needed to minimize and mitigate future events.

7.11.2.1 General Options

If the current NRC regulatory model is not the answer to a proactive approach to prevent or at least significantly mitigate future accidents, then other broad scope options for achieving this goal are possible. The items noted in the subsequent discussion are not necessarily complete but serve to illustrate the types of options that are available for minimizing the probability of a severe reactor accident.

The first option is the Swiss-German post-Fukushima Daiichi approach of abandoning nuclear power. This is essentially a default option that eventually eliminates nuclear power as an energy source. It could be a consistent worldwide option, but it imposes significant economic penalties.

Option 1 does not guarantee safety since plants continue to operate for a limited time before being decommissioned. In addition, the reactor's fuel requires disposition and subsequent regulatory attention. The problems associated with licensing the Yucca Mountain repository and the economic loss of electrical generating capacity suggest that this option presents significant challenges. However, it is achievable in the near term.

The second option is a modification of the first proposal. Option 2 is a gradual elimination of nuclear power plants by reducing their authorized power levels. A power reduction reduces the severity of an accident but does not necessarily reduce accident frequency. This option could be implemented gradually to permit other electrical generating options to be utilized, but an economic impact would result. The issues of decontamination, decommissioning, and spent fuel disposition remain as they did for the first option. As with Option 1, this approach is also achievable in the near term.

The third option is to continue to operate nuclear power plants using the current US regulatory framework. This approach has prevented additional major US accidents and could be justified on that basis. Option 3 would continue to incrementally improve safety by adopting the lessons learned from operational events and accidents (e.g., implementing the Fukushima Daiichi Task Force Recommendations) and the improved assessment approaches. It would also include the industry's flexible and diverse or FLEX strategy, which provides portable equipment for nuclear plants to maintain core cooling capability and electrical power during severe events. Following this approach, overall plant safety would improve, but it would not preclude the occurrence of new or previously unidentified events (e.g., events such as the Salem ATWS or Davis–Besse reactor vessel head erosion).

A fourth approach is similar to the previous option but replaces the Generation II reactors with Generation III designs after their licenses expire. The Generation III plants would be located in low population density areas. Although Option 4 requires the construction of additional transmission and generation facilities, it reduces the probability of a serious event since the Generation III plants utilize passive safety systems and have a lower core damage frequency than Generation II plants. The impact of a serious event affecting the public is reduced since the reactors would be located in low population density areas. As such, the disruption on nearby populations is minimized. This option improves safety over time, but in the near term, Option 4 is essentially Option 3.

A fifth approach restructures the regulatory authority to focus solely on reactor safety. This means that regulatory jurisdiction for hospitals, high- and low-level waste sites, universities, enrichment facilities, and commercial firms using or transporting radioactive material is transferred to the states/provinces/prefectures or other government agencies. Having a regulatory organization solely devoted to reactor safety has the benefit of not diluting the reactor safety mission with other aspects of radioactive materials and associated organizations.

A sixth approach involves abandoning the contention that the DID methodology is a basis for reactor safety. The DID systems are maintained but would not be the ultimate basis for reactor safety. Option 6 admits that events outside the original facility design basis occur, safety systems fail, and unanticipated and beyond design basis events occur. Using this approach, the current regulatory system continues to function to improve performance and safety and manage design basis events. However, beyond design basis events or conditions outside the facility design basis (e.g., a Fukushima Daiichi accident type) are managed from a new hardened emergency facility collocated with the existing reactor site. The hardened facility is the ultimate emergency response facility designed to operate and mitigate an accident in conditions encompassing worst-case events.

As envisioned by the French in its hard-core concept, the hardened facility provides control, power, and shutdown systems to protect and preserve the three fission product barriers. The new hard-core facility houses the requisite power and shutdown systems including pumps, water supplies, and power systems to ensure the core and SFP do not release radioactive material to the environment.

The concept requires a significant expenditure but adds a measure of safety beyond that provided by the currently accepted regulatory philosophy.

7.11.2.2 Specific Options and Future Directions

It is the author's view that none of these individual broad scope proposals is the answer to improving US nuclear regulation or forming the basis for international regulations. Future nuclear regulation must consider the current regulatory approach and evolve to a desired end state that includes elements of the aforementioned options.

Although there are numerous possible solutions for optimizing reactor safety, the following approach is proposed as an initial model subject to revision. It maintains the most positive aspects of current regulatory philosophy while transitioning to a more robust regulatory framework. This approach also has the potential to form the basis for an international system since it adopts elements of recommendations proposed by the Japanese, French, and US regulators.

The author proposes the following phased approach to improve safety at commercial power reactors. This approach includes near-term (<15 years), intermediate-term (15–30 years), and long-term (>30 years) transitions. The proposed near-term regulatory changes include:

1) Continue the operation of nuclear power facilities following the current NRC model. Improve reactor and regulatory performance by further developing the SOARCA and risk management proposals. This is essentially the current NRC model with planned improvements.
2) Limit the NRC regulatory role only to include power reactors. Regulatory responsibilities in other areas should be transferred to the states/provinces/prefectures or other agencies.
3) Utilize third-party groups such as the INPO, WANO, and IAEA to provide operational assessments of reactor performance and an independent reactor safety perspective. Require that national regulatory agencies evaluate and formally respond to INPO, WANO, and IAEA recommendations. National regulators would have the final decision to accept or reject these recommendations.
4) Facilitate the international exchange of utility personnel to broaden operating and maintenance experience.
5) Implement and expand the FLEX concept as an interim measure to enhance reactor safety.
6) Continue the development and licensing of Generation III and IV reactors.
7) Develop the French hard-core concept including appropriate design and cost analyses for all operating reactors. International participation and stakeholder groups should be involved in this process. Stakeholders should include groups that have previously expressed safety concerns (e.g., the UCS). The applicability of incorporating the FLEX approach to supplement the hard-core facility design should be determined.

8) Develop a consistent international regulatory approach using the NRC as the initial model. All nations with operating nuclear reactors or plans to initiate nuclear plant construction participate and industry groups (e.g., INPO and WANO) have a significant role in influencing development of the regulatory process.

These near-term actions advance existing concepts and form the basis for future international regulations. Developing nations are included in the near-term process to ensure that emerging nuclear power programs have a firm safety foundation.

Near-term Items 1–7 are significant steps, but Item 8 is crucial. Since a major accident affects all nuclear plants, it is essential that all nuclear reactors have a robust and sustainable regulatory foundation that provides the public and stakeholders an avenue to participate and influence regulatory issues in a positive manner.

The intermediate-term options assume that the near-term options have progressed and that the basis for international regulations has been achieved. With these assumptions, the intermediate-term actions include:

1) Incorporate the French hard-core concept into existing and future reactor designs.
2) Continue the design and licensing efforts for Generation IV reactors.
3) Use Generation III designs with hard cores for all new facilities and locate them in low population density areas. The licensing basis of these facilities includes a consistent international approach.
4) Implement the international regulatory approach developed in the near term.

The intermediate-term actions involve difficult choices, particularly incorporating the hard-core concept into new and existing facilities. If properly managed, the hard-core concept provides a methodology to build public and stakeholder consensus for nuclear reactors and add a degree of safety beyond that existing today.

If the near-term and intermediate-term actions are accomplished, the long-term goal of operating Generation IV reactors moves closer to reality. The major long-term action is to locate proposed Generation IV reactors with hard cores and their support facilities in remote, low population areas. This includes the Generation IV plants and their associated reprocessing and fuel fabrication facilities.

The implementation of Generation IV designs requires addressing significant issues (e.g., nuclear proliferation), but it enhances the use of uranium and plutonium resources and minimizes many of the issues associated with high-level waste storage.

This proposed operating and regulatory concept requires agreement of government, industry groups, the public, and stakeholder groups typically opposed to the current approach. Reconciling the viewpoints of these diverse groups is a very significant challenge. However, it increases the likelihood of consensus and fosters an open dialog with new safety ideas freely exchanged. This regulatory model also has the potential to enhance reactor safety performance over the current approach.

The facility operating and regulatory costs associated with the long-term approach are uncertain and require further analysis. However, given the aftermath of the Fukushima Daiichi accident, the cost issue should be included in discussions with stakeholders. These discussions should also review the merits of nuclear power in terms of global climate change and alternative power generating technologies.

7.11.2.3
Outlook

A review of US regulatory history suggests that major weaknesses have been minimized and improvements have occurred. However, the process has not been completely successful in eliminating significant events and programmatic weaknesses remain. An evaluation of the Fukushima Daiichi accident and subsequent recommendations in terms of the US regulatory history suggest an alternative licensing model is warranted. This model should evolve with time and incorporates international input. It also offers the potential for an international regulatory framework and minimizes opposition to the deployment of nuclear generating capacity.

7.12
Future Power Reactor Directions and Challenges

In the most basic terms of electrical generation, nuclear fission provides an energy source to boil water to produce steam that drives a turbine generator. As such, it must compete economically with other technologies that provide energy to boil water. Nuclear generation also faces public perceptions and concerns related to severe accidents and their associated economic and societal disruption. A challenging regulatory environment is also a unique aspect of nuclear generation.

There are significant issues confronting nuclear utilities. These issues include (i) the low cost of natural gas, (ii) pending facility and equipment upgrades to address weaknesses identified by analysis of the Fukushima Daiichi accident, (iii) possible facility modifications associated with seismic and flooding hazards, and (iv) a regulatory climate that is influenced by political considerations including mandates for renewable energy that negatively impact the economics of base load nuclear units.

From an economic perspective, nuclear generation is relatively inefficient with current light water technology having a peak thermal efficiency of about 34%. This efficiency is considerably less than the 60% or greater achieved with natural gas-fired power plants.

Nuclear construction is also a costly and time-consuming venture that is often complicated with litigation and regulatory issues. In addition, an advanced light water reactor has a $10 billion capital cost which limits their market to well-financed utilities with the capability to support a large technical staff. As a comparison, natural gas plants have significantly lower capital costs, smaller operating staffs, and shorter construction times. Natural gas facilities are available to a wide group of utilities, require industrial and trade levels of personnel expertise, and have minimal regulatory issues. Given the abundance of fuel and favorable economic and regulatory characteristics, natural gas-powered generation presents a significant challenge to future nuclear power deployment.

Financing costs in the 2005–2015 period have been at historically low levels. Future interest charges are expected to be significantly higher, and this increased cost will further dampen nuclear power plant construction. In fact, it may lead to the cancellation of planned units and delays in other reactor construction efforts. These delays will increase costs and further challenge nuclear development and sustainability.

Although there have been a number of evaluations of the Fukushima Daiichi accident, the financial impact of mandated facility changes on nuclear plant costs has yet to be fully evaluated. These regulatory changes may exceed the actions taken by utilities with respect to the availability of portable emergency equipment to improve the response to a loss-of-core-cooling event. Some BWRs may also be required to provide hardened, filtered vents to reduce the radiological release from a major nuclear accident. The unit cost of these BWR modifications lies in the $15–40 million range. This magnitude of these investments may force some utilities to retire their nuclear units.

Following the Fukushima Daiichi accident, the NRC ordered US nuclear plants to reevaluate seismic and flooding hazards against current requirements. If warranted by this evaluation, licensees were to update the facility design basis to address the updated flooding hazards. If the initial seismic calculations suggest the risk is substantially higher than previously determined, a more detailed seismic risk evaluation will be required. This evaluation would likely suggest modifications to ensure the facility could respond to operational and emergency events. These modifications are costly and could result in the shutdown of additional nuclear units.

In addition to these considerations, the government has yet to resolve issues associated with the storage of high-level nuclear waste. Attempts to advance the high-level waste repository at Yucca Mountain have been fraught with legal and technical challenges. Yucca Mountain licensing also involves a political component that cannot be resolved by technical means. The political aspect of nuclear power production and its influence on critical infrastructure such as the Yucca Mountain fuel repository is an additional impediment to US nuclear power advancement. In fact, it may be the most significant challenge to overcome and poses a serious threat to sustained nuclear development.

These challenges were somewhat mitigated in 2015 with the publication of the NRC's last two volumes of the Yucca Mountain safety evaluation report. However, the completion of the safety evaluation report does not represent an NRC decision to authorize construction. A final licensing decision is not possible until

DOE's supplemental environmental impact assessment is completed, the 300 intervenor contentions against the Yucca Mountain repository are adjudicated, and the commission has finalized its licensing review.

Many of the aforementioned issues have been addressed with Generation IV reactors. For example, their higher operating temperature improves efficiency. To successfully compete with other energy sources, Generation IV or alternative reactor designs should provide cost-competitive production, low operating costs, low capital costs, and short construction periods. Achieving these characteristics is challenging but can be achieved through (i) modular designs, (ii) reactor cores having long (10s of years) refueling cycles, (iii) fuels that produce less waste, (iv) reduced proliferation risk, (v) a design using innovative materials having the capability to retain fission products, and (vi) the capability to contain the reactor core and released fission products during a severe accident. However, the technical challenges noted in Chapter 2 must be overcome for Generation IV reactor technology to reach its projected potential.

It may be argued that these characteristics are impossible to achieve. A paradigm shift in nuclear power design and philosophy is necessary for these goals to be realized.

The nuclear industry and government regulators should emphasize a reactor operating cycle that has improved economic viability, enhanced safety, proliferation resistance, and elimination or significant reduction in nuclear waste production. Public and stakeholder groups should be involved in this paradigm change. This requires the development of new technologies and significant innovation. The task will not be easy, but it may be the only viable approach for sustaining nuclear power and eliminating issues that are limiting its further development and deployment. A national referendum on supporting the nuclear power option may also be required. Without political and strong public support, US nuclear power will continue to stagnate and fight for a sustained place in the nation's electrical generation structure.

There are inherent factors that have precluded this innovative approach. Significant, novel approaches to nuclear energy have not yet been developed because the industry has been forced to be risk averse. In addition, government policies, political maneuvers, and stakeholder legal action discourage innovation that advances nuclear reactor technology. It is difficult to forecast the future of nuclear generation, but significant challenges must be overcome for it to advance to its full potential.

Although the US outlook for nuclear power expansion must overcome the aforementioned challenges, the prospects for a global renaissance are more positive. In particular, the economic growth projections for Africa, Eastern Europe, Latin America, Middle East and South Asia, and South East Asia and the Pacific suggest a significant expansion for nuclear power. However, the development of a nuclear industry in less advanced nations must occur in a manner that ensures safe and dependable operations.

The fission reactor environment could be significantly perturbed by a breakthrough in fusion reactor technology. However, numerous issues must be overcome before fusion reactors achieve a level of development equivalent to Generation II nuclear reactors.

Problems

7.1 You are the senior health physicist at the Smelly Valley Nuclear Power Station, an advanced pressurized water reactor. Stakeholder groups are challenging a power increase licensing amendment. These groups are concerned with an elevated primary coolant source term. Plant management is evaluating a supplementary demineralizer system to reduce this source term and you are tasked with performing the radiological design review and ALARA evaluation to support resolution of the stakeholder's contention.

Data:

Component description:	Primary coolant demineralizer
Design configuration:	Stainless steel cylinder
Dimensions:	Height, 2 m; diameter, 1 m
Processing capability:	1000 l/min
Radionuclide efficiency:	99%
Routine demineralizer run time:	100 days
Routine demineralizer down time:	60 days
Influent ^{60}Co activity concentration:	70.3 Bq/ml
^{60}Co gamma constant:	3.1×10^{-4} mGy-m^2/h-MBq
^{60}Co half-life:	5.27 years
^{131}I gamma constant:	5.2×10^{-5} mGy-m^2/h-MBq
^{131}I half-life:	8.02 days
^{137}Cs gamma constant:	8.1×10^{-5} mGy-m^2/h-MBq
^{137}Cs half-life:	30.1 years

(a) Name four documents (e.g., federal regulations and facility documents) that are needed to perform the demineralizer ALARA evaluation.

(b) List and briefly describe four items that you should consider when evaluating the demineralizer from an ALARA perspective.

(c) Calculate the total activity in the demineralizer at the end of its run time and at the end of its down time. For the purpose of this question, ^{60}Co is the only radioisotope under consideration.

(d) Calculate the total absorbed dose rate 20 m above the demineralizer bed at the end of its down time. Ignore shielding provided by the demineralizer's bed, water, and steel shell.

(e) List four methods you could use to minimize the dose to plant personnel during maintenance of the demineralizer.

(f) The demineralizer design manual estimates a maximum loading of 750 TBq of ^{137}Cs and 500 TBq of ^{131}I following a failure of multiple fuel rods in a fuel assembly. What is the absorbed dose rate contribution from ^{131}I and ^{137}Cs at a centerline distance of 10 m above the demineralizer bed immediately following the maximum radionuclide loading? Ignore shielding provided by the demineralizer's bed, water, and steel shell.

(g) Using the data from Question (f), what absorbed dose rate is present after 1 year of decay?

(h) Based on the results of Question (f), how should the demineralizer cubicle be posted?

(i) Given the radiation levels noted in Question (f), what health physics controls should be imposed on resin removal (sluicing) from the demineralizer?

7.2 An intervenor group, Save the Land over Beautiful Seattle (SLOBS), has submitted a petition to the Nuclear Regulatory Commission to change the 10CFR20 radiation protection regulations from the ALARA concept to the threshold limit value (TLV) approach. You have been hired by the NRC to evaluate the SLOBS proposal, which has the following elements:

1. A daily effective dose limit defined as a TLV equivalent dose of 0.2 mSv/day.
2. An exemption to the TLV equivalent dose that allows a cumulative TLV equivalent dose of 20 mSv in a continuous 13-week period. This exemption is permitted once per calendar year.
3. Limits (1) and (2) also apply to the TLV equivalent dose to the eye.
4. TLV equivalent doses to the skin and extremities allow 10 times Limits (1) and (2).

(a) The ALARA principle is generally applied to radiation exposure controls, while chemical and microwave hazards are typically limited by TLVs. What is the basis for the difference between these two concepts?

(b) Is there a radiological basis for SLOBS element (1)?

(c) What effect will the SLOBS proposal have on work activities at commercial nuclear power plants performed by utility personnel?

(d) What effect will the SLOBS proposal have on work activities at commercial nuclear power plants performed by contractor personnel?

(e) How does the SLOBS proposal compare with the ICRP 103 recommendations?

7.3 As the newly appointed Radiation Protection Manager at Haven Unit 4 near San Francisco, you are responsible for all health physics activities at the advanced pressurized water reactor. Your task has been challenging since fuel damage resulted in elevated fission product activity in the reactor coolant system. The No. 3 waste gas decay tank is of concern since its ^{131}I concentration is well above release limits.

During the morning meeting, radiation air monitor alarms are reported in the auxiliary building near the waste gas decay tanks. The shift supervisor reports that a maintenance worker inadvertently breached the No. 3 waste gas decay tank while repairing an isolation valve.

Given the tank's ^{131}I activity, you order that nasal swabs be taken. The swabs are positive for the maintenance worker and indicate an intake of radioactive material has occurred. Accordingly, you direct that urine samples be obtained from the worker and whole-body counting be performed. The worker appears to have a positive result for ^{131}I in a spot urine sample. Due to an instrument failure, no air sample results are available.

Data:

1. From external thyroid counting, the following data are obtained for the maintenance worker:

Time postintake (d)	Thyroid activity (kBq)	IRF[a]
1	2500	0.133
7	2300	0.0995
10	1300	0.0751

a) Intake Retention Fraction (IRF) for inhalation of Class D (See next comment)^{131}I. This is the fraction of intake expected to be in the thyroid at the specified time postintake. Radioactive decay is included in these values.

2. For inhalation of Class D ^{131}I, the dose conversion factor for the thyroid is 2.9×10^{-7} Sv/Bq.
3. Approximately 75% of the ^{131}I is excreted from the body in the urine in 1–2 days with an effective halftime of about 6 h. The remaining 25% of ^{131}I is absorbed in the thyroid, reaching a maximum about 24 h postintake, and is excreted with an effective halftime of about 7 days.
4. The 10CFR20 organ dose weighting factor (w_T) for the thyroid is 0.03.
5. The Haven Facility's radiation protection program is governed by 10CFR20, which is based on ICRP 26.

(a) Given that you can choose *in vivo* or *in vitro* methods of analysis to perform bioassay, describe and discuss the optimal approach for this case. In your discussion, list two advantages and two disadvantages for the *in vitro* and *in vivo* methods of analysis as related to this event.

(b) How might your approach to bioassay change with time, given the metabolic model for iodine?

(c) Based on the thyroid counting data, what is your best estimate of the worker's intake?

(d) Assume that the intake was 5 MBq. What is the committed dose equivalent (CDE) to the thyroid for this intake? What is the committed effective dose equivalent (CEDE) for this intake? Assume that organs other than the thyroid make a negligible contribution to the CEDE. Have any regulatory limits been exceeded?

(e) Given the doses calculated in part (d), the NRC has fined your company \$250 000.00 for its poor radiological work practices. A subsequent investigation of the event reveals that the worker was involved with a terrorist group and that the waste gas decay tank was intentionally breached. The plant manager terminated the worker and the Federal Bureau of Investigation arrested him. You have been requested to prepare a radiological basis for appealing the NRC's fine. What is the basis for your appeal?

(f) A waste gas decay tank has been isolated for 3 years and will be released to the environment. What isotopes will dominate the release source term?

7.4 You are a health physicist employed by the state of Michigan. A large explosion occurred in Windsor, Canada, and a dirty bomb is suspected. The wind is blowing toward Detroit and you have been directed to report to the area where significant radioactive material has been deposited. When you arrive on the scene, first responders obtained gamma-ray spectra of the radioactive material, performed initial decontamination of individuals, and initiated the placement of rope barriers around contaminated areas. The radioactive material is extremely fine and has many of the characteristics of hot particles found at nuclear power plants.

Radioactive material was detected on individuals as they exited contaminated areas. As an initial estimate of the potential dose rate from these particles, technicians removed the radioactive material using tape and then measured the contact dose rate using an ion chamber with the end window open. Following that initial measurement, the particles are taken to the mobile gamma spectroscopy laboratory for analysis using a high-purity germanium detector.

Data:

1. Specifications for the air-filled ionization chamber vented to atmosphere used to monitor for contamination:

   Detector volume = 220 cm^3
   Chamber window = 7 mg/cm^2
   Beta shield = 1000 mg/cm^2

2. The ionization chamber was calibrated using a gamma-only source.
3. Gamma spectrum analysis indicates the following spectrum peaks for many of the detected microscopic particles:
   0.308, 0.511, 0.659, 0.819, 1.17 (dominant peak), 1.33 (dominant peak), and 2.50 MeV.

(a) What isotopes are suggested by the gamma-ray spectrum?

(b) When calculating the dose from a hot particle, what skin area and tissue depth are generally assumed for the purposes of estimating the skin dose?

(c) Is a dose measured by an ion chamber the "true" skin dose? State three reasons why or why not.

(d) The gamma spectrum reveals that the activity of a particular hot particle is due primarily to ^{60}Co. Given the seven peak energies observed in the spectrum, identify the most likely origin of each and describe the mechanism that causes each peak.

(e) What three follow-up actions would you take in the contaminated area upon discovery of these hot particles?

7.5 You are the Director of Radiation Safety for the city of Boston, Massachusetts. A truck hauling a tank of tritiated water (HTO) is being transported to a waste disposal facility during a December snowstorm. The truck leaves the highway and crashes into an industrial bakery. As a result of the accident, the tank ruptures and tritium is released into the facility. The governor has assigned you as the lead health physicist to investigate this event and characterize the doses received by the workers. You must interface with a number of stakeholder groups including Local 509 of the International Brotherhood of Pastry Workers.

During the previous week, an NRC information bulletin concerning a tritium intake at a government production facility was routed to you for action. Using the bulletin as a reference document, you direct that airborne monitoring and urine sampling for tritium be instituted for the bakery personnel.

Data:

Tritium Data:

HTO inhalation dose coefficient = 1.8×10^{-11} Sv/Bq

Half-life = 12.3 years

Biological Data for the Affected Workers:

Volume of free water within the whole body = 43 l

Mass of soft tissues within the whole body = 65 kg

Daily water loss (including urine) = 3 l/day

Daily urine loss = 1.4 l/day

(a) Which is more hazardous from a radiological perspective T_2 or HTO?

(b) Describe two airborne monitoring techniques that you would consider using to measure airborne tritium concentrations. Give one advantage and one disadvantage of each.

(c) A positive urine sample result of 500 dpm/ml is reported for a bakery worker. Since he was injured during the accident, the sample was obtained 60 days after the event. Assuming the metabolic model outlined in the problem statement, estimate the initial tritium intake.

(d) A baker has a single inhalation intake of 59.2 MBq of tritium (HTO). What is the individual's committed effective dose? Assume the inhalation dose coefficient is also applicable for skin absorption.

(e) What is the target organ for tritium in the HTO form?

7.6 You are the chairperson of the Radiation Protection Committee of a joint licensee-stakeholder review committee for the Hillary Gore Decommissioning Project (HGDP) where ^{90}Sr was used to manufacture radioisotope thermoelectric generators (RTGs) for offshore navigation buoys. The US

Nuclear Regulatory Commission licenses the facility. ICRP 103 terminology is utilized at the HGDP, but it follows 10CFR20 for its dose limits.

During a HGDP site visit, a worker inadvertently handles a contaminated drain valve. All radiation measurements were obtained with an ionization chamber.

Data:

1. The drain valve is the size of a baseball.
2. The beta dose rates are 8 Gy/h on contact with the drain valve and 2 Gy/h at 46 cm. The ionization chamber has a 7 mg/cm^2 end window. Beta correction factors are included in the absorbed dose values.
3. The gamma dose rates are 30 mGy/h on contact with the drain valve and 20 mGy/h at 46 cm. The ionization chamber used to obtain the measurements has a 300 mg/cm^2 end cap.
4. The monitoring instrument used for contact readings was protected by a plastic bag.
5. The worker hand-carried the drain valve for 2 min at 46 cm from his body.
6. The worker was wearing two pairs of rubber gloves, a set of coveralls, and a respirator with a hood.
7. Beta reduction relationship: $f_\beta = e^{-0.00435x}$
   where f_β is the beta reduction factor for $^{90}Sr/^{90}Y$ energies and x is the density thickness of the material in milligrams per square centimeter attenuating the radiation.
8. Ignore air attenuation.

Material	Density thickness (mg/cm^2)
Coveralls	29
One pair rubber gloves	39
Respirator facepiece	250
Plastic bag	15

(a) Calculate the equivalent dose to the lens of the eye. Disregard any field reduction by air and assume the distance to the eye is 46 cm.

(b) What are the 10CFR20 annual limits for the skin, lens of the eye, whole body, and extremities?

(c) At what tissue depth is the skin dose evaluated? What is the tissue at risk?

(d) At what tissue depth is the deep dose equivalent evaluated?

(e) A whole-body TLD with filters for skin and eye dose was worn on the chest under the coveralls during the incident. List four factors to consider when comparing the TLD dose to the calculated dose.

(f) A second worker involved in the event received a dose of 480 mSv to her hands when handling the drain valve. If her year to date total effective dose is 10 mSv, how much additional effective dose can be received for the remainder of the year.

(g) The HGDP site is planning to adopt the ICRP 103 dose recommendations including the 2011 Statement on Tissue Reactions. If the worker has a dose history summarized in the following table, how much effective dose can the individual receive in year 5 if ICRP 103 is adopted?

Year	Dose (mSv)		
	Effective	Skin	Eye
1	10	70	10
2	20	100	20
3	30	450	30
4	20	275	40

(h) Using the data from Question (g), how much eye dose can be received in year 5?

(i) Using the data from Question (g), how much skin dose can be received in year 5?

7.7 You supervise an in-house TLD system for occupationally exposed workers governed by a regulatory basis derived from ICRP 103. The TLD badge consists of two LiF chips of 235 mg/cm^2 thickness. Chip 1 is covered by 7 mg/cm^2 of plastic, and Chip 2 is shielded by 850 mg/cm^2 of lead and 150 mg/cm^2 of plastic. The TLD system is calibrated by exposing badges to known quantities of beta and gamma radiations and plotting the thermoluminescent (TL) reader output versus effective dose. Both the gamma and beta calibration curves are linear and pass through the origin. The gamma calibration curve indicates that 6000 TL units equals 5 mSv of gamma dose and the beta curve yields 750 TL units per 10 mSv of beta dose.

Data:

1. The control dosimeter reads 120 TL units on both Chips 1 and 2.
2. Both chips have the same beta and gamma sensitivity.
3. Upon heating, the TLD chip light output is:
   Chip 1 = 12 270 TL units.
   Chip 2 = 11 520 TL units.
4. The beta calibration curve for other tissue depths includes the following values:

Tissue depth (mg/cm^2)	Percentage of equivalent dose
7	100
100	50
300	25
500	10
1000	1

5. The gamma dose remains constant at tissue depths from 7 to 1000 mg/cm^2.

(a) Calculate the skin dose and effective dose for the exposed TLDs.

(b) Calculate the dose to the lens of the eye.

(c) Explain if any dose recommendations were exceeded. Justify your answer by stating the recommendation and identifying the source of the limits that you applied.

(d) A regulatory proposal is under review to limit the annual effective dose to 1 mSv. Would the doses calculated in the previous question exceed this limit? From a technical perspective, does this limit have merit?

(e) A second regulatory proposal is under review to abandon the linear-nonthreshold hypothesis and limit the annual effective dose to 50 mSv with a cumulative lifetime limit of 150 mSv. Would the doses calculated in the previous question exceed this limit? From a technical perspective, does this limit have merit?

(f) An employee works in a mixed radiation field, which includes beta particles, gamma photons, alpha particles, and thermal and mixed energy fast neutrons. The absorbed dose from external sources in the work environment was reported to be 30 μGy beta, 70 μGy gamma, 90 μGy thermal neutrons, and 25 μGy fast neutrons with an average energy of 10 MeV. Calculate the ICRP 60 equivalent dose in μSv.

(g) A 26-year-old male radiation worker has a lifetime effective dose of 0.32 Sv. Compare this worker's lifetime dose to the recommendations of NCRP 116.

(h) A radiation worker recorded the following effective doses over the past 4 years:

Year	Effective dose (mSv)
1	10
2	30
3	40
4	20

According to ICRP 103, what is the maximum effective dose allowed for the worker in year 5?

7.8 You are the Radiation Protection Director for Mega Drug of Tennessee's (MDT's) ^{131}I Medical Isotope Production Facility. A local stakeholder group Nashville United Together Strong (NUTS) is protesting and its president Dr Ima Moron is challenging the licensing basis assumptions and design basis for the facility. NUTS is concerned about facility releases contaminating farmland and a number of scenic ponds containing a rare species of tiger catfish. Dr Moron has provided a series of concerns that the MDT President will address at a public meeting. She requests a detailed technical assessment of the Dr Moron's concerns. NUTS has agreed to the meeting to resolve their concerns before they are submitted to the Nuclear Regulatory Commission as part of the facility's licensing proceedings.

Data:

^{131}I physical half-life = 8 days

^{131}I biological half-life (pond) = 15 days

^{131}I biological half-life (fish) = 21 days

(a) NUTS suggests that iodine concentrates in the atmosphere after it is released. Refute this assertion by listing mechanisms that reduce the concentration of airborne radioiodine during atmospheric transport.

(b) NUTS asserts that an individual continuously ingesting radioactive material at a constant rate reaches lethal dose levels since radioactive material is prevented from decaying by the unique human body chemistry. Demonstrate that radioactive material eventually achieves an equilibrium internal dose rate that depends on the effective half-life of the material.

(c) Dr Moron is concerned that the radioiodine concentration in pond water will increase without limit as it is released to a pond. Assuming that ^{131}I settles onto a pond at a steady-state rate, calculate the input rate to the pond's surface using the following information provided in the NUTS contention: the deposition velocity is 1×10^{-2} m/s, release rate is 1×10^{8} Bq/s, and atmospheric dispersion factor (χ/Q) is 1.8×10^{-7} s/m^3. NUTS asserts that the input rate of radioiodine into the pond is 550 Bq/m^2-s. Is their assertion valid?

(d) To further answer the concern noted in Question (c), assume the daily ^{131}I input rate to the pond is 0.5 Bq/m^2. Calculate the maximum steady-state concentration of ^{131}I in the pond assuming the pond surface is 100 m by 10 m and the pond depth is 1 m (average).

(e) Dr Moron is concerned that radioiodine will increase in fish without limit as it is released to a pond. To answer this concern, calculate the equilibrium concentration of radioactive iodine expected in fish. The following information utilized by NUTS to develop their assertion includes the daily water intake by fish (8×10^{-5} m^3/kg-day) and equilibrium activity in the pond (10 Bq/m^3).

(f) Question (e) assumed that all ^{131}I that entered the pond was available for concentration in the fish tissue. NUTS insists this is a valid assumption because the ^{131}I concentration in the pond water increases without limit. What factors contribute to the inaccuracy of the NUTS contention?

(g) Dr Moron suggests that the ^{131}I deposited in the pond is in a metastable state and therefore has a half-life much greater than determined in the scientific literature. He suggests that the facility's production process is altering ^{131}I to produce a long-lived metastable state. How do you counter this contention?

References

10CFR20 (2015) Standards for Protection Against Ionizing Radiation, National Archives and Records Administration, U.S. Government Printing Office, Washington, DC.

10CFR50 (2015) Domestic Licensing of Production and Utilization Facilities, National Archives and Records Administration, U.S. Government Printing Office, Washington, DC.

10CFR835 (2015) Occupational Radiation Protection, National Archives and Records Administration, U.S. Government Printing Office, Washington, DC.

Aerts, J.C.J.H., Lin, N., Botzen, W., Emanuel, K., and de Moel, H. (2013) Low-probability flood risk modeling for New York City. *Risk Anal.*, **33**, 772.

American Nuclear Society Report (2012) *ANS Committee Report: Fukushima Daiichi*, American Nuclear Society, LaGrange Park, IL.

Baverstock, K.F. and Stather, J.W. (eds) (1989) *Low Dose Radiation: Biological Bases of Risk Assessment*, Taylor and Francis, London.

Behling, U.H. and Hildebrand, J.E. (1986) *Radiation and Health Effects: A Report on the TMI-2 Accident and Related Health Studies*, GPU Nuclear Corporation, Middletown, PA.

Bevelacqua, J.J. (2003) Production equations in health physics. *Radiat. Prot. Manage.*, **20** (6), 9.

Bevelacqua, J.J. (2004a) Skin contaminations, internal intakes, and ALARA. *Radiat. Prot. Manage.*, **21** (1), 11.

Bevelacqua, J.J. (2004b) Muon colliders and neutrino dose equivalents: ALARA challenges for the 21st century. *Radiat. Prot. Manage.*, **21** (4), 8.

Bevelacqua, J.J. (2004c) Point source approximations in health physics. *Radiat. Prot. Manage.*, **21** (5), 9.

Bevelacqua, J.J. (2005a) An overview of the health physics considerations at a 21st century fusion power facility. *Radiat. Prot. Manage.*, **22** (2), 10.

Bevelacqua, J.J. (2005b) Internal dosimetry primer. *Radiat. Prot. Manage.*, **22** (5), 7.

Bevelacqua, J.J. (2009) *Contemporary Health Physics: Problems and Solutions*, 2nd edn, Wiley-VCH Verlag GmbH, Weinheim.

Bevelacqua, J.J. (2010a) *Basic Health Physics: Problems and Solutions*, 2nd edn, Wiley-VCH Verlag GmbH, Weinheim.

Bevelacqua, J.J. (2010b) Radiological dispersion devices and basic radiation science. *Phys. Teach.*, **48**, 302.

Bevelacqua, J.J. (2010c) Practical and effective ALARA. *Operational Radiation Safety. Health Phys.*, **98** (Suppl. 2), S39.

Bevelacqua, J.J. (2011) Conservatism in effective dose calculations for accident events involving fuel reprocessing waste tanks. *Health Phys.*, **101**, 48.

Bevelacqua, J.J. (2012a) Applicability of health physics lessons learned from the Three Mile Island Unit-2 accident to the Fukushima Daiichi accident. *J. Environ. Radioact.*, **105**, 6.

Bevelacqua, J.J. (2012b) Muon colliders and neutrino effective doses, Chapter 5 appearing in, in *Particle Physics* (ed E. Kennedy), InTech, p. 91. ISBN: 978-953-51-0481-0. http://www.intechopen.com/books/particle-physics/muon-colliders-and-neutrino-effective-doses (accessed 7 August 2015).

Bevelacqua, J.J. (2012c) Environmental radiation effects from muon and tau colliders and their impact on facility licensing. *J. Environ. Radioact.*, **113**, 123.

Bevelacqua, J.J. (2012d) The Fukushima Daiichi accident and its radiological impact on the environment. *Phys. Teach.*, **50**, 368.

Bevelacqua, J.J. (2013a) Nuclear regulation in the United States and a possible framework for an international regulatory approach. *Int. Nucl. Saf. J.*, **2** (1), 1. http://nuclearsafety.info/international-nuclear-safety-journal/index.php/INSJ/article/view/12/pdf_1 (accessed 7 August 2015).

Bevelacqua, J.J. (2013b) Methodology to ascertain core recriticality following a severe fission reactor accident. *Int. Nucl. Saf. J.*, **2** (1), 52. http://nuclearsafety.info/international-nuclear-safety-journal/index.php/INSJ/article/view/13/pdf_1 (accessed 7 August 2015).

Bevelacqua, J.J. (2014a) Health physics aspects of generation IV reactors. *Int. Nucl. Saf. J.*, **3** (1), 13. http://nuclearsafety.info/international-nuclear-safety-journal/index.php/INSJ/article/view/20/pdf_1 (accessed 7 August 2015).

Bevelacqua, J.J. (2014b) Radiation protection considerations for nuclear power expansion to nations with limited technical infrastructure. *Int. Nucl. Saf. J.*, **3** (3), 8. http://nuclearsafety.info/international-nuclear-safety-journal/index.php/INSJ/article/view/37/pdf (accessed 7 August 2015).

Bevelacqua, J.J. (2014c) Health physics considerations for the ITER. *Int. Nucl. Saf. J.*, **3** (3), 25. http://nuclearsafety.info/international-nuclear-safety-journal/index.php/INSJ/article/view/36/pdf (accessed 7 August 2015).

Brodsky, A. (2014) *Handbook for Survival: Saving Lives During Radiation Releases and Other Disasters*, Fideli Publishing, Inc., Martinsville, IN.

Bromet, E.J. (2014) Emotional consequences of nuclear power plant disasters. *Health Phys.*, **106** (2), 206.

Brumfiel, G. (2011a) The meltdown that wasn't. *Nature*, **471**, 417.

Brumfiel, G. (2011b) Fukushima set for epic clean-up. *Nature*, **472**, 146.

Brumfiel, G. and Cyranoski, D. (2011) Quake sparks nuclear crisis. *Nature*, **471**, 273.

Butler, D. (2011a) Radioactivity spreads in Japan. *Nature*, **471**, 355.

Butler, D. (2011b) Fukushima health risks scrutinized. *Nature*, **472**, 13.

Cohen, B.L. and Lee, I. (1979) A catalog of risks. *Health Phys.*, **36**, 707.

Doolittle, W.W., Bredvad, R.S., and Bevelacqua, J.J. (1992) An ALARA conscious hot particle control program. *Radiat. Prot. Manage.*, **9**, 27.

Doss, M. (2012) Shifting the paradigm in radiation safety. *Dose-Response*, **10**, 562.

EPA (1992) Manual of Protective Action Guides and Protective Actions for Nuclear Incidents, EPA-400-R-92-001, U. S. Environmental Protection Agency, Washington, DC

Fabrikant, J.I. (1981) Planning Guidance for Protection and Recovery Following Radiological Dispersal Device (RDD) and Improvised Nuclear Device (IND) Incidents. *Health Phys.*, **40**, 151.

Federal Register (2008) Friday, August 1, 2008, Notices, *Fed. Regist.*, **73** (149), 45029.

FEMA Program Manual (2013) *Radiological Emergency Preparedness*, Federal Emergency Management Agency, Washington, DC.

FNSA Report (2011) Complementary Safety Assessments of the French Nuclear Power Plants, French Nuclear Safety Authority, Paris.

Gollnick, D.A. (2011) *Basic Radiation Protection Technology*, 6th edn, Pacific Radiation Corporation, Altadena, CA.

Harada, K.H. *et al.* (2014) Radiation dose rates now and in the future for residents neighboring restricted areas of the Fukushima Daiichi Nuclear Power Plant. *Proc. Natl. Acad. Sci. U.S.A.*, **111**, E914–E923. www.pnas.org/cgi/doi/10.1073/pnas.1315684111 (accessed 7 August 2015).

IAEA (2000) *Climate Change and Nuclear Power*, Publication IAEA/PI/A72E, International Atomic Energy Agency, Vienna.

IAEA (2006a) *Effective Nuclear Regulatory Systems: Facing Safety and Security Challenges*, Publication STI/PUB/1272, International Atomic Energy Agency, Vienna.

IAEA (2006b) *Chernobyl's Legacy: Health, Environmental and Socio-economic Impacts and Recommendations to the Governments of Belarus, the Russian Federation and Ukraine*, The Chernobyl Forum: 2003–2005, Second revised version, Publication IAEA/PI/A.87 Rev.2 / 06-09181, International Atomic Energy Agency, Vienna.

IAEA Nuclear Energy (2014) *Experiences and Lessons Learned Worldwide in the Cleanup and Decommissioning of Nuclear Facilities in the Aftermath of Accidents (STI/PUB/1644)*, Series No. NW-T-2.7, International Atomic Energy Agency, Vienna.

IAEA Report (2009) *INES: The International Nuclear and Radiological Event Scale User's Manual 2008*, International Atomic Energy Agency, Vienna.

IAEA Report (2012) *Fukushima Daiichi Status Report 28 June 2012*, International Atomic Energy Agency, Vienna.

IAEA Safety (1992) *The Chernobyl Accident: Updating of INSAG-1*, Series No. 75 STI/PUB/913, International Atomic Energy Agency, Vienna.

IAEA-EPR (2013) *Actions to Protect the Public in an Emergency Due to Severe Conditions at a Light Water Reactor*, International Atomic Energy Agency, Vienna.

ICRP (1959) *Permissible Dose for Internal Radiation*, Publication 2, Pergamon Press, Oxford.

ICRP (1977) *Recommendations of the International Commission on Radiological Protection*, Publication 26, Pergamon Press, Oxford.

ICRP (1991) *Recommendations of the ICRP*, Publication 60, Pergamon Press, Oxford.

ICRP (1993) *Principles for Intervention for Protection of the Public in a Nuclear Emergency*, Publication 63, Pergamon Press, Oxford.

ICRP (2005) *Protecting People Against Radiation Exposure in the Event of a Radiological Attack*, Publication 96, Elsevier, Amsterdam.

ICRP (2007) *The 2007 Recommendations of the International Commission on Radiological Protection*, Publication 103, Elsevier, Amsterdam.

ICRP (2011) Statement 4825-3093-1464, Statement on Tissue Reactions April 21, http://www.icrp.org/docs/icrp%20statement%20on%20tissue%20reactions. pdf (accessed 7 August 2015).

ICRP (2013) *Radiological Protection in Geological Disposal of Long-lived Solid Radioactive Waste*, Publication 122, Elsevier, Amsterdam.

Intergovernmental Panel on Climate Change Report (2014) *Climate Change 2014: Impacts, Adaptation, and Vulnerability*, Cambridge University Press, Cambridge, New York. http://www.ipcc.ch/report/ar5/wg2/ (accessed 7 August 2015).

IRSN Report (2011) 22 March 2011 Information Report, Institut de Radioprotection et de Sûreté Nucléaire, Paris.

Kharecha, K.A. and Hansen, J.E. (2013) Prevented mortality and greenhouse gas emissions from historical and projected nuclear power. *Environ. Sci. Technol.*, **47** (9), 4889.

Knief, R.A. (1985) *Nuclear Criticality Safety*, American Nuclear Society, LaGrange Park, IL.

Lin, N., Emanuel, K., Oppenheimer, M., and Vanmarcke, E. (2012) Physically based assessment of hurricane surge threat under climate change. *Nat. Clim. Change*, **2**, 462.

Loewen E.P., (2012) President's special session: low-level radiation & its implications for the Fukushima recovery, in *2012 ANS Annual Meeting*, Chicago, IL, June 24–28, 2012, American Nuclear Society, LaGrange Park, IL.

Matsui, T. (2011) Deciphering the Measured Ratios of Iodine-131 and Cesium-137 at the Fukushima Reactors, http://arxiv.org/PS_cache/arxiv/pdf/1105/1105.0242v1.pdf(accessed 7 August 2015).

Mottram, M.E. and Hill, H.A. (1949) Radiation therapy of ringworm of the scalp. *Calif. Med.*, **70**, 189.

National Academy of Sciences (2012) Analysis of Cancer Risks in Populations Near Nuclear Facilities: Phase I. Report No. 13388, The National Academy Press, Washington, DC.

National Academy of Sciences (2013) Messaging for Engineering: From Research to Action. Report No. 13463, The National Academy Press, Washington, DC.

National Academy of Sciences (2014a) Lessons Learned from the Fukushima Nuclear Accident for Improving Safety of U.S. Nuclear Plants. Report No.18294, The National Academy Press, Washington, DC.

National Academy of Sciences (2014b) Best Practices for Risk- Informed Decision Making Regarding Contaminated Sites: Summary of a Workshop Series. Report No. 18747, The National Academies Press, Washington, DC.

National Academy of Sciences (2014c) An All-of-Government Approach to Increase Resilience for International Chemical, Biological, Radiological, Nuclear, and Explosive (CBRNE) Events. Report No. 18814, The National Academy Press, Washington, DC.

National Academy of Sciences (2014d) Analysis of Cancer Risks in Populations Near Nuclear Facilities: Phase 2 Pilot Planning.

Report No. 18968, The National Academy Press, Washington, DC.
National Academy of Sciences (2014e) The Science of Responding to a Nuclear Reactor Accident: Summary of a Symposium. Report No. 19002, The National Academy Press, Washington, DC.
National Research Council (1980) *Committee on the Biological Effects of Ionizing Radiation, The Effects on Populations of Exposure to Low Levels of Ionizing Radiation, BEIR III*, National Academy Press, Washington, DC.
National Research Council (1990) *The Health Effects of Exposure to Low Levels of Ionizing Radiation, BEIR V*, National Academy Press, Washington, DC.
National Research Council (2006) *Health Risks from Exposure to Low Levels of Ionizing Radiation, BEIR VII Phase 2*, National Academy Press, Washington, DC.
NCRP (1993) Limitations of Exposure to Ionizing Radiation. Report 116, NCRP Publications, Bethesda, MD.
NCRP (2001) Management of Terrorist Events Involving Radioactive Material. Report 138, NCRP Publications, Bethesda, MD.
NEA 3508(2002) Chernobyl: Assessment of Radiological and Health Impact -2002 Update of Chernobyl: Ten Years On, Nuclear Energy Agency, Paris.
Niehaus, R. (1983) The future role of risk assessment in nuclear safety. *IAEA Bull.*, **25**, 30.
Nuclear Energy Institute (2012) Diverse and Flexible Coping Strategies (FLEX) Implementation Guide. Report NEI 12-05, Revision B1, Nuclear Energy Institute, Washington, DC.
Orient, J.M. (2014) Fukushima and reflections on radiation as a terror weapon. *J. Am. Physicians Surg.*, **19**, 48.
Parmentola, J. (2013) Time for a game-changing nuclear technology. *Phys. Today*, **66** (2), 8.
Sandia National Laboratories (2014a) Evaluation of Options for Permanent Geologic Disposal of Spent Nuclear Fuel and High-Level Radioactive Waste in Support of a Comprehensive National Nuclear Fuel Cycle Strategy Volume I. Report SAND2014-0187P (Vol. I), Sandia National Laboratories, Albuquerque, NM.
Sandia National Laboratories (2014b) Evaluation of Options for Permanent Geologic Disposal of Spent Nuclear Fuel and High-Level Radioactive Waste in Support of a Comprehensive National Nuclear Fuel Cycle Strategy Volume II: Appendices. Report SAND2014-0189P (Vol. II), Sandia National Laboratories, Albuquerque, NM.
Schimmoller, B. (2013) Existential threats. *Power Eng.*, **117** (3), 12.
Schlosser, E. (2013) *Command and Control – Nuclear Weapons, the Damascus Accident, and the Illusion of Safety*, Penguin, New York.
Sinclair, W.K. (1981) Radiation protection: the NCRP guidelines and some considerations for the future. *Yale J. Biol. Med.*, **54**, 471.
Takahashi, S. (2014) *Radiation Monitoring and Dose Estimation of the Fukushima Nuclear Accident*, Springer, Tokyo.
Terje, A. and Renn, O. (2015) An evaluation of the treatment of risk and uncertainties in the IPCC reports on climate change. *Risk Anal.*, **35**, 701.
The National Diet of Japan Report (2012) The Official Report of the Fukushima Nuclear Accident Independent Investigation Commission, Fukushima Nuclear Accident Independent Investigation Commission, Tokyo.
Till, J.E. and Meyer, H.R. (eds) (1983) *Radiological Assessment: A Textbook on Environmental Dose Analysis*, NRC Report NUREG/CR-3332, NUREG, Washington, DC.
Tollefson, J. (2013) New York vs the Sea. *Nature*, **494**, 162.
Turner, J.E. (2007) *Atoms, Radiation, and Radiation Protection*, 3rd edn, Wiley-VCH Verlag GmbH, Weinheim.
U. S. Department of Energy (2009) Preliminary Report on Operational Guidelines Developed for Use in Emergency Preparedness and Response to a Radiological Dispersal Device Incident. DOE/HS-0001; ANL/EVS/TM/09-1, U.S. Department of Energy, Washington, DC.
U. S. Department of Energy Report (2014a) International Nuclear Safeguards Engagement Program, National Nuclear Security Administration, US Department of Energy, Washington, DC.

U. S. Department of Energy Report (2014b) Phase 1 – Radiological Release Event at the Waste Isolation Pilot Plant on February 14, 2014, US Department of Energy, Washington, DC.

U. S. Environmental Protection Agency (2013) PAG Manual, Protective Action Guides and Planning Guidance For Radiological Incidents, U.S. Environmental Protection Agency Draft for Interim Use and Public Comment, U. S. Environmental Protection Agency, Washington, DC.

U. S. Government Accountability Office (2014) Emergency Preparedness – Opportunities Exist to Strengthen Interagency Assessments and Accountability for Closing Capability Gaps. Report GAO-15-20, Government Printing Office, Washington, DC.

United States Government Accountability Office (2011) Nuclear Regulatory Commission Oversight of Underground Piping Systems Commensurate with Risk, but Proactive Measures Could Help Address Future Leaks. Report GAO-11-563, Government Accountability Office, Washington, DC.

U. S. Navy (2012) Occupational Radiation Exposure from U.S. Naval Nuclear Plants and Their Support Facilities. Report NT-12-2, Naval Nuclear Propulsion Program, Department of the Navy, Washington, DC.

USNRC Regulatory Guide 8.8 (1978) Information Relevant to Ensuring That Occupational Exposures at Nuclear Power Stations Will Be as Low as Reasonably Achievable (ALARA), Revision 3, U.S. Nuclear Regulatory Commission, Washington, DC.

USNRC Regulatory Guide (1979) Investigation into the March 28, 1979 Three Mile Island Accident by the Office of Inspection and Enforcement, NUREG-0600, U.S. Nuclear Regulatory Commission, Washington, DC.

USNRC Regulatory Guide (1980) Three Mile Island: A Report to the Commissioners and to the Public, NUREG/CR-1250, U.S. Nuclear Regulatory Commission, Washington, DC.

USNRC Regulatory Guide (1983) Generic Implications of ATWS Events at the Salem Nuclear Power Plant, NUREG-1000, U.S. Nuclear Regulatory Commission, Washington, DC.

USNRC Regulatory Guide (1989) Implications of the Accident at Chernobyl for Safety Regulation of Commercial Nuclear Power Plants in the United States, NUREG/CR-1251, U.S. Nuclear Regulatory Commission, Washington, DC.

USNRC Regulatory Guide 8.29 (1996) Instruction Concerning Risks from Occupational Radiation Exposure, Revision 1, U.S. Nuclear Regulatory Commission, Washington, DC.

USNRC Regulatory Guide (2008) Davis-Besse Reactor Pressure Vessel Head Degradation, NUREG/BR-0353. Revision 1, U.S. Nuclear Regulatory Commission, Washington, DC.

USNRC Regulatory Guide (2010a) A Short History of Nuclear Regulation, 1946–2009, NUREG/BR-0175, Revision 2, U.S. Nuclear Regulatory Commission, Washington, DC.

USNRC Regulatory Guide (2010b) Volume 1, Safety Evaluation Report Related to Disposal of High-Level Radioactive Wastes in a Geologic Repository at Yucca Mountain, Nevada: General Information, NUREG-1949, U.S. Nuclear Regulatory Commission, Washington, DC.

USNRC Regulatory Guide (2012a) State-of-the-Art Reactor Consequence Analyses (SOARCA) Report, NUREG-1935, U.S. Nuclear Regulatory Commission, Washington, DC.

USNRC Regulatory Guide (2012b) A Proposed Risk Management Regulatory Framework, NUREG-2150, U.S. Nuclear Regulatory Commission, Washington, DC.

USNRC Regulatory Guide (2014a) Volume 3, Safety Evaluation Report Related to Disposal of High-Level Radioactive Wastes in a Geologic Repository at Yucca Mountain, Nevada: Repository Safety After Permanent Closure, NUREG-1949, U.S. Nuclear Regulatory Commission, Washington, DC.

USNRC Regulatory Guide (2014b) Volume 4, Safety Evaluation Report Related to Disposal of High-Level Radioactive Wastes in a Geologic Repository at Yucca Mountain, Nevada: Administrative and Programmatic Requirements, NUREG-1949, U.S. Nuclear Regulatory Commission, Washington, DC.

USNRC Regulatory Guide (2015a) Volume 5, Safety Evaluation Report Related to Disposal of High-Level Radioactive Wastes in a Geologic Repository at Yucca Mountain, Nevada: Proposed Conditions on the Construction Authorization and Probable Subjects of License Specifications, NUREG-1949, U.S. Nuclear Regulatory Commission, Washington, DC

USNRC Regulatory Guide (2015b) Volume 2, Safety Evaluation Report Related to Disposal of High-Level Radioactive Wastes in a Geologic Repository at Yucca Mountain, Nevada: Repository Safety Before Permanent Closure, NUREG-1949, U.S. Nuclear Regulatory Commission, Washington, DC.

USNRC Report (2011) The Near-Term Task Force Review of Insights from the Fukushima Daiichi Accident, U.S. Nuclear Regulatory Commission, Washington, DC.

USNRC Report WASH-1400 (1975) Reactor Safety Study: An assessment of Accident Risks in U.S. Commercial Nuclear Power Plants, U.S. Nuclear Regulatory Commission, Washington, DC.

Vetter, R.J. (ed) (1990) The biological effects of low-dose radiation: a workshop. *Health Phys.*, **59**, 1.

Wang, Q. and Chen, X. (2012) Regulatory failures for nuclear safety – the bad example of Japan – implication for the rest of world. *Renewable Sustainable Energy Rev.*, **16**, 2610.

Weightman, M. (2011) *Japanese Earthquake and Tsunami: Implications for the UK Nuclear Industry – Interim Report*, Office for Nuclear Regulation, London.

World Nuclear Association Discussion Paper (2008) *Benefits Gained through International Harmonization of Nuclear Safety Standards for Reactor Designs*, World Nuclear Association, London.

Part VI
Solutions to Problems

Part VI of this book further defines and develops the material presented in Parts I–V. The problem solutions presented in Part VI illustrate many of the practical difficulties that will be encountered in twenty-first century health physics applications. Readers are strongly encouraged to examine carefully these solutions to gain the maximum benefit from this text.

The Part I solutions are general and provide an overview of possible black swan events and a preview of upcoming text material. These Chapter 1 problems are intended to stimulate consideration of issues and challenges that could be encountered by twenty-first century Health Physicists. These challenges are illustrated in considerably more detail in the Parts II–V problems and solutions that represent more probable events. The Chapters 2–7 solutions are based on the text material and included references. Solutions for Parts II–V are more detailed than the Part I solutions and amplify the concepts presented in this text.

Health Physics: Radiation-Generating Devices, Characteristics, and Hazards, First Edition. Joseph John Bevelacqua.

Solutions

Solutions for Chapter 1

1.1 The Hanford Tank Farms consist of 177 large underground tanks designed to store the radioactive and hazardous wastes generated during production of defense-related materials from the 1940s through the late 1980s. Over 190 million liters of wastes are stored in the Hanford Tank Farms.

The primary isotopes that have the potential for external exposure include ^{90}Sr, ^{90}Y, and ^{137}Cs. ^{239}Pu and ^{241}Am provide the dominant source terms for internal exposure. The total activity in the waste tanks is about 10 EBq.

A Tunguska-type event occurring within 1 km of the Hanford Tank Farms containing fuel reprocessing waste would mobilize and disperse radioactive material. The extent of mobilization and dispersion is uncertain, but the 15 MT detonation would likely transport a significant quantity of the waste over an area as large as the Tunguska-affected area.

The Hanford Site is isolated, but populated areas and a nuclear power plant are within tens of kilometers. These areas would be affected by the meteorite blast as well as the dispersed radioactivity. Health physics actions include isolation of the contaminated areas. The environment would be significantly affected with contamination of the Columbia River as well as neighboring agricultural and residential areas. The event would require significant resources for recovery. Specific effects and actions include the following:

Public effects:

1. Dispersion of fission products and actinides over a large area
2. Contamination of agricultural and residential areas
3. Contamination of the Columbia River and other surface waters
4. Contamination of human, plant, and animal populations
5. Injuries and fatalities
6. Property damage

Health Physics: Radiation-Generating Devices, Characteristics, and Hazards, First Edition. Joseph John Bevelacqua.

Health physics actions:

1. Evacuate and isolate contaminated areas and elevated radiation areas.
2. Characterize the extent, type, and magnitude of contamination in the affected areas.
3. Characterize the dose rates in the affected areas.
4. Sample agricultural food products including water, milk, meat, fruits, vegetables, and other foods to determine any required restrictions.
5. Establish safe levels of radioactive materials in food and water.
6. Establish reentry criteria for the public to return to their homes.
7. Decontaminate the affected areas.
8. Assess public internal intakes.
9. Assess public doses.
10. Initiate medical procedures to follow patients throughout their lives to provide care and determine health effects attributable to their exposure to the dispersed radioactive materials.
11. Decontaminate affected individuals.
12. Provide radiological assistance to medical personnel in dealing with contaminated/injured individuals.

1.2 Reactor site protection from toxic materials focuses on control room habitability. For example, the US Nuclear Regulatory Commission Regulatory Guide 1.78 (*Evaluating the Habitability of a Nuclear Power Plant Control Room during a Postulated Hazardous Chemical Release*) notes that the control room of a nuclear power plant should be protected from hazardous chemicals that may be discharged as a result of equipment failures, human errors, or events and conditions outside the control of the nuclear power plant. The design requirements of 10CFR50 (*Domestic Licensing of Production and Utilization Facilities*) requires that the facility design evaluates potential pathways for radioactivity and radiation that may lead to control room habitability problems and make necessary design provisions to preclude such problems.

Given these design requirements, it is likely that the facility control room is protected from this toxic gas release. Although the control room personnel will be protected, other station personnel could be significantly affected. Since the gas cloud covers the facility for an extended period, individuals outside the control room could be seriously injured or become fatalities. Loss of trained facility personnel would have a significant impact on facility operations.

Although the problem did not state any initial conditions, it is logical to assume the facility was operating normally prior to the toxic gas event. Minimum control room staffing is required by the license and a safe shutdown condition could be achieved by these personnel.

Any operational problems or equipment failures that would require support from other personnel could be impacted by the toxic gas release.

It is likely that a safe reactor shutdown will be accomplished, and there will be minimal health physics consequences. All fission product barriers can be protected by control room actions.

Communications with off-site organizations can be established and support personnel provided to the facility. There should be ample time to obtain specialized chemical protective equipment to perform any needed repair operations to maintain a safe reactor condition and continue core cooling. However, it is unclear if there is sufficient time to provide medical assistance to all injured personnel.

1.3 An extended (e.g., several weeks) power blackout event that occurs at a uranium enrichment facility using lasers and UF_6 gas as the working fluid deprives the facility of off-site power. Without off-site power, the facility must rely on emergency generators for electrical energy.

The NRC's Safety Evaluation Report for the laser-based uranium enrichment plant summarized in NUREG-2120 (*Safety Evaluation Report for the General Electric–Hitachi Global Laser Enrichment LLC Laser-Based Uranium Enrichment Plant in Wilmington, North Carolina*) notes that the facility's air permit requires diesel backup electrical generators. Given the extended power outage period, these generators will consume their available fuel and become inoperable. Since the region has also lost power, the fuel may not be replenished. Therefore, the facility could be in a station blackout condition.

Under normal operating conditions, the UF_6 in the enrichment process is a hot, pressurized gas. Upon loss of power, the gas cools and solidifies.

In spite of the loss of power, the facility's separation units operate as closed systems that confine the UF_6 working fluid. The loss of power leads to cooling of process components and the solidification of the UF_6 gas that limits its dispersibility. The UF_6 solidification occurs in the process vessels and feed transfer lines which further limits the spread of radioactive material.

The solidification of the UF_6 will significantly limit any environmental impact, and the dispersion of this material will likely be limited to the immediate processing areas. This facility may have challenges in reestablishing operations with the solidified UF_6, but the health physics consequences of the power outage should be manageable.

1.4 A dirty bomb or radiological dispersal device (RDD) is designed to spread radioactive material using conventional explosives or other means. The radiological properties of ^{32}P, ^{60}Co, and ^{131}I used in the RDD are summarized below.

Radiological characteristics of ^{32}P, ^{60}Co, and ^{131}I[a)]

Nuclide	Major radiation emitted		Half-life	Production modes
	Type	Energy (MeV)		
^{32}P	β^-	1.709 (max)	14.28 days	$^{31}P(n, \gamma)^{32}P$ $^{32}S(n, p)^{32}P$ $^{35}Cl(n, \alpha)^{32}P$
^{60}Co	β^-	0.318 (max)	5.271 years	$^{59}Co(n, \gamma)^{60}Co$ $^{60}Ni(n, p)^{60}Co$
	γ	1.1732 1.3325		$^{63}Cu(n, \alpha)^{60}Co$
^{131}I	β^-	0.606 (max)	8.023 days	Fission product
	γ	0.3645		$^{130}Te + n \rightarrow {}^{131}Te \xrightarrow{\beta^-} {}^{131}I$ $^{131}Xe(n, p)^{131}I$

a) The radiological hazard depends on the particle size and activity of the dispersed isotope.

The effectiveness of ^{32}P and ^{131}I in an RDD has been debated in the literature since their half-lives are short, and consequently their environmental impact is limited. Both ^{32}P and ^{131}I are internal and external radiation hazards. Under normal circumstances, ^{32}P is primarily an ingestion hazard, and ^{131}I can be inhaled, ingested, or absorbed through the skin (in certain chemical forms). However, ^{32}P would also be an inhalation hazard if dispersed by an RDD. Given the short half-lives of these radionuclides, entrance into the effected areas should be restricted to minimize their impact. Access should be permitted for search and rescue operations and required recovery activities.

By waiting 10 half-lives, the initial activity of ^{32}P and ^{131}I is reduced by a factor of about 1000. This minimizes their radiological impact. There may be areas of localized hot spots that can be decontaminated by recovery personnel. After 10 half-lives and localized decontamination efforts, the radiological effects from a ^{32}P or ^{131}I device should be minimal.

^{60}Co presents a more significant hazard with its 5.271-year half-life. With a long half-life, ^{60}Co contamination requires that the effected area be decontaminated to a level that is acceptable to stakeholders, property owners, and government officials. The contaminated area is primarily an external radiation hazard but presents an internal hazard if the ^{60}Co is inhaled or ingested. The approach to initial search and rescue and recovery activities is guided by radiological conditions. Following these activities, the contaminated areas are more thoroughly surveyed for their radiological hazard and physical damage. These surveys determine the optimum approach for the recovery of the contaminated areas and the optimum decontamination methods. Stakeholders should be involved in establishing acceptable residual contamination and radiation levels. These levels require considerable effort to establish since the various stakeholder groups will likely have divergent views of the final acceptable radiation and contamination levels.

1.5 The currents induced in transmission lines from a massive solar event can severely damage electrical equipment especially generators and transformers. A massive event causes induced currents in coils and cores, and the associated temperature increase can overload and severely damage transformers throughout an electrical distribution system. If a massive solar event struck the area surrounding a nuclear power reactor, the site would lose all off-site power. In a similar manner, the transformers and emergency electrical generators supplying power to the nuclear facility would be disrupted.

If the transformers and emergency electrical generators were disabled, the facility would be in a station blackout condition following the depletion of the station batteries. Given these conditions, the fission product barriers remain intact until the batteries are depleted. Without power, cooling water flow ceases and core temperature increases. If no backup power or supplementary cooling could be provided, the fuel fission product barrier would be breached.

If the event proceeded without power or cooling water, the fuel would melt and the second fission product barrier (reactor vessel and included piping) would also be breached. Breaching of the third barrier would depend on the construction of the containment and subsequent capability to provide cooling water or power.

The event could proceed as in the case of the Fukushima Daiichi accident where all three fission product barriers were lost. Given the emergency power and core cooling corrective actions mandated from the Fukushima Daiichi accident, it is likely that either supplemental power or cooling flow would be reestablished to mitigate the event.

If the transformers and generators remain available, the emergency diesel generators would provide power to the facility and maintain the safety systems needed to cool the core and protect the fission product barriers. The diesel generators would need to replenish their fuel supply during the month-long loss of off-site power event. However, it is likely that fuel replenishment would be accomplished and the fission product barriers protected.

1.6 A limited nuclear exchange has occurred between neighboring nations. Each nation has detonated three 250 kT ^{239}Pu fission devices over separate, heavily populated targets. The population group of interest lies outside these cities and at a distance removed from the immediate blast area. The radionuclides of concern include unfissioned weapons material and associated actinides (e.g., ^{238}Pu, ^{239}Pu, ^{240}Pu, ^{242}Pu, and ^{241}Am) and fission products (e.g., ^{89}Sr, ^{90}Sr, ^{103}Ru, ^{106}Ru, ^{131}I, ^{134}Cs, and ^{137}Cs). These isotopes reach the specified population group primarily through fallout from the radioactive plume.

These isotopes enter the food chain by contaminating soil, water, plants, and other food items. There are a number of pathways by which these materials reach the food chain including:

Fallout → water → ingestion by man
Fallout → water → ingestion by animals → milk → man
Fallout → crops → ingestion by animals → milk → man
Fallout → water → ingestion by animals → meat → man
Fallout → crops → ingestion by animals → meat → man
Fallout → grass → ingestion by animals → milk → man
Fallout → grass → ingestion by animals → meat → man
Fallout → crops → ingestion by man
Fallout → water → fish → ingestion by man
Fallout → inhalation by man
Fallout → inhalation by animals → meat → man
Fallout → inhalation by animals → milk → man

The absorbed dose from the consumption of contaminated food, water, and milk is minimized by taking a number of protective actions that include:

1. Feeding animals stored feed and water to limit the contamination of milk and meat.
2. Drinking water bottled prior to the nuclear exchange.
3. Consuming powdered milk processed prior to the nuclear exchange.
4. Consuming canned foods.
5. Importing food, water, and milk from areas unaffected by the nuclear exchange.
6. Establishing acceptable levels of contamination by isotope for foods for general consumption, drinking water, infant foods, and milk.
7. Limiting the distribution of foods for general consumption, drinking water, infant foods, and milk to items that meet the established acceptable levels of contamination by isotope.
8. Sheltering in place or evacuating the fallout area. This will depend on available resources and the condition of societal infrastructure.
9. Utilizing radioprotective chemicals (e.g., potassium iodide) to limit the intake of radioactive materials (e.g., iodine). These chemicals must be taken in a timely manner to be effective.

1.7 Failing to achieve a significant nuclear yield from an unsuccessful detonation is known as a *fizzle*. For the problem of interest, the fizzle involved a ^{235}U device. A fizzle has two generic categories. The first is a complete nuclear failure where the chemical explosives of the device disperse radioactive material near ground zero. This first category is essentially a ^{235}U dispersal device, and the primary isotope of concern is the fissile material. The second category of fizzle involves the fission of some of the ^{235}U, which generates a partial nuclear yield typically less than a few metric tons of TNT. Within this second category, ^{235}U and associated fission products produce a radiological hazard. The partial fission event is a significantly greater hazard than the first event category. It represents a combination of the effects of a ^{235}U dispersal devise and fission products produced by the limited nuclear yield.

The isotopes of concern for a weapon that does not achieve a significant nuclear yield are ^{235}U from the partially fissioned weapon and fission products including ^{90}Sr, ^{90}Y, ^{137}Cs, and ^{131}I. The fission products and unfissioned ^{235}U present internal and external dose challenges that must be addressed to permit reentry into the ground zero area. Limited activities of plutonium isotopes including ^{239}Pu will also be produced.

Reentry is governed by the dose rates and contamination levels. Since fission dose rates decrease by a factor of 10 during a 7-day period, reentry doses are significantly reduced by delaying reentry into the ground zero area. Earlier entry into ground zero could be achieved using robotic devices. Observing the ground zero area, determining absorbed dose rates, and mapping contamination levels can be achieved using aircraft or unmanned drones. These actions can be periodically performed to plan for manned entry into the ground zero area.

If human entry is required, it should await a reduction in dose rates. Appropriate protective clothing and respiratory protection are required. Given the presence of fission products and their associated beta–gamma radiation, coveralls and additional beta protection (e.g., fire gear) should be utilized to provide protection from contamination and the high-energy beta particles (e.g., $^{90}Sr/^{90}Y$). A self-contained breathing apparatus operating in pressure demand mode would provide maximum protection from internal deposition.

A number of parameters affect the absorbed dose rates and contamination levels near ground zero. Meteorological conditions that govern the dispersibility of the radioactive material beyond ground zero include the wind speed, associated Pasquill stability class, and precipitation. The dispersion of material is also dependent on the detonation height of the weapon. These factors change with time and have an impact on the specific health physics actions and requirements to reenter the ground zero area.

1.8 A research team has developed a cancer therapy technique using anti-^{12}C ions. There are a number of challenges and advantages of the proposed therapy protocol:

Challenges to deploying anti-^{12}C ions in cancer therapy:

1. Anti-^{12}C ions do not yet exist. Their creation presents a significant technological problem.
2. Once created, anti-^{12}C must be transported to the medical facility. It is unclear how anti-^{12}C will be confined to permit safe transport. Conventional antimatter confinement methods utilize electromagnetic fields within a research facility environment.
3. If administered as a beam, the anti-^{12}C must be collimated and accelerated to an appropriate energy to reach the tumor site. Collimation

and acceleration are significant operations to perform at a medical facility.

4. The anti-^{12}C ions need to be deposited at the tumor site. If administered as a beam, the ions will annihilate with surface tissue containing ^{12}C prior to reaching the tumor. This affects dose localization and enhances the irradiation of a healthy tissue.
5. If injected into the tumor site or transported there in another manner, annihilation outside the tumor is likely. The healthy tissue will be irradiated if the anti-^{12}C ions are not annihilated within the tumor.
6. Upon entering the body and irradiating the tumor site, the anti-^{12}C ions will annihilate with ^{12}C atoms to produce a variety of radiation types including pions and photons. Anti-^{12}C ions produce reasonable dose localization similar to ^{12}C heavy ion therapy, but the pions that result from the annihilation do not produce a tight absorbed dose profile and will irradiate the healthy tissue. The photons also irradiate the healthy tissue. Charged pion decay into muons must also be evaluated.

Positive aspects of anti-^{12}C ions in cancer therapy:

1. The photons produced during the annihilation event permit tracking the position of the anti-^{12}C ions. This facilitates determination of the location of the ions upon annihilation.
2. Significant energy deposition is possible if the ion annihilates within the tumor mass.
3. The technique could be useful for treating surface tumors since this minimizes the effects of annihilation of the ions prior to reaching the tumor.
4. The pions produced during the annihilation event can be localized at the tumor site if their kinetic energy is minimized. Low-energy anti-^{12}C ions facilitate the desired dose localization if these charged particles are delivered to the tumor site using an internal radiation-generating device.

Solutions for Chapter 2

2.1. (a) In formulating a criticality accident reentry plan, the primary considerations in developing recommendations are the worker's physical condition, plant's status, and radiological conditions. The requisite information is determined by evaluating the following items:

1. Determine if the worker is physically injured and the extent of the injuries. Medical personnel may be required if the injuries are severe.
2. Estimate the magnitude of the criticality event in terms of the number of fissions. The number of fissions is proportional to both the gamma and neutron absorbed doses.

3. Determine if the criticality event is terminated or ongoing. If the event was terminated, then the dose rates are decreasing. If the criticality is recurring, dose rates will likely increase.
4. Determine the location and time of the event. Planning an entry route and determining radiation levels depend on this information. All entry activities are to be performed in an ALARA manner.
5. The vessel containing the criticality and its status should be ascertained. If the vessel is not intact, then contamination is present and fission products have been released.
6. The likely radiation levels, airborne levels, and surface contamination levels should be determined. Installed radiation instrumentation, monitored from a remote location, supports this effort.
7. The likely toxicological concentrations are additional requisite data. Installed instrumentation, monitored from a remote location, supports this effort.
8. The availability and type of shielding should be determined.
9. Personnel locations relative to the site of the criticality event are needed information. The workers' position as a function of time determines their absorbed dose and the radiological hazards to a rescue team.
10. The time and duration of the criticality event affect the planned entry and exit routes. Significant dose rates are present after the event is terminated. This dose must be considered in planning subsequent actions.

Not all of this information is available, and sound judgment must be used to formulate your recommendations. Medical needs of the injured, rescue team safety, ALARA considerations, and the need for timely action are additional considerations in formulating the recommended rescue approach.

(b) The primary exposure pathways and radiation sources for three specific cases are requested. In general, the pathways depend on the condition of the vessel (i.e., intact or ruptured) containing the critical mass:

1. *Workers in the room at the time of the accident*:

   It is expected that the direct dose dominates the exposure considerations. Based on previous criticality events, workers near the tank receive large absorbed doses. For example, the 1999 Tokaimura criticality led to lethal doses in excess of 10 Gy. If the criticality has been terminated, a declining dose rate trend will be observed. If the criticality recurs, the dose rates will increase.

   For an intact vessel, the direct dose from the fission event includes prompt and delayed neutrons, fission gammas, activation gammas, activation betas, and bremsstrahlung. The individuals are also submerged in a noble gas and radioiodine cloud if the vessel provides a release pathway for these radionuclides.

For a ruptured vessel, the initial direct dose will be similar to the intact vessel case. The dose rates decrease since the critical geometry is lost following the vessel rupture. There will be a release of fission products including iodine and noble gases from the ruptured vessel. This produces skin, eye, and whole-body doses as well as the potential for an intake of radioactive materials. The source term includes particulates, noble gases, and iodine. Internal pathways include inhalation and ingestion of particulates and iodine in the released plume of radioactive material and skin absorption of iodine and tritium.

2. *Rescue workers assuming the criticality has stopped*:
   It is expected that the direct dose dominates the exposure considerations. Once the criticality is terminated, the dose rate will decrease, but the potential for a significant exposure remains.

   The radiological profile for the intact and ruptured vessel cases is similar to the description in the previous question. The dose pathways include submersion in a noble gas cloud, inhalation of particulates and iodine in the released plume of radioactive material, ingestion of particles and radioiodine, direct dose from the plume, and skin absorption of iodine and tritium. Respiratory protection decreases the internal intake, but the associated effective dose reduction should be evaluated against the increased direct dose caused by the inefficiencies created by personal protective equipment. Direct dose increases because respiratory protection decreases worker efficiency and extends task completion times.
3. *Other individuals within 0.1–10 km at the time of and following the incident*:
   Given the distance from the event, internal dose will dominate the exposure considerations. However, the direct dose must be monitored. If the criticality has ceased, a declining dose rate trend occurs. If the criticality recurs, the direct dose rates will increase.

   The direct dose rate depends on the quantity of shielding between the criticality location and the receptor. For a ruptured vessel, the direct dose is similar to the intact vessel case. There is also the potential for the release of fission products including iodine and noble gases from the vessel. This produces skin, eye, and whole-body doses as well as the potential for an intake of radioactive materials. The source term includes particulates, noble gases, and iodine. Dose pathways include (i) submersion in a noble gas cloud, (ii) inhalation of particulates and iodine in the released plume of radioactive material, (iii) ingestion of particles and radioiodine, (iv) direct dose, (v) skin absorption of iodine and tritium, (vi) radioactive material deposition on crops with subsequent ingestion, (vii) deposition on animal feed with subsequent ingestion of the animal's meat and milk, (viii) ground deposition and ground shine, (ix) immersion in contaminated water, (x) ingestion of contaminated water, and (xi) irrigation of crops with contaminated water and subsequent consumption.

(c) A method that could be used to quickly screen potentially irradiated persons near the criticality location is the collection and processing of the individual's dosimetry. Electronic dosimetry and self-reading pocket dosimetry provide immediate dose values. Thermoluminescent dosimetry is processed quickly if readers are available on-site.

Activation techniques are frequently used to rapidly screen individuals when exposure to a fast neutron field is suspected. Capture of fission neutrons by sodium atoms in the blood through the $^{23}Na(n, \gamma)^{24}Na$ reaction results in the formation of ^{24}Na which decays by photon emission. A Geiger–Müller detector placed near a large blood volume (i.e., under the armpit) detects the ^{24}Na photons. As a point of reference, a dose rate of about 10 μSv/h results from the acute exposure to 5 Gy of fast neutrons. Other activation sources such as rings and jewelry may also be counted, but the ^{24}Na activity provides a quick absorbed dose estimate. The ^{24}Na measurements must be performed in a low dose rate area.

(d) Medical interventions that could change the health outcome for an individual exposed to 7.5 Gy (whole body, deep dose) if administered during the first month following the incident include the following:

1. Place the victim in a sterile room to minimize infections since their ability to fight infection has been reduced.
2. Administer antibiotics to fight infection.
3. Administer fluids to minimize dehydration.
4. Evaluate the efficacy of bone marrow transplant therapy if a suitable donor or match is available.
5. Administer blood transfusions.
6. Administer hormones to assist lung tissue regeneration and maintain vital organ function.
7. Consider the use of radioprotective chemicals to mitigate the effects of the absorbed dose.

These interventions are intended to counter the acute radiation syndrome. The depletion of blood cells and damage to blood-forming organs and lung tissue are immediate concerns. These concerns manifest themselves as limited ability to fight infection, dehydration, deterioration of blood-forming organs, and deterioration of the lung and its ability to exchange gases. Vital organs should be monitored for possible radiation-induced detriment.

(e) Large acute radiation doses (e.g., from a criticality accident) are correctly expressed in units of Gy and not Sv. The absorbed dose (D) and dose equivalent (H) are related by the quality factor (Q) (ICRP 26):

$$H = DQ$$

Acute radiation effects are expressed in terms of absorbed dose instead of the dose equivalent because the quality factor is not defined for acute doses. Similar arguments apply to the radiation weighting factors used in ICRP 60 and 103.

2.2. (a) In this question, you are requested to calculate the committed dose equivalent ($H_{50,\mathrm{T}}$) to the specified organs and their respective committed effective dose equivalents (H_E):

1. Worker A's bone surface (BS) dose for an intake (I) of 5 ALI of ^{239}Pu and $w_\mathrm{T} = 0.03$:

$$H_{50,\mathrm{BS}} = \frac{I(^{239}\mathrm{Pu})}{\mathrm{ALI} - \mathrm{NS}(^{239}\mathrm{Pu})} 0.5\,\mathrm{Sv} = \frac{5\,\mathrm{ALI} - \mathrm{NS}(^{239}\mathrm{Pu})}{\mathrm{ALI} - \mathrm{NS}(^{239}\mathrm{Pu})} 0.5\,\mathrm{Sv} = 2.5\,\mathrm{Sv}$$

$$H_\mathrm{E} = w_{\mathrm{BS}} H_{50,\mathrm{BS}} = (0.03)(2.5\,\mathrm{Sv}) = 0.075\,\mathrm{Sv}$$

2. Worker B's thyroid (T) for an intake of 5 ALI of ^{131}I and $w_\mathrm{T} = 0.03$:

$$H_{50,\mathrm{T}} = \frac{I(^{131}\mathrm{I})}{\mathrm{ALI} - \mathrm{NS}(^{131}\mathrm{I})} 0.5\,\mathrm{Sv} = \frac{5\,\mathrm{ALI} - \mathrm{NS}(^{131}\mathrm{I})}{\mathrm{ALI} - \mathrm{NS}(^{131}\mathrm{I})} 0.5\,\mathrm{Sv} = 2.5\,\mathrm{Sv}$$

$$H_\mathrm{E} = w_\mathrm{T} H_{50,\mathrm{T}} = (0.03)(2.5\,\mathrm{Sv}) = 0.075\,\mathrm{Sv}$$

(b) In this part, you are to assess the physician's recommendation to remove the worker's thyroid to preclude the likelihood of thyroid cancer later in life. Although medical decisions are the responsibilities of a physician, you should provide relevant data for consideration by medical professionals.

A thyroid dose of 2.5 Gy is not sufficient to justify thyroid removal. Radiogenic thyroid cancer has a latency period of about 5 years, which suggests the worker should be monitored prior to radical surgery.

The NCRP reports absolute risk factors of 2.5×10^{-4} to 4.4×10^{-4} thyroid cancers per person-year-Gy for doses in the range of 0.06–15.0 Gy. Moreover, BEIR V notes that only about 10% of thyroid cancers are lethal and suggests a threshold of 2.5–5.0 Gy in animals. Although BEIR V notes a variety of risk models and coefficient values, the magnitude of the dose and risk coefficient, latency period, and lethality for thyroid cancer suggests observation, not removal, is the prudent course of action.

(c) Both workers develop solid tumor cancer 1 year later and are suing Ka-Boom Enterprises, claiming the cancers were caused by the spill. In court, the worker's attorneys claim that their clients received a dose that is five times the annual limit. The attorneys argue that it is likely that their cancers were caused by the spill.

Arguments to challenge the validity of this statement include the following:

1. There is a latency period for radiogenic cancers. BEIR V notes at least 5 years for thyroid cancer and at least 3 years for bone cancer.
2. The risk coefficients are low and do not support the contention that the cancers are radiogenic in origin. BEIR III quotes a risk coefficient of 27×10^{-4} sarcomas/person Gy for ^{224}Ra.
3. The organ doses are delivered over time, and adaptive response is likely to minimize the effect of the dose. This is particularly true in the case of ^{239}Pu where the dose is spread over 50 years.
4. Animal studies suggest a dose threshold for cancer incidence. The thyroid values are at the threshold lower limit of 2.5–5.0 Gy (BEIR V). Japanese atomic bomb survivor data suggests no excess cancers for doses in the 0–4 Gy range for low LET radiation. This data applies to Worker B's ^{131}I intake.
5. The dose values should be refined. A number of calculational improvements should be made in finalizing the dose assessment including:
   (a) The use of ICRP 60 or 103 methodology
   (b) The calculations should account for the worker's breathing pattern (e.g., nose vs. mouth) and use of a realistic particle size and shape rather than the 1.0 μm (sphere) ICRP 30 default value used to derive the listed ALI values

(d) On the day the spill occurred, the Worker A's physician administers the chelating agent DTPA. Chelation is appropriate for the ^{239}Pu intake because it binds divalent metals in the blood prior to uptake by the bone surfaces. Iodine is a halogen, not a metal, and chelating agents are less effective than saturation of the thyroid with a stable iodine compound such as KI.

Factors that determine the effectiveness of DTPA include:

1. *Timing of the initial dose*: The best results occur for prompt administration following the intake.
2. *Frequency of subsequent doses*: Aggressive, protracted therapy provides the best results.
3. *Concentration*: DTPA is ineffective at doses that are less than 10 μmol/kg.
4. *Long-term use in pregnant women*: Zn-DTPA is preferable since Ca-DTPA removes trace minerals. This same concern applies for long-term administration to male patients.

(e) The ICRP 66 Human Respiratory Tract Model is more sophisticated than the ICRP 30 model (see Appendix D). Improvements in the transuranic ICRP 66 lung model relative to the ICRP 30 model are as follows:

General changes:

1. The ICRP 66 model is based on the ICRP 60 formulation, and ICRP 30 is based on ICRP 26.
2. ICRP 66 uses the ICRP 89 age- and gender-specific biokinetic models, while ICRP 30 uses ICRP 23 Reference Man models.

3. ICRP 66 uses age- and gender-specific dose conversion factors, while ICRP 30 uses a single set for Reference Man.
4. ICRP 66 assumes a default particle size of 5 μm, while ICRP 30 uses a 1 μm value.
5. ICRP 66 uses a five-region model (ET_1, ET_2, BB, bb, and AI), while ICRP 30 uses a three-region model (NP, TB, and P).
6. ICRP 66 lung clearance is governed by model parameters and absorption to the blood as defined by F, M, and S types. ICRP 30 specifies pulmonary lung clearance is in terms of Class D, W, and Y materials.
7. ICRP 66 has a much more extensive treatment of the anatomy and physiology of the lung tissue.
8. ICRP 66 permits the consideration of a variety of particle shapes and sizes (0.6 nm to 100 μm), while ICRP 30 is limited to spherical particles between 0.2 and 10 μm.
9. ICRP 66 permits particles to transform from an initial classification (e.g., Type F) into another classification (Type M or S). ICRP 30 does not permit particles to change classification following their inhalation.

Specific changes:

1. ICRP 30 assumes that actinides remain at the initial bone surface location until eliminated from the body. This maximizes the bone surface dose but does not account for skeletal movement of actinides, bone growth, and clearance that are known to occur.
2. ICRP 30 assumes an intake model that includes a limited set of organs. Following intake and translocation to the blood, Pu is transported to the liver, bone, or soft tissue.
3. The Pu biokinetic model is summarized in ICRP 67 and is considerably more complex than the ICRP 30 model. ICRP 67's model includes the following tissues: massive soft tissue, blood, skeleton (cortical volume, cortical surface, cortical marrow, trabecular volume, trabecular surface, and trabecular marrow), liver (two compartments), kidneys, urine, bladder, GI contents, gonads, feces, and GI tract. A number of pathways with associated clearance and deposition parameters further define the ICRP 67 biokinetic model.

2.3. (a) In this question, you are to determine the neutron absorbed dose from the criticality event. To determine the absorbed dose, the following assumptions are made:

1. The criticality is treated as a point source. This is reasonable because the distance from the source is at least three times the maximum source dimension.
2. Radiation from the criticality is emitted isotropically.

3. Buildup is neglected within the polyethylene shield, tank, and air.
Unshielded neutron absorbed dose (D_{on}):

$$D_{\text{on}} = \frac{NYk}{4\pi r^2}$$

where

N	=	number of fission events = 1.0×10^{16} fissions
Y	=	neutron yield per fission = 3 n/fission
k	=	dose conversion coefficient = 0.002 mGy/h per 20 n/cm^2-s
r	=	distance from the source = 10.0 m = 1000 cm

With these values, the unshielded neutron absorbed dose is

$$D_{\text{on}} = \frac{(1.0 \times 10^{16}\ \text{fissions})\left(\frac{3\ \text{n}}{\text{fission}}\right)\left(\frac{0.002\ \text{mGy-cm}^2\text{-s}}{20\ \text{h-n}}\right)\left(\frac{1\ \text{h}}{3600\ \text{s}}\right)}{(4\pi)(1000\ \text{cm})^2}$$

$$= 66.3\ \text{mGy}$$

Shielded neutron absorbed dose (D_n):
The shielded neutron absorbed dose (D_{n}) is

$$D_{\text{n}} = f D_{\text{on}}$$

where f = shield's neutron attenuation factor = 0.005

$$D_{\text{n}} = (0.005)(66.3\ \text{mGy}) = 0.332\ \text{mGy}$$

(b) In this question, you are requested to determine the gamma absorbed dose from the criticality event. Using the assumptions from question (a), the unshielded gamma absorbed dose is
Unshielded gamma absorbed dose ($D_{o\gamma}$):

$$D_{\text{o}\gamma} = \frac{NYk}{4\pi r^2}$$

where

N	=	number of fission events = 1.0×10^{16} fissions
Y	=	gamma yield per fission = 8 γ/fission
k	=	dose conversion coefficient = 6.0×10^5 γ/cm^2-s per 10 mGy/h
r	=	distance from the source = 10.0 m = 1000 cm

With these values, the unshielded gamma absorbed dose is determined:

$$D_{\text{o}\gamma} = \frac{(1.0 \times 10^{16}\ \text{fissions})\left(\frac{8\gamma}{\text{fission}}\right)\left(\frac{10\ \text{mGy-cm}^2\text{-s}}{6.0\times 10^5\ h\text{-}\gamma}\right)\left(\frac{1\ \text{h}}{3600\ \text{s}}\right)}{(4\pi)(1000\ \text{cm})^2} = 29.5\ \text{mGy}$$

Shielded gamma absorbed dose (D_γ):

The shielded gamma absorbed dose (D_γ) is

$$D_\gamma = D_{o\gamma} e^{-\frac{\mu}{\rho}\rho t}$$

where

μ/ρ	=	mass attenuation coefficient for polyethylene = 0.073 cm^2/g
ρ	=	density of polyethylene = 1.5 g/cm^3
t	=	thickness of the polyethylene shield = 30 cm

$$D_\gamma = (29.5\,\text{mGy}) e^{-\left(\frac{0.073\,\text{cm}^2}{\text{g}}\right)\left(\frac{1.5\,\text{g}}{\text{cm}^3}\right)(30\,\text{cm})} = (29.5\,\text{mGy})(0.0374) = 1.10\,\text{mGy}$$

(c) Because the alarm set point is 5 mGy/h and the detector responds to 1/3500 of the actual gamma absorbed dose rate, the gamma absorbed dose rate at the detector (D_d) that will reach the set point is

$$D_d = (5\,\text{mGy/h})(3500) = 1.75 \times 10^4\,\text{mGy/h}$$

Since 1.0×10^{15} fissions yield a gamma absorbed dose of 20 mGy at 2.0 m, 1.0×10^{16} fissions produce an absorbed dose rate of 200 mGy at 2.0 m. Assuming that the 1.0×10^{16} fissions occur isotropically over the 1.0 ms transient, the average absorbed dose rate at 2 m (r_o) is

$$\dot{D}_o = \left(\frac{200\,\text{mGy}}{0.001\,\text{s}}\right)\left(\frac{3600\,\text{s}}{\text{h}}\right) = 7.20 \times 10^8\,\text{mGy/h at } 2.0\,\text{m}$$

The criticality normally occurs as a pulse that is not uniform in time. However, it is reasonable to assume that the detector responds in a manner that averages the absorbed dose rate ($\dot{D}$) over time. Since the criticality is adequately represented by a point source at the detector, the inverse square law is applicable:

$$\dot{D}r^2 = \text{constant},$$

where r is the distance from the criticality event. If d is the maximum distance where the detector alarms, the point source relationship yields

$$\dot{D}_d d^2 = \dot{D}_o r_o^2$$

$$d^2 = \left(\frac{\dot{D}_o}{\dot{D}_d}\right) r_o^2 \text{ or } d = \left(\frac{\dot{D}_o}{\dot{D}_d}\right)^{1/2} r_o$$

$$d = \left(\frac{7.20 \times 10^8\,\text{mGy/h}}{1.75 \times 10^4\,\text{mGy/h}}\right)^{1/2} (2.0\,\text{m}) = 406\,\text{m}$$

(d) Factors that affect criticality safety include:

1. Enrichment of the fissile isotope
2. Geometry of the fissile and surrounding materials

3. Mass of the fissile isotope
4. Presence and arrangement of moderator materials
5. Presence and arrangement of reflector materials
6. Presence and arrangement of absorber materials
7. Type of fissile material (e.g., ^{233}U, ^{235}U, and ^{239}Pu)
8. Form of the material (e.g., solid metal, solution, metal chips, and metal ribbons)
9. For suspensions, the particle size of the fissile material and its distribution within the moderator

2.4. (a) In this question, you are to determine the effective dose from ^{137}Cs photons to the worker. The problem assumes that the worker was standing at the center of the spill for 20 min and the dosimetric point of interest is 0.8 m above the spill. Any effects of self-shielding are to be neglected. A thin disk source approximates the effective dose rate from the spill geometry:

$$\dot{E} = \pi C_a \Gamma \ln \frac{r^2 + h^2}{h^2}$$

The effective dose (E) is just the product of the effective dose rate and the exposure time (t):

$$E = \dot{E}t = \pi C_a \Gamma t \ln \frac{r^2 + h^2}{h^2}$$

where

C_a	=	activity per unit area = A/a
A	=	^{137}Cs activity in the spill = 500 l × 1.85 × 10^4 MBq/l = 9.25 × 10^6 MBq
a	=	spill area = $\pi\, r^2$
r	=	spill radius = 5 m
a	=	(3.14)(5 m)^2 = 78.5 m^2
C_a	=	9.25 × 10^6 MBq/78.5 m^2 = 1.18 × 10^5 MBq/m^2
Γ	=	gamma constant for ^{137}Cs = 8.1 × 10^{-5} mGy-m^2/h-MBq
h	=	height above the spill = 0.8 m
t	=	exposure time = 20 min

Using these values, the effective dose is

$$E = (3.14)\left(1.18 \times 10^5\,\frac{\text{MBq}}{\text{m}^2}\right)\left(8.1 \times 10^{-5}\,\frac{\text{mGy-m}^2}{\text{h-MBq}}\right)$$
$$\times \left(20\text{ min}\,\frac{1\,\text{h}}{60\text{ min}}\right)\ln\frac{(5\,\text{m})^2 + (0.8\,\text{m})^2}{(0.8\,\text{m})^2}$$
$$= \left(10.0\,\text{mGy}\,\frac{1\,\text{mSv}}{1\,\text{mGy}}\right)(\ln(40.1)) = 37\,\text{mSv}$$

(b) In this question, you are requested to calculate the airborne radioactivity concentration of ^{90}Sr as measured by the lapel air sampler. The airborne concentration (C) of ^{90}Sr is determined from the lapel air sampler data:

$$C = \frac{A}{V}$$

where

A = ^{90}Sr activity deposited on the lapel air sampler filter $= 2 \times 10^6$ dpm
V = volume of air passing through the air sampler filter $= f\,t$
f = lapel air sampling rate $= 4$ l/min
t = sampling time $= 20$ min

$$C = \frac{\left(\left(\frac{2\times10^6\ \text{dis}}{\text{min}}\right)\left(\frac{1\ \text{min}}{60\ \text{s}}\right)\left(\frac{\text{Bq-s}}{\text{dis}}\right)\right)}{\left(4\,\frac{1}{\text{min}}\right)(20\ \text{min})} = 417\,\frac{\text{Bq}}{\text{l}}$$

(c) The committed effective dose ($E(50)$) to the worker from ^{90}Sr is determined from the intake I:

$$I = \frac{C(\text{BR})t}{(\text{PF})}$$

where

C = airborne concentration of ^{90}Sr $= 417$ Bq/l
BR = breathing rate $= 20$ l/min
t = exposure time $= 20$ min
PF = protection factor of the respirator $= 50$

$$I = \frac{\left(417\,\frac{\text{Bq}}{\text{l}}\right)\left(20\,\frac{\text{l}}{\text{min}}\right)(20\ \text{min})}{(50)} = 3.34 \times 10^3\ \text{Bq}$$

$E(50)$ is written in terms of the intake and the ^{90}Sr dose conversion factor (DCF):

$$E(50) = I(\text{DCF})$$

$$E(50) = (3.34 \times 10^3\ \text{Bq})\left(7.7 \times 10^{-8}\,\frac{\text{Sv}}{\text{Bq}}\right)\left(1000\,\frac{\text{mSv}}{\text{Sv}}\right) = 0.26\,\text{mSv}$$

(d) The letters AMAD are used for the activity median aerodynamic diameter.

(e) The AMAD is the diameter of an aerodynamic particle size distribution in which the total activity above and below this size are equal. A lognormal distribution of particle sizes is usually assumed, and this distribution is uniquely described by the geometric mean and geometric standard deviation. A lognormal distribution of radioactive particles can be characterized in terms of a number of associated parameters including the (i) count median diameter, (ii) surface area median diameter, (iii) mass median diameter (MMD), (iv) activity median aerodynamic diameter (AMAD), and (v) mass median aerodynamic diameter (MMAD).

Depending on the application, one of these parameters with the associated geometric standard deviation is used to describe the aerosol. For example, if mass is of interest, as is often the case in inhalation toxicology studies, then the MMD or MMAD can be used. For radioactive aerosols, the amount of radioactivity is the quantity of concern. Therefore, the AMAD is the quantity having dosimetric significance, and it should be used in health physics applications.

(f) If the spherical droplets have a specific gravity of 11.3 and a diameter of 5 μm, their AMAD is given by the relationship

$$\text{AMAD} = d_{\rho}\left(\frac{\rho}{\rho_{\text{unit}}}\right)^{1/2}$$

where

AMAD	=	activity median aerodynamic diameter
d_{ρ}	=	diameter of the spherical droplet having the specified density = 5 μm
ρ	=	density of the spherical drop = 11.3 g/cm^3
ρ_{unit}	=	unit density = 1 g/cm^3

$$\text{AMAD} = (5\,\mu\text{m})\left(\frac{11.3\,\frac{\text{g}}{\text{cm}^3}}{1.00\,\frac{\text{g}}{\text{cm}^3}}\right)^{1/2} = 16.8\,\mu\text{m}$$

The AMAD is also defined as the diameter of a unit density sphere that has the same settling velocity as the particle in question. The terminal settling velocity relationship given in the problem statement is

$$\frac{d_1^2\rho_1 g}{18\eta} = \frac{d_2^2\rho_2 g}{18\eta}$$

where

d_1	=	diameter of the particle having unit density = AMAD
ρ_1	=	unit density = 1 g/cm^3
d_2	=	particle diameter = 5 μm
ρ_2	=	particle density = 11.3 g/cm^3
g	=	acceleration due to gravity = 9.8 m/s^2
η	=	viscosity of air

By canceling the common terms, we obtain a simplified relationship for the particle's AMAD (d_1)

$$d_1^2\rho_1 = d_2^2\rho_2$$

$$d_1 = d_2 \left(\frac{\rho_2}{\rho_1} \right)^{1/2} = \text{AMAD}$$

Inserting the given parameters leads to the same result noted above:

$$\text{AMAD} = (5\,\mu\text{m}) \left(\frac{11.3\,\frac{\text{g}}{\text{cm}^3}}{1.00\,\frac{\text{g}}{\text{cm}^3}} \right)^{1/2} = 16.8\,\mu\text{m}$$

2.5. (a) The estimated intake for the worker is given by a simplified relationship since only data from a single measurement is provided:

$$I = \frac{\sum_{i=1}^{n} A_i f_i}{\sum_{i=1}^{n} f_i^2} = \frac{A}{f} = \frac{CV}{f}$$

where

I = estimated intake based on the methodology of NUREG/CR-4884
C = concentration of ^{137}Cs in the worker's urine = A_S/V_s
A_s = activity in analyzed portion of the urine sample = 15.9 kBq
V_s = volume of analyzed portion of the sample = 500 ml
C = 15.9 kBq/500 ml = 31.8 Bq/ml
V = daily urinary output = 1400 ml for men (assumed as the basis for the company's internal dosimetry model)
f = intake retention fraction at Day 20, the time the sample was obtained postintake = 2.59×10^{-3}

Using these values, the estimated intake is

$$I = \frac{CV}{f} = \frac{\left(31.8\,\frac{\text{Bq}}{\text{ml}}\right)(1400\,\text{ml})\left(\frac{1\,\text{MBq}}{10^6\,\text{Bq}}\right)}{(2.59 \times 10^{-3})} = 17.2\,\text{MBq}$$

(b) In this question, you are to determine the average concentration ($\overline{C}$) of ^{137}Cs in the air to which the worker was exposed. You are to assume the intake (I) for the worker was 55.5 MBq.

The rate of change of intake (I) as a function of the average air concentration is written in terms of a production equation:

$$\frac{\text{d}I}{\text{d}t} = Pe^{-kt} = \overline{C}(BR)e^{-kt}$$

where k = total removal rate for ^{137}Cs

$$k = \lambda + \frac{F}{V}$$

P = production term = $\overline{C}$(BR)
λ = radioactive decay constant for ^{137}Cs

$$\lambda = \frac{\ln 2}{T_{1/2}} = \left(\frac{\ln 2}{30.07\ \text{years}}\right)\left(\frac{\text{year}}{365\,\text{days}}\right)\left(\frac{\text{day}}{24\,\text{h}}\right) = 2.63 \times 10^{-6}/\text{h}$$

F/V = air turnover rate

$$\frac{F}{V} = \frac{1}{2\,\text{h}} = 0.5/\text{h}$$

Since radioactive decay is insignificant relative to the air turnover rate,

k	=	0.5/h
BR	=	worker's breathing rate = 1.2 m^3/h
t	=	time
T	=	exposure time = 30 min = 0.5 h

The intake is obtained by integrating the production equation

$$I(T) = \int_0^T \frac{\mathrm{d}I}{\mathrm{d}t}\mathrm{d}t = \overline{C}(\text{BR})\int_0^T e^{-kt}\,\mathrm{d}t = \frac{\overline{C}(\text{BR})}{k}(1 - e^{-kT})$$

where $I(T)$ is the intake (55.5 MBq).

The desired initial concentration is determined by algebraic solution of the integrated equation

$$\overline{C} = \frac{I(T)k}{(\text{BR})(1 - e^{-kT})} = \frac{(55.5\,\text{MBq})\left(\frac{0.5}{\text{h}}\right)}{\left(1.2\,\frac{\text{m}^3}{\text{h}}\right)\left(1 - e^{-\left(\frac{0.5}{\text{h}}\right)(0.5\,\text{h})}\right)} = 105\frac{\text{MBq}}{\text{m}^3}$$

(c) 3 In this question, you are to determine the percentages of the total activity that can be attributed to ^{234}U, ^{235}U, and ^{238}U, respectively. Naturally occurring uranium consists of the ^{234}U, ^{235}U, and ^{238}U isotopes. By mass, the natural distribution is

^{234}U: 0.013 g/mol-total U	$T_{1/2} = 2.5 \times 10^5$ years
^{235}U: 1.71 g/mol-total U	$T_{1/2} = 7.0 \times 10^8$ years
^{238}U: 236.4 g/mol-total U	$T_{1/2} = 4.5 \times 10^9$ years

The activity (A) is defined as follows:

$$A = \lambda N$$

where

λ	=	disintegration constant = $\ln(2)/T_{1/2}$
$T_{1/2}$	=	half-life
N	=	number of atoms = $(F N_a)/M$
N_a	=	Avogadro's number = 6.02×10^{23} atoms/mol-total U
M	=	molar mass of natural uranium
F	=	isotopic mass in natural material (g/mol-total U)

Given these relationships, the desired activity relationship is obtained:

$$A = \frac{\ln(2) N_a F}{M T_{1/2}}$$

In this equation, M is the sum of all constituent isotopes:

$$M = (0.013 + 1.71 + 236.4)\frac{\text{g}}{\text{mol-U}} = 238.1\frac{\text{g}}{\text{mol-U}}$$

This relationship permits the determination of the activities of the uranium isotopes:

^{234}U:

$$A = \frac{(0.693)(6.02 \times 10^{23}\,\text{atoms})\left(0.013\,\frac{\text{g}}{\text{mol-U}}\right)\left(1\,\frac{\text{dis}}{\text{atom}}\right)}{\left(238.1\,\frac{\text{g}}{\text{mol-U}}\right)(2.5 \times 10^{5}\,\text{years})}$$
$$= 9.11 \times 10^{13}\,\frac{\text{dis}}{\text{year}}$$

^{235}U:

$$A = \frac{(0.693)(6.02 \times 10^{23}\,\text{atoms})\left(1.71\,\frac{\text{g}}{\text{mol-U}}\right)\left(1\,\frac{\text{dis}}{\text{atom}}\right)}{\left(238.1\,\frac{\text{g}}{\text{mol-U}}\right)(7.0 \times 10^{8}\,\text{years})}$$
$$= 4.28 \times 10^{12}\,\frac{\text{dis}}{\text{year}}$$

^{238}U:

$$A = \frac{(0.693)(6.02 \times 10^{23}\,\text{atoms})\left(236.4\,\frac{\text{g}}{\text{mol-U}}\right)\left(1\,\frac{\text{dis}}{\text{atom}}\right)}{\left(238.1\,\frac{\text{g}}{\text{mol-U}}\right)(4.5 \times 10^{9}\,\text{years})}$$
$$= 9.20 \times 10^{13}\,\frac{\text{dis}}{\text{year}}$$

The total activity from all three isotopes is

$$A_T = 9.11 \times 10^{13}\text{dis/year} + 4.28 \times 10^{12}\text{dis/year} + 9.20 \times 10^{13}\text{dis/year} = 18.7 \times 10^{13}\text{dis /year}$$

The fractional amount (f) of each isotope is obtained by comparison to the total activity:

^{234}U: $f(^{234}\text{U}) = (9.11 \times 10^{13}$ dis/year)/(18.7 $\times 10^{13}$ dis/year) = 0.49.
^{235}U: $f(^{235}\text{U}) = (4.28 \times 10^{12}$ dis/year)/(18.7 $\times 10^{13}$ dis/year) = 0.02.
^{238}U: $f(^{238}\text{U}) = (9.20 \times 10^{13}$ dis/year)/(18.7 $\times 10^{13}$ dis/year) = 0.49.

(d) 2 The relative radiotoxicity (R) of the various enrichment technologies is estimated from the activity fraction values: $f_1(^{234}\text{U}) = 0.49$,

$f_2(^{235}\text{U}) = 0.02$, and $f_3(^{238}\text{U}) = 0.49$ for natural uranium:

$$R = \sum_i \frac{a_i}{b_i} f_i$$

where a_i is the enrichment of isotope i produced by the specified technology, b_i is the natural enrichment value for isotope i, f_i are the activity fractions calculated in the previous question for natural enrichment, and $i = 1$, 2, and 3 specify ^{234}U, ^{235}U, and ^{238}U, respectively. For natural uranium, $a_i = b_i$ and $R = 1.0$. Using the information provided in the problem statement ($b_1 = 0.0054$, $b_2 = 0.7204$, and $b_3 = 99.2742$), the R values relative to natural uranium are

$$R(\text{Centrifuge A}) = \left(\frac{1.0}{0.0054}\right) 0.49 + \left(\frac{5.0}{0.7204}\right) 0.02 + \left(\frac{94.0}{99.2742}\right) 0.49 = 91.3$$

$$R(\text{Centrifuge B}) = \left(\frac{2.0}{0.0054}\right) 0.49 + \left(\frac{10.0}{0.7204}\right) 0.02 + \left(\frac{88.0}{99.2742}\right) 0.49 = 182.$$

$$R(\text{SILEX}) = \left(\frac{0.1}{0.0054}\right) 0.49 + \left(\frac{8.0}{0.7204}\right) 0.02 + \left(\frac{91.9}{99.2742}\right) 0.49 = 9.75$$

$$R(\text{AVLIS}) = \left(\frac{0.1}{0.0054}\right) 0.49 + \left(\frac{50.0}{0.7204}\right) 0.02 + \left(\frac{49.9}{99.2742}\right) 0.49 = 10.7$$

$$R(\text{MLIS}) = \left(\frac{0.1}{0.0054}\right) 0.49 + \left(\frac{90.0}{0.7204}\right) 0.02 + \left(\frac{9.9}{99.2742}\right) 0.49 = 11.6$$

Based on these calculations, centrifuge technology B produces the highest-activity fuel.

(e) 4 This question deals with the recommendations of ANSI Z88.2, *Practices for Respiratory Protection*, which provides commentary for the use of supplied breathing air. The following items are reviewed against the recommendations of ANSI Z88.2:

1. Grade D breathing air specifications should be considered as the limits for compressed air of deteriorating quality. This statement is correct and is also referenced by the US Nuclear Regulatory Commission in NUREG-0041. The characteristics of Grade D breathing air are specified by the Compressed Gas Association that further recommends that Grade E air be supplied as a good practice. The characteristics of Grades D and E air are as follows:

Characteristic	Grade E	Grade D
Oxygen (vol%) balance is mainly nitrogen	Atmospheric (21)	Atmospheric (21)
Oxygen limits (vol%) synthesized air	19–23	19–23
Condensed hydrocarbons in mg/m^3 of gas at STP (max)	5	5
Carbon monoxide (ppm) max	10	20
Carbon dioxide (ppm) max	500	1000

2. The oxygen content of supplied breathing air shall be a minimum of 19.0% by volume. This statement is correct. The Grade D oxygen limit for synthesized air is 19–23%. However, the Bureau of Mines/NIOSH approvals require a minimum of 19.5% oxygen by volume.
3. Compressed oxygen may be used in supplied air or open-circuit self-contained breathing apparatus (SCBA) in which compressed air has previously been used. This statement is incorrect and is specifically excluded. Oxygen shall never be used with air line respirators.

Based on these statements, the correct answer is 4, that is, answers 1 and 2 are correct.

(f) Following Appendix A to 10CFR20, the appropriate match of protection factors and respirators is

1. Full facepiece, negative pressure mode, air-purifying respirator	(b) PF = 100
2. Full facepiece, pressure demand mode, self-contained breathing apparatus (SCBA)	(d) PF = 10 000
3. Half-mask facepiece, negative pressure mode, air-purifying respirator	(a) PF = 10
4. Full facepiece, powered air-purifying respirator	(c) PF = 1000

(g) Handheld instruments routinely used for the detection of uranium contamination on personnel as they leave contaminated areas include:

Instrument/radiation type detected	**Constraints/advantages/disadvantages**
Alpha scintillator ZnS(Ag)/alpha particles	Insensitive to humidity and pressure Less rugged than proportional counters Sensitive to gamma radiation (only rejects 0.1–0.2 mGy/h of photon radiation) About 35% efficient Extreme light sensitivity Requires window repairs if not carefully used Torn windows are very difficult to repair Radon may interfere with the uranium measurement

Instrument/radiation type detected	Constraints/advantages/disadvantages
Propane proportional counter/alpha particles	Window may be repaired with tape, but this diminishes the sensitive area About 50% efficient Less influenced by humidity than an air proportional counter Good gamma rejection up to 1 Gy/h Fast neutron interference Flammability hazard from propane Radon may interfere with the uranium measurement
Air proportional counter/alpha particles	Window may be repaired with tape, but this diminishes the sensitive area About 25% efficient Influenced by humidity Easy to use Good gamma rejection up to 1 Gy/h Fast neutron interference Sensitive to environmental factors particularly humidity Radon may interfere with the uranium measurement

2.6. (a) In this question, you are requested to calculate the committed effective dose equivalent a worker receives from a room air concentration of 185 MBq/m^3 as measured by workplace air monitoring. Tritium within the liquid waste cleanup system was initially a combination of HT and HTO. However, any HT that is released to the environment will eventually oxidize to form HTO. The conversion rate from the gas phase to HTO is accelerated by humidity, temperature, catalytic surfaces, discharges from electrical equipment, and other factors. Therefore, it is likely that the conversion of the HT to HTO is nearly complete since the maintenance evolution occurs after the tritium had been in residence for several weeks.

Assuming the tritium intake occurs in an HTO form, the committed effective dose equivalent is calculated from the exposure time and the specified concentration

$$H = \frac{N}{(2000\,\text{DAC-h})}0.05\,\text{Sv}$$

$$N = \frac{Ct}{C'}$$

where

H	=	committed effective dose equivalent from the tritium intake
N	=	number of DAC-h to which the worker is exposed
C	=	HTO concentration = 185 MBq/m^3
t	=	exposure duration = 1 h
C'	=	DAC (HTO) = 0.74 MBq/m^3-DAC

$$H = \frac{Ct}{(2000\,\text{DAC-h})C'} 0.05\,\text{Sv}$$

Using these values, the committed effective dose equivalent is determined:

$$H = \frac{(185\,\text{MBq/m}^3)(1\,\text{h})}{(2000\,\text{DAC-h})(0.74\,\text{MBq/m}^3\text{-DAC})}(0.05\,\text{Sv})\left(\frac{1000\,\text{mSv}}{\text{Sv}}\right)$$
$$= 6.3\,\text{mSv}$$

(b) The individual involved in the incident submits a postincident bioassay sample collected during the first 24 h. The results indicate a tritium concentration in urine of 1850 Bq/l. The committed effective dose equivalent (H) is obtained from the acute intake DCF based on urine tritium concentration (C). It is assumed that the committed effective dose equivalent is due to the acute exposure and not previous chronic occupational exposure:

$$H = C(\text{DCF})$$

$$H = (1850\,\text{Bq/l-urine})(7.57 \times 10^{-7}\ \text{mSv-l-urine/Bq})$$
$$= 1.40 \times 10^{-3}\,\text{mSv}$$

(c) The committed effective dose equivalent calculated from the urine concentration differs from the value obtained from the room air concentration. If the measurements and calculations were performed correctly, the likely sources of this discrepancy include the following:

1. The workplace air monitor location did not provide a representative measurement of the air concentration the worker experienced.
2. The urine concentration from the event is affected by chronic occupational exposure that was received by the worker before the event.
3. The worker's metabolism is not equivalent to that assumed in formulating the dose conversion factor and DAC values.
4. The tritium gas did not fully oxidize and includes both HT and HTO components.
5. The worker is not well represented by the dosimetric model.
6. The measured tritium air concentration may have been influenced by noble gas activity or residual gamma dose rates.
7. The 60 min exposure time was not accurate.

8. The tritium air sample was cross-contaminated.
9. The tritium urine sample was cross-contaminated.
10. The air sampling system malfunctioned.

(d) In this question, you are to identify two techniques that are used for tritium air monitoring and to specify one advantage and one disadvantage of each technique:

1. *Ion chamber tritium-in-air monitors*: The measurement of tritium in air presents special problems because the average energy of the beta particles is low (about 6 keV). Therefore, it is difficult to design a detector whose walls can be penetrated. Accordingly, tritium-contaminated air is pumped through the detector so that all the beta particle energy is converted to producing ion pairs inside the sensitive volume. Since external radiation also creates ion pairs, a second sealed detector is used to compensate for this source.

   The detector has the advantage of being convenient, and the tritium concentration is determined in real time. However, the instrument has a number of limitations. For example, any radioactive gas present in the air is measured as tritium and leads to a higher than actual reading. In addition, the gamma compensation is adequate only in relatively low gamma fields of about 100 μGy/h or less.

2. *Tritium bubbler*: The tritium bubbler is simple, accurate, and not affected by the gamma background or the presence of noble gases. The technique consists of bubbling air through clean water that collects the tritiated water vapor. Following collection, the tritium content of the water is analyzed. The bubbler consists of a pump, timer, flow gauge, and removable water jar containing about 100 ml of clean water. The bubbler fluid is then counted using liquid scintillation techniques to obtain the tritium air concentration.

   The tritium bubbler gives more accurate results than the ion chamber, but it is not as convenient. Although the technique is accurate, time must be allowed for sample preparation, counting, and processing. In addition, care must be taken to avoid cross-contamination of the water sample. Therefore, the technique does not provide real-time tritium air concentration information.

2.7. (a) Using a centrifuge designed for 5% ^{235}U is not appropriate for the intended operation of separating ^{239}Pu from ^{240}Pu. Plutonium delivers significantly more dose per unit intake than uranium. The contamination control measures appropriate for uranium enrichment may not be sufficient for plutonium. Radiological measures associated with contamination control must be more rigorous for the plutonium operation.

The engineering controls used for uranium enrichment must also be enhanced. In particular, the criticality controls must be significantly improved since the minimum masses for plutonium criticality are smaller than the corresponding uranium values. The external dose control measures must also be strengthened.

From a practical perspective, the centrifuge system may not be effective in separating the plutonium isotopes. The uranium enrichment approach involves the separation of ^{235}U and ^{238}U, which involves a difference of 3 mass units. Separation of ^{239}Pu and ^{240}Pu involves only 1 mass unit. The position of the product and tails withdrawal locations is not optimized for the separation of the plutonium isotopes.

(b) The decision to eliminate photons below 200 keV excludes the dominant low-energy photon radiation emitted from ^{239}Pu. Therefore, the whole-body counting technique is ineffective in detecting the low-energy ^{239}Pu photons. The fact that no positive counts have been observed does not indicate the radiation protection program is functioning properly.

(c) The activity (A) released during the 14-day period (t) is derived from a production equation (see Appendix B)

$$A(t) = \frac{P}{\lambda}(1 - e^{-\lambda t})e^{-\lambda t_{\text{decay}}}$$

where the production term (P) is the product of the release rate (F) and release concentration (C)

$$P = FC = \left(1\,\frac{\text{m}^3}{\text{s}}\right)\left(100\,\frac{\text{Bq}}{\text{m}^3}\right) = 100\,\frac{\text{Bq}}{\text{s}}$$

and λ is the disintegration constant determined from the ^{239}Pu half-life

$$\lambda = \frac{\ln 2}{T_{1/2}} = \left(\frac{\ln 2}{2.41 \times 10^4\,\text{years}}\right)\left(\frac{1\,\text{year}}{365\,\text{days}}\right)$$
$$= \frac{7.88 \times 10^{-8}}{\text{day}}\left(\frac{1\,\text{day}}{24\,\text{h}}\right)\left(\frac{1\,\text{h}}{3600\,\text{s}}\right) = \frac{9.12 \times 10^{-13}}{\text{s}}$$

Given the ^{239}Pu half-life, radioactive decay is negligible during the release time. Using these values, the activity released during the 14-day period is determined:

$$A(t) = \frac{P}{\lambda}(1 - e^{-\lambda t})e^{-\lambda t_{\text{decay}}} = \left(\frac{100\,\frac{\text{Bq}}{\text{s}}}{\left(\frac{9.12\times 10^{-13}}{\text{s}}\right)}\right)\left(1 - e^{-\left(\frac{7.88\times 10^{-8}}{\text{day}}\right)(14\,\text{days})}\right)$$
$$\times (1.0) = 1.21 \times 10^8\,\text{Bq}$$

Since the leak spreads particulate material uniformly over a $10\,\text{m}^2$ area (a), the concentration per unit area (C_A) is

$$C_A = \frac{A}{a} = \frac{1.21 \times 10^8\,\text{Bq}}{10\,\text{m}^2} = \frac{1.21 \times 10^7\,\text{Bq}}{\text{m}^2}$$

(d) Work activity in the contaminated area results in a resuspension factor (r) of 2×10^{-5}/m. The resulting airborne concentration (C) is

$$C = C_A r = \left(\frac{1.21 \times 10^7\,\text{Bq}}{\text{m}^2}\right)\left(\frac{2 \times 10^{-5}}{\text{m}}\right) = 242\frac{\text{Bq}}{\text{m}^3}$$

(e) A technician works for 8 h (t) in the air concentration derived in the previous question. If his breathing rate is 1.2 m^3/h (BR), the ^{239}Pu intake (I) is

$$I = C(\text{BR})t$$

$$I = \left(242\frac{\text{Bq}}{\text{m}^3}\right)\left(1.2\frac{\text{m}^3}{\text{h}}\right)(8\,\text{h}) = 2.32 \times 10^3\,\text{Bq}$$

(f) Since the ^{239}Pu is Class M with a dose conversion coefficient (e) of 4.7×10^{-5} Sv/Bq, the effective dose resulting from the intake is

$$E = eI = \left(4.7 \times 10^{-5}\,\frac{\text{Sv}}{\text{Bq}}\right)(2.32 \times 10^3\,\text{Bq})$$

$$= (0.11\,\text{Sv})\left(1000\,\frac{\text{mSv}}{\text{Sv}}\right) = 110\,\text{mSv}$$

(g) The consequences of this event (110 mSv) exceed the effective dose recommendations of ICRP 103 that specify 100 mSv over a 5-year period with no more than 50 mSv in any year.

2.8. (a) The time (t) required for the complete fission reaction of all PMA nuclei in the 1 cm^2 gamma-ray beam is given by the relationship

$$t = \frac{n}{R} = \frac{n}{n\sigma\phi} = \frac{1}{\sigma\phi}$$

where n is the density of atoms/cm^3, σ is the photoinduced reaction cross-section in b/atom, ϕ is the photon fluence in γ/cm^2-s, and R is the reaction rate in fissions/cm^3-s. Using the values in the problem statement, the time for transmutation of the 1 cm^2 area throughout the core depth is

$$t = \frac{1}{\sigma\phi} = \frac{1}{\left(1.0 \times 10^{-6}\,\frac{\text{b}}{\text{atom}}\right)\left(10^{-24}\,\frac{\text{cm}^2}{\text{b}}\right)\left(1.0 \times 10^{23}\,\frac{\gamma}{\text{cm}^2\text{-s}}\right)\left(\frac{\text{atom}}{\gamma}\right)}$$

$$= (1.0 \times 10^7\,\text{s})\left(\frac{1\,\text{h}}{3600\,\text{s}}\right)\left(\frac{1\,\text{day}}{24\,\text{h}}\right) = 116\,\text{days}$$

The time to irradiate the entire core area (T) is much longer:

$$T = \frac{(1.03\,\text{m})^2}{(1\,\text{cm}^2)}\left(\frac{100\,\text{cm}}{\text{m}}\right)^2(116\,\text{days})\left(\frac{1\,\text{year}}{365\,\text{days}}\right) = 3.37 \times 10^3\,\text{years}$$

This time is unrealistic and suggests that another approach be adopted. If it were possible to expand the beam to cover the entire core area or use multiple beams, a 116-day time for transmutation would be a reasonable production parameter. However, ignoring attenuation is not reasonable. The fluence is reduced as it passes through the core following the relationship

$$\phi(x) = \phi_o B e^{-\mu x}$$

where B is the buildup factor, μ is the core average linear attenuation coefficient (0.39/cm), and x is the penetration distance into the core. The attenuation is significant as noted in the following table:

x (cm)	μx	B	$e^{-\mu x}$	$\frac{\phi(x)}{\phi_o} = Be^{-\mu x}$
0.0	0.0	1.0	1.0	1.0
5.13	2.0	2.85	0.135	0.385
10.3	4.0	5.3	0.0183	9.70×10^{-2}
15.4	6.0	8.31	2.48×10^{-3}	2.06×10^{-2}
20.5	8.0	11.8	3.35×10^{-4}	3.95×10^{-3}
25.6	10.0	15.8	4.54×10^{-5}	7.17×10^{-4}
51.3	20.0	41.3	2.06×10^{-9}	8.51×10^{-8}
76.9	30.0	74.5	9.36×10^{-14}	6.97×10^{-12}
103	40.0	114	4.25×10^{-18}	4.85×10^{-16}

Since the transmutation time, increases as the position dependent flux decreases, the 116-day transmutation time $t(0)$ at the core surface $\phi(0)$ significantly increases as the beam penetrates the core $\phi(x)$ and is given by

$$t(x) = t(0)\frac{\phi(0)}{\phi(x)}$$

This attenuation makes the facility as outlined in the problem statement a poor transmutation approach. The approach is also limited because the beam is small (1 cm^2) and only irradiates a portion of the core.

(b) Process parameters that could be altered to improve transmutation performance are as follows:

1. Increase the gamma-ray energy to several hundred megaelectron-volts. The photofission cross-section will increase in comparison to the 1.25 MeV beam energy. The increase in beam energy also reduces the attenuation coefficient that permits better beam penetration.
2. Reduce the thickness of the core to permit better beam penetration. A core in the form of a thin sheet with a scanning photon beam would be more practical. The sheet geometry also minimizes the probability of an inadvertent criticality.
3. Increase the photon fluence to decrease the time required for transmutation of the PMA.
4. Increase the beam diameter to illuminate the entire core face.

(c) The neutron flux 10 m from an unshielded point source core is

$$\phi(x) = k\frac{S}{4\pi r^2}$$

where the beam delivers $3.0 \times 10^{24}\ \gamma/\text{s}$ to the core, the distance from the core (r) is 10 m, and 2.5×10^{-6} neutrons are produced for every incident photon (k). Using these values, the desired neutron flux is

$$\phi(x) = \left(2.5 \times 10^{-6}\ \frac{\text{n}}{\gamma}\right)\frac{\left(3.0 \times 10^{24}\ \frac{\gamma}{\text{s}}\right)\left(\frac{1\ \text{m}}{100\ \text{cm}}\right)^2}{(10\ \text{m})^2} = 7.50 \times 10^{12}\frac{\text{n}}{\text{cm}^2\text{-s}}$$

(d) The unattenuated effective dose at a location (r) 25 m from the center of the point source core is

$$E_\text{n}(r) = f\phi_\text{o}\left(\frac{r_\text{o}}{r}\right)^2$$

where the flux to dose conversion factor (f) is 25 μSv/h per 20 n/cm^2-s and the flux information was provided in the previous question. Using these values, the unshielded effective dose is

$$E_\text{n} = \left(\frac{25\ \mu\text{Sv/h}}{20\ \text{n/cm}^2\text{-s}}\right)\left(7.50 \times 10^{12}\ \frac{\text{n}}{\text{cm}^2\text{-s}}\right)\left(\frac{10.0\ \text{m}}{25.0\ \text{m}}\right)^2 = 1.50 \times 10^{12}\frac{\mu\text{Sv}}{\text{h}}$$

The concrete shielding (t) required to reduce the neutron effective dose to 10 μSv/h (E_std) is given by the relationship

$$E_\text{std} = E_\text{n}Be^{-\mu t}$$

where the neutron attenuation factor (μ) for concrete is 0.0576/cm and a buildup factor (B) of 60.2 was provided in the problem statement. Using these values, the required shielding thickness is obtained by solving the previous equation for t:

$$t = -\frac{1}{\mu}\ln\left(\frac{E_\text{std}}{BE_\text{n}}\right)$$

Using the previously defined parameters,

$$t = -\frac{1}{\left(\frac{0.0576}{\text{cm}}\right)}\ln\left(\frac{10\ \mu\text{Sv/h}}{(60.2)\,(1.50 \times 10^{12}\ \mu\text{Sv/h})}\right)$$

$$= \frac{29.8}{0.0576}\ \text{cm} = 517\ \text{cm} = 5.17\ \text{m}$$

Solutions for Chapter 3

3.1. (a) The waste in the tank has aged at least 40 years. ^{90}Sr, ^{90}Y, and ^{137}Cs present the dominant external radiation hazard. These isotopes dominate the source term at the Hanford Site Waste Tanks that have a constituency similar to that described in the problem statement.

(b) ^{239}Pu and ^{241}Am are the dominant internal radiation hazards. These isotopes dominate the source term at the Hanford Site Waste Tanks.

(c) RPP-13033, *Tank Farms Documented Safety Analysis*, discusses hazardous conditions that could produce an uncontrolled release of radioactive or hazardous material from these high-level waste tanks. These conditions include:

1. *Electrical sources* (e.g., battery banks, cable runs, power supplies, emergency diesel generators, transformers, motors, pumps, power tools, switch gear, transmission lines, underground wires, and facility wiring)
2. *Thermal sources* (e.g., electrical equipment, furnaces, boilers, heaters, steam lines, welding equipment, power sources, radioactive decay heat, exposed hot components, power tools, convective sources, solar radiation, and lighting components)
3. *Friction* (e.g., belts, bearings, fans, gears, motors, and power tools)
4. *Corrosives* (e.g., acids, caustics, other chemicals, decontamination solutions, and high-temperature waste)
5. *Rotational kinetic energy* (e.g., motors, pumps, power tools, and ventilation equipment)
6. *Linear kinetic energy* (e.g., motor vehicles, forklifts, dollies, carts, crane loads, and pressure vessel blowdown)
7. *Mass, gravity, and height* (e.g., human effort, stairs, lifts and cranes, slings, hoists, elevators, scaffold and ladders, pits and excavations, and vessels/tanks)
8. *Pressure and volume* (e.g., surge tanks, compressed gas bottles, pressure vessels, compressors, steam headers and lines, positive displacement pumps, and hydraulic systems)
9. *Explosives and pyrophoric materials* (e.g., chemicals, dusts, flammable gases, nitrates/nitrites, peroxides/hydrides, plutonium and uranium metal, sodium/phosphorus, and combustible vapors)
10. *Nuclear criticality* (e.g., fissile material including ^{235}U and ^{239}Pu)
11. *Flammable materials* (e.g., compressed gas bottles, packing materials, rags, gasoline, oil, paint solvent, diesel fuel, hydraulic fluids, grease, and chemicals)
12. *Hazardous materials* (e.g., metals, corrosives, and oxidizers)
13. *Ionizing radiation sources* (e.g., fissile material, radiography equipment, radioactive material, and radioactive sources)
14. *Uncontrolled chemical reactions* (e.g., reactions involving tank waste or added chemicals)

15. *External events* (e.g., aircraft impacts, explosions, fires, accidents at other sites, toxic materials, flammable liquids/gases, and explosive materials)
16. *Vehicles in motion* (e.g., aircraft, ground vehicles, and cranes)
17. *Natural phenomena* (e.g., earthquakes, floods, lightning, rain, snow, temperature extremes, wind, tornados, ashfall, tsunamis, and range fires)
18. *Terrorist attacks and sabotage* (e.g., external force ground and air attacks and insider actions)

(d) The design basis accidents create conditions for the mobilization of radioactive and toxic materials. Mobilization of the waste contaminates the air and ground above the underground tanks and can lead to both internal and external doses to the public and workers involved in tank waste operations or recovery activities. Design basis accidents associated with the underground waste tanks include:

1. *Flammable gas accidents*: This accident is a deflagration or detonation initiated by the ignition of flammable gases in the headspace of a waste tank. Flammable gases, primarily hydrogen, are produced in the waste tanks by radiolysis, organic decomposition, and corrosion. Ignition sources include the operation of installed equipment and manned activities.
2. *Nuclear criticality*: A criticality event must be considered because ^{235}U and ^{239}Pu are present. The criticality produces neutron and photon radiation that reaches the surface above the waste tank. It also generates a burst of energy that has the potential to mobilize the radioactive and toxic materials residing within the waste tank.
3. *Vacuum exhaust line rupture*: Vacuum methods are a common waste removal approach. These methods have an associated accident involving the release of waste aerosols to the environment following the rupture of a vacuum exhaust line during waste retrieval operations. In addition, some transport of waste aerosols into the vacuum exhaust line could occur during normal operations.
4. *Release from contaminated facility*: Facilities are required to support waste tank operations including waste transfers between tanks. Since these facilities mobilize and transfer tank waste, they have contaminated equipment and structures. For these facilities, the relevant accident is a flammable gas deflagration in a waste transfer structure that results in an uncontrolled release of radioactive and hazardous material. Fires, load drops, and compressed gas system failures in contaminated facilities supporting waste transfer operations could initiate this accident type.

5. *Tank failure due to excessive loads*: This accident is a waste tank structural failure caused by a load external to the tank, load drop, or internal load caused by waste storage. Structural failure of the waste tank creates a pathway for an airborne release and a mechanism for mobilizing the waste material.
6. *Tank failure due to vacuum or degradation*: A waste tank structural failure can be caused by excessive vacuum or structural degradation. Structural failure creates a pathway for an airborne release and a mechanism for mobilizing the waste material.
7. *Aboveground structure failure*: Aboveground structural failures include drops of contaminated equipment and other releases from contaminated aboveground structures. These failures can result in an uncontrolled release of radioactive and toxic materials.
8. *Mixing of incompatible material*: This accident is initiated by the addition of an incompatible material to tank waste that results in the release of radioactive and toxic materials. The incompatible material reacts with the tank waste to produce an energetic chemical reaction that mobilizes the waste.
9. *Waste transfer leak*: Waste transfer leaks occur in a variety of physical configurations. For example, a very narrow crack can form in a waste transfer line to produce a high-pressure aerosol spray of radioactive and toxic materials that is released into the air. A second example is a waste transfer leak that creates a subsurface or surface pool of tank waste. These pools have an enhanced potential for the mobilization of waste and contaminating areas beyond the pool boundary.
10. *Unplanned excavation/drilling*: This accident is initiated by excavating or drilling into an active or inactive liquid disposal site (e.g., waste tank or associated structure). The excavation mobilizes the waste material.
11. *External events*: An external event is assumed to generate sufficient energy to mobilize tank waste. For example, an aircraft directly impacts a waste tank. The aircraft penetrates the dome, and aviation fuel enters the tank and is ignited. Burning fuel causes the release of radioactive and hazardous aerosols and particulate material. Terrorist-induced events are included in this event type.
12. *Transportation accidents*: This postulated event involves handling and transfer activities incidental to transportation that could affect the waste tank and mobilize waste.
13. *Filtration failures leading to unfiltered releases*: A release of radioactive material can result from a HEPA filter failure. Filter failure can occur following high tank temperature or pressure, and the release of radioactive material occurs through the failed filter.

14. *Organic solvent fire*: Waste tanks with accumulated flammable organic material can experience a fire. These organic solvent fires include surface pools, solvent permeated salt cake burns, and organic vapor events.
15. *Tank bump*: This accident involves a rapid density change in a waste tank having a high decay heat load. The decay heat increases the waste temperature and eventually overpressurizes the tank headspace, which leads to a release of vapor and entrained waste. A tank bump accident occurs with the loss of tank cooling mechanisms. Tank waste boiling can also occur within this event type. The time to reach tank bump conditions depends on the decay heat level of radioactive materials residing within the tank.

3.2. (a) The primary responsibilities of the RCM include the following:

1. Perform timely calculations of the projected doses from facility releases of radioactive materials.
2. Ensure the radiation safety of the personnel at the EOF.
3. Formulate/update protective action recommendations (PARs).
4. Coordinate the movement of off-site sampling teams in order to assess the release.
5. Direct the overall radiological response to the event.
6. Determine if the use of thyroid blocking agents are warranted.
7. Discuss plant radiological conditions, dose projections, and PARs with utility emergency managers and senior state and federal officials.
8. Authorize emergency exposure requests and thyroid blocking agents.
9. Monitor the plant radiological status and coordinate response actions with on-site health physics personnel.
10. Ensure senior utility emergency managers are aware of changing plant radiological conditions.

(b) The three fission product barriers are (i) fuel pellets and fuel element cladding, (ii) reactor coolant system and included piping, and (iii) containment building.

(c) Fission product barrier status is normally categorized as:

1. *Intact*: The fission product barrier is capable of preventing the release of fission products to the environment. This is the normal condition.
2. *In jeopardy*: The barrier is currently intact, but plant conditions are such that the barrier will likely be breached.
3. *Breached*: The fission product barrier is not able to prevent the release of fission products to the environment.

(d) Based on plant data, the status of the three fission product barriers is as follows:

Fission product barrier	*Status*
Fuel pellets and fuel element cladding	The fuel barrier is breached as illustrated by the high activity in the letdown (primary) system
Reactor coolant system (RCS) and included piping	The RCS barrier is breached as evidenced by the high activity in the steam generator blowdown (secondary) sample and elevated Steam Generator "A" steam line dose rate. The breach is through the damaged steam generator tubes
Containment building	Containment integrity is in jeopardy because containment pressure is increasing. The primary to secondary leakage permits the possibility of a release to the environment if a relief valve lifts due to increases in secondary system pressure

(e) Factors affecting the calculation of off-site doses include:
1. Release magnitude and isotopic composition
2. Release rate
3. Meteorological conditions (atmospheric stability class, wind speed, and precipitation)
4. Effective release height
5. Release duration
6. Changing plant conditions due to repairs or equipment failures
7. Land topography
8. Receptor location

(f) The projected thyroid equivalent dose rate ($\dot{H}$) at 3.2 km, resulting from the ^{131}I released from the fuel, is affected by any partitioning from the secondary side of the plant to the environment. The ^{131}I release concentration (C) is determined from the steam generator blowdown sample (C_{BD}) and the iodine partitioning factor (ξ):

$$C = \xi C_{\mathrm{BD}}$$

$$C = (0.015)\left(2.33 \times 10^4\ \frac{\mathrm{Bq}}{\mathrm{cm}^3}\right) = 350\ \frac{\mathrm{Bq}}{\mathrm{cm}^3}$$

The release rate (Q) is obtained from the release concentration by assuming the release is through the atmospheric relief valve having a flow rate (F) of $1.4 \times 10^7\ \mathrm{cm}^3/\mathrm{s}$:

$$Q = CF = \left(1.4 \times 10^7\ \frac{\mathrm{cm}^3}{\mathrm{s}}\right)\left(350\ \frac{\mathrm{Bq}}{\mathrm{cm}^3}\right)$$

$$= 4.90 \times 10^9\ \frac{\mathrm{Bq}}{\mathrm{s}} = 4.90 \times 10^3\ \frac{\mathrm{MBq}}{\mathrm{s}}$$

The desired equivalent dose rate at 3.2 km delivered to the thyroid is

$$\dot{H} = \frac{\chi u}{Q} Q \frac{1}{u} \mathrm{DCF}$$

$$\dot{H} = \left(\frac{2.69 \times 10^{-3}}{\mathrm{m}^2}\right)\left(4.90 \times 10^3\,\frac{\mathrm{MBq}}{\mathrm{s}}\right)\left(\frac{\mathrm{h}}{25 \times 10^3\,\mathrm{m}}\right)$$
$$\times\left(\frac{3600\,\mathrm{s}}{\mathrm{h}}\right)\left(20.9\,\frac{\mathrm{Sv\text{-}cm^3}}{\mathrm{MBq\text{-}s}}\right)\left(\frac{1\,\mathrm{m}^3}{10^6\,\mathrm{cm}^3}\right)\left(\frac{3600\,\mathrm{s}}{\mathrm{h}}\right) = 0.143\,\frac{\mathrm{Sv}}{\mathrm{h}}$$

(g) Sheltering should be recommended at this time. Since a long-term release is not expected, evacuation is not warranted.

(h) A conservative estimate of the equivalent dose to the thyroid in the town is performed using the dose rate information calculated at the 3.2 km distance:

$$H = \dot{H}t = \left(0.143\,\frac{\mathrm{Sv}}{\mathrm{h}}\right)(0.25\,\mathrm{h}) = 0.0358\,\mathrm{Sv} = 35.8\,\mathrm{mSv}$$

Again, evacuation is not warranted. The population could not be evacuated in a 15 min period. Sheltering is the recommended protective action.

(i) The thyroid equivalent dose delivered in an 8 h period is

$$H = \dot{H}t = \left(0.143\frac{\mathrm{Sv}}{\mathrm{h}}\right)(8.0\,\mathrm{h}) = 1.14\,\mathrm{Sv}$$

The committed effective dose equivalent is determined from the relationship

$$H_{\mathrm{E}} = w_{\mathrm{T}}H = (0.03)(1.14\,\mathrm{Sv}) = 0.0342\,\mathrm{Sv} = 34.2\,\mathrm{mSv}$$

where w_{T} is the ICRP 26 thyroid tissue weighting factor. This dose warrants declaration of a General Emergency.

(j) Since the CEDE exceeds the 10 mSv PAG, an evacuation is warranted if it can be accomplished in a timely manner. The evacuation time study, usually documented in the facility's emergency plan, should be consulted to determine the time required to complete the evacuation protective action for the existing weather conditions. In addition, the equivalent dose to the thyroid exceeds the PAG of 50 mSv for KI administration, and it should be implemented to limit the thyroid equivalent dose.

3.3. (a) Particulate radon daughters are known to be present in the room at a beta concentration of 1.11×10^{-5} Bq/cm^3 with an effective half-life of ~27 min. You are requested to determine the count rate that should be observed on the monitor at 09:00 h. The expected count rate is determined by integrating the production relationship (see Appendix B)

$$\frac{\mathrm{d}A}{\mathrm{d}t} = Pe^{-\lambda t}$$
$$\int_0^A \mathrm{d}A = \int_0^T Pe^{-\lambda t}\mathrm{d}t$$
$$A(T) = \frac{P}{\lambda}\left(1 - e^{-\lambda T}\right)$$

where

A	=	activity deposited upon the filter paper
λ	=	total removal constant for the radon daughters or the effective removal constant = $\ln(2)/T_e$
T_e	=	effective half-life = 27 min
λ	=	$0.693/27\ \text{min} = 2.57 \times 10^{-2}/\text{min}$
T	=	sampling time = 1 h = 60 min
P	=	production rate = $C\,F\,e'$
C	=	air concentration of the radon daughters = 1.11×10^{-5} Bq/cm^3
F	=	monitor flow rate = 1 ft^3/min
e'	=	filter collection efficiency = 0.9

$$P = CFe'$$

$$P = \left(1.11 \times 10^{-5}\,\frac{\text{Bq}}{\text{cm}^3}\right)\left(1\,\frac{\text{ft}^3}{\text{min}}\right)\left(\frac{30.48\,\text{cm}}{\text{ft}}\right)^3(0.9) = 0.283\,\frac{\text{Bq}}{\text{min}}$$

With these parameter values, the activity collected on the filter between 08:00 and 09:00 h is determined:

$$A(T) = \frac{P}{\lambda}(1 - e^{-\lambda T})$$

$$A(T) = \frac{\left(0.283\,\frac{\text{Bq}}{\text{min}}\right)}{\left(\frac{2.57 \times 10^{-2}}{\text{min}}\right)}\left(1 - e^{-\left(\frac{2.57 \times 10^{-2}}{\text{min}}\right)(1\,\text{h})\left(\frac{60\,\text{min}}{\text{h}}\right)}\right)$$

$$= (11.0\,\text{Bq})(1 - 0.214) = (8.65\,\text{Bq})\left(\frac{\text{dis}}{\text{Bq-s}}\right)\left(\frac{60\,\text{s}}{\text{min}}\right) = 519\,\text{dpm}$$

The count rate (CR) from the collected activity is obtained from the counting efficiency (e)

$$\text{CR} = eA = \left(\frac{0.3\,c}{d}\right)(519\,\text{dpm}) = 156\,\text{cpm}$$

The measured count rate (CR_m) is the sum of the filter count rate and the background count rate (CR_B) of 70 cpm:

$$\text{CR}_\text{m} = \text{CR} + \text{CR}_\text{B} = 156\,\text{cpm} + 70\,\text{cpm} = 226\,\text{cpm}$$

(b) At 09:00 h, work begins in the room where the air monitor is located. At 09:45 h, the air in the room is contaminated following the premature detonation of the terrorist's explosive device. Over the next 10 min, the strip chart recorder shows that the average count rate has increased by 40 000 cpm. Based on this information, you are to determine the airborne concentration in the room.

Assuming that the release rate is constant over the 10 min period and radioactive decay is insignificant, the activity deposited on the filter (A)

is obtained from the following relationship

$$A = \left[\frac{\text{CR}}{ee'}\right]$$

where

CR	=	filter count rate during the 10 min period = 4.0×10^4 cpm
e'	=	filter collection efficiency = 0.9
e	=	counting efficiency = 0.3 cpm/dpm

$$A = \left[\frac{(4.0 \times 10^4\ \text{cpm})}{(0.9)\left(0.3\ \frac{\text{cpm}}{\text{dpm}}\right)}\right] = \left(1.48 \times 10^5\ \frac{\text{dis}}{\text{min}}\right)\left(\frac{1\ \text{min}}{60\ \text{s}}\right)\left(\frac{\text{Bq-s}}{\text{dis}}\right)$$
$$= 2.47 \times 10^3\ \text{Bq}$$

Since there is negligible decay during the sampling period, the average air concentration (C) is determined from the activity on the filter and the volume of air (V) drawn through it during the 10 min sampling period (t):

$$C = \frac{A}{V} = \frac{A}{Ft} = \frac{(2.47 \times 10^3\ \text{Bq})}{\left(1\ \frac{\text{ft}^3}{\text{min}}\right)\left(\frac{30.48\ \text{cm}}{\text{ft}}\right)^3 (10\ \text{min})} = 8.72 \times 10^{-3}\ \frac{\text{Bq}}{\text{cm}^3}$$

(c) In the part, you are requested to list advantages of a whole-body count over urine bioassay for assessing an intake of ^{137}Cs and ^{60}Co. The advantages include the following:

1. There is no need to wait for the collection of 24 h urine samples.
2. There is less dependency on worker action in collecting and returning urine.
3. ^{137}Cs and ^{60}Co are accurately measured with a whole-body counting system since they emit photons with energies of 662 keV and 1.25 MeV (average energy), respectively.
4. Immediate results are obtained with whole-body counting. Laboratory analysis and interpretation of the data are not required. The whole-body count provides an immediate indication of the intake.
5. Whole-body counts are easy to perform.
6. Whole-body counts provide an effective tool to monitor the internal deposition of radioactive materials as a function of time.
7. Whole-body counts are usually less expensive.
8. Whole-body counts have less dependency on human and quality errors (e.g., samples being lost, laboratory errors, and lack of vendor quality assurance).
9. Insoluble particulate chemical forms will not necessarily be eliminated through urine or may take longer to clear the body. The

whole-body count will detect the radioactive material independent of its clearance characteristics.

10. Complications arising from determining the intake pathway (i.e., inhalation or ingestion) for individuals preferentially using nose or mouth breathing are also avoided with whole-body counting.
11. The results are obtained immediately and can be used for subsequent action (e.g., medical intervention) if warranted.

(d) A count of the filter on a HPGe gamma spectroscopy system shows that the airborne radioactivity is due to 25% ^{60}Co and 75% ^{137}Cs. The effective dose for 1 h of exposure to an average measured air concentration of 74 Bq/cm^3 is determined from the intake (I):

$$I = C(\text{BR})t$$

where

C	=	concentration of nuclide in the air
$C(^{60}\text{Co})$	=	(0.25) (74 Bq/cm^3) = 18.5 Bq/cm^3
$C(^{137}\text{Cs})$	=	(0.75) (74 Bq/cm^3) = 55.5 Bq/cm^3
BR	=	breathing rate = 1.2 m^3/h
t	=	exposure time = 1 h

$$I(^{60}\text{Co}) = C(^{60}\text{Co})(\text{BR})t$$

$$I(^{60}\text{Co}) = \left(18.5\,\frac{\text{Bq}}{\text{cm}^3}\right)\left(\frac{100\,\text{cm}}{\text{m}}\right)^3\left(\frac{1.2\,\text{m}^3}{\text{h}}\right)(1\,\text{h})$$

$$= 2.22 \times 10^7\ \text{Bq}$$

$$I(^{137}\text{Cs}) = C(^{137}\text{Cs})(\text{BR})t$$

$$I(^{137}\text{Cs}) = \left(55.5\,\frac{\text{Bq}}{\text{cm}^3}\right)\left(\frac{100\,\text{cm}}{\text{m}}\right)^3\left(\frac{1.2\,\text{m}^3}{\text{h}}\right)(1\,\text{h})$$

$$= 6.66 \times 10^7\ \text{Bq}$$

The effective dose is determined from the dose conversion coefficient (e) values provided in the problem statement:

$$E = I(^{137}\text{Cs})e(^{137}\text{Cs}) + I(^{60}\text{Co})e(^{60}\text{Co})$$

$$E = (6.66 \times 10^7\ \text{Bq})\left(6.7 \times 10^{-9}\,\frac{\text{Sv}}{\text{Bq}}\right) + (2.22 \times 10^7\ \text{Bq})\left(7.1 \times 10^{-9}\,\frac{\text{Sv}}{\text{Bq}}\right)$$

$$= 0.45\,\text{Sv} + 0.16\,\text{Sv} = 0.61\,\text{Sv}$$

(e) Gamma spectroscopy screening of an air sample shows that ^{131}I is present. A technical support center worker is exposed for 4 h to the contaminated atmosphere and has an intake of 500 MBq of ^{131}I. The effective dose from this exposure is

$$E = I(^{131}\mathrm{I})e(^{131}\mathrm{I})$$
$$= (500\,\mathrm{MBq})\left(\frac{10^6\,\mathrm{Bq}}{\mathrm{MBq}}\right)\left(1.1\times10^{-8}\,\frac{\mathrm{Sv}}{\mathrm{Bq}}\right) = 5.5\,\mathrm{Sv}$$

(f) The Emergency Director's action was prudent. KI is effective if administered within a few hours of the intake. Given the calculated 5.5 Sv effective dose, medical personnel should be advised of the situation. Medical monitoring and consultation with TSC personnel should be subsequent actions.

3.4. (a) Assuming that the initial plume has passed, actions that could most significantly reduce the dose to the downwind population during the first week following the accident include:

1. Do not use locally produced foods (e.g., crops, milk, meat, and fish).
2. Do not use local water supplies.
3. Establish clean zones in areas that require public access.
4. Minimize travel in contaminated areas.
5. Assuming the population was relocated, allow reentry in an ALARA manner.
6. Fix plutonium in place with appropriate spray materials to permit limited use of land and facilities and to enhance the public health and welfare.
7. Establish an exclusion zone based on contamination levels.
8. Individuals not relocated should use dose reduction measures (e.g., scrub and/or flush surfaces, soak or plow soil, remove and dispose small hot spots, and spend more time indoors).
9. Perform bioassay to determine the extent of internal intakes to members of the public. Administer decorporation agents to minimize the internal dose. Decorporation approaches are addressed in Section 4.5.3.

(b) The intermediate phase of a nuclear emergency is defined as the period beginning after the source and release are brought under control and environmental measurements are available. These measurements are a key input for subsequent decisions. During the intermediate phase, decisions are made concerning the disposition of particular areas or properties from which persons have been evacuated. These areas will be either decontaminated and reoccupied or condemned and the occupants permanently relocated. The intermediate phase Protective Action Guidance are summarized in the following table:

Planning guidance and Protective Action Guides for radiological incidents		
Phase	**Protective action recommendation**	**Protective Action Guide or planning guidance**
Intermediate	Relocation of the public	20 mSv projected dose first year[a)] Subsequent years, 5 mSv/year projected dose
	Food interdiction	5 mSv/year projected dose, or 50 mSv/year, to any individual organ or tissue, whichever is limiting
	Limit emergency worker exposure	50 mSv/yr[a)]
	Reentry	Operational guidelines[b)] (stay times and concentrations) for specific activities

a) Projected dose is the sum of the effective dose from external radiation exposure (i.e., groundshine and cloudshine) and the committed effective dose from inhaled radioactive material.
b) See DOE/HS-0001; ANL/EVS/TM/09-(DOE 2009).

(c) The population effective doses incurred or projected are:
1. *Early phase*: 15–20 mSv
2. *Intermediate phase*: 13 mSv

The dose expected during the first-year postaccident is 13 mSv. This dose is incurred during the intermediate phase. Since the first-year dose is <20 mSv, the population may reenter the evacuated area, and dose reduction techniques should be applied to further reduce public doses. The other limits noted in the question (b) table must also be met.

(d) The relevant radiation data was provided in the problem statement:

Isotope	wt%	Selected radiation emissions	
		Photon energy (MeV)	Photon yield (%)
^{238}Pu	0.04	0.017	11
^{239}Pu	93.3	0.017	5
^{240}Pu	5.99	0.017	11
^{241,242}Pu	0.32	—	—
^{241}Am	0.30	0.017	37
		0.060	36
Pu (mixture)	100	a)	<5

a) Pu (mixture) energy represents multiple photons with energies >0.03 MeV.

As defined in the problem statement, the emission ratio of the 17–60 keV photons is ~2.5. The requested advantages are as follows:

1. *Advantages of the 17 keV calibration*:

1. It can detect plutonium preferentially over americium due to the large weight percentage of ^{239}Pu and the small weight percentage of ^{241}Am.
2. It can detect ^{239}Pu preferentially.
3. The 17 keV emission is 2.5 times the 60 keV emission.
4. It is more useful over contaminated surfaces that are smooth and provide minimal attenuation.

2. *Advantages of the 60 keV calibration*:
   1. It can determine the gross alpha contamination from ^{241}Am and the Pu-mixture if calibrated to the 60 keV peak.
   2. It can detect ^{241}Am preferentially if calibrated to the narrow ^{241}Am 60 keV peak vice the broader Pu mixture peak.
   3. It is more useful over contaminated surfaces that are nonuniform and provide attenuation of lower-energy photons.

(e) The choices of emission for the specified conditions are:

1. *Dry paved road*: The 17 keV energy is preferred because it has a higher emission ratio (2.5 times the 60 keV yield) and there is limited attenuation of the lower-energy photon on a dry, smooth surface.
2. *Agricultural field following an extended rain*: A wet agricultural field attenuates the lower-energy photons. Attenuation also increases because the water has a greater effect on the lower-energy photons. Therefore, the 60 keV photon calibration is preferred.

(f) The internal effective dose $E(50)$ from ground deposition of the Pu/Am mixture for an individual who walks for 1 h on soil contaminated at a level of 3.7 MBq/m^2 is determined from the relationship

$$E(50) = I(\mathrm{DCF})$$

where

I	=	intake of the Pu/Am mixture from inhalation
DCF	=	effective dose conversion factor for the inhalation of the Pu/Am mixture = 3.2×10^{-5} Sv/Bq inhaled

The intake is the product of the breathing rate (BR), air concentration (C), and exposure time (t):

$$I = C(\mathrm{BR})t$$

The air concentration is determined from the resuspension factor (r) and the surface contamination level (S):

$$C = Sr$$

$$C = \left(3.7 \times 10^{6}\ \frac{\mathrm{Bq}}{\mathrm{m}^2}\right)\left(\frac{1.0 \times 10^{-5}}{\mathrm{m}}\right) = 37\ \frac{\mathrm{Bq}}{\mathrm{m}^3}$$

With this result, the intake and effective dose are determined:

$$I = C(\mathrm{BR})t$$

$$I = \left(37\,\frac{\mathrm{Bq}}{\mathrm{m}^3}\right)\left(20\,\frac{\mathrm{l}}{\mathrm{min}}\right)(1\,\mathrm{h})\left(\frac{1\,\mathrm{m}^3}{1000\,\mathrm{l}}\right)\left(60\,\frac{\mathrm{min}}{\mathrm{h}}\right) = 44.4\,\mathrm{Bq}$$

Since the effective dose conversion coefficient provided in the problem is for the inhalation of the Pu/Am mixture,

$$E(50) = I(\mathrm{DCF})$$

$$E(50) = (44.4\,\mathrm{Bq})\left(3.2\times10^{-5}\,\frac{\mathrm{Sv}}{\mathrm{Bq}}\right)\left(1000\,\frac{\mathrm{mSv}}{\mathrm{Sv}}\right) = 1.4\,\mathrm{mSv}$$

(g) The external dose received by the individual is

$$E = C_{\mathrm{a}}(\mathrm{DCF})t$$

where C_{a} is the activity per unit area in the uniformly contaminated region, DCF is the external dose conversion coefficient, and t is the residence time in the contaminated area. Using the values in the problem statement,

$$C_{\mathrm{a}} = \frac{(2\,\mathrm{kg})\left(1000\,\frac{\mathrm{g}}{\mathrm{kg}}\right)}{(1000\,\mathrm{m}^2)} = 2\,\frac{\mathrm{g}}{\mathrm{m}^2}$$

$$\mathrm{DCF} = 2.8\times10^{-5}\,\frac{\mathrm{Sv\text{-}m}^2}{\mathrm{h\text{-}g}}$$

$$t = 8\,\mathrm{h}$$

determines the external dose

$$E = \left(2\frac{\mathrm{g}}{\mathrm{m}^2}\right)\left(2.8\times10^{-5}\,\frac{\mathrm{Sv\text{-}m}^2}{\mathrm{h\text{-}g}}\right)(8\,\mathrm{h})\left(\frac{1000\,\mathrm{mSv}}{\mathrm{Sv}}\right) = 0.45\,\mathrm{mSv}$$

(h) The NRC provides no specific emergency dose limits but utilizes the EPA values. These effective dose limits are 50 mSv/year (or greater under exceptional circumstances). Competent authority may approve doses to emergency workers above 50 mSv. In particular, the EPA specifies 100 mSv effective dose for the protection of property and 250 mSv for lifesaving activities. The worker's initial dose estimate of 48 mSv is below the EPA limits.

The default values are likely to be very conservative, and job-specific values will lead to a more realistic (lower) value. The worker's dose history should accurately reflect the effective dose. Therefore, specific accident details should be determined to refine the dose estimate. These details include:

1. Determining the specific dose rates and airborne concentrations at the worker's location as a function of time.
2. Determining the worker's residence time within the various dose rate and airborne concentration areas as a function of time.
3. Evaluating radiological data from the worker including nasal smears, whole-body counts, lung counts, thyroid counts, urine samples, and fecal samples.
4. Evaluating the worker's breathing zone air samples.
5. Obtaining bioassay data from the worker to refine the retention function.
6. Obtaining the dose results from TLD and other dosimetry worn by the worker during the accident.
7. Reviewing facility radiation monitor dose and air monitor contamination values and the trends in these data and correlating these data with the worker's location.
8. Reconstructing the worker's activities including their location and duration. This information should be evaluated in terms of the local radiological conditions and plant release data.
9. Documenting the timing and nature of radioprotective chemicals administered to the individual.
10. Evaluating the effectiveness of the workers personal protective equipment and any associated skin contamination.

These data should be noted and included in the final dosimetry report documenting the individual's effective dose.

3.5. (a) The actions that should be taken immediately in response to a generic radiological accident involving personal injury must consider the ALARA concept. These actions include the following:

Immediate response for medical concerns:

1. Lifesaving first aid and cardiopulmonary resuscitation (CPR) are the first priority.
2. Take a moment to familiarize yourself with the work area. Do not act too hastily without understanding the accident area and its associated hazards that could interfere with medical treatment.
3. Evacuate personnel to safe areas. The accounting of all personnel involved in the accident and an assessment of their medical condition are to be performed.
4. Prioritize medical treatment by first addressing the most serious injuries.
5. Utilize available facility medical staff and personnel trained in CPR and first aid.

Response for radiological conditions:

1. Stop the release by securing pumps and fans, closing isolation values and dampers, and securing ventilation systems. Take steps

to minimize the immediate spread of contamination. This may include isolating airborne and waterborne release pathways.

2. Warn facility personnel of the event and direct them to vacate the affected area.
3. Call for assistance. Notify facility emergency response personnel and activate the emergency response organization if warranted by the event.
4. Survey all affected personnel to assess the extent of external and internal contamination. Nasal swabs should be performed if facial contamination is detected. Whole-body counts are performed for positive nasal smears.
5. Identify injured, contaminated, and/or exposed personnel for immediate treatment. Ensure that medical assistance is provided to these accident victims.
6. Evacuate personnel from the affected areas.
7. Initiate actions to isolate the affected areas and begin recovery operations.
8. Collect and evaluate dosimetry.
9. Initiate a dose assessment for affected personnel.

Reentry and accident response:

1. Obtain survey instrumentation, protective clothing, and respiratory protection suitable to the accident conditions. These should be based on the facility's emergency procedures and emergency operation radiation work permits.
2. Utilize plant instrumentation to obtain a general indication of direct dose and airborne contamination levels.
3. Provide briefings to reentry and accident response personnel regarding the radiological and physical conditions in the facility.
4. Have health physics personnel accompany reentry and accident response teams to ensure that radiological hazards are bound by the anticipated conditions.
5. Reenter the accident area cautiously to locate the hazards. Initial entries will likely require respiratory protection (e.g., self-contained breathing apparatus) if the radiological conditions are unknown or suspected to be severe. Obtain radiological data to determine the radiation levels and set up air supplemental air samplers. Minimize time in the affected areas. Evaluate and document radiological data and post it for use in subsequent activities.
6. Interview witnesses to determine the cause of the event and to reconstruct the accident's sequence of events. Determine the radioactive materials that were released. Relay this information to attending physicians and emergency response personnel.
7. Establish liaison with outside authorities (e.g., state government and regulatory agencies).

8. Read dosimetry from the emergency response teams to characterize the external dose.
9. Perform appropriate bioassay for the emergency response teams to characterize their internal dose.

(b) For each of the five accident scenarios, the preferred bioassay monitoring techniques are as follows:

Accident no.	Isotope	Bioassay method
1	^{239}Pu	Lung and whole-body counting can detect the ^{239}Pu X-ray radiation. A detection activity of about 75 Bq in the lung is possible for a 2000 s count. Bone detection limits of about 185 Bq require 3000 s counts. Urinalysis is also an option since urine is a clearance pathway.
2	^{3}H	Urinalysis is a viable bioassay approach since tritium in the form of HTO behaves like water and is readily dispersed in the body's water. HTO is eliminated via urine and measured via liquid scintillation counting.
3	^{137}Cs	Whole-body counting is the most practical technique to detect internal ^{137}Cs depositions. The 662 keV photon is easily detected with whole-body counting techniques.
4	^{35}S	Intake retention functions suggest that either urine or feces are the most appropriate bioassay approach. The low-energy beta emission (167 keV) is analyzed using liquid scintillation counting. Depending on the solubility of the ^{35}S material being pipepetted, urinalysis will be a more esthetic choice. The reader should note that mouth pipepetting is not an acceptable practice and this accident can be easily avoided.
5	^{131}I	Whole-body counting or thyroid counting is the most appropriate bioassay technique for detecting ^{131}I. Iodine localizes in the thyroid and its photons are readily detected. Urinalysis is also an effective bioassay technique since the iodine is eliminated via several pathways including urine.

(c) Medical intervention techniques used to minimize the internal dose following an intake of radioactive material are divided into several general categories based on their protective actions. In order to understand these interventions, the total absorbed dose delivered to an organ (D) is defined by the relationship

$$D = \frac{AE}{m\lambda_{\text{eff}}} = \frac{AET_{\text{eff}}}{\ln(2)m}$$

where

A	=	activity deposited in the organ
E	=	energy emitted per disintegration of the radionuclide deposited in the organ
m	=	mass of the organ
λ_{eff}	=	effective disintegration constant of the radionuclide in the body
T_{eff}	=	effective half-life of the radionuclide in the body

The intervention agents either decrease A or T_{eff} to minimize the absorbed dose. As requested in the problem statement, the dose-savings categories and a brief description of the dose-savings principles of each include:

1. *Blocking or saturating agents that decrease A*: These agents saturate an organ with a stable isotope of the radionuclide deposited in the body, which minimizes the uptake of a radioisotope. For example, the administration of iodine (e.g., potassium iodide) saturates the thyroid, which minimizes the uptake of radioiodine and significantly reduces the absorbed dose.
2. *Chelating agents that decrease T_{eff}*: Chelation therapy involves administering a chemical compound that binds metal ions to form a soluble complex, which is readily excreted via urine. Increasing the solubility enhances the removal rate and decreases the absorbed dose. For example, greater than 50% of bone-deposited plutonium has been removed with DTPA.
3. *Physical intervention via stomach pump that decreases A*: The pump removes the contents of the stomach before radionuclides are absorbed via the alimentary tract. This action decreases the activity in the organ and decreases the absorbed dose.
4. *Physical intervention via lung lavage that decreases A*: Lung lavage removes material from the lung before it can be transferred to the blood or the alimentary tract. This action limits the activity in the lung and decreases the absorbed dose.
5. *Metabolism stimulation that decreases T_{eff}*: This approach refers to increasing the normal body response with the intent of increasing the clearance rate, which decreases the absorbed dose. The absorbed dose is proportional to the effective half-life of the material within the body. Tritium uptake mitigation utilizes an application of this technique. By forcing fluids or using agents to increase urine production (e.g., diuretics), the turnover of body water is increased.
6. Isotopic dilution that decreases the activity absorbed in the organ of interest and decreases the absorbed dose: Saturating the organ of interest with a nonradioactive isotope minimizes the radionuclide uptake to that organ and minimizes the organ dose. Isotopic dilution is illustrated by the use of potassium phosphates for minimizing the internal dose by minimizing the uptake of ^{32}P.
7. Ionic displacement that decreases A: With ionic displacement, radioactive material is displaced by a stable element having similar chemical properties. For example, an intake of ^{137}Cs may be mitigated using Prussian Blue (ferric ferrocyanide).

(d) 1. Chelation therapy following the inhalation of 5 ALI of ^{241}Am is appropriate. DTPA has been demonstrated to be effective in removing ^{241}Am from the bone even for long periods following deposition.

2. Lung lavage following the inhalation of 10 ALI of mixed fission products may not be appropriate. The most effective intervention technique depends on the specific mix of fission products. In addition, the particle size of the fission product aerosol and its retention within the lung affect its residence time, which may make the lavage unnecessary. Lavage will also be ineffective for radioiodine if it is performed after the iodine is transferred to the thyroid.

   As specific examples, the following techniques are superior to lung lavage and are less stressful to the individual:

$^{89}Sr/^{90}Sr$	Administer Sr or Ca or consider alginates
^{131}I	Administer KI or KIO_3 within 2 h of the intake
^{137}Cs	Administer Prussian Blue

(e) Assuming that the intakes associated with the five accidents scenarios are sufficiently high to warrant medical intervention, a specific intervention technique that is available for each accident and any special concerns or necessary precautions include the following:

1. In Accident 1, the ^{239}Pu intake is mitigated with chelation therapy using DTPA. The long-term use of chelation therapy tends to deplete the body of necessary trace elements. Early work with Ca-DTPA demonstrated these side effects after prolonged use. Zn-DTPA is less hazardous over the long term. The use of chelating agents for a declared pregnant worker requires special consideration due to the depletion of trace elements that can affect the health of the mother and development of the fetus. Other options noted by NCRP 161 include deferoxamine (DFOA) and ethylenediaminetetraacetic acid (EDTA).
2. In Accident 2, the ^{3}H intake is limited using diuretics or increasing the intake of fluids containing water. These techniques place a physiological stress on the body and must be assessed by a licensed physician since the urinary output will increase by about 50% with subsequent stress on the bladder and kidneys.
3. The Accident 3 ^{137}Cs intake is mitigated by administering Prussian Blue. Constipation is a side effect. Only limited use (less than 3 weeks) has been reported in humans.
4. NCRP 161 notes that sodium thiosulfate is a treatment option to mitigate the intake of ^{35}S. Lavage and purgatives could be considered. Since the material was ingested, evacuation of materials from the stomach by means of a gastric tube should be evaluated. Emetics or vomiting induction agents should also be considered.

Purgatives or laxatives may also be used. However, lavage, purgatives, or emetics place a physiological stress on the body and must be addressed by a physician prior to use.

5. The intake of ^{131}I can be effectively mitigated using blocking agents (e.g., KI or KIO_3) if these agents are administered in a timely manner following the intake. The use of these chemicals can lead to an allergic reaction. As such, a physician must examine the individual before prescribing their use. These blocking agents must be used within a few hours or they have minimal impact on reducing the iodine uptake by the thyroid. NCRP 161 notes KI is the preferred option. Other treatment options include propylthiouracil and methimazole.

3.6. (a) The likely cause of this event is a hydrogen explosion. The problem conditions noted that fuel temperatures were increasing and core damage had occurred. Given these conditions, the zirconium in the fuel cladding has reacted with water to produce hydrogen gas:

$$2\,\mathrm{Zr} + 2\,\mathrm{H_2O} \rightarrow 2\,\mathrm{ZrOH} + H_2 \uparrow$$

The hydrogen and fission products were released when the reactor pressure vessel and primary containment vessel were vented to the reactor building. Hydrogen is accumulated in the reactor building and detonated once it exceeded the lower flammability limit. Following the explosion, the fission products in the reactor building were released to the environment.

(b) The health physics consequences of the hydrogen explosion are significant. With fuel failure and venting, fission products are released to the environment. The explosion damages the reactor building upper floors, and building debris falls into the spent fuel pool. Falling debris could damage the fuel stored in the pool. The debris and explosion have the potential to affect the structural integrity of the pool. Loss of pool water reduces spent fuel cooling, leads to additional spent fuel damage, and releases fission products. The falling debris could also damage safety systems supporting the spent fuel pool.

The initial fuel damage will increase if core cooling is not reestablished. Without core cooling, the fuel will melt and breach the reactor pressure vessel and containment vessel. Fuel will then flow to the reactor building basement where it will react with the structural concrete and steel. The progression of the accident beyond this point is highly uncertain. If the fuel is not sufficiently cooled as it melts through the concrete base mat, it would enter the earth below the reactor and potentially interact with the soil and associated groundwater.

The fuel in the spent fuel pool is also at risk. Without cooling water, the pool will boil and fuel temperatures will increase. If sufficient spent fuel pool water inventory is unavailable to cool the fuel, it will suffer a fate similar to that outlined previously for the fuel comprising the reactor core.

The net result is additional hydrogen generation and the possibility of a subsequent explosion. Following the second hydrogen explosion, additional fission products are released to the environment. It is also possible that fuel particles are released to the environment.

Although these events are highly unlikely, their precedent was established by the Fukushima Daiichi accident. The extent of fuel damage was mitigated at Fukushima Daiichi because core cooling was reestablished. The scenario described in this problem is the Fukushima Daiichi accident without mitigation. However, the likelihood that personnel would not establish a suitable core cooling mechanism before it exceeded the Fukushima Daiichi consequence level is remote.

(c) The Fukushima Daiichi accident was a Level 7 event (Major Accident) which is the highest International Nuclear Event Scale classification. Since the postulated event is similar to the Fukushima Daiichi accident, it would be classified as an INES Level 7 event.

(d) The initial dose assessment is compared to the thresholds for initiating protective actions. For specificity, threshold dose values utilized by the NRC are used in answering this question:

1. Evacuation is warranted if the effective dose reaches 10 mSv.
2. Potassium iodide is recommended if the thyroid dose exceeds 50 mSv.

Distance from facility (km)	Thyroid equivalent dose (mSv)	Effective dose (mSv)
1[a), b)]	2000	1000
2[a), b)]	900	700
5[a), b)]	700	500
10[a), b)]	400	200
16[a), b)]	300	70
20	40	7
50	10	4
80	5	2

a) Evacuation criterion is met.
b) KI administration criterion is met.

Based on these results, evacuation is warranted to a distance of 16 km. However, the effective dose at 20 km is 70% of the evacuation PAG. Therefore, evacuation to 20 km should be discussed with the state. The administration of KI to the evacuated population is also warranted.

An evacuation is contingent on weather and travel conditions. If there are issues with safely evacuating the public, sheltering in place should be recommended. If sheltering is recommended, predistributed KI should be administered.

(e) During the early phase, the radionuclides most likely to affect the food chain and the affected foods and associated media are as follows:

Affected food or medium	Dominant radionuclides
Beef	^{131}I and ^{137}Cs
Fish	^{131}I and ^{137}Cs
Fruits	^{131}I and ^{137}Cs
Milk	^{131}I and ^{137}Cs
Rice	^{131}I and ^{137}Cs
Seafood	^{131}I and ^{137}Cs
Soil	^{90}Sr, ^{131}I, and ^{137}Cs
Tea	^{131}I and ^{137}Cs
Vegetables	^{131}I and ^{137}Cs
Water	^{131}I and ^{137}Cs

The tabulated results are based on the Fukushima Daiichi accident. During the intermediate phase, the ^{131}I will have decayed and ^{137}Cs will be the dominant radionuclide. At Fukushima Daiichi, ^{137}Cs dominated the intermediate phase with limited ^{90}Sr deposition.

(f) Based on experience at the TMI and Fukushima Daiichi accidents, the reactor building is heavily contaminated and the dose rates are elevated. Initial characterization efforts should be accomplished using available installed radiation and air monitors. This instrumentation may be unavailable or inaccurate due to the accident damage.

Robotic techniques are used to perform general area radiation surveys. The results of the initial robotic inspections determine the scope of subsequent human entries. Human entries are used for further reactor building characterization. Accessible areas should have air and surface contamination levels determined by isotope. The accessible areas are used as staging locations for further characterization. Telescoping detectors and TLD strings dropped into elevated radiation areas add direct radiation data. Each entry adds radiological data that are used to plan subsequent entries and further characterization.

Following radiological characterization, areas are decontaminated to facilitate additional access. Decontamination efforts include reducing surface and airborne contamination levels and lowering the radiological source term. This includes draining contaminated systems and flushing lines and tanks. The process is slow and labor-intensive. However, sound ALARA principles lead to an effective recovery effort.

(g) There are a number of methods to assess if a criticality is in progress. If the nuclear instruments are undamaged, a rapid increase in neutron count rate is an indication of a criticality event. This indication should be confirmed with reactor coolant liquid and gas samples. An increase in fission product activity would be a strong indication of a criticality event. The presence of ^{131}I and short-lived noble gases would be obvious indications of new fission activity. These nuclides would have decayed from previous reactor operation due to their short half-lives.

A criticality can also be verified by comparing the ratio of isotopes such as $^{131}I/^{137}Cs$, $^{131}I/^{134}Cs$, and $^{131}I/^{90}Sr$. These ratios are dependent on the criticality duration, core operating history, initial fuel composition, fuel burnup, and sample location.

3.7. (a) The three fission product barriers associated with the MSR design are the fuel and salt coolant eutectic mixture, reactor vessel and included piping, and containment building. Based on the release, the reactor vessel and containment designs are poor and not equivalent to licensed Generation II and III reactor designs. These barriers have a number of apparent weaknesses.

The fuel and salt eutectic mixture was drained to the fuel storage tanks following a reactor vessel weld failure. This failure indicates a design weakness and a quality issue. The liquid fuel storage tanks ruptured which suggests additional quality and design issues.

Failure of the containment following liquid fuel storage tank rupture is also indicative of a basic design flaw. The sequence of weld failure, tank failure, and containment failure suggests significant design and licensing issues. This event sequence should have been precluded through the design and regulatory review. These failures indicate a complete failure of the MSR regulatory process. Issues are also associated with performance of the fuel fission product barrier and its radionuclide retention properties.

The fuel barrier has a liquid state and is not as robust as conventional fuel pellet plus cladding designs. The effectiveness of the MSR fuel design depends on the capability of the liquid fuel to retain fission products under accident conditions. From a health physics perspective, this represents another weakness. Based on the activity available for release, the fission product retention capability of the fuel plus salt eutectic mixture is poor.

(b) The ^{131}I ground-level concentration (χ) at plume centerline at a location 10 km downwind from the facility is obtained from the Pasquill–Gifford equation for a gas:

$$\chi = \frac{Q}{\pi \sigma_y \sigma_z u} \exp\left[-\frac{1}{2}\left(\frac{y^2}{\sigma_y^2} + \frac{h^2}{\sigma_z^2}\right)\right]$$

where Q = iodine release rate

$$Q = \frac{1}{\mathrm{DF}} \frac{A}{t}$$

DF = iodine decontamination factor for the containment release pathway = 10^3
A = released activity = 1300×10^{15} Bq
t = release duration = 72 h

$$Q = \frac{1}{10^3} \frac{1300 \times 10^{15}\,\text{Bq}}{72\,\text{h}} = 1.81 \times 10^{13}\,\frac{\text{Bq}}{\text{h}}$$

σ_y = horizontal standard deviation = 300 m
σ_z = vertical standard deviation = 50 m
u = mean wind speed = 11 km/h
y = crosswind distance = 0
h = effective release height = 25 m

Using these values, the ground-level concentration is determined:

$$\chi = \frac{\left(1.81 \times 10^{13}\,\frac{\text{Bq}}{\text{h}}\right)}{\pi(300\,\text{m})(50\,\text{m})\left(11 \times 10^3\,\frac{\text{m}}{\text{h}}\right)} \exp\left[-\frac{1}{2}\left(\frac{(0\,\text{m})^2}{(300\,\text{m})^2} + \frac{(25\,\text{m})^2}{(50\,\text{m})^2}\right)\right]$$

$$= \left(3.49 \times 10^4\,\frac{\text{Bq}}{\text{m}^3}\right)(0.882) = 3.08 \times 10^4\,\frac{\text{Bq}}{\text{m}^3}$$

(c) The ratio of ^{133}Xe/^{131}I ground-level concentrations at 100 km from the facility is

$$\frac{\chi(^{133}\text{Xe})}{\chi(^{131}\text{I})} = \frac{\frac{Q(^{133}\text{Xe})}{\pi\sigma_y\sigma_z u}\exp\left[-\frac{1}{2}\left(\frac{y^2}{\sigma_y^2}+\frac{h^2}{\sigma_z^2}\right)\right]}{\frac{Q(^{131}\text{I})}{\pi\sigma_y\sigma_z u}\exp\left[-\frac{1}{2}\left(\frac{y^2}{\sigma_y^2}+\frac{h^2}{\sigma_z^2}\right)\right]}$$

$$= \left(\frac{Q\left(^{133}\text{Xe}\right)}{Q(^{131}\text{I})}\right) = \frac{\left[\frac{A(^{133}\text{Xe})}{\text{DF}(^{133}\text{Xe})t}\right]}{\left[\frac{A(^{131}\text{I})}{\text{DF}(^{131}\text{I})t}\right]} = \frac{\left[\frac{A(^{133}\text{Xe})}{\text{DF}(^{133}\text{Xe})}\right]}{\left[\frac{A(^{131}\text{I})}{\text{DF}(^{131}\text{I})}\right]}$$

$$= \frac{\left(\frac{3700 \times 10^{15}\,\text{Bq}}{1.0}\right)}{\left(\frac{1300 \times 10^{15}\,\text{Bq}}{10^3}\right)} = 2.85 \times 10^3$$

(d) An individual inhales the uniform concentration (C) determined in question (b) for 5 h (t) with a breathing rate (BR) of 1.2 m^3/h. The associated intake (I) is

$$I = C(\text{BR})t$$

$$I = \left(3.08 \times 10^4\,\frac{\text{Bq}}{\text{m}^3}\right)\left(1.2\,\frac{\text{m}^3}{\text{h}}\right)(5\,\text{h}) = 1.85 \times 10^5\,\text{Bq}$$

(e) Since the effective dose conversion factor is 1.1×10^{-8} Sv/Bq, the effective dose to the individual receiving the ^{131}I intake is

$$E = (1.85 \times 10^5\,\text{Bq})\left(1.1 \times 10^{-8}\,\frac{\text{Sv}}{\text{Bq}}\right)\left(\frac{1000\,\text{mSv}}{\text{Sv}}\right) = 2.0\,\text{mSv}$$

(f) Based on the release of fission products, the severity of this MSR event is comparable to the Chernobyl source term (see Section 3.3.2). The ^{131}I doses were reduced because a decontamination factor of 1000 was assumed. At Chernobyl, much of the core inventory was released because no containment was incorporated into its design.

(g) The iodine source term can be reduced by spraying a chemical into the containment atmosphere to scavenge the iodine and minimize its release. In a pressurized water reactor, a sodium hydroxide solution is sprayed into the containment atmosphere to reduce pressure and scavenge iodine. Iodine volatilization is minimized in a basic pH solution. The specific chemical additive will be determined once the MSR fuel is optimized for fission product retention.

Charcoal filters should also be used to reduce the iodine source term. Although the release location is not specified, penetrations (e.g., purge exhaust valves) should include isolation valves to limit the release. These release points should be routed through filters or have a separate filtration system to minimize the iodine release.

The noble gas source term cannot be minimized. There is no effective approach to scavenge noble gases. The reactor accidents at Three Mile Island, Chernobyl, and Fukushima Daiichi released large noble gas activities.

The best way to avoid the release of fission products is to reduce the accident's probability of occurrence. The weld and containment failures suggest that the facility's design basis is flawed.

(h) The decontamination factors appear to be reasonable, but the source terms are speculative. In addition, no iodine removal methods are defined in the problem statement. Credible iodine decontamination factor values are 10^2 for spent fuel pool mechanical fuel damage and secondary system relief valve releases and 10^4 for releases through the condenser. The TMI-2 iodine decontamination factor was about 10^6 and involved a complex release pathway: primary coolant → open pressurizer relief valve → reactor coolant drain tank → containment building sump → auxiliary building sump → auxiliary building waste gas system → station vent filters → environment. It is likely that radioiodine will be reduced as demonstrated by the effective use of sodium hydroxide, water scavenging, and charcoal filters at Three Mile Island. The decontamination factor of unity for noble gases is appropriate based on previous operating experience.

3.8. (a) The types of demographic information that should be included in the facility's emergency plan are:

1. Population within the 16 km Plume Exposure Pathway by county and municipality
2. Surface area of each county and municipality within the Plume Exposure Pathway
3. Population density within the Plume Exposure Pathway by county and municipality
4. Cumulative total population and population density by distance from the facility (e.g., 0–3.2, 0–8, and 0–16 km)
5. Total surface area and area per sector by distance from the facility (e.g., 0–3.2, 3.2–8, 8–16, and 0–16 km)
6. Distance and direction of municipalities within the Plume Exposure Pathway
7. List of schools within the Plume Exposure Pathway, their location, distance, and direction from the facility, and enrollment
8. List of hospitals within the Plume Exposure Pathway, their location, distance, and direction from the facility, and patient capacity
9. Percent of land use by county (e.g., agricultural, business, forest and woodland, industrial, livestock, pasture, public, urban, and other use)
10. Distance and direction of major industries within the Plume Exposure Pathway
11. Maps with major roads, railroads, rivers, and topographical features
12. Plume Exposure Pathway map showing county boundaries and population centers
13. Plume Exposure Pathway sector map with population information by distance
14. 80 km Ingestion Pathway map showing county boundaries and population centers
15. Plume Exposure Pathway evacuation time estimates

(b) The general types of emergency action levels that should be included in the emergency plan are:

1. Reactor pressure vessel level
2. Reactor pressure vessel pressure
3. Reactor power
4. Containment pressure and atmospheric composition
5. Torus temperature
6. Torus water level
7. Reactor coolant system integrity
8. Fuel integrity
9. Radiation effluent monitor readings
10. Radioactive materials control including fuel handling
11. Actual or projected off-site integrated doses
12. Status of control room indications
13. Electrical power availability

14. Plant equipment and engineered safety features status
15. Natural hazards and conditions (e.g., earthquakes, water intake levels, and high winds)
16. Man-made hazards (e.g., aircraft crash, explosion, toxic gas, flammable gas, and turbine rotor failure)
17. Fires
18. Security and sabotage
19. Fission product barrier status

(c) Based on the emergency action level (off-site effective dose exceeds 0.1 mSv but less than 0.5 mSv or exceeds 0.5 mSv but less than 2.5 mSv child thyroid equivalent dose), the projected 0.15 mSv effective dose warrants declaration of an Alert.

(d) Based on the emergency action level (off-site effective dose $\geq$0.5 mSv but less than 10 mSv or $\geq$2.5 mSv but less than 50 mSv child thyroid equivalent dose), the 44.5 mSv child thyroid equivalent dose warrants declaration of a Site Area Emergency.

(e) Based on the emergency action level (off-site effective dose $\geq$10 mSv or $\geq$50 mSv child thyroid equivalent dose), the 30 mSv effective and 100 mSv child thyroid equivalent doses warrant declaration of a General Emergency.

(f) If the Fukushima Daiichi accident occurred in the United States, a General Emergency classification would be warranted. This classification is based on a number of EALs including the following:
1. *Reactor pressure vessel level*: The reactor pressure vessel water level decreased beyond the depth where the fuel was uncovered leading to melting and the loss of all fission product barriers.
2. *Containment pressure and atmospheric composition*: The hydrogen concentration achieved an explosive mixture and detonated.
3. *Reactor coolant system integrity*: The loss of coolant accident occurred with a failure of the emergency core cooling system following the loss of all power. This led to the failure of all fission product barriers.
4. *Actual or projected off-site integrated doses*: The loss of fission product barriers led to off-site doses that exceeded the General Emergency criteria.
5. *Plant equipment and engineered safety features status*: Decay heat removal capability was interrupted that resulted in the loss of all fission product barriers.
6. *Fission product barrier status*: All barriers were lost.
7. *Electrical power*: All normal and emergency power sources were unavailable. A station blackout condition existed including the loss of station battery capacity.

The reader should note that each EAL noted in question (b) does not have a General Emergency entry. The EAL's construction depends on the facility design. The author's solution is based on the EALs for a US reactor using a design similar to the Fukushima Daiichi Generation II boiling water reactors with a Mark I containment.

Solutions for Chapter 4

4.1. (a) Based on a 1 m^3 air sample at the downwind location, you are to calculate the ^{239}Pu airborne activity in Bq/m^3 correcting for the contribution from ^{222}Rn progeny. The sample counts (R) is a combination of the short-lived Rn and long-lived Pu contributions:

$$R = R_{\mathrm{Pu}} + R_{\mathrm{Rn}}$$

Two sample counts were made:

Count no.	Time of count (min)	Gross counts	Count time (min)
1	0	500	10
2	60	360	10

The net count rate (n) is the gross count rate (g) minus the background count rate (b):

$$R_{\mathrm{n}} = R_{\mathrm{g}} - R_{\mathrm{b}}$$

where

$$R_{\mathrm{b}} = \frac{180\,\text{counts}}{60\ \text{min}} = 3\,\text{cpm}$$

Using these two counts, a set of coupled equations is written for net count rate:

$$R_1 = R_{\mathrm{Pu}} + R_{\mathrm{Rn}} \quad (1)$$

$$R_2 = R_{\mathrm{Pu}} + R_{\mathrm{Rn}} e^{-\lambda_{\mathrm{Rn}} t} \quad (2)$$

where

R_1	=	total net count rate for the first sample count
	=	500 counts/10 min − 3 cpm = 50 cpm − 3 cpm = 47 cpm
R_2	=	total net count rate for the second sample count
	=	360 counts/10 min − 3 cpm = 36 cpm − 3 cpm = 33 cpm
R_{Pu}	=	initial Pu count rate
R_{Rn}	=	initial radon progeny count rate
λ_{Rn}	=	radon progeny disintegration constant
	=	$\ln(2)/T_{1/2} = 0.693/30\ \text{min} = 0.0231/\text{min}$
t	=	time of count 2 after count 1 = 60 min

In view of the long half-life of ^{239}Pu (24 100 years) relative to the Rn progeny (30 min) and the time involved in the sampling and counting, the decay of ^{239}Pu is insignificant. Using these values, Eqs. (1) and (2) become

$$47\,\text{cpm} = R_{\text{Pu}} + R_{\text{Rn}} \quad (3)$$

$$33\,\text{cpm} = R_{\text{Pu}} + R_{\text{Rn}} e^{-(0.0231/\,\text{min})(60\,\text{min})} \quad (4)$$

$$33\,\text{cpm} = R_{\text{Pu}} + 0.25\,R_{\text{Rn}} \quad (5)$$

Subtracting Eq. (5) from Eq. (3),

$$(47 - 33)\text{cpm} = 0.75\,R_{\text{Rn}}$$

$$R_{\text{Rn}} = (14\,\text{cpm})/0.75 = 18.7\,\text{cpm}$$

Using Eq. (3),

$$R_{\text{Pu}} = 47\,\text{cpm} - 18.7\,\text{cpm} = 28.3\,\text{cpm}$$

The airborne concentration (C) of ^{239}Pu is obtained from the Pu count rate on the filter (R_{Pu})

$$C = \frac{R_{\text{Pu}}}{e_{\text{f}} e_{\text{c}} e_{\text{d}} V g}$$

where

R_{Pu}	=	^{239}Pu count rate on the filter = 28.3 cpm
V	=	volume of air collected = 1 m^3
e_{f}	=	filter alpha self-absorption = 0.4
e_{c}	=	filter collection efficiency = 0.8
e_{d}	=	detector efficiency for alpha = 0.3 cpm/dpm
g	=	fraction of the filter area monitored by the detector

$$g = \frac{A_{\text{d}}}{A_{\text{f}}}$$

A_{d}	=	detector active area = 60 cm^2
A_{f}	=	active filter area with uniform activity distribution = 500 cm^2

$$g = \frac{60\,\text{cm}^2}{500\,\text{cm}^2} = 0.12$$

With these values, the ^{239}Pu air concentration is determined:

$$C = \frac{R_{\text{Pu}}}{e_{\text{f}} e_{\text{c}} e_{\text{d}} V g} = \frac{(28.3\,\text{cpm})(1\,\text{min}/60\,\text{s})\left(\frac{1\,\text{Bq-s}}{\text{dis}}\right)}{(0.4)(0.8)(0.3\,\text{cpm/dpm})(1\,\text{m}^3)(0.12)} = 40.9\,\text{Bq/m}^3$$

(b) The LLD (in cpm) for this counting system at the 95% confidence interval is

$$\mathrm{LLD} = 3.29\sqrt{\frac{R_\mathrm{b}}{t_\mathrm{g}} + \frac{R_\mathrm{b}}{t_\mathrm{b}}} + \frac{2.71}{t_\mathrm{g}}$$

where

R_b	=	background count rate = 3 cpm
t_b	=	background count time = 60 min
t_g	=	sample count time = 10 min

Using these values, the LLD is determined:

$$\mathrm{LLD} = 3.29\sqrt{\frac{3\,\mathrm{cpm}}{10\ \mathrm{min}} + \frac{3\,\mathrm{cpm}}{60\ \mathrm{min}}} + \frac{2.71\,\mathrm{c}}{10\ \mathrm{min}}$$
$$= (3.29)(0.592\,\mathrm{cpm}) + 0.271\,\mathrm{cpm} = 2.22\,\mathrm{cpm}$$

(c) In this part, you are requested to calculate the committed effective dose $E(50)$ to a person standing at the sampler location. $E(50)$ is the product of the intake and the dose conversion factor:

$$E(50) = I\,e(50)$$

where

I	=	intake = C BR t
C	=	average inhaled air concentration of $^{239}Pu = 20\ Bq/m^3$
BR	=	breathing rate = $1.2\ m^3/h$
t	=	exposure period = 4 h
$e(50)$	=	^{239}Pu committed effective dose conversion factor per unit inhalation intake = 4.7×10^{-5} Sv/Bq

Using these parameters, the intake and $E(50)$ are determined:

$$I = C(\mathrm{BR})t = (20\,\mathrm{Bq/m^3})(1.2\,\mathrm{m^3/h})(4\,\mathrm{h}) = 96\,\mathrm{Bq}$$

$$E(50) = Ie(50) = (96\,\mathrm{Bq})(4.7 \times 10^{-5}\,\mathrm{Sv/Bq}) = 4.5 \times 10^{-3}\,\mathrm{Sv} = 4.5\,\mathrm{mSv}$$

(d) Approaches to improve the dose estimate for off-site individuals include the following:

a. Perform bioassay (e.g., lung counts, urinalysis, and fecal analysis) to better characterize the intake.
b. Determine if the individual is a nose or mouth breather.
c. Determine the particle size and shape of the inhaled material.
d. Apply ICRP 66 methodology to account for the particle size and shape of the ^{239}Pu aerosol.

e. Improve the estimate of the air concentration as a function of the individual's position.
f. Account for any respirator use.
g. Use realistic meteorology to improve the estimate of the local air concentration.
h. Determine the position of the individuals relative to the plume centerline.
i. Account for plume lofting and particulate fallout in estimating the local air concentration.
j. Obtain ground deposition data to verify the assumed air concentration.
k. Account for meteorology including rainout or washout.
l. Obtain air concentrations as a function of time.
m. Collect all available sample data to verify the air concentration and intake values.
n. If available, use personal lapel air sampler data.
o. Use all air sample data including samples using larger air volumes than the 1 m^3 snapshot of the air concentration.
p. Derive a more accurate resuspension factor to better characterize the airborne concentration.
q. Determine the individual's actual retention function.
r. Determine the intake material's type (i.e., F, M, or S).

(e) Possible methods to reduce the effective dose to individuals from brush fires or other high resuspension events include the following:

Prior to the event:
(a) Remove excess brush and vegetation.
(b) Control vegetation growth with pesticides and chemicals.

During the event:
(a) Use respiratory protection (SCBAs) when entering affected areas.
(b) Fight fires from the air when possible.
(c) Fight fires from an upwind location.
(d) Evacuate firefighters and the general public as warranted by radiological or safety considerations.
(e) Use chelating agents in a timely manner for highly exposed individuals.
(f) Shelter/evacuate personnel and the public as warranted by radiological conditions.

After the event:
(a) Use respiratory protection (SCBAs) when entering affected areas.
(b) Spray areas with a fixodent to minimize blowing dust and burned vegetation.
(c) Shelter/evacuate workers and the general public as warranted by radiological or safety considerations.

(d) Use chelating agents in a timely manner for highly exposed individuals.
(e) Spray areas with water to minimize dust generation.
(f) Reseed the area with grass following the fire.
(g) Control access to contaminated areas.
(h) Fence and post affected areas.
(i) Remove excess brush and vegetation.
(j) Control vegetation growth with pesticides and chemicals.
(k) Remove contaminated materials (e.g., vegetation, soil, and debris) to minimize future intakes and dispersal of radioactive material.

4.2. (a) The current (I) generated by an exposure rate ($\dot{X}$) of 1 R/h is given by the relationship

$$I = \rho V \frac{T_{\text{STP}}}{T} \frac{P}{P_{\text{STP}}} \dot{X}$$

where

ρ	=	density of air at STP = 1.29 kg/m^3
V	=	detector volume = 235.5 cm^3
T	=	temperature = 20 °C = (20 + 273) K = 293 K
P	=	pressure = 720 mmHg
T_{STP}	=	standard temperature = 0 °C = 273 K
P_{STP}	=	standard pressure = 760 mmHg

Using these input values, the current is determined:

$$I = (1.29\,\text{kg/m}^3)(235.5\,\text{cm}^3)\left(\frac{1\,\text{m}^3}{10^6\,\text{cm}^3}\right)\left(\frac{273\,\text{K}}{293\,\text{K}}\right)\left(\frac{720\,\text{mmHg}}{760\,\text{mmHg}}\right)(1\,\text{R/h})$$
$$x\left(\frac{2.58\times10^{-4}\,\text{C}}{\text{kg-R}}\right)\left(\frac{1\,\text{h}}{3600\,\text{s}}\right)\left(\frac{\text{A-s}}{\text{C}}\right) = 1.92\times10^{-11}\,\text{A}$$

(b) A measurement of 12.6 R/h is taken on a hot day of 35 °C and 740 mmHg. The exposure rate at standard conditions is obtained from the relationship used in the previous part:

$$I = \rho V \frac{T_{\text{STP}}}{T} \frac{P}{P_{\text{STP}}} \dot{X}$$

From the problem statement, the currents for the observed and standard conditions are the same, which permits us to write

$$I = \rho V \frac{T_{\text{STP}}}{T_1} \frac{P_1}{P_{\text{STP}}} \dot{X}_1 = \rho V \frac{T_{\text{STP}}}{T_2} \frac{P_2}{P_{\text{STP}}} \dot{X}_2$$

$$\frac{P_1 \dot{X}_1}{T_1} = \frac{P_2 \dot{X}_2}{T_2}$$

where

$\dot{X}_1$	=	exposure rate at 35 °C and 740 mmHg = 12.6 R/h
T_1	=	temperature on a hot day = 35 °C = (35 + 273) K = 308 K
P_1	=	pressure on a hot day = 740 mmHg
$\dot{X}_2$	=	exposure rate at standard conditions
T_2	=	temperature for standard conditions = 0 °C = (0 + 273) K = 273 K
P_2	=	pressure for standard conditions = 760 mmHg

Using these input values, the exposure rate at standard conditions is determined:

$$\dot{X}_2 = \dot{X}_1\left(\frac{P_1}{P_2}\right)\left(\frac{T_2}{T_1}\right) = (12.6\,\text{R/h})\left(\frac{740\,\text{mmHg}}{760\,\text{mmHg}}\right)\left(\frac{273\,\text{K}}{308\,\text{K}}\right)$$
$$= 10.9\,\text{R/h}$$

(c) Since the crack width is smaller than the ion chamber length, an estimate of the true exposure rate is made by determining the detector volume irradiated by the beam. Assuming electronic equilibrium, 0 °C, and 760 mmHg, the desired exposure rate is determined.

Ignoring beam scattering in the detector, the detector volume (V') irradiated through the 1 cm crack is

$$V' = Ah = \pi r^2 h'$$

where

h'	=	width of the beam traversing the detector = 1 cm
r	=	radius of the detector = 5 cm

The true exposure rate ($\dot{X}_\text{T}$) is written in terms of the measured exposure rate ($\dot{X}$) of 20 mR/h, the detector volume (V), and the volume of the detector irradiated by the beam:

$$\dot{X}_\text{T} = \dot{X}\left(\frac{V}{V'}\right) = \dot{X}\left(\frac{\pi r^2 h}{\pi r^2 h'}\right) = \dot{X}\left(\frac{h}{h'}\right)$$
$$= (20\,\text{mR/h})\left(\frac{3.0\,\text{cm}}{1.0\,\text{cm}}\right) = 60.0\,\text{mR/h}$$

where h is the detector length (3.0 cm). This solution assumes that the wall is sufficiently thick to attenuate the radiation emitted postdetonation.

4.3. (a) The average ^{32}P contamination level of 250 dpm/100 cm^2 and highest measured absorbed dose rates of 0.05 mGy/h suggest most of the affected area can be considered a cold zone as defined in NCRP 165. Immediate attention should be directed toward search and rescue teams and providing medical treatment to survivors. These teams should have health physics support (e.g., emergency response personnel or Medical Reserve Corps volunteers) or instrumentation to access the elevated radiation areas in an ALARA manner. However, the initial lifesaving activities are not expected to be significantly limited by the radiological conditions.

The first responders should be briefed on measures to minimize their external and internal doses.

Following search and rescue activities and accounting for all casualties, the area should be controlled and no entries permitted. Authorized activities include fixing contamination in place and decontamination, which limit the likelihood of radioactive material spreading to additional areas. However, the short half-life of ^{32}P (14.3 days) suggests that unrestricted access should be possible within a few months. After this time, final clearance surveys and any localized decontamination can be accomplished before returning the undamaged areas to unrestricted use.

Subsequent actions are governed by the economic importance of the affected area and its value to the city. If the contaminated area is not vital to city services or has a limited economic value, it could be sequestered for a few months until the ^{32}P activity decays to levels that are acceptable to all stakeholder groups. Temporary relocation of business offices permits economic activity to continue until the ^{32}P contamination decays to the acceptable level.

(b) The initial radiation and contamination levels are within the NCRP 165 cold zone criteria. However, the radiological conditions inside the collapsed tunnel are uncertain and a total ^{60}Co activity greater than 500 TBq is possible.

Moving debris requires heavy equipment, and these activities merit careful planning. Initial radiological control actions include spraying water or chemical mixtures to control dust and shielding construction equipment. These actions limit the spread of contamination and permit activities to proceed in a controlled manner. Debris removal must not jeopardize the tunnel's structural integrity so shoring may be needed to prevent the collapse of excavated areas.

Careful planning and radiological input should be incorporated into each tunnel excavation phase. High dose rates and contamination levels are possible since the size of the RDD is not definitively known.

Contaminated debris requires packaging and shipping following the requisite 49CFR requirements. Burial space allotments and acceptance of the material requires additional attention.

Initial efforts focus on search and rescue and initial recovery efforts. These activities should proceed with deliberate speed while recognizing the need for ALARA planning to minimize recovery worker doses.

The tunnel exit areas should be covered to minimize the spread of ^{60}Co contamination. Each end of the tunnel becomes a staging area for recovery activities, and should be staffed with health physics resources to ensure a timely response to emerging radiological conditions.

Recovery workers should receive focused radiological training to ensure the hazards are understood. There will be a strong desire for immediate rescue operations. Health physics personnel should support this desire but must ensure that ALARA measures are implemented throughout the recovery effort. A careful balance must be struck between good radiological practices and the need for a rapid effort to rescue survivors.

(c) The first responders recovered the injured and determined the initial radiation levels. Controlling access and establishing radiological boundaries for the affected area are the next actions. In addition, the injured and first responders should be evaluated to determine if actions are required to mitigate any internal depositions and to determine the associated equivalent and effective doses.

Following access control and a more complete radiological characterization, measures should be taken to minimize the spread of contamination and to fix radioactive material in place. Since the radioactive material is fuel reprocessing waste that contains ^{90}Sr and ^{137}Cs, the problem is long term. Decontamination plans, reentry criteria, and acceptable radiation and contamination levels must be established. Stakeholders should be included in establishing the radiological criteria and priorities for decontaminating the affected areas.

(d) A number of initial actions are required. The injured must be treated and evacuated from the contaminated areas. Following medical treatment, their radiological status should be determined. Additional contamination management actions are outlined in Table 4.17.

The emergency response personnel should also be evaluated to determine their internal and external doses. Actions should be taken to mitigate radionuclide intakes above the clinical decision guide values.

The Protective Action Guides of Table 3.7 should be implemented and worker doses controlled using the guidelines of Table 4.11. The NCRP 138 guidelines (see Tables 4.12 and 4.13) should also be reviewed for applicability.

Since this event involved the National Command Authority, health physics activities including restricting access will be coordinated with various agencies including the Department of Homeland Security, Secret Service, and Federal Bureau of Investigation. After security requirements are met, the radiological conditions are determined and appropriate access controls for recovery activities are established.

Given the contamination levels, health physics support during search and rescue operations is also anticipated. Since plutonium is involved, care must be exercised to minimize internal intakes and the spread of contamination while supporting the various initial actions. Spraying areas to fix the contamination should be considered until decontamination plans are developed. Implementing the requisite health physics practices will be challenging given the security requirements associated with the National Command Authority.

4.4. (a) In this question, you are to determine the ^{137}Cs inhalation intake to a person present at the concert during the 8 h period that the air was contaminated. The intake depends on the average air concentration inhaled by the individual and is obtained from the deposition rate of ^{137}Cs on the ground.

The ^{137}Cs deposition rate (r_d) is the product of the average air concentration (C) and the deposition velocity (v):

$$r_d = Cv$$

The released activity is derived from the production equation

$$A(t) = \frac{P}{\lambda}(1 - e^{-\lambda t})$$

where the production term (P) depends on the area (S) of the deposited activity

$$P = r_d S = CvS$$

Using these equations,

$$A(t) = \frac{CvS}{\lambda}(1 - e^{-\lambda t})$$

The ground contamination per unit area C_S is

$$C_S = \frac{A(t)}{S} = \frac{CvS}{\lambda S}(1 - e^{-\lambda t}) = \frac{Cv}{\lambda}(1 - e^{-\lambda t})$$

The average air concentration is obtained from algebraic manipulation of this equation

$$C = \frac{C_S \lambda}{v(1 - e^{-\lambda t})}$$

where

C	=	average ^{137}Cs air concentration inhaled by the individual during the 8 h time period
C_S	=	measured ^{137}Cs deposition on the soil = 518 kBq/m^2
v	=	deposition velocity = 0.002 m/s
t	=	release time = 8 h
λ	=	^{137}Cs disintegration constant

$$\lambda = \left(\frac{0.693}{30.07\,\text{years}}\right)\left(\frac{1\,\text{year}}{365\,\text{days}}\right)\left(\frac{1\,\text{day}}{24\,\text{h}}\right) = \frac{2.63 \times 10^{-6}}{\text{h}}$$

Using the data, the average air concentration is determined:

$$C = \frac{\left(518 \times 10^3 \frac{\text{Bq}}{\text{m}^2}\right)\left(\frac{2.63 \times 10^{-6}}{\text{year}}\right)}{\left(0.002 \frac{\text{m}}{\text{s}}\right)\left(3600 \frac{\text{s}}{\text{h}}\right)\left(1 - e^{-\frac{(2.63 \times 10^{-6})(8\,\text{h})}{\text{h}}}\right)} = 8.99 \times 10^3 \frac{\text{Bq}}{\text{m}^3}$$

The careful reader will note that the production equation could have been simplified since λt is small which permits approximating the exponential by the first two terms of its power series expansion

$$C_S = \frac{Cv}{\lambda}(1 - e^{-\lambda t}) \approx \frac{Cv}{\lambda}(1 - (1 - \lambda t)) = \frac{Cv}{\lambda}(\lambda t) = Cvt$$

$$C = \frac{C_S}{vt} = \frac{\left(518 \times 10^3 \frac{\text{Bq}}{\text{m}^2}\right)}{\left(0.002 \frac{\text{m}}{\text{s}}\right)(8\,\text{h})\left(3600 \frac{\text{s}}{\text{h}}\right)} = 8.99 \times 10^3 \frac{\text{Bq}}{\text{m}^3}$$

This was not done to further emphasize the use of the production equation and its functional form derived in Appendix B.

The intake is determined using the average air concentration and the following relationship

$$I = C(\text{BR})t$$

where

I	=	intake of ^{137}Cs resulting from the 8 h inhalation
BR	=	0.8 m^3/h

$$I = \left(8.99 \times 10^3 \frac{\text{Bq}}{\text{m}^3}\right)\left(0.8 \frac{\text{m}^3}{\text{h}}\right)(8\,\text{h}) = 57.5 \times 10^3\,\text{Bq} = 57.5\,\text{kBq}$$

(b) The effective dose (E) delivered to a person present for the entire concert is the product of the intake and effective dose conversion factor $e(50)$:

$$E = Ie(50) = (57.5 \times 10^3\,\text{Bq})\left(4.6 \times 10^{-9} \frac{\text{Sv}}{\text{Bq}}\right)$$
$$= 2.6 \times 10^{-4}\,\text{Sv} = 0.26\,\text{mSv}$$

(c) In this question, you are requested to determine the quantity of ^{137}Cs released from the device based on an air concentration of 11.1 kBq/m^3. The city park air concentration is on the plume centerline and was obtained at ground level.

The total ^{137}Cs released activity (q) is defined by the relationship

$$q = Qt$$

where t is the release time (8 h) and Q is the uniform ^{137}Cs release rate, which is obtained from the dispersion equation

$$\chi = \frac{Q}{2\pi\sigma_y\sigma_z u} e^{-\left(\frac{y^2}{2\sigma_y^2} + \frac{H^2}{2\sigma_z^2}\right)}$$

where

χ	=	ground-level concentration at the concert site = 11.1 kBq/m^3
u	=	wind speed = 5 m/s
σ_y	=	horizontal dispersion coefficient = 205 m
σ_z	=	vertical dispersion coefficient = 120 m
H	=	effective release height = 40 m
y	=	crosswind distance = 0

These values simplify the dispersion equation

$$\chi = \frac{Q}{2\pi\sigma_y\sigma_z u} e^{-\frac{H^2}{2\sigma_z^2}}$$

The release rate is obtained by solving the dispersion equation for Q:

$$Q = 2\pi\sigma_y\sigma_z u\chi e^{+\frac{H^2}{2\sigma_z^2}}$$

and the total ^{137}Cs released is

$$q = Qt = 2\pi\sigma_y\sigma_z u\chi t e^{+\frac{H^2}{2\sigma_z^2}}$$

$$q = (2)(3.14)(205\,\text{m})(120\,\text{m})\left(5\,\frac{\text{m}}{\text{s}}\right)\left(11.1\times 10^3\,\frac{\text{Bq}}{\text{m}^3}\right)(8\,\text{h})$$
$$\times\left(\frac{3600\,\text{s}}{\text{h}}\right)e^{+\frac{(40\,\text{m})^2}{2(120\,\text{m})^2}} = (2.47\times 10^{14}\,\text{Bq})(1.06) = 2.62\times 10^{14}\,\text{Bq}$$

(d) In order to confirm the release scenario and reduce the uncertainty in the release estimate, additional information is required. This information includes:

1. The wind speed and direction as a function of time during the release
2. Available air samples and direct dose readings (e.g., environmental TLDs) as a function of time and position during the release
3. The actual release rate as a function of the release time
4. Ground deposition values as a function of position
5. The terminal settling velocity and deposition velocity to correct for effective stack height and assess the deposition, respectively
6. Knowledge of the meteorological classification as a function of the release time
7. The particle size distribution of the released ^{137}Cs aerosol

8. The actual release time
9. The plume geometry and location as a function of time
10. The ^{137}Cs release type (i.e., F, M, or S)
11. The aerosol particle shape

(e) Tasks that should be routinely performed to ensure the quality of environmental counting system measurements include the following:
1. Periodically determine the background count rate.
2. Perform functional tests to verify operability.
3. Perform a battery check to verify the backup power supply.
4. Verify the calibration date.
5. Perform source checks.
6. Count known samples to verify system operability and reliability.
7. Verify the instrument efficiency curve using standard sources.
8. Verify peak positions occur at the proper channel numbers and with the correct heights when counting standard samples.
9. Check all electronic settings prior to using the counting system.
10. Verify the counter's operation by performing a chi-square test with a standard source in a fixed position.
11. Use a control chart to verify expected system performance.
12. Determine the background count of a blank sample before each sample count or series of sample counts.

4.5. (a) The event scenario is similar to the Fukushima Daiichi accident. Based on this correspondence, the largest activity releases are noble gases (e.g., krypton and xenon) and radioiodine. The particulate release is dominated by ^{137}Cs. Smaller amounts of ^{134}Cs, ^{103}Ru, ^{106}Ru, ^{89}Sr, and ^{90}Sr are also released. Minimal quantities of ^{241}Am, ^{238}Pu, and ^{239}Pu are released off-site, but these radionuclides may be found on-site.

(b) Based on the initial dose assessment summary, a General Emergency should be declared. The lower limit protective action guideline for the effective dose (10 mSv) and child thyroid equivalent dose (50 mSv) for KI administration have been reached at distances up to 32 km from the facility. A General Emergency declaration is also warranted based on the loss of all fission product barriers, degrading facility conditions, and inability to control the release of radioactive material to the environment.

(c) Under current the US guidance, evacuations are issued at distances to 16 km of the facility. However, given the magnitude of the projected dose, evacuation within 32 km of the facility should be recommended. With the expectation of variable winds, the evacuation should encompass all 360° around the facility. The use of potassium iodide should also be recommended. Given the magnitude of the projected doses, KI should be issued to residents within 32 km of the facility.

(d) Following the USEPA guidance, emergency doses above 250 mSv may be authorized to prevent loss of life and massive spread of destruction. EPA rules require that the worker be fully aware of the risks associated with their exposure. The 2013 EPA *Protective Action Guides and Planning Guidance for Radiological Incidents* specifies that the dose limit include the sum of the external dose equivalent and committed effective dose to adults from direct exposure and internal deposition during the emergency activity. Based on concern for the well-being of the fetus, the declared pregnant worker should be eliminated from consideration. Since the remaining workers are qualified for the required tasks, the maintenance manager should select the team from this group.

It would be prudent to advise the pregnant worker of your concerns. The risks to the fetus and their scientific basis should be clearly explained. NCRP 174 *Preconception and Prenatal Radiation Exposure: Health Effects and Protective Guidance* provides relevant data to support the discussion with the worker.

(e) The briefing for the emergency task should outline the scope of activity to be performed and the associated radiological conditions. Required protective clothing and respiratory protection should be discussed. Facility areas to be avoided should be specifically addressed.

The access and egress routes, anticipated physical and radiological conditions within the plant, the known extent of physical damage, and possible hazards created by the explosions should be addressed. Given the uncertain conditions, a discussion of emergency escape routes, contingencies, and recovery options for potential problems should be performed.

Health physics personnel should accompany the repair team to verify the assumed radiological conditions. If significant deviations from the assumed conditions are encountered, the team should exit the area and regroup. All actions should be taken in an ALARA manner.

These items should be discussed during the prejob briefing supporting the task. The briefing should also address appropriate dosimetry and review communications equipment, respiratory protection use and limitations, operation of special equipment, and any unique items identified in the ALARA review.

A brief discussion of the health physics consequences of the anticipated doses should be part of the task briefing. Given anticipated levels of radioiodine, KI should be administered prior to initiating the repair operation. The early administration of KI will saturate the thyroid and limit the thyroid's uptake of radioiodine.

The workers should be provided the opportunity to ask questions, offer alternative approaches, and express any concerns. All questions and concerns must be satisfactorily addressed. However, the importance of the task must be emphasized as well as the need to restore core cooling in an expeditious manner.

(f) External dosimetry indicated an average effective dose of 800 mSv, which is consistent with the 1000 mSv estimate. Whole-body counting suggests an average thyroid dose of 400 mSv that is much less than the anticipated 6000 mGy dose. The thyroid dose estimate was based on the available air concentration data and did not consider KI administration. The worker's thyroid equivalent dose based on the whole-body count is credible since KI was administered and limited the thyroid uptake.

(g) A physician asks you to assist him in addressing the emergency team to discuss any anticipated health effects. Your discussion is guided by their dosimetry that indicates an average effective dose of 800 mSv and whole-body counting that suggests an average thyroid equivalent dose of 400 mSv.

Prior to meeting with the workers, a review of the acute radiation syndrome should be provided to the physician. The 400 mSv thyroid equivalent dose will not lead to observable symptoms or health effects. This dose is less than the annual occupational dose limit in 10CFR20. Periodic whole-body or thyroid counts are recommended to further refine the thyroid equivalent dose estimate and to ensure that it is credible.

The whole-body doses are more significant. The 800 mSv effective dose suggests that detectable deterministic effects will be observable. Given a threshold of about 50 mGy, a detectable increase in chromosome aberrations may be observable. Nausea, vomiting, and diarrhea have a threshold of about 1 Gy. Given an effective dose of 0.8 Sv, these effects could be observed in some of the exposed workers. The prodromal phase symptoms of nausea and vomiting can occur with an onset at 1–24 h after exposure. However, the workers do not face an immediate life-threatening situation.

The physician should also discuss other possible effects that could occur (e.g., blood cell concentrations and depression of sperm counts). Other health detriments including the long-term cancer risk should also be addressed.

The workers should be monitored on an outpatient basis for several weeks following the exposure. Following the initial examinations, a periodicity for long-term monitoring should be established. The physician should present the medical briefing and answer emergency team questions. As the Radiation Protection Manager, you should be present since you know the workers and can answer any plant-specific issues and assist the physician in answering the worker's questions.

(h) Given the severity of the accident and the 32 km evacuation radius, crops and food items should be initially restricted. These restrictions should be lifted after sampling data verifies that the individual food items are safe for consumption. Verification must be expedited to ensure the area recovers from the event in a timely manner.

Detailed radiological surveys of air, water, and soil will be accomplished to characterize land use restrictions. As part of these surveys, crops, milk, water, fish, and other foods will be sampled to define the intake of radionuclides by plant and animal life. Given the extent of the release, these samples will be required within the 80 km ingestion pathway.

Guidance for restricting food consumption is provided by the EPA and summarized in Table 3.7. Additional food consumption guidance from the IAEA, DHS, NCRP, and Province of Ontario, Canada, is summarized in Tables 3.8, 4.10, 4.12, and 4.14, respectively. Easing restrictions on food consumption should be performed in a timely manner to minimize the economic impact of the accident and to speed recovery from its aftermath. Stakeholders must be advised of any issues, the basis for the issue, applicable guidance, and activities being taken to return the area to a normal condition. Issues raised by stakeholders should be addressed in a timely and complete manner.

(i) If feasible, farm animals should be relocated. If this is not practical, they should be fed stored feed and water. Minimizing the intake of deposited radioactive material in farm animals limits subsequent restrictions regarding the use of these animals or their food products.

Samples will be periodically taken and compared to established limits for milk and meat. Initially, concern for the contaminated milk and meat forms a natural restriction on their consumption. Survey results govern further restrictions. Since these restrictions have a significant economic impact, their basis must be clearly communicated to stakeholders.

(j) Future land use depends on the levels of contamination, spectrum of isotopes contaminating the land, and established criteria for allowable radioactive material concentrations in foods. The depth and contamination profile in soil is also an important consideration.

EPA guidance is provided for reducing the effects of soil contamination. The EPA provides simple dose reduction techniques including soaking or plowing soil and removing soil from locations where radioactive materials have concentrated. Final land use depends on the initial contamination levels and effectiveness of dose reduction methods. Stakeholders should be involved in future land use decisions and in establishing acceptable radiological criteria. The limits noted in question (h) would be an important input in establishing these criteria.

4.6. (a) Since the surface dose rates are due to fallout, the time delay for rescue of the trapped city workers is determined from the fallout relationship

$$\dot{D}(t) = \dot{D}(1)t^{-1.2}$$

where $\dot{D}(1)$ is the absorbed dose rate measured at 1-day postdetonation (1 Gy/h), $\dot{D}(t)$ is the dose rate at the recovery site at the time of rescue operations, and t is the time postdetonation for the rescue operation to begin. The dose rate $\dot{D}(t)$ is determined from the estimated rescue time, limiting rescue dose (0.1 Gy), and time to complete the task (24 h):

$$\dot{D}(t) = \frac{D(t)}{t} = \frac{0.1\,\text{Gy}}{24\,\text{h}} = 4.17 \times 10^{-3}\,\frac{\text{Gy}}{\text{h}}$$

Using this dose rate and solving the fallout equation for t lead to the desired time for recovery operations to be initiated:

$$\dot{D}(t) = \dot{D}(1)t^{-1.2}$$

$$\frac{\dot{D}(t)}{\dot{D}(1)} = t^{-1.2}$$

$$\ln\left(\frac{\dot{D}(t)}{\dot{D}(1)}\right) = \ln(t^{-1.2}) = -1.2\ln(t)$$

$$\ln(t) = -\frac{1}{1.2}\ln\left(\frac{4.17 \times 10^{-3}\,\text{Gy/h}}{1\,\text{Gy/h}}\right) = 4.57$$

$$e^{\ln(t)} = t = e^{4.57}d = 96.5\,\text{days}$$

The rescue operations could begin at 96.5-days postdetonation to meet the 0.1 Gy absorbed dose limit criteria established in the problem statement. Since the shelter has a supply of 4 months of food and water, this time delay is acceptable. Of course, this assumes the individuals can withstand the stress of an extended stay in the shelter. In addition, the dose received during the city employee's stay in the shelter is not a significant concern:

$$D_{\text{shelter}} = \left(0.3 \times 10^{-6}\,\frac{\text{Gy}}{\text{h}}\right)(96.5\,\text{days})\left(24\,\frac{\text{h}}{\text{day}}\right)\left(1000\,\frac{\text{mGy}}{\text{Gy}}\right)$$
$$= 0.695\,\text{mGy}$$

(b) The downwind distance (d) for 0.1 Gy/h to be reached is obtained from the relationships summarized in Table 4.6

$$d = 38\,W^{0.45}\,\text{km}$$

where W is the weapons yield in kT (100 kT). Using this yield, the desired downwind distance is

$$d = 38(100)^{0.45}\,\text{km} = 302\,\text{km}$$

(c) For a 50 km/h wind speed (v), the downwind distance increases by a factor F given by the relationship (Eq. 4.6)

$$F = 1 + \frac{v - 24\,\text{km/h}}{96\,\text{km/h}} = 1 + \frac{(50 - 24)\,\text{km/h}}{96\,\text{km/h}} = 1.27$$

$$d = (1.27)(302\,\text{km}) = 384\,\text{km}$$

(d) The emergency planning basis assumed a 10 kT yield. The actual detonation yield of 100 kT affects the emergency planning basis in the following ways:

1. The damage zone caused by the blast and subsequent shock and thermal effects is considerably larger. Consequently, assumed emergency response facilities and personnel may be unavailable.
2. The radiation levels are higher and the affected areas significantly larger than assumed in the 10 kT planning basis. Assumed emergency response criteria and requirements will likely be inadequate to meet the increased demand for radiological services.
3. The number of casualties is significantly larger and will overwhelm the assumed resources and response capabilities noted in the 10 kT emergency response plan.
4. The increased damage and more severe radiological conditions significantly complicate search and rescue and recovery operations.
5. Essential resources including hospitals, emergency response equipment, and emergency response personnel will likely be unavailable due to the increased damage area attributed to the larger nuclear detonation.
6. The effects of the larger weapon's yield may compromise assumed evacuation routes and evacuation centers.
7. Emergency notification and communications systems and their supporting infrastructure are negatively affected by the larger weapon's yield.
8. Additional resources beyond those assumed in the emergency plan are required to address the increased devastation. With the 10 kT planning basis, there are no provisions to mobilize these resources and effectively utilize them in emergency operations.

4.7. (a) Common radionuclides, source activities, and their conventional uses are provided in the following table:

Radionuclide	Typical source activity	Conventional uses
^{3}H	10s of kBq to 100s of MBq	Research Self-illuminating dials and signs Nuclear fusion research
^{32}P	10s of kBq to 10s of MBq	Research Medical therapy
^{35}S	10s of kBq to 10s of MBq	Research
^{60}Co	100s of GBq to 10s of TBq	Industrial radiography Cancer therapy and irradiators
^{90}Sr	10s of kBq to 100s of TBq	Medical therapy Process control gauges Radioisotope thermoelectric generators
^{131}I	10s of kBq to 10s of TBq	Medical imaging and therapy
^{137}Cs	100s of GBq to 10s of TBq	Therapy sources Blood irradiators Industrial radiography Soil density gauges Well logging
^{192}Ir	10s to 100s of GBq	Radiography Radiation therapy
^{226}Ra	10s of kBq to 10s of TBq	Cancer therapy (legacy sources) Self-luminous products
^{238}Pu	100s of MBq to 10s of GBq	Calibration sources Neutron generators
^{241}Am	100s of MBq to 10s of GBq	Hydrocarbon content determination Smoke detectors Soil moisture determination Well logging
^{252}Cf	100s of MBq to 10s of GBq	Neutron generators

Solution derived from NCRP 161 II, Management of Persons Contaminated with Radionuclides: Scientific and Technical Bases (2008).

The applicability of ^{3}H, ^{32}P, ^{35}S, and ^{131}I for radiological terrorism can be debated. All of these radioactive materials are readily available, but their radiological characteristics (i.e., short half-life and/or low disintegration energy) are less effective for nefarious purposes.

(b) The methods for direct measurement of internally deposited photon-emitting radionuclides are influenced by the photon energy. Direct measurements must ensure that all external contamination has been removed. Individuals wear clean clothing prior to the measurement. All jewelry, watches, rings, glasses, and other personal accessories are removed. Individuals should also be questioned regarding their prior receipt of radioactive materials from medical treatment including imaging and therapy procedures. The following table outlines methods for the direct measurement of gamma-emitting radionuclides in the body and their associated detectors:

Radionuclide type	Method	Counting geometry	Detectors
Gamma-emitting radionuclides ($E > 100\,\text{keV}$)	Whole-body counting	Stationary detector arrays	NaI(Tl) in a shielded or partially shielded facility
Gamma-emitting radionuclides	Whole-body counting	Stationary detector arrays	HPGe in a shielded room
Gamma-emitting radionuclides (low-energy photons)	Organ specific (e.g., lung) or specific body location (e.g., skull) counting	Detector positioned over desired organ or body part with care to note radionuclides in adjacent organs	HPGe thin detectors in a shielded room Thin NaI(Tl) detectors or phoswich detectors (thin NaI(Tl) plus CsI(Tl) duel detector) in a shielded room
Radioiodine or ^{99m}Tc	Thyroid	Detector positioned over the thyroid	HPGe thin detectors in a shielded room (for radioiodine) Planar germanium detector or thin NaI(Tl) detector in a shielded room (for low-energy radiation from radioiodine) Thick NaI(Tl) detector in a shielded room or with minimal shielding in emergency situations

Solution derived from NCRP 161 II, Management of Persons Contaminated with Radionuclides: Scientific and Technical Bases (2008).

(c) Radionuclides corresponding to the gamma-ray peaks noted in the Drill-02A Spectrum are:

Energy (MeV)	Nuclide
0.0136	^{239}Pu
0.0711	^{201}Tl
0.140	^{99m}Tc
0.186	^{226}Ra
0.316	^{192}Ir
0.364	^{131}I
0.662	^{137}Cs/^{137m}Ba
1.17	^{60}Co
1.33	^{60}Co

(d) The recommended bioassay sample type and analysis method for assessing the internal deposition of the radionuclides determined in the previous question are:

Radionuclide	Bioassay sample type	Analysis method
^{99m}Tc	a)	Gamma spectroscopy
^{60}Co	a)	Gamma spectroscopy
^{131}I	a)	Gamma spectroscopy
^{137}Cs/^{137m}Ba	a)	Gamma spectroscopy
^{192}Ir	a)	Gamma spectroscopy
^{201}Tl	a)	Gamma spectroscopy
^{226}Ra	a)	Alpha spectroscopy Proportional counter
^{239}Pu	a)	Alpha spectroscopy Thermal ionization mass spectrometry

a) Urine, feces, and whole-body counting can be used. In most cases, whole-body counting is performed. In many cases, urine is preferable to feces as a bioassay approach. For water-soluble radioactive materials, urine is a more convenient bioassay sampling approach than feces. However, multiple bioassay types are utilized if the intake is significant.

(e) The internal deposition of ^{241}Am is quantified from measurement of alpha activity in excreta or from external photon detection. ^{241}Am has a half-life of 433 years. Alpha energies attributed to ^{241}Am are 5.49 MeV @ 85% and 5.44 MeV @ 13%. The principal photons emitted from ^{241}Am are 25 and 60 keV.

Measurements of ^{241}Am in excreta usually involve radiochemical separation followed by alpha spectroscopy. Detectors are calibrated and located to estimate ^{241}Am in the total body, lungs, or skeleton. Skeletal measurements are feasible following a large intake by positioning the detectors near the skull since it will contain 10–20% of the total ^{241}Am that accumulates in this system.

(f) The following types of information should be maintained for all medical and emergency response personnel involved in the radiological terrorist event:

1. Worker's name, address, company title, and contact information
2. Brief description of the incident and the individual's involvement in the response
3. Physical injuries
4. Date and time of the radiation exposure
5. Date and time of entry into and exit from radiologically controlled areas
6. Location of areas entered, the associated stay time, and radiological conditions encountered (e.g., dose rates and contamination levels by radiation type)

7. Radionuclides and activity estimate of the source to which the individual was exposed
8. Strength of sources to which the individual was exposed
9. Personal contamination level, radionuclides comprising the contamination, and body location and extent (e.g., area) of contaminated skin
10. Decontamination method and its effectiveness including radionuclides comprising the remaining contamination, body location, and extent (e.g., area) of contaminated skin
11. Postdecontamination levels
12. Personal protective equipment worn
13. Respiratory protection worn including type and associated protection factor
14. Type of radiation exposure (e.g., external radiation, external contamination, and internal contamination)
15. Contamination measurements (e.g., nasal counts, wound counts, and whole-body counts)
16. Personal dosimetry worn and associated pre- and postentry values
17. List of bioassay samples taken and time of sampling
18. Results of whole-body counts including radionuclide, deposited activity, and body location
19. Results of any *in vivo* or *in vitro* bioassay measurements
20. Description of clinical symptoms (e.g., nausea, vomiting, diarrhea, headache, and fatigue)
21. Medical treatment provided and names of attending medical personnel
22. Radioprotective chemicals administered, name of administering physician, quantities administered, administration effectiveness, and associated bioassay approach documenting the results
23. Assessment of the intake and associated dose calculations including all supporting calculations and data
24. Name of personnel performing the intake and dose calculations, reviewer names, and individual approving the dose calculations
25. Title and qualifications of all individuals associated with the intake and internal dose calculations
26. Assessment of external doses and associated calculations
27. Names of health physics personnel and their duties associated with the worker's records and activities
28. Personal statements or comments made by the affected individuals

4.8. (a) ^{60}Co is commonly available and used in industrial and medical applications. It could have been stolen from a medical center or a variety of industrial facilities. A list of possible sources is provided in Table 4.8.

^{210}Po is not commonly used. It is used in PoBe (α, n) neutron sources and devices that eliminate static electricity in machinery associated with a variety of industrial processes such as paper rolling, manufacturing sheet plastics, and spinning synthetic fibers. Static eliminators typically contain from one to tens of gigabecquerels of ^{210}Po. Brushes containing ^{210}Po are also used to remove accumulated dust from photographic films and camera lenses. ^{210}Po has been used as a heat source in satellites and was also used in each of the Soviet rovers deployed on the surface of the Moon to keep their internal components warm during the lunar nights. It has also been used as a clandestine poison.

(b) The intakes of ^{60}Co and ^{210}Po are determined from the relationship

$$I = \frac{\sum_{i=1}^{n} \mathrm{IRF}_i A_i}{\sum_{i=1}^{n} \mathrm{IRF}_i^2}$$

where the intake retention fractions (IRFs) and measured activities (A) are provided in the problem statement. The intakes of ^{60}Co and ^{210}Po are based on the three urine samples at 1, 5, and 10 days postexposure:

$$I(^{60}\mathrm{Co}) = \frac{[(2.59\times10^{-2})(1.05\,\mathrm{MBq}) + (2.26\times10^{-3})(0.12\,\mathrm{MBq}) + (1.27\times10^{-3})(0.065\,\mathrm{MBq})]}{(2.59\times10^{-2})^2 + (2.26\times10^{-3})^2 + (1.27\times10^{-3})^2}$$

$$= \frac{0.0275\,\mathrm{MBq}}{6.78\times10^{-4}} = 40.6\,\mathrm{MBq}$$

The ^{60}Co intake exceeds the adult CDG value of 15 MBq:

$$I(^{210}\mathrm{Po}) = \frac{[(4.67\times10^{-4})(0.15\,\mathrm{kBq}) + (6.34\times10^{-4})(0.22\,\mathrm{kBq}) + (5.78\times10^{-4})(0.20\,\mathrm{kBq})]}{(4.67\times10^{-4})^2 + (6.34\times10^{-4})^2 + (5.78\times10^{-4})^2}$$

$$= \frac{3.25\times10^{-4}\,\mathrm{kBq}}{9.54\times10^{-7}} = 341\,\mathrm{kBq}$$

The ^{210}Po intake exceeds the adult CDG value of 110 kBq. For nuclides other than iodine, the CDGs for the pregnant women are one-fifth of the adult values. Since the woman is pregnant, the CDG values must be appropriately reduced for a proper medical assessment of the intake and the need for subsequent action.

(c) The effective dose (E) from ^{60}Co is obtained from the effective dose coefficient (e):

$$E(^{60}\mathrm{Co}) = eI = \left(1.0\times10^{-8}\,\frac{\mathrm{Sv}}{\mathrm{Bq}}\right)(40.6\times10^{6}\,\mathrm{Bq}) = 0.41\,\mathrm{Sv}$$

The effective dose from ^{210}Po is obtained in a similar manner:

$$E(^{210}\text{Po}) = eI = \left(3.3 \times 10^{-6}\,\frac{\text{Sv}}{\text{Bq}}\right)(341 \times 10^{3}\,\text{Bq}) = 1.1\,\text{Sv}$$

The total effective dose is

$$E_{\text{total}} = E(^{60}\text{Co}) + E(^{210}\text{Po}) = 0.41\,\text{Sv} + 1.1\,\text{Sv} = 1.5\,\text{Sv}$$

(d) The decision to induce labor is a medical option. This action removes the fetus from the mother's blood supply that can transfer radioactive material to the unborn child. This action eliminates the source of radioactive material supplied by the mother and minimizes the fetal dose.

(e) The effective doses are delivered in a relatively short period given the estimated biological half-lives. These doses are determined by the effective half-lives:

$$T_{\text{eff}} = \frac{T_{\text{p}}T_{\text{b}}}{T_{\text{p}} + T_{\text{b}}}$$

$$T_{\text{eff}}(^{60}\text{Co}) = \frac{(5.27\,\text{year})\left(365\,\frac{\text{days}}{\text{year}}\right)(40\,\text{days})}{(5.27\,\text{year})\left(365\,\frac{\text{days}}{\text{year}}\right) + (40\,\text{days})} = 39.2\,\text{days}$$

$$T_{\text{eff}}(^{210}\text{Po}) = \frac{(138\,\text{days})(10\,\text{days})}{(138\,\text{days}) + (10\,\text{days})} = 9.32\,\text{days}$$

The effective dose rate delivered at time (T) is given in terms of the initial dose rate and effective half-life:

$$\dot{E}(t) = \dot{E}(0)e^{-\lambda_{\text{eff}}t}$$

The effective dose as a function of time is obtained by integration:

$$\int_0^T \dot{E}(t)\text{d}t = \int_0^T \dot{E}(0)e^{-\lambda_{\text{eff}}t}\,\text{d}t$$

$$E_{\text{total}}(T) = \frac{\dot{E}(0)}{\lambda_{\text{eff}}}(1 - e^{-\lambda_{\text{eff}}T})$$

where $\frac{\dot{E}(0)}{\lambda_{\text{eff}}}$ is the effective dose calculated in question (c).

For ^{60}Co:

$$E(T) = E(^{60}\text{Co})(1 - e^{-\lambda_{\text{eff}}T})$$

$$E(10\,\text{days}) = 0.41\,\text{Sv}\left(1 - e^{-\frac{\ln 2}{39.2\,\text{days}}10\,\text{days}}\right) = 0.066\,\text{Sv}$$

For ^{210}*Po*:

$$E(T) = E(^{210}\text{Po})(1 - e^{-\lambda_{\text{eff}}T})$$

$$E(10\,\text{days}) = 1.1\,\text{Sv}\left(1 - e^{-\frac{\ln 2}{9.32\,\text{days}}10\,\text{days}}\right) = 0.58\,\text{Sv}$$

The total effective dose is the sum of the ^{60}Co and ^{210}Po contributions:

$$E(10\,\text{days}) = 0.066\,\text{Sv} + 0.58\,\text{Sv} = 0.65\,\text{Sv}$$

Although the ^{60}Co and ^{210}Po effective doses are not immediately life threatening, it is likely the physician will elect to use decorporation therapy since the Clinical Decision Guide values were exceeded.

Possible biological effects of this exposure are suggested by the basis for deriving the CDG values:

1. A 0.25 Sv (50-year effective dose) is defined for consideration of stochastic effects. Using ICRP 103 values, this represents a 1.3% lifetime risk of fatal cancer attributed to the internal deposition.
2. A 30-day RBE-weighted absorbed dose value of 0.25 Gy-Eq for consideration of deterministic effects to bone marrow.
3. A 30-day RBE-weighted absorbed dose value of 1 Gy-Eq for consideration of deterministic effects to the lungs.

It should be noted that for nuclides other than iodine, the CDGs for children and pregnant women are defined as one-fifth of the adult values.

Specific organs at risk from ^{210}Po are the bone marrow, kidneys, liver, and lungs. The ^{60}Co deposition can affect the lung.

(f) The infant dose depends on the quantity of radioactive material transferred from the mother. These values were derived from whole-body counting data. The effective dose (E) is determined using the infant's dose conversion factors (e) and intake (I) values:

$$E(^{60}\text{Co}) = eI = \left(4.2 \times 10^{-8}\,\frac{\text{Sv}}{\text{Bq}}\right)(5.0 \times 10^{6}\,\text{Bq}) = 0.21\,\text{Sv}$$

$$E(^{210}\text{Po}) = eI = \left(1.5 \times 10^{-5}\,\frac{\text{Sv}}{\text{Bq}}\right)(3.8 \times 10^{4}\,\text{Bq}) = 0.57\,\text{Sv}$$

$$E_{\text{total}} = E(^{60}\text{Co}) + E(^{210}\text{Po}) = 0.21\,\text{Sv} + 0.57\,\text{Sv} = 0.78\,\text{Sv}$$

Since the ^{60}Co and ^{210}Po intakes exceed the infant CDG values, the physician will evaluate the need for medical intervention.

(g) Decorporation agents appropriate for this situation include:

^{60}*Co*: dimercaptosuccinic acid (DMSA), diethylenetriamine pentaacetate (DTPA) which is preferred (NCRP 161), ethylenediaminetetraacetic acid (EDTA), and *N*-acetyl-L-cysteine (NAC)

^{210}Po: British Anti-Lewisite (BAL) which is preferred (NCRP 161), DMSA, and penicillamine

Solutions for Chapter 5

5.1. (a) Based on the problem information, you are requested to calculate the absorbed dose delivered to an individual residing 1.0 m from the patient. The absorbed dose to the exposed individual (D_∞) is determined from the relationship provided with this question:

$$D_\infty = \left(\frac{34.6\,\Gamma\, Q_0}{(100\text{ cm})^2}\right)\begin{pmatrix} E_1 T_p (0.8)(1 - e^{-\ln 2(0.33\text{ days})/T_p}) + E_2 F_1 T_{1\text{eff}} e^{-\ln 2(0.33\text{ days})/T_p} \\ + E_2 F_2 T_{2\text{eff}} e^{-\ln 2(0.33\text{ days})/T_p} \end{pmatrix}$$

where

F_1 = extrathyroid uptake fraction = 0.93
F_2 = thyroidal uptake fraction = 0.07
E_1 = occupancy factor for the first 8 h = 0.05
E_2 = occupancy factor from the first 8 h to total decay = 0.10
Γ = absorbed dose rate constant for $^{131}I = 5.2 \times 10^{-8}$ Gy-m^2/MBq-h
Q_0 = administered activity = 7400 MBq
T_p = physical half-life of ^{131}I = 8.04 day

For the thyroid administration, the following effective half-life values apply:

$T_{1\text{eff}}$ = 0.32 day
$T_{2\text{eff}}$ = 7.3 days

Using these values, the absorbed dose to the exposed individual is

$$D_\infty = \left((34.6)\left(\frac{\frac{5.2\times10^{-8}\text{ Gy-m}^2}{\text{h - MBq}}}{(100\text{ cm})^2}\right)(7400\text{ MBq})\right)\left(\frac{100\text{ cm}}{\text{m}}\right)^2$$
$$\times \begin{pmatrix} (0.05)(8.04\text{ days})(0.8)(1 - e^{-\ln 2(0.33\text{ days})/8.04\text{ days}}) \\ + (0.10)(0.93)(0.32\text{ days})(e^{-\ln 2(0.33\text{ days})/8.04\text{ days}}) \\ + (0.10)(0.07)(7.3\text{ days})(e^{-\ln 2(0.33\text{ days})/8.04\text{ days}}) \end{pmatrix}$$

$$D_\infty = \left(1.33 \times 10^{-2}\,\frac{\text{Gy}}{\text{h}}\right)[(0.322\text{ days})(1 - 0.972)$$
$$+ (0.0298\text{ days})(0.972) + (0.0511\text{ days})(0.972)](24\text{ h/day})$$
$$= (0.028\text{ Gy})\left(1000\,\frac{\text{mGy}}{\text{Gy}}\right) = 28\text{ mGy}$$

The second part of this question requests that you make a determination regarding the need for written safety instructions to the patient. Following 10CFR35.75(b), the licensee shall provide the released individual with instructions, including written instructions, regarding actions needed to maintain doses to other individuals as low as reasonably achievable if the total effective dose equivalent (TEDE) to any other individual is likely to exceed 1 mSv. Since the dose exceeds 1 mSv, written safety instructions are required.

(b) Assuming that the patient cannot be treated as an outpatient, two restrictions that would allow you to release her from the hospital are as follows:
1. The licensee may authorize the release from its control of any individual who has been administered radiopharmaceuticals or permanent implants containing radioactive material if the total effective dose equivalent to any individual from exposure to the released individual is not likely to exceed 5 mSv [10CFR35.75(a)].
2. The licensee shall provide the released individual with instructions, including written instructions, on actions needed to maintain doses to other individuals as low as reasonably achievable if the total effective dose equivalent is likely to exceed 1 mSv [10CFR35.75(b)].

(c) In this question, you are to outline three general requirements she could apply to minimize dose to members of her family. These are the general requirements:
1. Maintain the distance from other persons, including separate sleeping relationships.
2. Minimize time in proximity to family members.
3. Take precautions to reduce the spread of radioactive material.
4. Avoid breast-feeding to minimize the dose to small children.

This would preclude family members from using the same clothing, bedding, bath towels, and eating utensils. Meal preparation should include contamination control considerations.

(d) The patient has a 15-month-old child at home. If the child is breast-feeding and the dose to the child could exceed 1 mSv, then additional instruction should be provided to the mother regarding:
1. Guidance on the interruption or discontinuation of breast-feeding.
2. Information on the consequences of failure to follow this guidance.

The age of the child also suggests that items 1–4 of question (c) be emphasized. The mother's need to hold and care for the child must be addressed in view of the necessary radiological restrictions. Care should be exercised to minimize the ^{131}I intake by the 15-month-old child. The mother should also clearly understand the time period of these restrictions.

(e) If the patient had a 15-year-old child at home, the mother should explain to the teenager why the restrictions are necessary and their duration. Items 1–3 of question (c) would apply.

5.2. (a) The chief potential advantage of using a negative K meson beam versus a ^{60}Co beam for treating a tumor is the delivery of a localized dose at high LET. In this problem, the capture of a negative K meson at the end of its range results in the transfer of 130.6 MeV to the tumor from protons (60.0 MeV) and heavy fragments (70.6 MeV). This energy is transferred via scattering or nuclear interactions to the atoms at the tumor site. The particles interacting with the tumor have high LET, and the charged particles have short ranges, which selectively deposits their energy at the target location. By comparison, ^{60}Co irradiates the tumor and healthy tissue.

(b) 1 The dominant energy-loss mechanism for light charged particles (e.g., electrons) in matter is the collision of these particles with atomic electrons. These collisions do not appreciably deflect the path in the tissue of a heavy charged particle like a charged K meson, which is much more massive than the electron. With an electron or positron beam, the particles have the same mass as the atomic electrons, and so large-angle deflections and beam spreading occur.

Elastic nuclear collisions through the Coulomb force have a much smaller cross-section, but these collisions involve multiple small-scattering angles, which spread the beam. For heavy charged particles, multiple scattering is a significant effect.

For negative K mesons, multiple Coulomb scattering by atomic nuclei is the dominant effect (Response 1). Response 5 would apply for electron or positron beams, but it is not the best general answer for a heavy charged particle.

(c) 3 Neutrons, photons, and muons are the penetrating radiation types generated with the accelerator running. Residual photons are created from the decay of activation products. Neutrons are produced from negative K meson–nuclear reactions. K mesons decay into muons. Given their interaction characteristics, neutrons, photons, and muons present the dominant radiation hazard from the negative K meson beam.

(d) 1 Induced beta and gamma activity is present in the treatment area when the accelerator is not operating. These radiation types arise from activated materials that decay by beta and gamma emission.

(e) 2 This energy difference is required to overcome the nuclear binding energy in order to fragment the oxygen nucleus.

(f) 1 A negative K meson decays into a negative muon, which then decays into an electron. Neutron and pion production from proton interactions with the target are also expected. The primary K^- and μ^- decay modes are

$$K^- \to \mu^- + \bar{\upsilon}_\mu$$

$$\mu^- \to e^- + \upsilon_\mu + \bar{\upsilon}_e$$

(g) The distance traveled by the negative K meson in the tissue is expressed in terms of the standard particle range relationship

$$R = \rho t$$

where

R	=	range = 11.5 g/cm^2
t	=	physical distance traveled in the tissue
ρ	=	density of the material (soft tissue) = 0.95 g/cm^3

$$t = \frac{R}{\rho} = \frac{11.5\,\frac{\text{g}}{\text{cm}^2}}{0.95\,\frac{\text{g}}{\text{cm}^3}} = 12.1\,\text{cm}$$

(h) *Assumptions*:

1. Protons and heavier nuclear fragments have ranges less than 4.0 cm and stop in the sphere.
2. Neutrons and gamma photons lose negligible energy in the sphere when compared to protons and heavy fragments.

Based upon these assumptions the energy deposited into the 4.0 cm radius sphere of water per stopped negative K meson is derived from the radiation types produced in the interaction:

Reaction products and average energies from capture of a stopped negative K meson by ^{16}O

Emitted particle	Average kinetic energy per capture (MeV)	Energy absorbed into a 4 cm radius sphere
Fast neutrons	215	0.00
Protons	60.0	60.0
Heavy fragments	70.6	70.6
Gamma rays	21.2	0.00
Total	366.8	130.6

The absorbed dose is defined in terms of energy deposited per unit mass. The mass of material (M) contained within the 4.0 cm radius (r) sphere of water having a density (ρ) is given by

$$M = \rho V = \frac{4}{3}\pi r^3 \rho = \frac{4}{3}\pi (4\,\text{cm})^3 \left(1.0\,\frac{\text{g}}{\text{cm}^3}\right) = 268\,\text{g}$$

The absorbed dose D is obtained from its definition, the energy (E) deposited per unit mass (m), and the appropriate conversion factors

$D = \frac{E}{m}$:

$$D = \frac{(130.6\,\text{MeV})\left(1.6 \times 10^{-13}\,\frac{\text{J}}{\text{MeV}}\right)\left(1000\,\frac{\text{g}}{\text{kg}}\right)\left(\frac{\text{Gy-kg}}{\text{J}}\right)}{(268\,\text{g})}$$

$$= 7.80 \times 10^{-11}\,\text{Gy}$$

(i) The assumption that charged particles stop in the 4.0 cm radius sphere is accurate. A 60 MeV proton has a range of about 3 cm in water, and the charged fragment ranges are less than the proton range. The following additional data would be needed to refine the dose estimate:

1. A more accurate neutron and gamma-ray energy spectrum is needed. In addition, the double differential cross-section in terms of energy loss and angle is needed for hydrogen and oxygen.
2. The attenuation coefficients, mean free path values, and relevant buildup factors, for neutrons and gamma rays in water at the energies of interest.
3. The stopping power for low-energy negative K mesons in water.
4. The negative K meson macroscopic cross-section in water for low energies.
5. The incident fluence for low-energy negative K mesons incident on the water sphere.

Items 3–5 are derived from the dose deposited by a charged particle in the tissue. For a tissue volume irradiated by a parallel beam of particles, the absorbed dose (D) as a function of penetration distance x is given by

$$D(x) = \frac{1}{\rho}\left(-\frac{\mathrm{d}E}{\mathrm{d}x}\right)\Phi(x)$$

where ρ is the density of the material (water) attenuating the negative K meson, $-\mathrm{d}E/\mathrm{d}x$ is the stopping power, and Φ is the negative K meson fluence. The particle fluence varies with penetration distance according to the relationship

$$\Phi(x) = \Phi(0)\exp(-\mu x)$$

where $\Phi(0)$ is the entrance fluence and μ is the macroscopic reaction cross-section (linear attenuation coefficient). The linear attenuation coefficient is defined as

$$\mu = n\sigma$$

where n is the number of atoms of absorbing material per unit volume and σ is the total microscopic reaction cross-section for the negative K meson–water interaction.

(j) Dose localization is greatest for ^{12}C since it has a narrow Bragg peak at the end of its range. The negative pions and negative K mesons have broader dose distributions within the tissue. These three radiation types deposit some energy outside the tumor volume since they incorporate external beams, which deliver dose to the healthy tissue between the skin surface and the tumor volume. Dose to the healthy tissue also arises from beam misdirection, scattering, and a dose profile extending beyond the tumor volume.

5.3. (a) In this question, you are requested to calculate the ^{131}I activity administered to the patient to deliver the prescribed 70 Gy dose. The patient has a thyroid mass that is five times the mass of the Reference Man thyroid (i.e., 100 g/20 g = 5). A mass correction for S is needed since dose is defined on a per unit mass basis. Assuming the thyroid functions as in Reference Man except for its mass, an effective S factor for the patient can be determined:

$$S_{\text{patient}} = \frac{S_{\text{Reference Man}}}{5} = S$$

where $S_{\text{Reference Man}}$ = Reference Man absorbed dose per unit cumulated activity (S factor) for the thyroid as source and target organ = 1.57×10^{-3} mGy/MBq-s:

$$S = \frac{1.57 \times 10^{-3}\ \frac{\text{mGy}}{\text{MBq-s}}}{5} = 3.14 \times 10^{-4}\ \frac{\text{mGy}}{\text{MBq-s}}$$

The thyroid absorbed dose D is determined using the Medical Internal Radiation Dose formulation (see Appendix D):

$$D = \tilde{A}S$$

where $\tilde{A}$ = total cumulated activity:

$$\tilde{A} = \frac{f_2 A}{\lambda_e} = 1.44 f_2 A T_{\text{eff}}$$

f_2	=	thyroid uptake in the patient = 0.6
A	=	administered activity
T_{eff}	=	effective half-life = 5 days
D	=	administered dose = 70 Gy

In terms of the aforementioned quantities,

$$D = 1.44 f_2 A T_{\text{eff}} S$$

The desired activity is obtained by algebraic manipulation:

$$A = \frac{D}{1.44 f_2 T_{\text{eff}} S}$$

$$A = \frac{(70\ \text{Gy})\left(\frac{1000\ \text{mGy}}{\text{Gy}}\right)\left(\frac{1\ \text{day}}{24\ \text{h}}\right)\left(\frac{1\ \text{h}}{3600\ \text{s}}\right)}{(1.44)(0.6)(5\ \text{days})\left(3.14 \times 10^{-4}\ \frac{\text{mGy}}{\text{MBq-s}}\right)} = 597\ \text{MBq}$$

(b) In this part, you are requested to calculate the cumulative external dose to his spouse under the following conditions:

- Sleeping arrangements: distance (r) is 1 m.
- The thyroid is the only source of exposure.
- Time spent in the vicinity (1 m) of the spouse, over a period of 24 h = 8 h.
- Specific gamma-ray dose constant (Γ) at 1 m = 5.2×10^{-5} mSv/h-MBq.
- Assume the patient was administered 1480 MBq (A).

It is reasonable to treat the thyroid as a point source and to neglect attenuation from the thyroid and intervening tissue. This is also necessary since no attenuation data are provided. Using the point source approximation, the effective dose rate ($\dot{E}$) from the thyroid uptake is

$$\dot{E}(t) = \frac{Af_2\Gamma}{r^2} e^{-\lambda_{\text{eff}} t}$$

and the cumulative effective dose is obtained by integration:

$$E = \int_0^\infty \dot{E}(t)\,dt = \frac{Af_2\Gamma}{r^2} \int_0^\infty e^{-\lambda_{\text{eff}} t}\,dt = \frac{Af_2\Gamma}{\lambda_{\text{eff}} r^2}$$

To perform this problem rigorously, one should perform the integration over 8 h periods. This is tedious and is reasonably approximated by

$$E_{\text{Spouse}} = \frac{E}{3}$$

because the spouse meets the assumed conditions only 8 h in a 24 h period. Therefore,

$$E_{\text{Spouse}} = \frac{Af_2\Gamma}{3\lambda_{\text{eff}} r^2}$$

$$E_{\text{Spouse}} = \frac{(1480\,\text{MBq})(0.6)\left(5.2 \times 10^{-5}\,\frac{\text{mSv-m}^2}{\text{MBq-h}}\right)}{(3)\left(\frac{0.693}{5\,\text{days}}\right)\left(\frac{1\,\text{day}}{24\,\text{h}}\right)(1\,\text{m})^2} = 2.7\,\text{mSv}$$

(c) Assuming the dose equivalent to the patient's spouse is 2.5 mSv, you are to determine if the licensee is in compliance with the radiation protection limits of 10CFR35 if the patient is released from the hospital immediately after the administration.

Following 10CFR35.75, a licensee may not authorize release from confinement for medical care any patient administered a radiopharmaceutical until a number of requirements are met. These include the following:

> 10CFR35.75(a) *A licensee may authorize the release from its control of any individual who has been administered unsealed byproduct material or implants containing byproduct material if the total effective dose equivalent to any other individual from exposure to the released individual is not likely to exceed 5 mSv (0.5 rem).*

Since the dose equivalent to the patient's spouse is 2.5 mSv, the licensee is in compliance with the radiation limits of 10CFR35 even if the patient is released from the hospital immediately after administration. The current revision of NUREG-1556, Vol. 9, *Consolidated Guidance about Materials Licenses: Program-Specific Guidance about Medical Licenses* provides methods to calculate the dose to other individuals. It also contains tables of activities not likely to cause doses exceeding 5 mSv.

> 10CFR35.75(b) *A licensee shall provide the released individual, or the individual's parent or guardian, with instructions, including written instructions, on actions recommended to maintain doses to other individuals as low as is reasonably achievable if the total effective dose equivalent to any other individual is likely to exceed 1 mSv (0.1 rem). If the total effective dose equivalent to a nursing infant or child could exceed 1 mSv (0.1 rem) assuming there were no interruption of breast-feeding, the instructions must also include—*
>
> 1. *Guidance on the interruption or discontinuation of breast-feeding; and*
> 2. *Information on the potential consequences, if any, of failure to follow the guidance*

Since the effective dose to each infant is estimated to be 1.25 mSv, the requirements of 10CFR35.75(b) were not met. She should have been provided written instructions in maintaining the dose to her children as low as reasonably achievable. In addition, no breast-feeding instructions were provided to the nursing mother.

The dose to the patient's spouse also exceeds 1 mSv, and no written instructions were provided for maintaining this dose ALARA. These failures place the hospital in conflict with 10CFR35.

> 10CFR35.75 (c) *A licensee shall maintain a record of the basis for authorizing the release of an individual in accordance with §35.75(a).*

The medical facility performed calculations to determine the dose to the husband and infants. These calculations should have been documented and would form the basis for release. However, as noted in 10CFR35.75(d), release instructions should have been provided and documented. No instructions were provided.

> 10CFR35.75 (d) *The licensee shall maintain a record of instructions provided to a breast-feeding female in accordance with § 35.75(b).*

Although instructions were not provided to the patient, the problem does not provide any information regarding records. If the instructions were not produced, the licensee would not be in compliance with the requirements of 10CFR35.75(d).

(d) General precautionary measures that should be provided to a patient upon release from the hospital reflect contamination control and dose minimization considerations. These measures include direction to:

a. Maximize the distance between the patient and other individuals (spouse, children, and friends) to minimize their external dose.
b. Sleep in a separate bed to minimize ^{131}I cross-contamination from body fluids (e.g., sweating) and external dose.
c. Use separate tableware, bedding, and linen.
d. Avoid contact with small children.
e. Avoid contact with pregnant women.
f. Minimize physical contact with other people.
g. Wash her own dishes, clothing, and linen.
h. Use separate bathrooms and showers.
i. Avoid food preparation for consumption by members of the household.
j. Use infant formula and discontinue breast-feeding. This measure will be evaluated by the physician in consultation with the medical physicist.

The standardized set of instructions and the length of their applicability should be reviewed with the patient and her family. Any questions should be fully answered.

5.4. (a) Using the 125 kVp X-ray spectrum data, the equivalent concrete thickness for the control panel wall is to be determined. The problem provides X-ray data for the existing wall whose composition is unknown. In general, the variation in kerma rate as a function of shield thickness (x) is written as

$$K(x) = K_o B e^{-\mu x} = K_o e^{-\mu x}$$

where

$K(x)$	=	the shielded kerma from the X-ray system at the control panel due to the presence of the control panel wall whose composition is not known
	=	0.0125 mGy for a 1 min exposure at 10 mA-min and 125 kVp
K_o	=	unattenuated X-ray kerma rate at the control panel determined from survey data
B	=	buildup factor = 1.0 per the problem statement
x	=	thickness of the existing control panel wall (composition unknown) = 15 cm
μ	=	linear attenuation coefficient for the unknown wall material

The equivalent concrete wall thickness is determined from the unattenuated 30 cm survey data (180 mGy at 5 mA-min and 125 kVp) and the attenuated dose 0.1 mGy (10 mA-min for 8 min) through the 15 cm wall at a distance of 385 cm (370 cm + 15 cm). The desired unshielded value at the control panel is determined from a point source approximation by correcting for X-ray beam current and distance

$$K_o(385\,\text{cm}) = (180\,\text{mGy})\left(\frac{30\,\text{cm}}{385\,\text{cm}}\right)^2\left(\frac{10\,\text{mA- min}}{5\,\text{mA- min}}\right) = 2.19\,\text{mGy}$$

This information now permits the attenuation coefficient for the unknown wall material (μ_1) to be determined by algebraic manipulation of the kerma relationship

$$\mu_1 = -\frac{1}{x}\ln\frac{K(x)}{K_o} = -\frac{1}{(15\,\text{cm})}\ln\left(\frac{0.0125\,\text{mGy}}{2.19\,\text{mGy}}\right) = \frac{0.344}{\text{cm}}$$

The equivalent concrete thickness is obtained by comparing the kerma attenuation relationships for the existing control panel wall of unknown composition (1) and equivalent concrete wall (2):

$$K(x) = K_o B_1 e^{-\mu_1 x_1} = K_o B_2 e^{-\mu_2 x_2}$$

since $B_1 = B_2 = 1$, this equation is significantly simplified:

$$\mu_1 x_1 = \mu_2 x_2$$

The linear attenuation coefficient for concrete is obtained from the 125 kVp X-ray HVL given in the problem statement:

$$\mu_2 = \frac{\ln 2}{\text{HVL}} = \frac{0.693}{2.0\,\text{cm}} = \frac{0.347}{\text{cm}}$$

The equivalent concrete thickness of the existing wall is

$$x_2 = \left(\frac{\mu_1}{\mu_2}\right)x_1 = \left(\frac{\frac{0.344}{\text{cm}}}{\frac{0.347}{\text{cm}}}\right)(15.0\,\text{cm}) = 14.9\,\text{cm}$$

(b) Assuming the maximum activity loading, the workload for the RCB system is

$$W = At$$

where

W	=	workload in units of TBq-min/week
A	=	RCB ^{192}Ir activity = 0.37 TBq
t	=	maximum exposure time per week for the RCB system
	=	(8 patients/week)(4 min/patient) = 32 min/week

With these values, the workload is

$$W = (0.37\,\text{TBq})\left(32\,\frac{\text{min}}{\text{week}}\right) = 11.8\,\frac{\text{TBq- min}}{\text{week}}$$

A second definition for the workload is obtained by using the kerma rate constant (Γ) for ^{192}Ir (1.1×10^{-4} mGy-m^2/MBq-h):

$$W' = \left(11.8\,\frac{\text{TBq- min}}{\text{week}}\right)\left(1.1 \times 10^{-4}\,\frac{\text{mGy-m}^2}{\text{MBq-h}}\right)\left(\frac{1\,\text{h}}{60\,\text{min}}\right) \times \left(\frac{10^6\,\text{MBq}}{\text{TBq}}\right) = 22\,\frac{\text{mGy-m}^2}{\text{week}}$$

(c) The additional workload (W') requires that shielding be added to the control panel wall before using the RCB system. The additional weekly kerma at the control panel is given by

$$\dot{K}_{\text{RCB}} = \frac{W'}{d^2}e^{-\mu x}$$

where

$\dot{K}_{\text{RCB}}$ = additional weekly kerma due to RCB operations
W' = RCB workload = 22 mGy-m^2/week
μ = linear attenuation coefficient for ^{192}Ir = ln(2)/HVL = 0.693/4.0 cm
= 0.173/cm
x = existing wall thickness based only on the operation of the X-ray system = 15.0 cm of unknown material
d = distance between the source and the control panel
= 370 cm + x = 370 cm + 15.0 cm = 385.0 cm

These values determine the additional weekly kerma due to operation of the RCB system:

$$\dot{K}_{\text{RCB}} = \frac{W'}{d^2}e^{-\mu x} = \frac{22\,\frac{\text{mGy-m}^2}{\text{week}}}{\left[(385\,\text{cm})\left(\frac{1\,\text{m}}{100\,\text{cm}}\right)\right]^2}e^{-\left(\frac{0.173}{\text{cm}}\right)(15.0\,\text{cm})}$$

$$= \left(1.48\,\frac{\text{mGy}}{\text{week}}\right)(0.0746) = 0.11\,\frac{\text{mGy}}{\text{week}}$$

In addition to the RCB dose component, the control panel also experiences radiation from the X-ray (superficial) therapy system. The weekly X-ray kerma from the system is given by

$$\dot{K}_{\text{X-ray}} = \frac{\dot{K}(x)W_{\text{X-ray}}}{I}$$

where

$\dot{K}_{X\text{-ray}}$	=	weekly X-ray kerma rate at the control panel
$\dot{K}(x)$	=	kerma rate at the control panel area due to the X-ray beam (0.0125 mGy for a 1 min exposure)
	=	(0.0125 mGy)/(1 min)
	=	0.0125 mGy/min
I	=	X-ray beam current = 10 mA
$W_{X\text{-ray}}$	=	weekly workload of the superficial therapy system
	=	750 mA-min/week

With these values, the weekly X-ray kerma is determined:

$$\dot{K}_{X\text{-ray}} = \frac{\left(0.0125\,\frac{\text{mGy}}{\text{min}}\right)\left(750\,\frac{\text{mA-min}}{\text{week}}\right)}{(10\,\text{mA})} = 0.94\,\frac{\text{mGy}}{\text{week}}$$

The total weekly kerma rate is the sum of the X-ray and RCB contributions:

$$\dot{K}_{\text{total}} = \dot{K}_{X\text{-ray}} + \dot{K}_{\text{RCB}}$$

$$\dot{K}_{\text{total}} = 0.94\,\frac{\text{mGy}}{\text{week}} + 0.11\,\frac{\text{mGy}}{\text{week}} = 1.1\,\frac{\text{mGy}}{\text{week}}$$

This total weekly kerma exceeds the design limit of 1.0 mGy/week. Following NCRP 151, an additional HVL (4.0 cm) of concrete should be added to reduce the weekly exposure to the 1.0 mGy/week limits specified in the problem statement. The additional shielding accommodates the RCB system addition and its associated ^{192}Ir source. Additional shielding may also be advisable based upon future research plans and ALARA considerations.

The reader should note that the effects of shielding from the patient's body and structures within the room have not been considered. This shielding would reduce the 1.1 mGy/week value to below the 1.0 mGy/week design limit. A more detailed analysis would include the fixed components in the room that would provide shielding. The NCRP provides guidance in using the preshielding concept for imaging facilities but does not apply this concept to therapy facilities. Credible occupancy factors should also be determined as part of the shielding analysis.

(d) This question concerns the calculation of the shielding requirements for the treatment room ceiling. Shielding is required to address the occasional exposure of technicians who access the roof to perform radiation surveys during patient treatments. For these considerations, the use factor and the occupancy factor could each be assigned a conservative value of 1.0. The technicians are radiation workers who are occupationally exposed. NCRP 49 notes that for occupationally exposed persons, the occupancy factor is usually assumed to be unity. Moreover, occupationally exposed workers are assumed to spend their entire work day in radiologically controlled areas. Since the fraction of time a worker spends in any given location cannot be predicted, that person can occupy any accessible area for an extended time unless there is an exclusion mechanism in place.

A more refined assignment notes that the ceiling pathway involves scattered radiation. Scattered radiation is assigned a use factor of unity. The problem notes occasional occupancy that would justify an assignment of 1/16 following NCRP 49. NCRP 151 updates the NCRP 49 report for therapy facilities. Although it does not specifically address the situation of this problem, Table B.1 of this report could be used to justify a value of 1/20. Unless a time motion study is performed either the less conservative values of 1/16 or 1/20 could be justified based on discussions with technicians and their maximum anticipated access of the area. This initial assignment should be verified by a subsequent evaluation. If required, the space above the ceiling could be locked and access restricted. Worker access could also be limited by key or key card control measures.

(e) Precautions that could have prevented the patient from leaving the hospital with the ^{192}Ir source inside her body include the following:

1. A radiation survey of the patient should be performed prior to her release from the treatment room. This survey should also include the source shield and the treatment area to ensure the radioactive material is properly stored and is not outside its shield.
2. A radiation detector with both audible and visible alarms should be located near the exit from the treatment room. This instrument would have detected the source within the patient if it were properly maintained and calibrated and if the technicians properly responded when it alarmed.
3. A source accountability log should be established. Removal of ^{192}Ir sources from the storage area should be logged with the technician signing their name to indicate the sources removed and its identification number. Upon removal from the patient, the sources must be logged and returned to the storage area and documented through a sign-off. For added security, a second signature by a supervisor who verifies log out and return would add a supporting administrative control. Although this control is not as rigorous as physical measurements of the patients, it enhances source accountability and personnel responsibility.

5.5. (a) Assuming that radiological hazards to medical personnel while treating the injury are minimal, the health physicist should provide the following radiological information about the incident and individual to the physician:

1. The isotope or isotopes that are the source of the contamination.
2. The radiation types associated with the contamination (e.g., gamma, beta, and alpha). The associated radiation energies should also be presented.
3. The parts of the body that are contaminated, the size of the contaminated area, and the contamination levels.
4. The dose rates on contact and at 30 cm from the body surface.
5. The radiological hazard to the patient in terms of the dose delivered to the skin. In most cases, medical treatment should not be delayed for decontamination. Patient injuries should receive priority attention.
6. The risk to the hospital staff. ALARA measures, such as the use of bed shields, may be appropriate.
7. Appropriate contamination control measures for containing body fluids and tissues. A health physics technician should accompany the patient and provide assistance as warranted.
8. Possible skin decontamination methods that could be utilized and the damage they could impose on the skin.
9. Estimate of internal intakes by isotope.
10. Dose estimate from internal intakes of radioactive material.

(b) Considerations concerning the patient's risk from removing stubborn contamination by using a radical technique are as follows:

1. The skin and tissue doses should be calculated to determine if the radical technique is warranted. In most cases, there will be little risk if the contamination remains in place. However, the final treatment decision resides with the physician. The health physicist provides technical recommendations for consideration by the physician.
2. The radical technique may drive the contamination into the underlying tissues where the clearance mechanisms are slower than the skin's normal turnover rate.
3. The technique may force the contamination into the systemic circulation and produce a deposition in other organs.
4. Radical methods such as surgical removal of the tissue create long-term effects that may not be warranted on the basis of the radiological risk.
5. The radical technique could create an airborne hazard, and the radioactive material might be inhaled or ingested by the medical staff.

(c) If the dose is limited to 1.0 Gy, the level of long-lived contamination in counts per minute that could be left on the skin is calculated based on the following factors and assumptions:

1. Since no radioactive decay information is provided, assume that the only removal mechanism is due to sloughing. The effective removal rate of the initial radioactive material deposited on the skin is the sloughing (biological) fractional removal rate constant (k) of 0.05/day.
2. The skin contamination is uniformly distributed over an area (S) of 50 cm^2.
3. The total integrated dose D due to the decay of all the activity is given by

$$D = \int_0^\infty \dot{D}\,\mathrm{d}t = \int_0^\infty A_S(t)(\mathrm{DRF})\,\mathrm{d}t = \int_0^\infty A_S(0)e^{-kt}(\mathrm{DRF})\,\mathrm{d}t$$
$$= A_S(0)(\mathrm{DRF})\int_0^\infty e^{-kt}\,\mathrm{d}t = \frac{A_S(0)(\mathrm{DRF})}{k}$$

where

D	=	total dose = 1.0 Gy
$A_S(0)$	=	initial activity deposited per unit area
DRF	=	dose rate factor = 1.35×10^{-3} mGy-cm^2/Bq-h
k	=	sloughing fractional removal rate constant = 0.05/day

4. The initial activity per unit area is defined in terms of the measured pancake probe measurement

$$A_S(0) = \frac{R}{eS}$$

where

R	=	frisker count rate (cpm)
e	=	frisker efficiency = 0.1 counts/dis
S	=	probe area = 15 cm^2

The initial activity equation is solved for the desired frisker count rate:

$$R = A_S(0)eS$$

The total dose equation is used to recast the initial activity in terms of the information provided in this problem:

$$A_S(0) = \frac{kD}{(\mathrm{DRF})}$$

When this equation is combined with frisker count rate equation, the frisker count rate is obtained in terms of the available data:

$$R = \frac{kDeS}{(\mathrm{DRF})}$$

$$R = \frac{\left(\frac{0.05}{\text{day}}\right)\left(\frac{1\ \text{day}}{24\ \text{h}}\right)(1\ \text{Gy})\left(\frac{1000\ \text{mGy}}{\text{Gy}}\right)\left(0.1\ \frac{\text{c}}{\text{dis}}\right)(15\ \text{cm}^2)}{\left(1.35\times 10^{-3}\ \frac{\text{mGy-cm}^2}{\text{Bq-h}}\right)\left(\frac{\text{Bq-s}}{\text{dis}}\right)\left(\frac{1\ \text{min}}{60\ \text{s}}\right)}$$

$$= 1.39\times 10^5\ \text{cpm}$$

This count rate value represents the skin contamination level (cpm) that could be left on the skin such that the physician-imposed 1.0 Gy limiting skin dose is not exceeded.

5.6. (a) The relative hazards can be expressed in terms of the radiation weighting factor. This factor derived from ICRP 103 is provided in parenthesis following the specific radiation type: photons (1), electrons (1), muons (1), protons (2), charged pions (2), alpha particles (20), ^{12}C nuclei (20 is specified for heavy ions), ^{16}O nuclei (20 is specified for heavy ions), ^{56}Fe nuclei (20 is specified for heavy ions), thermal neutrons (5), antiprotons (not specified in ICRP 103), and anti-^{12}C nuclei (not specified in ICRP 103).

Antiprotons and protons interact to produce a variety of radiation types including charged pions and photons. Since antiprotons and protons have the same mass, have the same absolute value of their charge, and produce charged pions upon annihilation, a radiation weighting factor of 2 is a reasonable first-order approximation. Based on their heavy ion character, anti-^{12}C ions would be assigned an initial radiation weighting factor of 20.

(b) The antiprotons annihilate protons in the tissue to provide a variety of reaction products. Proton–antiproton annihilation events result in an average of 4–5 charged and neutral pions of mean energy about 400 MeV being produced plus 3 high-energy gamma rays. For example, reactions of the following type will occur:

$$p + \overline{p} \rightarrow 3\pi^+ + 3\pi^- + \gamma' s$$

(c) The charged pions do not preferentially deposit their energy in a small volume, but their dose localization is better than achieved by photons and neutrons. The charged pion Bragg peak is much broader than the corresponding peak from protons and heavy ions. Photon and neutrons deposit their energy over an extended range.

The photons offer the possibility of tracking the radiation profile in a manner analogous to imaging using radioactive materials. Tomographic techniques would need to be developed to relate the detected photons to their generation location.

(d) Protons and heavy ions have narrow Bragg peaks, which would facilitate dose localization. Photons and neutrons have broad dose deposition profiles. The other radiation types have broader dose deposition profiles as a function of penetration depth into the tissue than protons and heavy ions. Antimatter will produce pions and photons and also produce a broader deposition profile than protons and heavy ions.

(e) For a volume irradiated by a parallel beam of protons, the absorbed dose (D) as a function of penetration distance x is given by

$$D(x) = \frac{1}{\rho}\left(-\frac{dE}{dx}\right)\Phi(x)$$

where ρ is the density of the material attenuating the heavy ion, $-dE/dx$ is the stopping power, and Φ is the proton fluence. The proton fluence varies with penetration distance according to the relationship

$$\Phi(x) = \Phi(0)\exp(-\mu x)$$

where $\Phi(0)$ is the fluence as it exits the internal radiation-generating device and μ is the macroscopic reaction cross-section (linear attenuation coefficient). The linear attenuation coefficient is defined as

$$\mu = n\sigma$$

where n is the number of atoms of absorbing material per unit volume and σ is the total microscopic reaction cross-section for the proton tissue interaction.

Each of the parameters identified earlier are required to perform a dose assessment. The calculation is complex because the cross-section, stopping power, and attenuation coefficients are energy and angle dependent. Since the generated radiation type (e.g., protons for this problem) continuously loses energy as it penetrates the tissue, each distance has unique parameter values that must be calculated. Calculation of the energy and angle-dependent parameters (e.g., differential scattering cross-sections) is time consuming from a computational perspective.

5.7. (a) ^3H is a low-energy beta emitter with a maximum energy (E) of 0.0186 MeV and a 12.3-year half-life. It has a range (R) of

$$\begin{aligned} R &= 412E^{1.265-0.0954\ln(E)}\ \mathrm{mg/cm^2} \\ &= 412(0.0186)^{1.265-0.0954\ln(0.0186)}\ \mathrm{mg/cm^2} \\ &= 0.586\ \mathrm{mg/cm^2} \end{aligned}$$

Assuming blood vessels have a unit density, the physical range (t) of the tritium beta particles is

$$\begin{aligned} t &= \frac{R}{\rho} = \frac{(0.586\ \mathrm{mg/cm^2})(1\ \mathrm{g}/1000\ \mathrm{mg})}{(1\ \mathrm{g/cm^3})}\left(\frac{1\ \mathrm{m}}{100\ \mathrm{cm}}\right)\left(\frac{10^6\ \mu\mathrm{m}}{\mathrm{m}}\right) \\ &= 5.86\ \mu\mathrm{m} \end{aligned}$$

Tritium has the range to penetrate only a portion of the 20 μm arteriole wall thickness, and it must be removed from the body after it has imparted sufficient damage to the tissue. The 12.3-year physical half-life suggests that the material be removed from the body before the microsphere degrades, releases tritium or HTO into the body, and irradiates the healthy tissue. Since it does not have sufficient energy to damage the entire arteriole wall, it is not an appropriate vascular disruption agent. In addition, a method must be developed to deliver ^{3}H to the arteriole wall. Given its limited range, placing the ^{3}H into a microsphere would not be a priority option.

^{32}P is a high-energy beta emitter with a maximum energy of 1.71 MeV and a 14.3-day half-life. It has a range of

$$\begin{aligned} R &= 412\,E^{1.265-0.0954\ \ln(E)}\ \text{mg/cm}^2 \\ &= 412\,(1.71)^{1.265-0.0954\ \ln(1.71)}\ \text{mg/cm}^2 \\ &= 790\,\text{mg/cm}^2 \end{aligned}$$

Assuming blood vessels have a unit density, the physical range of the ^{32}P beta particles is

$$t = \frac{(790\,\text{mg/cm}^2)(1\,\text{g}/1000\,\text{mg})}{(1\,\text{g}/\text{cm}^3)} \left(\frac{1\,\text{m}}{100\,\text{cm}}\right) \left(\frac{10^6\,\mu\text{m}}{\text{m}}\right) = 7900\,\mu\text{m}$$

^{32}P radiation has sufficient range to penetrate the arteriole wall but also irradiates the healthy tissue. Moreover, its bremsstrahlung photons irradiate the tissue well beyond the arteriole wall. It has been used in tumor blood vessel irradiation studies. The 14.3-day physical half-life is reasonable for the intended application, but the ^{32}P radiation is not localized in the arteriole wall.

^{60}Co has a half-life of 5.27 years and emits a 0.318 MeV maximum energy beta particle and two photons with energies of 1.17 and 1.33 MeV. The photons will easily penetrate the arteriole wall and irradiate the healthy tissue. Therefore, it is not an appropriate radionuclide for the research.

^{125}I has a half-life of 59.4 days and emits a low-energy photon (0.0355 MeV). Dose localization is better than achieved with the higher-energy photons, but radiation is deposited beyond the arteriole wall.

^{201}Tl has a half-life of 3.04 days and emits two low-energy photons (0.135 and 0.167 MeV). The short half-life is desirable, but the photons have sufficient energy to penetrate beyond the arteriole wall and irradiate the healthy tissue.

^{252}Cf has a 2.65-year half-life and emits neutrons via spontaneous fission, alpha particles having energies of 6.08 and 6.12 MeV, and low-energy photons (0.0434 and 0.100 MeV). The range of the alpha particles (in air) is

$$R_{\text{air}} = (1.24E - 2.62)\,\text{cm} = (1.24(6.1) - 2.62)\ \text{cm} = 4.94\,\text{cm}$$

The range in the tissue with unit density is obtained from the Bragg–Kleeman rule:

$$R_{\text{Tissue}} = \frac{\rho_{\text{air}}}{\rho_{\text{Tissue}}} \left(\frac{M_{\text{Tissue}}}{M_{\text{air}}}\right)^{1/2} R_{\text{air}}$$

Using the molecular weight (M) and density of air (ρ) at standard temperature and pressure simplifies the result:

$$\begin{aligned}
\rho_{\text{air}} &= \text{density of air} = 1.293 \times 10^{-3}\,\text{g/cm}^3 \text{ at STP} \\
\rho_{\text{water}} &= \text{density of water} = 1.0\,\text{g/cm}^3 \\
M_{\text{air}} &= \text{molecular weight of air} = 14.5\,\text{g/mol} \\
M_{\text{tissue}} \approx M_{\text{water}} &= \text{molecular weight of water} = 18\,\text{g/mol}
\end{aligned}$$

$$\begin{aligned}
R_{\text{Tissue}} &= \left(\frac{1.293 \times 10^{-3}\,\frac{\text{g}}{\text{cm}^3}}{1.0\,\frac{\text{g}}{\text{cm}^3}}\right)\left[\frac{\left(18\,\frac{\text{g}}{\text{mol}}\right)}{\left(14.5\,\frac{\text{g}}{\text{mol}}\right)}\right]^{1/2}(4.94\,\text{cm}) \\
&= (7.12 \times 10^{-3}\,\text{cm})\left(10^4\,\frac{\mu\text{m}}{\text{cm}}\right) = 71.2\,\mu\text{m}
\end{aligned}$$

The alpha particles penetrate the arteriole wall and deposit energy about 50 μm into tumor cells beyond the arteriole wall. The neutron and photon radiation resulting from spontaneous fission and the 2.65-year half-life are negative attributes of this radionuclide since these radiation types irradiate the healthy tissue well beyond the arteriole wall.

Based on these comments, none of the listed nuclides is appropriate for the research. However, the results suggest that low-energy beta-, low-energy gamma-, and pure alpha-emitting radionuclides are candidates for further study. It should be noted that ^{32}P and ^{90}Y have been utilized in tumor vascular disruption applications.

(b) The following are desirable characteristics for the radionuclide incorporated into the nanoparticles to facilitate tumor blood vessel destruction:
1. The nuclide should have a short half-life.
2. The particle's range should be on the order of 20 μm to deliver maximum absorbed dose to the arteriole wall and limit the dose delivered to the healthy tissue.
3. The nanoparticle should have the capability to preferentially attach to the wall of an arteriole supplying blood to the tumor.
4. Radionuclides emitting low-energy photons, low-energy beta particles, and alpha particles appear to be the optimum radiation types for the desired application.
5. The nanoparticle design should facilitate removal from the body following completion of the therapy application.

(c) The dose rate from an activity (A) injected into the patient is obtained from a point isotropic source approximation:

$$\dot{D} = \frac{k}{4\pi r^2} A(0) e^{-\lambda t}$$

where

r	=	distance to the outer arteriole wall = 20 μm (assuming nanoparticle attachment to the interior wall)
	=	20.0×10^{-4} cm
k	=	^{113}I dose conversion coefficient = 364 pGy-cm^2/α
$A(0)$	=	the initial activity
λ	=	^{113}I disintegration constant = ln 2/T$_{1/2}$(^{113}I)
$T_{1/2}(^{113}\text{I})$	=	half-life of ^{113}I = 5.9 s
t	=	decay time

The total absorbed dose is obtained by integrating the dose rate with respect to time:

$$D = \int_0^{\infty} \dot{D}\text{d}t = \int_0^{\infty} \frac{k}{4\pi r^2} A(0) e^{-\lambda t} \text{d}t = \frac{k}{4\pi r^2} \frac{A(0)}{\lambda}$$

The absorbed dose (D) required to disrupt the arteriole wall is provided in the problem statement (100 Gy). This relationship is solved for the activity of material attached to the arteriole wall:

$$A(0) = \frac{4\pi r^2 D\lambda}{k} = \frac{4\pi (20.0 \times 10^{-4}\,\text{cm})^2 (100\,\text{Gy}) \left(\frac{\ln 2}{5.9\,\text{s}}\right)\left(\frac{\text{Bq-s}}{\text{dis}}\right)}{\left(364 \times 10^{-12}\,\frac{\text{Gy-cm}^2}{\alpha}\right)\left(\frac{\alpha}{\text{dis}}\right)}$$

$$= 1.62 \times 10^6\,\text{Bq}$$

The decay-corrected activity A_{DC} accounting for the time between milking the ^{113}I generator and injection (1 min) and attachment time following injection (1 min) is

$$A(0) = A_{\text{DC}}\, e^{-\lambda T}$$

where $T = 2$ min. Solving for A_{DC} leads to the desired activity

$$A_{\text{DC}} = A(0)\, e^{+\lambda T} = (1.62 \times 10^6\,\text{Bq})\, e^{+\left(\frac{\ln 2}{5.9\,\text{s}}\right)(2\,\text{min})\left(\frac{60\,\text{s}}{\text{min}}\right)} = 2.15 \times 10^{12}\,\text{Bq}$$

(d) Upon milking, the unshielded gamma absorbed dose rate 1 cm from the ^{113}I point isotropic source is

$$\dot{D} = \frac{A\Gamma}{r^2}$$

where A = activity = 2.15×10^{12} Bq, r = 1 cm, and Γ = gamma emission constant which is obtained from the relationship

$$\Gamma = 1.2 \times 10^{-13}\,\frac{\text{Gy-m}^2}{\text{Bq-h}} \sum_i E_i(MeV)\, Y_i$$

$$= 1.2 \times 10^{-13}\,\frac{\text{Gy-m}^2}{\text{Bq-h}}\,(0.463\,(1.0) + 0.622\,(1.0))$$

$$= 1.3 \times 10^{-13}\,\frac{\text{Gy-m}^2}{\text{Bq-h}}$$

Using these values, the unshielded absorbed dose rate is determined:

$$\dot{D} = \frac{A\Gamma}{r^2} = \frac{(2.15 \times 10^{12}\,\text{Bq})\left(1.3 \times 10^{-13}\,\frac{\text{Gy-m}^2}{\text{Bq-h}}\right)}{(0.01\,\text{m})^2} = 2.8 \times 10^3\,\text{Gy/h}$$

(e) ALARA provisions that should be implemented for milking the ^{113}I generator and producing the nanoparticles are as follows:

1. All operations should be performed remotely.
2. Draw the ^{113}I generator sample directly into a shielded container.
3. Since the nanoparticles are formed quickly, the production operation should be performed within the shielded container.
4. Provide training to staff involved in the operation to emphasize ALARA measures to minimize dose.
5. Practice the transfer operation using surrogate materials to ensure proficiency and minimize worker doses.

ALARA provisions that should be implemented for administering the isotope to the patient include the following:

1. Use a shielded vial for administering the radioactive material.
2. Utilize shielded bed shields to minimize staff dose.
3. Transfer the material from the generator to the patient using a shielded transport container.
4. Use shielded vests and face shields to minimize dose.
5. Provide training to staff involved in the operation to emphasize ALARA measures.
6. Practice the administration operation using surrogate materials to ensure proficiency.

The dose rate at the time ($t = 1$ min) of administration is

$$\dot{D} = \dot{D}(0)e^{-\lambda t} = \left(2.8 \times 10^3\,\frac{\text{Gy}}{\text{h}}\right) e^{-\left(\frac{\ln 2}{5.9\,\text{s}}\right)(1\,\text{min})\left(60\,\frac{\text{s}}{\text{min}}\right)} = 2.4\,\frac{\text{Gy}}{\text{h}}$$

The external doses are high and staff doses should be carefully evaluated and additional ALARA measures implemented. The use of robotic techniques to limit staff doses should also be investigated.

(f) The methodology used in the initial calculations is not sufficiently rigorous to finalize a microsphere design because the ^{113}I alpha particle range was not verified. Figure 5.8 illustrates a more rigorous calculation that determined the alpha particle energy needed to penetrate the arteriole wall. This figure illustrates that alpha particles having energies less than 3 MeV will not penetrate the target wall. The research group should have performed these calculations before proceeding. For a successful approach, the nuclides summarized in Table 5.14 and the required ranges summarized in Figure 5.8 should be used to select the candidate microsphere radionuclide.

5.8. (a) Assuming an isotropic emission of neutrons, the direct neutron fluence rate (flux) at a distance 3 m North of the target is given by the relationship

$$\phi = \frac{S}{4\pi r^2}$$

where

ϕ	=	neutron fluence rate (n/cm^2-s)
S	=	neutron emission source strength (n/s)
	=	$I\,Y\,k$
I	=	beam current = (200 μA) (1.0×10^{-6} A/μA) = 2.0×10^{-4} A
Y	=	neutron yield in the target/incident electron = 0.001 n/e
k	=	conversion factor = (1 C/s-A)(1 e/1.6×10^{-19} C)
S	=	(2.0×10^{-4} A)(0.001 n/e)(1 C/s-A)(1 e/1.6×10^{-19} C)
	=	1.25×10^{12} n/s
r	=	distance from the target = 3.0 m × 100 cm/m = 300 cm

These values complete the specification of the flux

$$\phi = \frac{(1.25\times10^{12}\,\text{n/s})}{(4\pi)(300\,\text{cm})^2} = 1.11\times10^{6}\,\frac{\text{n}}{\text{cm}^2\text{-s}}$$

(b) The reaction being described is ^{23}Na(n, γ)^{24}Na which is governed by the thermal neutron flux. For a thermal neutron fluence rate of 2×10^7 n/cm^2-s at one of the concrete walls, the activity of ^{24}Na in 1 cm^3 of concrete at saturation (assuming that the accelerator can run continuously) is given by the activation production relationship of Appendix B:

$$A_{\text{sat}} = N\sigma\phi$$

where A_{sat} is the saturation activity (Bq), N is the number of atoms in the target, σ is the microscopic cross-section (b/atom), and ϕ is the activating the flux (n/cm^2-s). The problem requests the specific activity (A'_{sat}) in Bq/cm^3 which is obtained from the previous relationship

$$A'_{\text{sat}} = \frac{A_{\text{sat}}}{V} = \frac{N\sigma\phi}{V} = n\sigma\phi$$

where n is the number of atoms per unit volume. The desired specific activity is obtained in terms of the parameters provided in the problem statement by rewriting this equation in terms of the macroscopic cross-section (μ):

$$\mu = n\sigma$$

$$A'_{\text{sat}} = n\sigma\phi = \mu\phi$$

A final algebraic manipulation using the density (ρ) provides the saturation activity per unit volume in terms of the information provided in the problem statement:

$$A'_{\text{sat}} = \mu \phi = \rho \frac{\mu}{\rho} \phi$$

where

ρ	=	number of grams of ^{23}Na per cm^3 of concrete = 0.012 g/cm^3 (100% of natural sodium is ^{23}Na)
μ/ρ	=	cross-section for the ^{23}Na(n, γ)^{24}Na reaction = 0.0139 cm^2/g
ϕ	=	thermal fluence rate = 2.0 × 10^7 n/cm^2-s

Using these values, the specific saturation activity concentration of ^{24}Na is

$$A'_{\text{sat}} = \left(0.012 \frac{\text{g}}{\text{cm}^3}\right)\left(0.0139 \frac{\text{cm}^2}{\text{g}}\right)\left(2.0 \times 10^7 \frac{\text{n}}{\text{cm}^2\text{-s}}\right)\left(1 \frac{\text{dis}}{\text{n}}\right) \times \left(\frac{\text{Bq-s}}{\text{dis}}\right) = 3.34 \times 10^3 \frac{\text{Bq}}{\text{cm}^3}$$

(c) After an irradiation time of 8 h, the ^{24}Na activity in 1 cm^3 of the concrete wall is

$$A'(t) = A'_{\text{sat}}(1 - e^{-\lambda t_{\text{irr}}})$$

where

A'_{sat}	=	saturation specific activity
λ	=	^{24}Na disintegration constant = 0.693/T$_{1/2}$ = 0.693/15.0 h = 4.62 × 10^{-2}/h
t_{irr}	=	irradiation time = 8 h

With these values, the ^{24}Na specific activity after 8 h of accelerator irradiation is

$$A'(t) = \left(3.34 \times 10^3 \frac{\text{Bq}}{\text{cm}^3}\right)\left(1 - e^{-\left(\frac{4.62 \times 10^{-2}}{\text{h}}\right)(8\,\text{h})}\right) = 1.03 \times 10^3 \frac{\text{Bq}}{\text{cm}^3}$$

(d) After an irradiation time of 8 h followed by a decay time of 8 h, the ^{24}Na activity in 1 cm^3 of a concrete wall is

$$A'(t) = A'_{\text{sat}}(1 - e^{-\lambda t_{\text{irr}}})e^{-\lambda t_{\text{d}}}$$

Using the values from the previous question, $\boldsymbol{A'(t)}$ becomes

$$A'(t) = \left(3.34 \times 10^3 \frac{\text{Bq}}{\text{cm}^3}\right)\left(1 - e^{-\left(\frac{4.62 \times 10^{-2}}{\text{h}}\right)(8\,\text{h})}\right) e^{-\left(\frac{4.62 \times 10^{-2}}{\text{h}}\right)(8\,\text{h})}$$

$$= 712 \frac{\text{Bq}}{\text{cm}^3}$$

(e) The ratio of saturation activities of ^{42}K and ^{24}Na is

$$\frac{A'_{sat}(^{42}\text{K})}{A'_{sat}(^{24}\text{Na})} = \frac{\rho(^{41}\text{K})\,a(^{41}\text{K})\,\frac{\mu}{\rho}(^{41}\text{K(n, }\gamma)^{42}\text{K})\,\phi}{\rho(^{23}\text{Na})\,a(^{23}\text{Na})\,\frac{\mu}{\rho}(^{23}\text{Na(n, }\gamma)^{24}\text{Na})\,\phi} = \frac{\rho(^{41}\text{K})\,a(^{41}\text{K})\,\frac{\mu}{\rho}(^{41}\text{K(n, }\gamma)^{42}\text{K})}{\rho(^{23}\text{Na})\,a(^{23}\text{Na})\,\frac{\mu}{\rho}(^{23}\text{Na(n, }\gamma)^{24}\text{Na})}$$

where a is the abundance of the specific isotope:

$$\frac{A'_{sat}(^{42}\text{K})}{A'_{sat}(^{24}\text{Na})} = \frac{\left(0.008\,\frac{\text{g}}{\text{cm}^3}\right)(0.0677)\left(1.22\times10^{-3}\,\frac{\text{cm}^2}{\text{g}}\right)}{\left(0.012\,\frac{\text{g}}{\text{cm}^3}\right)(1.0)\left(1.39\times10^{-2}\,\frac{\text{cm}^2}{\text{g}}\right)} = 3.96\times10^{-3}$$

(f) It would be inappropriate to use a bare, unmodified BF_3 proportional counter to measure the neutron flux at the North wall inside the accelerator room because the LINAC has a pulsed output. Since the proportional counter detects individual ^{10}B(n, α) neutron events, the LINAC's pulse rate and pulse width and the detector's resolving time may be such that the BF_3 counter detects the neutron pulse repetition frequency. Thus, the detector would count pulses and not neutron flux at the North wall. BF_3 counters can be built to measure neutrons in a pulsed field, but modifications to the basic circuitry are required. The following paper describes the issues associated with pulsed field measurements: M. Caresana *et al.*, A neutron detector for pulsed fields: Preliminary measurements, *Progress in Nuclear Science and Technology* **4,** 725 (2014).

The 25 MeV electrons produce photons, thermal neutrons, and fast neutrons. Detecting the neutron flux requires selectively determining both thermal and fast components while excluding the photon contribution to the BF_3 detector.

The BF_3 detector measures thermal neutrons via the ^{10}B(n, α) reaction. Therefore, the detector requires a modification to measure the fast neutron output from the accelerator. This modification includes the addition of a material with a large thermal neutron capture cross-section to remove the thermal neutrons and hydrogenous material to thermalize the remaining fast neutrons to permit their detection. The BF_3 detector will also require pulse height discrimination to distinguish between the neutron and gamma radiation types.

The geometry at the North wall and scattered radiation inside the accelerator room suggest that it may be difficult to discriminate between the various radiation types. This issue is also complicated by scattered radiation, which alters the energy of the neutron and gamma spectrum.

(g) After the initial 8 h run, your survey indicates that the measurements for the East, West, and North walls are the same. Immediately after the 8 h run, ^{24}Na, ^{42}K, and ^{59}Fe are produced. The activity of these thermal neutron activation products are as follows:

Activation product saturation specific activity:

^{24}Na:

$$A'_{\text{sat}}(^{24}\text{Na}) = \rho(^{23}\text{Na})\,a(^{23}\text{Na})\,\frac{\mu}{\rho}(^{23}\text{Na(n, }\gamma)^{24}\text{Na})\,\phi$$

$$= 3.34 \times 10^{4}\,\frac{\text{Bq}}{\text{cm}^3}\ \text{(From question (b))}$$

^{42}K:

$$A'_{\text{sat}}(^{42}\text{K}) = \rho(^{41}\text{K})\,a(^{41}\text{K})\,\frac{\mu}{\rho}(^{41}\text{K(n, }\gamma)^{42}\text{K})\,\phi$$

$$= \left(0.008\,\frac{\text{g}}{\text{cm}^3}\right)\left(2.0 \times 10^{7}\,\frac{\text{n}}{\text{cm}^2\text{-s}}\right)(0.0677)$$

$$\times \left(1.22 \times 10^{-3}\,\frac{\text{cm}^2}{\text{g}}\right)\left(1\,\frac{\text{dis}}{\text{n}}\right)\left(1\,\frac{\text{Bq-s}}{\text{dis}}\right) = 13.2\,\frac{\text{Bq}}{\text{cm}^3}$$

^{59}Fe:

$$A'_{\text{sat}}(^{59}\text{Fe}) = \rho(^{58}\text{Fe})\,a(^{58}\text{Fe})\,\frac{\mu}{\rho}(^{58}\text{Fe(n, }\gamma)^{59}\text{Fe})\phi = \left(0.018\,\frac{\text{g}}{\text{cm}^3}\right)$$

$$\times \left(2.0 \times 10^{7}\,\frac{\text{n}}{\text{cm}^2\text{-s}}\right)(0.0031)\left(3.01 \times 10^{-5}\,\frac{\text{cm}^2}{\text{g}}\right)$$

$$\times \left(1\,\frac{\text{dis}}{\text{n}}\right)\left(1\,\frac{\text{Bq-s}}{\text{dis}}\right) = 3.36 \times 10^{-2}\,\frac{\text{Bq}}{\text{cm}^3}$$

Activation product specific activity after 8 h of accelerator operation: The specific activity after 8 h is given by the relationship

$$A'(t) = A'_{\text{sat}}\,(1 - e^{-\lambda t_{\text{irr}}})$$

^{24}Na:

$$A'_{\text{sat}}(^{24}\text{Na}) = 1.03 \times 10^{3}\,\frac{\text{Bq}}{\text{cm}^3}\ \text{(From question (c))}$$

^{42}K:

$$\lambda = \frac{0.693}{T_{1/2}} = \frac{0.693}{12.4\,\text{h}} = \frac{5.59 \times 10^{-2}}{\text{h}}$$

$$A'(^{42}\text{K}) = \left(13.2\,\frac{\text{Bq}}{\text{cm}^3}\right)\left(1 - e^{-\left(\frac{5.59\times10^{-2}}{\text{h}}\right)(8\ \text{h})}\right) = 4.76\,\frac{\text{Bq}}{\text{cm}^3}$$

^{59}Fe:

$$\lambda = \frac{0.693}{T_{1/2}} = \frac{0.693}{45.6\ \text{days}} = \frac{1.52 \times 10^{-2}}{\text{day}}$$

$$A'(^{59}\text{Fe}) = \left(3.36 \times 10^{-2}\,\frac{\text{Bq}}{\text{cm}^3}\right)\left(1 - e^{-\left(\frac{1.52\times10^{-2}}{\text{day}}\right)(8\,\text{h})\left(\frac{1\ \text{day}}{24\,\text{h}}\right)}\right)$$

$$= 1.70 \times 10^{-4}\,\frac{\text{Bq}}{\text{cm}^3}$$

During testing, more scattering is expected since this operation is designed to verify beam performance and is not focused on scattered radiation. Testing is a short-duration activity and will produce additional large-angle scattering that will not occur during treatment operations. During treatment operations, most of the beam will strike the North wall since the physical interactions occur primarily in the beam direction.

Based on the testing results, the wall activity is dominated by ^{24}Na, a short-lived activation product. The thermal neutron fluence rates incident on all the walls will be similar because multiple scattering of the fast neutrons produces thermal neutrons. This results in a relatively isotropic thermal neutron distribution within the accelerator near the target. Accordingly, the measurements for the East, West, and North walls are essentially the same.

The other source of activity is from induced photonuclear reactions, which produce ^{22}Na with a 2.60-year half-life, ^{38}K with a 7.63 min half-life, and ^{55}Fe with a 2.75-year half-life. These reactions occur primarily in the forward beam direction at the North wall.

Based on the problem statement, the North wall does not have increased activity. This suggests these photonuclear reactions are not significant immediately after the 8 h initial accelerator run. After the 8 h run, the ^{22}Na and ^{55}Fe sources have not had sufficient time to accumulate and affect the North wall activity. ^{38} K will reach saturation but rapidly decays since it has a short 7.63 min half-life.

(h) After 5 years of operation, the accelerator is shut down and surveyed with a shielded pancake GM probe. The North wall shows activation on your survey, but the East and West walls do not. This result may be understood based on the discussion of the previous question. The North wall is preferentially irradiated because the photonuclear reactions occur preferentially in the beam direction, which irradiates the North wall.

After the 2-week shutdown, the short-lived isotopes ^{24}Na (15.0 h), ^{42}K (12.4 h), and ^{38}K (7.63 min) will not significantly contribute to the total activity. ^{59}Fe contributes minimal activity as noted in question (g). However, the 5 years of accelerator operation will permit the accumulation of ^{22}Na (2.60 years) and ^{55}Fe (2.75 years), and these isotopes occur predominantly in the North wall. These two isotopes lead to the elevated activity measured in the North wall.

A 25 MeV electron beam leads to bremsstrahlung production in the beam direction. The fact that the target is thin minimizes large-angle scattering, promotes the production of photons in the forward (North wall) direction, and minimizes their production at 90° (East and West walls). For these reasons, the North wall is expected to have an elevated level of activation.

Solutions for Chapter 6

6.1. (a) This question requires the calculation of the ^{238}U deposition rate to edible parts of plants from direct application of overhead irrigation. The deposition rate (r_d) is

$$r_d = \frac{C_W I_R T_v r_v}{Y_v}$$

where

C_W	=	^{238}U concentration in groundwater used for overhead irrigation = 1.85 Bq/l
I_R	=	irrigation rate = 2.5 l/m^2-day
T_v	=	translocation factor, transfer of radionuclides from plant surfaces to edible parts for nonleafy vegetables = 0.1
r_v	=	fraction of deposited activity retained on plant surfaces = 0.25
Y_v	=	plant yield (nonleafy vegetables) = 4 kg plant wet weight/m^2

Using these parameters, the deposition rate is determined:

$$r_d = \frac{\left(1.85\,\frac{\text{Bq}}{\text{l}}\right)\left(2.5\,\frac{\text{l}}{\text{m}^2\text{-day}}\right)(0.1)(0.25)}{\left(4\,\frac{\text{kg(plant wet)}}{\text{m}^2}\right)} = 0.0289\,\frac{\text{Bq}}{\text{kg(plant wet)-day}}$$

(b) This question requests that you assume a daily direct irrigation deposition rate to the edible parts of plants of 0.037 Bq/kg-day for ^{238}U and calculate the ^{238}U concentration in the plants at the end of the growing season (from direct deposition (dd) only). The ^{238}U concentration in the plants (C_p^{dd}) is calculated using the production equation concept of Appendix B with an associated production term (P):

$$P = T_v r_v r_d$$

$$C_p^{dd} = P \int_0^T e^{-\lambda t}\,dt$$

$$C_p^{dd} = T_v r_v r_d \int_0^T e^{-\lambda t}\,dt = T_v r_v r_d \frac{(1 - e^{-\lambda T})}{\lambda}$$

where

T_v	=	translocation factor, transfer of radionuclides from plant surfaces to edible parts for nonleafy vegetables = 0.1
r_v	=	fraction of deposited activity retained on plant surfaces = 0.25
r_d	=	daily direct irrigation deposition rate to the edible parts of plants for ^{238}U = 0.037 Bq/kg plant wet-day
λ	=	effective weathering and decay constant of uranium on plant surfaces = 0.12/day
T	=	crop growing period = 90 days

Using these values, the ^{238}U concentration in the plants is determined:

$$C_{\mathrm{p}}^{\mathrm{dd}} = (0.1)(0.25)\left(0.037\,\frac{\mathrm{Bq}}{\mathrm{kg\,(plant\,wet)\text{-}day}}\right)\frac{(1 - e^{-(0.12/\mathrm{day})(90\,\mathrm{days})})}{0.12\,/\mathrm{day}}$$

$$= 7.71 \times 10^{-3}\,\frac{\mathrm{Bq}}{\mathrm{kg(plant\,wet)}}$$

(c) In this question, you are to assume an equilibrium concentration of ^{238}U in the soil of 7.77 Bq/kg and calculate the plant concentration at the end of the growing season as a result of root uptake and resuspension (rur). The desired concentration due to root uptake and resuspension ($C_{\mathrm{p}}^{\mathrm{rur}}$) is

$$C_{\mathrm{p}}^{\mathrm{rur}} = C_{\mathrm{p}}^{\mathrm{soil}}\left(\frac{B}{W_{\mathrm{W-d}}} + \frac{T_{\mathrm{v}}\,r_{\mathrm{v}}}{M_{\mathrm{L}}}\right)$$

where

$C_{\mathrm{p}}^{\mathrm{soil}}$	=	equilibrium concentration of ^{238}U in the soil = 7.77 Bq/kg soil
B	=	soil-to-plant concentration factor (nonleafy vegetables dry weight basis) = 0.012 $\frac{\mathrm{kg(soil)}}{\mathrm{kg(plant\,dry)}}$
$W_{\mathrm{W-d}}$	=	wet-to-dry weight conversion factor = 0.25 $\frac{\mathrm{kg(plant\,wet)}}{\mathrm{kg(plant\,dry)}}$
M_{L}	=	mass-loading factor for resuspension to edible proportions (also termed the crop external contamination factor) = 0.1 $\frac{\mathrm{kg(plant\,wet)}}{\mathrm{kg(soil)}}$
T_{v}	=	translocation factor, transfer of radionuclides from plant surfaces to edible parts for nonleafy vegetables = 0.1
r_{v}	=	fraction of deposited activity retained on plant surfaces = 0.25

Using these values, the desired plant concentration from root uptake and resuspension is determined:

$$C_{\mathrm{p}}^{\mathrm{rur}} = \left(7.77\,\frac{\mathrm{Bq}}{\mathrm{kg\,(soil)}}\right)\left(\frac{0.012\,\frac{\mathrm{kg(soil)}}{\mathrm{kg(plant\,dry)}}}{0.25\,\frac{\mathrm{kg(plant\,wet)}}{\mathrm{kg(plant\,dry)}}} + \frac{(0.1)(0.25)}{\left(0.1\,\frac{\mathrm{kg(plant\,wet)}}{\mathrm{kg(soil)}}\right)}\right)$$

$$= 2.32\,\frac{\mathrm{Bq}}{\mathrm{kg(plant\,wet)}}$$

(d) Assuming a uranium concentration of 0.0296 Bq/kg plant wet weight ($C_{\mathrm{p}}^{\mathrm{dd}}$) from direct deposition and a concentration of 0.0592 Bq/kg plant wet weight ($C_{\mathrm{p}}^{\mathrm{rur}}$) from root uptake and resuspension, the effective dose to an individual from 1 year of produce consumption is to be determined. The total plant concentration on a wet weight basis $C_{\mathrm{p,total}}$ is the sum of the two intake pathways:

$$C_{\text{p, total}} = C_{\text{p}}^{\text{dd}} + C_{\text{p}}^{\text{rur}} = 0.0296\,\frac{\text{Bq}}{\text{kg(plant wet)}} + 0.0592\,\frac{\text{Bq}}{\text{kg(plant wet)}}$$

$$= 0.0888\,\frac{\text{Bq}}{\text{kg(plant wet)}}$$

The effective dose ($\dot{E}$) to an individual (per year) is determined from the total plant concentration:

$$\dot{E} = C_{\text{p, total}}\,Q(\text{DCF})$$

where

Q	=	consumption rate of produce = 50 kg (wet weight)/year
DCF	=	^{238}U (Type F) ingestion dose conversion factor = 4.4×10^{-8} Sv/Bq

Using these input values, the effective dose to an individual from 1 year of consuming produce is

$$\dot{E} = \left(0.0888\,\frac{\text{Bq}}{\text{kg (plant wet)}}\right)\left(50\,\frac{\text{kg (plant wet)}}{\text{year}}\right)\left(4.4 \times 10^{-8}\,\frac{\text{Sv}}{\text{Bq}}\right)$$

$$= \left(2.0 \times 10^{-7}\,\frac{\text{Sv}}{\text{year}}\right)\left(1.0 \times 10^{6}\,\frac{\mu\text{Sv}}{\text{Sv}}\right) = 0.20\,\frac{\mu\text{Sv}}{\text{year}}$$

(e) This question requires you to list factors that may influence plant uptake of uranium. These factors include:

1. Soil moisture content
2. Precipitation
3. Irrigation methods
4. Solar loading
5. Plant type
6. Plant growth rate
7. Ambient temperature
8. Chemical form of the uranium in the soil
9. Soil nutrients
10. Meteorology including wind speed and cloud cover
11. Soil type
12. Soil porosity
13. Use and type of fertilizers

(f) Other exposure pathways from the terrorist dispersal device that are not considered previously include:

1. Dust inhalation
2. Dust ingestion
3. Water runoff direct consumption
4. Water runoff → plants → animal feed → milk → man
5. Water runoff → plants → animal feed → meat → man
6. Water runoff → animal consumption → milk → man

7. Water runoff → animal consumption → meat → man
8. Direct consumption of contaminated groundwater
9. Water runoff → plants → man
10. Dust inhalation by animals → milk → man
11. Dust inhalation by animals → meat → man
12. Dust ingestion by animals → milk → man
13. Dust ingestion by animals → meat → man
14. Plants → animal feed → milk → man
15. Plants → animal feed → meat → man

(g) This question requests that you list methods of determining the concentration of uranium in the body and an advantage and disadvantage of each method. These methods include:

Uranium bioassay methods

Method	Advantage	Disadvantage
Urinalysis	Well-established method ICRP models provide the formalism to convert the intake into an effective dose Intake retention fractions are available to determine the intake	Requires worker participation and return of samples Long-term follow-up and evaluation are usually necessary to obtain reasonable accuracy Monitoring frequency depends on the solubility type of the uranium material Less likely to detect insoluble materials than fecal analysis
Fecal analysis	More likely to detect insoluble materials than urinalysis or *in vivo* counting ICRP models provide the formalism to convert the intake into an effective dose Intake retention fractions are available to determine the intake	Requires worker participation and return of samples Long-term follow-up and evaluation are usually necessary to obtain reasonable accuracy Esthetic and administrative challenges Monitoring frequency depends on the solubility type of the uranium material
In vivo monitoring/lung counting	No reliance on workers to return samples ICRP models provide the formalism to convert the intake into an effective dose Intake retention fractions are available to determine the intake	Detection of uranium X-rays is more difficult than higher-energy photon measurements Long-term follow-up and evaluation are usually necessary to obtain reasonable accuracy Monitoring frequency depends on the solubility type of the uranium material Sensitivity varies depending on the isotopic mixture

6.2. (a) The steady-state indoor radon concentration in the first floor living space is determined from a production equation:

$$\frac{dA'}{dt} = P' e^{-kt}$$

where

$\frac{dA'}{dt}$	=	time rate of change of activity per unit area
P'	=	production term
	=	constant radon flux into home per unit area = 0.074 Bq/m^2-s = J_i
k	=	total removal rate constant = $\lambda + F/v$
λ	=	physical decay constant for ^{222}Rn = ln(2)/$T_{1/2}$
$T_{1/2}$	=	^{222}Rn half-life = 3.82 days
λ	=	(0.693/3.82 days) (1 day/24 h) = 7.56×10^{-3}/h
F/V	=	ventilation removal constant = 0.5/h
V	=	free air volume of the home = $S\,H$
S	=	building area = 200 m^2
H	=	building room height = 2.5 m
V	=	200 m^2 × 2.5 m = 500 m^3
k	=	7.56×10^{-3}/h + 0.5/h = 0.508/h
t	=	time

The production relationship is determined by integrating over time:

$$dA' = P' e^{-kt}\, dt$$

$$A' = \int_0^T P' e^{-kt}\, dt = \frac{P'}{k}(1 - e^{-kT})$$

In the problem, steady-state conditions are implied ($T \to \infty$):

$$A' = \frac{P'}{k}$$

The activity A is the product of A' and S:

$$A = A'\, S$$

The concentration (C) is just the ratio of the activity (A) and free space volume (V):

$$C = \frac{A}{V} = \frac{A'\,S}{V} = \frac{P'\,S}{k\,V}$$

$$C = \frac{\left(0.074\,\frac{\text{Bq}}{\text{m}^2\text{-s}}\right)(200\,\text{m}^2)\left(3600\,\frac{\text{s}}{\text{h}}\right)}{\left(\frac{0.508}{\text{h}}\right)(500\,\text{m}^3)} = 210\,\frac{\text{Bq}}{\text{m}^3}$$

(b) In this part, you are to determine the Number of Working Level Months (NWLMs) per year assuming an occupancy factor (F) of 0.7 and a ^{222}Rn concentration of 0.518 Bq/l (C). The working level unit is defined in terms of English units, and the concentration is converted into those units:

$$C = \left(0.518\,\frac{\text{Bq}}{\text{l}}\right)\left(\frac{1\,\text{Ci}}{3.7\times10^{10}\,\text{Bq}}\right)\left(\frac{10^{12}\,\text{pCi}}{\text{Ci}}\right) = 14\,\frac{\text{pCi}}{\text{l}}$$

The desired annual exposure is determined using the relationship

$$\text{NWLM} = \frac{C_{\text{WL}}\,F}{k}$$

where

k	=	conversion factor
	=	$\left(\frac{170\,\text{WL-h}}{\text{WLM}}\right)\left(\frac{1\,\text{day}}{24\,\text{h}}\right)\left(\frac{1\,\text{year}}{365\,\text{days}}\right)$
F	=	occupancy factor = 0.7
C_{WL}	=	working level (WL) concentration
	=	$C\,(F_{\text{eq}})\,K$
C	=	^{222}Rn concentration = 14 pCi/l
F_{eq}	=	equilibrium factor = 0.4
K	=	conversion factor = 1 WL/(100 pCi/l) at 100% equilibrium
C_{WL}	=	(14 pCi/l) (0.4) (1 WL-l/100 pCi) = 0.056 WL

$$\text{NWLM} = \frac{(0.056\,\text{WL})(0.7)}{\left(\frac{170\,\text{WL-h}}{\text{WLM}}\right)\left(\frac{1\,\text{day}}{24\,\text{h}}\right)\left(\frac{1\,\text{year}}{365\,\text{days}}\right)} = 2.02\,\frac{\text{WLM}}{\text{year}}$$

(c) Sources of uncertainty in the application of the results from epidemiological studies of populations of underground miners to health effects in the general population include the following:

a. Many miners are heavy smokers which is not indicative of the general public.
b. Miners are subjected to dust, smoke, and diesel fumes that are inconsistent with the environment of a home.
c. The concentration of radon in the home is less than that in mines.
d. Miners inhaled dust containing uranium ores.
e. There are lifestyle differences between the miners and home owners.
f. The equilibrium factor in mines and homes are not the same.
g. The unattached fraction in mines and homes are not the same.
h. The general health of miners and home dwellers differ.
i. The diet of miners and home dwellers differ.
j. The population of miners and the general public population groups are not the same. For example, these groups do not have the same age and sex characteristics.
k. The occupancy factors for the home and miners are not the same.

(d) Sources of uncertainty in the dosimetry model for the respiratory tract as applied to risk estimates for radon exposures include:

a. Equilibrium factor
b. Unattached fraction
c. Radon and daughter concentrations
d. Dose conversion factors as a function of age
e. Metabolic model parameters as a function of age
f. Breathing mode (i.e., mouth breather vs. nose breather)
g. Breathing rate for the individual (e.g., resting, light activity, and exertion)
h. Shape and particle size of the inhaled aerosol
i. Receptor's age, gender, ethnic origin, and body size
j. Aerodynamic and thermodynamic diameters of the inhaled aerosols
k. Particle size distribution of the inhaled aerosol
l. Confounding factors including smoking, chemical pollutants, and environmental factors
m. Clearance models, removal efficiencies, and model parameters
n. Occupancy factor
o. Metabolic parameters and transfer coefficients
p. Assumed model and uncertainties in modeling parameters
q. Respiratory tract compartment specification and characteristics
r. Particle types in the initial and transformed states
s. Individual's state of health

(e) Methods to reduce the radon entry into a home or building are as follows:

a. Increase the air turnover rate.
b. Seal foundation and wall cracks.
c. Cover sumps.
d. Ventilate the soil gas.
e. Remove high radon bearing soil and replace with clean soil.
f. Use building materials with low radon emanation rates.
g. Use city water versus well water.
h. Use basement fans to remove high radon concentration air.

(f) 4 The best estimate of the water to air transfer factor for the reduction in concentration of radon in water (in Bq/l) to the indoor air concentration (in Bq/l) is ~10 000 to 1 reduction (i.e., a 10 000 Bq/l water concentration to a 1 Bq/l air concentration).

6.3. (a) In this question, you are to calculate the absorbed dose to the fetus. The fetal dose is determined from both X-ray and fluoroscopy procedures.

X-ray exposure: The trauma series of X-rays include one each of the head/neck, chest, abdomen, pelvis, and lumbar, thoracic, and cervical spine. All X-ray projections are anterior–posterior (AP). Following standard practice (see Kereiakes and Rosenstein, Handbook of

Radiation Doses in Nuclear Medicine and Diagnostic X-rays, CRC Press, Inc., Boca Raton, FL (1980)), the following facts are noted for the performed X-ray procedures:

Chest	Minimal fetal dose contribution expected (*)
Pelvic	Fetal dose contribution ($i = 1$)
Head/neck	Minimal fetal dose contribution expected (*)
Abdomen	Fetal dose contribution ($i = 2$)
Lumbar spine	Fetal dose contribution ($i = 3$)
Thoracic spine	Minimal fetal dose contribution expected (*)
Cervical spine	Minimal fetal dose contribution expected (*)

Examinations annotated with a "*" are excluded because they are physically removed from the fetal location. The included procedures ($i = 1$, 2, and 3) directly irradiate the fetus.

Using the tabulated data and selected procedures permits the calculation of the fetal absorbed dose from X-rays ($D_{\text{X-ray}}$):

$$D_{\text{X-ray}} = \sum_{i=1}^{3} \text{ESE}_i f\,k = f\,k \sum_{i=1}^{3} \text{ESE}_i$$

where

ESE_i	=	entrance skin exposure for the ith procedure
i	=	procedure index
ESE_1	=	pelvic AP X-ray ESE = 0.79×10^{-4} C/kg
ESE_2	=	abdominal AP X-ray ESE = 1.08×10^{-4} C/kg
ESE_3	=	lumbar spine X-ray ESE = 1.40×10^{-4} C/kg
f	=	fraction of ESE contributing to the fetal dose = 0.45
k	=	conversion factor = 34 Gy-kg/C

Using these values, the X-ray contribution to the fetus is determined:

$$D_{\text{X-ray}} = (0.45)\left(34\,\frac{\text{Gy-kg}}{\text{C}}\right)(0.79 + 1.08 + 1.40)$$
$$\times 10^{-4}\,\frac{\text{C}}{\text{kg}}\left(1000\,\frac{\text{mGy}}{\text{Gy}}\right) = 5.00\,\text{mGy}$$

Fluoroscopy exposure: The fetal dose from the fluoroscopy procedure is obtained from the relationship

$$D_{\text{fluoro}} = \text{ESE}_{\text{fluoro}}\,I f\,t$$

where

D_{fluoro}	=	fluoroscopy absorbed dose contribution to the fetal dose
$\text{ESE}_{\text{fluoro}}$	=	fluoroscopy entrance skin exposure = 14.9 mGy/mA-min
I	=	fluoroscopy current = 2 mA
t	=	patient fluoroscopy exposure time = 4 min
f	=	fraction of ESE contributing to the fetal dose = 0.45

Using these values, the fluoroscopy contribution to the fetus is determined:

$$D_{\text{fluoro}} = \left(14.9\,\frac{\text{mGy}}{\text{mA-min}}\right)(2\,\text{mA})(4\ \text{min})(0.45) = 53.6\,\text{mGy}$$

The total fetal absorbed dose (D) is the sum of the X-ray and fluoroscopy components:

$$D = D_{\text{X-ray}} + D_{\text{fluoro}} = 5.00\,\text{mGy} + 53.6\,\text{mGy} = 58.6\,\text{mGy}$$

(b) Assuming that the radiation dose calculated in question (a) was 35 mGy, the advice based on NCRP 174, *Preconception and Prenatal Radiation Exposure: Health Effects and Protective Guidance*, that you would give the woman's physician regarding terminating the pregnancy or letting it proceed includes the following:
   1. The risk to the fetus for deterministic effects (e.g., fetal loss, malformations, growth retardation, and mental retardation) from prenatal exposure is generally low. At absorbed doses to the fetus below 0.1 Gy, there may be no increased risk based on mammalian animal studies.
   2. For the third to fifth week postconception, the minimum human acute lethal dose for the embryo is estimated to be in the range of 0.25–0.50 Gy. This dose range is derived from animal studies.
   3. The 35 mGy dose is below the no-adverse-effect level for the induction of birth defects.
   4. For the third to fifth week postconception, greater than 0.5 Gy is needed for the induction for major malformations.
   5. The growth retardation no-adverse-level is in the range of 0.2–0.5 Gy.

   Based on the NCRP 174 information, the calculated dose is below the thresholds for observable fetal health effects. Therefore, the NCRP 174 compilation suggests the pregnancy should proceed normally.

(c) Information necessary to determine the risk of injury to the fetus in this incident includes:
   1. Time of the examination postconception
   2. Type of procedure performed (e.g., X-ray, fluoroscopy, or CT)
   3. Beam quality of the X-ray machine or exposure parameters for fluoroscopy

4. Tube potential peak kilovoltage of the machine for each examination configuration
5. Beam current for each examination configuration
6. Exposure time for each examination configuration
7. Beam size
8. Source-skin distance for each examination configuration
9. Entrance Skin Exposure for each examination configuration
10. Exam orientation (e.g., AP, PA, and LAT) for each examination configuration
11. Fetal dose conversion coefficients for the entrance skin exposure configurations
12. Survey data and machine parameters for the X-ray and fluoroscopy examinations
13. Actual technique factors (mA-s) for each examination configuration

(d) During subsequent communications with the woman's physician, you are informed that she appeared to have skin burns. The patient dose may be viewed in terms of five generic factors:

- X-ray source factors
- Patient/geometry factors
- Image intensifier factors
- Computer factors.
- Administrative factors

A discussion of these factors provides a possible cause for the skin burns.

As a point of reference, fluoroscopy dose rates are normally in the range of 10–50 mGy/min. In the high output mode, the dose rates are between 100 and 200 mGy/min. Skin injuries vary with cumulated dose. Temporary epilation occurs at about 3 Gy. Erythema occurs at about 6 Gy. Between 15 and 20 Gy, moist desquamation, dermal necrosis, and secondary ulceration occur.

X-ray source factors:

1. *X-ray tube output*: As the X-ray output (mGy/min) increases, the dose increases.
2. *Tube current (mA)*: An increase in current leads to a larger dose.
3. *Activating pedal operation for the fluoroscopy unit*: The pedal could be equipped with high output or low output mode options. Patient dose is significantly reduced by intermittent pedal operation.
4. *Beam energy*: As the tube voltage (kVp) and beam filtration decrease, the skin dose increases.
5. *Beam filtration*: If the beam filtration increases, the dose decreases. The beam also hardens with tube filtration and becomes more penetrating. As the tube filtration increases, the dose decreases.

6. *Tube head and collimator leakage*: An increase in leakage causes the secondary barrier dose to increase.
7. *Automatic brightness control (ABC)*: The ABC senses light output from the screen phosphor. If the light output is low, the X-ray output will increase which subsequently increases the dose to maintain the desired brightness.
8. *Software*: The software interpreting the input instructions could contain errors within the code.
9. *Number of images*: As the number of radiographs increases, the dose increases. The radiologist may have taken more radiographs than anticipated.
10. *Scattering field size*: The patient dose is composed of primary and scattered radiation. As the field size increases, the dose increases. The increase in field size adds an additional scattering component that increases the dose.

Patient/geometry factors:

1. *Patient thickness/density*: The patient's body characteristics affect the delivered dose.
2. *Source/patient distance*: If all other factors are constant, increasing distance decreases patient dose.
3. *Patient/image intensifier distance*: Image quality and associated patient dose are affected by the distance between the patient and the image intensifier.
4. *Attenuation*: Postpatient attenuation includes the tabletop and scatter rejection grids. The use of a scatter rejection grid placed below the patient increases the patient dose. The magnitude of the dose (D) increase is proportional to the bucky factor (BF):

$$D(\text{with grid}) = \text{BF} \times D(\text{no grid})$$

Image intensifier factors:

1. *Fluoroscopy gain*: As the gain decreases the patient dose increases. The gain reflects the operation of the image intensifier. The total gain is the product of the acceleration gain and the minification gain. The acceleration gain is due to the accelerating potential, and the minification gain = (input diameter/output diameter)^2.
2. *Magnification*: Increasing the magnification of the image intensifier increases the dose. If half the size is imaged, the dose increases by a factor of 4.
3. *Digital acquisition*: An increase in digital acquisition increases the dose.

4. *Run duration*: As the cinema run duration increases, the dose increases.
5. *Number of runs*: The dose increases with the number of cinema runs.
6. *Frame length*: The frame length may have been longer than assumed. As the frame length increases, the dose increases.
7. *Beam width*: The beam width may be different than assumed. The F# is defined as the ratio of the focal length to the beam width. As the F# increases, the light gathering decreases, the lens speed decreases, and the dose increases. The F# is usually fixed, but it controls the brightness on some older units.

Computer factors:

1. *Recording sensitivity*: Digital recording sensitivity offers a postprocessing capability that can significantly lower the dose.
2. *Pixel size*: The use of a smaller pixel size increases the dose. A smaller pixel size means that fewer photons enter the pixel. Accordingly, there is more noise in the image, and additional dose is required to obtain a clear image that meets the radiologist's requirements.
3. *Frame averaging*: Frame averaging reduces the current (mA) or exposure (mA-s) which reduces the dose.

Administrative factors:

1. *Procedures*: The lack of operating procedures has the potential to increase the dose especially for inexperienced personnel.
2. *Safety awareness*: The actions of personnel that lack radiation safety awareness or concern for patient and staff exposures will lead to increased dose.
3. *Training*: The lack of effective training increases doses. The "See one, do one, teach one" approach increases doses.
4. *Maintenance and testing*: Proper machine preventive maintenance, calibration, periodic servicing, and knowledge of the equipment status lead to lower doses.
5. *Radiation monitoring*: Equipment should be monitored by a qualified radiological physicist. Ignoring this factor increases dose.

(e) Machine parameters that affect fetal radiation exposure from CT, X-ray, or fluoroscopy include:

1. Field size
2. Type of procedure performed
3. Beam quality of the X-ray machine or exposure parameters for fluoroscopy
4. Tube potential (kVp) of the machine
5. Beam current

6. Exposure time
7. Beam size
8. Source-skin distance
9. Entrance Skin Exposure
10. Exam orientation (e.g., AP, PA, and LAT)
11. Beam quality (HVL)

Since the patient was involved in an air crash, the records of the accident should be reviewed to determine if the burns are associated with a fire occurring after the crash. If the burns are attributable to the fluoroscopy examination, the event should be reviewed to minimize the probability of its recurrence.

The 35 mGy absorbed dose from question (b) is insufficient to produce skin burns. If the burns are found to be attributable to fluoroscopy, the dose assessment and exam information should be carefully reviewed to resolve the inconsistency between the calculated skin dose and observed skin burns.

6.4. (a) *Assumptions*:

1. A point source approximation is appropriate for this source–receptor configuration. Since the sources are much smaller than the receptor distance, this is a reasonable approximation.
2. The sources offer minimal self-shielding and attenuation of the ^{60}Co activity.
3. There is minimal energy degradation of the ^{60}Co photons as they traverse the distance between the source and the point of interest.
4. The crane cab provides minimal shielding. This assumption is necessary because no crane data are provided.

The absorbed dose rate from a point source shielded by a material is given by the relationship

$$\dot{D}(r) = \left(\frac{A\Gamma}{r^2}\right) B e^{-\mu x}$$

where

$\dot{D}(r)$	=	absorbed dose rate at distance r from the point source
A	=	source activity
Γ	=	gamma constant for $^{60}Co = 3.1 \times 10^{-4}$ mGy-m^2/h-MBq
r	=	distance from the point source
$B(\mu x)$	=	gamma-ray buildup factor
μ	=	linear attenuation coefficient
x	=	shielding thickness

Since the source activity is not known, the underwater data must be used to determine the dose rate in the crane cab. In this problem, two cases are presented. The first involves the absorbed dose rate at a distance of 1.0 m (r_1) when the sources are shielded by 100 cm (x_1) of water. In Case 1 the following parameters are given:

$\dot{D}_1(r_1)$	=	30 mGy/h
r_1	=	1.0 m
x_1	=	1.0 m
μ_1	=	^{60}Co attenuation coefficient for water = 0.0707/cm
$\mu_1 x_1$	=	(100 cm) (0.0707/cm) = 7.07
$B_1(7.07) \approx B_1(7)$	=	15.8

The second case involves the irradiation of a worker a distance of 10.0 m in air when the sources are submerged in 1.0 m of water. In Case 2, the following parameters are given:

r_2	=	11 m
x_2^{air}	=	10.0 m
x_2^{water}	=	1.0 m
μ_2^{air}	=	attenuation coefficient for air = 7.75×10^{-5}/cm
μ_2^{water}	=	attenuation coefficient for water = 0.0707/cm
$\mu_2 x_2$	=	$\mu_2^{\text{air}} x_2^{\text{air}} + \mu_2^{\text{water}} x_2^{\text{water}}$
$\mu_2 x_2$	=	(1000 cm)(7.75×10^{-5}/cm) + (100 cm)(0.0707/cm)
	=	0.0775 + 7.07 = 7.15

The buildup factor for air is about unity. A more refined value is obtained by interpolation between $\mu x = 0$ and 0.5

$$B_{2\,\text{air}}(\mu x) = B_0 + \frac{\mu x - \mu_0 x_0}{\mu_2 x_2 - \mu_0 x_0}(B_2 - B_0)$$
$$= 1.0 + \frac{0.0775 - 0.0}{0.5 - 0.0}(1.47 - 1.00) = 1.07$$

The buildup factor for water was found in Case 1 to be 15.8. The total buildup factor for Case 2 is a complex calculation that can be conservatively approximated to first order using the relationship

$$B_2 = B_2^{\text{air}} + B_2^{\text{water}} = 1.07 + 15.8 = 16.9$$

These parameters and assumptions permit the dose to the cab to be determined. This determination is possible because $A\Gamma$ is a constant. Solving for $A\Gamma$ in the point source equation

$$\dot{D}(r) = \left(\frac{A\Gamma}{r^2}\right) B e^{-\mu x}$$

$$A\Gamma = \frac{\dot{D}(r) r^2\, e^{+\mu x}}{B}$$

In these equations, μx represents the total attenuation. In Case 1, 1.0 m of water provides the attenuation. The Case 2 attenuation is provided by 1.0 m of water and 10.0 m of air.

The values for $A\Gamma$ from Cases 1 and 2 are equated to determine $\dot{D}_2(r_2)$:

$$\frac{\dot{D}_1(r_1)r_1^2\, e^{+\mu_1 x_1}}{B_1} = \frac{\dot{D}_2(r_2)r_2^2\, e^{+\mu_2 x_2}}{B_2}$$

The desired dose rate is obtained by solving for $\dot{D}_2$ and using the available data:

$$\dot{D}_2(r_2) = \frac{\dfrac{\dot{D}_1(r_1)r_1^2\, e^{+\mu_1 x_1}}{B_1}}{\dfrac{r_2^2\, e^{+\mu_2 x_2}}{B_2}} = \dot{D}_1(r_1)\left(\frac{r_1}{r_2}\right)^2 \left(\frac{B_2}{B_1}\right) \frac{e^{+\mu_1 x_1}}{e^{+\mu_2 x_2}}$$

$$\dot{D}_2(r_2) = \left(30\,\frac{\text{mGy}}{\text{h}}\right)\left(\frac{1.0\,\text{m}}{11.0\,\text{m}}\right)^2 \left(\frac{16.9}{15.8}\right)\left(\frac{e^{7.07}}{e^{7.15}}\right) = 0.245\,\frac{\text{mGy}}{\text{h}}$$

(b) Using the notation from the previous problem, the absorbed dose rate in the crane cab when the sources are lifted to the water surface is obtained from the relationship

$$\dot{D}_2(r_2) = \frac{\dfrac{\dot{D}_1(r_1)r_1^2\, e^{+\mu_1 x_1}}{B_1}}{\dfrac{r_2^2\, e^{+\mu_2 x_2}}{B_2}} = \dot{D}_1(r_1)\left(\frac{r_1}{r_2}\right)^2 \left(\frac{B_2}{B_1}\right) \frac{e^{+\mu_1 x_1}}{e^{+\mu_2 x_2}}$$

where

r_1	=	1.0 m (water)
$\mu_1 x_1$	=	7.07
B_1	=	15.8
r_2	=	10 m (air)
$\mu_2 x_2$	=	0.0775
B_2		1.07

$$\dot{D}_1(r_1) = 30\,\frac{\text{mGy}}{\text{h}}$$

$$\dot{D}_2(r_2) = \left(30\,\frac{\text{mGy}}{\text{h}}\right)\left(\frac{1.0\,\text{m}}{10.0\,\text{m}}\right)^2 \left(\frac{1.07}{15.8}\right)\left(\frac{e^{7.07}}{e^{0.0775}}\right) = 22.1\,\frac{\text{mGy}}{\text{h}}$$

(c) The ^{137}Cs activity in the small rod is determined following the methodology and notation of the previous problem:

$$\dot{D}(r) = \left(\frac{A\Gamma}{r^2}\right) B e^{-\mu x}$$

A point source approximation is used to approximate the source because no dimensions are provided and the rod is described in the problem statement as being small. From the point source equation, the activity A is determined by algebraic manipulation:

$$A = \frac{\dot{D}(r)\, r^2 e^{+\mu x}}{\Gamma B}$$

where

$\dot{D}(r) = 100\,\text{mGy/h}$
$r = 30\,\text{cm}$
$x = 30\,\text{cm of air}$
$\mu = 1.00 \times 10^{-4}/\text{cm}$
$\mu x = (1.00 \times 10^{-4}/\text{cm})\,(30\,\text{cm}) = 3.00 \times 10^{-3}$
Since $\mu x \ll 1$, $B(\mu x) \approx 1$
Γ = ^{137}Cs gamma constant, which is approximated by the relationship

$$\Gamma = \frac{1}{2} E Y$$

which yields Γ in units of R-m^2/h-Ci

E = gamma-ray energy (MeV) = 0.662 MeV
Y = gamma-ray yield = 0.851

$$\Gamma = \frac{1}{2}(0.662)(0.851)\left(\frac{\text{R-m}^2}{\text{h-Ci}}\right)\left(\frac{0.877\,\text{rad}}{\text{R}}\right)\left(\frac{1\,\text{Gy}}{100\,\text{rad}}\right)$$
$$\times\left(\frac{1000\,\text{mGy}}{\text{Gy}}\right)\left(\frac{1\,\text{Ci}}{3.7 \times 10^4\,\text{MBq}}\right) = 6.7 \times 10^{-5}\,\frac{\text{mGy-m}^2}{\text{MBq-h}}$$

Using these values the activity of the small rod is determined:

$$A = \frac{\dot{D}(r)\,r^2 e^{+\mu x}}{\Gamma B}$$

$$A = \frac{\left(100\,\frac{\text{mGy}}{\text{h}}\right)(0.3\,\text{m})^2\,e^{+0.003}}{\left(6.7 \times 10^{-5}\,\frac{\text{mGy-m}^2}{\text{MBq-h}}\right)(1.0)} = 1.4 \times 10^5\,\text{MBq}$$

(d) The dose rate from a 3.0 m long, thin-walled, 1.0 cm diameter pipe is 9 mGy/h at 1 m from the midpoint. In this question, you are requested to calculate the activity per unit length of the pipe.
Assumptions:
1. The problem configuration will be sufficiently represented by a line source. This is reasonable based on the geometry.
2. There is no attenuation by the pipe wall.
3. There is no attenuation by any material residing within the pipe.
4. There is no energy degradation as the ^{60}Co photons penetrate the pipe material, pipe wall, and air. This permits the use of the ^{60}Co gamma constant.

The absorbed dose rate is given by the line source relationship provided in Appendix C:

$$\dot{D} = \frac{C_L \Gamma \theta}{w}$$

which can be solved for the concentration per unit length (C_L):

$$C_L = \frac{\dot{D} w}{\Gamma \theta}$$

where

$\dot{D}$	=	absorbed dose rate = 9 mGy/h
w	=	perpendicular distance from the line source = 1.0 m
Γ	=	^{60}Co gamma-ray constant = 3.1×10^{-4} mGy-m^2/h-MBq
θ	=	included angle

$$\tan\frac{\theta}{2} = \frac{1.5\,\text{m}}{1.0\,\text{m}} = 1.5$$

$$\frac{\theta}{2} = \tan^{-1}(1.5) = 56.3° \,\frac{\pi}{180°} = 0.982$$

$$\theta = 1.96$$

Using these values, the activity per unit length is determined:

$$C_L = \frac{\dot{D} w}{\Gamma \theta} = \frac{\left(9.0\,\frac{\text{mGy}}{\text{h}}\right)(1.0\,\text{m})}{\left(3.1 \times 10^{-4}\,\frac{\text{mGy-m}^2}{\text{MBq-h}}\right)(1.96)} = 1.5 \times 10^4\,\frac{\text{MBq}}{\text{m}}$$

(e) In order to determine the required shielding, the source activity must be determined. To facilitate this calculation, the following assumptions are made:

1. A point source approximation is valid for the small source.
2. The small source size provides minimal self-shielding of the ^{60}Co source.
3. There is no energy degradation as the ^{60}Co photons traverse the air and added lead shielding.
4. The attenuation of the air is negligible.

The absorbed dose rate from an unshielded point source is given by

$$\dot{D}(r) = \frac{A\Gamma}{r^2}$$

where

A	=	activity of ^{60}Co residing within the source
$\dot{D}(r)$	=	1.5 mGy/h
Γ	=	^{60}Co gamma-ray constant = 3.1×10^{-4} mGy-m^2/h-MBq
r	=	distance from the source where the measured absorbed dose rate is obtained = 1.0 m

With these values, the activity residing within the source is obtained:

$$A = \frac{\dot{D}(r)\, r^2}{\Gamma}$$

$$A = \frac{\left(1.5\,\frac{\text{mGy}}{\text{h}}\right)(1.0\,\text{m})^2}{\left(3.1 \times 10^{-4}\,\frac{\text{mGy-m}^2}{\text{MBq-h}}\right)} = 4.8 \times 10^3\,\text{MBq}$$

The dose rate a distance of 30 cm (r) from the source when shielded by lead is

$$\dot{D}(r) = \left(\frac{A\Gamma}{r^2}\right) B e^{-\mu x}$$

where

B	=	buildup factor for lead
μ	=	linear attenuation coefficient for lead = 0.679/cm
X	=	thickness of lead shielding to yield an absorbed dose rate that does not exceed 1.0 mGy/h at 30 cm from the source

The unshielded absorbed dose rate at 30 cm must be calculated to determine how much lead is required:

$$\dot{D}(r) = \frac{A\Gamma}{r^2} = \frac{(4.8 \times 10^3\,\text{MBq})\left(3.1 \times 10^{-4}\,\frac{\text{mGy-m}^2}{\text{MBq-h}}\right)}{(0.3\,\text{m})^2} = 17\,\frac{\text{mGy}}{\text{h}}$$

The absorbed dose must be reduced by a factor of 17 to meet the 1 mGy/h requirement. As a first estimate of the required shielding thickness, ignore buildup to determine the number (N) of half-value layers (HVL) required for a factor of 17 reduction in the absorbed dose rate. The required attenuation is

$$2^N = 17$$

$$N \ln(2) = \ln(17)$$

$$N = \frac{\ln(17)}{\ln(2)} = 4.09$$

The HVL thickness is obtained from the ^{60}Co attenuation coefficient:

$$\text{HVL} = -\frac{\ln(0.5)}{\mu} = \frac{0.693}{\frac{0.679}{\text{cm}}} = 1.02\,\text{cm}$$

Two lead blankets would be 2 × 2.5 cm thick or 5.0 cm thick, which does not include the PVC covering. A consideration of attenuation suggests that about two blankets would give the desired reduction $\left(\frac{1}{2}\right)^5$. However, this does not consider buildup. Therefore, as a first guess, calculate the attenuation for three blankets to account for buildup.

For three blankets, the equivalent lead thickness is:

$$x = 3\,(2.5\,\text{cm}) = 7.5\,\text{cm}$$
$$\mu x = (0.679/\text{cm})\,(7.5\,\text{cm}) = 5.09$$

$B(5.09)$ is reasonably approximated by $B(5) = 10.0$. Buildup for air is used which is consistent with standard practice. This is also conservative since the buildup factor for air is larger than the corresponding lead buildup factor. Using these values, the shielded absorbed dose rate corresponding to three blankets covering the source is

$$\dot{D}(r) = \left(\frac{A\Gamma}{r^2}\right) B e^{-\mu x}$$

$$\dot{D}(r) = \left(\frac{(4.8 \times 10^3\,\text{MBq})\left(3.1 \times 10^{-4}\,\frac{\text{mGy-m}^2}{\text{MBq-h}}\right)}{(0.3\,\text{m})^2}\right)(10)\,e^{-\left(\frac{0.679}{\text{cm}}\right)(7.5\ \text{cm})}$$

$$= \left(17\,\frac{\text{mGy}}{\text{h}}\right)(10)(6.14 \times 10^{-3}) = 1.0\,\frac{\text{mGy}}{\text{h}}$$

Within the range of accuracy of the gamma constant, the three-lead-blanket case is the desired result. By adding a fourth blanket, the absorbed dose would be reduced below the desired 1 mGy/h value and would accommodate the 15–20% inherent inaccuracy of the gamma constant.

(f) With law enforcement support meet with the concerned stakeholders and suggest a visit to the crash site. Prior to the site visit, provide the stakeholders basic instruction on the operation of portable radiation instrumentation including their response to ^{60}Co and ^{137}Cs sources. With law enforcement safety oversight have the stakeholders perform a survey of the crash site to confirm the contention that the area is free of radioactive material.

6.5. (a) State and/or local government response actions are derived from the Three Mile Island Emergency Response Plan. Actions appropriate for Alert, Site Area Emergency, and General Emergency classifications are provided.

Alert:

1. Provide fire, rescue, ambulance, and security assistance as required.
2. Prepare for reclassification or closeout of the emergency.
3. Notify elected officials of the emergency conditions.

4. Augment resources and bring the primary response centers and Emergency Broadcast System to a standby status.
5. Place key emergency response personnel including monitoring teams and associated communications systems in a standby status.
6. Place route alert teams on an advanced state of readiness.
7. As warranted, provide confirmatory off-site radiation monitoring.
8. Review resource needs and supplement as warranted.
9. Provide radiation monitoring results to the utility, Department of Energy, and other organizations and jointly evaluate these data.

Site Area Emergency:

1. Verify all Alert actions have been implemented.
2. Activate public notification systems and provide periodic status updates.
3. If sheltering near the reactor site is warranted, utilize public notification systems within at least 2 mi of the site to announce this protective action.
4. Issue notification of the declaration of a Site Area Emergency to the risk counties.
5. Verify that the risk counties have disseminated public information materials regarding protective actions.
6. Verify that the risk counties have sounded their emergency sirens.
7. Augment resource and response capability by activating Emergency Operations Centers.
8. Dispatch key emergency personnel including radiation monitoring teams.
9. Alert additional emergency personnel (e.g., those needed for evacuation) to standby status and dispatch personnel (e.g., first aid, vehicle response, and fuel services) to their duty stations.
10. Verify that the risk counties have notified county and local government heads, key staff, emergency response forces, volunteer organizations, schools, hospitals, nursing homes, businesses, and industry of the incident and the possible need for protective actions.
11. Place the Emergency Broadcast System on standby status and activate as necessary.
12. Issue dosimetry, KI, and radiation survey instrumentation to emergency personnel.
13. Place reception and mass care centers on standby status.
14. Continuously evaluate radiological and plant information from the utility and radiological information from monitoring teams.

15. Based on plant and radiological updates, issue updated protective actions.
16. Provide radiation monitoring results to the utility, Department of Energy, and other organizations and jointly evaluate these data.
17. Recommend placing milk-producing animals within 2 miles of the site on stored feed and periodically assess the need to extend this distance.
18. Evaluate the need for the sheltering of animals and the collection and disposal of contaminated farm products.
19. Provide press briefings. Some of these should be held jointly with other organizations including the utility, NRC, and DOE.
20. Establish a press office to serve as the principal point of contact with utility and government communications organizations. The press office will facilitate information exchange with these and other communications organizations.
21. Maintain Site Area Emergency status until emergency closeout or reclassification.

General Emergency:

1. Provide assistance as requested.
2. Activate the process for public notification of this emergency classification and provide periodic updates.
3. Direct risk counties to activate their warning sirens.
4. Utilize the Emergency Broadcast System as a medium for public information dissemination and protective action updates.
5. Review and update protective action bulletins as warranted.
6. Fully activate all emergency response personnel and response activities.
7. Evaluate and assign resources to support risk county needs.
8. Dispatch key emergency response personnel including radiation monitoring teams and support and other personnel to their assigned duty stations.
9. Ensure status information is provided to state and county services that will be needed to support protective actions including evacuation.
10. Advise fire and ambulance organizations of the emergency condition in effect and have them staff their duty stations.
11. Provide radiation monitoring results to the utility, Department of Energy, and other organizations and jointly evaluate these data.
12. Continuously evaluate radiological and plant information from the utility and radiological information from monitoring teams.
13. Based on plant and radiological updates, issue updated protective actions.

14. Recommend placing milk-producing animals within 10 miles of the site on stored feed and periodically assess the need to extend this distance.
15. Evaluate the need for the sheltering of animals and the collection and disposal of contaminated farm products.
16. Provide press briefings. Some of these should be held jointly with other organizations including the utility, NRC, and DOE.
17. Maintain General Emergency status until emergency closeout or reclassification.

(b) The utility has escalated to a Site Area Emergency after failure of the fuel fission product barrier. However, no release is in progress, and the utility projects no release will occur. The utility recommends sheltering in place as the protective action recommendation. Given these conditions, sheltering is currently appropriate. However, failure of the fuel fission product barrier is a significant concern. The condition of the remaining fission product barriers and any associated releases must be carefully monitored to determine if evacuation or issuance of KI is warranted.

(c) Question (b) notes that fuel failure has occurred. This suggests that fission products are available for release. With an airborne release in progress, additional fission product barrier failures have occurred. Given these circumstances, noble gases and radioiodine are the dominant concerns. Specific radionuclides available for release include:

Noble gases: ^{85}Kr, ^{88}Kr, ^{133}Xe, ^{133m}Xe, ^{135}Xe, and ^{135m}Xe

Radioiodine: ^{129}I, ^{131}I, and ^{133}I

Cesium and strontium: ^{89}Sr, ^{90}Sr, ^{134}Cs, ^{136}Cs, ^{137}Cs, and ^{138}Cs

Activation products: ^{3}H, ^{58}Co, and ^{60}Co

Transuranics: Unless there were significant breaches of the reactor coolant system and containment building, very limited quantities of transuranics will be released off-site. This was true at the TMI-2 and Fukushima Daiichi accidents.

The aforementioned list of radionuclides is consistent with the accidents that involved pressurized and boiling water reactors.

(d) The lower-limit Protective Action Guides have not been exceeded (10 mSv effective dose and 50 mSv child thyroid equivalent dose for KI administration). However, with a failed fission product barrier, sheltering should be maintained. Preparations for evacuation are warranted since the fuel fission product barrier is breached. Appropriate state, county, and local governments should be placed on standby for a possible evacuation if the radiological situation degrades.

(e) The reactor coolant system is breached and fuel severely damaged. Containment radiation monitors are off-scale high, and a utility field teams report a direct radiation measurement of 175 mSv at 3.2 km and 10 mSv at 16 km from the facility. The utility recommends sheltering in place as a protective action.

The utility's recommendation is incorrect. The protective action guidelines have been exceeded. Given the meteorological conditions (i.e., variable winds) and the 24 h release duration estimate, evacuation of the 16 km emergency planning zone should be recommended to the Governor. Administration of KI should also be considered given the nature of the release and loss of all fission product barriers. To verify the need for KI administration, dose projections or field team measurements should be obtained for the thyroid dose.

6.6. (a) The absorbed dose (D) received by the passengers and crew during a typical 2 h Utopian Air flight from Los Angeles to LEO-1 is determined by the integrated 2 h fluence (Φ), the dose conversion factor (F), the aircraft shell thickness (t), and attenuation coefficient (μ):

$$D = \sum_{i=1}^{2} \Phi_i F_i e^{-\mu_i t}$$

where $i = 1$ and 2 define the proton (p) and heavy ion (HI) source terms, respectively. The absorbed dose is determined from the values provided in the problem statement:

$$\begin{aligned} D &= \left(3 \times 10^5 \frac{\text{p}}{\text{cm}^2}\right)\left(\frac{3000\,\text{pGy-cm}^2}{\text{p}}\right)\left[\frac{1\text{Gy}}{10^{12}\text{pGy}}\right] e^{-\left(\frac{0.2}{\text{cm}}\right)(3\text{ cm})} \\ &\quad + \left(4 \times 10^4 \frac{\text{HI}}{\text{cm}^2}\right)\left(\frac{7000\,\text{pGy-cm}^2}{\text{HI}}\right)\left[\frac{1\text{Gy}}{10^{12}\text{pGy}}\right] e^{-\left(\frac{0.35}{\text{cm}}\right)(3\text{ cm})} \\ &= 4.94 \times 10^{-4}\text{Gy} + 0.980 \times 10^{-4}\text{Gy} \\ &= (5.92 \times 10^{-4}\text{Gy})\left(1000\,\frac{\text{mGy}}{\text{Gy}}\right) \\ &= 0.592\,\text{mGy} \end{aligned}$$

(b) The absorbed dose (D) received by the passengers and crew from a massive solar particle event during the 2 h Utopian Air flight from Los Angeles to LEO-1 is determined using the methodology provided in the previous question:

$$\begin{aligned} D &= \left(5 \times 10^9 \frac{\text{p}}{\text{cm}^2}\right)\left(\frac{4500\,\text{pGy-cm}^2}{\text{p}}\right)\left[\frac{1\text{Gy}}{10^{12}\text{pGy}}\right] e^{-\left(\frac{0.15}{\text{cm}}\right)(3\text{ cm})} \\ &\quad + \left(6 \times 10^8 \frac{\text{HI}}{\text{cm}^2}\right)\left(\frac{9500\,\text{pGy-cm}^2}{\text{HI}}\right)\left[\frac{1\text{Gy}}{10^{12}\text{pGy}}\right] e^{-\left(\frac{0.30}{\text{cm}}\right)(3\text{ cm})} \\ &= 14.3\,\text{Gy} + 2.32\,\text{Gy} = 16.6\,\text{Gy} \end{aligned}$$

(c) The calculated absorbed dose during the massive solar event is of considerable concern. In all likelihood, the aircraft will receive warning of the extreme radiological conditions prior to the event. It would be logical to assume that this warning would preclude the launch and eliminate the hazard.

If the event occurred after launch, onboard radiation monitors would indicate a radiation hazard. Emergency procedures provide dose reduction methods that include (i) aborting the launch trajectory and returning to earth and (ii) altering the planned trajectory to change the orbit or return to the atmosphere.

It would also be desirable to provide a shielded enclosure to minimize crew and passenger dose. The shielding thickness needs to consider the bounding solar event, launch weight limitations, and the number of passengers and crew.

(d) Equivalent aluminum shielding provided by the hotel (4 cm) and shelter (15 cm) yields 19 cm total aluminum attenuation. Since the integrated fluence values are based on 2 h of exposure, they must be increased to account for the actual event duration. The doses received by the passengers and crew within the shielded LEO-1 emergency shelter for the 5 h event is given by the relationship

$$D = \begin{bmatrix} \left(5\times10^{9}\,\frac{\text{p}}{\text{cm}^2}\right)\left(\frac{4500\,\text{pGy-cm}^2}{\text{p}}\right)\left[\frac{1\text{Gy}}{10^{12}\text{pGy}}\right]e^{-\left(\frac{0.15}{\text{cm}}\right)(19\ \text{cm})} \\ +\left(6\times10^{8}\,\frac{\text{HI}}{\text{cm}^2}\right)\left(\frac{9500\,\text{pGy-cm}^2}{\text{HI}}\right)\left[\frac{1\,Gy}{10^{12}pGy}\right]e^{-\left(\frac{0.30}{\text{cm}}\right)(19\,\text{cm})} \end{bmatrix}\left(\frac{5\,\text{h}}{2\,\text{h}}\right)$$

$$= (1.30\,\text{Gy} + 0.019\,\text{Gy})(2.5) = 3.30\,\text{Gy}$$

(e) Given the vast distance between the Sun and the earth, the fluence reaching the spacecraft is essentially the same as the fluence reaching the earth's outer atmosphere. The unattenuated dose near the earth and outside the atmosphere is $D = \sum_{i=1}^{2} \Phi_i F_i$

$$D = \begin{bmatrix} \left(5\times10^{9}\,\frac{\text{p}}{\text{cm}^2}\right)\left(\frac{4500\,\text{pGy-cm}^2}{\text{p}}\right)\left[\frac{1\,\text{Gy}}{10^{12}\text{pGy}}\right] \\ +\left(6\times10^{8}\,\frac{\text{HI}}{\text{cm}^2}\right)\left(\frac{9500\,\text{pGy-cm}^2}{\text{HI}}\right)\left[\frac{1\,\text{Gy}}{10^{12}\text{pGy}}\right] \end{bmatrix}\left(\frac{5\,\text{h}}{2\,\text{h}}\right)$$

$$= (22.5\,\text{Gy} + 5.70\,\text{Gy})(2.5) = 70.5\,\text{Gy}$$

However, the radiation is significantly attenuated by the earth's atmosphere:

$$D = D_{\text{proton}} + D_{\text{Heavy Ion}}$$

$$D = \left[(22.5\,\text{Gy})\,2^{\left(-\frac{25\text{km}}{2\text{km}}\right)} + (5.70\,\text{Gy})2^{\left(-\frac{25\,\text{km}}{1.5\,\text{km}}\right)}\right]\left(\frac{1000\,\text{mGy}}{\text{Gy}}\right)(2.5)$$

$$= (3.88\,\text{mGy} + 0.0548\,\text{mGy})(2.5) = 9.84\,\text{mGy}$$

(f) Assuming the radiological conditions are constant throughout the event, the effective dose is written in terms of the absorbed dose (D_R) from radiation of type (R) and the radiation weighting factor (w_R). Using the values from question (e) for a 5 h event and the w_R values in the problem statement, the effective dose (E) is

$$E = \sum_{R=1}^{2} w_R D_R = \left[(3.88\,\mathrm{mGy})\left(2\frac{\mathrm{mSv}}{\mathrm{mGy}}\right) + (0.0548\,\mathrm{mGy})\left(20\frac{\mathrm{mSv}}{\mathrm{mGy}}\right)\right](2.5)$$

$$E = (8.86\,\mathrm{mSv})(2.5) = 22.2\,\mathrm{mSv}$$

Since this effective dose occurs over a 5 h period, the effective dose over the 4-week duration of the solar event is

$$E_{\mathrm{Total}} = \left(\frac{22.2\,\mathrm{mSv}}{5\,\mathrm{h}}\right)(4\,\mathrm{weeks})\left(\frac{7\,\mathrm{days}}{\mathrm{week}}\right)\left(\frac{24\,\mathrm{h}}{\mathrm{day}}\right)\left(\frac{1\,\mathrm{Sv}}{1000\,\mathrm{mSv}}\right) = 2.98\,\mathrm{Sv}$$

(g) The situation is grave and the massive solar event is well beyond the LEO-1 design basis. A comparison with Table 6.2 suggests the current event is more severe than the 1859 Carrington flare having an integrated fluence of 18.8×10^9 protons/cm^2. The massive event's proton (p) fluence is

$$\Phi_{\mathrm{p}} = \left(\frac{5 \times 10^9\,\frac{\mathrm{p}}{\mathrm{cm}^2}}{2\,\mathrm{h}}\right)(28\,\mathrm{days})\left(24\frac{\mathrm{h}}{\mathrm{day}}\right) = 1.68 \times 10^{12}\,\frac{\mathrm{p}}{\mathrm{cm}^2}$$

which is about 90 times larger than the Carrington event. Survival of the individuals at LEO-1 is enhanced if dose reduction measures are implemented in a timely manner.

You should advise the Utopian Air Flight Operations Director that the total dose from the event will produce deterministic effects at LEO-1 over the 4-week period. Direct the residents of LEO-1 to remain in the shelter until further notice. Additional shielding (e.g., food stores, construction materials, available metal plates, and portable water shields) should be placed around the emergency shelter to minimize the projected doses. You should perform a dose assessment to determine the dose reduction from these shielding additions.

Other emergency actions to be evaluated include:

1. Assessing if a change in the orbit of LEO-1 is feasible
2. Administering radioprotective agents
3. Determining if an evacuation of LEO-1 is feasible
4. Performing a dose assessment to determine the best option to minimize LEO-1 doses
5. Implementing medical measures to minimize deterministic effects that may occur following exposure of the LEO-1 inhabitants
6. Obtaining continuous radiation levels and particle fluence values from LEO-1

Ground personnel should be advised of measures to minimize their effective dose including sheltering in an underground location. Your dose projections should also be communicated to senior corporate management for transmission to government officials.

Clear communication with the all Utopian Air employees is important. The risks from the event should be presented in a clear, logical manner. Dialog should be encouraged, and all employee concerns should be addressed.

6.7. (a) The radionuclide composition of the meteorite can be determined using a portable HPGe detector. The detector provides the gamma-ray spectrum emitted by the meteorite and the peak energies. These data and the associated count rates are used to determine the radionuclide composition and an estimate of their activity.

(b) Given the estimated 1.0 m meteorite diameter, the dose rate at 100 m is determined using a point source approximation:

$$\dot{D}_1 r_1^2 = \dot{D}_2 r_2^2$$

$$\dot{D}_2 = \dot{D}_1 \frac{r_1^2}{r_2^2} = \left(10 \, \frac{\text{mGy}}{\text{h}}\right) \left(\frac{1000\,\text{m}}{100\,\text{m}}\right)^2 = 1000\,\text{mGy/h} = 1\,\text{Gy/h}$$

Based on the calculation, the State Police should not move to 100 m from the crater. Additional planning is required to permit access to the vicinity of the impact crater.

(c) A number of techniques can be used to approach the crater. If the isotopes forming the meteorite are short-lived, radioactive decay facilitates a decrease in dose rates. If the half-lives are long, decay is not a viable option.

For long-lived radionuclides, the use of shielding permits closer access to the crater. The crater could also be observed and photographed by aircraft or drones. If there is a national security need, satellite imagery could also be utilized. Bulldozers could push earth around the crater to reduce the dose and permit the placement of closed circuit television cameras and radiation detectors. These instruments would permit remote viewing and monitoring. Robotic techniques present another observational possibility.

(d) The 2 s HPGe gamma scan suggests the following isotopes are present in the meteorite: ^{238}U (0.0496 MeV), ^{239}Pu (0.0516 MeV), ^{137}Cs/^{137m}Ba (0.66 MeV), and ^{60}Co (1.17 and 1.33 MeV). Only ^{238}U is naturally occurring. ^{60}Co (neutron activation product) and ^{137}Cs (fission product) are produced from neutron reactions. ^{239}Pu is produced from neutron capture in ^{238}U and subsequent beta decays. Other isotopes could be present that did not appear in the initial scan and would provide insight into the origin of the meteorite. A more detailed analysis is required to determine its origin and composition. However, based on the initial gamma scan, the meteorite does not have a natural origin, and its constituent radionuclides could have been produced by a fission device.

(e) The estimated activity (A) of ^{60}Co and ^{137}Cs in the meteorite is obtained from the relationship

$$A = \frac{\mathrm{CR}(E)}{t\,Y(E)\,e'(E)}$$

where A is the activity of the desired isotope, CR(E) is the count rate in a given photopeak energy, t is the count time, $Y(E)$ is the yield of the isotope leading to a photon of energy E, and $e'(E)$ is the efficiency of the detector at energy E at a distance of 100 m. Since the detector is being used in a nonstandard geometry, the efficiency must be corrected for distance

$$e'(E) = \left(\frac{r_{\text{standard scan}}}{r_{100\,\text{m scan}}}\right)^2 e(E) = \left(\frac{0.1\,\text{m}}{100\,\text{m}}\right)^2 e(E) = 10^{-6}\,e(E)$$

where $e(E)$ is the efficiency for the standard source–detector distance of 10 cm.

A(^{60}Co, 1.17 MeV photopeak): The problem statement provides the requisite parameters values:

$$\begin{aligned} \mathrm{CR} &= 5.5 \times 10^8 \text{ counts} \\ t &= 2\,\text{s} \\ Y(E) &= 1.0 \end{aligned}$$

$$e'(E) = 10^{-6}\,e(E) = 0.0093 \times 10^{-6}\,\frac{\text{counts}}{\text{dis}}$$

$$A = \frac{\mathrm{CR}(E)}{t\,Y(E)\,e'(E)} = \frac{(5.5 \times 10^8\,\text{counts})\left(\frac{\text{Bq-s}}{\text{dis}}\right)}{(2\,\text{s})(1.0)\left(0.0093 \times 10^{-6}\,\frac{\text{counts}}{\text{dis}}\right)} = 3.0 \times 10^{16}\,\text{Bq}$$

A(^{137}Cs): The ^{137}Cs activity is obtained from the data in the problem statement:

$$\begin{aligned} \mathrm{CR} &= 1.6 \times 10^8 \text{ counts} \\ t &= 2\,\text{s} \\ Y(E) &= 0.851 \end{aligned}$$

$$e'(E) = 10^{-6}\,e(E) = 0.022 \times 10^{-6}\,\frac{\text{counts}}{\text{dis}}$$

$$A = \frac{\mathrm{CR}(E)}{t\,Y(E)\,e'(E)} = \frac{(1.6 \times 10^8\,\text{counts})\left(\frac{\text{Bq-s}}{\text{dis}}\right)}{(2\,\text{s})(0.851)\left(0.022 \times 10^{-6}\,\frac{\text{counts}}{\text{dis}}\right)} = 4.3 \times 10^{15}\,\text{Bq}$$

(f) The absorbed dose rate at a distance of 1 km from the impact location is obtained from the point source approximation:

$$\dot{D}(^{60}\mathrm{Co}) = \frac{A\Gamma}{r^2} = \frac{(3.0 \times 10^{16}\,\mathrm{Bq})\left(\frac{1\,\mathrm{MBq}}{10^6\,\mathrm{Bq}}\right)\left(3.1 \times 10^{-4}\,\frac{\mathrm{mGy\text{-}m^2}}{\mathrm{MBq\text{-}h}}\right)}{(1000\,\mathrm{m})^2}$$
$$= 9.3\,\mathrm{mGy/h}$$

$$\dot{D}(^{137}\mathrm{Cs}) = \frac{A\Gamma}{r^2} = \frac{(4.3 \times 10^{15}\,\mathrm{Bq})\left(\frac{1\,\mathrm{MBq}}{10^6\,\mathrm{Bq}}\right)\left(0.81 \times 10^{-4}\,\frac{\mathrm{mGy\text{-}m^2}}{\mathrm{MBq\text{-}h}}\right)}{(1000\,\mathrm{m})^2}$$
$$= 0.35\,\mathrm{mGy/h}$$

$$\dot{D}_{\mathrm{Total}} = \dot{D}(^{60}\mathrm{Co}) + \dot{D}(^{137}\mathrm{Cs}) = 9.3\,\mathrm{mGy/h} + 0.35\,\mathrm{mGy/h} = 9.7\,\mathrm{mGy/h}$$

6.8. (a) Since the warehouse personnel have no radiological training, your instructions should be very basic. Considering the level of radiological knowledge of the warehouse personnel, direct the trucking manager to take the following actions:

1. Evacuate all personnel from the warehouse.
2. Secure all heating, ventilation, and air-conditioning systems.
3. Close and lock all warehouse windows, doors, and entryways.
4. Move all personnel from the immediate vicinity of the warehouse.
5. Establish an exclusion zone around the warehouse.
6. Have security personnel ensure no personnel enter the exclusion zone.
7. Request police assistance to control access to the area.
8. Notify city and county emergency management officials of the situation and that state resources are in route.
9. Inform the manager when you expect to arrive and the resources that you will deploy to support recovery of the radioactive materials.
10. If the warehouse has an emergency response plan, request that it be implemented.

(b) Although the bill of lading notes that the package was Radioactive Yellow III and contains ^{60}Co sources, no shipping papers were provided to you. These may be available at the event location (see question (c)). Therefore, the radiological information is likely incomplete, and other radionuclides and radiation types may be present.

You should bring instrumentation to determine the isotopes present and to characterize the radiation and contamination levels within the warehouse. Accordingly, a portable HPGe instrument should be used to determine which photon-emitting radionuclides are present. In addition, handheld survey instruments to detect beta – gamma, alpha, and neutron radiation should be part of your response kit. It is prudent to have additional instrumentation present because the radioactive material package may contain radioactive materials in addition to the ^{60}Co sources.

(c) The Source 1 activity (A) is obtained from the production equation for activation (Appendix B):

$$A = N\sigma\phi(1 - e^{-\lambda t_{irr}})e^{-\lambda t_d}$$

$$N = \frac{(10\,\text{g})\left(6.02 \times 10^{23}\,\frac{\text{atoms}}{\text{mol}}\right)}{\left(\frac{59\,\text{g}}{\text{mol}}\right)} = 1.02 \times 10^{23}\,\text{atoms}$$

$$\sigma = \left(\frac{37\,\text{b}}{\text{atom}}\right)\left(1.0 \times 10^{-24}\,\frac{\text{cm}^2}{\text{b}}\right) = 3.7 \times 10^{-23}\,\frac{\text{cm}^2}{\text{atom}}$$

$$\phi = 1.0 \times 10^{10}\,\frac{\text{n}}{\text{cm}^2\text{-s}}$$

$^{59}Co(n, \gamma)^{60}Co$ activation is induced by thermal neutrons.

$$t_{irr} = \text{irradiation time} = 10\,\text{years}$$
$$t_d = \text{decay time} = 0.5\,\text{year}$$

$$\lambda = \frac{\ln 2}{T_{1/2}} = \frac{0.693}{5.27\,\text{years}} = \frac{0.131}{\text{year}}$$

Given these values, the Source 1 activity is determined:

$$\begin{aligned} A &= (1.02 \times 10^{23}\,\text{atoms})\left(3.7 \times 10^{-23}\,\frac{\text{cm}^2}{\text{atom}}\right) \\ &\times \left(1.0 \times 10^{10}\,\frac{\text{n}}{\text{cm}^2\text{-s}}\right)\left(\frac{1\,\text{dis}}{\text{n}}\right)\left(\frac{\text{Bq-s}}{\text{dis}}\right) \\ &\times \left[1.0 - e^{-\left(\frac{0.131}{\text{year}}\right)(10\,\text{years})}\right] e^{-\left(\frac{0.131}{\text{year}}\right)(0.5\,\text{year})} \\ &= 2.58 \times 10^{10}\,\text{Bq} = 2.58 \times 10^4\,\text{MBq} \end{aligned}$$

(d) The absorbed dose rate at a point 2.0 m from Source 1 in air is obtained from the point source approximation (see Appendix C). Assuming no attenuation within the particle and air, the absorbed dose rate is

$$\begin{aligned} \dot{D} = \frac{A\Gamma}{r^2} &= \frac{(2.58 \times 10^4\,\text{MBq})\left(3.1 \times 10^{-7}\,\frac{\text{Gy-m}^2}{\text{MBq-h}}\right)\left(\frac{1000\,\text{mGy}}{\text{Gy}}\right)}{(2.0\,\text{m})^2} \\ &= \frac{2.0\,\text{mGy}}{\text{h}} \end{aligned}$$

(e) The absorbed dose rate at a point 0.3 m from Source 2 is obtained from the point source relationship

$$\dot{D} = \frac{A\Gamma}{r^2} = \frac{(1.1 \times 10^{12}\,\text{Bq})\left(\frac{1\,\text{MBq}}{10^6\,\text{Bq}}\right)\left(3.1 \times 10^{-7}\,\frac{\text{Gy-m}^2}{\text{MBq-h}}\right)}{(0.3\,\text{m})^2} = \frac{3.8\,\text{Gy}}{\text{h}}$$

(f) The line source equation (see Appendix C) is utilized for this question:

$$\dot{D} = \frac{C_L \Gamma \theta}{w}$$

where

C_L	=	activity per unit length = 3.0 TBq/10 m = 0.3 TBq/m
W	=	distance from the line source = 2.0 m
θ	=	included angle

$$\tan\theta = \frac{10.0\,\text{m}}{2\,\text{m}} = 5.0$$

$$\theta = \tan^{-1}(5.0) = (78.7°)\left(\frac{\pi}{180°}\right) = 1.37$$

Using these values, the absorbed dose rate is determined:

$$\dot{D} = \frac{\left(3.0\times10^{11}\,\frac{\text{Bq}}{\text{m}}\right)\left(\frac{1\,\text{MBq}}{1.0\times10^6\,\text{Bq}}\right)\left(3.1\times10^{-7}\,\frac{\text{Gy-m}^2}{\text{MBq-h}}\right)(1.37)\left(\frac{1000\,\text{mGy}}{\text{Gy}}\right)}{(2.0\,\text{m})}$$

$$= 64\,\frac{\text{mGy}}{\text{h}}$$

(g) The absorbed dose rate at a distance of 10.0 m above the centerline of the thin disk source (see Appendix C) is determined from the relationship

$$\dot{D} = \pi C_a \Gamma \ln\frac{R^2+h^2}{h^2}$$

where

C_a	=	activity per unit area = (A)/(π R^2) = (15.1 TBq)/(π (10 m)^2)
	=	0.0481 TBq/m^2 = 4.81 × 10^{10} Bq/m^2
R	=	spill radius = 10.0 m
h	=	distance above the spill = 10.0 m

$$\dot{D} = \pi\left(4.81\times10^{10}\,\frac{\text{Bq}}{\text{m}^2}\right)\left(\frac{1\,\text{MBq}}{1.0\times10^6\,\text{Bq}}\right)$$

$$\times\left(3.1\times10^{-7}\,\frac{\text{Gy-}\,\text{m}^2}{\text{MBq-h}}\right)\ln\frac{(10.0\,\text{m})^2+(10.0\,\text{m})^2}{(10.0\,\text{m})^2}$$

$$\times\left(\frac{1000\,\text{mGy}}{\text{Gy}}\right) = 32\,\frac{\text{mGy}}{\text{h}}$$

(h) The boundary distance is determined using a point source approximation and the results of question (e):

$$\dot{D}_1 r_1^2 = \dot{D}_2 r_2^2$$

$$r_1 = \left(\frac{\dot{D}_2}{\dot{D}_1}\right)^{0.5} r_2$$

$$r_1 = \left(\frac{3.8\,\text{Gy/h}}{1\times10^{-5}\,\text{Gy/h}}\right)^{0.5}(0.3\,\text{m}) = 185\,\text{m}$$

Solutions for Chapter 7

7.1. (a) Documents that are needed to perform the demineralizer ALARA evaluation include:

1. Current and planned System Design Descriptions for the Demineralizer project that includes flow rates, materials of construction, resin addition and removal pathways, and removal efficiency by isotope
2. Current and planned piping and instrumentation drawings for the demineralizer system
3. Current and planned demineralizer operating procedures
4. Title 10, Code of Federal Regulations, Part 20, *Standards for Protection Against Radiation*
5. Final Safety Analysis Report
6. Technical Specifications
7. Radiation survey records for the demineralizer and adjacent areas
8. Previous ALARA evaluations for demineralizer evolutions including resin addition and removal (sluicing) operations
9. Radiation Work Permits for previous demineralizer system activities including surveillance, maintenance, testing, resin addition, and resin removal
10. Work packages and supporting documents for the demineralizer installation

(b) Items that should be considered when evaluating the demineralizer from an ALARA perspective include:

1. The contamination levels residing both inside and outside the demineralizer systems including associated piping and instrument lines. These contamination levels govern the use of protective clothing and respiratory protection that influence the time required to perform the individual job tasks supporting the demineralizer addition.
2. The estimated times to complete the various work package activities. These times multiply the expected dose rates to get the dose for the various activities.
3. The estimated number of personnel required to perform the various work package activities. The number of personnel is used to establish the collective dose for the task.
4. The dose rates (by radiation type) projected for the various work package activities. The dose rates when multiplied by the associated times yield the dose for completion of the various tasks supporting the demineralizer job.
5. The anticipated location of proposed shielding installations, their configurations, and materials of construction are additional input data. This information permits an assessment of the anticipated

dose to install the shielding and the dose savings once the shielding is installed. Shielding removal is also considered in the dose assessment.

6. The access routes to the demineralizer cubicle and associated systems are important in assessing the installation time. The installation dose as well as the expected dose during routine, abnormal, and emergency operations should be evaluated.
7. The estimated dose rates that are anticipated for long-term (postinstallation) demineralizer operations are needed to assess the long-term radiological impacts of the new system.
8. The types of valves have a significant impact on operational exposures. The use of motor operators, air-operated valves, valve extension handles, and reach rods should be considered as a means of reducing operator doses.
9. The location of installed radiation and air contamination monitors minimize routine health physics surveys.
10. The use of remote viewing devices (e.g., closed circuit TV cameras, periscopes, and viewing windows) minimizes the need for operator entries into high radiation areas for routine rounds and surveillance activities.

(c) The total activity in the demineralizer at (i) the end of its run time and (ii) the end of its down time is to be determined. For the purpose of this question, ^{60}Co is the only radioisotope under consideration.
The activity buildup in the demineralizer is given by the production relationship (see Appendix B)

$$A(t) = \frac{CFe}{\lambda}(1 - e^{-\lambda T})e^{-\lambda t_d}$$

where

$A(t)$	=	demineralizer activity as a function of time
C	=	influent concentration of ^{60}Co = 70.3 Bq/ml
F	=	demineralizer flow rate = 1000 l/min
e	=	demineralizer removal efficiency = 0.99
λ	=	^{60}Co disintegration constant = ln(2)/$T_{1/2}$
$T_{1/2}$	=	^{60}Co half-life = 5.27 year
λ	=	0.693/5.27 years = 0.131/year
T	=	demineralizer run time = 100 days
t_d	=	decay time = 60 days

Using these values, the total activity present in the demineralizer at the end of its run time ($t_d = 0$) is

$$A_e(t) = \frac{\left(70.3\,\frac{\text{Bq}}{\text{ml}}\right)\left(\frac{1000\,\text{ml}}{\text{l}}\right)\left(1000\,\frac{\text{l}}{\text{min}}\right)(0.99)}{\left(\frac{0.131}{\text{year}}\right)\left(\frac{1\,\text{year}}{365\,\text{days}}\right)\left(\frac{1\,\text{day}}{24\,\text{h}}\right)\left(\frac{1\,\text{h}}{60\,\text{min}}\right)}$$
$$\times\left(\frac{1\,\text{MBq}}{10^6\,\text{Bq}}\right)\left(1-e^{-\left(\frac{0.131}{\text{year}}\right)(100\,\text{days})\left(\frac{1\,\text{year}}{365\,\text{days}}\right)}\right)$$
$$= (2.79\times 10^8\,\text{MBq})(1-0.965) = 9.77\times 10^6\,\text{MBq}$$

The total activity present in the demineralizer at the end of its run time and at the end of its down time ($t_d = 60$ days) is given by the relationship

$$A(t) = \frac{CFe}{\lambda}(1-e^{-\lambda T})e^{-\lambda t_d} = A_e(t)e^{-\lambda t_d}$$

$$A(t) = (9.77\times 10^6\,\text{MBq})\,e^{-\left(\frac{0.131}{\text{day}}\right)(60\,\text{days})\left(\frac{1\,\text{year}}{365\,\text{days}}\right)}$$
$$= (9.77\times 10^6\,\text{MBq})(0.979) = 9.56\times 10^6\,\text{MBq}$$

(d) This question requests that you calculate the absorbed dose rate 20 m above the demineralizer bed at the end of its down time. The distance from the demineralizer bed is greater that three times the largest source dimension. Therefore, a point source approximation is appropriate and provides a result within 1% of the disk source solution. Any shielding from the demineralizer bed, water, and demineralizer shell are to be ignored. The absorbed rate is

$$\dot{D}(r) = \frac{A\Gamma}{r^2}$$

where

$\dot{D}(r)$	=	absorbed dose rate from the demineralizer bed at the end of its down time
R	=	distance above the demineralizer at the point of interest = 20 m
A	=	demineralizer activity = 9.56×10^6 MBq
Γ	=	gamma constant for ^{60}Co = 3.1×10^{-4} mGy-m^2/h-MBq

$$\dot{D}(r) = \frac{(9.56\times 10^6\,\text{MBq})\left(3.1\times 10^{-4}\,\frac{\text{mGy-m}^2}{\text{h-MBq}}\right)}{(20.0\,\text{m})^2} = 7.4\,\frac{\text{mGy}}{\text{h}}$$

(e) Methods that could be used to minimize the dose to plant personnel during maintenance of the demineralizer include the following:

1. Use shielding in high dose areas or where justified by the ALARA evaluation.
2. Limit the demineralizer activity to keep dose rates low.
3. Practice the various demineralizer tasks using mockups and experienced personnel.
4. Provide training to the crews performing the various demineralizer tasks.

5. Use low cobalt alloys to minimize ^{60}Co deposition on demineralizer piping and in the demineralizer bed.
6. Utilize low dose rate waiting areas during maintenance tasks.
7. Use glove bags to minimize internal depositions.
8. Decontaminate system components (pumps and valves) prior to maintenance.
9. Flush system components to minimize contamination levels.
10. Use remote tools or robotics wherever practical.
11. Maintain primary system chemistry to ensure activity deposits on core surfaces instead of primary system and interfacing piping.
12. Perform demineralizer maintenance at the end of the down time period to maximize the decay of short-lived radionuclides and minimize the dose rates.
13. Components residing in elevated radiation areas should be removed, decontaminated, and repaired in low dose rate areas.
14. Sluice resin from the demineralizer and flush the system before performing maintenance in proximity to the vessel.

(f) Given the size of the demineralizer and the distance of interest, the dose rate contribution from ^{131}I and ^{137}Cs is determined using a point source approximation:

$$\dot{D} = \dot{D}(^{137}\text{Cs}) + \dot{D}(^{131}\text{I})$$

$$= \frac{A(^{137}\text{Cs})\,\Gamma(^{137}\text{Cs})}{r^2} + \frac{A(^{131}\text{I})\,\Gamma(^{131}\text{I})}{r^2}$$

$$= \frac{(750 \times 10^{12}\,\text{Bq})\left(\frac{1\,\text{MBq}}{1{\times}10^6\,\text{Bq}}\right)\left(8.1 \times 10^{-5}\,\frac{\text{mGy-m}^2}{\text{h - MBq}}\right)\left(\frac{\text{Gy}}{1000\,\text{mGy}}\right)}{(10\,\text{m})^2}$$

$$+ \frac{(500 \times 10^{12}\,\text{Bq})\left(\frac{1\,\text{MBq}}{1{\times}10^6\,\text{Bq}}\right)\left(5.2 \times 10^{-5}\,\frac{\text{mGy-m}^2}{\text{h - MBq}}\right)\left(\frac{\text{Gy}}{1000\,\text{mGy}}\right)}{(10\,\text{m})^2}$$

$$= 0.61\,\frac{\text{Gy}}{\text{h}} + 0.26\,\frac{\text{Gy}}{\text{h}} = 0.87\,\frac{\text{Gy}}{\text{h}}$$

(g) After 1 year of decay, the dose rate is

$$\dot{D} = \left(0.61\,\frac{\text{Gy}}{\text{h}}\right)\left(e^{-\frac{(0.693)(1\,\text{year})}{30.1\,\text{years}}}\right) + \left(0.26\,\frac{\text{Gy}}{\text{h}}\right)\left(e^{-\frac{(0.693)(1\,\text{year})\left(365\,\frac{\text{day}}{\text{year}}\right)}{(8.02\,\text{days})}}\right)$$

$$= \left(0.61\,\frac{\text{Gy}}{\text{h}}\right)(0.977) + \left(0.26\,\frac{\text{Gy}}{\text{h}}\right)(2.01 \times 10^{-14}) = 0.60\,\frac{\text{Gy}}{\text{h}}$$

(h) The demineralizer cubicle should be posted as a very high radiation area (VHRA). A VHRA is defined in 10CFR20 as an area, accessible to individuals, in which radiation levels could result in an individual receiving an absorbed dose in excess of 5 Gy in an hour at a distance of 1 m from a radiation source or 1 m from any surface that the radiation penetrates. Given the dose rate calculated in question (f) and (g), the

VHRA criteria is met by assuming the applicability of a point source relationship

$$\dot{D}(1\,\text{m}) \approx \dot{D}(10\,\text{m})\left(\frac{10\,\text{m}}{1\,\text{m}}\right)^2 = \left(0.60\,\frac{\text{Gy}}{\text{h}}\right)(100) = 60\,\frac{\text{Gy}}{\text{h}}$$

(i) Resin sluicing is a routine activity, and this operation does not require access to the interior of the demineralizer cubicle. Given the elevated dose rates, an ALARA review should be performed to determine the batch size for resin transfers and to determine dose rates near resin transfer lines and in radioactive waste processing areas where the resin is deposited into a high-integrity container. The ALARA review determines the number of batches, the number of waste containers, supplemental shielding requirements, and additional health physics controls.

The sluicing activity is optimally performed on back shifts when minimal staff is present. Surveys and management walkthroughs verify that all high radiation areas are posted. Sufficient health physics resources should be utilized to ensure personnel understand that high radiation areas are not to be accessed during resin transfer operations. Supervision of the activity by operations and health physics management is recommended. A thorough prejob briefing is also advisable.

7.2. (a) The ALARA principle is applied to ionizing radiation to limit the risks of stochastic effects (e.g., cancer and hereditary disease), which are assumed to have no threshold in the current regulatory formulation. In the TLV approach, the deterministic effects of concern have a threshold or total exposure that must be exceeded for detrimental effects to be observed. Threshold Limit Values are used to prevent detrimental deterministic effects associated with hazardous materials or nonionizing radiation sources.

The TLV is the maximum concentration or magnitude of an agent (e.g., chemical) to which workers can be exposed for a fixed time (normally 8 h/day) without developing a physiological impairment. The TLVs are normally based on 95% confidence level values.

ALARA is based on maintaining exposures to values as low as reasonably achievable and presumes the validity of the linear-nonthreshold dose response model. Observable radiation-related stochastic effects are limited by setting regulatory limits.

(b) SLOBS limit (1) is based on the threshold acute absorbed dose (*D*) for observing any blood changes that is ~0.05 Gy. Using this value for an annual limit leads to the SLOBS element (1) proposal

$$\dot{D} = \left(0.05\,\frac{\text{Gy}}{\text{year}}\right)\left(\frac{1000\,\text{mGy}}{\text{Gy}}\right)\left(\frac{1\,\text{year}}{50\,\text{weeks}}\right)\left(\frac{1\,\text{week}}{5\,\text{days}}\right)\left(\frac{1\,\text{mSv}}{1\,\text{mGy}}\right)$$

$$= 0.2\,\frac{\text{mSv}}{\text{day}}$$

The 0.05 Sv value is also the maximum annual TEDE allowed by the ICRP 26 recommendations and the maximum annual effective dose in a 5-year period per the ICRP 60 and ICRP 103 recommendations (see Table D.6).

(c) Utility staff doses could be accommodated by the SLOBS proposal. Worker annual doses are typically in the range of 3–5 mSv/year. Doses are typically higher during refueling and maintenance outages but in most cases would be manageable with the element (2) exemption.

(d) Contract workers often perform high dose and specialized activities, and some of these workers could exceed an equivalent dose of 20 mSv in a continuous 13-week period. There is also a potential legal challenge to the proposal since it would limit the earning potential of a subset of contractor workers when compared to the flexibility provided by the current 10CFR20 dose limit of 50 mSv/year TEDE.

(e) The ICRP 103 recommendations summarized in Table D.6 allow an average of 20 mSv/year effective dose over a 5-year period. The cumulative effective dose during the 5-year period is 100 mSv with a maximum of 50 mSv in any year. The SLOBS proposal allows a maximum dose of 50 mSv in a year and is inconsistent with the ICRP 103 effective dose recommendations.

Since the ICRP 103 eye dose has the same limits as the effective dose, it is also inconsistent with the SLOBS proposal. The skin dose limit is consistent with ICRP 103. However, a continuous 13-week skin and extremity dose restriction is not part of the ICRP 103 recommendations.

The ICRP 103 recommendations also include a fetal dose (for declared pregnant workers) of 1 mSv. The SLOBS proposal did not address fetal dose.

7.3. (a) In this question, you are asked to describe and discuss the optimal bioassay approach for the ^{131}I intake. Advantages and disadvantages for the *in vivo* (within the body) and for *in vitro* (outside the body) methods of analysis as related to this case are also requested.

Optimum approach: *In vivo* whole-body or thyroid counting is the preferred approach. Whole-body counters and thyroid counters readily detect the photons emitted by ^{131}I. These *in vivo* counts require minimum participation by the worker and can be quickly repeated. The methods of analyzing the data are well established and accepted.

In vivo:

Advantages:

1. The method is quick and requires little effort by the worker.
2. An immediate determination of the intake is obtained from the thyroid activity and intake retention fraction.
3. Detectors are used to isolate and count the thyroid.
4. The method is well established and defensible.

5. The measurements are directly related to the accumulated thyroid activity and dose.

Disadvantages:

1. The location of the radioactive material within the body may not be clearly defined unless a shielded detector is used. A whole-body count detects the contamination but not necessarily its exact location.
2. Quantification of the intake is complicated by differences between the subject and Reference Man (ICRP 23) in terms of body organ sizes, chest wall thickness, organ shape and weight, metabolic and physiological characteristics, and the absorption and scattering characteristics of the individual.
3. Inaccuracies are introduced if the individual has external contamination.
4. ^{131}I may be simulated by other radionuclides, including some radon daughters.
5. The method requires the presence of the worker during each whole-body or thyroid count.
6. Internal contamination in the gastrointestinal tract or respiratory system, particularly within a few days of intake, leads to inaccuracies if these depositions are interpreted as a thyroid uptake.

In vitro:

Advantages:

1. The method is well established and defensible.
2. The input parameters (intakes and intake retention fractions) are tabulated and readily available. They are based on ICRP 26 and supporting methodology that forms the basis for NRC regulations in the United States.
3. Urine is a medium that is readily obtained and analyzed.
4. The exposed individual need not be present during the analysis of the 24 h urine samples.

Disadvantages:

1. Samples require time for collection and additional time for preparation and analysis.
2. The method depends on the worker to participate and properly implement the required sampling protocol.
3. The method does not provide an immediate quantification of the intake. Analysis of the sample with a subsequent model-dependent calculation is required.
4. The use of standard models leads to inaccurate results unless the individual's anatomy and physiology resembles and responds in a manner assumed in the Reference Man formulation.
5. Excreta must be collected over time.

6. The collected sample only represents the excretion and not necessarily the thyroid burden. This is particularly important if a thyroid-blocking agent (e.g., KI) was used.

(b) In this question, you are requested to describe how your approach to bioassay might change as a function of time, given the metabolic model for iodine. For most low-level intakes, only one method is utilized. This is typically whole-body or thyroid counting. The intake addressed in this problem is more significant and merits additional monitoring. Within the scope of the specified model, the following initial and later phase commentary is provided:

Initial phase: Approximately 75% of the ^{131}I is excreted from the body in the urine in 1–2 days with an effective halftime of about 6 h. In the early phase, whole-body counting is the easiest approach. Given the specified model, most of the activity is excreted via urine. Therefore, urinalysis during the first few days would be helpful in verifying the intake. However, limited data is obtained since a 24 h void is the usual protocol.

Later phase: The remaining 25% of ^{131}I is trapped in the subject's thyroid, reaching a maximum about 24 h postintake, and is excreted with an effective halftime of about 7 days. After the ^{131}I localizes in the thyroid, whole-body or thyroid counting will be performed. Periodic urine sampling should also be performed to verify the intake estimates.

In vivo counting should be performed throughout the monitoring period since it is easy and requires a minimum of effort. As noted earlier, periodic urine samples also establish the individual's retention function and verify the *in vivo*-based intake.

(c) Based on the available thyroid counting data, the best estimate of the subject's intake is provided by the methodology of NUREG/CR-4884. This methodology and the following data permit determination of the intake:

i	Time postintake (days)	Thyroid activity [*A*(*i*)] (kBq)	IRF(*i*)
1	1	2500	0.133
2	7	2300	0.0995
3	10	1300	0.0751

$$I = \frac{\sum_{i=1}^{3} A(i)\,\mathrm{IRF}(i)}{\sum_{i=1}^{3} [\mathrm{IRF}(i)]^2}$$

$$I = \frac{(0.133)(2500\,\text{kBq}) + (0.0995)(2300\,\text{kBq}) + (0.0751)(1300\,\text{kBq})}{(0.133)^2 + (0.0995)^2 + (0.0751)^2}$$
$$= \frac{659\,\text{kBq}}{0.0332} = 1.98 \times 10^4\,\text{kBq} = 19.8\,\text{MBq}$$

(d) Assuming that the intake was 5 MBq and that no organs other than the thyroid make a significant contribution to the committed effective dose equivalent (CEDE), you are requested to determine the committed dose equivalent (CDE) to the thyroid and the CEDE. A determination if any regulatory limits have been exceeded is also requested.
CDE: The CDE is determined from the intake

$$\text{CDE} = H_{50,\text{T}} = I(\text{DCF})$$

where

I = intake of $^{131}I = 5\,\text{MBq}$
DCF = dose conversion factor $= 2.9 \times 10^{-7}$ Sv/Bq

$$\text{CDE} = H_{50,\text{T}} = (5\,\text{MBq})\left(\frac{10^6\,\text{Bq}}{\text{MBq}}\right)\left(2.9 \times 10^{-7}\,\frac{\text{Sv}}{\text{Bq}}\right) = 1.5\,\text{Sv}$$

CEDE:

$$\text{CEDE} = w_\text{T} H_{50,\text{T}}$$

where

w_T = thyroid organ weighting factor (ICRP 26) = 0.03
CDE = thyroid committed dose equivalent $= H_{50,\text{T}} = 1.5$ Sv

$$\text{CEDE} = (0.03)(1.5\,\text{Sv}) = 0.045\,\text{Sv} = 45\,\text{mSv}$$

Regulatory limits: The Regulatory Dose Limits for both 10CFR20 (NRC Licensees) and 10CFR835 (DOE Licensees) are based on ICRP 26. The organ dose limit (including the thyroid) is 0.5 Sv CDE. The organ dose of 1.5 Sv exceeds the regulatory limit for the thyroid. The CEDE limit is 0.05 Sv (50 mSv), and it was not exceeded.

(e) From Part (d), the regulatory limit for the committed dose equivalent to the thyroid was exceeded and the NRC fined the operating utility for this violation. Since the dose resulted from an individual participating in a terrorist attack, you have been requested to prepare a radiological basis for appealing the NRC's fine.

The NRC dose limits cited in question (d) are applicable since the worker was occupationally exposed. The operating utility is required to certify the worker's fitness for duty (FFD) under 10CFR26.4 (*FFD*

Program Applicability to Categories of Individuals). In particular, the utility failed to ensure that the worker met the access authorization requirements summarized in 10CFR26. This section requires that the utility conduct background investigations or psychological assessments to make access authorization determinations. These assessments and investigations failed to determine that the worker could commit radiological sabotage. Therefore, there is no basis for appealing the fine for exceeding the thyroid dose limit. Additional regulatory action is likely because of weaknesses in the utility's FFD program.

Any argument regarding the worker's radiological performance is invalid because of his status as a maintenance worker. He was subject to the occupational dose limits, and the organ dose limit was exceeded.

(f) The radioactive materials stored in a waste gas decay tank are primarily noble gases and radioiodine. After 3 years, the dominant isotope is ^{85}Kr. All other noble gases and ^{131}I have decayed to insignificant levels.

7.4. (a) Based on the dominant peaks, the major isotope in the spectrum is ^{60}Co that has 1.17 and 1.33 MeV photons with a yield of 1.0 for each. The other peaks are related to the primary 1.17 and 1.33 MeV peaks and are addressed in question (d).

(b) When calculating the skin dose from a hot particle, an area of 10 cm^2 and tissue depth of 7 mg/cm^2 are generally utilized. In 1999, NCRP 130 published this updated guidance that has been adopted by the NRC. However, ICRP 103 (see Table 7.9) recommends the skin dose be averaged over 1 cm^2 regardless of the area exposed.

(c) The dose measured by an ion chamber is not the "true" skin dose. True skin dose is measured at the depth of the basal cell layer (7 mg/cm^2). This dose is not the same as the ion chamber dose:

1. The ion chamber dose represents the dose averaged over the detector volume (220 cm^3) which is not equivalent to the dose delivered at the depth of the basal cell layer.
2. The true skin dose is measured in the tissue, but the ionization chamber is measuring the exposure in air. These are not the same quantities.
3. The meter was calibrated with a gamma-only source. Its response will not represent the true skin dose that is a beta plus gamma dose. For hot particles, the beta dose is normally the dominant contribution. ^{60}Co emits a 318 keV beta particle that is a significant skin dose contributor.
4. An open window measurement includes beta plus gamma contributions. The actual beta dose rates and gamma exposure rates are determined from a combination of open window (OW) and closed

window (CW) readings:

$$\dot{D}_{\text{gamma}} = \text{CW}$$

$$\dot{D}_{\text{beta}} = k(\text{OW} - \text{CW})$$

where k is a beta correction factor that depends on the size of the source and the distance the detector resides from the source.

5. The ionization chamber yields a dose at a density thickness that does not correspond to the basal cell depth.
6. The skin dose from a point source particle falls off as $1/r^2$. The geometry is not the same for the ionization chamber and the particle residing on the skin.
7. The hot particle only illuminates a portion of the detector volume that depends on the measurement distance. This limited illumination volume affects the measured dose rate.

(d) The gamma spectrum reveals that the activity of a particular hot particle is due primarily to ^{60}Co. The seven peak energies, the most likely origin of each, and the mechanism that causes each peak are described in the following discussion.

To answer this question, the reader must recall that the primary radiation types emitted by ^{60}Co are:

Gamma	1.17 MeV @ 100%
	1.33 MeV @ 100%
Beta	0.318 MeV @ 100%

0.308 MeV peak: The 1.33 MeV photon creates an electron–positron pair. These positrons annihilate electrons and yield two 0.511 MeV photons. The 0.308 MeV peak is the double escape peak for the 1.33 MeV photopeak. If both the annihilation photons escape from the detector, a peak with an energy of $1.33\,\text{MeV} - 2 \times 0.511\,\text{MeV}$ occurs at 0.308 MeV.

0.511 MeV peak:The 0.511 MeV peak is an annihilation peak. Photons above the pair production threshold (1.022 MeV) produce a positron–electron pair. The positron annihilates an electron with the production of two 0.511 MeV photons. Both primary photons (1.17 and 1.33 MeV) contribute to the 0.511 MeV peak.

0.659 MeV peak: The 0.659 MeV peak is the single escape peak for the 1.17 MeV photopeak. If one of the annihilation photons escapes from the detector, a peak with energy $1.17 - 0.511\,\text{MeV}$ occurs at 0.659 MeV.

0.819 MeV peak: The 0.819 MeV peak is the single escape peak for the 1.33 MeV photopeak. If one of the annihilation photons escapes from the detector, a peak with an energy of $1.33 - 0.511\,\text{MeV}$ occurs at 0.819 MeV.

1.17 MeV *peak*: The 1.17 MeV peak is a photopeak. It arises from the ^{60}Co nucleus beta decaying to an excited ^{60}Ni energy level, which subsequently decays (i.e., the 2.5057 MeV 4^+ level deexcites to the 1.3325 MeV 2^+ level in ^{60}Ni).

1.33 MeV peak: The 1.33 MeV peak is a photopeak. It arises from the ^{60}Co nucleus beta decaying to an excited ^{60}Ni energy level, which subsequently decays (i.e., the 1.3325 MeV 2^+ level transitions to the 0.0 MeV 0^+ ground state in ^{60}Ni).

2.50 MeV peak: The 2.50 MeV peak is a sum peak. It is caused by the coincident detection of the 1.17 and 1.33 MeV photons. As noted previously, these are the dominant ^{60}Co photopeaks.

(e) Follow-up actions that should be initiated for the contaminated area upon discovery of these hot particles include the following:

1. Review the health physics controls for entry into the contaminated area. Ensure that emergency response personnel performing radiological surveys are familiar with the instrument response to a hot particle. This response will be a sharp needle deflection as the detector passes over the particle with a rapid return to background levels. Distributed contamination presents a more uniform instrument response. To minimize skin dose, hot particles should be removed from the body upon detection.
2. Ensure all radiological and medical personnel are informed of the NCRP 130 recommendations for hot particles found on skin, the eye, in the respiratory system, and in the alimentary tract.
3. Workers exiting the contaminated area should perform a detailed frisk to ensure particles are detected and removed.
4. Increase the survey frequency of workers to detect contamination and minimize their skin doses.
5. Implement a hot particle control kit to facilitate the removal of radioactive material from contaminated individuals.
6. Implement an increased glove change-out frequency.
7. Utilize tacky mats or carpet at the step-off-pad to the contaminated area to minimize the transport of the particles beyond the contaminated areas.
8. Advise and train workers of the hot particle hazard and of the controls implemented to preclude unanticipated exposures.
9. Add an exit automated frisking monitor to speed personnel screening from the contaminated area.
10. Provide additional friskers and portal monitors to enhance detection of the particles.
11. Implement procedural controls to enhance the detection and removal of hot particles.
12. Train workers, particularly health physics technicians, in skin decontamination techniques to enhance removal of hot particles.

13. Spray the contaminated area to fix the contamination and minimize the spread of the radioactive material.
14. Have adequate health physician resources to assist in contamination detection and removal.

7.5. (a) HTO presents a greater radiological hazard than T_2. It is a water analog that is incorporated into the body and irradiates the tissue. T_2 is a gas that is inhaled and then exhaled and provides a short-term dose. T_2 can be converted into HTO over time

$$2T_2 + O_2 + 2H_2O \rightarrow 4HTO$$

As an illustration of the radiological hazards of T_2 and HTO, the ICRP inhalation dose coefficients of tritium gas and tritiated water are 1.8×10^{-15} and 1.8×10^{-11} Sv/Bq, respectively.

(b) Airborne monitoring techniques for tritium are as follows:

1. Pass the sampled air through a collection chamber surrounded by cooling liquid nitrogen or cold water.
   Advantages:
   - Well-established technique based on condensation of water vapor containing HTO.
   - Samples are easily prepared for liquid scintillation counting (LSC).
   - Not affected by gamma-ray or noble gas backgrounds.

   Disadvantages:
   - Liquid nitrogen requires replenishment.
   - Liquid nitrogen presents an industrial safety hazard.
   - A cooling water supply must be available in the sampling area.
   - Fine ice particles flake and cause a loss in collection efficiency.
2. Pass the sampled air through a desiccant or other absorbent material.
   Advantages:
   - Technique is easy to perform.
   - Not affected by gamma-ray or noble gas backgrounds.

   Disadvantages:
   - Desiccant replenishment is required.
   - Desiccant must be properly stored, so it is not partially saturated.
   - Collected sample must be prepared for LSC. Heating or chemical processing of the desiccant is required.
   - Saturation of the desiccant decreases the collection efficiency.
3. Diffuse the sampled air in a known volume of water.
   Advantages:
   - Bubbler technique is easy to perform.
   - Samples are easily prepared for liquid scintillation counting.
   - Not affected by gamma-ray or noble gas backgrounds.
   - High collection efficiency for HTO.

Disadvantages:

- Accuracy depends on atmospheric conditions (e.g., humidity, temperature, and pressure).
- Low collection efficiency for T_2 gas.

4. Portable tritium-in-air monitor that pumps tritium-contaminated air through an ionization chamber.

   Advantages:

   - Convenient, handheld monitor such as SCINTREX Tritium-in-air Monitor Model 209C.
   - Subsequent LSC is not required.

   Disadvantages:

   - External gamma radiation interferes with the measurement.
   - Functions most effectively in a low gamma-ray background area.
   - Noble gases also interfere with the measurement.

For techniques 1, 2, and 3, the resultant air samples are subsequently prepared and counted in a liquid scintillation counter to determine the airborne tritium concentration.

(c) The 60-day urine sample may be used to determine the initial intake through the relationship

$$C(t) = C(0)e^{-\lambda_e t}$$

where

$C(t)$	=	urine concentration = 500 dpm/ml at 60 days
$C(0)$	=	initial urine concentration assuming uniform mixing of body water
λ_e	=	effective disintegration constant = $\lambda_p + \lambda_b$
λ_p	=	physical disintegration constant = $0.693/T_{1/2}$
$T_{1/2}$	=	physical half-life = 12.3 years
λ_p	=	$(0.693/12.3\ \text{years}) \times (1\ \text{year}/365\ \text{days}) = 1.54 \times 10^{-4}/\text{day}$
λ_b	=	biological removal rate = F/V
F	=	daily water turnover rate = 3 l/day
V	=	free water volume = 43 l
λ_b	=	(3 l/day)/(43 l) = 0.0698/day
λ_e	=	$6.98 \times 10^{-2}/\text{day} + 1.54 \times 10^{-4}/\text{day} = 0.070/\text{day}$

With these results, $C(0)$ is determined:

$$C(0) = C(t)e^{+\lambda_e t} = \left(\frac{500\,\text{dpm}}{\text{ml}}\right) e^{+\left(\frac{0.070}{\text{day}}\right)(60\,\text{days})} = 3.33 \times 10^4\,\frac{\text{dpm}}{\text{ml}}$$

The intake (I) is

$$I = C(0)V$$

$$= \left(\frac{3.33 \times 10^4\,\text{dpm}}{\text{ml}}\right)(43\,\text{l})\left(\frac{1000\,\text{ml}}{\text{l}}\right)$$

$$= \left(\frac{1.43 \times 10^9\,\text{dis}}{\text{min}}\right)\left(\frac{1\ \text{min}}{60\,\text{s}}\right)\left(\frac{\text{Bq-s}}{\text{dis}}\right)\left(\frac{1\,\text{MBq}}{10^6\,\text{Bq}}\right) = 23.8\,\text{MBq}$$

(d) The committed effective dose $E(50)$ from the tritium intake is determined from the relationship

$$E(50) = Ief$$

where

I = inhalation intake = 59.2 MBq
e = inhalation dose coefficient = 1.8×10^{-11} Sv/Bq
f = factor accounting for tritium skin absorption (ICRP 30) = 1.5

$$E(50) = (59.2 \times 10^6\,\text{Bq})\left(1.8 \times 10^{-11}\,\frac{\text{Sv}}{\text{Bq}}\right)(1.5) = 1.6 \times 10^{-3}\,\text{Sv} = 1.6\,\text{mSv}$$

(e) The target organ for HTO is the whole body.

7.6 (a) The attenuation to the lens of the eye is determined from the intervening materials between the source and the eye:

Air	0 mg/cm^2
Respirator facepiece	250 mg/cm^2
Depth of the lens of the eye	300 mg/cm^2
Total attenuation	550 mg/cm^2

At 46 cm, the 2 Gy/h ionization chamber measurement is available, but it must be corrected for attenuation. The beta correction required by the window (7 mg/cm^2) and the plastic bag (15 mg/cm^2) is

$$\dot{D}' = \left(2.00\,\frac{\text{Gy}}{\text{h}}\right) e^{0.00435(7+15)} = 2.20\,\frac{\text{Gy}}{\text{h}}$$

This absorbed dose rate is the unattenuated beta dose in air. To obtain the eye dose, the 2.20 Gy/h absorbed dose rate is reduced to account for the respirator facepiece and depth of the lens of the eye. The absorbed dose rate at the lens of the eye is derived from the gamma and beta contributions at 46 cm:

$$\dot{D}_{\text{gamma}} = 20.0\,\frac{\text{mGy}}{\text{h}}$$

$$\dot{D}_{\text{beta}} = \left(2.20\,\frac{\text{Gy}}{\text{h}}\right) e^{-(0.00435)(550)} = 0.201\,\frac{\text{Gy}}{\text{h}} = 201\,\frac{\text{mGy}}{\text{h}}$$

The gamma attenuation is expected to be minimal, and no reduction factor is applied. This is a necessary assumption since no gamma attenuation data is supplied in the problem statement.

Since the radiation weighting factors for gamma and beta radiation types are unity, 1 Gy = 1 Sv. The equivalent dose rate to the lens of the eye is

$$\dot{H} = \dot{H}_{\text{gamma}} + \dot{H}_{\text{beta}} = w_{\text{gamma}}\dot{D}_{\text{gamma}} + w_{\text{beta}}\dot{D}_{\text{beta}}$$
$$= \left(20.0\,\frac{\text{mGy}}{\text{h}}\right)\left(\frac{1\,\text{mSv}}{1\,\text{mGy}}\right) + \left(201\,\frac{\text{mGy}}{\text{h}}\right)\left(\frac{1\,\text{mSv}}{1\,\text{mGy}}\right)$$
$$= 20.0\,\frac{\text{mSv}}{\text{h}} + 201\,\frac{\text{mSv}}{\text{h}} = 221\,\frac{\text{mSv}}{\text{h}}$$

The equivalent dose to the lens of the eye is the product of the exposure time (2 min) and equivalent dose rate:

$$H = \dot{H}t = \left(221\,\frac{\text{mSv}}{\text{h}}\right)(2\text{ min})\left(\frac{1\,\text{h}}{60\text{ min}}\right) = 7.37\,\text{mSv}$$

(b) The 10CFR20 requested annual limits are:

Organ	Annual limit (mSv) (10CFR20)
Skin	500
Lens of the eye	150
Whole body	50
Extremities	500

(c) Skin dose is evaluated at a tissue depth of 7 mg/cm^2. The tissue at risk is the basal cell layer.

(d) The deep dose equivalent is evaluated at a tissue depth of 1000 mg/cm^2.

(e) A whole-body TLD with filters for skin and eye dose was worn on the chest under the coveralls during the incident. Factors to consider when comparing the TLD dose to the calculated dose include the following:

1. The TLD has a directional response. No directionality was considered in the calculation. This correction applies to beta and gamma contributions.
2. The protective clothing (one set of coveralls) provides additional beta attenuation (29 mg/cm^2) that should be considered. This attenuation is estimated from the problem data:

$$f_\beta = \exp(-0.00435 \times 29)$$
$$= 0.88 \text{ or a 12\% beta correction}$$

3. The air provides additional attenuation:

$$x = 1.293\,\text{mg/cm}^3 \times 46\,\text{cm} = 59.5\,\text{mg/cm}^2$$
$$f_\beta = \exp(-0.00435 \times 59.5)$$
$$= 0.77 \text{ or a 23\% beta correction}$$

4. The 2 min exposure is approximate, and uncertainty in the time introduces an error in the estimate of equivalent dose.
5. Bremsstrahlung has been ignored.
6. Photon scattering has not been included.
7. Photon buildup and attenuation are ignored.
8. The actual transport distance to the eye probably differed from a constant 46 cm.
9. The energy response of the TLD must be considered.
10. A reenactment of the event should be performed and videotaped to refine the calculation and better define the calculational parameters.

(f) The facility is dominated by beta–gamma radiation fields that uniformly expose the whole body and hands. She had already received 10 mSv effective and extremity dose before the event which provided an additional 480 mSv to her hand. Since the limiting dose to the hands is 500 mSv, the worker can receive an additional 10 mSv extremity dose for the remainder of the year (500-480-10 mSv). Therefore, the worker is limited to an additional 10 mSv extremity dose for the year. Since the radiation fields uniformly irradiate the whole body and extremities, the effective dose is also limited to 10 mSv. The ICRP 26 dose recommendations are provided in Table D.6.

(g) Following ICRP 103, the effective dose is limited to 100 mSv averaged over 5 years. A maximum effective dose of 50 mSv can be received in any 1 year. The total effective dose (E) for years 1–4 is

$$E = \sum_{i=1}^{4} E_i = 10\,\text{mSv} + 20\,\text{mSv} + 30\,\text{mSv} + 20\,\text{mSv} = 80\,\text{mSv}$$

Therefore, the worker is limited to 20 mSv for year 5. The ICRP 103 effective dose recommendations are provided in Table D.6.

(h) In its 2011 Statement on Tissue Reactions, the ICRP recommends the same limits for the eye and effective dose or 100 mSv over a 5-year period. The total eye dose (H_{eye}) received during years 1–4 is

$$H_{\text{eye}} = \sum_{i=1}^{4} H_{i\,\text{eye}} = 10\,\text{mSv} + 20\,\text{mSv} + 30\,\text{mSv} + 40\,\text{mSv} = 100\,\text{mSv}$$

Since 100 mSv eye dose was received in years 1–4, no additional eye dose can be received in year 5.

(i) In ICRP 103, the dose to the skin, hands, and feet is limited to 500 mSv/year. Therefore, the worker can receive 500 mSv in year 5. There is no cumulative skin dose recommendation in ICRP 103.

7.7. (a) Chip1 is shielded by 7 mg/cm^2, and Chip 2 is shielded by a total of 1000 mg/cm^2. Given the problem data, Chip 1 provides the beta plus gamma dose and Chip 2 provides the gamma dose. Accordingly, the beta dose is derived from the difference of the output of the two chips.

$$\begin{aligned}\text{Gamma calibration factor (GCF)} &= 6000\ \text{TL units}/5\ \text{mSv}\\ \text{GCF} &= 1200\ \text{TL units/mSv}\\ \text{Beta calibration factor (BCF)} &= 750\ \text{TL units}/10\ \text{mSv}\\ \text{BCF} &= 75.0\ \text{TL units/mSv}\end{aligned}$$

$$\begin{aligned}\text{Gamma dose} &= (\text{Chip 2} - \text{Control Chip 2})/\text{GCF}\\ &= (11\,520 - 120)\ \text{TL units}/(1200\ \text{TL units/mSv})\\ &= 9.5\ \text{mSv}\end{aligned}$$

Since both chips have the same control dosimeter output, the beta dose relationship is simplified.

$$\begin{aligned}\text{Beta dose} &= (\text{Chip 1} - \text{Chip 2})/\text{BCF}\\ &= (12\,270 - 11\,520)\ \text{TL units}/(75.0\ \text{TL units/mSv})\\ &= 10.0\ \text{mSv}\end{aligned}$$

The beta dose varies as a function of depth into the tissue. This variation is provided in the problem statement as the fraction (f) of the entrance beta dose equivalent as a function of depth. Using these data the effective dose and skin dose are:

$$\begin{aligned}\text{Effective dose} &= \text{beta dose } (1000\ \text{mg/cm}^2) + \text{gamma dose } (1000\ \text{mg/cm}^2)\\ &= f\ (1000\ \text{mg/cm}^2) \times \text{beta dose} + \text{gamma dose}\\ &= 0.01 \times 10.0\ \text{mSv} + 9.5\ \text{mSv} = 9.6\ \text{mSv}\\ \text{Skin dose} &= \text{beta dose } (7\ \text{mg/cm}^2) + \text{gamma dose } (7\ \text{mg/cm}^2)\\ &= f\ (7\ \text{mg/cm}^2) \times \text{beta dose} + \text{gamma dose}\\ &= 1.0 \times 10.0\ \text{mSv} + 9.50\ \text{mSv} = 19.5\ \text{mSv}\end{aligned}$$

(b) The correct depth to evaluate dose to the lens of the eye is 300 mg/cm^2. The beta equivalent dose at 300 mg/cm^2 is 25% of the dose evaluated at 7 mg/cm^2.

$$\begin{aligned}\text{Eye dose} &= \text{beta dose } (300\ \text{mg/cm}^2) + \text{gamma dose } (300\ \text{mg/cm}^2)\\ &= f\ (300\ \text{mg/cm}^2) \times \text{beta dose} + \text{gamma dose}\\ &= 0.25 \times 10.0\ \text{mSv} + 9.5\ \text{mSv}\\ &= 12.0\ \text{mSv}\end{aligned}$$

(c) No ICRP 103 recommendations were exceeded. The effective dose was less than the 20 mSv/year recommendation. Lens of the eye and skin doses were also below the recommended 20 and 500 mSv annual values, respectively.

(d) The regulatory proposal to limit the annual effective dose to 1 mSv would be exceeded. From a technical perspective, this limit has no merit. It is well below the threshold for any observed biological effect and is a fraction of the typical annual background effective dose. It represents a more restrictive requirement than current limits based on the linear-non-threshold hypothesis. It is an example of supralinearity that was rejected in the BEIR VII report.

(e) The second regulatory proposal to abandon the linear-non-threshold hypothesis and limit the annual effective dose to 50 mSv with a cumulative lifetime limit of 150 mSv has more merit than the approach noted in the previous question. The 50 mSv is the same as the maximum ICRP 103 recommendation, which also allows 100 mSv during a 5-year period. The 150 mSv lifetime limit is only 50% larger than the ICRP 5-year recommendation. The lifetime limit is considerable less than would be allowed under ICRP 103.

The 150 mSv lifetime limit would not pose an issue for many radiation workers, but would likely be exceeded by some groups (e.g., power reactor outage contractors) and a small percentage of other radiation workers. However, abandoning the linear-non-threshold approach would have positive benefits in terms of the public's perception and their associated fears of radiation. Given improving technology and a general decreasing effective dose trend, this approach should be thoroughly reviewed.

Possible modifications would include a larger lifetime limit (e.g., 300 mSv) or allowing for extenuating circumstances by expanding the Planned Special Exposure category used in US Regulations. In addition, some older radiation workers have cumulative effective doses that exceed 150 mSv. An exception for their existing doses would be required for implementation of the proposed limits.

(f) The ICRP equivalent dose (H_T) is given by the relationship

$$H_\mathrm{T} = \sum_\mathrm{T} w_\mathrm{R} D_\mathrm{T,R}$$

This sum is illustrated with the following table:

Radiation type (R)	**w_R (μSv/μGy) × $D_{T,R}$ (μGy)**	**$H_T = w_R\, D_{T,R}$ (μSv)**
Beta	(1 μSv/μGy) (30 μGy)	= 30 μSv
Gamma	(1 μSv/μGy) (70 μGy)	= 70 μSv
Thermal neutrons	(5 μSv/μGy) (90 μGy)	= 450 μSv
10 MeV neutrons	(10 μSv/μGy) (25 μGy)	= 250 μSv
Sum		= 800 μSv

(g) A 26-year-old male radiation worker had a lifetime effective dose of 0.32 Sv. NCRP 116 recommends that the cumulative occupational exposure be limited to $10 \times N$ mSv, where N is the worker's age in years. Since the worker is 26 years old, he exceeds the NCRP 116 recommendation by $0.32 - 0.26 = 0.06$ Sv. Based on NCRP recommendations, the annual dose should be limited to 10 mSv instead of 50 mSv for workers who exceed their lifetime limit. This restriction is removed when the cumulative occupational dose limit is no longer exceeded.

(h) The worker has a cumulative effective dose in years 1–4 of

$$E = (10 + 30 + 40 + 20)\ \mathrm{mSv} = 100\ \mathrm{mSv}$$

Since ICRP 103 recommends a dose of 100 mSv averaged over 5 years, the worker should receive no exposure during the fifth year.

7.8. (a) Mechanisms by which airborne iodine is reduced during atmospheric transport include:
1. Radioactive decay.
2. Diffusion or dispersion.
3. Precipitation removal (rainout or washout).
4. Gravitational settling.
5. Ground contact.

These mechanisms are well established and negate the NUTS contentions associated with the atmosphere's capability to concentrate ^{131}I. There are no credible mechanisms for the atmosphere to concentrate radioiodine.

(b) The effective dose (E) resulting from an activity (A) of radioactive material deposited in the body is written in terms of a dose conversion factor (DCF):

$$E = A(\mathrm{DCF})$$

The activity is determined from a production equation (see Appendix B) that includes the constant rate of intake P and effective removal rate (k) of the isotope from the body

$$\frac{\mathrm{d}A}{\mathrm{d}t} = Pe^{-kt}$$

where

$$P = C(\mathrm{BR})$$

and C is the inhaled concentration of radioactive material and BR is the individual's breathing rate. The effective removal rate includes contributions from physical decay and biological removal of the radioactive material from the body. Since the solution of the production equation is derived in Appendix B, it is provided without further discussion. Given the production equation, the effective dose is

$$E = \frac{C(\mathrm{BR})(\mathrm{DCF})}{k}[1 - e^{-kT}]$$

where T is the time the individual is exposed to the airborne concentration of radioactive material. This equation clearly illustrates that the dose does not continually increase, but reaches an equilibrium value (E_{eq}) as $T \to \infty$:

$$E_{\mathrm{eq}} = \frac{C(\mathrm{BR})(\mathrm{DCF})}{k}$$

Therefore, the NUTS contention that the effective dose continues to increase because radioactive material does not decay in the body is flawed. There are no credible mechanisms to prevent radioactive material from decaying within the human body.

(c) The input or deposition rate (r_d) of ^{131}I onto the pond's surface is written as

$$r_d = v\chi = vQ\frac{\chi}{Q}$$

where

$$v = \text{deposition velocity} = 1 \times 10^{-2}\ \text{m/s}$$
$$Q = \text{release rate} = 1 \times 10^{8}\ \text{Bq/s}$$
$$\chi/Q = \text{atmospheric dispersion factor} = 1.8 \times 10^{-7}\ \text{s/m}^3$$

Using these values, the input rate to the pond surface is determined:

$$r_d = \left(1 \times 10^{-2}\ \frac{\text{m}}{\text{s}}\right)\left(1.0 \times 10^{8}\ \frac{\text{Bq}}{\text{s}}\right)\left(1.8 \times 10^{-7}\ \frac{\text{s}}{\text{m}^3}\right) = 0.18\ \frac{\text{Bq}}{\text{m}^2\text{-s}}$$

The NUTS value of 550 Bq/m^2-s is a factor of about 3000 too large.

(d) The maximum steady-state concentration (C_{eq}) of ^{131}I in the pond is obtained from a production equation where the production term (P) is

$$P = r_d S$$

where

$$r_d = \text{daily } ^{131}\text{I input rate to the pond is } 0.5\ \text{Bq/m}^2\text{-day}$$
$$S = \text{pond surface area} = 100\ \text{m} \times 10\ \text{m} = 1000\ \text{m}^2$$

$$A(T) = \frac{P}{k}[1 - e^{-kT}]$$

At equilibrium or steady-state conditions,

$$A_{eq} = \frac{P}{k}$$

and

$$C_{eq} = \frac{A_{eq}}{V} = \frac{1}{V}\frac{P}{k} = \frac{1}{V}\frac{r_d S}{k}$$

The pond volume (V) and effective removal term (k) are

$$V = Sd$$

$$k = \frac{\ln 2}{T^{p}_{1/2}} + \frac{\ln 2}{T^{b}_{1/2}}$$

where

$$d = \text{pond depth} = 1\ \text{m (average)}$$
$$T^{p}_{1/2} = {}^{131}\text{I physical half-life} = 8\ \text{days}$$
$$T^{b}_{1/2} = {}^{131}\text{I biological half-life (pond)} = 15\ \text{days}$$

Using these values, the volume, effective removal term, and equilibrium concentration are determined:

$$V = (1000\,\text{m}^2)(1\,\text{m}) = 1000\,\text{m}^3$$

$$k = \frac{\ln 2}{8\,\text{days}} + \frac{\ln 2}{15\,\text{days}} = \frac{0.133}{\text{day}}$$

$$C_{\text{eq}} = \frac{\left(0.5\,\frac{\text{Bq}}{\text{m}^2\text{-day}}\right)(1000\,\text{m}^2)}{(1000\,\text{m}^3)\left(\frac{0.133}{\text{day}}\right)} = 3.76\,\frac{\text{Bq}}{\text{m}^3}$$

Since physical decay and biological removal mechanisms are present in the pond, the water concentration reaches an equilibrium value and does not increase without limit as alleged by NUTS.

(e) The equilibrium concentration of radioactive iodine per unit mass (m) of fish (C_{f}) is determined from the production equations summarized in Appendix B:

$$A(T) = \frac{P}{k}[1 - e^{-kT}]$$

$$C_{\text{f}} = \frac{A(T)}{m} = \frac{P}{km}[1 - e^{-kT}]$$

where P/m is the production term per unit mass of fish given by

$$\frac{P}{m} = C_{\text{eq}} I$$

and C_{eq} is the ^{131}I equilibrium activity in the pond. I is the daily intake of pond water by the fish per unit mass of fish. The equilibrium concentration in the fish is

$$C_{\text{f-eq}} = \frac{P}{km} = \frac{C_{\text{eq}} I}{k}$$

$C_{\text{f-eq}}$ is determined using the information provided in the problem statement:

I	=	daily water intake by fish = 8×10^{-5} m^3/kg-day
$T^{\text{b}}_{1/2}$	=	^{131}I biological half-life (pond) = 21 days
C_{eq}	=	equilibrium activity in the pond = 10 Bq/m^3

$$k = \frac{\ln 2}{8\,\text{days}} + \frac{\ln 2}{21\,\text{days}} = \frac{0.120}{\text{day}}$$

$$C_{\text{f-eq}} = \frac{\left(10\,\frac{\text{Bq}}{\text{m}^3}\right)\left(8 \times 10^{-5}\,\frac{\text{m}^3}{\text{kg-day}}\right)}{\left(\frac{0.120}{\text{day}}\right)} = 6.67 \times 10^{-3}\,\frac{\text{Bq}}{\text{kg}}$$

(f) Inaccuracies of the stakeholder assumptions include the following:

1. Water evaporation increases the ^{131}I concentration in the pond.

2. Rain and snow decrease the ^{131}I concentration in the pond.
3. Settling of the ^{131}I decreases the concentration in the pond.
4. The NUTS model assumes that all ^{131}I incorporated into the fish stays there. The fish's excretion function reduces its radioiodine concentration and returns ^{131}I to the pond.
5. The ^{131}I removal by other food-chain members is ignored and decreases the radioiodine concentration in the pond water.
6. Physical and chemical removal of ^{131}I by the pond decreases the concentration in pond water.
7. Bottom feeding by fish may concentrate ^{131}I. However, the concentration does not increase without limit.
8. Changes in pond pH may increase the evolution of ^{131}I from the pond and decrease its concentration.
9. Changes in atmospheric pressure and temperature affect evaporation rates that alter the ^{131}I concentration in pond water.
10. Changes in the temperature of the pond affect evaporation rates that will alter the ^{131}I concentration.
11. Chemical changes in the pond can lead to ^{131}I precipitation, which decreases the soluble concentration in pond water.
12. Radioactive decay decreases the ^{131}I concentration. The short physical half-life suggests a rapid decrease over time after the production term is terminated.
13. Iodine is reactive and combines with other pond elements that affect its intake by biota.

The results of the previous question and the aforementioned list of effects suggest that the concentration of ^{131}I in fish will be less than alleged by NUTS. The net effect of these items is a decrease in the radioiodine concentration as a function of time. Some bioaccumulation may occur, but it should be off-set by the aforementioned considerations. In any event, a continuous increase is not credible.

(g) The intervenor is unlikely to accept the wealth of scientific literature. Therefore, with the intervenor present, have an independent agent (e.g., university professor or another scientific source that NUTS trusts) obtain a sample of radioiodine from the facility and count it to verify the 8-day half-life.

After completing the documentation of your responses to the stakeholder's concerns, publish the results in the report requested by the MDT President. Submit this report to the NRC if NUTS files its contentions with the regulator. With company concurrence, present this report to the public and invite peer review from credible scientific sources. Since the facility will be a neighbor for years, it is important to establish credibility within the community, insure that sound science is injected into public discussions, and vigorously challenge pseudoscientific allegations.

It is important to thoroughly refute the NUTS allegations and answer all stakeholder questions to establish trust in the facility staff and its operations. An open dialog with the stakeholders is needed throughout the lifetime of the facility to establish and maintain a positive relationship.

Part VII
Appendices

The eight appendices in Part VII provide supporting data and commentary that supplement and further develop the concepts presented in Chapters 1–7. These appendices should be consulted as they are referenced in the text or as needed by the reader.

Health Physics: Radiation-Generating Devices, Characteristics, and Hazards, First Edition. Joseph John Bevelacqua.

A
Selected Data on Radionuclides of Health Physics Interest

A.1
Introduction

Although there are over 2500 known radionuclides, it is important to become familiar with the fundamental characteristics of those systems commonly encountered in the radiation protection field. These characteristics include the decay mode, type of radiation emitted, energy of the emitted radiation, half-life, and production mode.

Table A.1 outlines the fundamental characteristics of selected radionuclides. These radionuclides include those emphasized in the American Board of Health Physics Examination Preparation Guide.

Nuclear systems move toward stability through a number of modes noted in Table A.1. These include alpha, beta, and gamma decay. Electron capture, internal conversion, positron emission, and spontaneous fission (SF) are additional nuclear deexcitation mechanisms.

A.2
Alpha Decay

Almost all naturally occurring alpha emitters are heavy elements. Alpha decay becomes the dominant decay mode for proton (neutron)-rich nuclides with $A \geq 160$ ($\geq$211). This decay mode occurs preferentially in the ^{232}Th, ^{235}U, and ^{238}U natural decay series. In the heaviest known transuranic nuclear systems, alpha emission competes with spontaneous fission as the dominant decay mode. Other decay modes (e.g., beta decay) occur but are usually not the dominant decay mechanism.

A.3
Beta Decay

In beta decay, a nucleus emits an electron and an antielectron neutrino. These particles arise from the decay of a neutron into a proton in an unstable nuclear system.

Health Physics: Radiation-Generating Devices, Characteristics, and Hazards, First Edition. Joseph John Bevelacqua.

Table A.1 Fundamental characteristics of commonly encountered radionuclides[a)–c)].

Nuclide	Major radiation emitted		Half-life	Production modes
	Type	Energy (MeV)		
^{3}H	β^-	0.018591 (max)	12.32 years	$^{2}H(n, \gamma)^{3}H$
				$^{3}He(n, p)^{3}H$
				$^{6}Li(n, \alpha)^{3}H$
				$^{10}B(n, 2\alpha)^{3}H$
				Spallation of atmospheric nuclides induced by cosmic rays
^{7}Be	γ	0.4776	53.3 days	$^{10}B(p, \alpha)^{7}Be$
	ε			Spallation of atmospheric nuclides induced by cosmic rays
^{11}C	β^+	0.960 (max)	20.36 min	$^{12}C(\gamma, n)^{11}C$
	γ	0.511		$^{12}C(n, 2n)^{11}C$
	ε			$^{14}N(p, \alpha)^{11}C$
^{13}N	β^+	1.190 (max)	9.97 min	$^{12}C(p, \gamma)^{13}N$
	γ	0.511		$^{13}C(p, n)^{13}N$
				$^{14}N(\gamma, n)^{13}N$
				$^{14}N(n, 2n)^{13}N$
				$^{16}O(p, \alpha)^{13}N$
^{14}C	β^-	0.157 (max)	5715 years	$^{13}C(n, \gamma)^{14}C$
				$^{14}N(n, p)^{14}C$
				$^{17}O(n, \alpha)^{14}C$
^{15}O	β^+	1.72 (max)	2.037 min	$^{12}C(\alpha, n)^{15}O$
	γ	0.511		$^{14}N(p, \gamma)^{15}O$
	ε			$^{16}O(\gamma, n)^{15}O$
				$^{16}O(n, 2n)^{15}O$
^{16}N	β^-	4.27 (max)	7.13 s	$^{15}N(n, \gamma)^{16}N$
		10.44 (max)		$^{16}O(n, p)^{16}N$
	γ	6.129		$^{19}F(n, \alpha)^{16}N$
		7.115		
	α	1.85		
^{18}F	β^+	0.635 (max)	1.8293 h	$^{18}O(p, n)^{18}F$
	γ	0.511		$^{19}F(n, 2n)^{18}F$
	ε			$^{16}O(^{3}He, p)^{18}F$
^{22}Na	β^+	0.546 (max)	2.604 years	$^{19}F(\alpha, n)^{22}Na$
	γ	0.511		$^{23}Na(n, 2n)^{22}Na$
		1.2745		$^{23}Na(\gamma, n)^{22}Na$
	ε			

Table A.1 (Continued)

Nuclide	Major radiation emitted		Half-life	Production modes
	Type	Energy (MeV)		
^{24}Na	β^-	1.391 (max)	14.97 h	^{23}Na(n, γ)^{24}Na
	γ	1.3686		^{24}Mg(n, p)^{24}Na
		2.7540		^{27}Al(n, α)^{24}Na
^{32}P	β^-	1.709 (max)	14.28 days	^{31}P(n, γ)^{32}P
				^{32}S(n, p)^{32}P
				^{35}Cl(n, α)^{32}P
^{35}S	β^-	0.1674 (max)	87.2 days	^{34}S(n, γ)^{35}S
				^{35}Cl(n, p)^{35}S
^{40}K	β^-	1.33	1.25×10^9 years	Naturally occurring
	γ	1.4608		
^{41}Ar	β^-	1.198 (max)	1.83 h	^{40}Ar(n, γ)^{41}Ar
	γ	1.2936		^{41}K(n, p)^{41}Ar
				^{44}Ca(n, α)^{41}Ar
^{55}Fe	γ	0.126	2.75 years	^{54}Fe(n, γ)^{55}Fe
	ε			^{58}Ni(n, α)^{55}Fe
				^{56}Fe(γ, n)^{55}Fe
^{58}Co	β^+	0.474 (max)	70.88 days	^{57}Co(n, γ)^{58}Co
	γ	0.511		^{59}Co(n, 2n)^{58}Co
		0.8108		^{58}Ni(n, p)^{58}Co
	ε			
^{60}Co	β^-	0.318 (max)	5.271 years	^{59}Co(n, γ)^{60}Co
	γ	1.1732		^{60}Ni(n, p)^{60}Co
		1.3325		^{63}Cu(n, α)^{60}Co
^{65}Zn	β^+	0.325 (max)	244.0 days	^{64}Zn(n, γ)^{65}Zn
	γ	0.511		^{66}Zn(n, 2n)^{65}Zn
		1.1155		
	ε			
^{85}Kr	β^-	0.687 (max)	10.76 years	Fission product
	γ	0.514		^{84}Kr(n, γ)^{85}Kr
^{90}Sr	β^-	0.546 (max)	28.8 years	Fission product
				^{88}Sr + n → ^{89}Sr + n → ^{90}Sr
^{90}Y	β^-	2.281 (max)	2.669 days	Fission product
				^{90}Sr daughter
				^{89}Y(n, γ)^{90}Y
				^{90}Zr(n, p)^{90}Y
				^{93}Nb(n, α)^{90}Y

(continued overleaf)

Table A.1 *(Continued)*

Nuclide	Major radiation emitted		Half-life	Production modes
	Type	Energy (MeV)		
^{99m}Tc	e^-	0.0022 0.1427	6.008 h	$^{98}Tc(n, \gamma)^{99m}Tc$
	γ	0.1405		$^{98}Mo(n, \gamma)^{99}Mo \xrightarrow{\beta^-} {}^{99m}Tc$
^{125}I	γ	0.03549	59.4 days	Fission product
	ε			$^{124}Xe + n \rightarrow {}^{125}Xe \xrightarrow{\beta^+} {}^{125}I$
	e^-			
^{129}I	β^-	0.15 (max)	1.57×10^7 years	Fission product
	γ	0.0396		$^{128}Te + n \rightarrow {}^{129}Te \xrightarrow{\beta^-} {}^{129}I$
^{131}I	β^-	0.606 (max)	8.023 days	Fission product
	γ	0.3645		$^{130}Te + n \rightarrow {}^{131}Te \xrightarrow{\beta^-} {}^{131}I$
^{133}Xe	β^-	0.346 (max)	5.243 days	Fission product
	γ	0.08099		$^{132}Xe(n, \gamma)^{133}Xe$
^{137}Cs	β^-	0.514 (max)	30.07 years	Fission product
	γ	0.6617		$^{136}Xe + n \rightarrow {}^{137}Xe \xrightarrow{\beta^-} {}^{137}Cs$
^{201}Tl	γ	0.1353 0.1674	3.043 days	$^{203}Tl(p, 3n)^{201}Pb \xrightarrow{\beta^+} {}^{201}Tl$
	ε			
^{214}Pb	β^-	0.67 (max) 0.73 (max)	27 min	^{238}U decay series
	γ	0.242 0.2952 0.3519		
^{214}Bi	β^-	1.51 (max) 1.54 (max) 3.27 (max)	19.9 min	^{238}U decay series
	γ	0.6093 1.1203 1.7645		
	α	5.450 5.513		
^{214}Po	γ	0.799	163.7 μs	^{238}U decay series
	α	7.6869		
^{218}Po	γ	0.510	3.10 min	^{238}U decay series
	α	6.0024		

Table A.1 (*Continued*)

Nuclide	Major radiation emitted		Half-life	Production modes
	Type	Energy (MeV)		
^{220}Rn	γ	0.5497	55.6 s	^{232}Th decay series
	α	6.2882		
^{222}Rn	γ	0.510	3.8235 days	^{238}U decay series
	α	5.4895		
^{226}Ra	γ	0.1862	1599 years	^{238}U decay series
	α	4.602		
		4.7844		
^{232}Th	γ	0.06381	1.4×10^{10} years	Naturally occurring
		0.14088		
	α	3.947		
		4.012		
	SF			
^{238}U	γ	0.0496	4.468×10^{9} years	Naturally occurring
	α	4.147		
		4.197		
	SF			
^{239}Pu	γ	0.0516	2.41×10^{4} years	$^{238}\text{U} + \text{n} \rightarrow$
	α	5.105		$^{239}\text{U} \xrightarrow{\beta^-} {}^{239}\text{Np} \xrightarrow{\beta^-} {}^{239}\text{Pu}$
		5.144		
		5.156		
	SF			
^{241}Am	γ	0.0595409	432.7 years	$^{239}\text{Pu} + \text{n} \rightarrow {}^{240}\text{Pu} + \text{n} \rightarrow$
		0.0263–0.955		$^{241}\text{Pu} \xrightarrow{\beta^-} {}^{241}\text{Am}$
	α	5.4430		
		5.4857		
	SF			
^{252}Cf	γ	0.0434 0.1002	2.646 years	Multiple neutron capture from a variety of nuclides (e.g., ^{238}U, ^{239}Pu, and ^{244}Cm)
	α	6.0756		
		6.1181		
	SF			

a) Baum *et al.* (2010).
b) Electron capture (ε).
c) Conversion electron (e^-).

Beta decay predominates in systems with excess neutrons (e.g., fission products). This decay mode is an efficient method to move the unstable nucleus toward the line of stability.

A.4
Gamma Emission

Gamma emission is a common nuclear decay mode. The emission of photons reduces the energy of an excited nucleus and permits it to reach its ground state or facilitates its decay to a more stable nuclear system.

A.5
Internal Conversion

Internal conversion is a process that transfers the energy of an excited nuclear system to an atomic electron. The electron, usually in the K or L shell, is ejected from the atom. This process competes with gamma emission.

A.6
Electron Capture

Orbital electron capture competes with positron emission to move a nucleus with excess protons toward the line of stability. A nucleus with excess protons captures an orbital electron, usually from the K shell. The result of this capture is the conversion of a proton into a neutron and the emission of an electron neutrino.

A.7
Positron Emission

Positron emission occurs in systems with excess protons (e.g., accelerator products). The decay results in the conversion of the proton into a neutron within the nucleus with the emission of a positron and electron neutrino. Competition between positron emission and electron capture is governed by the specific nuclear systems and their energy level structures.

A.8
Spontaneous Fission

Spontaneous fission is a decay mode of some heavy nuclear systems that splits the nucleus into two intermediate mass fragments and several neutrons. Because the maximum in the binding energy per nucleon curve occurs near $A = 56$ (^{56}Fe), nuclides with A greater than about 100 are theoretically unstable with respect to spontaneous fission. However, measurable spontaneous fission rates are only observed in nuclei with $A > 230$. This occurs because higher energies are required for fission product emission through the Coulomb barrier. For very heavy nuclei, spontaneous fission becomes an important decay mode. spontaneous fission produces a variety of radiation types including fission fragments, neutrons, gamma rays, beta particles, positrons, and neutrinos.

References

Baum, E.M., Ernesti, M.C., Knox, H.D., Miller, T.R., and Watson, A.M. (2010) *Nuclides and Isotopes, Chart of the Nuclides*, 17th edn, Knolls Atomic Power Laboratory, Bechtel Marine Propulsion Corporation, Schenectady, NY.

Bevelacqua, J.J. (2008) *Health Physics in the 21st Century*, Wiley-VCH Verlag GmbH, Weinheim.

Bevelacqua, J.J. (2009) *Contemporary Health Physics, Problems and Solutions*, 2nd edn, Wiley-VCH Verlag GmbH, Weinheim.

Bevelacqua, J.J. (2010) *Basic Health Physics, Problems and Solutions*, 2nd edn, Wiley-VCH Verlag GmbH, Weinheim.

Brookhaven National Laboratory, National Nuclear Data Center www.nndc.bnl.gov (accessed 08 August 2015).

Firestone, R.B., Frank Chu, S.Y., and Baglin, C.M. (1998) *Table of Isotopes*, 8th edn, John Wiley & Sons, Inc., New York.

ICRP (2008) Nuclear decay data for dosimetric calculations, ICRP Report No. 107. *Ann. ICRP*, **38** (3), 1.

Magill, J., Pfennig, G., Dreher, R., and Sóti, Z. (2012) *Karlsruhe Nuclide Chart*, 8th edn, Karlsruher Institut für Technologie, Karlsruhe, http://www.nucleonica.com (accessed 08 August 2015).

Shleien, B., Slaback, L.A. Jr., and Birky, B.K. (1998) *Handbook of Health Physics and Radiological Health*, 3rd edn, Lippincott, Williams, and Wilkins, Philadelphia, PA.

Tuli, J.K. (ed.) (2015) Nuclear Data Sheets. The Nuclear Structure and Decay Data in Nuclear Data Sheets is also Available in the Evaluated Nuclear Structure Data File (ENDSF), http://www.nndc.bnl.gov/ensdf (accessed 14 August 2015).

Tuli, J.K. (2004) *Nuclear Wallet Cards for Radioactive Nuclides*, Brookhaven National Laboratory, Upton, NY.

Tuli, J.K. (2011) *Nuclear Wallet Cards*, 8th edn, Brookhaven National Laboratory, Upton, NY.

Turner, J.E. (2007) *Atoms, Radiation, and Radiation Protection*, 3rd edn, Wiley-VCH Verlag GmbH, Weinheim.

B
Production Equations in Health Physics

B.1
Introduction

The mathematical framework for the accumulation or production of radioactive material is governed by linear differential equations having exponential solutions. In many health physics applications, the solutions of these equations have a similar mathematical framework that leads to a common structure in the resultant relationships. This structure produces a common solution type even if additional constraints are imposed.

For example, the assumption that radioactive material enters a system at a constant rate and is removed at a different rate leads to a set of production equations that describe a broad class of phenomena encountered by health physicists. Equations governing activation, buildup of radioactive material in a filter or demineralizer, deposition of material on a surface from a radioactive plume, and release of material into a room are examples of phenomena described in a consistent manner by production equations. This appendix describes production equations and their applications in a variety of health physics areas.

B.2
Theory

In health physics applications, the time rate of change of the radioactive material activity in a system is described by first-order linear differential equations that have exponential solutions. Since exponential forms appear throughout the field, it is expected that phenomena describing the accumulation of radioactive material have a similar mathematical structure. This appendix refers to these structures as production equations.

In order to define a general form of production relationship, consider the time rate of change of activity $\dot{A}$ associated with the continuous introduction of a radionuclide into a system or structure. For a given radionuclide

$$\dot{A} = \frac{dA}{dt} = Pe^{-Kt} \tag{B.1}$$

Health Physics: Radiation-Generating Devices, Characteristics, and Hazards, First Edition. Joseph John Bevelacqua.

where P is the production term or the rate at which activity is introduced into the system (e.g., a room, accelerator target, or filter paper), K is the total removal rate of the radionuclide from the system, and t is the time from the start of production. In order to simplify the equation resulting from the integration of Eq. (B.1), it is assumed that P is constant. The production term has units of activity per unit time (Bq/s). Examples of the production term for a variety of physical phenomena are provided in Table B.1.

When using Eq. (B.1), it is important to realize that the production equation is applied separately for each radionuclide of interest. The quantities P and K depend on the radionuclide half-life as well as its physical and chemical properties.

The total removal rate (K) has numerous components. The most common components are derived from radioactive decay (λ), biological removal (λ_b), or ventilation (λ_v). Explicit forms for these removal terms are

$$\lambda = \frac{\ln 2}{T_{1/2}} \tag{B.2}$$

$$\lambda_b = \frac{\ln 2}{T^b_{1/2}} \tag{B.3}$$

$$\lambda_v = \frac{F}{V} \tag{B.4}$$

where $T_{1/2}$ is the physical half-life, $T^b_{1/2}$ is the biological half-life, F is the fluid flow rate (e.g., ventilation flow rate) of the system, and V is the free fluid volume (e.g., free air volume of a room) of the system. The total removal rate

$$K = \lambda + \lambda_b + \lambda_v + \ldots \tag{B.5}$$

is the sum of the individual removal rates as they apply to the problem of interest. Not all terms in Eq. (B.5) appear in each application. The specific application of removal rates is addressed in subsequent discussion.

Equation (B.1) is integrated with respect to time from $t = 0$ to $t = T$ where the time T is the end of the production interval:

$$\int_0^T \dot{A}\mathrm{d}t = \int_0^T \frac{\mathrm{d}A}{\mathrm{d}t}\mathrm{d}t = \int_0^{A(T)} \mathrm{d}A = A(T) = \int_0^T Pe^{-Kt}\mathrm{d}t = P\int_0^T e^{-Kt}\mathrm{d}t \tag{B.6}$$

In Eq. (B.6), we assume that no activity is initially present in the system ($A(0) = 0$ at $t = 0$). Using this condition leads to the result

$$A(T) = \frac{P}{K}(1 - e^{-KT}) \tag{B.7}$$

Equation (B.7) provides a relationship describing the buildup of activity during the time that the production term is active. For $KT \gg 1$, the system activity reaches its maximum or equilibrium value. Accordingly, Eq. (B.7) is written as

$$A(\infty) = A_{eq} = \frac{P}{K} \tag{B.8}$$

Table B.1 Examples of production terms in health physics applications.

Physical phenomena	P (Bq/s)	Definition of terms (units)
Activation of material in an accelerator	$N\sigma\phi\lambda$	N = number of target atoms of the nuclide being activated (atoms) σ = microscopic activation cross-section for the specific activation reaction (b/atom or cm^2/atom) ϕ = activating flux of a beam of particles (particles/cm^2-s) λ = radioactive disintegration constant (1/s)
Activation of material in a reactor	$N\sigma\phi\lambda$	N = number of target atoms of the nuclide being activated (atoms) σ = microscopic activation cross-section for the specific activation reaction (b/atom or cm^2/atom) ϕ = activating flux of neutrons (neutrons/cm^2-s) λ = radioactive disintegration constant (1/s)
Deposition of radioactive material in a demineralizer bed	CFe	C = influent activity concentration of an isotope entering the demineralizer (Bq/m^3) F = flow rate of fluid through the demineralizer (m^3/s) e = isotope specific removal efficiency of the demineralizer bed
Deposition of radioactive material in a filter	CFe	C = influent activity concentration of an isotope entering the filter (Bq/m^3) F = flow rate of fluid through the filter (m^3/s) e = isotope specific removal efficiency of the filter
Surface deposition from a radioactive plume	wS	w = ground deposition rate (Bq/m^2-s) S = surface area of the deposition (m^2)
Inhalation of radioactive material	Cr	C = air concentration of radioactive material (Bq/m^3) r = breathing rate (m^3/s)
Surface deposition from a leaking radioactive fluid	CF	C = activity concentration of the isotope in the fluid leaking onto the surface (Bq/m^3) F = leak rate of the fluid onto the surface (m^3/s)
Airborne entry of ^{222}Rn into a home	CF	C = air concentration of ^{222}Rn entering the home (Bq/m^3) F = air infiltration rate entering the home (m^3/s)
Release of radioactive material from a stack	CF	C = air concentration of radioactive material being released (Bq/m^3) from a stack F = stack flow rate (m^3/s)
Release of radioactive material into a room	Q	Q = release rate of airborne radioactive material into the room (Bq/s)
Intake rate of a radioactive liquid into an organism	Cg	C = concentration of a radionuclide in a liquid (Bq/m^3) g = liquid consumption per unit time (m^3/s)
Intake rate of a radioactive solid (e.g., food) into an organism	Cg	C = concentration of a radionuclide in a solid (Bq/kg) g = solid consumption per unit time (kg/s)

The saturation or equilibrium activity is the maximum activity that can be achieved in the system.

If T is defined as the time the production term is active and t is the time after the production ceases, Eq. (B.7) is rewritten to describe the activity variation following the production interval and during the subsequent decay period:

$$A(T) = \frac{P}{K}(1 - e^{-KT})e^{-kt} \tag{B.9}$$

where k is the total removal rate postproduction, that is, during the decay time t. As a matter of specificity, $t = 0$ corresponds to the time that production ceases.

B.3
Examples of Production Equations

A number of examples are provided to illustrate the utility of the general production equation approach. These examples include the (i) activation of a target by an accelerator beam or reactor neutron source, (ii) buildup of activity on a filter or demineralizer, (iii) accumulation of activity in a pond, and (iv) release of radioactive material into a room.

B.3.1
Activation

Activation is a process described by the reaction C(c, d)D during which radiation of type c strikes a target nucleus C and produces residual nucleus D and radiation of type d. Examples of activation reactions include ^{59}Co(n, γ)^{60}Co, ^{16}O(n, p)^{16}N, ^{27}Al(n, α)^{24}Na, and ^{3}H(p, n)^{3}He.

Using the generalized production equation (Eq. (B.9)) and the production term from Table B.1 leads to a relationship that describes the activity in the target as a function of time:

$$A = N\sigma\phi(1 - e^{-\lambda T})e^{-\lambda t} \tag{B.10}$$

where N, σ, and ϕ are defined in Table B.1. For nongaseous products, the removal rates (K and k) are equal to the physical decay constant (λ). T is the irradiation time, which is the time the target is irradiated by the accelerator's beam or the time the material to be activated is exposed to the reactor's neutron fluence rate (flux). The time after the reactor is shut down or the accelerator beam is terminated is t. The steady state (saturation) or equilibrium activity is $N\sigma\phi$.

The application of Eq. (B.10) is further illustrated by considering the activation of ^{59}Co by thermal neutrons. In this example, N is the number of ^{59}Co atoms in the target, σ is the microscopic cross-section for the ^{59}Co(n$_{\text{thermal}}$, γ)^{60}Co reaction, ϕ is the number of thermal neutrons per cm^2-s, and λ is the ^{60}Co decay constant. Equation (B.10) is applied separately for each activated species.

B.3.2
Demineralizer Activity

Ion exchange is a process used in a variety of nuclear facilities to reduce the radioactive material concentration in water by removing radioactive ions and replacing them with nonradioactive ions. The device in which the ion exchange occurs is commonly called a *demineralizer*.

The activity that accumulates within a demineralizer bed is also obtained from Eq. (B.9) and Table B.1:

$$A(T) = \frac{CFe}{\lambda}(1 - e^{-\lambda T})e^{-\lambda t} \tag{B.11}$$

Equation (B.11) is also to be applied individually for each isotope trapped in the demineralizer bed. In Eq. (B.11), C, F, and e are defined in Table B.1, λ is the physical decay constant of the trapped material, T is the time the demineralizer is on-line (i.e., inlet valve is open) and removing radioactivity from the influent stream, and t is the time after the demineralizer is no longer in service (i.e., inlet valve is closed). For the demineralizer application, the total removal rate is just the physical decay constant.

Equation (B.11) also applies to filters. The saturation activity for both filters and demineralizers is CFe/λ.

B.3.3
Surface Deposition

The deposition of radioactive material onto a surface from an airborne plume is also described by a production equation. Again, using Table B.1 and Eq. (B.9), the activity deposited onto a surface is

$$A = \frac{wS}{K}(1 - e^{-KT})e^{-kt} \tag{B.12}$$

and w and S are defined in Table B.1. The removal rates k and K are discussed in the following text. Equation (B.12) is used to illustrate the versatility of the production equation methodology.

Assuming the continuous release of radioactive material from a plume, an expression for the equilibrium activity that is removed from the plume and deposited on a surface of area S is written as

$$A_{\text{eq}} = \frac{wS}{K} \tag{B.13}$$

If it is also assumed that the material deposits on the surface of a stationary body of water, such as a pond, then Eq. (B.13) still applies and

$$k = K = \lambda + \lambda_{\text{b}} \tag{B.14}$$

where λ_{b} is the biological removal rate from the pond.

If the radionuclide deposited onto the surface of the pond is also soluble in the pond water and instantaneous mixing of the radionuclide within the pond occurs, then the equilibrium concentration of the radionuclide in the pond water (C_{eq}) is determined from the relationship

$$C_{eq} = \frac{A_{eq}}{V} \tag{B.15}$$

where V is the volume of water in the pond.

Using Eqs. (B.13) and (B.15), the equilibrium concentration of a radionuclide in a pond is determined:

$$C_{eq} = \frac{wS}{KV} \tag{B.16}$$

The production concept can be extended to calculate the equilibrium concentration in an organism, such as a fish, living in the pond. The equilibrium activity concentration per unit mass (Bq/kg) in the fish ($C_{eq\text{-fish}}$) is written as

$$C_{eq\text{-fish}} = \frac{IC_{eq}}{K'} \tag{B.17}$$

where I is the intake of pond water by the fish (m^3/kg(fish)-s) and K' is the total removal rate of the isotope from the fish:

$$K' = \lambda + \lambda'_b \tag{B.18}$$

where λ'_b is the biological removal rate from the fish. A careful examination of Eq. (B.18) indicates that the term $C_{eq}I$ is P per unit mass of the fish.

B.3.4
Release of Radioactive Material into a Room

The release of airborne radioactive material into a room is obtained from Eq. (B.9) and Table B.1:

$$A = \frac{Q}{K}(1 - e^{-KT})e^{-kt} \tag{B.19}$$

where removal of radioactive material includes both physical decay and ventilation terms:

$$K = k = \lambda + \frac{F}{V} \tag{B.20}$$

In Eq. (B.20), the ventilation rate is assumed constant during the production and postproduction periods. Typically, Q is the product of the release flow rate (m^3/s) and the release concentration (Bq/m^3).

B.4
Alternative Derivation of the Production Equation

Students readily understand Eq. (B.1), but it is not the only mathematical connection to the production equation summarized in Eq. (B.7). A more fundamental derivation equates the rate of change of activity in a system to the difference between source and removal terms:

$$\dot{A} = \frac{dA}{dt} = P - KA \tag{B.21}$$

or

$$\frac{dA}{P - KA} = dt \tag{B.22}$$

Equation (B.22) can be integrated assuming no activity is present at $T = 0$:

$$\int_0^{A(T)} \frac{dA}{P - KA} = \int_0^T dt \tag{B.23}$$

which has the solution

$$-\frac{1}{K}\ln|P - KA|\,|_0^{A(T)} = T \tag{B.24}$$

Equation (B.24) is simplified to the form

$$\ln\left(\frac{P - KA\,(T)}{P}\right) = -KT \tag{B.25}$$

Exponentiation of both sides of Eq. (B.25) yields

$$1 - \frac{K}{P}A(T) = e^{-KT} \tag{B.26}$$

which simplifies to the expected production equation

$$A(T) = \frac{P}{K}(1 - e^{-KT}) \tag{B.27}$$

Equation (B.1) is the derivative of Eq. (B.27), which makes Eqs. (B.1) and (B.27) equivalent mathematical formulations. However, the decay dynamics are best illustrated by the solution sequence illustrated in this section.

B.5 Conclusions

The use of production equations provides a unified explanation for a variety of phenomena encountered in health physics applications. The specific application determines the P, K, and k values, but the form of the equation remains the same. The use of production equations greatly simplifies the understanding of a variety of health physics concepts that appear to involve dissimilar phenomena.

References

Bevelacqua, J.J. (2003) Production equations in health physics. *Radiat. Prot. Manage.*, **20** (6), 9.

Bevelacqua, J.J. (2008) *Health Physics in the 21st Century*, 2nd edn, Wiley-VCH Verlag GmbH, Weinheim.

Bevelacqua, J.J. (2009) *Contemporary Health Physics, Problems and Solutions*, 2nd edn, Wiley-VCH Verlag GmbH, Weinheim.

Bevelacqua, J.J. (2010) *Basic Health Physics, Problems and Solutions*,

2nd edn, Wiley-VCH Verlag GmbH, Weinheim.

Cember, H. and Johnson, T.A. (2008) *Introduction to Health Physics*, 4th edn, McGraw-Hill Medical, New York.

Turner, J.E. (2007) *Atoms, Radiation, and Radiation Protection*, 3rd edn, Wiley-VCH Verlag GmbH, Weinheim.

C
Key Health Physics Relationships

C.1
Introduction

A number of physical relationships are frequently utilized in health physics applications. This appendix provides a summary of important relationships encountered in a number of areas including external dosimetry, electromagnetic theory, classical mechanics, quantum mechanics, ionizing radiation, and nonionizing radiation.

The production equations that describe a range of health physics applications, including activation, filtration, and plume fallout, are addressed in Appendix B. Internal dosimetry relationships and commentary are provided in Appendix D.

The equations provided in Appendices B–D represent a set of key health physics relationships that are utilized throughout the field. Applications of many of these equations are provided in Chapters 1–7 problems.

C.2
Notation and Terminology

Within this appendix, the following notation is used:

A	=	Source activity
		Laser beam area
		Hot particle activity
B	=	Magnetic induction
		Buildup factor
C	=	Capacitance
C_a	=	Activity per unit area
C_L	=	Activity per unit length
C_v	=	Activity per unit volume
D	=	Displacement current
		Divergence angle

Health Physics: Radiation-Generating Devices, Characteristics, and Hazards, First Edition. Joseph John Bevelacqua.

DF	=	Duty factor
		Hot particle dose factor
DRCF	=	Dose rate conversion factor
E	=	Electric field strength
		Effective dose
		Energy
		Irradiance
$\dot{E}$	=	Effective dose rate
E_o	=	Rest energy
F	=	Force
GB	=	Gaussian beam
H	=	Effective dose (used to avoid confusion when the energy appears in an equation defining the effective dose)
		Magnetic field strength
		Radiant exposure
$\dot{H}$	=	Effective dose rate
I	=	Current
		Measured radiation quantity (e.g., absorbed dose, effective dose, energy fluence, equivalent dose, exposure, flux, and kerma)
J	=	Current density
L	=	Inductance
M	=	Magnetization
MPE	=	Maximum permissible exposure
N	=	Number of atoms
OD	=	Optical density
P	=	Polarization
		Power
		Pressure
PRF	=	Pulse repetition frequency
PW	=	Pulse width
Q	=	Heat
		Release rate
R	=	Ideal gas constant
		Resistance
		Radius of disk source
S	=	Source strength
		Poynting vector
STP	=	Standard temperature and pressure
T	=	Half-life
		Kinetic energy
		Temperature
V	=	Voltage
		Volume
		Potential energy
W	=	Work
Y	=	Yield
Z	=	Impedance

a	=	Acceleration
		Aperture radius
		Area
c	=	Speed of light
		Specific heat
		Heat capacity
d	=	Daughter
		Distance from laser aperture
e	=	Energy stored in an electric field per unit volume
		Energy (used to avoid confusion when the energy appears in an equation defining a related quantity such as the radiant exposure)
h	=	Release height
		Distance from disk source (on axis)
		Energy stored in a magnetic field per unit volume
		Planck's constant
		$\hbar = \frac{h}{2\pi}$
i	=	Summation index
k	=	Conversion factor (value depends on the selected units and the particular relationship)
l	=	Angular momentum
m	=	Mass
m_o	=	Rest mass
n	=	Number of moles
		Number of atoms per unit volume
p	=	Momentum
		Parent
q	=	Charge
	=	Total released activity
r	=	Radius of circular orbit
		Radius of laser beam
		Distance from radiation source
s	=	Distance
t	=	Shield thickness
		Time
		Thickness of disk source
u	=	Mean wind speed
v	=	Velocity
x	=	Shield thickness
$\times$	=	Vector cross product $[\vec{A} \times \vec{B} = \|\vec{A}\|\|\vec{B}\| \cos\theta]$
y	=	Cross wind distance
w	=	Perpendicular distance from line source
$\vec{\nabla}$	=	Gradient operator
Γ	=	Absorbed dose factor or gamma constant (Gy-m^2/MBq-h)
ΔE	=	Uncertainty in energy or the width of an energy level
Δt	=	Uncertainty in time or the lifetime of an energy level
Δp	=	Uncertainty in momentum

Δx	=	Uncertainty in position
ΔT	=	Change in temperature
β	=	Velocity relative to the speed of light = v/c
γ	=	Lorentz factor
ε	=	Permittivity
ε_{o}	=	Permittivity of free space
θ	=	Angle between the two vectors involved in the cross product
		Included angle that the point of interest makes with the ends of a line source
μ	=	Attenuation coefficient
		Permeability of a medium
$\frac{\mu}{\rho}$	=	Mass attenuation coefficient
$\frac{\mu_{\mathrm{en}}}{\rho}$	=	Mass energy absorption coefficient
μ_{o}	=	Permeability of free space
μ_{en}	=	Energy absorption coefficient
ν	=	Frequency
λ	=	Disintegration constant
		Wavelength
ρ	=	Density
		Charge density
		Gas density in an ionization chamber
		Physical density
σ	=	Microscopic cross-section
σ_y	=	Horizontal standard deviation
σ_z	=	Vertical standard deviation
χ	=	Concentration of radioactive material in a plume
χ/Q	=	Dispersion factor (s/m^3)
$\chi u/Q$	=	Dispersion factor (1/m^2)
ψ	=	Wave function
ω	=	Angular frequency
$\vec{a}$	=	An arrow over a variable indicates it is a vector quantity

C.3 Key Relationships

C.3.1 Activation

See Appendix B for a discussion of production equations.

C.3.2 Activity

$$A = \lambda N$$

$$A(t) = A(0)e^{-\lambda t}$$

$$A_{\mathrm{d}}(t) = \frac{\lambda_{\mathrm{d}} A_{\mathrm{p}}(0)}{\lambda_{\mathrm{d}} - \lambda_{\mathrm{p}}}(e^{-\lambda_{\mathrm{p}} t} - e^{-\lambda_{\mathrm{d}} t})$$

C.3.3
Attenuation

$$I(x) = I(0)Be^{-\mu x}$$

$$B(\mathrm{Fe}) = 1 + \mu x \quad \text{for small} \quad \mu x \quad \text{values}$$

$$B(\mathrm{Pb}) = 1 + \mu x/3 \quad \text{for small} \quad \mu x \quad \text{values}$$

C.3.4
Duty Factor

$$\mathrm{DF} = \frac{I_{\text{average}}}{I_{\text{peak}}} = \frac{P_{\text{average}}}{P_{\text{peak}}} = \mathrm{PW} \times \mathrm{PRF}$$

C.3.5
External Dosimetry

Dose – point source

$$\dot{E} = \frac{A\Gamma}{r^2}$$

$$\dot{H} = \frac{S}{4\pi r^2}\left(\frac{\mu_{\text{en}}}{\rho}\right)\sum_i E_i Y_i$$

Dose – line source

$$\dot{E} = \frac{C_{\mathrm{L}}\Gamma\theta}{w}$$

Dose – thin disk source

$$\dot{E} = \pi C_{\mathrm{a}}\Gamma \ln\left(\frac{R^2 + h^2}{h^2}\right)$$

Dose – thick disk source

$$\dot{E} = \frac{\pi C_{\mathrm{v}}\Gamma(1 - \exp(-\mu t))}{\mu} \ln\left(\frac{R^2 + h^2}{h^2}\right)$$

Gamma constant or dose factor

$$\Gamma = k\sum_i E_i Y_i$$

Hot particle absorbed dose

$$D = \frac{A(\mathrm{DF})t}{a}$$

Ionization chamber dose–current relationship

$$I = \rho V \frac{T_{\mathrm{STP}}}{T} \frac{P}{P_{\mathrm{STP}}} \dot{H}$$

C.3.6
Internal Dosimetry

See Appendix D.

C.3.7
Dispersion Relationships

Dispersion theory – Pasquill–Gifford equation for a gas release

$$\chi = \frac{Q}{\pi \sigma_y \sigma_z u} \exp\left[-\frac{1}{2}\left(\frac{y^2}{\sigma_y^2} + \frac{h^2}{\sigma_z^2}\right)\right]$$

Dispersion

$$\dot{E} = Q\left(\frac{\chi u}{Q}\right)\frac{1}{u}(\text{DRCF})$$

$$\dot{E} = Q\left(\frac{\chi}{Q}\right)(\text{DRCF})$$

C.3.8
Electromagnetic Relationships

Constants

Permittivity of free space: $\varepsilon_o = 8.854 \times 10^{-12}$ F/m
Permeability of free space: $\mu_o = 4\pi \times 10^{-7}$ N/A^2
Speed of light: $c = 3.0 \times 10^8$ m/s $= (\varepsilon_o\, \mu_o)^{-1/2}$
Impedance of free space: $Z = (\mu_o/\varepsilon_o)^{1/2} = 376.7$ ohms
Charge: 1 C = 1 A-s
Potential: 1 V = 1 J/C
Magnetic field: 1 T = 1 N/A-m = 1.0×10^4 gauss

Capacitance

$$C = \frac{q}{V}$$

Constitutive equations

$$\vec{D} = \varepsilon_o \vec{E} + \vec{P}$$

$$\vec{H} = \frac{\vec{B}}{\mu_o} - \vec{M}$$

Constitutive equations in a linear medium

$$\vec{D} = \varepsilon \vec{E}$$

$$\vec{H} = \frac{\vec{B}}{\mu}$$

Current

$$I = \frac{q}{t}$$

Electric field strength

$$E = \frac{F}{q}$$

Energy

$$E = qV$$

Energy stored in an electromagnetic field per unit volume

$$e = {}^1\!/_2 \varepsilon_0 E^2$$
$$h = {}^1\!/_2 \mu_0 H^2$$

Forces:
Electric force

$$\vec{F} = q\vec{E}$$

Magnetic force

$$\vec{F} = q\vec{v} \times \vec{B} = q|\vec{v}||\vec{B}| \sin\theta$$
$$\vec{B} = \mu\vec{H}$$

Lorentz force

$$\vec{F} = q(\vec{E} + \vec{v} \times \vec{B})$$

Impedance (alternating current)

$$V = ZI$$
$$V = V_0 \sin \omega t$$
$$Z = \left[R^2 + \left(\omega L - \frac{1}{\omega C}\right)\right]^{1/2}$$

Ohm's law (direct current)

$$V = IR$$

Power

$$P = IV = I^2 R$$

Poynting vector

$$\vec{S} = \vec{E} \times \vec{H} = |\vec{E}||\vec{H}| \sin\theta$$

Maxwell equations

$$\vec{\nabla}\cdot\vec{D} = \rho$$

$$\vec{\nabla}\times\vec{H} - \frac{\partial\vec{D}}{\partial t} = \vec{J}$$

$$\vec{\nabla}\cdot\vec{B} = 0$$

$$\vec{\nabla}\times\vec{E} + \frac{\partial\vec{B}}{\partial t} = 0$$

C.3.9 Mechanics Relationships

Angular momentum

$$l = mvr$$

Centrifugal force

$$F = \frac{mv^2}{r}$$

Force

$$F = ma$$

Heat

$$Q = mc\Delta T$$

Ideal gas

$$PV = nRT$$

Kinetic energy

$$T = \frac{1}{2}mv^2 = \frac{p^2}{2m}$$

Momentum

$$\vec{p} = m\vec{v} = \gamma m_o\vec{v}$$

Total energy

$$E = mc^2 = \gamma m_o c^2$$

$$E^2 = p^2c^2 + m_o^2c^4 = (m_o c^2 + T)^2$$

Relativistic mass

$$m = m_o\gamma$$

Relativistic notation

$$\beta = \frac{v}{c}$$

$$\gamma = \frac{1}{\sqrt{1-\beta^2}}$$

Rest energy

$$E_o = m_o c^2$$

Wavelength

$$c = \nu\lambda$$

Work

$$W = Fs$$

C.3.10 Relationships

Gaussian beam radius

$$r_{GB} = \left(a^2 + \frac{d^2 D^2}{4}\right)^{1/2}$$

Irradiance

$$E = \frac{P}{A}$$

Nominal Ocular Hazard Distance (NOHD)

$$\mathrm{NOHD} = \frac{2}{D}\sqrt{\frac{P}{\pi(\mathrm{MPE})} - a^2}$$

Optical density

$$\mathrm{OD} = \log_{10}\left[\frac{H}{\mathrm{MPE}}\right] = \log_{10}\left[\frac{E}{\mathrm{MPE}}\right]$$

Radiant exposure

$$H = \frac{e}{A}$$

C.3.11 Production Equations

See Appendix B.

C.3.12 Quantum Mechanics

Schrödinger equation

$$\left(-\frac{\hbar^2}{2m}\nabla^2 + V\right)\psi = E\psi$$

Uncertainty relationships

$$\Delta E \Delta t \geq \hbar$$

$$\Delta p \Delta x \geq \hbar$$

References

Bevelacqua, J.J. (2003) Production equations in health physics. *Radiat. Prot. Manage.*, **20** (6), 9.

Bevelacqua, J.J. (2005) Internal dosimetry primer. *Radiat. Prot. Manage.*, **22** (5), 7.

Bevelacqua, J.J. (2008) *Health Physics in the 21st Century*, Wiley-VCH Verlag GmbH, Weinheim.

Bevelacqua, J.J. (2009) *Contemporary Health Physics, Problems and Solutions*, 2nd edn, Wiley-VCH Verlag GmbH, Weinheim.

Bevelacqua, J.J. (2010) *Basic Health Physics, Problems and Solutions*, 2nd edn, Wiley-VCH Verlag GmbH, Weinheim.

Cember, H. and Johnson, T.A. (2008) *Introduction to Health Physics*, 4th edn, McGraw-Hill Medical, New York.

Halliday, D., Resnick, R., and Walker, J. (2014) *Fundamentals of Physics Extended*, 10th edn, John Wiley & Sons, Inc., Hoboken, NJ.

Turner, J.E. (2007) *Atoms, Radiation, and Radiation Protection*, 3rd edn, Wiley-VCH Verlag GmbH, Weinheim.

D
Internal Dosimetry

D.1
Introduction

Internal dosimetry can be overwhelming and sometimes confusing because there is a wealth and diversity of models and terminology. The International Commission on Radiological Protection (ICRP) and Medical Internal Radiation Dose (MIRD) methodologies are the most commonly used internal dosimetry models. The various ICRP and MIRD models are similar in terms of their assumptions and defining equations. This similarity is obscured by differing terminology and notation. These differences contribute to the confusion and can limit a full understanding of these models. Emphasizing the definition of absorbed dose and using this definition to illustrate the MIRD and ICRP terminology and notation minimize the confusion.

Contemporary internal dosimetry models began with the single compartment models of ICRP 2, 10, and 10A (ICRP 2/10/10A). The MIRD methodology and ICRP 26 and 30 (ICRP 26/30) developed the concept of source and target organs. ICRP 60 and supporting publications including ICRP 66 (ICRP 60/66/30) continued to refine the internal dosimetry methodology. In this appendix, the notation ICRP 2/10/10A, ICRP 26/30, and ICRP 60/66/30 is used to refer to the defining internal dosimetry publications and major supporting documents. Additional refinement is provided as part of the 2007 ICRP recommendations ICRP 103 and supporting publications ICRP 66 and ICRP 100 (ICRP 103/66/100).

This appendix presents the essential elements of the most frequently utilized internal dosimetry models. The presentation begins by defining the key elements of the MIRD and ICRP models in terms of the absorbed dose. With the key elements established, the MIRD and ICRP methodologies are presented in additional detail.

D.2
Overview of Internal Dosimetry Models

As an introduction to the MIRD and ICRP internal dosimetry models, the absorbed dose rate following an intake of radioactive material is calculated. If an

Health Physics: Radiation-Generating Devices, Characteristics, and Hazards, First Edition. Joseph John Bevelacqua.

isolated (single compartment) organ having a mass m contains an activity $q(t)$ of a radionuclide that emits a radiation type of energy E per disintegration, then the initial absorbed dose rate ($\dot{D}_o$) to this organ is

$$\dot{D}_o = k\frac{q(0)E}{m} \tag{D.1}$$

where k is a conversion factor and $q(0)$ is the initial activity in the organ. If $q(0)$ is expressed in μCi, E in MeV/disintegration, and m in grams, then

$$k = 2.13\,(\text{rad/h})\,(\text{g-dis/MeV-}\mu\text{Ci}) \tag{D.2}$$

Equation (D.2) is presented in English units because it has historical roots and is still used in the literature in the United States.

The dose rate as a function of time t is written in terms of the initial absorbed dose rate

$$\dot{D}(t) = \dot{D}_o \exp(-\lambda_{\text{eff}} t) \tag{D.3}$$

where λ_{eff} is the effective removal rate from the organ and

$$\lambda_{\text{eff}} = \lambda_p + \lambda_b \tag{D.4}$$

In Eq. (D.4), λ_p is the physical removal rate (disintegration constant) and λ_b is the biological removal rate. The removal rates are related to their respective half-lives (T) through the relationship

$$\lambda = \frac{\ln(2)}{T} \tag{D.5}$$

The absorbed dose (D) is the integral of the absorbed dose rate with respect to time. Equations (D.1) and (D.3) lead to an expression for the absorbed dose

$$D = \int_0^T \dot{D}(t)\text{d}t = \int_0^T \dot{D}_o \exp(-\lambda_{\text{eff}} t)\text{d}t = \int_0^T k\frac{q(0)E}{m} \exp(-\lambda_{\text{eff}} t)\text{d}t \tag{D.6}$$

Equation (D.6) simplifies by recognizing that only the activity in the organ varies with time:

$$D = \frac{kE}{m}\int_0^T q(0)\exp(-\lambda_{\text{eff}} t)\text{d}t \tag{D.7}$$

Equation (D.7) can be compared to the basic equations for internal dose within the MIRD (Eq. (D.8)) and ICRP (Eq. (D.9)) methodologies:

$$\overline{D} = \widetilde{A} S \tag{D.8}$$

$$H_{50,T} = 1.6 \times 10^{-10}\frac{\text{Sv-g}}{\text{MeV}} U_S \text{SEE} \tag{D.9}$$

where $\overline{D}$ is the mean absorbed dose delivered to an organ, $\widetilde{A}$ is the total cumulated activity, S is the mean dose per unit cumulated activity, $H_{50,T}$ is the 50-year committed dose equivalent (CDE), U_S is the number of transformations in the source organ over 50 years, and SEE is the specific effective energy. The constant 1.6×10^{-10} in Eq. (D.9) is the product of conversion factors 1.6×10^{-13} J/MeV and 1000 g/kg, and its units include the definition that a Sv is equivalent to a J/kg.

Table D.1 Comparison of the MIRD and ICRP models.

Equation (D.7) term	Corresponding quantities	
	MIRD	ICRP
D	$\overline{D}$	$H_{50,T}$
kE/m	S	$1.6 \times 10^{-10} \frac{\text{Sv-g}}{\text{MeV}}$ SEE
T	∞	50 year
$\int_0^T q(0)\exp(-\lambda_{\text{eff}}t)\text{d}t$	$\widetilde{A}$	U_S

A comparison of Eq. (D.7) to Eqs. (D.8) and (D.9) leads to the explicit identifications summarized in Table D.1. Table D.1 illustrates that the ICRP and MIRD methodologies are essentially equivalent. With the exception of terminology, the major difference is in the upper limit of integration in Eq. (D.7) (i.e., $T = 50$ years for the ICRP and $T = \infty$ for MIRD).

Equations (D.7)–(D.9) and the comparisons of Table D.1 illustrate the inherent consistency of the internal dosimetry models. With this consistency established, model-specific details are presented. These aspects should be periodically reviewed with regard to Table D.1 to simplify and unify the presented concepts.

D.3 MIRD Methodology

The Committee on MIRD of the Society of Nuclear Medicine developed a methodology to perform radiation absorbed dose calculations. These calculations assess the absorbed dose associated with the administration of radiopharmaceuticals for medical studies including imaging, therapy, and metabolic applications.

The MIRD technique is a computational methodology that facilitates absorbed dose calculations for specified target organs from radioactive decays that occur in source organs. The source organs contain the radioactive material, and the dose is calculated in the target organs. The target and source organs can be the same tissue. In subsequent discussion, the terms tissue and organ are used interchangeably.

To specify the MIRD methodology, it is necessary to define several terms. The mean energy emitted per transition (Δ), in Gy-kg/Bq-s, is the product of the mean particle energy (E) in megaelectronvolts or joules and the number of particles emitted per nuclear transformation (n):

$$\Delta = KEn \tag{D.10}$$

where K is a conversion factor. Within the MIRD methodology, particles were originally defined to be photons, beta particles, or positrons. These are the radiation types used most frequently in nuclear medicine procedures. Recent work expands this set to include alpha particles. Given their limited range, alpha radionuclides selectively deposit their energy in diseased tissue and minimize the dose to healthy tissue.

The cumulated activity or the total number of nuclear transitions occurring within the source organ from time $t = 0$ to time T is

$$\tilde{A} = \int_0^T A(t)\mathrm{d}t \tag{D.11}$$

The activity as a function of time is

$$A(t) = A(0)\exp(-\lambda_{\mathrm{eff}}t) \tag{D.12}$$

Using Eq. (D.12), the cumulated activity is simplified if the MIRD upper integration limit of $T = \infty$ is selected. In this case, the total cumulated activity is

$$\tilde{A}(\infty) = \frac{A(0)}{\lambda_{\mathrm{eff}}} = \frac{A(0)T_{\mathrm{eff}}}{\ln(2)} = 1.44T_{\mathrm{eff}}A(0) \tag{D.13}$$

where

$$T_{\mathrm{eff}} = \frac{T_{\mathrm{p}}T_{\mathrm{b}}}{T_{\mathrm{p}} + T_{\mathrm{b}}} \tag{D.14}$$

The initial activity in the organ, $A(0)$, is related to the intake activity $q(0)$:

$$A(0) = f_2 q(0) \tag{D.15}$$

where f_2 is the fraction of the intake reaching the organ of interest.

The total energy emitted by the source organ is the product of Δ and the cumulated activity. However, only a fraction (f) of this energy is deposited in the target organ, which is the location of interest in the dose calculation. With these quantities and knowledge of the mass of the target organ (m), the mean absorbed dose $\overline{D}$ is

$$\overline{D} = \frac{\tilde{A}\Delta f}{m} \tag{D.16}$$

The MIRD methodology also defines the specific absorbed fraction (F):

$$F = \frac{f}{m} \tag{D.17}$$

where f is the energy absorbed by the target divided by the energy emitted by the source. The specific absorbed fraction represents the mean target dose per unit energy emitted by the source. Therefore, the mean absorbed dose is

$$\overline{D} = \tilde{A}\Delta F \tag{D.18}$$

The MIRD methodology defines the mean dose to the target (T) per unit cumulated activity in the source (S) in mGy/MBq-s:

$$S(T \leftarrow S) = \frac{\Delta f}{m} = \Delta F \tag{D.19}$$

Equations (D.18) and (D.19) yield the expected MIRD dose relationship

$$\overline{D} = \widetilde{A}\ S(T \leftarrow S) \tag{D.20}$$

In Eq. (D.20), most of the metabolic factors are contained in the $\widetilde{A}$ term, which depends on the uptake and subsequent biological elimination of the radiopharmaceutical by the source organ. The S factor represents the physical decay characteristics of the radionuclide, the range of the emitted radiations, and the organ size and configuration. If a standard anatomy is utilized, S can be calculated and tabulated for a variety of radionuclides and source–target combinations. MIRD Pamphlet No. 11 provides a convenient tabulation of these S factors.

D.4
ICRP Methodology

The ICRP internal dosimetry models are based in part on evolving assessments of the biological effects of ionizing radiation (BEIR). These assessments affect the selection of the model's organs/tissues and their associated weighting factors. The radionuclide transport, metabolic processes, and organ models determine the calculated doses that lead to recommendations regarding occupational exposures. These ICRP model aspects are reviewed in subsequent sections of this appendix.

The specific ICRP recommendations are incorporated into national and international regulations. ICRP 2/10/10A is the basis for the 29CFR1910 radiation protection regulations utilized by the Occupational Safety and Health Administration. The US Department of Energy uses the dose limits in ICRP 26 and the formalism of ICRP 60/66/30 in its 10CFR835 regulations. ICRP 26/30 is the basis for the US Nuclear Regulatory Commission's ionizing radiation regulations (10CFR20). Current international regulations utilize ICRP 60/66/30. The 2007 ICRP 103/66/100 recommendations replaced the ICRP 60/66/30 recommendations.

D.5
Biological Effects

The ICRP models should be viewed in their historical context. The models continue to evolve and incorporate available data regarding the biological effects of ionizing radiation.

A portion of the scientific basis for ICRP 26/30, ICRP 60/66/30, and ICRP 103/66/100 is summarized in Table D.2. ICRP 26/30 are based in part on the BEIR III Report. In BEIR III, the dose–response relationships for both solid tumors and leukemia are defined to have a linear quadratic (LQ) dose–response relationship:

$$f(d) = ad + bd^2 \tag{D.21}$$

Table D.2 Comparison of the basis for recent ICRP models.

ICRP model	Basis	Dose–response relationship[a)]		Risk model
		Solid tumors	Leukemia	
26/30	BEIR III	LQ	LQ	Absolute
60/66/30	BEIR V	L	LQ	Relative
103/66/100	BEIR VII	L	LQ	Various[b)]

a) (L) linear; (LQ) linear quadratic.
b) See Table D.4.

where $f(d)$ is the effect of the radiation dose, d is the effective dose, and a and b are the risk coefficients. BEIR III based its preferred age-specific cancer model on the absolute (additive) risk model

$$r(d) = r_o + f(d)g(\beta) \tag{D.22}$$

where $r(d)$ is the number of cancers of a specific type in the population group, r_o is the natural incidence of the specific cancer type, and $g(\beta)$ is the excess risk function that contains the time dependence of these effects.

BEIR V forms a portion of the basis for ICRP 60/66/30. In BEIR V, the dose–response model is linear (L) for solid tumors:

$$f(d) = cd \tag{D.23}$$

and LQ for leukemia. In Eq. (D.23), c is a risk coefficient. In contrast to BEIR III, BEIR V uses a relative (multiplicative) risk model:

$$r(d) = r_o[1 + f(d)g(\beta)] \tag{D.24}$$

Both BEIR III and BEIR V assume the dose–response models have no threshold. That is any dose no matter how small has an effect (detriment).

There are significant differences between the BEIR III and BEIR V reports. Table D.3 illustrates the variation in both leukemia and nonleukemia (solid tumor) cancer risk estimates. The solid tumors include respiratory, digestive, breast, and other cancer types. For leukemia, BEIR V leads to a factor of 4–5 greater risk. A similar increase of about 3–5 occurs for nonleukemia cancers if BEIR III and V relative risk models are compared.

Considerably larger factors of 11–19 occur for nonleukemia cancers if the BEIR III absolute risk model is compared to BEIR V's relative risk model. BEIR VII supports a combination of absolute and relative risk models, and it is compared to BEIR III and BEIR V in Table D.4.

The BEIR VII Report is consistent with BEIR V. The key elements of BEIR VII and their comparison with BEIR III and BEIR V are summarized in Table D.4.

The BEIR VII total cancer morality and leukemia risk estimates from radiation exposure have not changed significantly from BEIR V. BEIR VII's risk estimates are based on expanded epidemiological data including cancer incidence data and

Table D.3 Lifetime cancer risk estimates (deaths per 100 000 persons)[a].

Cancer type	Continuous lifetime exposure 1 mGy/year		Instantaneous exposure 0.1 Gy	
	Male	Female	Male	Female
Leukemia				
BEIR III	15.9	12.1	27.4	18.6
BEIR V	70	60	110	80
BEIR V/BEIR III	4.4	5.0	4.0	4.3
Nonleukemia				
BEIR III (absolute)	24.6	42.4	42.1	66.5
BEIR III (relative)	92.9	118.5	192	213
BEIR V (relative)	450	540	660	730
BEIR V/BEIR III (relative)	4.8	4.6	3.4	3.4
BEIR V/BEIR III (absolute)	18.3	12.7	15.7	11.2

a) Derived from Bevelacqua (2009).

Table D.4 BEIR III, V, and VII comparison.

Parameter/quantity	BEIR III (1980a)	BEIR V (1990)	BEIR VII (2006)
Dose–response model – solid tumors	LQ[a]	L[a]	L
Dose–response model – leukemia	LQ	LQ	LQ
Preferred risk model	Absolute	Relative	Various[b),c)]
Dosimetry system[d]	T65D	DS86	DS02
DDREF[e] (range)	—	2–10	1.1–2.3
DDREF (adopted)	—	—	1.5 for linear models

a) (L) linear; (LQ) linear quadratic.
b) For solid cancers other than lung, breast, and thyroid, the preferred risk model is a weighted average (on a logarithmic scale) of relative and absolute risk models with relative risk given a weight of 0.7 and absolute risk a weight of 0.3. These weights are reversed for lung cancer. The preferred breast cancer model is based on the absolute risk model. The preferred thyroid cancer model is based on the relative risk model.
c) For leukemia the preferred risk model is a weighted average (on a logarithmic scale) of relative and absolute risk models with relative risk given a weight of 0.7 and absolute risk a weight of 0.3.
d) T65D: Tentative 1965 Dosimetry; DS86: Dosimetry System 1986; DS02: Dosimetry System 2002.
e) Dose and dose rate effectiveness factor.

15 years of additional mortality follow-up for the Japanese atomic bomb survivors. Studies involving occupational and environmental exposure were evaluated but not utilized in BEIR VII.

In formulating its risk models, the BEIR VII Report used the revised Dosimetry System 2002 (DS02) for atomic bomb survivors as a portion of the basis for evaluation of the dependence of risk on dose. The risk models were developed from atomic bomb survivor and medical therapy patient data.

BEIR VII also reviewed the dose–response model and its functional dependence, the emergence of hormesis as a positive consequence of the radiation dose, and the existence of a threshold for radiation-induced effects. According to BEIR VII, the updated molecular and cellular data from studies of radiation exposure do not support the postulate that low doses of low-LET radiation are more harmful than predicted by the linear-nonthreshold (LNT) model. That is, the contention that the dose–response curve exhibits supralinearity is not supported. In addition, the updated molecular and cellular data from studies of radiation exposure do not support hormesis. BEIR VII reaffirms the LNT hypothesis and concludes there is cellular level evidence for the LNT approach. Thresholds were considered but not endorsed as representing the best scientific view of low-dose risk.

BEIR VII also noted that other effects were observed to exist but were too small to definitively quantify. In particular, BEIR VII concluded that the genetic risks of low-dose, low-LET radiation are very small compared to the baseline frequencies of genetic disease. In addition, a dose–response for noncancer mortality in atomic bomb survivors has been demonstrated, but data are not sufficient to determine if this effect exists at low doses and dose rates. BEIR VII does not provide risk estimates for noncancer mortality.

Reports such as BEIR VII are important because they refine the internal dosimetry models and affect the risk estimates. Consequently, the conclusions of BEIR VII carry significant weight and ideally are clear, unambiguous, and widely accepted. Appendix H reviews the impact of the BEIR reports and their associated assumptions on radiation protection regulations.

D.6
ICRP 26/30 and ICRP 60/66/30 Terminology

ICRP 26/30 and ICRP 60/66/30 utilize different terminology to describe similar quantities. The ICRP 103/66/100 and ICRP 60/66/30 terminology is consistent. Table D.5 summarizes the terminology appropriate to each model. The specific terms are defined in subsequent sections of this appendix.

Table D.5 Terminology utilized in recent ICRP models.

ICRP model	Terminology	
	Organ dose	Whole body dose
26/30	Committed dose equivalent ($H_{50,T}$)	Effective dose equivalent[a] (H_E)
60/66/30	Equivalent dose (H_T)	Effective dose (E)
103/66/100	Equivalent dose (H_T)	Effective dose (E)

a) US regulations use the term committed effective dose equivalent.

D.7
ICRP 26 and ICRP 60 Recommendations

Prior to reviewing specific ICRP internal dose formalism, the ICRP 26 and ICRP 60 recommendations are outlined. The ICRP recommendations are based on the following two general principles that radiation protection guidance should (i) prevent the occurrence of clinically significant radiation-induced deterministic effects and (ii) limit the risk of stochastic effects to a reasonable level.

The National Council on Radiation Protection and Measurement (NCRP) also adopts these two general principles. In addition, the NCRP recommends that risk be limited over a working lifetime to be no greater than the risk of accidental death in a safe industry.

The deterministic effects have a threshold. The term *deterministic effect* was introduced in ICRP 60. Deterministic effects include erythema, cataracts, impairment of fertility, and depletion of blood-forming cells in bone marrow. These effects only occur in irradiated individuals. By keeping the dose below the threshold for the deterministic effect, the detriment is eliminated. With deterministic effects, the severity of the effect varies with dose. ICRP 26 refers to deterministic effects as nonstochastic effects.

Stochastic effects include cancer and hereditary effects. These effects occur in the general population as well as in irradiated individuals. The probability of a stochastic effect increases with increasing dose without threshold.

With ICRP 26, these recommendations are implemented by limiting the effective dose equivalent and CDE and by establishing stochastic and nonstochastic annual limits on intakes (ALIs). Considering the purpose of this appendix, the applicable ICRP recommendations are summarized in Table D.6. In Table D.6, the deep dose equivalent, eye dose equivalent, and skin dose equivalent are evaluated at depths of 1000, 300, and 7 mg/cm^2, respectively.

Table D.6 Applicable ICRP 26, ICRP 60, and ICRP 103 recommendations.

Dose recommendation	Dose (mSv)		
	ICRP 26	ICRP 60	ICRP 103
Annual	50[a)]	50 maximum[b)]	50 maximum[b)]
Cumulative	None	100 over 5 years[b)] 20/year average[b)]	100 over 5 years[b)] 20/year average[b)]
Eye	150[c)]	150[d)]	100 over 5 years[d),e)] 20/year average[d),c)]
Skin, hands, and feet	500[c)]	500[d)]	500[d)]

a) Effective dose equivalent.
b) Effective dose.
c) Committed dose equivalent.
d) Equivalent dose.
e) Modified by ICRP Statement on Tissue Reactions (2011).

In ICRP 60, the restrictions on effective dose are sufficient to ensure the avoidance of deterministic effects in all body tissues except the lens of the eye and the skin. The limits for the eye and skin preclude deterministic effects. Therefore, only a stochastic ALI is needed in the ICRP 60/66/30 internal dosimetry formulation.

The ICRP 60 and 103 recommendations are similar. As noted in Table D.6, the ICRP 103 eye dose recommendation was subsequently modified to account for recent epidemiological evidence. These data suggest that some tissue reaction effects having a very late manifestation might have lower threshold doses than previously considered. For the lens of the eye, the ICRP considers the absorbed dose threshold to be 0.5 Gy.

D.7.1
Calculation of Internal Dose Equivalents Using ICRP 26/30

Internal dose equivalents are calculated in a variety of ways. These include the use of the annual limit on intake, derived air concentration (DAC), and SEE and U_S values.

Within the ICRP 26/30 methodology, the stochastic and nonstochastic recommendations for internal dose equivalents are developed in terms of the ALI. Following ICRP 26/30, the ALI is defined to be the largest value of intake that satisfies both of the inequalities of Eqs. (D.25) and (D.26). In Eqs. (D.25) and (D.26), $\mathrm{ALI_S}$ is the stochastic ALI and $\mathrm{ALI_{NS}}$ is the nonstochastic ALI:

$$\mathrm{ALI_S} \sum_T w_T H'_{50,T} \leq 0.05\,\mathrm{Sv} \quad \text{for stochastic effects} \tag{D.25}$$

$$\mathrm{ALI_{NS}} H'_{50,T} \leq 0.5\,\mathrm{Sv} \quad \text{for non-stochastic effects} \tag{D.26}$$

where w_T is the ICRP 26/30 organ/tissue weighting factor and $H'_{50,T}$ is the dose per unit intake (Sv/Bq) which yields the correct units for the ALI. The organ/tissue weighting factors for ICRP 26/30 and 60/66/30 are summarized in Table D.7.

ICRP 26/30 form the basis for the current US regulations embodied in 10CFR20 for US Nuclear Regulatory Commission licensees and the 10CFR835 dose limits for US Department of Energy licensees. The 10CFR20 regulations require the calculation of individual organ doses (i.e., the CDE) and the committed effective dose equivalents (CEDEs). The dose limits are based on the risk to the various organs/tissues included in the ICRP 26/30 model. The CDE and CEDE are calculated in terms of the intake (I) and ALI values as follows:

$$\mathrm{CDE} = H_{50,T} = \frac{I}{\mathrm{ALI_{NS}}} 0.5\,\mathrm{Sv} = 1.6 \times 10^{-10}\,\frac{\mathrm{Sv\text{-}g}}{\mathrm{MeV}} U_S \mathrm{SEE}(T \leftarrow S) \tag{D.27}$$

$$\mathrm{CEDE} = H_E = \sum_T w_T H_{50,T} = \frac{I}{\mathrm{ALI_S}} 0.05\,\mathrm{Sv} \tag{D.28}$$

Equations (D.27) and (D.28) can also be rewritten in terms of the DAC:

$$\mathrm{DAC} = \frac{\mathrm{ALI}}{2400\,\mathrm{m^3}} \tag{D.29}$$

Table D.7 Weighting factors for the ICRP 26 and 60 models.

Organ or tissue	ICRP 26/30	ICRP 60/66/30
Gonads	0.25	0.20
Breast	0.15	0.05
Red bone marrow	0.12	0.12
Lung	0.12	0.12
Thyroid	0.03	0.05
Bone surfaces	0.03	0.01
Stomach	—	0.12
Colon	—	0.12
Esophagus	—	0.05
Bladder	—	0.05
Skin	—	0.01
Liver	—	0.05
Remainder	0.30[a]	0.05[b]

a) Five highest other organs.
b) Adrenals, brain, small intestine, spleen, kidneys, muscle, pancreas, upper large intestine, thymus, and uterus.

D.7.2
Calculation of Equivalent and Effective Doses Using ICRP 60/66/30

Within the ICRP 60/66/30 formalism, new dose terminology is introduced including the equivalent dose and the effective dose. The equivalent dose (H_T) is defined as

$$H_T = \sum_R w_R D_{T,R} \tag{D.30}$$

where w_R is the radiation weighting factor and $D_{T,R}$ is the average absorbed dose in tissue T due to radiation of type R. The ICRP 60/66/30 radiation weighting factors are provided in Table D.8.

The effective dose (E) is defined as

$$E = \sum_T w_T H_T \tag{D.31}$$

Using Eq. (D.30), the effective dose is written as

$$E = \sum_R w_R \sum_T w_T D_{T,R} = \sum_T w_T \sum_R w_R D_{T,R} \tag{D.32}$$

Within ICRP 60/66/30, only one ALI is required. The committed effective dose $E(50)$ is written as

$$E(50) = \frac{I}{\text{ALI}} 0.02\,\text{Sv} = \sum_{T=1}^{12} w_T H_T(50) + w_{\text{remainder}} \frac{\sum_{T=13}^{22} m_T H_T(50)}{\sum_{T=13}^{22} m_T} \tag{D.33}$$

Table D.8 ICRP 60 radiation weighting factors[a)].

Type and energy range[b)]	Radiation weighting factor
Photons (all energies)	1
Electrons and muons (all energies)[c)]	1
Neutrons	
<10 keV	5
10–100 keV	10
>100 keV to 2 MeV	20
>2–20 MeV	10
>20 MeV	5
Protons, other than recoil protons (>2 MeV)	5
Alpha particles, fission fragments, and heavy nuclei	20

a) All values relate to the radiation incident on the body or, for internal sources, emitted from the source.
b) The choice of values for other radiation types is discussed in Appendix A, ICRP 60.
c) Excluding Auger electrons emitted from nuclei bound to DNA.

where $H_T(50)$ is the committed equivalent dose, m_T is the mass of the remainder tissue, and $w_{\text{remainder}} = 0.05$. In Eq. (D.33), the first sum is over the 12 organs/tissues with assigned weighting factors (see Table D.7) and the second sum is over the 10 remainder organs/tissues (i.e., adrenals, brain, small intestine, spleen, kidneys, muscle, pancreas, upper large intestine, thymus, and uterus). Equation (D.33) is applicable whenever one of the 12 organs with assigned weighting factors has the largest committed equivalent dose. In the exceptional case in which one of the remainder organs receives a committed equivalent dose in excess of the highest committed equivalent dose in any of the 12 organs for which a weighting factor is assigned, a weighting factor of 0.025 is applied to that remainder organ or tissue. A weighting factor of 0.025 is also assigned to the average dose in the rest of the remainder, and in the exceptional case the $E(50)$ equation has the form

$$E(50) = \sum_{T=1}^{12} w_T H_T(50) + 0.025 H_{T'}(50) + 0.025 \frac{\sum_{T=13}^{22} m_T H_T(50) - m_{T'} H_{T'}(50)}{\sum_{T=13}^{22} m_T - m_{T'}} \tag{D.34}$$

where $m_{T'}$ is the mass of the remainder tissue or organ in which the committed equivalent dose is calculated to be higher than in any of the 12 specified tissues/organs with assigned weighting factors and $H_{T'}(50)$ is the committed equivalent dose in that remainder tissue/organ.

The careful reader will note that the first term in Eq. (D.33) contains no ALI subscript since the ICRP 60/66/30 formulation only utilizes a stochastic ALI. The 0.02 Sv (20 mSv) multiplier is a direct consequence of Table D.6 cumulative effective dose recommendation.

D.8 ICRP 103/66/100 Methodology

ICRP 103 documents the 2007 Recommendations of the ICRP, and it replaces the ICRP 60 recommendations. The ICRP 103 recommendations incorporate the results of the BEIR VII report and supporting DS02. In addition, the LNT assumption is integral to ICRP 103. The 2007 ICRP recommendations incorporate the ICRP 100 human alimentary tract model (HATM) and the human respiratory tract model (HRTM) presented in ICRP 66.

The HRTM and HATM are replacement models for the predecessor models of ICRP 30. The ICRP 30 models were developed for the calculation of occupational doses. An improvement of both the HATM and the HRTM is their capability to calculate doses from intakes by children as well as male and female adults. Therefore, the new models are applicable to environmental as well as occupational exposures. Another important development in the HATM and HRTM is the calculation of doses to target regions containing cells considered susceptible to cancer induction.

D.8.1 Radiation Effects, Tissue Weighting Factors, and Radiation Weighting Factors

The LNT hypothesis is a fundamental assumption used in the formulation of radiation protection approaches including ICRP 103. It is the basis for the averaging and summing of doses, the effective dose concept, the collective dose concept, individual dosimetry, and keeping dose records.

ICRP 103 revised the tissue weighting factors and significant revisions occurred for the breast, gonads, and treatment of remainder tissues. The ICRP 103 tissue weighting factors are provided in Table D.9. Table D.9 compares the ICRP 103 values with values from other ICRP and UNSCEAR approaches. For the purposes of radiological protection, the tissue weighting factors are assumed valid for both sexes and modeled age groups.

Biological and dosimetric considerations were evaluated in the reappraisal of the ICRP 60 radiation weighting factors. Most relative biological effectiveness values were derived from high doses and were extrapolated to the low-dose regime appropriate for radiological protection purposes. The w_R values for photons, beta particles, and alpha particles remain the same as the ICRP 60 values. The proton radiation weighting factor was revised because the ICRP 60 value was judged to be a significant overestimate of the biological effectiveness. ICRP 103 also included a charged pion radiation weighting factor having a value of 2. For neutrons, the w_R values are energy dependent and based on the continuous function

$$w_R = \begin{cases} 2.5 + 18.2e^{-[\ln(E_n)]^2/6}, & E_n < 1\,\text{MeV} \\ 5.0 + 17.0e^{-[\ln(2E_n)]^2/6}, & 1 \le E_n \le 50\,\text{MeV} \\ 2.5 + 3.25e^{-[\ln(0.04E_n)]^2/6}, & E_n > 50\,\text{MeV} \end{cases} \tag{D.35}$$

Table D.9 Tissue weighting factors for various models.

Tissue	ICRP 26 (1977)	UNSCEAR (1988)	ICRP 60 (1991)	ICRP 103 (2007)
Gonads	0.25	—	0.20	0.08
Breast	0.15	0.05	0.05	0.12
Bone marrow (red)	0.12	0.17	0.12	0.12
Lung	0.12	0.17	0.12	0.12
Thyroid	0.03	—	0.05	0.04
Bone surfaces	0.03	—	0.01	0.01
Stomach	—	0.18	0.12	0.12
Colon	—	0.09	0.12	0.12
Esophagus	—	0.04	0.05	0.04
Bladder	—	0.05	0.05	0.04
Ovary	—	0.03	—	—
Skin	—	—	0.01	0.01
Liver	—	—	0.05	0.04
Multiple myeloma	—	0.03	—	—
Brain	—	—	—	0.01
Salivary glands	—	—	—	0.01
Remainder	0.30	0.19	0.05[a)]	0.12[b)]

a) The ICRP 60 remainder tissues are the adrenals, brain, small intestine, spleen, kidneys, muscle, pancreas, upper large intestine, thymus, and uterus.
b) The ICRP 103 remainder tissues are the adrenals, extrathoracic (ET) region, gall bladder, heart, kidneys, lymphatic nodes, muscle, oral mucosa, pancreas, prostate (♂), small intestine, spleen, thymus, and uterus/cervix (♀).

where E_n is the neutron kinetic energy. The ICRP 103 radiation weighting factors are summarized in Table D.10.

D.8.2
Sex Averaging

The effective dose is calculated from the equivalent dose or from coefficients of the effective dose or equivalent dose. These coefficients are represented by convention as e and h for the effective dose and equivalent dose, respectively.

Given the uncertainties in the dose assessment, the ICRP calculated a single value for effective dose for both sexes. Therefore, the tissue weighting factors of Table D.10 are sex-averaged values for all tissues and organs including the male and female breast, testis, and ovary. The effective dose (E) is calculated from the equivalent dose (H) for tissue T of the Reference Male (H_T^{M}) and Reference Female (H_T^{F}), including the remainder tissues using the equation

$$E = \sum_{T=1}^{27} w_T \left[\frac{H_T^{\mathrm{M}} + H_T^{\mathrm{F}}}{2} \right] \tag{D.36}$$

Table D.10 ICRP 103 recommended radiation weighting factors[a].

Radiation type	Radiation weighting factor (w_R)
Photons	1
Electrons and muons	1
Protons and charged pions	2
Alpha particles, fission fragments, and heavy ions	20
Neutrons	b)

a) All values relate to the radiation incident on the body or, for internal sources, emitted from the incorporated radionuclide(s).
b) A continuous function of the neutron energy (see Eq. (D.35)) defines the weighting factor.

where the sum includes organs with assigned weighting factors ($T = 1-14$) and remainder tissues ($T = 15-27$).

The equivalent dose to the tissues of the remainder of the Reference Male and Reference Female is computed as the arithmetic mean

$$H^{\mathrm{M}}_{\text{remainder}} = \frac{1}{13}\sum_{T=15}^{27} H^{\mathrm{M}}_T \quad \text{and} \quad H^{\mathrm{F}}_{\text{remainder}} = \frac{1}{13}\sum_{T=15}^{27} H^{\mathrm{F}}_T \tag{D.37}$$

where the sum is over the 13 tissues of the remainder listed in footnote b of Table D.9. For radiation protection purposes, the effective dose is based on the mean dose to the organs/tissues in the ICRP 89 Reference Male and Reference Female. Following ICRP 103, the weighting factors are mean values representing an average over many individuals of both sexes. For males, remainder tissues are the adrenals, extrathoracic region, gall bladder, heart, kidneys, lymphatic nodes, muscle, oral mucosa, pancreas, prostate, small intestine, spleen, and thymus. For females, the sum excludes the prostate and includes the uterus/cervix.

D.8.3 Assessment of Occupational Dose

The total effective dose for demonstrating compliance with dose limits and constraints is defined in terms of the committed effective dose $E(50)$ from internal exposure and the personal dose equivalent from external exposure $H_{\mathrm{p}}(10)$:

$$E \cong H_{\mathrm{p}}(10) + E(50) \tag{D.38}$$

and

$$E(50) = \sum_j e_{j,\text{inh}}(50) I_{j,\text{inh}} + \sum_j e_{j,\text{ing}}(50) I_{j,\text{ing}} \tag{D.39}$$

where the personal dose equivalent is evaluated at a depth of 10 mm, $e_{j,\text{inh}}(50)$ is the committed effective dose coefficient for inhalation intakes of radionuclide j, $I_{j,\text{inh}}$ is the inhalation activity intake of radionuclide j, $e_{j,\text{ing}}(50)$ is the committed effective dose coefficient for ingestion intakes of radionuclide j, and $I_{j,\text{ing}}$ is the ingestion activity intake of radionuclide j. The $e(50)$ values are sex-averaged effective dose coefficients for the intake of specified radionuclides.

D.9 Human Respiratory Tract Model (HRTM)

In 1994, the ICRP recommended an improved model for the respiratory tract. The model includes a detailed description for aerosol deposition in the respiratory tract, particle clearance behavior, and the absorption of materials from different regions of the respiratory tract into the blood. ICRP 66 extended the ICRP 30 dosimetric model to include both workers and members of the public. Characteristic breathing rates for various groups of people (males, females, children, and adults) and ages were also defined in ICRP 66. ICRP 30 was based on the ICRP 23 Reference Man Philosophy. ICRP 66 utilizes the gender- and age-specific approach of ICRP 89.

ICRP 66 further subdivides the ICRP 30 regions and models the lymph nodes as connections to each region. A comparison of the modeled regions in ICRP 30 and ICRP 66 is provided in Table D.11.

Table D.11 Respiratory tract region comparison.

Formulation		Anatomy modeled
ICRP 30	**ICRP 66**	
Nasopharyngeal region	Extrathoracic region (ET)	
	ET_1	Anterior nose
	ET_2	Posterior nasal passages
		Larynx
		Pharynx
		Mouth
Tracheobronchial region	Bronchial region (BB)	Trachea
		Bronchi
	Bronchiolar region (bb)	Bronchioles
		Terminal bronchioles
Pulmonary region	Alveolar–interstitial region (AI)	Respiratory bronchioles
		Alveolar ducts and sacs
		Alveoli
		Interstitial connective tissue
Lymph	a)	

a) All ICRP 66 regions contain lymphatic tissue or components of it. Fluid collected in the interstitial connective tissue is collected in lymph capillaries, from which it flows and drains via lymph nodes (LNs). LN_{ET} drains the extrathoracic region. LN_{TH} is located in the BB region and drains the BB, bb, and AI regions.

D.9.1 Absorption

The ICRP 66 respiratory tract model includes absorption rates that vary with time following deposition in the lung. If experimental data are available, specific absorption behavior is used to obtain the most realistic assessment of the systemic uptake.

For the cases in which only a generic knowledge of the material is available, ICRP 66 adopts a methodology that is similar to the Class D, W, and Y assumptions used in ICRP 30. ICRP 66 adopts the use of three types of absorption (Types F, M, and S).

Type F absorption behavior corresponds to the ICRP 30 Class D. For materials that are only moderately soluble, Type M is defined. This solubility lies roughly between the ICRP 30 Class W and Class Y absorption. The most slowly absorbed materials are classified as Type S.

D.9.2 Particle Sizes

ICRP 66 extended the ICRP 30 range of particle size (0.2–10 μm) to particles of atomic dimensions (0.6 nm diameter) through aerosols of course particles (100 μm AMAD (activity median aerodynamic diameter)). The default occupational particle size in ICRP 66 is 5 μm as compared with the 1 μm size in ICRP 30. ICRP 66 more realistically treats the effects of particle density, particle shape, and their absorption from the different regions of the respiratory tract. Additional comparisons between ICRP 30 and ICRP 66 are provided in Table D.12.

Table D.12 Comparison of the ICRP 30 lung and ICRP 66 human respiratory tract models.

Model parameter/quantity	ICRP 30	ICRP 66
Biokinetic model basis	Reference man (ICRP 23)	Age-specific models (ICRP 89)
Dose coefficients availability	Reference man only	Age-specific dose coefficients
Dosimetry capability	Occupational only	Occupational and environmental
Clearance	Pulmonary clearance (Class D, W, and Y)	Blood absorption (Type F, M, and S) Translocation from initial to transformed state (Type F, M, and S)
Physiology basis	Reference man (ICRP 23)	Updated physiology/transport models (ICRP 89)
Dose recommendation basis	ICRP 26	ICRP 103
Default occupational particle size	1 μm	5 μm
Particle size range	0.2–10 μm particles	0.6 nm to 100 μm particles
Particle shapes allowed	Spherical only	User-defined particle shapes and sizes

D.9.3
Additional Model Details

The ICRP 66 HRTM model contains considerable flexibility in terms of the character and shape of the inhaled particle, the transformation of aerosol characteristics once it is deposited in the lung, and in the translocation of material within the lung. In the ICRP 66 model, a radioactive aerosol is deposited in the respiratory tract into the particles in initial state (PIS) subsystem and into the ET_1 region. The ICRP 66 model assumes that most compartments in the PIS subsystem have a mirror compartment in the particles in transformed state (PTS) subsystem. The PTS subsystem permits a particle entering the lung with a given type classification (e.g., Type M) to be altered by interaction with the lung environment and be transformed into another type (e.g., Type F or S). The transformed aerosol would behave differently in the PTS subsystem than it did in the PIS subsystem.

There are 13 specific compartments in the PIS and PTS subsystems. These compartments are labeled as 1 (AI_1), 2 (AI_2), 3 (AI_3), 4 (bb_1), 5 (bb_2), 6 (bb_{seq}; seq = sequestered), 7(BB_1), 8 (BB_2), 9 (BB_{seq}), 10 (LN_{TH}), 11 (ET_2), 12 (ET_{seq}), and 13 (LN_{ET}).

The PIS subsystem also includes the ET_1 compartment that connects to the environment. Both PIS and PTS Compartment 11 (ET_2) connect to the gastrointestinal (GI) tract. The model allows the following transport pathways: $AI_1 \rightarrow bb_1$, $AI_2 \rightarrow bb_1$, $AI_3 \rightarrow bb_1$, $AI_3 \rightarrow LN_{TH}$, $bb_1 \rightarrow BB_1$, $BB_1 \rightarrow ET_2$, $ET_2 \rightarrow$ GI tract, $ET_1 \rightarrow$ environment (PIS only), $bb_2 \rightarrow BB_1$, $BB_2 \rightarrow ET_2$, $BB_{seq} \rightarrow LN_{TH}$, $bb_{seq} \rightarrow LN_{TH}$, $ET_{seq} \rightarrow LN_{ET}$, PIS $\rightarrow$ body fluids, PTS $\rightarrow$ body fluids, and PIS $\rightarrow$ PTS.

From each PIS compartment, the material is transferred into the body fluids at an absorption rate s_p. As warranted, it is also simultaneously transferred from the PIS block at a rate s_{pt} to the corresponding PTS compartment (i.e., the material is transferred from PIS compartment i to PTS compartment i where $i = 1-13$). Each numbered compartment in PIS has a counterpart in PTS. In each PTS compartment, the isotope is transferred at a constant rate s_t into the body fluids.

As an example, the total transfer rate (K) for Compartment 2 (AI_2) in PIS is

$$K(AI_2 \text{ in PIS}) = k_{PIS}(2,4) + s_{pt} + s_p \quad \text{(D.40)}$$

where $k_{PIS}(2,4)$ is the transfer rate from PIS Compartment 2 (AI_2) to PIS Compartment 4 (bb_1), s_p is the transfer rate from PIS to the body fluids, and s_{pt} is the transfer rate from PIS to PTS. A number of factors including the chemical form of the inhaled radioactive isotope determine the transfer rates (s_{pt}, s_p, and s_t).

D.10
Human Alimentary Tract Model (HATM)

ICRP 100 presents a revised dosimetric model for the human alimentary tract that replaces the ICRP 30 ingestion model. The model is fully consistent with the anatomical and physiological data given in ICRP 89. ICRP 100 provides examples of radionuclide behavior and doses to alimentary tract regions.

The revised HATM model consists of the following regions: mouth, esophagus, stomach, small intestine, right colon, left colon, and rectosigmoid. These sections are modeled as hollow tubes, extending from the pharynx to the anus, and are comprised of four concentric layers. From the lumen outward, these layers are the mucosa, submucosa, muscularis, and adventitia or serosa.

The HATM includes the following processes:

1) *Entry of a radionuclide into the oral cavity by ingestion or into the esophagus following mechanical clearance from the respiratory tract.* After entering the oral cavity, sequential transfer occurs through the esophagus, stomach, small intestine, and segments of the colon, followed by elimination as feces.
2) *Radionuclide deposition and retention on or between the teeth and return to the oral cavity.*
3) *Deposition and retention in the oral mucosa or walls of the stomach and intestines.*
4) *Transfer from the oral mucosa or walls of the stomach and intestines back into the lumenal contents or into blood (absorption).*
5) *Transfer from various secretory organs or blood into the contents of certain segments of the alimentary tract (secretion).*

First-order kinetic processes are assumed as the basis for the HATM. Although this is a considerable simplification of the complex processes involved in the transfer of material through the lumen of the alimentary tract, it is judged by the ICRP to provide a reasonably accurate representation of the mean residence time of a radionuclide in each tract segment.

For computational purposes, each parameter value of the model is represented by a transfer coefficient that describes the rate of outflow of a substance from a compartment. The transfer coefficient or rate constant is defined as the instantaneous fraction of the contained substance leaving the compartment per unit time.

A comparison of the ICRP 30 gastrointestinal tract and ICRP 100 HATM is provided in Table D.13. This table compares a number of the ingestion model characteristics including their absorption characteristics, biokinetic basis, dose coefficient basis, entry points, and transit times.

D.10.1 Absorption to Blood

Retention and absorption to blood are permitted in all regions except the esophagus, but the possibility of lymphatic transport is excluded. For most radionuclides, absorption occurs in the small intestine, and limited or no information is available for retention in alimentary tract tissues or absorption into the blood from other HATM regions.

Table D.13 Comparison of the ICRP 30 gastrointestinal tract and ICRP 100 human alimentary tract models.

Model parameter/quantity	ICRP 30	ICRP 100
Absorption of radionuclides into blood	Occurs only in the small intestine	Absorption occurs in all compartments of the HATM with the exception of the esophagus, but the possibility of lymphatic transport is excluded
Applicability	Occupational doses	Occupational and environmental doses
Biokinetic model basis	Reference man (ICRP 23)	Provides age-dependent parameter values based on ICRP 89 for the alimentary tract regions and associated transit times for the movement of materials through these regions For adults, gender-dependent parameter values are given for dimensions and transit times
Dose coefficient availability	Reference man	Age- and gender-specific dose coefficients
Dose recommendation basis	ICRP 26	ICRP 103
Dosimetry calculations	Doses are calculated to the walls and contents of the modeled tissues	Doses are calculated to targets for cancer induction (stem cells) in the oral cavity, esophagus, stomach, small intestine, right colon, left colon, and rectosigmoid ICRP 103 tissue weighting factors are included for the stomach, colon, and esophagus. The small intestine and oral mucosa are included in the remainder
Point of entry into alimentary tract	Stomach	Oral cavity, esophagus, and stomach
Transit times	Independent of age, gender, and the type of material ingested	Provides age- and gender-specific transit times for all segments of the HATM For the upper segments of the HATM (oral cavity, esophagus, and stomach) provides material-specific transit times

While absorption occurs predominantly in the small intestine, the HATM includes absorption in the oral cavity, stomach, or the three segments of the colon. Absorption from the oral cavity is modeled as the transfer from the oral cavity contents to the oral mucosa and then direct transfer from the oral mucosa to blood. Absorption from other segments of the HATM is modeled as the transfer from the contents to the wall of that segment, followed by transfer to blood in the portal vein with the potential for direct uptake by the liver prior to entry into the general circulation. The model does not include the slow transfer from the alimentary tract to the blood via the lymphatic system, which is not expected to significantly contribute to the total absorption.

The ICRP 103 fractional absorption (f_A) values replace the f_1 values of ICRP 30. The f_A value defines the total absorption to blood in the HATM and represents the fraction of the material entering the alimentary tract that is absorbed in the absence of radioactive decay or endogenous input to the tract. It is defined by the sum of the fractions of the material entering the alimentary tract (f_i) absorbed in all of the regions (i) of the alimentary tract:

$$f_A = \sum_i f_i \tag{D.41}$$

In most situations, data is only available on the total absorption of an element to the blood with no information on regional absorption, and the ICRP 100 model assumes that this absorption takes place entirely from the small intestine (SI). In this case, the fractional transfer from the small intestine to blood, f_{SI}, is equal to f_A. If an element is absorbed from the stomach (ST) as well as from the small intestine (SI), f_A is

$$f_A = f_{ST} + f_{SI} \tag{D.42}$$

In order to perform calculations, transfer coefficients describing the uptake from the HATM to the blood are required. In the absence of retention in the walls, teeth, and oral mucosa, the transfer coefficient ($\lambda_{i,B}$), describing uptake to the blood (B) from compartment i of the HATM, is defined as

$$\lambda_{i,B} = \frac{f_i \lambda_{i,i+1}}{1 - \sum_i f_i} \tag{D.43}$$

where f_i is the fraction of the intake assumed to be absorbed from compartment i, and $\lambda_{i,i+1}$ is the HATM transfer coefficient for movement of material from compartment i to compartment $i+1$ of the alimentary tract. For the most common situation involving uptake only from the small intestine, the transfer coefficient for uptake from the small intestine to blood ($\lambda_{SI,B}$) is

$$\lambda_{SI,B} = \frac{f_{SI} \lambda_{SI,RC}}{1 - f_{SI}} \tag{D.44}$$

where $\lambda_{SI,RC}$ is the coefficient for transfer from the small intestine to the right colon.

A second example is more complex with blood uptake occurring from both the small intestine and stomach (S). In this case, the transfer coefficients to blood are

$$\lambda_{ST,B} = \frac{f_{ST} \lambda_{ST,SI}}{1 - f_{ST}} \tag{D.45}$$

$$\lambda_{SI,B} = \frac{f_{SI} \lambda_{SI,RC}}{1 - f_{SI} - f_{ST}} \tag{D.46}$$

Similar expressions, based on Eq. (D.43), can be written for the uptake from other compartments of the HATM.

D.10.2
Dose Calculations

The HATM dosimetry approach evaluates the dose to target regions within the mucosal lining of the alimentary tract. Doses are calculated in the epithelial stem cells, which are usually the targets for cancer induction.

D.10.3
Model Dependence

Equations (D.25)–(D.39) and Tables D.4 and D.7–D.10 illustrate the model dependence of the various ICRP internal dosimetry formulations. The selection of defined tissues is determined by their model-dependent risk. This risk is derived from the set of model assumptions, selected biological effects data, and associated doses to produce these effects. An examination of Table D.9 and the differences in the number of listed tissues, their associated weighting factors, and the treatment of the remainder illustrate the evolving nature of the ICRP internal dosimetry models.

References

10CFR20 (2015) Standards for Protection Against Radiation, National Archives and Records Administration, U. S. Government Printing Office, Washington, DC.

10CFR835 (2015) Occupational Radiation Protection, National Archives and Records Administration, U. S. Government Printing Office, Washington, DC.

29CFR1910 (2015) Occupational Safety and Health Standards, National Archives and Records Administration, U. S. Government Printing Office, Washington, DC.

Bevelacqua, J.J. (2004) Skin contaminations, internal intakes, and ALARA. *Radiat. Prot. Manage.*, **21** (1), 11.

Bevelacqua, J.J. (2005) Internal dosimetry primer. *Radiat. Prot. Manage.*, **22** (5), 7.

Bevelacqua, J.J. (2008) *Health Physics in the 21st Century*, Wiley-VCH Verlag GmbH, Weinheim.

Bevelacqua, J.J. (2009) *Contemporary Health Physics, Problems and Solutions*, 2nd edn, Wiley-VCH Verlag GmbH, Weinheim.

Bevelacqua, J.J. (2010a) *Basic Health Physics, Problems and Solutions*, 2nd edn, Wiley-VCH Verlag GmbH, Weinheim.

Bevelacqua, J.J. (2010b) Radiological dispersion devices and basic radiation science. *Phys. Teach.*, **48**, 302.

Bevelacqua, J.J. (2010c) Practical and effective ALARA. *Oper. Radiat. Saf., Health Phys.*, **98** (Suppl. 2), S39.

Bevelacqua, J.J. (2011) Conservatism in effective dose calculations for accident events involving fuel reprocessing waste tanks. *Health Phys.*, **101**, 48.

Bevelacqua, J.J. (2012) Applicability of health physics lessons learned from the Three Mile Island Unit-2 accident to the Fukushima Daiichi accident. *J. Environ. Radioact.*, **105**, 6.

ICRP (1959) *Permissible Dose for Internal Radiation*, Publication 2, Pergamon Press, Oxford.

ICRP (1968) *Evaluation of Radiation Doses to Body Tissues from Internal Contamination due to Occupational Exposure*, Publication 10, Pergamon Press, Oxford.

ICRP (1971) *The Assessment of Internal Contamination Resulting from Recurrent or Prolonged Uptakes*, Publication 10A, Pergamon Press, Oxford.

ICRP (1975) *Reference Man: Anatomical, Physiological, and Metabolic Characteristics*, Publication 23, Pergamon Press, Oxford.

ICRP (1977) *Recommendations of the International Commission on Radiological Protection*, Publication 26, Pergamon Press, Oxford.

ICRP (1979) *Limits for Intakes of Radionuclides by Workers*, Publication 30, Pergamon Press, Oxford.

ICRP (1991) *Recommendations of the ICRP*, Publication 60, Pergamon Press, Oxford.

ICRP (1994) *Human Respiratory Tract Model for Radiological Protection*, Publication 66, Pergamon Press, Oxford.

ICRP (2003) *Basic Anatomical and Physiological Data for Use in Radiological Protection: Reference Values*, Publication 89, Elsevier, Amsterdam.

ICRP (2006) *Human Alimentary Tract Model for Radiological Protection*, Publication 100, Elsevier, Amsterdam.

ICRP (2007) *The 2007 Recommendations of the International Commission on Radiological Protection*, Publication 103, Elsevier, Amsterdam.

ICRP (2011) Statement on Tissue Reactions, Ref 4825-3093-1464, Approved by the Commission on April 21, 2011, http://www.icrp.org/docs/icrp%20statement%20on%20tissue%20reactions.pdf (accessed 8 August 2015).

Loevinger, R., Budinger, T.F., and Watson, E.E. (1988) *MIRD Primer for Absorbed Dose Calculations*, The Society of Nuclear Medicine, New York.

National Research Council (1980a) Committee on the Biological Effects of Ionizing Radiation, *The Effects on Populations of Exposures to Low Levels of Ionizing Radiation, BEIR III*, National Academy Press, Washington, DC.

National Research Council (1980b) *The Effects on Populations of Exposures to Low Levels of Ionizing Radiation (BEIR IV)*, Committee on the Biological Effects of Ionizing Radiation, National Academy Press, Washington, DC.

National Research Council (1990) *The Health Effects of Exposures to Low Levels of Ionizing Radiation, BEIR V*, National Academy Press, Washington, DC.

National Research Council (2006) *Health Risks from Exposure to Low Levels of Ionizing Radiation, BEIR VII Phase 2*, National Academy Press, Washington, DC.

NCRP (1993) Limitation of Exposure to Ionizing Radiation. Report No. 116, National Council on Radiation Protection and Measurement, Bethesda, MD.

Snyder, W.S., Ford, M.R., Warner, G.G., and Watson, S.B. (1975) *"S", Absorbed Dose per Unit Cumulated Activity for Selected Radionuclides and Organs*, MIRD Pamphlet No. 11, The Society of Nuclear Medicine, New York.

Tubiana, M. and Aurengo, A. (2005) Dose-effect relationship and estimation of the carcinogenic effects of low doses of ionizing radiation: the Joint Report of the Académie des Sciences (Paris) and the Académie Nationale de Médecine. *Int. J. Low Radiat.*, **2** (3/4), 1.

United Nations Scientific Committee on the Effects of Atomic Radiation (UNSCEAR) (1988) *Sources, Effects and Risks of Ionizing Radiation, United Nations Committee on the Effects of Atomic Radiations, 1988 Report to The General Assembly*, United Nations, New York.

United Nations Scientific Committee on the Effects of Atomic Radiation (UNSCEAR) (2000) *Sources and Effects of Ionizing Radiation: United Nations Scientific Committee on the Effects of Atomic Radiation (UNSCEAR) 2000 Report to the General Assembly with Scientific Annexes*, United Nations, New York.

United Nations Scientific Committee on the Effects of Atomic Radiation (UNSCEAR) (2001) *Hereditary Effects of Radiation: UNSCEAR 2001 Report to the General Assembly, with Scientific Annex*, United Nations, New York.

United Nations Scientific Committee on the Effects of Atomic Radiation (UNSCEAR) (2005) *United Nations Scientific Committee on the Effects of Atomic Radiation*, United Nations, New York.

United Nations Scientific Committee on the Effects of Atomic Radiation (UNSCEAR) (2008) *United Nations Scientific Committee on the Effects of Atomic Radiation: UNSCEAR 2006 Report to the General Assembly, with Scientific Annexes*, United Nations, New York.

United Nations Scientific Committee on the Effects of Atomic Radiation (UNSCEAR) (2010) *United Nations Scientific Committee on the Effects of Atomic Radiation: UNSCEAR 2008 Report to the General Assembly, with Scientific Annexes*, United Nations, New York.

United Nations Scientific Committee on the Effects of Atomic Radiation (UNSCEAR) (2013) *United Nations Scientific Committee on the Effects of Atomic Radiation: UNSCEAR 2013 Report to the General Assembly, with Scientific Annexes*, United Nations, New York.

E
Health Physics-Related Computer Codes

E.1
Overview

This appendix summarizes a selected listing of computer codes and supporting data used in health physics applications. The listing only represents a sample of the broad scope of available models. Additional codes are noted in the publications listed in the reference section.

The codes summarized in this appendix are utilized in a variety of health physics applications encompassing the topics addressed in this book. This appendix contains a brief summary of the code, its applications, and a web address that provides additional information.

Code selection is based on the author's experience with a variety of models and an assessment regarding the applicability of the listed software to topics covered in this book. Omissions are based on space limitations and do not reflect on the value of any software package.

E.2
Code Descriptions

There are large ensembles of computer codes that facilitate calculations involving the fuel cycle, power reactor, medical, environmental, and regulatory health physics areas. Codes are also available to assess a variety of accident- and terrorist-induced events and nuclear weapons effects. Additional codes offer the capability to evaluate the radiation profiles resulting from medical imaging and therapy procedures. A subset of these codes that are often utilized in studies related to the aforementioned topics is addressed in subsequent discussions.

E.2.1
CAP-88 (http://www.epa.gov/radiation/assessment/CAP88/index.html)

The CAP-88 (Clean Air Act Assessment Package-1988) computer model is a set of computer programs, databases, and associated utility programs for estimation of dose and risk from radionuclide emissions to air. CAP-88 is written in FORTRAN77 and allows the user to complete dose and risk assessment calculations using a personal computer.

The code uses a modified Gaussian plume model to estimate the average dispersion of radionuclides released from up to six sources. These sources are either elevated stacks or uniform area sources. Plume rise is calculated assuming either momentum or buoyancy-driven effects.

CAP-88 computes radionuclide concentrations in air, rates of deposition on ground surfaces, concentrations in food, and intake rates to people from ingestion of food produced in the assessment area. Estimates of the radionuclide concentrations in produce, milk, and meat consumed by humans are made by linking the output of the atmospheric transport models with the US Nuclear Regulatory Commission (NRC) Regulatory Guide 1.109 terrestrial food chain models.

E.2.2
COMPASS (http://orise.orau.gov/environmental-assessments-health-physics/resources/marssim.aspx)

The COMPASS software is designed to facilitate the use of Multi-Agency Radiation Survey and Site Investigation Manual (MARSSIM) and assists the user in designing final status radiological surveys. COMPASS has a number of contaminant decay series, which include uranium ore, processed natural uranium, enriched uranium, depleted uranium, ^{230}Th, ^{232}Th, and ^{226}Ra. The published NRC screening values for surface soil and building surfaces are also included in the software. COMPASS interfaces with the decontamination and decommissioning program Visual Sampling Plan (VSP) to merge the systematic MARSSIM planning of COMPASS with the graphical capabilities of VSP.

E.2.3
COMPLY (http://www.epa.gov/radiation/assessment/comply.html#download)

COMPLY is a computational model that calculates the effective dose equivalent from radionuclides released into the atmosphere from stacks and vents. It can be utilized to demonstrate compliance with the National Emission Standards for Hazardous Air Pollutants in 40CFR61, Subparts H and I. Atmospheric concentrations are estimated using a Gaussian plume model, and the code also includes building wake effects.

E.2.4
DCAL (http://www.epa.gov/radiation/assessment/dcal.html#download)

DCAL (Dose and Risk Calculation) is a computer model that calculates the tissue dose and associated health risk from intakes of radionuclides or exposure to radionuclides present in environmental media. The system incorporates contemporary biokinetic and dosimetric data and models. DCAL uses the metabolic models from International Commission on Radiological Protection (ICRP) Publications 68 and 72 and data from ICRP Publications 23 and 89 to calculate the dose per unit intake for over 800 radionuclides. The code determines risk based in US Environmental Protection Agency models.

E.2.5
DWUCK/CHUCK/MERCURY (https://rsicc.ornl.gov/codes/psr/psr5/psr-546.html)

The Distorted-Wave University of Colorado Kunz (DWUCK)/(Coupled-Channels University of Colorado Kunz) CHUCK/MERCURY codes are nuclear models used to calculate differential and total reaction cross-sections using the distorted-wave Born approximation. This code package contains DWUCK-4, DWUCK-5/MERCURY, and CHUCK-3.

DWUCK-4 calculates the differential scattering cross-section in a zero-range approximation, which is most useful for light projectiles. The incoming and outgoing waves may be in any combination of spin 0, $^1/_2$, or 1 particles. CHUCK-3 is used for coupled-channel calculations to evaluate nuclear scattering amplitudes and differential collision cross-sections. Both nonrelativistic and relativistic kinematics are available. DWUCK-5 is used for heavier ions and more complex nuclear reactions. This code is directly applicable to the calculation of heavy ion cross-sections used in cancer therapy applications.

E.2.6
EGS Code System (http://rcwww.kek.jp/research/egs/egs5.html)

EGS5 (Electron Gamma Shower) is a Monte Carlo code that simulates the transport of electrons and photons in arbitrary geometries. It was originally developed at the Stanford Linear Accelerator Center for high-energy physics applications. With the help of the National Research Council of Canada and the High Energy Research Organization in Japan, EGS was extended to apply to lower-energy applications. EGS5 is applicable to the kiloelectronvolt to teraelectronvolt energy range. The EGS5 code has been extensively benchmarked for medical physics applications.

E.2.7
ENDF (http://www.nndc.bnl.gov/exfor3/endf00.htm)

The Evaluated Nuclear Data File (ENDF) library includes a nuclear reaction database containing evaluated (recommended) cross-sections, spectra, angular distributions, fission product yields, photo-atomic, and thermal scattering data. The emphasis of the data set is neutron-induced reactions, and data were analyzed to produce recommended libraries for national (United States, European, Japanese, Russian, and Chinese) nuclear data projects. All data are stored in the internationally adopted format (ENDF-6).

E.2.8
FLUKA (http://www.fluka.org/)

FLUKA is a fully integrated Monte Carlo simulation package. It has applications in high-energy physics; engineering; shielding, detector, and telescope design; cosmic ray studies; dosimetry; medical physics; and radiobiology.

E.2.9
GENII-LIN (http://www-rsicc.ornl.gov/codes/ccc/ccc7/ccc-728.html)

The GENII-LIN code package provides a platform for calculating environmental doses. It generalizes the original Hanford Dosimetry code used to estimate potential radiation doses to individuals or populations from both routine and accidental releases of radionuclides to air or water and residual contamination from spills or decontamination operations.

GENII-LIN includes the capability to evaluate annual, committed, and accumulated doses for acute and chronic releases. The dose from various exposure pathways is also calculated. These pathways include direct aquatic exposure (e.g., swimming, boating, and fishing), soil dose from buried and surface sources, and air doses from semi-infinite and finite cloud models. Inhalation and ingestion pathways are included.

The release scenarios include (i) an acute release to air or water from ground level or elevated sources, (ii) a chronic release to air or water from ground level or elevated sources, and (iii) contamination of soil or surfaces. GENII-LIN 2.0 incorporates the internal dosimetry models recommended by ICRP 72 and the radiological risk estimating procedures of the US Environmental Protection Agency Federal Guidance Report 13.

E.2.10
HOTSPOT (http://www-rsicc.ornl.gov/codes/mis/mis0/mis-009.html)

The HOTSPOT codes provide a first-order approximation of the radiological effects associated with the atmospheric release of radioactive materials. Four programs (PLUME, EXPLOSION, FIRE, and RESUSPENSION) facilitate a downwind dose assessment following the release of radioactive material resulting from a continuous or puff release, explosive release, fuel fire, or an area contamination event. Additional programs address the release of plutonium, uranium, and tritium and provide an initial accident assessment of the effects of a nuclear detonation.

E.2.11
IDD–SAM (www.bevelacquaresources.com)

The Internal Device Dose–Simulation and Modeling (IDD–SAM) Program is a FORTRAN code developed by Bevelacqua Resources. IDD–SAM calculates absorbed dose within a three-dimensional Cartesian lattice. Stopping powers are determined using Bethe's formulation, and energy-dependent cross-sections are obtained from Shen's parameterization or the DWUCK/CHUCK/MERCURY code package. The model permits user-defined source locations, radiation types, output energies, and radiation-type fluence values. This code is currently proprietary and may be publicly released. Release is contingent upon publication of planned research, code documentation and development, and finalization of licensing agreements.

E.2.12
IMBA (www.hpa-radiationservices.org.uk/services/imba)

The Integrated Modules for Bioassay Analysis (IMBA) is a set of software modules for internal dosimetry that implements the biokinetic and dosimetric models currently recommended by the ICRP. IMBA replaced LUDEP (Lung Dose Evaluation Program) that modeled the ICRP 66 Human Respiratory Tract Model.

IMBA performs basic internal dosimetry calculations. Output is provided in tabular and graphical formats. For standard calculations, the ICRP default values are selected from built-in databases. The user can enter individual parameter values for more detailed calculations.

E.2.13
ISO-PC (https://rsicc.ornl.gov/codes/ccc/ccc6/ccc-636.html)

ISO-PC is a kernel integration code used to perform general shielding calculations utilizing a limited set of geometries. It is based on the ISOSHLD-II program. ISO-PC calculates dose rates from X-rays, gamma rays, and bremsstrahlung radiation. The following source shield geometries are included: (i) point source with slab shields, (ii) line source with slab shields, (iii) sphere with spherical or slab shields, (iv) truncated cone with slab shields, (v) infinite slab or infinite plane with infinite slab shields, (vi) flat disk with slab shields, (vii) cylinder with cylindrical or slab shields, (viii) end of a cylinder with slab shields, (ix) rectangular solid with slab shields, and (x) cylindrical shell sources with cylindrical and slab shields. Hundreds of radionuclides are incorporated into the ISO-PC source library.

E.2.14
JENDL (http://wwwndc.tokai-sc.jaea.go.jp/jendl/jendl.html)

The Japanese Evaluated Nuclear Data Library (JENDL) provides standard technical information for fast breeder reactors, thermal reactors, fusion reactors, shielding calculations, and other applications. JENDL-3.3 (2002) contains neutron-induced reaction data for 337 nuclides in the neutron energy range from 10^{-5} eV to 20 MeV.

E.2.15
LISE++ (http://lise.nscl.msu.edu/lise.html)

LISE++ was developed to calculate the transmission and yields of nuclear fragments produced and collected in a spectrometer. The code also calculates the passage of an ion through matter including its energy deposition, stopping power, and range including straggling. Considerable flexibility is available to accommodate heavy ion interactions with all elements in the periodic table and with common materials. A variety of two-dimensional plots facilitates visualization of the program's calculations. Projectile fragmentation, fusion–evaporation, fusion–fission, Coulomb fission, and abrasion–fission are included in LISE++ as production reaction mechanisms to simulate interactions at energies above the Coulomb barrier. The code can also be configured to simulate heavy ion interactions with bone and soft tissue.

E.2.16
MACCS2 (http://www.nrc.gov/about-nrc/regulatory/research/comp-codes.html)

MACCS2 is the successor code of the NRC's MELCOR Accident Consequence Code Systems (MACCSs) code. The code is based on the straight-line Gaussian plume model, and it evaluates doses and health risks from the accidental dispersion of radioactive material to the environment. MACCS2 models accident-related effects and impacts including (i) atmospheric transport and deposition under time-variant meteorology, (ii) short-term and long-term mitigative actions and exposure pathways, (iii) deterministic and stochastic health effects, and (iv) economic costs.

E.2.17
MARS (http://www-ap.fnal.gov/MARS/)

MARS is a Monte Carlo code for the simulation of three-dimensional hadronic and electromagnetic cascades. Its applications include muon, heavy ion, and low-energy neutron transport in accelerators; detector development and evaluation; spacecraft shielding design; and a variety of shielding applications. MARS is applied to energies spanning the electronvolt to 100 TeV range.

E.2.18
MCNP (http://mcnp-green.lanl.gov/index.html)

MCNP is a general-purpose Monte Carlo N-Particle code used for neutron, photon, electron, or coupled neutron/photon/electron transport. Applications include radiation protection and dosimetry, radiation shielding, radiography, medical physics, nuclear criticality safety, detector design and analysis, well logging, accelerator target design, fission and fusion reactor design, and decontamination and decommissioning.

E.2.19 MCNPX (http://mcnpx.lanl.gov/)

MCNPX is a three-dimensional, time-dependent, and general-purpose Monte Carlo radiation transport code for modeling radiation interactions in a wide variety of situations. MCNPX stands for Monte Carlo N-Particle Extended. It extends the capabilities of MCNP to many particle types and over a wide energy range. MCNPX is applicable to a diverse set of applications including earth orbit and planetary space radiation evaluations, oil exploration, nuclear medicine, nuclear safeguards, accelerator applications, and nuclear criticality safety.

E.2.20 MICROSHIELD® (http://www.radiationsoftware.com/mshield.html)

MicroShield® is a photon shielding and dose assessment program. It has applications in the health physics, waste management, radiological design, and radiological engineering. MicroShield® has a relatively simple input format and is based on the ISOSHLD-II program.

E.2.21 MICROSKYSHINE® (http://www.radiationsoftware.com/mskyshine.html)

MicroSkyshine® calculates the photon dose from sky scattered gamma radiation, and its method of solution is based on the use of "beam functions" for a point source as developed for the US NRC. The MicroSkyshine® code has been used to evaluate conformance with US Regulations (e.g., a portion of the 10CFR50, Appendix A radiological requirements and 40CFR190 fuel cycle exposure criteria). Typical applications include photon scattering in boiling water reactor turbine buildings, radioactive waste storage facilities, and waste disposal sites.

E.2.22 MIDAS (http://www.absconsulting.com/midas.cfm)

Meteorological Information and Dose Assessment System (MIDAS)-NU models the atmospheric dispersion of releases of radioactive materials during routine and accident conditions. The code has a graphical user interface that facilitates data entry to define accident conditions. MIDAS-NU also provides critical protective action information during an incident. The model has the capability to provide real-time emergency dose assessment, routine operation 10CFR50 Appendix A environmental analysis, radiological effluent assessment, and automatic meteorological and radiological data collection.

E.2.23 **MULTIBIODOSE (http://www.multibiodose.eu/software.html)**

MULTIBIODOSE is a Java software program created to calculate doses from laboratory data. Up to five retrospective (biological and physical) assays are combined to yield a single estimate for each individual exposed to ionizing radiation from a large-scale radiation accident or event involving mass casualty situations.

E.2.24 **OLINDA/EXM (http://olinda.vueinnovations.com/olinda)**

The Organ Level Internal Dose Assessment/Exponential Modeling (OLINDA/EXM) code calculates organ doses and effective doses in nuclear medicine studies from systemically administered radiopharmaceuticals. OLINDA/EXM also performs regression analysis for user-supplied biokinetic data. This code is an upgrade of the MIRDOSE code that was widely used in the radiopharmaceutical industry and research community for medical internal dose calculations.

E.2.25 **PRESTO (http://www.epa.gov/radiation/assessment/presto.html)**

The Prediction of Radiological Effects Due to Shallow Trench Operations (PRESTO) code evaluates radiation exposure from contaminated soil layers. PRESTO simulates the transport of radionuclides in air, surface water, and groundwater pathways and calculates the dose for ingestion, inhalation, immersion, and external source pathways. The model calculates the maximum annual committed effective dose to a critical population group and cumulative fatal health effects and genetic effects to the general population. The following scenarios are included: (i) near-surface disposal trench containing low-level radioactive waste and/or naturally occurring or accelerator-produced radioactive material, (ii) residual radionuclides remaining in soil layers after cleanup of a contaminated site, (iii) agricultural land application of technologically enhanced naturally occurring radioactive material (TENORM) waste, and (iv) stripped land reclamation with applied TENORM waste.

E.2.26 **RADTRAD (http://www-rsicc.ornl.gov/codes/ccc/ccc8/ccc-800.html)**

The Radionuclide Transport, Removal, and Dose (RADTRAD) code estimates doses from a nuclear power reactor release at off-site locations and in the control room. It can also be used to estimate dose attenuation due to modification of a facility or accident sequence.

RADTRAN has source terms to describe the fission product release from the reactor coolant system from power and test reactors. The model includes spray and natural deposition that reduce the quantity of radioactive material transported through the containment. Radioactive material transport includes pathways between buildings, from buildings to the environment, or into control rooms through high-efficiency particulate air filters, piping, or other conduits.

E.2.27
RASCAL (http://www-rsicc.ornl.gov/codes/ccc/ccc7/ccc-783.html)

The Radiological Assessment System for Consequence Analysis (RASCAL) code evaluates releases from nuclear power plants, spent fuel storage pools and casks, fuel cycle facilities, and radioactive material handling facilities. RASCAL is designed to be used by the NRC in the independent assessment of dose projections during the response to radiological emergencies. The code has the capability to calculate power reactor source terms, the airborne transport of radioactive materials using both Gaussian plume and puff models, and the associated doses. Fuel cycle events that can be addressed using RASCAL include uranium fires and explosions, criticality accidents, and isotopic releases from transportation and materials events.

E.2.28
RESRAD (http://web.ead.anl.gov/resrad/home2/)

RESRAD is a family of computer models designed to estimate radiation doses and excess lifetime cancer risk to a chronically exposed on-site resident from residual radioactive materials. The model includes the following environmental pathways: direct exposure; ingestion of aquatic foods, plants, meat, milk, soil, and water; and inhalation of particulates and radon. Sources are modified to account for radioactive decay and in growth, leaching, erosion, and mixing.

E.2.29
SCALE 5 (http://www-rsicc.ornl.gov/codes/ccc/ccc7/ccc-725.html)

The Standardized Computer Analyses for Licensing Evaluation (SCALE) system was developed for the US NRC to provide a method of analysis for nuclear fuel facilities and package designs. The system has the capability to perform criticality safety, shielding, radiation source term, spent fuel depletion/decay, and heat transfer analyses.

The criticality safety analysis sequence (CSAS) control module calculates the neutron multiplication factor for one-dimensional (1-D) (XSDRNPM S) and multidimensional (KENO V.a) system models. It also has the capability to perform criticality searches (optimum, minimum, or specified values of k_{eff}) on geometry dimensions or nuclide concentrations in KENO V.a.

The SAS2H module uses ORIGEN S to perform a 1-D fuel depletion analysis. This module can be used to characterize spent fuel and generate source terms.

Four shielding analysis sequence (SAS) codes are provided. General 1-D shielding problems are analyzed using XSDRNPM S. Shielding analysis using the

MORSE SGC Monte Carlo code is available. The SAS4 module is used to perform a Monte Carlo shielding analysis for cask-type geometry. The QADS module analyzes three-dimensional gamma-ray shielding problems via the point kernel code QAD CGGP.

The thermal analysis module HTAS1 performs a two-dimensional thermal analysis for a specific class of spent fuel casks during normal, fire, and postfire conditions.

E.2.30
SKYSHINE-KSU (http://www-rsicc.ornl.gov/codes/ccc/ccc6/ccc-646.html)

SKYSHINE-KSU was developed at Kansas State University to form a comprehensive system for calculating gamma-ray scattering from air. It includes the SKYNEUT 1.1, SKYDOSE 2.2, and MCSKY 2.3 codes plus the DLC-0188/ZZ-SKYDATA library.

SKYNEUT evaluates neutron and neutron-induced secondary gamma-ray skyshine doses from an isotropic, point, neutron source collimated by three simple geometries. These geometries are an open silo; a vertical, perfectly absorbing wall; and a rectangular building. The source may emit monoenergetic neutrons or neutrons with a spectrum of energies.

SKYDOSE evaluates the gamma-ray skyshine dose from an isotropic, monoenergetic, point gamma-photon source collimated by three simple geometries. These are a source in a silo, a source behind an infinitely long, vertical, perfectly absorbing wall, and a source in a rectangular building. In all three geometries, an optional overhead slab shield may be specified.

MCSKY evaluates the gamma-ray skyshine dose from an isotropic, monoenergetic, point gamma source collimated either into a vertical cone or into a vertically oriented structure with an N-sided polygon cross-section. An overhead laminate shield composed of two different materials is assumed.

E.2.31
SPAR (http://www-rsicc.ornl.gov/codes/ccc/ccc2/ccc-228.html)

SPAR (Stopping Powers and Ranges) is a legacy code that computes stopping powers and ranges for muons, pions, protons, and heavy ions in any nongaseous medium for energies up to several hundred gigaelectronvolts. The original code may require modification to run with conventional FORTRAN compilers.

E.2.32
TRACE (http://www-rsicc.ornl.gov/codes/psr/psr4/psr-481.html)

The TRAC/RELAP Advanced Computational code is a state-of-the-art model designed to consolidate and extend the capabilities of NRC's legacy safety codes. TRACE has the capability to analyze large and small break loss of coolant system accidents and other reactor transients in both pressurized and boiling water reactors. The capability exists to model thermal hydraulic phenomena in both one- and three-dimensional spaces.

E.2.33
VARSKIN (http://www-rsicc.ornl.gov/codes/ccc/ccc7/ccc-781.html)

The VARSKIN computer code facilitates the calculation of skin dose from radioactive contamination. Skin dose from both beta and gamma sources is obtained from radioactive contamination residing on the skin or clothing. VARSKIN computes the dose at a user-specified skin depth or skin volume, with point, disk, cylindrical, spherical, or slab (rectangular) sources. Dose calculations from multiple sources can also be performed.

E.2.34
VSM (http://www.doseinfo-radar.com/RADARSoft.html)

The Visual Monte Carlo Program (VSM) is an external dose Monte Carlo simulator. VSM is a computer model that simulates the irradiation of the human body by external sources. The code uses a Yale University voxel phantom and Monte Carlo techniques to simulate the emission of photons. VSM includes the following sources: point photon, ground, cloud, and X-ray source. The code transports the photons through the human body phantom and calculates the dose to each body region.

E.2.35
VSP (http://vsp.pnnl.gov/)

The VSP code provides guidance in decontamination and decommissioning planning to determine the appropriate number of samples and the sample locations. VSP provides statistical solutions to sampling design, mathematical and statistical algorithms, and a visual interface. The code can merge the MARSSIM planning capabilities of COMPASS with the graphical features of VSP.

E.3
Code Utilization

Computer code users need to exercise caution in using any numerical algorithm. Users must clearly understand the limitations and capability of a code to address the problem of interest. This caution is more encompassing than the old adage "GARBAGE IN–GARBAGE OUT." It involves the interpretation of results and understanding the inherent limits, assumptions, and methodology of the code package.

As an example, a series of shielding design calculations performed by the author are cited. The problem involved the scattering of photons through a complex shielding arrangement. Before utilizing MCNP, a scaling calculation based on data from a similar problem was performed. The next step was to perform a hand calculation using Rockwell's methodology. Next, a deterministic shielding code was run to provide a refined calculation. Finally, MCNP was utilized.

At each step, differences from the previous step were evaluated and assessed for credibility. If MCNP had been run without the other steps and without any internal benchmarking, how would a user know if the results were credible? Issues could include input/geometry errors, misinterpreting MCNP caution or error flags, applying MCNP to a problem that was outside its zone of applicability, or encountering a previously unidentified code error. In the case cited, all codes and hand calculations provided a consistent solution that suggested a reasonable degree of confidence in the MCNP results. The message to any code user is to be cautious and to perform internal benchmarking to improve confidence in the results.

E.4 Code Documentation

Many programmers write new codes and do not sufficiently document the models, solution methods, convergence criteria, and test cases used to verify the results. It is quite important for new codes to be thoroughly documented and for the users of existing codes to thoroughly understand the limitations of the models that are utilized. This can be a challenge for codes written over a number of years with multiple authors. A code user must exercise considerable caution when utilizing software that they did not develop.

References

10CFR50 (2015) Domestic Licensing of Production and Utilization Facilities, National Archives and Records Administration, U.S. Government Printing Office, Washington, DC.

40CFR61 (2015) National Environmental Standards for Hazardous Air Pollutants, National Archives and Records Administration, U.S. Government Printing Office, Washington, DC.

40CFR190 (2015) Radiation Protection Standards for Nuclear Power Operations, National Archives and Records Administration, U.S. Government Printing Office, Washington, DC.

Ainsbury, E.A. *et al.* (2014) Multibiodose radiation emergency categorization software. *Health Phys.*, **107**, 83.

ANSI ANSI/ANS-6.6.1-1987 (R1998). (2013) *Calculation and Measurement of Direct and Scattered Gamma Radiation from LWR Nuclear Power Plants*, American Nuclear Society, Lagrange Park, IL.

Bevelacqua, J.J. (2008) *Health Physics in the 21st Century*, Wiley-VCH Verlag GmbH, Weinheim.

Bevelacqua, J.J. (2010) Feasibility of using internal radiation-generating devices in radiotherapy. *Health Phys.*, **98**, 614.

Bevelacqua, J.J. (2012) Angular absorbed dose dependence of internal radiation- generating devices in radiotherapy. *Health Phys.*, **102**, 2.

ICRP (1975) *Reference Man: Anatomical, Physiological, and Metabolic*

Characteristics, ICRP Publication 23, Pergamon Press, Oxford.
ICRP (1994) *Dose Coefficients for Intakes of Radionuclides by Workers*, ICRP Publication 68, Pergamon Press, Oxford.
ICRP (1995) *Age-dependent Doses to the Members of the Public from Intake of Radionuclides – Part 5 Compilation of Ingestion and Inhalation Coefficients*, ICRP Publication 72, Pergamon Press, Oxford.
ICRP (2003) *Basic Anatomical and Physiological Data for Use in Radiological Protection: Reference Values*, ICRP Publication 89, Elsevier, Amsterdam.
International Atomic Energy Agency Nuclear Data Section www-nds.iaea.org/ (accessed 08 August 2015).
Japan Atomic Energy Agency Nuclear Data Center http://wwwndc.jaea.go.jp/codes/index.html (accessed 08 August 2015).
National Atmospheric Release Advisory Center https://narac.llnl.gov/ (accessed 08 August 2015).
National Nuclear Data Center – Brookhaven National Laboratory www.nndc.bnl.gov/ (accessed 08 August 2015).
NCRP (2003) Radiation Protection for Particle Accelerator Facilities. NCRP Report No. 144, National Council on Radiation Protection and Measurements, Bethesda, MD.
Organization for Economic Co-operation and Development, Nuclear Energy Agency Data Bank Computer Program Services http://www.oecd-nea.org/dbprog/ (accessed 08 August 2015).
Radiation Safety Information Computational Center (RSICC) www-rsicc.ornl.gov/ (accessed 08 August 2015).
Rockwell, T. III, (1956) *Reactor Shielding Design Manual*, McGraw-Hill Book Company, Inc., New York.
US Environmental Protection Agency (1999) Cancer Risk Coefficients for Environmental Exposure to Radionuclide: Updates and Supplements. Federal Guidance Report No. 13, U.S. Environmental Protection Agency, Washington, DC.
USNRC (2000) Multi-Agency Radiation Survey and Site Investigation Manual (MARSSIM). USNRC Regulatory Guide NUREG-1575, Revision 1, U.S. Nuclear Regulatory Commission, Washington, DC.

F
Systematics of Charged Particle Interactions with Matter

F.1
Introduction

The interaction of ions with matter is described by well-known relationships. These relationships provide the ion's stopping power, range, and dosimetric information over a span of energies and are provided in Chapter 5. This appendix provides an overview of the associated calculations and radiation protection considerations that are important in a variety of health physics-related areas. These areas include cancer therapy using a variety of ions, accelerator transmutation of high-level radioactive waste, and the emerging space tourism industry.

A description of the angular dependence of charged particle interactions is provided in Appendix G. Appendix E summarizes computer codes that provide numerical algorithms for charged particle and associated effective dose calculations. Applicable codes are references in subsequent discussion.

F.2
Overview of External Radiation Sources

Charged particle interactions with matter are important health physics considerations. These interactions are encountered in cancer therapy applications, during low earth orbit tourist excursions, and during major solar particle events. Charged particle interactions affect fuel cycle accelerator facility designs that are being developed to transmute minor actinides. A range of energies and diversity of radiation types are encountered in these applications.

F.2.1
Cancer Therapy

Cancer therapy applications are discussed in Chapter 5. Proton, electron, and heavy ion cancer therapy applications utilize charged particle energy deposition characteristics to preferentially localize absorbed dose within a tumor volume. By carefully selecting the ion and its energy, the absorbed dose delivered to healthy

Health Physics: Radiation-Generating Devices, Characteristics, and Hazards, First Edition. Joseph John Bevelacqua.

Table F.1 Range of protons and electrons in water[a].

Kinetic energy (MeV)	Range (g/cm^2)	
	Protons	Electrons
0.01	0.00003	0.0002
0.1	0.0001	0.0140
1	0.002	0.430
10	0.118	4.88
100	7.57	32.5
1000	321	101

a) Turner (2007).

tissue can be minimized. Some charged particle beams (e.g., heavy ions) can also be tracked to ensure proper absorbed dose location using positron emission tomography.

The stopping power for electrons used in therapy applications is about 2 MeV/cm in tissue and about twice this value in bone. For electron energies below 1 MeV, the maximum effective dose occurs near the skin surface. As the electron energy increases from 4 to 20 MeV, the shape of the effective dose curve shifts from a surface peak to a broader plateau extending into tissue. Beyond 20 MeV, the plateau expands and additional tissue is irradiated.

Protons have a range that varies with energy. Proton beams produce a relatively low constant deposition profile that terminates in a narrow Bragg peak at the end of the particle's range. As a matter of reference, Table F.1 summarizes the range of electrons and protons in water as a function of energy. Chapter 5 provides a similar table for heavy ions.

Various computer models are used to evaluate therapy approaches and to assess the associated doses. These codes include EGS5 for photon and electron interaction assessments; FLUKA, MCNP (Monte Carlo N-Particle), and MCNPX (Monte Carlo N-Particle Extended) for radiation transport studies; IDD-SAM (Internal Device Dose–Simulation and Modeling) for internal radiation-generating device calculations; OLINDA/EXM (Organ Level Internal Dose Assessment/Exponential Modeling) for organ doses and effective dose studies from systemically administered radiopharmaceuticals; and SPAR (Stopping Powers and Range) for stopping powers and range calculations.

F.2.2
Accelerator Transmutation of High-Level Waste

The partitioning and transmutation of high-level nuclear were addressed in Chapter 2. Accelerators are a key approach to minimize the effects of minor actinides (e.g., Np, Am, and Cm) in waste storage facilities.

Accelerator health physics issues are primarily associated with shielding the generated radiation to ensure the applicable effective dose standards and requirements are met. The transmutation of minor actinides as a method to process high-level waste is an emerging accelerator application. The accelerated particles may include protons and heavy ions. Neutrons produced by charged particle interactions are also being considered for transmutation applications.

A variety of shielding and radiation transport codes are utilized to assess the impact of various radiation types on the effective dose and determine the shielding requirements to meet the associated regulatory requirements. These codes include DWUCK-4 (Distorted-Wave University of Colorado Kunz), DWUCK-5/MERCURY, and CHUCK-3 (Coupled-Channels University of Colorado Kunz) for cross-section determination; EGS5 for photon and electron interaction assessments; FLUKA, MCNP, and MCNPX for radiation transport studies; ISO-PC, MicroShield®, MicroSkyshine®, and SKYSHINE-KSU for shielding calculations; and SPAR for stopping powers and range calculations.

F.2.3 Space Tourism

Chapter 6 provides an overview of the radiological implications of the emerging space tourism industry. External radiation in low earth orbit involves ultrarelativistic energies, is often more diverse than a transmutation accelerator's environment, and presents a complex dosimetry challenge. Radiation encountered in space tourism applications arises from radiation trapped by the earth's electromagnetic field, galactic cosmic rays, and solar particle events involving a variety of radiation types including photons, electrons, protons, and heavy ions. For photon radiation, scattering and attenuation reduce the photon fluence as it penetrates the spacecraft shielding.

With electrons, the density builds to an equilibrium value inside the shield such that the electron fluence rises to a maximum and then decreases with increasing depth into the shield. Electron backscatter increases the surface fluence and is considered in the shielding analysis. The depth of the maximum fluence increases with increasing electron energy. With electrons, the primary particles slow down in the shield and produce high energy deposition values per unit length as they reach their maximum range. For depths beyond the maximum range, the electron fluence decreases very rapidly to a value of only a few percent of the maximum value. Similar comments apply to electrons that penetrate the shield and reach tissue. Proton and heavy ion interactions are addressed in the next section.

Numerous shielding and radiation transport codes are available to evaluate various radiation types and their impact on space tourists in low earth orbit. These codes determine the effective dose and associated shielding requirements to limit the radiological hazard during suborbital and orbital flights. Applicable codes include DWUCK-4, DWUCK-5/MERCURY, and CHUCK-3 for cross-section determination; FLUKA, MCNP, and MCNPX for radiation transport studies; ISO-PC, MicroShield®, MicroSkyshine®, and SKYSHINE-KSU for shielding calculations; and SPAR for stopping powers and range calculations.

F.3
Tissue-Absorbed Dose from a Heavy Ion or Proton Beam

Given the various external sources that can interact with tissue, it is important to have a well-established methodology to calculate the absorbed dose. For a tissue volume irradiated by a beam of particles, the absorbed dose (D) as a function of penetration distance x is given by

$$D(x) = \frac{1}{\rho}\left(-\frac{\mathrm{d}E}{\mathrm{d}x}\right)\Phi(x) \tag{F.1}$$

where ρ is the density of the tissue type attenuating the ion, $-\mathrm{d}E/\mathrm{d}x$ is the stopping power, and Φ is the ion fluence. The particle fluence varies with penetration distance according to the relationship

$$\Phi(x) = \Phi(0)\exp(-\mu x) \tag{F.2}$$

where $\Phi(0)$ is the entrance fluence and μ is the macroscopic reaction cross-section (linear attenuation coefficient). The linear attenuation coefficient is defined as

$$\mu = n\sigma \tag{F.3}$$

where n is the number of atoms of absorbing material per unit volume and σ is the total microscopic reaction cross-section for the heavy ion–tissue interaction.

In principle, the dose distribution from each ion in the beam is summed to obtain the total absorbed dose distribution. However, in performing this sum the absorbed dose must be modified by an energy-dependent radiation weighting factor or relative biological effectiveness value.

The actual dosimetry situation involved in space tourism situations is more complex than assumed in Eqs. (F.1)–(F.3). In particular, the heavy ion beam is shielded by spacecraft structures prior to impinging on tissue. For that case, the solar or cosmic fluence is modified to account for the attenuation of the ion beam. In addition, secondary particle fluence is generated from interactions of the primary particles and spacecraft structures.

F.4
Determination of Total Reaction Cross-Section

Equation (F.3) uses the total microscopic reaction cross-section to obtain the total macroscopic reaction cross-section. The microscopic reaction cross-section is obtained from data parameterizations or the use of nuclear optical model codes such as DWUCK or MERCURY. These codes are described in Appendix E.

Parametric models fit available cross-section data using established relationships including trends in nuclear radii, reaction kinematics, and energy dependence. The optical model codes require parameterization of the entrance and exit channels, nuclear structure information for the transferred particles, spectroscopic information, and specification of kinematic information related to the reaction under investigation. Each of these approaches has its inherent shortcomings, and these must be clearly understood. The best practice is to use measured data. However, the use of models is often required because a complete set of cross-sections are often not available.

F.5
Calculational Considerations

Calculations using the codes summarized in Appendix E are well established and will not be repeated. These calculations become more difficult if topics outside the applicability scope of these codes are attempted.

In general, the calculation of absorbed doses is efficiently computed using Eqs. (F.1)–(F.3) if cross-section parameterizations are available. Without these parameterizations, cross-sections are determined as a function of energy and angle using computer codes such as DWUCK or MERCURY (see Appendix E). The standard codes of Appendix E will not contain all necessary cross-sections if unique heavy ions and associated energies are selected for evaluation as a therapy protocol.

Determination of the microscopic cross-section is more challenging when applicable codes (e.g., DWUCK or MERCURY) are called as a subroutine for each energy decrement and required angle as the ions lose energy and scatter in tissue. For an energetic heavy ion penetrating tissue, a three-dimensional dose distribution calculation requires hours to days to complete using a top flight personal computer. These calculations are greatly simplified after the cross-sections are parameterized or input into the Appendix E codes. This simplification can be accomplished after ions and energies are selected for the proposed therapy protocol.

References

Bethe, H. (1930) Zur Theorie des Durchgangs schneller Korpuskularstrahlung durch Materie. *Ann. Phys. (Leipzig)*, **5**, 325.

Bevelacqua, J.J. (2005) Systematics of heavy ion radiotherapy. *Radiat. Prot. Manage.*, **22** (6), 4.

Bevelacqua, J.J. (2008) *Health Physics in the 21st Century*, Wiley-VCH Verlag GmbH, Weinheim.

Bevelacqua, J.J. (2009) *Contemporary Health Physics, Problems and Solutions*, 2nd edn, Wiley-VCH Verlag GmbH, Weinheim.

Bevelacqua, J.J. (2010a) *Basic Health Physics, Problems and Solutions*, 2nd edn, Wiley-VCH Verlag GmbH, Weinheim.

Bevelacqua, J.J. (2010b) Feasibility of using internal radiation-generating devices in radiotherapy. *Health Phys.*, **98**, 614.

Bevelacqua, J.J. (2012) Angular absorbed dose dependence of internal radiation- generating devices in radiotherapy. *Health Phys.*, **102**, 2.

Bevelacqua, J.J. (2014) Tumor vascular disruption using various radiation types. *PeerJ*, **2**, e320. 10.7717/peerj.320, https://peerj.com/articles/320/ (accessed 8 August 2015).

Bevelacqua, J.J., Stanley, D., and Robson, D. (1977) Sensitivity of the reaction $^{40}Ca(^{13}C, ^{14}N)^{39}K$ to the $^{39}K+p$ form factor. *Phys. Rev.*, **C15**, 447.

Charlton, L.A. (1973) Specific new approach to finite-range distorted-wave born approximation. *Phys. Rev.*, **C8**, 146.

Charlton, L.A. and Robson, D. (1973) MERCURY. Technical Report No. 5, Florida State University, Tallahassee, FL.

Kraft, G. (2000) Tumor therapy with heavy charged particles. *Prog. Part. Nucl. Phys.*, **45**, S473.

Kunz, P. D. (1986) Computer Code DWUCK, University of Colorado, Bolder, CO, available as part of Radiation Safety Information Computational Center Computer Code Collection: Code Package PSR-235 DWUCK, Nuclear Model Computer Codes for Distorted Wave Born Approximation and Coupled Channel Calculations, Oak Ridge National Laboratory, Oak Ridge, TN, http://rsicc.ornl.gov (accessed 8 August 2015).

Marmier, P. and Sheldon, E. (1969) *Physics of Nuclei and Particles*, vol. **1**, Academic Press, New York.

Radiation Safety Information Computational Center Computer Code Collection (1985) Calculation of Stopping Powers and Ranges for Muons, Charged Pions, Protons, and Heavy Ions, Code Package CCC-228 SPAR, Oak Ridge National Laboratory, Oak Ridge, TN, http://rsicc.ornl.gov (accessed 8 August 2015).

Shen, W.Q., Wang, B., Feng, J., Zhan, W.L., Zhu, Y.T., and Feng, E.P. (1989) Total reaction cross section for heavy-ion collisions and its relation to the neutron excess degree of freedom. *Nucl. Phys. A*, **491**, 130.

Turner, J.E. (2007) *Atoms, Radiation, and Radiation Protection*, 3rd edn, Wiley-VCH Verlag GmbH, Weinheim.

Weber, U.(1996) Volumenkonforme Bestrahlung mit Kohlenstoffionen. PhD thesis, Universität Gh Kassel, Germany.

G
Angular Absorbed Dose Dependence of Heavy Ion Interactions

G.1
Introduction

The angular absorbed dose dependence of an ion interacting with a medium is a complex calculation that requires the determination of the energy- and angle-dependent differential cross-section. The calculations are time consuming because the ion loses energy as it penetrates the media and the differential cross-section must be calculated at each energy increment and angle for the particular medium that it encounters. This appendix provides an overview of the angular absorbed dose calculations that are encountered in cancer therapy applications that utilize protons and heavy ions.

G.2
Basic Theory

For a tissue volume irradiated by a beam of ions of a given energy (E), the absorbed dose ($D(r, \theta)$) as a function of penetration distance r into tissue at an angle θ relative to the beam direction is obtained from the relationship

$$D(r,\theta) = \frac{1}{\rho}\left(-\frac{\mathrm{d}E}{\mathrm{d}r}\right)\Phi(r,\theta) \tag{G.1}$$

where standard spherical coordinates (r, θ, ϕ) are used, ρ is the density of the material (e.g., tissue, tumor, or other structure) attenuating the ion, $-\mathrm{d}E/\mathrm{d}r$ is the stopping power, and Φ is the ion fluence, which is dependent on r and θ. For specificity, the ranges of the spherical coordinates are $0 \le r \le \infty$, $0 \le \theta \le \pi$, and $0 \le \phi \le 2\pi$.

The particle fluence varies with tissue penetration depth and angle according to the relationship

$$\Phi(r,\theta) = \Phi(0,0)\exp(-\Sigma(\theta)r) \tag{G.2}$$

where $\Phi(0,0)$ is the entrance fluence into tissue at 0° relative to the beam direction and $\Sigma(\theta)$ is the angular macroscopic cross-section for a given nuclear reaction at

Health Physics: Radiation-Generating Devices, Characteristics, and Hazards, First Edition. Joseph John Bevelacqua.

energy E defined as

$$\Sigma(\theta) = n\frac{d\sigma(\theta)}{d\Omega} \tag{G.3}$$

In Eq. (G.3), n is the number of target atoms per cm^3, $d\sigma(\theta)/d\Omega$ is the microscopic differential cross-section (cm^2/atom-sr) for the reaction of interest, and $d\Omega$ is the spherical coordinate area element ($r^2 \sin(\theta) d\theta\, d\phi$). Given these cross-section model definitions, $\Sigma(\theta)$ has units of 1/cm-sr.

The angular macroscopic reaction cross-section is not readily obtained since there are no relationships analogous to the total reaction cross-section parameterization of Shen *et al.* In principle, this requires the calculation of the complete set of elastic and inelastic (including transfer) reaction differential scattering cross-sections for ions incident on the various tissue constituents.

In Eq. (G.3), the differential cross-section $d\sigma(\theta)/d\Omega$ is usually obtained from a distorted-wave Born approximation (DWBA) code such as DWUCK4 or MERCURY. The calculation of the differential cross-section for each location is required to obtain the three-dimensional absorbed dose distribution. To calculate the three-dimensional absorbed dose distribution, the cross-section is determined for each induced reaction and energy decrement as the tissue is traversed by the ion. Additional commentary regarding the models used to calculate differential cross-sections is provided in Appendix E.

G.3
Differential Scattering Cross-Section

DWBA codes calculate the differential scattering cross-section for a general transfer or elastic scattering reaction: $A(a, b)B$. Optical potentials are used to define scattering in the entrance (projectile (a) plus target nucleus (A)) and exit (ejectile (b) plus residual nucleus (B)) channels and are derived by parameterizing scattering data in terms of well-defined interaction strengths, radii, and diffuseness values.

An illustration of the process used to select optical potentials is provided by a discussion of the low-energy proton interactions with water. A survey of the literature for low-energy proton-induced reactions on ^{1}H and ^{16}O reveals that the total cross-section is dominated by $^{1}H(p, p)^{1}H$ and $^{16}O(p, p)^{16}O$ elastic scattering. Other reactions (e.g., $^{16}O(p, n)^{16}F$, $^{16}O(p, t)^{14}O$, and $^{16}O(p, \alpha)^{13}N$), which have smaller cross-sections, can be omitted from the discussion without a significant perturbation on the results. Accordingly, the use of $p + {}^{1}H$ and $p + {}^{16}O$ elastic scattering cross-sections is sufficient to determine the essential features of the angular dependence of the absorbed dose.

Although cross-section simplification occurs for lower-energy protons, general heavy ion water interactions must include all naturally occurring isotopes (i.e., ^{1}H, ^{2}H, ^{3}H, ^{16}O, ^{17}O, and ^{18}O) and their elastic, inelastic, and transfer reactions. Unless a limited set of reactions dominate (e.g., $^{1}H(p, p)^{1}H$ and $^{16}O(p, p)^{16}O$ for low-energy protons interacting with water), the calculations become complex and require evaluation of all viable nuclear interactions.

The subsequent discussion uses the center-of-mass (cm) system to facilitate comparison with published data. The differential cross-sections are largest in the beam direction ($\theta_{cm} = 0°$) and generally decrease as the cm angle increases. Differential cross-sections tend to have this general angular dependence for both elastic, inelastic, and transfer reactions. In addition, the cross-sections become more pronounced in the beam direction as the ion energy increases. These general trends in the differential cross-section are significant characteristics that influence the angular dependence of the absorbed dose distribution.

G.4
Model Calculations

Model calculations can be performed using a zero-range (ZR) approximation between the coordinates of the incoming and outgoing waves. For the purposes of this appendix, zero-range distorted-wave Born approximation (ZRDWBA) calculations are a credible approximation for reactions involving light ions (e.g., $^{1}H(p, p)^{1}H$, $^{16}O(p, p)^{16}O$, $^{16}O(p, n)^{16}F$, $^{16}O(p, d)^{15}O$, $^{16}O(p, t)^{14}O$, and $^{16}O(p, {}^{3}He)^{14}N$). Finite range effects can be important for reactions involving heavier ions (e.g., $^{16}O(p, {}^{6}Li)^{11}C$, $^{16}O(p, {}^{7}Li)^{10}C$, $^{16}O(p, {}^{8}Be)^{9}B$, and $^{16}O(p, {}^{7}B)^{10}Be$). Specific examples of the use and application of these codes are provided in this appendix's references.

References

Baron, N., Leonard, R.F., and Lind, D.A. (1969) Elastic scattering of 21-MeV protons from nitrogen-14, oxygen-16, argon-40, nickel-58, and tin-116. *Phys. Rev.*, **160**, 978.

Bevelacqua, J.J. (1982) Center-of-mass effects in the $^{12}C(p, t)^{10}C$ reaction at 80 MeV. *Rev. Mex. Fis.*, **28**, 573.

Bevelacqua, J.J. (2005) Systematics of heavy ion radiotherapy. *Radiat. Prot. Manage.*, **22**, 4.

Bevelacqua, J.J. (2010) Feasibility of using internal radiation-generating devices in radiotherapy. *Health Phys.*, **98**, 614.

Bevelacqua, J.J. (2012) Angular absorbed dose dependence of internal radiation-generating devices in radiotherapy. *Health Phys.*, **102**, 2.

Bevelacqua, J.J. and Prewett, S.V. (1979) Form factor effects in the $^{18}O(p, t)^{16}O$ reaction. *Can. J. Phys.*, **57**, 1484.

Bevelacqua, J.J. and Prewett, S.V. (1981) Zero-range distorted-wave born approximation analysis of (p, t) transfer reactions. *Can. J. Phys.*, **59**, 231.

Charlton, L.A. (1973) Specific new approach to finite-range distorted-wave born approximation. *Phys. Rev.*, **C8**, 146.

Charlton, L.A. (1975) Finite range calculations of light-ion two-nucleon transfer reactions. *Phys. Rev.*, **C12**, 351.

Charlton, L.A. and Robson, D. (1973) MERCURY. Technical Report No. 5, Florida State University, Tallahassee, FL.

Davis, J.C., Anderson, J.D., Grimes, S.M., and Wong, C. (1973) Proton-proton final-state interactions in the $^{2}H(p, n)^{2}p$ reaction. *Phys. Rev.*, **C8**, 863.

Fabrici, E., Micheletti, S., Pignanelli, M., Resmini, F.G., DeLeo, R., D'Erasmo, G., and Panteleo, A. (1980) Proton elastic scattering on light nuclei: II. Nuclear structure effects. *Phys. Rev.*, **C21**, 844.

Hurd, J.R., James, D.R., Bevelacqua, J.J., and Medsker, L.R. (1979) (^{7}Li, ^{8}Be) reaction on ^{40}Ca. *Phys. Rev.*, **C20**, 1208.

International Atomic Energy Agency (1998) *Handbook for Calculations of Nuclear Reaction Data: Reference Input Parameter Library*, IAEA-TECDOC-1034, IAEA, Vienna.

Jarmie, N. and Jett, J.H. (1976) Proton-proton elastic scattering at 19.700 MeV. *Phys. Rev.*, **C13**, 2554.

Jarmie, N., Jett, J.H., Detch, J.L., and Hutson, R.L. (1971) Proton-proton elastic scattering from 9.6 to 13.6 MeV. *Phys. Rev.*, **C3**, 10.

Kunz, P.D. (1986) Radiation Safety Information Computational Center Computer Code Collection: Code Package P00546 DWUCK/CHUCK, Nuclear Model Computer codes for Distorted Wave Born Approximation and Coupled Channel Calculations, Oak Ridge, TN, Oak Ridge National Laboratory, rsicc.ornl.gov(accessed 8 August 2015).

Madey, R., Flanders, B.S., Anderson, B.D., Baldwin, A.R., Bertozzi, W., Chittrakarn, T., Comfort, J.R., Finn, J.M., Foster, C.C., Hyde-Wright, C., Kelly, J.J., Lebo, C., Orihara, H., Pairsuwan, W., and Watson, J.W. (1997) 0 to 0 transition in the ^{16}O(p, n)^{16}F reaction at 79 MeV. *Phys. Rev.*, **C56**, 3210.

Marmier, M. and Sheldon, E. (1969) *Physics of Nuclei and Particles*, vol. I and II, Academic Press, New York.

van Oers, W.T.H. and Cameron, J.M. (1969) Optical-model analyses of p-^{16}O elastic scattering from 23–53 MeV. *Phys. Rev.*, **184**, 1061.

Olver, F.W.J., Lozier, D.W., Boisvert, R.F., and Clark, C.W. (2010) *NIST Handbook of Mathematical Functions*, Cambridge University Press, Cambridge.

Perey, C.M. and Perey, F.G. (1976) Compilation of phenomenological optical model parameters 1954–1975. *At. Data Nucl. Data Tables*, **17**, 1.

Shen, W.Q., Wang, B., Feng, J., Zhan, W.L., Zhu, Y.T., and Feng, E.P. (1989) Total reaction cross section for heavy-ion collisions and its relation to the neutron excess degree of freedom. *Nucl. Phys.*, **A491**, 130.

Turner, J.E. (2007) *Atoms, Radiation, and Radiation Protection*, 3rd edn, Wiley-VCH Verlag GmbH, Weinheim.

H
Basis for Radiation Protection Regulations

H.1
Overview

In Chapter 7, a discussion of the basis for future radiation protection regulations is presented. These regulations are currently based on the linear-nonthreshold (LNT) hypothesis, but this approach is not universally accepted and issues have been raised regarding its acceptability. One of the concerns with current regulatory models is their LNT basis derived from high-dose and dose rate data (e.g., atomic bomb and medical therapy) extrapolated in a linear manner to low doses. Other data (e.g., occupational and environmental) are excluded even though the dosimetry is good and the exposed groups are large and well defined. In addition to the inclusion of all dosimetric data, the new regulations should consider a variety of dose–response models including those that do not rely on the LNT hypothesis.

This appendix provides supporting information that forms a portion of the technical basis for current radiation protection regulations. This basis includes the LNT approach and the selection and modeling of dose–response models, risk models, excess risk functions, risk coefficients, biological detriments, and the dosimetry associated with these detriments. This appendix also illustrates the influence of these models in assessing radiation risk to workers.

The motivation for establishing a new regulatory basis is outlined in this appendix. Options for a new regulatory format are summarized in Chapter 7.

H.2
Risk

Radiation is one of the most thoroughly studied agents associated with a biological detriment. These detriments are quantified in terms of stochastic and nonstochastic effects and their associated health risks. The risk (R) is often quantified in terms of a risk coefficient (r) expressing excess radiation-induced effects per unit radiation dose (D). Accordingly, the risk of the radiation exposure is often determined from the LNT relationship

$$R = rD \tag{H.1}$$

Health Physics: Radiation-Generating Devices, Characteristics, and Hazards, First Edition. Joseph John Bevelacqua.

Table H.1 Ionizing radiation risk coefficient summary[a].

Year	Report	Risk coefficient ($\times 10^{-2}$) (RIE/Sv)
1972	BEIR I	1
1977	ICRP 26	2
1980	BEIR III	2
1985	EPA NESHAP	4
1988	NRC BRC policy	5
1990	BEIR V	8
1991	ICRP 60	7
2006	BEIR VII	5[b]
2007	ICRP 103	6

a) Bevelacqua (2010).
b) Excess cancer deaths extracted from BEIR VII data.

where the dose is the radiation exposure received and under evaluation. The risk coefficient depends on the data under evaluation and the underlying modeling assumptions. Risk estimates are also influenced by the radiation characteristics (e.g., dose, dose rate, fractionalization, and radiation type), biological characteristics (e.g., age, sex, genetic background, and nature of the tissue or organ), and the approach to the analysis (e.g., dose–response model, projection model, and risk model).

In view of these factors, it is not surprising that there is considerable variance in risk estimates. For example, the ICRP-26 (International Commission on Radiological Protection) risk coefficient is 2×10^{-2} radiation induced effects (RIE)/Sv, while BEIR V with its 8×10^{-2} RIE/Sv coefficient yields a larger characterization of the risk. A summary of risk coefficients derived from major studies is provided in Table H.1.

Equation (H.1) is often applied carelessly. This equation is most valid for a large ensemble of subjects (10 000–100 000) who have each received at least 0.1 Gy of acute radiation exposure.

The total risk coefficient (r) is the sum of the risk coefficients for the organs or tissues (T) composing the modeled human body:

$$r = \sum_T r_T \tag{H.2}$$

Table D.9 summarizes the various organs that are assumed in the ICRP 26, UNSCEAR 88 (United Nations Scientific Committee on the Effects of Atomic Radiation), ICRP 60, and ICRP 103 formulations. The formulations do not contain the same organs or level of organ risk. This table also provides the values of the tissue weighting factors (w_T) for these models:

$$w_T = \frac{r_T}{r} \tag{H.3}$$

where the weighting factor is a dimensionless number with a value between zero and unity.

Tables H.1 and D.9 illustrate the modeling variations encountered in the risk estimates. The ICRP 60 and 103 models include more organs with specified organ weighting factors than the ICRP 26 formulation and also include a set of specified organs to be included in the remainder. An examination of Table D.9 illustrates the model dependence (i.e., number of organs and assigned weighting factors) of the various ICRP internal dosimetry formulations. For example, the tissue weighting factor for the gonads changed significantly in the ICRP 26 (0.25), ICRP 60 (0.20), and ICRP 103 (0.08) formulations.

The risk coefficients summarized in Table H.1 and the associated tissue weighting factors summarized in Table D.9 are derived from an assessment of the number of radiation-induced effects per unit dose. These assessments require that the source of the measured effect be determined and directly related to the radiation dose. This assessment utilizes basic epidemiological principles, which are briefly outlined in the next section of this appendix.

H.3
Basic Epidemiology

Studies of radiation risk utilize epidemiological input that requires a sample size dependent on the magnitude of the radiation exposure. The sample size, required for statistically meaningful results, is 5×10^4, 5×10^8, and 5×10^{12} individuals for acute absorbed doses of 100, 1, and 0.01 mGy, respectively. These values illustrate how the size of the required exposed group varies with the absorbed dose. Meaningful results are possible for larger exposures, but the population size required for typical annual occupational or environmental exposures is prohibitive.

The BEIR VII Committee did not know whether dose rates of gamma or X-rays of about 1 mGy/year are detrimental to humans. Somatic effects at these doses would be masked by environmental and other factors that produce the same types of health effects as ionizing radiation. Therefore, assessments of the impact of doses on the order of magnitude of 1 mGy or less are not practical from a statistical perspective.

Epidemiological studies must also consider a number of factors including sex, age, time since exposure, and the age at exposure. Accurate studies are also of long duration since time is required to follow the exposed population and control group.

The number of cancers expected in a cohort (E) of exposed individuals is given by the sum

$$E = \sum_{x} c(x)\, r(x) \tag{H.4}$$

where $r(x)$ is the annual incidence (morbidity) per person at age x per year and $c(x)$ is the sum of all years spent by cohort members at age x. Once the expected incidence is determined, the number of excess cancers (ECs) is readily obtained:

$$EC = O - E \tag{H.5}$$

where O is the observed cancer incidence in the risk or exposed population.

The excess cancers per population year per incident exposure (Z) is given by

$$Z = \frac{(O - E)}{N} \tag{H.6}$$

The quantity N has the units of population year-Gy:

$$N = d_i \; y_i \tag{H.7}$$

where d_i is the dose to the ith group and y_i is the number of years that the ith group is observed. Therefore, Z is expressed in excess cancers per population year-Gy.

With these definitions, commonly utilized epidemiology terms can be defined. The relative risk (RR), standard mortality ratio (SMR), and excess relative risk (ERR) are defined as

$$\mathrm{RR} = \frac{O}{E} \tag{H.8}$$

$$\mathrm{SMR} = 100\mathrm{RR} \tag{H.9}$$

and

$$\mathrm{ERR} = \mathrm{RR} - 1 \tag{H.10}$$

H.4
Dose–Response Relationships

Dose–response relationships describe how an effect varies with dose. Currently, the two most popular dose–response relationships are the linear and linear quadratic models. These models are discussed in Appendix D.

The dose–response models utilized in the BEIR reports are zero threshold approaches. This hypothesis leads to the suggestion that detrimental health effects exist at very low doses. The linear extrapolation from high-dose and dose rate data to low doses is open to challenge and addressed in subsequent discussion.

Uncertainties associated with the extrapolation from high-dose and dose rate data to the low-dose region are alleviated by utilizing the complete set of available radiation dosimetry data. The complete set of radiation data includes a wealth of information including occupational data from power reactors, Department of Energy (DOE) weapons complex facilities and national laboratories, universities, medical facilities, and commercial facilities utilizing radioactive materials. Environmental data are also tabulated from high-dose rate areas of the world. In addition, low-dose imaging data and other diagnostic medical data are available.

Utilization of the complete set of dosimetry data could significantly improve the justification for the functional dependence of the dose–response relationship. These relationships should consider thresholds and functions more diverse than the assumed linear or linear quadratic relationships.

H.5 Risk Models

There are two general types of models that are traditionally used in assessing risk. These are the absolute and the relative risk models that were reviewed in Appendix D.

As noted in Table H.2, the BEIR models have been applied to a variety of cancer types. An example of their application to leukemia and nonleukemia cancers illustrates the conclusions drawn by the BEIR Committees in attempting to assess radiation risk. This is illustrated by summarizing a portion of the BEIR V report and its conclusions regarding the relative risk model.

These conclusions are specific and well defined but are based on high-dose and dose rate data. As noted previously, their extrapolation to low doses using LNT models is open to challenge.

The assumed functional forms of the absolute and relative risk models and the typical exponential and step functions used in the excess risk function are not unique. Other functional relationships should be investigated with the utilization of complete dosimetry data sets.

The large lifetime cancer risk uncertainties illustrated in Table D.3 also suggest that the investigation of other functional forms for the dose–response model,

Table H.2 BEIR V preferred relative risk model.[a)]

Cancer type	Dose–response model	Comments
Leukemia	Linear quadratic	Minimum latency of 2 years
Breast	Linear	Highest risk in women under age 15 at the time of exposure
		Risk is low for women if exposed after age 40
Respiratory	Linear	Minimum latency of 10 years
		Risk decreases with time after exposure
		Relative risk for females is twice that for males
Digestive	Linear	About seven times the risk if exposure occurs at age 30 or less
		Risk does not change with time post exposure
Other	Linear	Contributes significantly to total risk
		No age or sex effects have been noted
		Insufficient data to permit detailed modeling

a) BEIR V does not support the absolute risk model.

risk model, and excess risk function is warranted. This investigation should also include thresholds and all available dosimetry data.

H.6
BEIR VII Uncertainties

Although BEIR VII does not provide an excess cancer risk coefficient, a public risk coefficient for excess cancer deaths (ecd) can be developed from the report's data. An illustration of the uncertainties involved in the BEIR VII analysis is provided by developing this risk coefficient.

Using BEIR VII data, the number of ecd from exposure to 0.1 Gy to males is 410 (200, 830) ecd in an exposed population of 100 000 individuals. The values in parenthesis are the 95% confidence intervals. For females, the number of excess deaths from exposure to 0.1 Gy is 610 (300, 1200) ecd in an exposed population of 100 000. These distributions are broad and indicate the uncertainties encountered in the BEIR VII analysis. When considered in the historical context of the BEIR III and V reports summarized in Table D.3, a view of data uncertainty is provided. This uncertainty suggests that many functional forms could be used to fit the available data. Limiting the analysis of evaluated data sets to linear and linear quadratic models with no thresholds is open to challenge. In addition, restricting the analysis to absolute and relative risk models or combinations of these models is also overly restrictive. Other functional forms, the existence of thresholds, and a variety of risk models should be evaluated to ensure that radiation protection regulations are based on an unbiased analysis.

The public risk coefficient for all cancers is obtained by averaging over age and sex (410 ecd + 610 ecd)/2 which produces a value of 510 ecd. This data can be used to obtain a corresponding risk coefficient:

$$r = \frac{\left(\frac{410\,\text{ecd}+610\,\text{ecd}}{2}\right)}{(100\,000\,\text{persons})\ (0.1\,\text{Gy/person})}\frac{1\,\text{Gy}}{1\,\text{Sv}} = 5\times10^{-2}\,\text{ecd/Sv} \qquad \text{(H.11)}$$

As a comparison, BEIR V derived a value of 695 ecd/100 000 persons exposed to 0.1 Gy (no dose and dose rate effectiveness factor (DDREF) utilized). This is again averaged over males and females ((660 ecd + 730 ecd)/2 = 695 ecd). If the BEIR VII DDREF is applied to the BEIR V data, 695 ecd/1.5 provides a value of 463 ecd. Using the methodology illustrated by Eq. (H.11) and keeping one significant figure lead to a public ecd risk coefficient for BEIR V of 5×10^{-2} ecd/Sv. Therefore, BEIR V, BEIR VII, and ICRP 60 have the same excess cancer risk coefficient of 5×10^{-2} ecd/Sv. This calculation illustrates the consistency of these reports.

Reports such as BEIR VII are important because they refine the internal dosimetry models and develop risk estimates. Consequently, conclusions of BEIR VII carry significant weight and ideally are clear, unambiguous, and widely accepted.

The LNT hypothesis is an expedient regulatory model, but it is not universally accepted. For example, a number of professional organizations including two

Table H.3 Doubling dose.

Report	Doubling dose (Sv)
BEAR[a] (1956)	0.05 – 1.0
BEIR I (1972)	0.20 – 1.0
BEIR III (1980)	0.50 – 2.5
BEIR V (1990)	<1.0
BEIR VII (2006)	1.0

a) BEAR: Biological Effects of Atomic Radiation

French academies have challenged the BEIR VII Report's conclusion regarding the LNT hypothesis. Issues associated with the LNT hypothesis are summarized in subsequent discussion.

One of the weaknesses of the BEIR VII approach is not fully utilizing the wealth of available dosimetry data (e.g., occupational and environmental). With the inclusion of all dosimetric data, thresholds, hormesis, and various functional forms for the dose – response and risk models should be rigorously evaluated to determine the validity of the LNT hypothesis.

H.7 Doubling Dose

The qualitative relationship between radiation dose and the probability of a mutation is often described in terms of the doubling dose. The doubling dose is the radiation dose that would lead to a doubling of the natural mutation rate. Table H.3 summarizes the doubling dose from the Biological Effects of Atomic Radiation (BEAR) and from BEIR I, III, V, and VII Reports. A doubling dose of about 1.0 Sv appears to a consistent value from the reports summarized in Table H.3.

H.8 Probability of Causation

A consequence of the LNT hypothesis is that its assumptions are replicated in derivative work. One of the LNT derivatives is the concept of Probability of Causation (PC), which is used to assess if an effect is attributable to radiation exposure.

US Public Law 97-414, the Orphan Drug Act of 1984, directed the Secretary of Health and Human Services to construct radioepidemiological tables providing the probability that certain cancers could result from prior exposure to ionizing radiation. The probability of causation is defined as a number that represents the probability that a given cancer, in a specific tissue, has been caused by a previous exposure or series of exposures to a carcinogenic agent such as ionizing radiation.

The PC tables are based upon the BEIR III report. NIH Publication No. 85-2748 established the groundwork for the PC concept for radiogenic tumors. The original PC tables are outdated because BEIR III has been superseded by BEIR VII.

The PC has the form

$$\mathrm{PC} = \frac{R}{(1+R)} \tag{H.12}$$

where R is the relative excess.

In the case of a single exposure of short duration to an individual representative of a population group, the relative excess is given by

$$R = FTK \tag{H.13}$$

In this equation, F is the exposure factor that characterizes the dependence of R on the radiation dose to the risk organ. The use of effective dose from TLD (thermoluminescent dosimetry) packages is not appropriate because absorbed dose in tissue is the desired quantity. The appropriate value of F is defined as a function of absorbed tissue dose (D), measured in centigray or radian. The factors T and K are defined in the subsequent discussion.

The specific functional form for F depends on the radiation quality and cancer site. For example, consideration of ^{224}Ra irradiating the bone and leading to bone cancer results in the simple relationship

$$F_{\mathrm{Bone}} = D \tag{H.14}$$

for high-linear energy transfer (LET) alpha radiation. For low-linear energy transfer radiation, the values of F for thyroid, breast, and other cancers are

$$F_{\mathrm{thyroid}} = D \tag{H.15}$$

$$F_{\mathrm{Bresat}} = D \tag{H.16}$$

and

$$F_{\mathrm{other}} = D + \left(\frac{1}{116}\right) D^2 \tag{H.17}$$

The second factor (T) in the definition of relative excess represents the relative likelihood that a cancer induced at age A_1 will be diagnosed after Y years. For diagnosis times between Y and $Y+1$ years, Y is utilized in the computation. Under the constant relative risk model, which is used for cancers other than leukemia and bone cancer, T depends only on Y and has a value that increases with Y. For $Y = 0-4$ years, $T = 0$ and it rises to a value of unity for $Y \geq 10$ years. T values of 0.25, 0.5, and 0.75 occur at about $Y = 6$, 7, and 8 years, respectively.

The constant relative risk model has not been assumed to hold for bone cancer and leukemia. For these two cancer types, T is a conditional probability, which assumes that the cancer has been caused by an exposure at age A_1 and will be diagnosed Y years later. For these cases, T is calculated as the lognormal probability that a cancer is detected between years Y and $Y+1$ after exposure at age A_1. The PC tables compile T for the various forms of cancer.

The final factor defining the relative excess is K, and it provides the dependence of R on age and baseline cancer incidence for persons of age A_2 and sex (S) for exposure at age A_1:

$$K = K(A_1, A_2, \text{and } S) \tag{H.18}$$

The PC tables include both human and animal data. Smoking data is also included, but prior medical exposure is not included.

The reader has by now drawn the conclusion that the PC concept is not precise. A qualitative estimate of the uncertainties in the PC estimate is illustrated by a few examples. If the PC is calculated to be 2% or less, the true PC could be as large as 7% even if an accurate knowledge of all the input parameters is known. If the PC is within the 5–10% range, the true PC could lie within the 1–30% range. Finally, if the PC is calculated to be a least 20%, the true PC could be in the 5–40% range.

A final complication of the PC concept lies in its ties to the BEIR III methodology. The differences between BEIR III and BEIR VII suggest that a review of the current PC approach and its underlying assumptions is in order.

H.9
Energy Employees Occupational Illness Compensation Program Act

The PC concept has been revised, and it is now used as the basis for determining the legal standard for resolving radiation-related claims associated with the Energy Employees Occupational Illness Compensation Program Act of 2000 (EEOICPA). As outlined in 42CFR81, 42CFR82, and 42CFR83, EEOICPA models incorporate relevant epidemiology, BEIR reports, and ICRP reports available at the time of its enactment.

The methods for calculating internal dose from the intake of radioactive material use the ICRP 66 human respiratory tract model and the ICRP 30 ingestion model. In addition, supporting radionuclide data (e.g., ICRP 56, 67, and 68) are utilized in the internal dose assessment. The EEOICPA permits calculational methods to be updated to reflect new reports and science as they become available. For example, ICRP 100 provides the human alimentary tract model that updates the ICRP 30 ingestion model.

The EEOICPA established that a lump-sum payment and medical benefits can be awarded as compensation to covered employees suffering from designated illnesses (e.g., cancer resulting from radiation exposure) incurred as a result of their performance of duty for the DOE and designated contractors. Under EEOICPA, an employee seeking compensation for cancer is eligible if the cancer has a 50% or greater probability of being caused by radiation doses incurred in the performance of duty or the employee is included in a specified cohort.

The risk models address a number of cancer types, and most types of radiation exposure are relevant to employees covered by the EEOICPA. These models include the employee's cancer type, year of birth, year of cancer diagnosis, exposure information, and the dose received from gamma radiation, X-rays, alpha radiation, beta radiation, and neutrons. In addition, the risk model for lung cancer includes the worker's smoking history and radon exposure, and the risk model for skin cancer incorporates race and ethnicity. None of the risk models explicitly includes exposure to other occupational, environmental, or dietary carcinogens. Models incorporating chemical agents have not yet been developed.

Although it is appropriate for an organization to compensate workers for harm incurred through employment, the PC concept is also tied to the LNT hypothesis. The PC results are inherently limited by the issues previously outlined for the LNT approach.

H.10
Future Dose Limits

The aforementioned discussion involves risk and its characterization. This characterization and the establishment of dose limits, risk coefficients, tissue weighting factors, and PC tables are a direct consequence of the ICRP, NCRP (National Council on Radiation Protection and Measurements), and other scientific organizations and their assessment of radiation and its associated detrimental health effects. These recommendations and the characterization of risk are reflected in national regulations that govern radiation protection activities. The regulations provide defined dose limits that are derived from the risk estimates.

Radiation protection regulations are currently based on the hypothesis that any radiation dose may result in a health detriment including cancer and hereditary effects. As noted previously, this hypothesis is based on the premise that health effects observed at high doses are also present at low doses. There are significant issues in demonstrating the validity of this premise. The current regulatory basis also assumes that detrimental effects occur in a linear, direct relationship with the dose delivered to an individual. These two assumptions form the basis for the LNT hypothesis, for estimating health effects. The LNT hypothesis is briefly examined in subsequent discussion. In particular, arguments supporting and challenging the LNT hypothesis are briefly presented.

H.10.1
LNT Hypothesis

The LNT hypothesis presumes that radiation-induced detriment is linearly proportional to dose even in the limit of zero exposure. It assumes that any dose, no matter how small, produces detrimental health effects.

A corollary to the LNT hypothesis is the introduction of the collective dose assumption. Collective dose is the sum of individual doses in an exposed group and is a method for quantifying dose in a population group. This assumption presumes that small doses to large populations can be summed to predict a set of calculated health effects that are representative of the population risk.

Collective dose often overstates the presumed risk and equivalent collective doses do not imply equivalent risk. For example, a large dose to members of a small group is not equivalent to a small dose to members of a large group, even if the collective doses are the same. For groups in which individual lifetime doses are less than 100 mSv above background, collective dose is a speculative and uncertain measure of risk. It should not be used for estimating the health risks to an exposed population.

The LNT hypothesis and opposing viewpoints cannot be sufficiently addressed in the limited space allotted to this appendix. Accordingly, only salient arguments supporting and opposing the LNT hypothesis are outlined. Additional commentary is provided in Chapter 7.

H.10.1.1
Arguments Supporting the LNT Hypothesis

At low and intermediate doses (10 mGy to 1 Gy), Brenner notes that mutation and chromosome aberration induction data are consistent with a linear dose–response relation. NCRP 136 supports this view: "although other dose-response relationships for the mutagenic and carcinogenic effects of low-level radiation cannot be excluded, no alternate dose-response relationship appears to be more plausible than the linear-nonthreshold model on the basis of present scientific knowledge."

At lower doses, biophysical arguments are used by Brenner to justify the LNT hypothesis. These arguments include:

1) Tumors are largely of monoclonal origin.
2) Ionizing radiation produces sufficient damage in a cell to initiate oncogenesis.
3) As the dose of ionizing radiation decreases, fewer cells are damaged by more than one radiation track. This results in a proportional decrease in the number of cells in which this damage occurs. The proportional decrease remains valid at very low doses.

The proportional decrease in the limit of zero dose forms the basis for the LNT hypothesis. If this proportionality can be unambiguously demonstrated as doses approach zero, the LNT hypothesis becomes the LNT Law.

H.10.1.2
Arguments against the LNT Hypothesis

Section 7.10.3 provides a number of physiological perspectives that challenge the LNT hypothesis. These challenges include the LNT focus on DNA damage and mutations, but these are not the only factors affecting the onset and propagation of cancer. The LNT approach also ignores the effect of the human immune system response. Finally, the large variability in cancer rates is ignored by not specifying a threshold. Raabe presents additional arguments against the LNT hypothesis.

Raabe notes that the development of a radiation-induced malignant tumor is not the result of a single random interaction of the ionizing radiation with an isolated cell. He offers the following arguments against the LNT hypothesis and suggests that major revisions of methodology and standards are needed:

1) The cancer risk associated with ionizing radiation exposure is a nonlinear function of the lifetime average dose rate to the affected tissues.
2) Cancer risk exhibits a virtual threshold at low lifetime average dose rates.
3) Cumulative radiation dose is not an accurate or appropriate measure of cancer risk, but it is useful for describing the virtual threshold for various exposures.
4) High-dose rate atomic bomb survivor data from Hiroshima and Nagasaki cannot be used to estimate cancer risk from ionizing radiation exposures over long times and at low dose rates.

Based on these considerations, currently accepted ionizing radiation models should be reevaluated to assess the validity of LNT estimates of ionizing radiation cancer risk. Other arguments offered by the Health Physics Society (HPS) suggest that the LNT hypothesis is an oversimplification. The LNT approach can be rejected for specific cancer types (e.g., bone cancer and chronic lymphocytic leukemia). In addition, significant heritable genetic damage has not been observed in human studies. The effects of various biological mechanisms (e.g., DNA repair and adaptive response) on the induction of cancers and genetic mutations as a function of dose and dose rate have not been thoroughly investigated. These mechanisms do not appear to be credibly modeled by the LNT hypothesis.

H.10.2
Threshold Dose limits

The credibility of the LNT hypothesis is challenged by the observation that radiogenic health effects have not been consistently demonstrated below 100 mSv. Primary cancers have been observed in humans only at doses exceeding 50 – 100 mSv delivered at high dose rates. Below this threshold, estimates of radiation detriment are speculative. As noted previously, risk estimates in exposed populations are based on epidemiological studies of well-defined groups (e.g., the Japanese atomic bomb survivors and medical therapy patients) exposed to relatively high doses delivered at high dose rates. Adverse health effects have not been observed in individuals exposed to chronic doses less than 100 mSv.

In its radiation risk in perspective position statement, the HPS concluded that risk estimates should be limited to individuals receiving a dose of 50 mSv in 1 year or a lifetime dose of 100 mSv. This dose is in addition to natural background. Below these doses, risk estimates should not be performed. In addition, the HPS recommends that expressions of risk should only be qualitative and presented as a range of values based on uncertainties. This range of uncertainty values should include the inability to detect any increased health detriment, which acknowledges that zero health effects are a credible outcome.

These risk assessments can be used as a regulatory basis to select from a group of options associated with work involving radiation exposure. This risk assessment approach can be applied to a variety of radiological work activities including the selection of methods to remediate sites contaminated with radioactive material, disposition of slightly radioactive material, recovery options from a reactor accident, transport of radioactive material, and selection of decontamination end-state criteria.

H.10.3
Radiation Carcinogenesis

Raabe notes that ionizing radiation carcinogenesis is not a linear function of cumulated dose. Moreover, it is not a stochastic single-cell phenomenon. It is a whole-organ process that is dependent on a variety of factors including the lifetime average dose to the sensitive organ cells. As a collective process, the arguments of Doss suggest a whole-body response including the importance of the human immune system. The elimination of a single-cell effect and influence of collective body defense mechanisms suggest the LNT response model is an oversimplification of the onset and development of carcinogenesis.

H.11
Future Regulations

The discussion presented in this appendix suggests that other viable regulatory bases for radiation protection merit consideration. These approaches should investigate thresholds and utilize alternative dose–response and risk models. To evaluate these proposals, all radiation data must be assessed and radiation protection regulations should not be based solely on high-dose and high-dose rate data linearly extrapolated to low doses. The new regulatory basis should also investigate the need to incorporate dose and dose rate effectiveness factors, adaptive response, and hormesis. New regulations should explicitly include risk and utilize risk estimates to select the optimum radiological work approach. The risk-based process would supplement the traditional ALARA (as low as reasonably achievable) regulatory philosophy used to minimize worker doses.

References

10CFR20 (2015) Standards for Protection Against Radiation, Code of Federal Regulations Title 10, Part 20, National Archives and Records Administration, U.S. Government Printing Office, Washington, DC.

10CFR835 (2015) Occupational Radiation Protection, Code of Federal Regulations Title 10, Part 835, National Archives and Records Administration, U.S. Government Printing Office, Washington, DC.

42CFR81 (2015) Guidelines for Determining the Probability of Causation under the Energy Employees Occupational Illness Compensation Program Act of 2000, Code of Federal Regulations Title 42, Part 81, National Archives and Records Administration, Washington, DC.

42CFR82 (2015) Methods for Radiation Dose Reconstruction under the Energy Employees Occupational Illness Compensation Program Act of 2000, Code of Federal

Regulations Title 42, Part 82 National Archives and Records Administration, Washington, DC.

42CFR83 (2015) Procedures for Designating Classes of Employees as Members of the Special Exposure Cohort under the Energy Employees Occupational Illness Compensation Program Act of 2000, Code of Federal Regulations Title 42, Part 83, National Archives and Records Administration, Washington, DC.

ANS (2001) American Nuclear Society Background Information for Position Statement 41 (2001) *Health Effects of Low-Level Radiation*, American Nuclear Society, LaGrange Park, IL.

Bevelacqua, J.J. (2009) *Contemporary Health Physics, Problems and Solutions*, 2nd edn, Wiley-VCH Verlag GmbH, Weinheim.

Bevelacqua, J.J. (2010) *Basic Health Physics, Problems and Solutions*, 2nd edn, Wiley-VCH Verlag GmbH, Weinheim.

Brenner, D.J. and Raabe, O.G. (2001) Topics under debate-is the linear-no-threshold hypothesis appropriate for use in radiation protection? *Radiat. Prot. Dosim.*, **97**, 279.

Brown, C.C. and Chu, K.C. (1982) *Approaches to Epidemiologic Analysis of Prospective and Retrospective Studies, Epidemiology: Risk Assessment*, SIAM, Philadelphia, PA.

Cember, H. and Johnson, T.A. (2008) *Introduction to Health Physics*, 4th edn, McGraw-Hill Medical, New York.

Doss, M. (2012) Shifting the paradigm in radiation safety. *Dose-Response*, **10**, 562.

Health Physics Society Position Statement PS010-2, Radiation Risk in Perspective, Health Physics Society, McLean, VA (2012).

ICRP (1977) *Recommendations of the International Commission on Radiological Protection*, Publication 26, Pergamon Press, Oxford.

ICRP (1979) *Limits for Intakes of Radionuclides by Workers*, Publication 30, Pergamon Press, Oxford.

ICRP (1990) *Age-dependent Doses to Members of the Public from Intake of Radionuclides-Part 1*, Publication 56, Pergamon Press, Oxford.

ICRP (1991) *Recommendations of the ICRP*, Publication 60, Pergamon Press, Oxford.

ICRP (1994a) *Human Respiratory Tract Model for Radiological Protection*, Publication 66, Pergamon Press, Oxford.

ICRP (1994b) *Age-dependent Doses to Members of the Public from Intake of Radionuclides–Part 2*, Publication 67, Ingestion Dose Coefficients, Pergamon Press, Oxford.

ICRP (1994c) *Dose Coefficients for Intakes of Radionuclides by Workers*, Publication 68, Pergamon Press, Oxford.

ICRP (1995) *Age-dependent Doses to Members of the Public from Intake of Radionuclides–Part 4 Inhalation Dose Coefficients*, Publication 71, Pergamon Press, Oxford.

ICRP (2007) *The 2007 Recommendations of the International Commission on Radiological Protection*, Publication 103, Elsevier, Amsterdam.

National Academy of Sciences, Committees on the Biological Effects of Atomic Radiation (1956) *Biological Effects of Atomic Radiation*, National Academy Press, Washington, DC.

National Academy of Sciences-National Research Council, Advisory Committee on the Biological Effects of Ionizing Radiation (1972) *Biological Effects of Ionizing Radiation*, National Academy Press, Washington DC.

National Research Council (1980) *The Effects on Populations of Exposures to Low Levels of Ionizing Radiation (BEIR III)*, Committee on the Biological Effects of Ionizing Radiation, National Academy Press, Washington, DC.

National Research Council (1990) *The Health Risks from Exposure to Low Levels of Ionizing Radiation, BEIR V*, National Academy Press, Washington, DC.

National Research Council (2006) *Health Risks from Exposure to Low Levels of Ionizing Radiation, BEIR VII Phase 2*, National Academy Press, Washington, DC.

NCRP (1997) Uncertainties in Fatal Cancer Risk Estimates Used in Radiation Protection. Report No. 126, NCRP Publications, Bethesda, MD.

NCRP (2001) Evaluation of the Linear-Nonthreshold Dose-Response Model for Ionizing Radiation. Report No. 136 NCRP Publications, Bethesda, MD.

NIH (1985) *Report of the National Institutes of Health Ad Hoc Working Group to Develop Radioepidemiological Tables*, Publication No. 85-2748, National Institutes of Health, Bethesda, MD.

Orient, J.M. (2014) Fukushima and reflections on radiation as a terror weapon. *J. Am. Physicians Surg.*, **19**, 48.

Oversight Committee on Radioepidemiological Tables (1984) *Assigned Share for Radiation as a Cause of Cancer – Review of Radioepidemiological Tables Assigning Probability of Causation (Final Report)*, National Academy Press, Washington, DC.

Raabe, O.R. (2010) Concerning the health effects of internally deposited radionuclides. *Health Phys.*, **98**, 515.

Raabe, O.R. (2011) Toward improved ionizing radiation safety standards. *Health Phys.*, **101**, 84.

Raabe, O.R. (2012) Ionizing Radiation Carcinogenesis, in *Current Topics in Ionizing Radiation Research*, Chapter 15 (ed M. Nenoi), InTech, Rijeka, pp. 299–348.

Raabe, O.R. (2014) Concerning radiation carcinogenesis. *Health Phys.*, **107**, 571.

Tubiana, M. and Aurengo, A. (2005) Dose-effect relationship and estimation of the carcinogenic effects of low doses of ionizing radiation: the Joint Report of the Académie des Sciences (Paris) and the Académie Nationale de Médecine. *Int. J. Low Radiat.*, **2** (3/4), 1.

Turner, J.E. (2007) *Atoms, Radiation, and Radiation Protection*, 3rd edn, Wiley-VCH Verlag GmbH, Weinheim.

Index

*** There are more than 100 occurrences of this index entry. Only the first instance (page number) from each chapter or the page reference that defines the term is listed**

Health Physics: Radiation-Generating Devices, Characteristics, and Hazards, First Edition. Joseph John Bevelacqua.

q

r

t